Organosilicon Chemistry III

Edited by N. Auner and J. Weis

 WILEY-VCH

Organosilicon Chemistry III

From Molecules to Materials

Edited by Norbert Auner and Johann Weis

WILEY-VCH

Weinheim · Berlin · New York · Chichester · Brisbane · Singapore · Toronto

CHem

085190½2

Prof. Dr. N. Auner
Institut für Chemie
Fachinstitut für Anorganische
und Allgemeine Chemie
Humboldt-Universität zu Berlin
Hessische Straße 1–2
D-10115 Berlin
Germany

Dr. J. Weis
Wacker-Chemie GmbH
Geschäftsbereich S
Werk Burghausen
Johannes-Heß-Straße 24
D-84489 Burghausen
Germany

This book was carefully produced. Nevertheless, authors, editors and publisher do not warrant the information contained therein to be free of errors. Readers are advised to keep in mind that statements, data, illustrations, procedural details or other items may inadvertently be inaccurate.

Editorial Director: Dr. Anette Eckerle
Production Manager: Hans-Jochen Schmitt

Library of Congress Card No. applied for.

A catalogue record for this book is available from the British Library.

Deutsche Bibliothek Cataloguing-in-Publication Data:
Organosilicon chemistry : from molecules to materials / ed. by
Norbert Auner und Johann Weis. - Weinheim ; New York ; Basel ;
Cambridge ; Tokyo : VCH
    3 (1997)
    ISBN 3-527-29450-3

© WILEY-VCH Verlag GmbH, D-69469 Weinheim (Federal Republic of Germany), 1998

Printed on acid-free and chlorine-free paper.

Printing: betz-druck, D-64291 Darmstadt
Bookbinding: Wilh. Osswald, D-67433 Neustadt
Printed in the Federal Republic of Germany

Preface

This volume summarizes the lectures and poster contributions of the *III. Münchner Silicontage* that were held in April 1996. This symposium as well as the two predecessors (in 1992 and 1994) were again jointly organized by the *Gesellschaft Deutscher Chemiker* and *Wacker-Chemie GmbH*. The number of participants from industry and university, especially of students and young scientists, was again pleasing and is convincing evidence for the great interest in this meeting; in addition it was appreciation for the effort of the organizers. Moreover the book reviews of *Organosilicon Chemistry II - From Molecules to Materials* from all over the world have encouraged us to continue the series with the current issue.

The volumes of *Organosilicon Chemistry* are not considered to be textbooks in a common sense which should help students to pass basic examinations. These contributions from internationally renowned experts and researchers in a fascinating part of the rapidly growing field of main group chemistry describes current trends in organosilicon chemistry and provides summaries of the latest knowledge in this area.

However, in order to facilitate students and "non-silicon" scientists an easier access to the ongoing research on the basis of the relevant historical background, we decided to split this volume in two parts, each with a comprehensive introduction, one on molecular and one on polymer and solid state (organo)silicon chemistry.

During the *I. Münchner Silicontage* the *Wacker Silicon-Preis* was awarded to the two pioneers of silicone chemistry - Prof. Dr. Richard Müller and Prof. Dr. Eugene Rochow on the occasion of the 50th anniversary of the "Direct Process". In the course of the *II. Münchner Silicontage* this award was conferred on Prof. Dr. Edwin Hengge, Technische Universität Graz, for his fundamental work in polysilane chemistry. At the *III. Münchner Silicontage* Prof. Dr. Hubert Schmidbaur was honoured by presenting the *Wacker Silicon-Preis* for his outstanding contributions in the field of the synthesis and characterization of organosilicon "molecules and materials".

Right in the middle of the editorial phase of this volume, we were sad to hear of the unexpected decease of our friend Prof. Dr. Edwin Hengge. With him the organosilicon community has lost a passionate lecturer and researcher, unforgettable for his integrity and charming personality. Therefore it is the editors special wish that this volume shall keep alive the memory of an excellent organosilicon chemist!

Collecting and publishing these papers it is our main intention to encourage students and young scientists to focus on organosilicon chemistry and to continue the work in the future! We want to light a beacon - outstanding success in the last few years should not deceive us, that there are still a lot of challenging problems to be solved in the future: this includes basic research as well as the development of new materials.

In August 1997

Prof. Dr. Norbert Auner   *Dr. Johann Weis*

Acknowledgment

First of all we would like to thank the numerous authors for their intense cooperation, which made this overview of current organosilicon chemistry possible. The tremendous work load to achieve the attractive layout of this volume was mainly performed by Dr. Claus-Rüdiger Heikenwälder and Dr. Mathias Kersten. We thank both for their admirable engagement!

Prof. Dr. Norbert Auner
Humboldt-Universität
zu Berlin

Dr. Johann Weis
Wacker-Chemie GmbH
München

Contents

I Fascinating Organosilicon Compounds

Introduction 1
N. Auner, G. Fearon, J. Weis

More Compelling Evidence that Silicon is Better Than Carbon: The Thermal 17
Isomerization of Olefins to Carbenes
T. J. Barton, J. Lin, S. Ijadi-Maghsoodi, M. D. Power, X. Zhang, Z. Ma, H. Shimizu,
M. S. Gordon

NMR and Quantum Chemical Characterization of Silicon-Substituted Carbocations 25
H.-U. Siehl, B. Müller, O. Malkina

Matrix Isolation Studies of the Reactions of Silicon Atoms 31
G. Maier, H. P. Reisenauer, H. Egenolf

Cycloaddition Reactions of Dimethylaminomethylsilylene with Dienes and Heterodienes 36
S. Meinel, J. Heinicke

Do Unsubstituted Silacyclobutadienes Exist? 39
G. Maier, H. P. Reisenauer, J. Jung, A. Meudt, H. Pacl

A Thermally Stable Silylene: Reactivity of the Bis(amino)silylene $\overline{Si[\{N(CH_2tBu)\}_2C_6H_4\text{-}1,2]}$ 44
B. Gerhus, P. B. Hitchcock, M. F. Lappert

Pyrido[b]-1,3,2λ^2-diazasilole: The First Stable Unsymmetrical Silylene 50
A. Oprea, J. Heinicke

A New Route to Silaheterocycles: Heterobutadiene Cycloaddition 53
H. H. Karsch, P. Schlüter

Base Coordination: A Way to Nucleophilic Silylenes? 58
J. Belzner

Isoelectronic Replacement of Si by P$^+$: A Comparative Study of the Structures of the 65
Spirocyclic E^{II} Compounds E[C(PMe$_2$)$_2$(X)]$_2$ (E = Si, Ge, Sn; X = PMe$_2$,SiMe$_3$) and
a Novel Spirocyclic 10 e-Phosphorus Cation(P^{III}) P[C(PPh$_2$)$_2$(SiMe$_3$)]$_2$$^+$
H. H. Karsch, E. Witt

The Main Group Carbonyls RLi–CO and R$_2$Si–CO: An Ab Initio Study 70
M. Tacke

Unprecedented Multistep Reactions of Decamethylsilicocene, $(Me_5C_5)_2Si$:, with CO_2, CS_2, ...76
COS, RNCS (R = Me, Ph), with CF_3CCCF_3, and with $HMn(CO)_5$
P. Jutzi, D. Eikenberg

Rearrangement of Bis(hypersilyl)silylene and Related Compounds – An Unusual Way to 82
Three-Membered Rings
K. W. Klinkhammer

Oxidation of Silenes and Silylenes: Matrix Isolation of Unusual Silicon Species 86
W. Sander, M. Trommer, A. Patyk

New Silaheterocycles: Formation and Properties 95
E. Kroke, M. Weidenbruch

Cycloaddition Reactions of 1,1-Dichloro-2-neopentyl-1-silen with Monoterpenes 101
C.-R. Heikenwälder, N. Auner

Silaspirocycles as Precursors for a 2-Silaallene 106
B. Goetze, B. Herrschaft, N. Auner

Catalytic Carbon-Carbon Hydrogenation of Silicon-Functionalized Olefins 113
H.-U. Steinberger, N. Auner

Dieno- and Enophilicity of Sila-, Germa-, and Stannaethenes 117
N. Wiberg, S. Wagner

Iminosilanes and Silaamidides: Synthesis and Reactions 120
J. Niesmann, A. Frenzel, U. Klingebiel

Metastable Compounds Containing Silicon-Phosphorus and Silicon-Arsenic Multiple Bonds: 126
Syntheses, Structures and Reactivity
M. Driess, S. Rell, U. Winkler, H. Pritzkow

Silole and Germole Dianions and their Dilithium Derivatives – Are they Aromatic? 144
T. Müller, Y. Apeloig, H. Sohn, R. West

Supersilylmetal Compounds 152
N. Wiberg, K. Amelunxen, H. Nöth, A. Appel, M. Schmidt, K. Polborn

Trialkylsilyl Substituted Homobimetallic Phosphanides of the Alkaline Earth Metals as 157
well as Zinc
M. Westerhausen, B. Rademacher, M. Hartmann, M. Wieneke, M. Digeser

The Tris(trimethylsilyl)silyl Substituent: An Old Hat With A New Feather 162
A. Heine, L. Lameyer, D. Stalke

Functionalized Trisilylmethanes and Trisilylsilanes as Precursors of a New Class of 172
Tripodal Amido Ligands
M. Schubart, B. Findeis, H. Memmler, L. H. Gade

Methoxy-bis[tris(trimethylsilyl)silyl]methane: The First Geminal Di(hypersilyl) Compound 178
E. Jeschke, T. Gross, H. Reinke, H. Oehme

The Use of the Tris(trimethylsilyl)silyl Group in Stabilization of Low Valent Gallium 182
Compounds
W. Köstler, G. Linti

Synthesis, Structure, and Reactions of Tris(trimethylsilyl)silyl Gallanes and Gallates 189
H. Urban, R. Frey, G. Linti

Novel Pathways in the Reactions of Vinylsilanes with Lithium Metal 195
A. Maercker, K. Reider, U. Girreser

New Organosilicon Reagents : Synthesis, Structure, and Reactivity of 206
(Lithiomethyl)(aminomethyl)silanes
B. C. Abele, C. Strohmann

(Phenylthiomethyl)silanes as New Bifunctional Assembling Ligands for the Construction 211
of Heterometallic Complexes
M. Knorr, S. Kneifel, C. Strohmann

Synthesis of 1,3-Disilacyclobutanes, 1,3-Digermacyclobutanes, and 1-Germa-3-sila- 217
cyclobutanes with New 1,3-Dimetallated Organoelement Building Blocks
C. Strohmann, E. Wack

Trialkylsilyldiazomethane Derivatives: Wonderful Chemical Building Blocks 223
G. Bertrand

A New Route to Silaheterocycles: Nucleophilic Aminomethylation 237
H. H. Karsch, K. A. Schreiber

Infrared and Raman Spectra, ab initio Calculations, and Rotational Isomerism of 241
Methylated Disilanes
K. Schenzel, A. Jähn, M. Ernst, K. Hassler

1,2-Di-*tert*-butyltetrafluorodisilane: A Highly Fluxional Molecule 248
R. Zink, K. Hassler, N. W. Mitzel, B. A. Smart, D. W. H. Rankin

Amino-Substituted Disilanes by Reductive Coupling 254
S. Mantey, J. Heinicke

Multifunctional Disilane Derivatives 257
H. Stüger, P. Lassacher, E. Hengge

New Transition Metal Substituted Oligosilanes 262
W. Palitzsch, U. Böhme, G. Roewer

Regiospecific Chlorination and Oxygenation of Pentahydridodisilanyl Complexes of Iron 267
and Ruthenium
S. Möller, H. Jehle, W. Malisch, W. Seelbach

Inter- and Intramolecular Oxidative Addition of Si–H Bonds 271
R. Karch, H. Gilges, U. Schubert

Novel Synthetic Approach to Molybdenum-Silicon Compounds: Structures and 275
Reactivities
P. Jutzi, S. H. A. Petri

Unexpected Reactivity of Bis-1,2-[(bromodiphenyl)methyl]-1,1,2,2-tetramethyldisilane 281
F. Pillong, O. Schütt, C. Strohmann

Trichlorosilane/Triethylamine – An Alternative to Hexachlorodisilane in Reductive 286
Trichlorosilylation Reactions?
L.-P. Müller, A. Zanin, J. Jeske, P. G. Jones, W.-W. du Mont

Disproportionation of Chloromethyldisilanes using Lewis Base Heterogeneous Catalysts – 291
A Way to Influence the Polymer Structure
T. Lange, N. Schulze, G. Roewer, R. Richter

Supersilylated Bromodisilanes, Cyclotri-, and Cyclotetrasilanes 296
N. Wiberg, H. Auer, Ch. M. M. Finger, K. Polborn

Syntheses and ^{29}Si NMR Spectra of Halogenated Trisilanes and Cyclopentasilanes 301
K. Hassler, W. Köll, U. Pöschl

Chlorination of Methylphenyloligosilanes: Products and Reactions 307
C. Notheis, E. Brendler, B. Thomas

Selective Hydrogenation of Methylchlorooligosilanes 312
U. Herzog, G. Roewer

Electrochemical Formation of Cyclosilanes 317
S. Graschy, C. Grogger, E. Hengge

Undecamethylcyclohexasilanyl Derivatives of Tin(IV) and Lead(IV) 322
F. Uhlig, U. Hermann, K. Klinkhammer, E. Hengge

New Results in Cyclosilane Chemistry: Siloxene-like Polymers 327
A. Kleewein, U. Pätzold, E. Hengge, S. Tasch, G. Leising

Stepwise Synthesis of Functional Polysilane Dendrimers 333
C. Marschner, E. Hengge

Synthesis and Reactivity of Novel Polysilynes and Branched Copolysilanes 337
W. Uhlig

Unusual Polyhedra by Lithiation of Silazanes 342
G. Becker, S. Abele, U. Eberle, G. Motz, W. Schwarz

Isomeric Halosilylhydroxylamines: Preparation and Thermal Rearrangements 348
R. Wolfgramm, U. Klingebiel

Reactions of Hydridosilylamides 353
K. Junge, N. Peulecke, K. Sternberg, H. Reinke, E. Popowski

Silylhydrazines: Precursors for Rings, Hydrazones and Pyrazolones 358
C. Drost, U. Klingebiel, H. Witte-Abel

Products from Multiple Insertion Reactions between Diisocyanates and Antiheteroaromatic 364
1,4-Bis(trimethylsilyl)-1,4-dihydropyrazine
T. Sixt, F. M. Hornung, A. Ehlend, W. Kaim

Some Surprising Chemistry of Sterically Hindered Silanols 369
P. D. Lickiss

Silanetriols: Preparation and Their Reactions 376
R. Murugavel, A. Voigt, M. G. Walawalkar, H. W. Roesky

Silsesquioxanes as Crown Ether Analogs 395
U. Dittmar, H. C. Marsmann, E. Rikowski

Azomethine-Substituted Organotrialkoxysilanes and Polysiloxanes 400
F. Mucha, G. Roewer

On the Reaction of (tBu$_2$SnO)$_3$ with Organochlorosilanes. Simple Formation of 403
[(tBu$_2$SnO)$_2$(tBu$_2$SiO)]
J. Beckmann, K. Jurkschat, D. Schollmeyer

Silanols and Siloxanes Substituted with the Chiral Iron Fragments Cp(OC)(RPh$_2$P)Fe 407
[R = Ph, (H)(Me)(Ph)C(Me)N]
W. Malisch, M. Neumayer, K. Perneker, N. Gunzelmann, K. Roschmann

Si–H Functionalized Ferrio-Trisiloxanes $C_5R_5(OC)_2Fe$–Si(Me)(OSiMe$_2$H)$_2$ (R = H, Me) 412
J. Reising, W. Malisch, R. Lankat

Novel Siloxy-Bridged Di-, Tri-, and Tetranuclear Metal Complexes from Ferrio- and 415
Tungsten-Silanols
W. Malisch, J. Reising, M. Schneider

The 2-Dimethylaminomethyl-4,6-dimethylphenyl Substituent: A New Intramolecular 418
Coordinating System with High Steric Demand
U. Dehnert, J. Belzner

Reaction Behaviour of Hypervalent Silanes 423
H. Lang, E. Meichel, M. Weinmann, M. Melter

Investigations of Nucleophilic Substitution at Silicon: An Unprecedented Equilibrium 429
between an Ionic and a Covalent Chlorosilane
D. Schär, J. Belzner

Ligand Exchange Mechanism in Novel Hexacoordinate Silicon Complexes 435
D. Kost, S. Krivonos, I. Kalikhman

Ligand Exchange *via* Coordinative Si–N Bond Cleavage and Pseudorotation in Neutral 446
Pentacoordinate Silicon Complexes
I. Kalikhman, D. Kost

Phosphine Coordination to Silicon Revisited 452
G. Müller, M. Waldkircher, A. Pape

Novel Sila-Phospha-Heterocycles and Hypervalent Silicon Compounds with Phosphorus 460
Donors
H. H. Karsch, R. Richter, E. Witt

Germanium Analogues of Zwitterionic Spirocyclic λ^5Si-Silicates 466
J. Heermann, R. Tacke, P. G. Jones

II Silicon Based Materials

Introduction 471
N. Auner, G. Fearon, J. Weis

The Direct Process to Methylchlorosilanes: Reflections on Chemistry and Process 478
Technology
B. Pachaly, J. Weis

On the Nature of the Active Copper State and on Promoter Action in Rochow Contact 484
Masses
H. Ehrich, D. Born, J. Richter-Mendau, H. Lieske

On the Acid- and Base- Catalyzed Reactions of Silanediols and Siloxanediols in Water 496
H. Kelling, W. Rutz, K. Busse, C. Wendler, D. Lange

Trace Analysis of Mono- and Trifunctional Groups in Polydimethylsiloxanes Using 500
Reaction Headspace GC
J. Graßhoff

Synthesis and Investigation of the Surface Active Properties of New Silane Surfactants 504
S. Stadtmüller, K.-D. Klein, K. Köppen, J. Venzmer

Carbohydrate-Modified Siloxane Surfactants: The Effect of Substructures on the 510
Wetting Behaviour on Non Polar Solid Surfaces
R. Wagner, L. Richter, Y. Wu, J. Weißmüller, J. Reiners, K.-D. Klein, D. Schaefer,
S. Stadtmüller

Synthesis and Application of ω-Epoxy-Functionalized Alkoxysilanes 515
G. Sperveslage, K. Stoppek-Langner, J. Grobe

Diffuse Reflectance IR and Time-of-Flight SIMS Investigation of Methoxysilane 520
SA-Layers on Silica and Alumina
K. Stoppek-Langner, K. Meyer, A. Benninghoven

Ether-Substituted Triethoxy- and Diethoxymethylsilanes: Precursors for Hydrophilic, ...526
Elastic Consolidants for Natural Stones
R. Fabis, C. Zeine, J. Grobe

Organosilicon Compounds for Stone Impregnation – Long-Term Effectivity and 531
Weathering Stability
C. Bruchertseifer, S. Brüggerhoff, K. Stoppek-Langner, J. Grobe, M. Jursch, H.-J. Götze

Studies on the Regioselectivity of the Hydroformylation with Alkenylalkoxysilanes 538
M. Wessels, J. Grobe

Novel Precursors for Inorganic-Organic Hybrid Materials 543
S. Kairies, K. Rose

Mesomorphic Properties of Poly(diphenylsiloxane) 550
B. R. Harkness, M. Tachikawa, I. Mita

Applications of Silicone Elastomers for Electrical and Electronic Fields 555
M. Takahashi

Cyclic Liquid Crystalline Siloxanes: Chemistry and Applications 566
F.-H. Kreuzer, N. Häberle, H. Leigeber, R. Maurer, J. Stohrer, J. Weis

Modified Polydimethylsiloxanes with Fluorescent Properties 587
B. Strehmel, C. W. Frank, W. Abraham, M. Garrison

Sensitized Cationic Photocrosslinking of α,ω-Terminated Disiloxanes: Cation Formation 594
in Nonpolar Media
U. Müller, A. Kunze, Ch. Herzig, J. Weis

Cationic Photoinitiators for Curing Epoxy-Modified Silicones 605
C. Priou

Photoconductivity in Polysilylenes: Doping with Electron Acceptors 617
A. Eckhardt, V. Herden, W. Schnabel

Functionalized Polycarbosilanes as Preceramic Materials 622
S. Back, H. Lang, M. Weinmann, W. Frosch

Novel Polyorganoborosilazanes for the Synthesis of Ultra-High Thermal Resistant Ceramics 628
L. M. Ruwisch, W. Dressler, S. Reichert, R. Riedel

Precursors for Silicon-Alloyed Carbon Fibers 632
J. Dautel, W. Schwarz

One-Pot Syntheses of Poly(diorganylsilylene-*co*-ethynylene)s 638
W. Habel, A. Moll, P. Sartori

Localization Phenomena of Photogenerated Charge Carriers in Silicon Structures: 643
From Organosilicon Compounds to Bulk Silicon
T. Wirschem, S. Veprek

Functionalized Structure-Directing Agents for the Direct Synthesis of Nanostructured 649
Materials
P. Behrens

Novel Aspects of the Chemical Modification of Silica Surface 670
V. Tertykh

Microporous Thermal Insulation: Theory, Properties, Applications 682
H. Katzer, J. Weis

Nitridosilicates – High Temperature Materials with Interesting Properties 691
W. Schnick, H. Huppertz, T. Schlieper

Author Index 705

Subject Index 711

List of Contributors

Prof. Dr. Yitzhak Apeloig
Department of Chemistry
Technion – Israel Institute of Technology
Technion City
32000 Haifa
Israel

Prof. Dr. T. Barton
Department of Chemistry
Ames Laboratory (U.S. Department of Energy)
Iowa State University
Ames, Iowa 50011
USA

Prof. Dr. Gerd Becker
Institut für Anorganische Chemie
Universität Stuttgart
Pfaffenwaldring 55
D-70550 Stuttgart
Germany

Prof. Dr. Peter Behrens
Institut für Anorganische Chemie
Ludwigs-Maximilians-Universität München
Meiserstraße 1
D-80333 München
Germany

Dr. Johannes Belzner
Institut für Organische Chemie
Georg-August-Universität Göttingen
Tammannstraße 2
D-37077 Göttingen
Germany

Prof. Dr. Guy Bertrand
Laboratoire de Chimie de Coordination du CNRS
205 route de Narbonne
F-31077 Toulouse Cedex
France

Dr. Uwe Böhme
Institut für Anorganische Chemie
Technische Universität Bergakademie Freiberg
Leipziger Straße 29
D-09596 Freiberg
Germany

Dr. Joachim Dautel
Institut für Anorganische Chemie
Universität Stuttgart
Pfaffenwaldring 55
D-70569 Stuttgart
Germany

Matthias Driess
Lehrstuhl für Anorganische Chemie I
Fakultät für Chemie
Ruhr-Universität Bochum
Universitätsstraße 150
D-44801 Bochum
Germany

Prof. Dr. W.-W. du Mont
Institut für Anorganische und Analytische Chemie
Technische Universität Braunschweig
Hagenring 30
D-38106 Braunschweig
Germany

Dr. Gordon Fearon
Central Research and Development
Dow Corning Corporation
Midland, MI, 48686-09994
USA

Dr. Lutz H. Gade
Institut für Anorganische Chemie
Universität Würzburg
Am Hubland
D-97074 Würzburg
Germany

Dr. Barbara Gehrhus
School of Chemistry and Molecular Sciences
University of Sussex
Brighton BN1 9QJ
United Kingdom

Dr. Jürgen Graßhoff
Hüls Silicone GmbH
Analytical Department
D-01612 Nünchritz
Germany

Prof. Dr. Joseph Grobe
Anorganisch-Chemisches-Institut
Westfälische Wilhelms-Universität
Wilhelm-Klemm-Straße 8
D-48149 Münster
Germany

Dr. Christa Grogger
Institut für Anorganische Chemie
Technische Universität Graz
Stremayrgasse 16
A-8010 Graz
Austria

Dr. B. R. Harkness
Dow Corning Asia Ltd.
Research Center
603 Kishi, Yamakita
Kanagawa 258-01
Japan

Priv. Doz. Dr. Karl Hassler
Institut für Anorganische Chemie
Technische Universität
Stremayrgasse 16
A-8010 Graz
Austria

Prof. Dr. Joachim Heinicke
Institut für Anorganische Chemie
Ernst-Moritz-Arndt-Universität Greifswald
Soldtmannstraße 16
D-17487 Greifswald
Germany

Prof. Dr. Klaus Jurkschat
Fachbereich Chemie der Universität Dortmund
Lehrstuhl für Anorganische Chemie II
D-44221 Dortmund
Germany

Prof. Dr. Peter Jutzi
Fakultät für Chemie
Universität Bielefeld
Universitätsstraße 25
D-33615 Bielefeld
Germany

Dipl.-Ing. Alois Kleewein
Institut für Anorganische Chemie
Technische Universität Graz
Stremayrgasse 16 / IV
A-8010 Graz
Austria

Prof. Dr. Wolfgang Kaim
Institut für Anorganische Chemie
Universität Stuttgart
Pfaffenwaldring 55
D-70550 Stuttgart
Germany

Prof. Dr. Uwe Klingebiel
Institut für Anorganische Chemie
Georg-August-Universität Göttingen
Tammannstraße 4
D-37077 Göttingen
Germany

Prof. Dr. Hans Heinz Karsch
Anorganisch-Chemisches Institut
Technische Universität München
Lichtenbergstraße 4
D-85747 Garching
Germany

Dr. Karl Wilhelm Klinkhammer
Institut für Anorganische Chemie
Universität Stuttgart
Pfaffenwaldring 55
D-70550 Stuttgart
Germany

Dr. Hans Katzer
Wacker-Chemie GmbH
Geschätsbereich S-Werk Kempten
Max-Schaidhauf-Straße 25
D-87437 Kempten
Germany

Dr. Michael Knorr
Institut für Anorganische Chemie
Universität des Saarlandes
Postfach 1150
D-66041 Saarbrücken
Germany

Prof. Dr. Hans Kelling
Fachbereich Chemie
Universität Rostock
Buchbinderstraße 9
D-18051 Rostock
Germany

Prof. Dr. Daniel Kost
Department of Chemistry
Ben Gurion University of the Negev
Beer-Sheva 84105
Israel

Dr. F.-H. Kreuzer
Consortium für elektrochemische Industrie GmbH
Corporate Research Company of
Wacker-Chemie GmbH
Zielstatterstraße 20
D-81379 München
Germany

Prof. Dr. Adalbert Maercker
Institut für Organische Chemie
Universität Siegen
D-57068 Siegen
Germany

Prof. Dr. Heinrich Lang
Institut für Chemie
Lehrstuhl für Anorganische Chemie
Technische Universität Chemnitz
Straße der Nationen 62
D-09107 Chemnitz
Germany

Prof. Dr. Günther Maier
Institut für Organische Chemie
Justus-Liebig-Universität Gießen
Heirich-Buff-Ring 58
D-35392 Gießen
Germany

Dr. Paul D. Lickiss
Department of Chemistry
Imperial College of Science
Technology and Medicine
London SW7 2AY
United Kingdom

Prof. Dr. Wolfgang Malisch
Institut für Anorganische Chemie
Universität Würzburg
Am Hubland
D-97074 Würzburg
Germany

Prof. Dr. Heiner Lieske
Institut für Angewandte Chemie Berlin-
Adlershof e.V.
Abteilung Katalyse
Rudower Chaussee 5
D-12484 Berlin
Germany

Dr. Christoph Marschner
Institut für Anorganische Chemie
Technische Universität Graz
Stremayrgasse 16
A-8010 Graz
Austria

Dr. Gerald Linti
Institut für Anorganische Chemie
der Universität (TH) Karlsruhe
Engesserstraße Geb. 30.45
D-76128 Karlsruhe
Germany

Prof. Dr. Heinrich Christian Marsmann
Fachbereich Chemie und Chemietechnik
Universität-GH Paderborn
Warburger Straße 100
D-33095 Paderborn
Germany

Prof. Dr. Gerhard Müller

Fakultät für Chemie der Universität Konstanz

Universitätsstraße 10

D-78464 Konstanz

Germany

Dr. Thomas Müller

Fachinstitut für Anorganische und Allgemeine

Chemie

Humboldt-Universität zu Berlin

Hessische Straße 1-2

D-10115 Berlin

Germany

Dr. U. Müller

Institut für Organische Chemie

Martin-Luther-Universität Halle-Wittenberg

Geusaer Straße

D-06217 Merseburg

Germany

Christina Notheis

Institut für Analytische Chemie

Technische Universität Bergakademie Freiberg

Leipziger Straße 29

D-09599 Freiberg

Germany

Prof. Dr. Hartmut Oehme

Fachbereich Chemie

Abteilung Anorganische Chemie

Universität Rostock

Buchbinderstraße 9

D-18051 Rostock

Germany

Dr. B. Pachaly

Wacker-Chemie GmbH

Geschäftsbereich S-Werk Burghausen

Johannes-Heß-Straße 24

D-84489 Burghausen

Germany

Prof. Dr. Eckhard Popowski

Fachbereich Chemie

Universität Rostock

Buchbinderstraße 9

D-18051 Rostock

Germany

Dr. C. Priou

Rhone-Poulenc Chimie

Silicone Division

1 rue des frères Perret, B.P. 22

F-69191 Saint Fons cedex

France

Dr. Ralf Riedel

Fachbereich Materialwissenschaft

Fachgebiet Disperse Feststoffe

Technische Hochschule Darmstadt

Petersenstraße 23

D-64287 Darmstadt

Germany

Prof. Dr. Herbert W. Roesky

Institut für Anorganische Chemie der Universität

Göttingen

Tammannstraße 4

D-37077 Göttingen

Germany

Prof. Dr. Gerhard Roewer
Institut für Anorganische Chemie
Technische Universität Bergakademie Freiberg
Leipziger Straße 29
D-09596 Freiberg
Germany

Dr. Klaus Rose
Fraunhofer-Institut für Silicatforschung
Neunerplatz 2
D-97082 Würzburg
Germany

Prof. Dr. Wolfram Sander
Lehrstuhl für Organische Chemie II
Ruhr-Universität Bochum
D-44780 Bochum
Germany

Prof. Dr. Peter Sartori
Anorganische Chemie
Gerhard Mercator Universität-Gesamthochschule-
Duisburg
Lotharstraße 1
D-47048 Duisburg
Germany

Prof. Dr. Wolfram Schnabel
Hahn-Meitner-Institut Berlin GmbH
Bereich Physikalische Chemie
Glienicker Straße 100
D-14109 Berlin
Germany

Prof. Dr. Wolfgang Schnick
Laboratorium für Anorganische Chemie
Universität Bayreuth
Universitätsstraße 30
D-95440 Bayreuth
Germany

Prof. Dr. Hans-Ullrich Siehl
Abteilung für Organische Chemie der
Universität Ulm
D-89069 Ulm
Germany

Prof. Dr. U. Schubert
Institut für Anorganische Chemie
Technische Universität Wien
Getreidemarkt 9
A-1060 Wien
Austria

Dr. Stefan Stadtmüller
Th. Goldschmidt AG
Goldschmidtstraße 100
D-45127 Essen
Germany

Prof. Dr. Dietmar Stalke
Institut für Anorganische Chemie
Universität Würzburg
Am Hubland
D-97074 Würzburg
Germany

Dr. Karl Stoppek-Langner
Anorganisch-Chemisches-Institut
Westfälische Wilhelms-Universität
Wilhelm-Klemm-Straße 8
D-48149 Münster
Germany

Prof. Dr. Reinhold Tacke
Institut für Anorganische Chemie
Bayerische Julius-Maximilians-Universität zu
Würzburg
Am Hubland
D-97074 Würzburg
Germany

Dr. B. Strehmel
Fachinstitut für Physikalische und Theoretische
Chemie
Humboldt Universität zu Berlin
Bunsenstraße 1
D-10117 Berlin
Germany

Dr. Masaharu Takahashi
Silicone-Electronics Materials Research Center
Shin-Etsu Chemical Co., Ltd.
1-10, Hitomi, Matsuida
Gunma 379-02
Japan

Priv.-Doz. Dr. Carsten Strohmann
Institut für Anorganische Chemie
Universität des Saarlandes
Postfach 1150
D-66041 Saarbrücken
Germany

Prof. Dr. Valentin Tertykh
Department of Chemisorption
Institute of Surface Chemistry
National Academy of Sciences
Prospekt Nauki 31
252022 Kiev
Ukraina

Dr. Harald Stüger
Institut für Anorganische Chemie
Technische Universität Graz
Stremayrgasse 16
A-8010 Graz
Austria

Dr. Frank Uhlig
Lehrstuhl für Anorganische Chemie II
Universität Dortmund
Otto-Hahn-Straße 6
D-44227 Dortmund
Germany

Dr. Matthias Tacke
University College Dublin
Department of Chemistry
Belfield
Dublin 4
Ireland

Dr. Wolfram Uhlig
Laboratorium für Anorganische Chemie
Eidgenössische Technische Hochschule Zürich
ETH-Zentrum
CH-8092 Zürich
Switzerland

Prof. Dr. Stan Veprek

Institut für Chemie Anorganischer Materialien
Technische Universität München
Lichtenbergstraße 4
D-85747 Garching
Germany

Dr. R. Wagner

Max-Planck-Institut für Kolloid-und
Grenzflächenforschung
Rudower Chaussee 5
D-12489 Berlin
Germany

Prof. Dr. Manfred Weidenbruch

Fachbereich Chemie
Universität Oldenburg
Carl-von-Ossietzky-Straße 9
D-26111 Oldenburg
Germany

Prof. Dr. Matthias Westerhausen

Institut für Anorganische Chemie
Ludwig-Maximilians-Universität München
Meisterstraße 1
D-80333 München
Germany

Prof. Dr. Nils Wiberg

Institut für Anorganische Chemie
Ludwig-Maximilians-Universität München
Meisterstraße 1
D-80333 München
Germany

Editors:

Prof. Dr. Norbert Auner
Fachinstitut für Anorganische und Allgemeine
Chemie
Humboldt-Universität zu Berlin
Hessische Straße 1-2
D-10115 Berlin
Germany

Dr. Johann Weis
Wacker-Chemie GmbH
Geschäftsbereich S
Werk Burghausen
Johannes-Heß-Straße 24
D-84489 Burghausen
Germany

PART I

FASCINATING ORGANOSILICON COMPOUNDS

INTRODUCTION

Norbert Auner
Humboldt-Universität
Berlin, Germany

Gordon Fearon
Dow Corning Corporation
Midland, MI, USA

Johann Weis
Wacker-Chemie GmbH
München, Germany

To understand Group 14 - and especially organosilicon - chemistry some comparisons between silicon and carbon have to be considered. There are two major properties that distinguish silicon from carbon. Silicon atoms are about 50 % larger than carbon atoms and this increased size will have some ramifications and consequences, such as lower barriers to silicon-element bond rotations and less stable π-bonds. Furthermore, the smaller Pauling electronegativity of silicon results in differently polar silicon-element bonds compared to carbon and thus will change its reactivity and enable reactions not possible in carbon chemistry.

The common environments exhibited for carbon and silicon show that silicon differs from carbon by its strongly reduced ability to form multiple bonds in comparison to its congener carbon but also in its capacity to form stable derivatives with more than four bonds. Carbon exhibits a maximum of four single covalent bonds, derivatives with more (e.g. five) nearest neighbors only exist in 'non classical (carbon bridged) ions', in a few carbon gold complexes and in some organometallic carbide compounds. The lack of multiple bonds seriously hampers synthetic strategies in silicon chemistry compared to the numerous possibilities available to the organic chemists. But this lack is counterbalanced by the ease of formation of silicon hypervalent species, the low activation energies for nucleophilic substitution at silicon compared to carbon and the numerous reaction pathways at tetrahedral silicon centers that are not accessible to carbon.

Although over years a hybridization state has often been used to describe the expanded octet geometries of silicon - this is by no means the exclusive bonding view for higher coordinate species - one has to take into account that the participation of d-orbitals in the description of bonding in silicon compounds was a subject of continuing debate and that nowadays an alternative explanation using multicenter bonding has been accepted.

Besides other organizing principles, compounds of silicon might be ordered in terms of oxidation states and coordination numbers at the silicon centers. Tetravalent silicon mostly exhibits the oxidation state +4 (e.g. SiX_4, X = halogen, H, organo group, SiO_2, Si_3N_4, SiC etc.), in Ca_2Si however it is -4. Divalent silicon in the oxidation state +2 is found in compounds such as $:SiX_2$, SiO or SiS which are available only at high temperatures and can be identified spectroscopically in the gas phase at low pressure and/or in low temperature matrices after condensation on a cold surface (T~10K). As examples for tri- and monovalent silicon with positive oxidation numbers +3 and +1 compounds SiX_3 and SiX are named, which are formed competitively to silylenes $:SiX_2$ by photolysis and/or thermolysis of stable precursors SiX_4. Negative oxidation states -3, -2 and -1 with tri-, di-, and monovalent silicon are represented in silicides, as exemplified by $BaMg_2Si_2$, CaSi, and $CaSi_2$.

In its compounds the silicon center is surrounded by at least one up to ten neighbors. Some representative specimens in various coordination numbers are listed below and discussed later in this introduction.

Coordination number	Compound
1	$Si≡O$, $Si≡S$
2	$O=Si=O$, $Me_3SiN=Si=NSiMe_3$, $tBuCH_2CH=Si=CHCH_2tBu$ $HSi≡N$, $PhSi≡N$; $X_2Si:$
3	$R_2Si=Y$, $·SiX_3$, R_3Si^+
4	SiX_4, SiO_2
5	$XSi(1,2-O_2C_6H_4)_2^-$ (X = Ph, F) $C_9H_6NOSiR_2X$ (X = halogen, SO_3CF_3) (see page 10) $X_3Si(o-C_6H_4CH_2NMe_2)$
6	SiF_6^{2-} $X_2Si(o-C_6H_4CH_2NMe_2)_2$
7	$XSi(o-C_6H_4CH_2NMe_2)_3$ (X = H, F)
8	Mg_2Si $H_2Si\{o,o-C_6H_3(CH_2NMe_2)\}_2$
10	$(C_5Me_5)_2Si:$ (see page 9)

This list of compounds with silicon in the whole range of its coordination numbers and environments, including their synthesis and the investigation of their chemistry gives an impressive overview of organosilicon research currently going on worldwide.

Such developments require different synthetic methods to form silicon-carbon bonds in a laboratory as well as an easy availability of organo and halogen functionalized silanes on a preparative or even industrial scale.

First experiments in this field have been done by Friedel and Crafts who synthesized Et_4Si from reaction of Et_2Zn with $SiCl_4$ at 140°C in a sealed tube. Later Pape reacted a mixture of $SiCl_4$, Zn and propyl iodide and isolated Pr_4Si. The first methods to link silicon and carbon were explored at the turn of this century and included basically the transfer of an organic group from an active organometallic reagent to $SiCl_4$ or $Si(OEt)_4$. The first breakthrough came after 1900, when Grignard had published his discovery of the reagents named after him: Kipping isolated a mixture of ethylchlorosilanes Et_nSiCl_{4-n} (n = 0-4) from $SiCl_4$ and EtMgI. The major 'scientific revolution' occured a little bit more than 50 years ago, when Müller and Rochow - independently from each other - developed the 'Direct Process' for the large scale production of halide functional organosilanes. From the copper catalyzed reaction between silicon metal and chlorinated hydrocarbons a wide variety of organochlorosilanes - especially of Me_2SiCl_2, the precursor for the production of polydimethylsiloxanes - became easily available on an industrial scale. This discovery is on one hand the birthday of the organosilicon industry and on the other hand it has been the starting shot for incredible research activities in academia in a very broad and basic sense.

While the second part of this volume deals with new silicon based materials (polymers) and industrial application (see Introduction, page ...), the first part is concerned with the description of defined molecules which might be precursors for new materials one day in the future. These 'molecules' include unsaturated silicon compounds as well as silyl cations, silylenes, small and big cycles and heterosilacycles, di- and oligosilanes - both cyclic and linear - and silyl metal compounds. Even single silicon atoms seem to have a big synthetic potential - at least in a matrix! Some of these 'molecules' will now be considered in closer detail and a wide list of reviewing articles appropriate to the broad field of organosilicon chemistry will be presented at the end of this introduction into a world of fascinating compounds.

Unsaturated silicon compounds

As mentioned before, carbon and silicon mostly differ in their ability to from multiple p_π-p_π bonds E=Y with suitable partners (E = C, Si; Y = element of group 14 to 16). While the p orbital overlap in compounds >C=Y is sufficient to yield stable multiple bonded species, this overlap is strongly reduced in the case of silicon (classical double bond rule of Pitzer and Mulliken). Consequently, under comparable conditions the equivalents of many unsaturated monomeric compounds of carbon, such as $H_2C=CH_2$, $R_2C=O$ or CO_2 are silicon single bonded polymeric products, e.g. polysilanes $(-H_2Si-SiH_2-)_n$, silicones $(-R_2Si-O-)_n$ and silicon dioxide $(SiO_2)_n$.

p_π-p_π unsaturated silicon compounds, such as Si≡E (E = O, S), are formed from saturated organosilicon precursors only in the gas phase at very low pressure and high temperatures. Under these conditions they are stable. The same holds for compounds >Si=Y (Y = Si<, C<, Ge<, N-, P-, O, S) which - because of their extraordinary high reactivity - usually dimerize or polymerize under normal conditions or might be trapped by suitable trapping agents (polar reagents HX, polar C=C or C=X species or dienes). This is shown exemplarily for silene formation using the two main synthetic approaches, namely the gas phase pyrolysis of stable precursors (e.g. silacyclobutanes) or the salt elimination of organolithio precursors at low temperatures and in solution, and for general trapping reactions of >Si=Y species.

$$R_2Si\ \square \quad \xrightarrow[- H_2C=CH_2]{\Delta} \quad R_2Si=CH_2 \quad \longrightarrow \quad 1/2 \quad R_2Si \diamond SiR_2$$

$$\underset{X \quad\; Li}{R_2Si-CR_2} \quad \xrightarrow{- LiX} \quad R_2Si=CR_2 \quad \longrightarrow \quad 1/2 \quad R_2Si \underset{C}{\overset{C}{\diamond}} SiR_2$$

$$\underset{Cl \quad\;\; Li}{R_2Si-CHCH_2tBu} \quad \xrightarrow{- LiCl} \quad R_2Si=CHCH_2tBu \quad \longrightarrow \quad 1/2 \quad R_2Si \diamond SiR_2$$

$$\uparrow LitBu$$

$$R_2(Cl)Si-CH=CH_2 \qquad\qquad R = organo\ group; \quad X = halogen$$

$$\begin{array}{c} | \\ -Si-Y \\ | \quad | \\ ML_n \end{array} \quad \xleftarrow[+ ML_n]{complex\ formation}$$

$$Me_3N \rightarrow \overset{|}{\underset{|}{Si}}=Y \quad \underset{+ Me_3N\ (Y \neq Si)}{\overset{base\ addition}{\rightleftharpoons}}$$

$$\begin{array}{c} | \\ -Si-Y \\ | \quad | \\ HO \quad H \end{array} \quad \xleftarrow[+ HO\text{-}H]{insertion}$$

$$\begin{array}{c} | \\ -Si-Y \\ | \quad | \\ H_2C \quad\;\; H \\ \; \\ HC=CH_2 \end{array} \quad \xleftarrow[+ H_2C=CH\text{-}CH_3]{ene\text{-}reaction}$$

$$\boxed{\overset{|}{\underset{|}{Si}}=Y}$$

$$\xrightarrow[+ R_2C=N=N]{[2+1]\ cycloadd.} \quad \begin{array}{c} | \\ -Si-Y \\ | \quad | \\ N \\ \;\; N=CR_2 \end{array}$$

$$\xrightarrow[+ O=CPh_2]{[2+2]\ cycloadd.} \quad \begin{array}{c} | \\ -Si-Y \\ | \quad | \\ O-CPh_2 \end{array}$$

$$\xrightarrow[+ R_3SiN=N=N]{[2+3]\ cycloadd.} \quad \begin{array}{c} | \\ -Si-Y \\ | \quad | \\ R_3SiN \quad N \\ \;\; N \end{array}$$

$$\xrightarrow[+ H_2C=CH\text{-}CH=CH_2]{[2+4]\ cycloadd.} \quad \begin{array}{c} | \\ -Si-Y \\ | \quad | \\ H_2C \quad CH_2 \\ \; \\ HC=CH \end{array}$$

Investigations during the past years show that unsaturated silicon compounds are stabilized by protection of the p_π-p_π bond with very bulky substituents against consecutive reactions. Thus, metastable Si=Y-π-compounds (Y = Si, C, N, P, As, S, Se) could be fully charaterized by NMR and/or X-Ray crystallography under normal conditions in condensed phase or in the solid state. Some examples are given below.

$$\text{Mes}_2\text{Si}=\text{SiMes}_2 \qquad \text{Dsi}_2\text{Si}=\text{SiDsi}_2 \qquad \text{Tbt(Mes)Si}=\text{Si(Mes)Tbt}$$

four-membered Si ring with $t\text{BuMe}_2\text{Si}$ substituents

$$(\text{Me}_3\text{Si})_2\text{Si}=\text{C(OSiMe}_3)(\text{CEt}_3) \qquad \text{Me}_2\text{Si}=\text{C(SiMe}t\text{Bu}_2)(\text{SiMe}_3)$$

$$(\text{Me}_3\text{Si})(\text{Me}_2t\text{BuSi})\text{Si}=\text{C (adamantyl)} \qquad \text{Mes}_2\text{Si}=\text{C(CH}_2\text{--}t\text{Bu})(\text{H})$$

$$t\text{Bu}_2\text{Si}=\text{N--Si}t\text{Bu}_3 \qquad \text{Mes}_2\text{Si}=\text{P--Mes*} \qquad \text{Is}_2\text{Si}=\text{As--Si}i\text{Pr}_3 \qquad \text{Is(Tbt)Si}=\text{S} \qquad \text{Is(Tbt)Si}=\text{Se}$$

Me	=	CH_3	
Et	=	C_2H_5	
iPr	=	CHMe_2	
tBu	=	CMe_3	
Mes	=	$2,4,6\text{-Me}_3\text{C}_6\text{H}_2$	

Is	=	$2,4,6\text{-}i\text{Pr}_3\text{C}_6\text{H}_2$
Mes*	=	$2,4,6\text{-}t\text{Bu}_3\text{C}_6\text{H}_2$
Dsi	=	$(\text{Me}_3\text{Si})_2\text{CH}$
Tbt	=	$2,4,6\text{-(Dsi)}_3\text{C}_6\text{H}_2$

$$R^1R^2\text{Si}=\text{C}=\text{C (biphenyl } R^3,R^4,R^5)$$

$$R^1R^2\text{Si}=\text{C}=\text{C}R^3R^4$$

R^1 / R^2	$R^3 / R^4 / R^5$
Mes* / ^{1}ad	iPr / OMe / H
Mes* / tBu	Me / Me / Me
tBu / tBu	H / H / H

R^1 / R^2	R^3 / R^4
Is / Is	Ph / tBu
Mes* / tBu	Ph / tBu
tBu / tBu	Ph / Ph

Furthermore, surprisingly stable 1-silaallenes $R_2Si=C=CR_2$ [1] and even a tetrasilabutadiene $Is_2Si=SiIs–SiIs=SiIs_2$ [2] could be isolated and analyzed by single crystal X-ray analysis. From these investigations it is obvious that the Si=Y-π-bond is roughly shortened by about 10% compared to the corresponding Si-Y single bonds.

Y	C	Si	N	P	As	O	S	Se
d (Si–Y)[a] [pm]	187	235	172	225	236	163	214	228
d (Si=Y)[a] [pm]	170	215	157	206	216	151[*]	194	209[*]

*) estimated values

a) the given values deviate by ~± 5 pm depending on the substituent pattern at Si and Y

Although there is no question about the kinetic stabilization of Si=Y-π-bonds the extent of a thermodynamic stabilization is still an ongoing debate. Silabenzenes were produced by pyrolysis of suitable precursors and have been characterized in the gas phase by PE spectroscopy; they dimerize at comparably low temperatures, even with bulky *tert*-butyl groups at silicon (~-100°C). On the other hand only very recently a 2-silanaphthalene was characterized by X-ray crystallography [3].

While the parent silene $H_2Si=CH_2$ and its methyl substituted derivatives $Me(H)Si=CH_2$ and $Me_2Si=CH_2$ show an extraordinary high reactivity and were only characterized in the gas phase (life time for $H_2Si=CH_2$ at r.t. 30±2 ms [4]), these silenes could be stabilized by reducing the silicon Lewis acidity by coordination of dimethylether as a Lewis base. In consequence, the donor adducts $R^1R^2Si=CH_2 \cdot O(CD_3)_2$ (R^1, R^2 = H, Me; R^1 = H, R^2 = Me) have been detected at about -130°C by NMR spectroscopic methods [5]. A very similar donor coordination stabilizes the monomeric silanone $tBu_2Si=O \cdot thf$; donor free $R_2Si=O$-species, the monomer building units for silicones, were not found so far.

Silyl Cations

The synthesis and the structural analysis of silicon unsaturated compounds impressively demonstrate that tricoordinate silicon species are able to exist under 'normal' conditions. Since more than 50 years organosilicon research tries to answer the question if this might be true for silyl cations R_3Si^+, in which silicon also is surrounded by three neighbors. Those compounds seem to meet the requirements because silicon is more electropositive than the carbon homologue and therefore should accommodate a positive charge more efficiently. And in fact, calculations confirm that $[H_3Si]^+$ is thermodynamically much more stable than $[H_3C]^+$.

In good experimental agreement silyl cations are formed in the gas phase without any problems. In addition, they were proposed to participate in reactions of organosilicon compounds as plausible, but highly reactive and short lived species. However, there was - until very recently - no clear spectroscopic proof for the existance of R_3Si^+ type derivatives in the condensed phase or in the solid state. The obvious reason for that is found in the extremely strong desire of silyl cations to react with Lewis bases to complete the silicon's coordination sphere. This tendency is proved by the identification of numerous stable, but highly coordinate silyl cations.

The identification of an almost unlimited stable silanorbornyl cation shows that in principle a C=C double bond might be nucleophilic enough to π-stabilize a Si^+ center [6].

Numerous investigations, especially by the groups of Lambert, Olah and Reed demonstrate that the verification of a silyl cation requires counteranions and solvents with a nucleophilicity as low as possible and bulky substituents at the silicon center. For thermodynamic stabilisation of silyl cations amino-, alkylthio-, or alkoxy- substituents NR_2, SR or OR are recommended, using their ability for back bonding into the empty p orbital at silicon.

Based on these synthetic strategies the following compounds were synthesized and published during the past years; their physical properties and spectroscopic parameters were, however, discussed very controversially, and, the skeptics seem to be right:

$$R_3Si^+ \; OClO_3^- \qquad vs. \qquad R_3Si\text{---}OClO_3 \qquad (R = Ph, MeS, EtS, iPrS)$$

$$Et_3Si^+ \; [B(C_6F_5)_4]^- \; / \; toluene \qquad vs. \qquad \left[Me\text{---}\underset{H}{\overset{SiEt_3}{\bigoplus}} \right] [B(C_6F_5)_4]^-$$

The interpretation of physical and spectroscopic features shows the 'non positively' charged organosilicon compounds to be the true products; in the case of Reeds compounds $[R_3Si]^+[CB_9Br_5H_5]^-$ and $[R_3Si]^+[CB_{11}X_6H_6]^-$ (X = Cl, Br, I) the silicon 'cation' coordinates to the halogen substituent of the carborane anion (Si^+ character: 55 - 70%). Obviously the three ethyl groups in $[Et_3Si\cdot toluene]^+$ are too small to protect the Si^+ center from nucleophilic solvent attack. But the replacement by mesityl substituents, using Mes_3SiH as precursor for silyl cation formation, prevents the cleavage of the Si-H bond by trityl cations as reaction partner; evidently these groups are too bulky. The synthetic detour, which - at last - led to the verification of a pretty stable Mes_3Si^+ by Lambert is shown in the following reaction scheme [7]:

The ^{29}Si NMR signal of Mes$_3$Si$^+$ at 225.5 ppm (in C$_6$D$_6$) mainly proves that now the positive charge is located at the silicon center. Instead, for the silanorbornyl cation a ^{29}Si chemical shift of 87.7 ppm is recorded in d$_8$-toluene. Running the NMR spectra of both species in acetonitrile as solvent (δ ^{29}Si = 37.0 and 35.9 ppm, respectively), it is elegantly demonstrated that this nucleophile is - in contrast to aromatic hydrocarbons - small enough to coordinate at the Si$^+$ centers. As expected the positive charges are then transferred from silicon to nitrogen.

The developments in the silicon cation field during the past years impressively show that long term activities in the synthesis and characterization of these compounds have come to a successful end. Now the synthetic and catalytic potential of these highly Lewis acidic species has to be explored.

This concluding statement might be transferred without any restrictions to the history of the development of the silylene chemistry, the sila analogs of the carbenes :CR$_2$. Again, there are main milestones describing the efforts in this area: first, silylenes were discussed as highly reactive intermediates in gas phase reactions of suitable silyl precursors, then followed by the spectroscopic characterization in the gas phase or in low temperature matrices. In a third period, these compounds were kinetically and/or thermodynamically stabilized and characterized in condensed phase and in the solid state, even by single crystal X-ray analysis.

The parent compound, H$_2$Si:, is formed reacting gaseous silicon (heating elemental silicon up to 1500-1700°C) with hydrogen. At liquid helium temperature it can be condensed and characterized spectroscopically to be a triplet silylene ^{3}SiH$_2$ which - in part - is transformed into singlet silylene, ^{1}SiH$_2$. The latter is metastabile at 10 K and shows a non-linear geometry (C$_{2v}$-symmetry; r$_{SiH}$ = 152 pm, $\measuredangle$ HSiH = 92.8°). Inorganic derivatives which have been investigated over some decades very intensively are especially the dihalogenated silylenes X$_2$Si: (X = F, Cl, Br; available from the pyrolysis of SiX$_4$ in the presence of elemental silicon), which under normal conditions are unstable. The same is true for organic derivatives R$_2$Si: (R = Me, Et etc.); again, these species were characterized in the gas phase and isolated in low temperature matrices. It should be mentioned that during the last years the academic and industrial interest was focussed on 'mixed' derivatives RXSi:, e.g. Me(Cl)Si:, because those compounds are discussed to play an important role in the course of the 'Direct Process' (silylene-catalyst interactions).

The following two schemes summarize synthetic methods for the synthesis of organosubstituted silylenes and describe their reactivity.

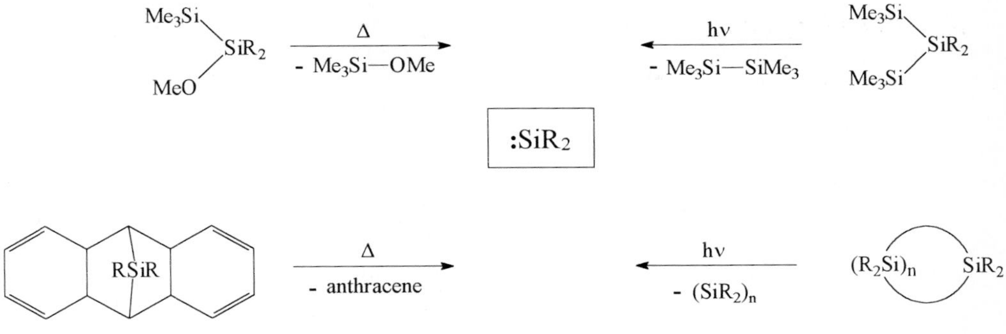

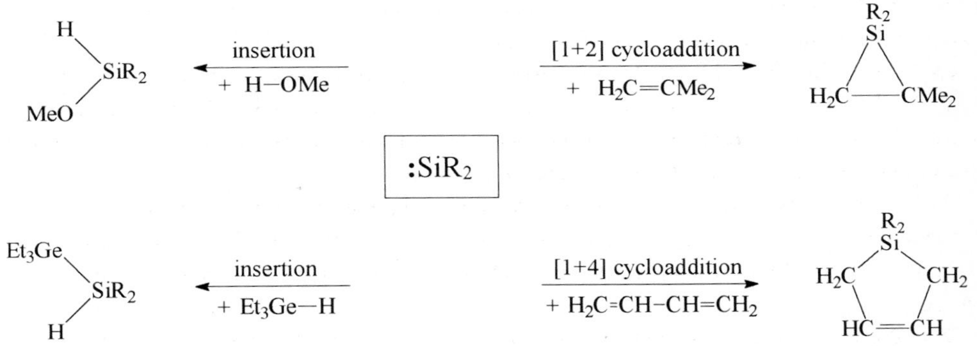

Organosubstituted silylenes are colored compounds which - in most cases - are only metastable against di-, oligo- or polymerization at low temperatures. This metastability again increases with the steric bulk at the silicon centers: Me_2Si: already polymerizes above 10 K to yield $(Me_2Si)_n$, Mes_2Si: requires temperatures of more than 77 K to dimerize to the disilene $Mes_2Si=SiMes_2$. Excellent research during the last years shows that even silylenes could be stabilized and analyzed by X-ray crystallography under normal conditions. These derivatives are listed below:

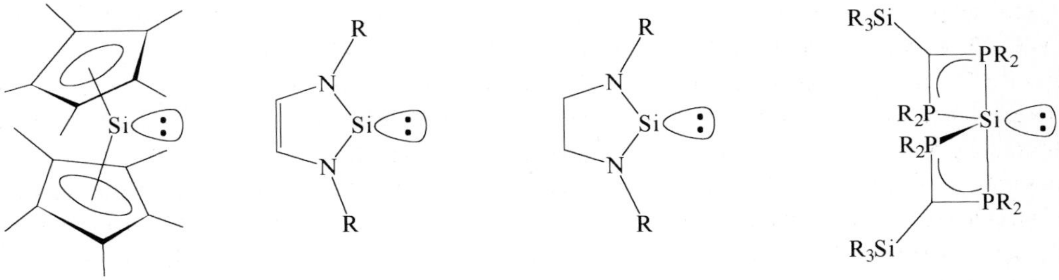

While silicocene is π-base stabilized by the pentamethylcyclopentadienyl substituents, the stabilization in the silicon analog of the Arduengo carbene is mainly caused by the C=C-π system including the lone pairs at nitrogen ('aromatic' 6π system). This is basically proved by the reduced stability of the carbon saturated diazasilacyclopentane. In contrast, in the tetraphosphasilaspirocyclus the silicon is stabilized by the free p lone pairs at phosphorus. These kinds of silylene stabilization require - very similar to donor stabilized compounds >Si=Y·D - the ability of coordination sphere expansion at silicon. This feature is not only limited to silicon in low coordination numbers but will be found even at tetravalent silicon centers. This is the reason that nucleophilic S_N2 substitution reactions at silicon occur rapidly with increase in the coordination number. This is basically shown by the facile hydrolysis of $SiCl_4$ with 'strong' silicon chlorine bonds, while the carbon analog CCl_4 does not react with water. In continuation to the modelling of the pathway of an S_N2 reaction, which was first performed over 15 years ago, only very recently the 'trajectory' of a bimolecular nucleophilic substitution reaction at silicon could be 'shown' by X-ray crystallographic investigations on a series of closely related five-coordinate complexes based on quinoline ligands. Each complex contains an internal 'nucleophile' which undergoes an intramolecular interaction with the silicon atom. Thus, each structure represents a 'frozen snapshot' of the substitution at a particular point on the modelled reaction profile.

X	F	Cl	Br	SO$_3$CF$_3$
d (Si-X) [pm]	168	232	265	276
d (Si-X)$_{typ}$ [pm]	156	203	217	150
d (O→Si) [pm]	206	194	185	174
% Si-X extension	29	67	109	126

To describe the degree to which the Si-X bond is 'stretched' by the incoming nucleophile, a percentage Si-X extension term might be defined: 0 % extension is that of a typical Si-X bond and 100 % extension is the sum of the crystal ionic radii for Si and X. As can be concluded from these data, the degree of Si-X bond stretching increases with strengthening nucleophilic attack. In the case of the triflate derivative the stretching is great enough that the Si-X distance is more characteristic of an inter-ion contact distance than a covalent interaction [8].

Generally the expansion of the coordination sphere at tetravalent silicon centers SiX_4 with electronegative substituents X (X = halogen, O, N etc.) occurs with donors D (D = halide ions, ethers, amines, water, phosphanes etc.) by the formation of cationic, anionic and neutral adducts $SiX_4·D$, $SiX_4·2D$ and even compounds of the composition $SiX_4·3D$ and $SiX_4·4D$. As expected, the silicon donor bond length increases with increasing coordination number at silicon. Exemplarily, this will be demonstrated for a very well investigated series of compounds with markedly different Si-N (donor)bonds (the number in brackets refers to the coordination number at silicon, the bond length is given in pm).

d $(Si^{(3)}=N)$	157
d $(Si^{(3)}\leftarrow N)$	199
d $(Si^{(4)}=N)$	157-162
d $(Si^{(4)}-N)$	164-178
d $(Si^{(5)}\leftarrow N)$	190-235
d $(Si^{(6)}\leftarrow N)$	260-280
d $(Si^{(7)}\leftarrow N)$	298-304
d $(Si^{(8)}\leftarrow N)$	290-312

From this table it can be concluded that - as mentioned before - even unsaturated tetravalent silicon compounds $X_2Si=Y$ (e.g. $X_2Si=NR\cdot D$) and $Y=Si=Y$ (e.g. $L_nM=Si=ML_n$, M = metal, L = ligand) and divalent species $:SiX_2$ will expand their coordination sphere, very often accompanied by a stabilization of the π subunit. Thus, from literature adducts of the types $X_2SiY\cdot D$, $SiY_2\cdot 2D$, $SiX_2\cdot D$ and $SiX_2\cdot 2D$ are known; their general structural features together with those of coordination numbers up to seven at silicon are listed below.

Summarizing this very basic introduction of current results and trends in organosilicon chemistry the authors avoided to cite the numerous original papers. Only those results published during the term of this year or being in press or in preparation are listed. All the other relevant literature is in the mean time collected in excellent reviews on the different topics of organosilicon chemistry. Some of those are given below to facilitate the reader to go deeper into this fast growing area of chemistry.
At last we want to thank Professors Dr. N. Wiberg and J. Y. Corey for the permission to cite their contributions to textbooks and to use some figures and formulae.

References

[1] M. Trommer, G. E. Miracle, D. R. Powell, R. West, *Organometallics* **1997**, in press.

[2] M. Weidenbruch, S. Wilhelm, W. Saak, G. Henkel, *Angew. Chemie* **1997**, *109*, in press.

[3] N. Tokitoh, K. Wakita, R. Okazaki, S. Nagase, P. v. Ragué Schleyer, H. Jiao, *J. Am. Chem. Soc.* **1997**, *119*, 6951.

[4] S. Bailleux, M. Bogey, J. Breidung, J. Demaison, H. Bürger, M. Senzlober, W. Thiel, F. Faigar, J. Pola, *J. Chem. Phys.* **1997**, *106*, 10016.

[5] N. Auner, J. Grobe, T. Müller, H. Rathmann, to be published.

[6] H.-U. Steinberger, T. Müller, C. Maercker, P. v. Ragué Schleyer, *Angew. Chemie* **1997**, *109*, 667; *Angew. Chemie Int. Ed. Engl.* **1997**, *36*, 626.

[7] J. B. Lambert, Y. Zhao, *Angew. Chemie* **1997**, *109*, 389; *Angew. Chemie Int. Ed. Engl.* **1997**, *36*, 400.

[8] A. R. Bassindale, D. J. Parker, P. G. Taylor, B. Herrschaft, N. Auner, to be published.

General Overviews:

H. Schmidbaur, *"Kohlenstoff und Silicium"*, *Chemie in unserer Zeit* **1967**, *1*, 184.

H. Bürger, *"Anomalien in der Chemie des Siliciums - wie ähnlich sind homologe Elemente"*, *Angew. Chemie* **1973**, *85*, 519; *Angew. Chemie Int. Ed. Engl.* **1973**, *12*, 474.

L. Vikov, L.S. Khaikin, *"Stereochemistry of Compounds Containing Bonds between Si, P, Cl, and N or O"*, *Topics Curr. Chem.* **1975**, *53*, 25.

R. Janoschek, *"Kohlenstoff und Silicium - wie verschieden können homologe Elemente sein?"*, *Chemie in unserer Zeit* **1988**, *21*, 128.

H. Bock, *"Grundlagen der Siliciumchemie: Molekülzustände Silicium enthaltender Verbindungen"*, *Angew. Chemie* **1989**, *101*, 1659; *Angew. Chemie Int. Ed.* **1989**, *28*, 1627.

J. Y. Corey, *"Historical overview and comparison of silicon with carbon"*, in: *"The chemistry of organic silicon compounds"* (Eds.: S. Patai and Z. Rappoport), John Wiley & Sons, **1989**, p. 1.

Y. Apeloig, *"Theoretical aspects of organosilicon compounds"*, in: *"The chemistry of organic silicon compounds"* (Eds.: S. Patai and Z. Rappoport), John Wiley & Sons, **1989**, p. 57.

W. S. Sheldrick, *"Structural chemistry of organic silicon compounds"*, in: *"The chemistry of organic silicon compounds"* (Eds.: S. Patai and Z. Rappoport), John Wiley & Sons, **1989**, p. 227.

R. J. P. Corriu, C. Guerin, J. E. Moreau, *"Dynamic stereochemistry at silicon"*, in: *"The chemistry of organic silicon compounds"* (Eds.: S. Patai and Z. Rappoport), John Wiley & Sons, **1989**, p. 305.

L. Birkofer, O. Stuhl, *"General synthetic pathways to organosilicon compounds"*, in: *"The chemistry of organic silicon compounds"* (Eds.: S. Patai and Z. Rappoport), John Wiley & Sons, **1989**, p. 655.

A. R. Bassindale, P. G. Taylor, *"Acidity, basicity and complex formation of organosilicon compounds"*, in: *"The chemistry of organic silicon compounds"* (Eds.: S. Patai and Z. Rappoport), John Wiley & Sons, **1989**, p. 809.

A. G. Brook, *"The photochemistry of organosilicon compounds"*, in: *"The chemistry of organic silicon compounds"* (Eds.: S. Patai and Z. Rappoport), John Wiley & Sons, **1989**, p. 965.

R. Tacke, H. Linoh, *"Bioorganosilicon Chemistry"*, in: *"The chemistry of organic silicon compounds"* (Eds.: S. Patai and Z. Rappoport), John Wiley & Sons, **1989**, p. 1143.

Divalent Silicon Compounds:

W. H. Atwell, D. R. Weyenberg, *"Zwischenverbindungen des zweiwertigen Siliciums"*, Angew. Chem. **1969**, *81*, 485; *Angew. Chem. Int. Ed. Engl.* **1969**, *8*, 469.

J. L. Margrave, P. W. Wilson, *"SiF₂, a Carbene Analogue. Its Reactions and Properties"*, Acc. Chem. Res. **1971**, *4*, 145.

J. L. Margrave, K. G. Sharp, P. W. Wilson, *"The Dihalides of Group IV B Elements"*, Fortschr. Chem. Forsch. **1972**, *26*, 1.

P. L. Timms, *"Low Temperature Condensation of High Temperature Species as a Synthetic Method"*, Adv. Inorg. Radiochem. **1972**, *14*, 121.

H. Bürger, R. Eujen, *"Low-Valent Silicon"*, Topics Curr. Chem. **1974**, *50*, 1.

C. S. Liv, T.-L. Hwang, *"Inorganic Silylenes. Chemistry of Silylene, Dichlorsilylene and Difluorsilylene"*, Adv. Inorg. Radiochem. **1985**, *29*, 1.

R. West, M. Denk, *Pure Appl. Chem.* **1996**, *68*, 785.

Trivalent Silicon Compunds, Silyl Cations:

R. J. P. Corriu, M. Henner, *"The Siliconium Ion Question"*, J. Organomet. Chem. **1974**, *74*, 1.

J. B. Lambert, W. J. Schulz, Jr., *"Trivalent silyl ions"*, in: *"The chemistry of organic silicon compounds"* (Eds.: S. Patai and Z. Rappoport), John Wiley & Sons, **1989**, p. 1007.

P. D. Lickiss, *J. Chem. Soc. Dalton Trans.* **1992**, 1333.

J. B. Lambert, L. Kania, S. Zhang, *Chem. Rev.* **1995**, *95*, 1191.

C. Maercker, J. Kapp, P. v. Ragué Schleyer, *"The Nature of Organosilicon Cations and Their Interactions"*, in: *Organosilicon Chemistry II – From Molecules to Materials* (Eds.: N. Auner, J. Weis), VCH, Weinheim, **1996**, p. 329.

P. v. Ragué Schleyer, *"Germanyl and Silyl Cations – Free at last"*, Science **1997**, *275*(5296), 39.

J. Belzner, *"Frei oder nicht frei , das ist die Frage – Silyl- und Germylkationen in kondensierter Phase"*, Angew. Chemie **1997**, *109*, 1330; *Angew. Chemie Int. Ed. Engl.* **1997**, *36*, 1277.

Unsaturated Silicon Compounds:

L. F. Gusel`nikov, N. S. Nametkin, *"Formations and Properties of Unstable Intermediates Containing Multiple $p_\pi p_\pi$ Bonded 4 B Metals"*, Chem. Rev. **1979**, *79*, 529.

P. Jutzi, *"Die klassische Doppelbindungsregel und ihre vielen Ausnahmen"*, Chemie in unserer Zeit **1981**, *15*, 149.

N. Wiberg, *"Unsaturated Compounds of Silicon and Group Homologues. Unsaturated Silicon and Germanium Compounds of the Types $R_2E=C(SiR_3)_2$ and $R_2E=N(SiR_3)$ (E=Si, Ge)"*, J. Organomet. Chem. **1984**, *273*, 141.

G. Raabe, J. Michl, *"Multiple Bonding to Silicon"*, Chem. Rev. **1985**, *85*, 419.

A. G. Brook, K.M. Baines, *"Silenes"*, Adv. Organomet. Chem. **1986**, *25*, 1.

R. West, *"Chemie der Silicium-Silicium-Doppelbindung"*, Angew. Chem. **1987**, *99*, 1231; Angew. Chem. Int. Ed. Engl. **1987**, *26*, 1201.

G. Raabe, J. Michl, *"Multiple bonds to silicon"*, in: *"The chemistry of organic silicon compounds"* (Eds.: S. Patai and Z. Rappoport), John Wiley & Sons, **1989**, p. 1015.

R. S. Grev, *"Structure and Bonding in the Parent Hydrides and Multiply Bonded Silicon and Germanium Compounds: from MH_n to $R_2M=MR_2$ and $RM\equiv MR$"*, Adv. Organomet. Chem. **1991**, *33*, 125.

T. Tsumuraya, S. A. Batcheller, S. Masamune, *"Verbindungen mit SiSi-, GeGe- und SnSn-Doppelbindungen sowie gespannte Ringsysteme mit Si-, Ge-, und Sn-Gerüsten"*, Angew. Chem. **1991**, *103*, 916; Angew. Chem. Int. Ed. Engl. **1991**, *30*, 902.

M. Drieß, *"Mehrfachbindungen zwischen schweren Hauptgruppenelementen?"*, Chemie in unserer Zeit **1993**, *27*, 141.

Reviews on multiple bonded silicon compounds Si=Y *see:* Adv. Organomet. Chem. **1995**, *39*, (Eds. R. West, F. A. Stone).

Silanes:

E. Hengge, *"Siloxen und schichtförmig gebaute Siliciumverbindungen"*, Fortschr. Chem. Forsch. **1967**, *9*, 145.

G. Schott, *"Oligo- und Polysilane und ihre Derivate"*, Fortschr. Chem. Forsch. **1967**, *9*, 60.

B. J. Aylett, *"Silicon Hydrides and their Derivatives"*, Adv. Inorg. Radiochem. **1968**, *11*, 249.

G. Urry, *"Systematic Synthesis in Polysilane Series"*, Acc. Chem. Res. **1970**, *3*, 306.

J. E. Drake, Ch. Riddle, *"Volatile Compounds of Hydrides of Silicon and Germanium with Elements of Group V and VI"*, Quart. Rev. **1970**, *24*, 263.

E. Wiberg, E. Amberger, *Hydrides of the Elements of Main Group I-IV*, Elsevier, Amsterdam, **1971**.

A. Weiß, G. Beil, H. Meyer, *"The Topochemical Reaction of $CaSi_2$ to a Two-Dimensional Subsiliceous Acid $Si_6H_3(OH)_3$ (= Kautzky`s Siloxene)"*, Z. Naturforsch. **1979**, *34b*, 25.

A. Sekiguchi, H. Sakurai, *"Cage and Cluster Compounds of Silicon, Germanium and Tin"*, Adv. Organomet. Chem. **1995**, *37*, 1.

Halogen Substituted Compounds of Silicon:

E. Hengge, *"Inorganic Silicon Halides"*, in: *Halogen Chemistry Vol. II* (Ed.: V. Gutmann), Academic Press, London **1967**, *2*, 169.

M. Schmeisser, P. Voss, *"Darstellung und chemisches Verhalten von Siliciumsubhalogeniden"*, *Fortschr. Chem. Forsch.* **1967**, *9*, 165.

D. Naumann, *Fluor und Fluorverbindungen*, Steinkopff, Darmstadt, **1980**, p. 78.

Substitution Mechanisms at Si Centers:

L. H. Sommer, *Stereochemistry, Mechanism and Silicon*, McGraw-Hill, New York, **1965**.

R. J. P. Corriu, C. Guerin, *"Nucleophilic Displacement at Silicon. Stereochemistry and Mechanistic Implications"*, *J. Organomet. Chem.* **1980**, *198*, 231.

R. J. P. Corriu, C. Guerin, *"Nucleophilic Displacement at Silicon: Recent Developments and Mechanistic Implications"*, *Adv. Organomet. Chem.* **1982**, *20*, 265.

A. R. Bassindale, P. G. Taylor, *"Reaction mechanism of nucleophilic attack at silicon"*, in: *"The chemistry of organic silicon compounds"* (Eds.: S. Patai and Z. Rappoport), John Wiley & Sons, **1989**, p. 839.

J. Chojnowski, W. Stanczyk, *"Dissoziative Pathways in Substitution: Silicon Cations R_3Si^+, $R_3Si^+ \leftarrow Nu$, and Silene-Type Species $R_2Si=X$ as Intermediates"*, *Adv. Organomet. Chem.* **1990**, *30*, 243.

Hypervalent Silicon Compounds:

S. N. Tandura, M. G. Vorokov, N. V. Alekseev, *"Molecular and Electronic Structure of Penta- and Hexacoordinate Silicon Compounds"*, *Topics Curr. Chem.* **1986**, *131*, 99.

P. G. Harrison, *"Silicon, Germanium, Tin and Lead"*, *Comprehensive Coord. Chem.* **1987**, *3*, 183.

C. Cumit, R. J. P. Corriu, C. Reye, J. C. Young, *"Reactivity of Penta- and Hexacoordinate Silicon Compounds and their Role as Reaction Intermediates"*, *Chem. Rev.* **1993**, *93*, 1371.

R. J. P. Corriu, J. C. Young, *"Hypervalent silicon compounds"*, in: *"The chemistry of organic silicon compounds"* (Eds.: S. Patai and Z. Rappoport), John Wiley & Sons, **1989**, p. 1241.

Silyl and Silylen Metal Complexes:

F. Höfler, *"Silicon-Transition-Metal Compounds"*, *Topics Curr. Chem.* **1974**, *50*, 129.

J. L. Speier, *"Homogenous Catalysis of Hydrosilation by Transition Metals"*, *Adv. Organomet. Chem.* **1979**, *17*, 407.

B. J. Aylett, *"Some Aspects of Silicon-Transition-Metal Chemistry"*, *Adv. Inorg. Radiochem.* **1982**, *25*, 1.

W. Petz, *"Transition Metal Complexes with Derivatives of Divalent Silicon, Germanium, Tin, and Lead as Ligand"*, *Chem. Rev.* **1986**, *86*, 1019.

P. G. Harrison, T. Kikabbai, *"Silicon, Germanium, Tin and Lead"*, *Compr. Coord. Chem.* **1987**, *2*, 15.

T. D. Tilley, *"Transition-metal silyl derivatives"*, in: *"The chemistry of organic silicon compounds"* (Eds.: S. Patai and Z. Rappoport), John Wiley & Sons, **1989**, p. 1415.

M. F. Lappert, R. S. Rowe, *"The Role of Group 14 Elements Carbene Analogues in Transition Metal Chemistry"*, *Coord. Chem. Rev.* **1990**, *100*, 267.

Ch. Zybill, *"The Coordination Chemistry of Low Valent Silicon"*, *Topics Curr. Chem.* **1991**, *190*, 1.

Ch. Zybill, H. Handwerker, H. Friedrich, *"Sila-Organometallic Chemistry on the Basis of Multiple Bonding"*, *Adv. Organomet. Chem.* **1994**, *36*, 229.

Silicides:

B. Atonsson, T. Lundström, S. Rundquist, *Borides, Silicides, Phosphides*, Methuen, London, **1969**.

More Compelling Evidence That Silicon is Better Than Carbon: The Thermal Isomerization of Olefins to Carbenes [1]

Thomas J. Barton*, Jibing Lin, Sina Ijadi-Maghsoodi, Martin D. Power, Xianping Zhang, Zhongxin Ma, Hideaki Shimizu, Mark S. Gordon

Department of Chemistry, Ames Laboratory (U.S. Department of Energy)
Iowa State University
Ames, Iowa 50011, U.S.A.
Tel.: Int. code + (515)294-2770 – Fax: Int. code + (515)294-4456
E-Mail: barton@ameslab.gov

Keywords: Rearrangement / Silicon / Carbene

Summary: Through a variety of thermolytic, kinetic, and theoretical studies, it is established that certain vinyl silanes can undergo thermally-induced isomerization without intermediacy of ß-silyl carbenes which actually turn out to be the transition state for the concerted vinylsilane-to-vinylsilane rearrangement. The situation is quite different for the all-carbon analogs where 1,2-alkyl migration produces a carbene considerably lower in energy than the transition state.

For more than 20 years we have been intrigued with the facility with which silicon undergoes various intramolecular migrations. This ability to bounce around on a molecular framework is certainly one of the endearing features which separates the chemistry of silicon from that of its more staid neighbour in the periodic chart, carbon. One of the more severe challenges to which silicon has been put is 1,2-migration over two carbons pi bonded to one another. Thus we have through the years investigated migrations of silyl acetylenes [2], silyl ketones [3], and silyl thioketones [4] and have accumulated considerable evidence that with sufficient thermal encouragement silicon will migrate in a 1,2-fashion over these pi frameworks (Eqs. 1 and 2).

$$R_3Si-C \equiv CR' \xrightarrow{\Delta} \quad :C = C \underset{R'}{\overset{SiR_3}{\diagup}}$$

Eq. 1. Ref. [2].

$$\underset{\substack{HSi \\ R_2}}{\overset{\substack{R_2 \\ HSi}}{\diagdown}} C = C = X \xrightarrow{\Delta} \quad \underset{\substack{HSi \\ R_2}}{\overset{\cdot\cdot}{C}} - C \underset{X}{\overset{\substack{R_2 \\ SiH}}{\diagup}}$$

Eq. 2. Ref. [3, 4].

The ultimate test would be to have silicon migrate on a simple, unstrained olefin (Eq. 3). With the exception of a few, very strained olefins, such as l-norbornene [5] (Eq. 4), the thermal rearrangement of an olefin to a carbene is unknown in carbon chemistry.

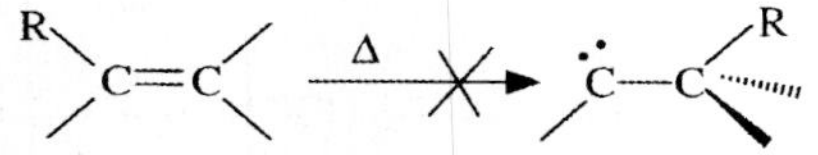

Eq. 3.

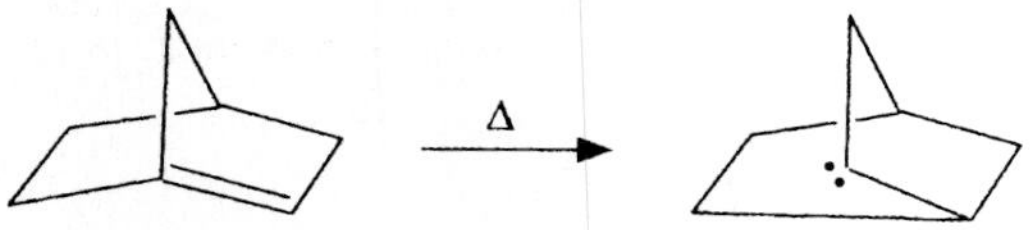

Eq. 4. Ref. [5].

However we rationalized that using silicon as a migrating group could result in a unique stabilization of the energy surface of olefin isomerization. This rationalization was based on silicon's well known ability to stabilize both α-carbanions and β-carbocations. Thus a hypothetical "dual-stabilized" zwitterion would be produced by a 90° twist of a vinyl silane, and a following 1,2-shift of Silicon would produce a singlet carbene possessed of the same hyperconjugative stabilization as in a β-silyl cation (Eq. 5).

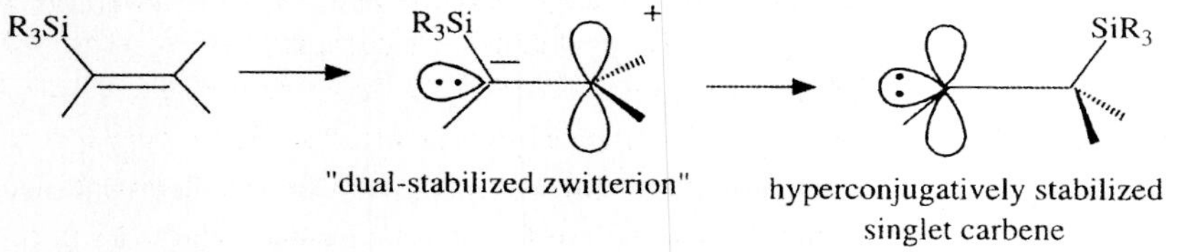

"dual-stabilized zwitterion"   hyperconjugatively stabilized
                                              singlet carbene

Eq. 5.

As satisfying as was this rationalization for the facile rearrangement of vinyl silanes to β-silyl carbenes, we unfortunately could produce no experimental results in our laboratory in need of this explanation despite our best efforts to provide enticing opportunities for this thermal rearrangement of vinyl silanes. Indeed in a last-ditch effort to add some validity to this picture of attached silicon aiding the breaking of a C=C double bond we undertook an extensive study of gas-phase, thermal cis-trans isomerizations of alkyl- and silyl-substituted olefins (Eqs. 6 and 7) which dissapointingly produced no rate enhancements that could not simply be accounted for by steric differences.

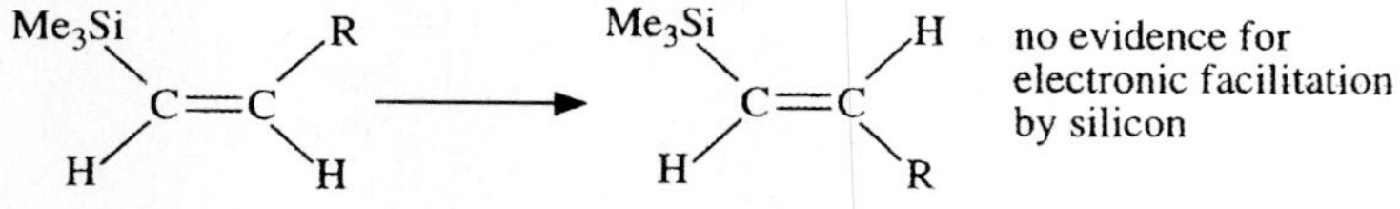

no evidence for
electronic facilitation
by silicon

Eq. 6.

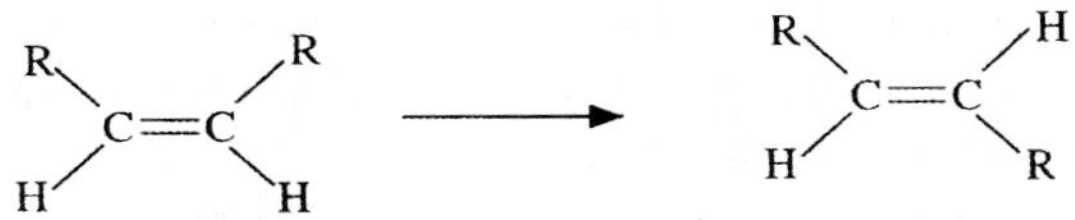

Eq. 7.

There was however a possible case of thermal migration of a silicon on a simple olefin in the largely overlooked 1985 report by Conlin [6] that α-methylenesilacyclobutane **1** cleanly isomerized to a mixture of silacyclopentenes **3** and **4** (Eq. 8). Although no mechanistic studies were performed, Conlin boldly proposed that this reaction proceeded through the intermediacy of carbene **2** which, if correct, would make this the sole example of a thermal isomerization of an untwisted olefin to a carbene.

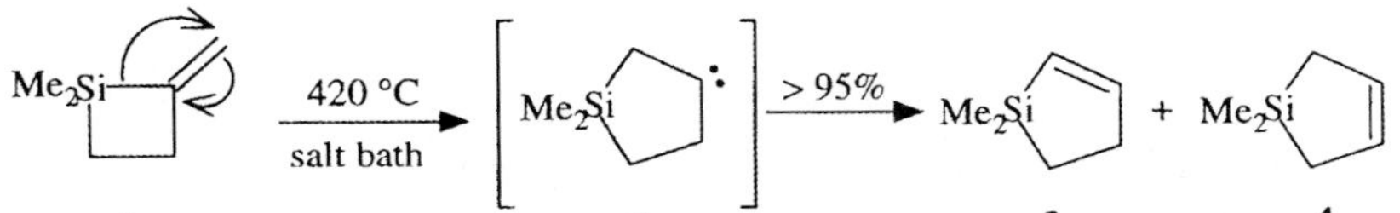

Eq. 8.

This report by Conlin raised a lot of questions in my mind, but these questions remained dormant until in a futile attempt to prepare α,α-silylenevinylene polymers we accidentally synthesized 1,3-di-methylene-1,3-disilacyclobutane **5**. This serindipitious synthesis allowed us to establish a two-case generality for the isomerization of methylenesilacyclobutanes since gas-phase pyrolysis of **5** cleanly and solely produced methylenedisilacyclopentene **7** (Eq. 9). This isomerization was kinetically followed in a stirred-flow reactor to afford Arrhenius parameters that clearly revealed this to be a concerted reaction. (Well, maybe not that clearly since you wouldn't bet your life on a log A of 12.5, but there certainly aren't any 54 kcal/mol sigma bonds in **5**.)

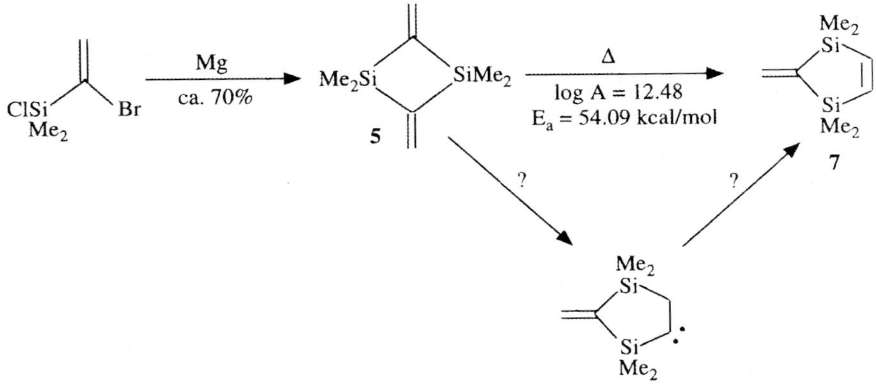

Eq. 9

In addition to the key question as to whether carbene **6** is an intermediate in this isomerization, was whether or not the ring expansion was unique to silicon systems. To answer this latter question we laboriously synthesized the all-carbon analog of **5**, **8**, and established that it does not perform the analogous thermal isomerization to cyclopentene **10** (Eq. 10). Indeed **8** does nothing until ca. 550°C where it begins an apparent molecular explosion into a myriad of products.

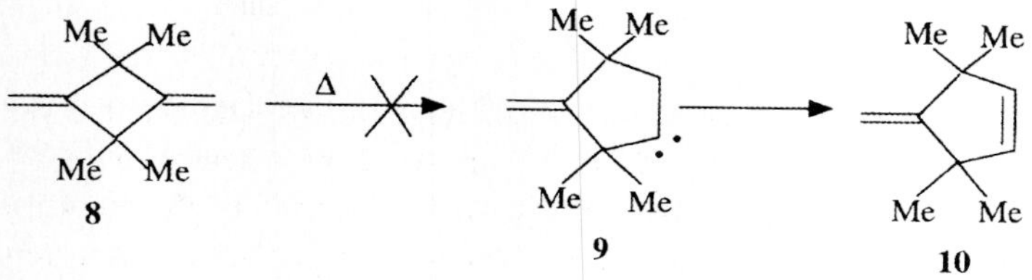

Eq. 10.

So why does silicon do it and carbon not? The answer is not found in the energetic differences between the two four-membered rings and their corresponding carbenes since *ab initio* calculations revealed that on this basis the isomerization of **8** to carbene **9** would be favoured by more than 10 kcal/mol (Eqs. 11 and 12).

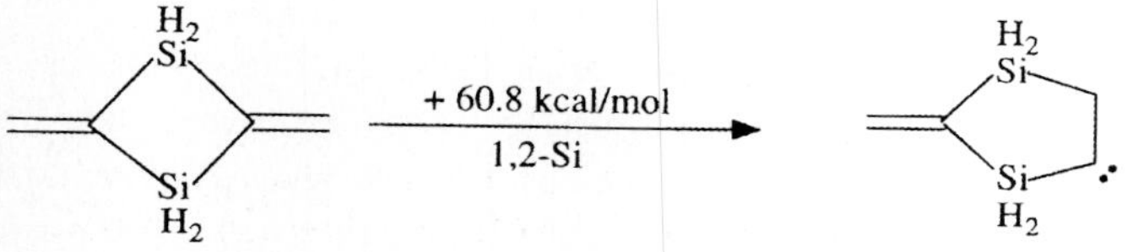

Eq. 11.

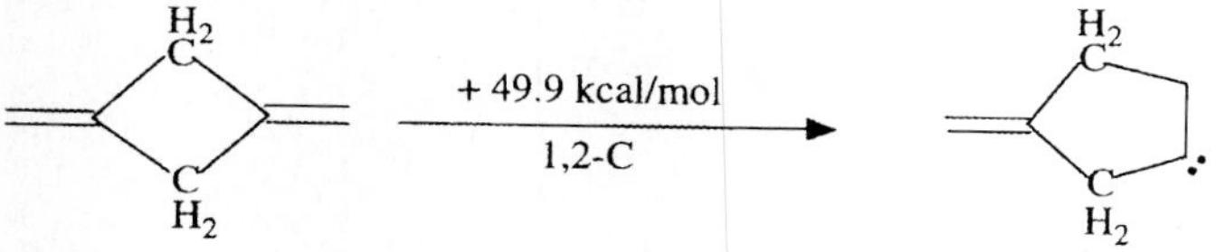

Eq. 12.

Likewise ring-strain arguments favoured rearrangement of **8** instead of **5** since (Eqs. 13 and 14) expanding the all-carbon ring to either the carbene or the ultimate cyclopentene released considerably more strain energy than for the silicon system.

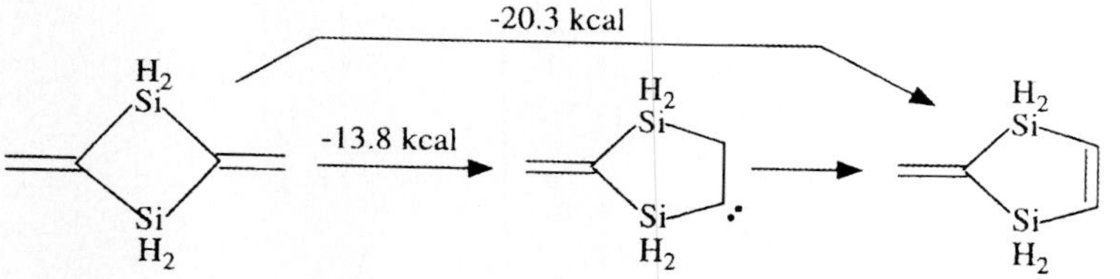

Eq. 13.

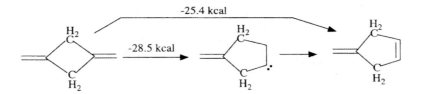

Eq. 14.

So what's going on? Everything says that the experimental results for **5** and **8** are reversed from what should be expected on energetic grounds. The answer comes from theoretical examination of the transition states for the two processes. At the HF/6-31G(d) level the transition state for the silacycle is slightly (<1 kcal) above the carbene but introduction of electron correlation to the MP2/6-31G(d) level reverses the relative energies so that the carbene becomes the transition state and isomerization from the 4- to the 5-membered ring is seen to be actually concerted (Fig. 1).

The situation is very different for the all-carbon system where even at the MP2 level of calculation the bridged transition state remains a considerable 24 kcal above the ring-expanded carbene (Fig. 1).

These results are not unique to these ring expansions as is clearly revealed in the theoretical results displayed in Eqs. 15 and 16 which also reveal that olefin-to-carbene isomerization by alkyl migration is an unreasonable expectation without the added benefit of some element of unusual strain.

HF/6/31G(d) (kcal/mol)	74.1	68.5
MP2/6-31G(d)	68.5	71.1
MP4/6-311G(d,p)	63.4	68.1

Eq. 15.

HF	91.3	70.7
MP2	88.6	84.8
MP4	84.9	78.8

Eq. 16.

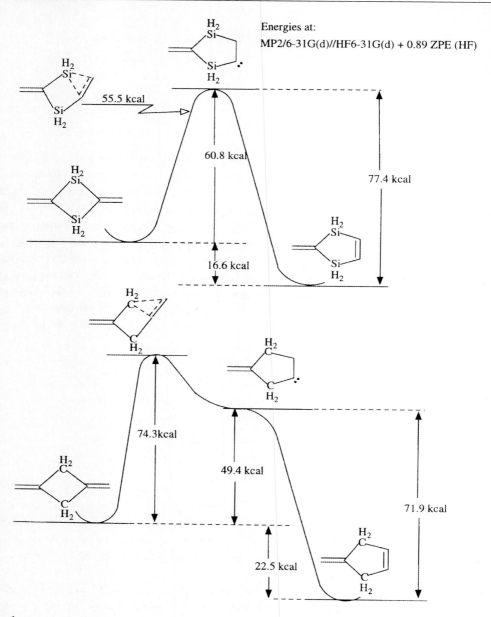

Fig. 1.

Finally we returned to the questions surrounding the isomerization of **1** to **3**. Gas-phase kinetic studies again revealed a concerted isomerization. Here we would base that conclusion more on the low value of log A (11.3) rather than E_{act} (47.5 kcal/mol) since it was not obvious that ring C-C homolysis would not be ca. 50 kcal/mol. Indeed homolysis of a ring C–C bond is what we would have predicted *a priori* for the first thermal event of **1** based on the established behaviour of dimethylsilacyclobutane [7] and methylenecyclobutane [8] (Eqs. 17 and 18).

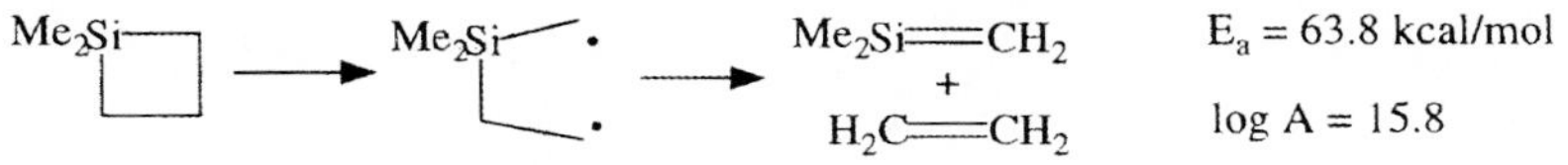

Me2Si
Me2Si
Me2Si=CH2
+
H2C=CH2
Ea = 63.8 kcal/mol
log A = 15.8

Eq. 17.

Ea = 63.3 kcal/mol
log A = 15.68

Eq. 18.

Of course it was possible that **1** actually was undergoing ring opening to a 1,4-biradical, but that this process was undetected since it was not proceeding on to the expected silene and allene. That this is indeed the case was established by deuterating **1** on the exocyclic carbon and observing a very clean equilibration of deuterium between the vinyl and allylic positions when thermolysis of **1-D** was carried out slightly below temperatures required for ring expansion to **3** (eqn. 19). This equilibration was so clean that it was possible to perform a kinetic analysis using ^{2}H NMR to obtain Arrhenius parameters completely consistent with a homolytic ring opening.

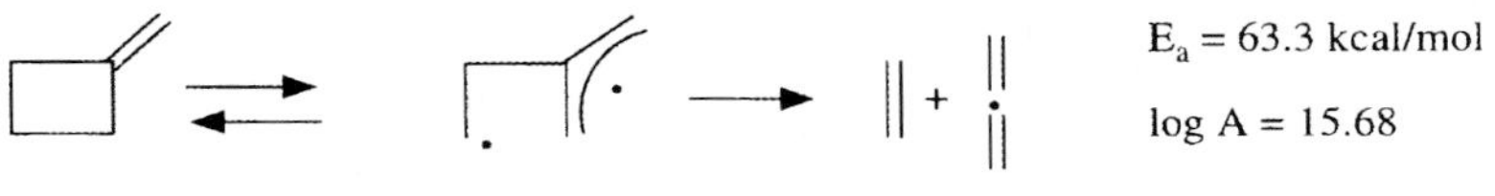

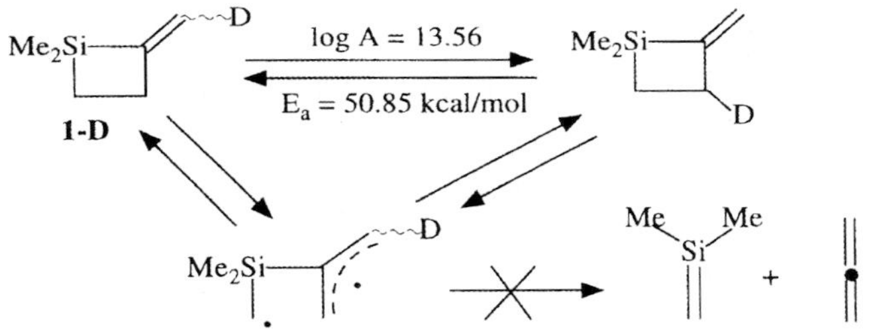

Me2Si
D
log A = 13.56
Ea = 50.85 kcal/mol
1-D
Me2Si
D
Me2Si
D
Me
Me
Si
+

Eq. 19.

Thus a nearly complete picture of the energetics of the thermolysis of **1** (Fig. 2), missing only the barrier of the unobserved decomposition of the 1,4-diradical to silene and allene, is now available. This picture provides a reminder with regard to the dangers of reliance on bond strengths and energies of activation while ignoring entropy change, in the prediction of thermal reactions - the lowest E_{act} is not always the best deal.

In closing I suppose an explanatory note on my semifacetious title is called for, if for no other reason than to smooth the feathers of my organic breathren. True beauty is indeed defined in the eyes of the beholder, and to this author the far more complex thermochemistry inevitably encountered for organosilicon compounds relative to their all-carbon analogues is beautiful. If mechanistic complexity enchants you not, the conclusions of the title reverse.

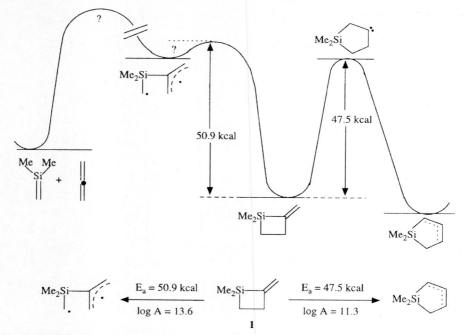

Fig. 2. Energetic profile of the thermal isomerizations of **1**.

Acknowledgments: Ames Laboratory is operated for the *U.S. Department of Energy by Iowa State University* under Contract No. W-7405-ENG-82. This work was supported by the *Director for Energy Research, Office of Basic Energy Sciences.* T. J. B. and M. S. G. would also like to acknowledge the *National Science Foundation* for their support of portions of this work. The calculations reported here were performed on IBM RS 6000 workstations generously provided by *Iowa State University.*

References:

[1] The work briefly presented herein is all described in far greater detail in: *J. Am. Chem. Soc.* **1995**, *117*, 11695.

[2] a) T. J. Barton, B. L. Groh, *Organometallics* **1985**, *4*, 575.
b) T. J. Barton, B. L. Groh, *J. Org. Chem.* **1985**, *50*, 158.

[3] T. J. Barton, B. L. Groh, *J. Am. Chem. Soc.* **1985**, *107*, 7221.

[4] T. J. Barton, G. C. Paul, *J. Am. Chem. Soc.* **1987**, *109*, 5292.

[5] T. J. Barton, M. H. Yeh, *Tetrahedron Lett.* **1987**, *28*, 6421.

[6] R. H. Conlin, H. B. Huffaker, Y. Kawk, *J. Am. Chem. Soc.* **1985**, *107*, 731.

[7] T. J. Barton, G. Marquardt, J. A. Kilgour, *J. Organomet. Chem.* **1974**, *85*, 317; C. M. Golino, R. D. Bush, P. On, L. H. Sommer, *J. Am. Chem. Soc.* **1975**, *97*, 8678; P. B. Valkovich, T. I. Ito, W. P. Weber, *J. Org. Chem.* **1974**, *39*, 3543; R. T. Conlin, M. Namavari, J. S. Chickos, R. Walsh, *Organometallics* **1989**, *8*, 168.

[8] J. P. Chesick, *J. Phys. Chem.* **1961**, *65*, 2170.

NMR and Quantum Chemical Characterization of Silicon-Substituted Carbocations

Hans-Ullrich Siehl*, Bernhard Müller
Abteilung für Organische Chemie
Universität Ulm
D-89069 Ulm, Germany
Tel./Fax: Int. code + (731)5022800
E-mail: ullrich.siehl@chemie.uni-ulm.de

Olga Malkina
Computing Center
Slovak Academy of Sciences
SK-84236 Bratislava, Slovakia

Keywords: Carbocations / Silyl effect / ^{13}C NMR / DFT Calculations

Summary: The protonation of E-1-*p*-anisyl-2-triisopropylsilyl-ethene with FSO₃H leads to the formation of *syn*- and *anti*-1-*p*-anisyl-2-triisopropylsilyl-ethyl cations. The experimental ^{13}C NMR chemical shifts and energy barrier for the *syn/anti*-isomerization are compared with ab initio DFT quantum chemical calculations of chemical shifts and energies.

Introduction

NMR spectroscopy has evolved as *the* most important experimental method for the direct study of structure and dynamics of carbocations in solution. Quantum chemical methods have developed as indispensable tools to complement experimental results. Despite great interest in ab initio calculations of experimentally observabed molecular properties, calculations of NMR chemical shifts and coupling constants have become routine only very recently. The development of the IGLO- and GIAO-methods for chemical shift calculations has been a major breakthrough in this area of quantum chemistry. The most recent developments are the implementation of electron correlation methods into the GIAO-method to achieve higher accuracy [1] and the application of density functional theory (DFT) methods for the GIAO- [2] and IGLO- [3] methods to allow the calculation of larger molecules.

A silyl substituent can stabilize the positive charge in a carbocation if the spatial arrangement allows overlap of the C–Si σ-bond and the vacant p-orbital at the C⁺-carbon. The hyperconjugative interaction of a β-silyl group, as shown in the MO- and VB-structures in Fig.1, is called the β-silyl effect.

ß-Trialkylsilyl-substituted carbocations are particularly important reactive intermediates in chemical reactions in polar media. Due to the high affinity of silicon for fluorine and oxygen, the facile formation of five-coordinate Si-intermediates or transition states, and the polar nature of the carbon-silicon bond, the C–Si bond is prone to facile cleavage. Silicon-substituted carbocations are, therefore, not generally accessible as persistent species in superacid solution.

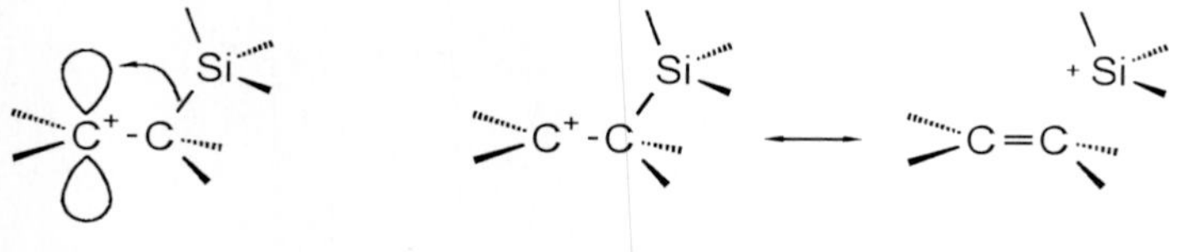

Fig. 1.   β-Silyl effect in MO representation and VB *no-bond* resonance structures.

The *β*-Silyl Effect in Benzyl Cations

The protonation of substituted styrenes generally leads to sequential oligomerization and polymerization. Recently, we have generated and characterized by NMR spectroscopy formerly elusive ß-trialkylsilyl α-aryl-substituted vinyl cations by protonation of suitable substituted alkynes using matrix co-condensation techniques [4, 5]. When similar carefully controlled experimental conditions for protonation are applied to silyl-substituted styrenes with bulky trialkylsilyl groups the corresponding β-silyl-substituted benzyl cations are formed as the only products observed by NMR spectroscopy at low temperature in solution [5]. Protonation of E-1-*p*-anisyl-2-triisopropylsilyl-ethene with FSO$_3$H yields a mixture of *syn*- and *anti*-1-p-anisyl-2-triisopropylsilyl-ethylcations (Fig. 2).

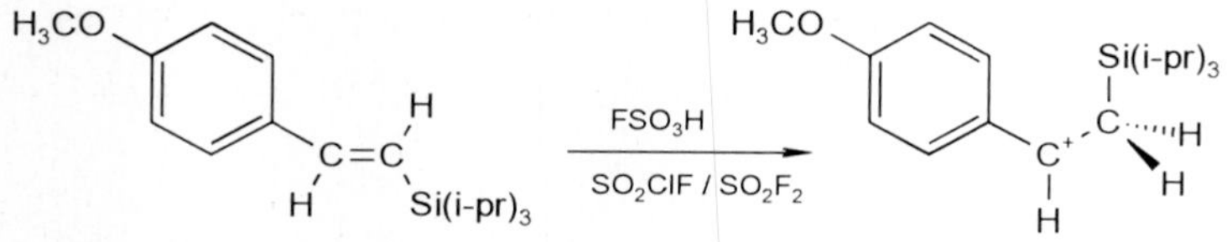

Fig. 2.   Formation of β-silyl-benzyl cations by protonation of silyl-substituted styrenes.

Below -120°C interconversion of the *syn*- and *anti*-isomers (Fig. 3) is slow on the NMR time scale. The two ortho carbons and the two meta carbons, respectively, are not equivalent in the *syn*- and *anti*-isomers. The aryl-methine signals form two sets of unequal intensity. Each set consists of four lines. One set is for the C$_2$ and C$_6$ *ortho* and C$_3$ and C$_5$ *meta* carbons in the *anti*-isomer and another set consists of signals of lower intensity, due to the C$_{2'}$ and C$_{6'}$ *ortho* and C$_{3'}$ and C$_{5'}$ *meta* carbons in the *syn*-isomer.

At higher temperature the 8 signals for the aromatic methine carbons show kinetic line broadening (Fig. 4). Upon warming above the coalescence temperature of about -110°C four lines for the aromatic methine carbons with decreasing line width are observed, until at a temperature of about -70°C fast decomposition takes place. From the line shape analysis of the dynamic NMR spectra the energy barrier for the isomerization is obtained as $\Delta G^{\neq} = 7.5$ kcal mol^{-1}.

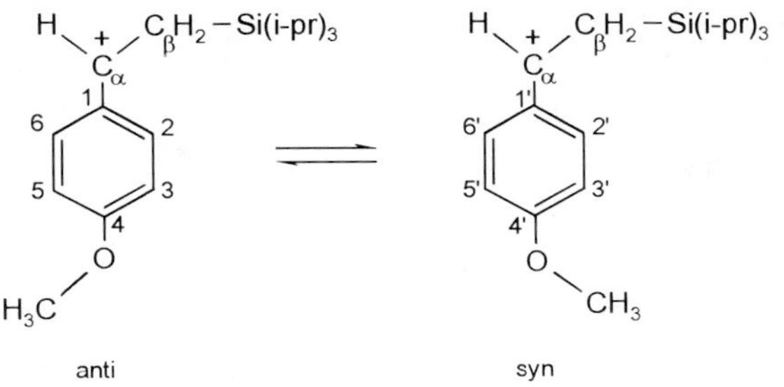

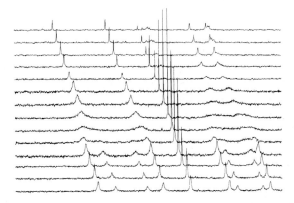

Fig. 3. Interconversion of *syn-* and *anti*-1-*p*-anisyl-2-triisopropylsilyl-ethylcations.

Fig. 4. 100 MHz ^{13}C NMR (aryl CH-signals, range 100-160 ppm) of *syn-* and *anti*-1-*p*-anisyl-2-triisopropylsilyl ethyl cations between -140 and -70°C.

Quantum chemical calculations with methods of density functional theory (DFT) were used for geometry optimization of the *syn-* and *anti*-1-*p*-anisyl-2-SiH$_3$ substituted ethyl cations, which serve as close models for the experimentally observed cations with alkyl groups at silicon. We have shown that varions alkyl groups at silicon have no significant impact on the structure and stabilization of α-aryl β-silyl-substituted carbocations. The structures were optimized using the B3LYP hybrid method with the 6-31G(d) basis set. The *syn* isomer is calculated to be 0.3 kcal mol^{-1} higher in energy than the *anti*-isomer. The optimized transition state for the *anti/syn*-isomerization is 10.3 kcal·mol^{-1} higher in energy than the *anti*-isomer. In the transition state structure the oxygen-methyl bond of the *p*-methoxy group is oriented perpendicular to the plane of the aryl ring and conjugation of the OCH$_3$ group with the aromatic π-system is not possible. The energy difference between the transition state and the minimum structures is a measure for the barrier of the methoxy group rotation around the C$_4$–oxygen bond. The calculated energy difference is in satisfactory agreement with the experimentally determined energy barrier of $\Delta G^{\neq} = 7.5$ kcal mol^{-1} (-100°C). The rotation around the C$_{ipso}$–C$^+$ bond in benzyl type cations is known to be higher in energy.

The energy barrier for C$_{para}$–OCH$_3$ rotation is a measure of the relative electron demand of the carbocation center. The better a β-substituent stabilizes the positive charge, the less delocalization of positive charge into the aryl ring is necessary and thus the lower the barrier for the rotation of the *para*-methoxy group. The corresponding rotational barriers in the parent 1-*p*-anisyl methyl cation, the 1-*p*-cation are $\Delta G^{\neq} = 12$; 10.6; 8.9, and 8 kcal mol^{-1}, respectively. The relative height of the rotational barriers in substituted 1-*p*-anisylmethyl cations (Fig. 5) shows that a –CH$_2$SiR$_3$ group at the C$^+$ carbon stabilizes the positive charge much better than an alkyl group.

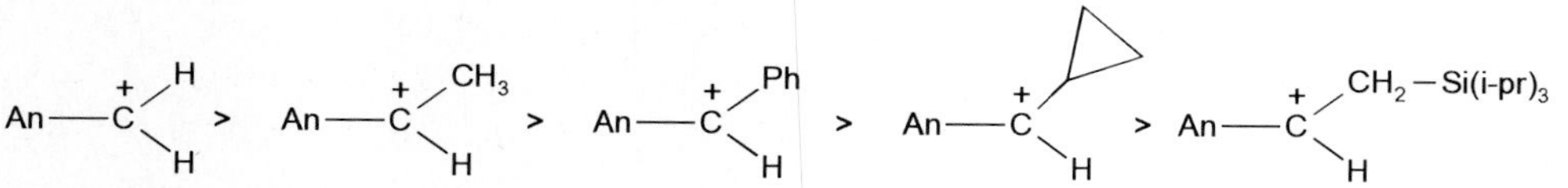

$$\text{An—}\overset{+}{\underset{H}{\overset{H}{C}}} \;>\; \text{An—}\overset{+}{\underset{H}{\overset{CH_3}{C}}} \;>\; \text{An—}\overset{+}{\underset{H}{\overset{Ph}{C}}} \;>\; \text{An—}\overset{+}{\underset{H}{\overset{\triangle}{C}}} \;>\; \text{An—}\overset{+}{\underset{H}{\overset{CH_2-Si(i\text{-}pr)_3}{C}}}$$

Fig. 5. Relative barriers for C$_{para}$–OCH$_3$ rotation in substituted 1-*p*-anisylmethyl cations.

β-Silyl stabilization in this type of benzyl cations is about as efficient as the effect of an α-phenyl group or an α-cyclopropyl ring. This is in accord with a comparison of the ^{13}C NMR chemical shifts of the *para*-carbon in 1-*p*-anisylmethyl cations An–C$^+$(H)R, with R = CH$_3$, CH$_2$CH$_3$, and CH$_2$Si(iPr)$_3$, which are $\delta_{Cpara} = 185.84$ ppm, 181.93 ppm, and 178.84 ppm, respectively.

Calculations of the NMR chemical shifts for the B3LYP/6-31G(d) geometry of the *syn*- and *anti*-1-*p*-anisyl-2-SiH$_3$-ethyl cations were performed with the GIAO-DFT method, using the BLYP functional with 6-31G(d) basis sets, the DFT/SCF hybrid method B3LYP with 6-31G(d,p) basis sets, and the SOS-DFPT(IGLO) approach with PW91 functional and IGLO III basis sets. The calculated NMR chemical shifts together with the experimental shifts for the *anti*- and the *syn*-isomer are shown in Table 1.

Table 1. Calculated and experimentally observed ^{13}C NMR chemical shifts for the *syn*- and *anti*-1-*p*-anisyl-2-triisopropylsilyl ethyl cation; calculated δ for TMS.

anti	C$_1$	C$^+$	C$_\beta$	C$_6$	C$_2$	C$_5$	C$_3$	C$_4$	OCH$_3$
GIAO BLYP/ 6-31G(d)[a]	120.8	185.7	29.0	142.0	127.9	105.5	115.6	167.3	53.5
GIAO B3LYP/ 6311G(d,p)[b]	136.9	206.5	37.1	158.8	144.0	118.9	131.5	187.1	61.8
PW91 IGLO III[c]	136.8	198.3	39.9	155.1	140.7	118.3	129.8	182.9	63.4
expt.[d]	132.2	205.3	43.2	152.2	138.1	115.0	123.9	178.8	58.9

syn	C$_1$	C$^+$	C$_\beta$	C$_{6'}$	C$_{2'}$	C$_{5'}$	C$_{3'}$	C$_4$	OCH$_3$
GIAO BLYP/ 6-31G(d)[a]	120.6	185.7	28.8	138.6	131.0	114.1	107.0	167.2	53.4
GIAO B3LYP/ 6311G(d,p)[b]	136.8	206.6	36.9	155.2	147.3	129.8	120.4	187.0	61.7
PW91 IGLO III[c]	136.6	198.34	38.8	151.7	143.5	128.1	119.9	183.1	63.2
expt.[d]	132.2	205.3	43.2	148.5	141.8	122.0	116.7	178.8	58.9

[a] 186.0 ppm – [b] 183.9 ppm – [c] 183.5 ppm – [d] 100 MHz, T = -126°C in SO$_2$ClF/SO$_2$F$_2$, external capillary TMS: δ = 0 ppm.

The deviations from experiment are quite large (Δ = +5 to +20 ppm) for the BLYP/6-31G(d) shift calculations. At B3LYP/6-311G(d,p) level of theory a better agreement is achieved, the maximum difference from the experimental values is -8.3 ppm. The SOS DFT IGLO III method yields, overall, the smallest deviations. The deviations are at least in part due to the approximation using a model structure. The calculated differences between the chemical shifts for the signals of the two isomers allow, however, unequivocal assignment of the experimental signals even at the lowest level of calculation. The good agreement between calculated and observed chemical shifts confirms the interpretation of the experimental data and gives confidence in the computational results. DFT-based ab initio calculations of chemicals shifts are feasible for this size of carbocations molecules and, as shown here, are very useful for signal assignment by interfacing experimental and computational investigations.

Variable Demand for β-Silicon Stabilization in Benzyl Cations

In β-silyl substituted benzyl cations with different α-aryl substituents the stabilizing effect of a β-silyl group depends on the electron donating ability of the α-aryl group. The better the α-aryl substituent can stabilize the positive charge, the lower is the need for β-σ-hyperconjugative stabilization by the silyl group and the *no-bond* resonance structure becomes less important (Fig. 6).

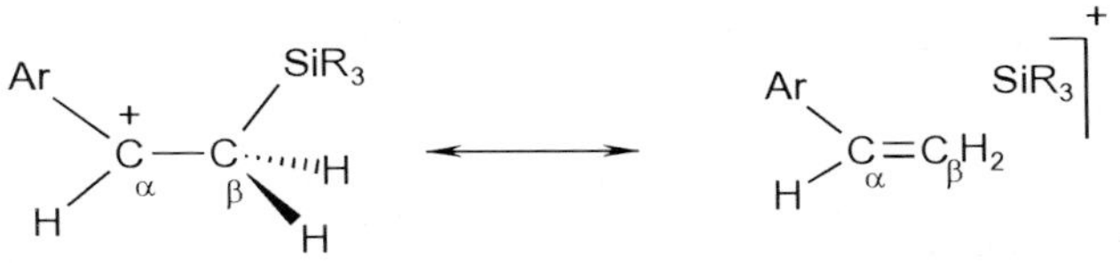

Fig. 6.   β-Silyl hyperconjugative stabilization in benzyl cations.

The ^{29}Si NMR chemical shifts in the α-phenyl-, α-tolyl-, α-*p*-anisyl-, and α-ferrocenyl-substituted β-silyl ethyl cations, shown in Fig. 7, are 66.34 ppm, 56.92 ppm, 38.88 ppm, and 23.48 ppm, respectively. ^{29}Si NMR shifts are thus suitable for monitoring the electron demand in β-silyl substituted carbocations.

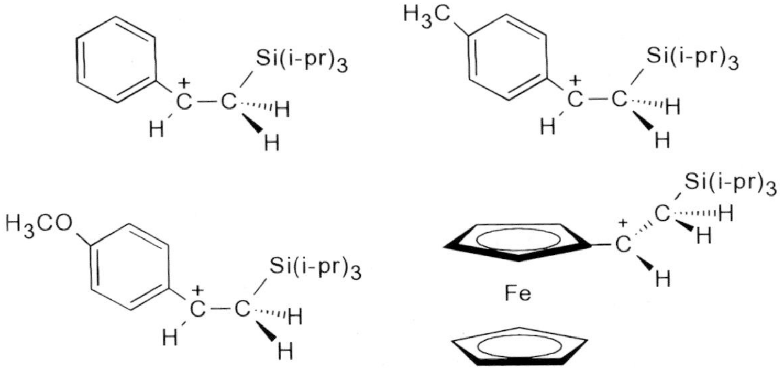

Fig. 7.   β-Silyl-substituted ethyl cations with various α-aryl groups.

Acknowledgement: This work was supported by the *Deutsche Forschungsgemeinschaft* and the *Fonds der Chemischen Industrie.* Financial support from the Slovak Grant Agency VEGA (grant No. 2/3008/97) is gratefully acknowledged.

References:

[1]   J. F. Stanton, J. Gauss, H.-U. Siehl, *Chem. Phys. Lett.* **1996**, *262*, 183; and references cited therein.

[2]   K. Wolinski, J. F. Hilton, P. Pulay, *J. Am. Chem. Soc.* **1990**, *112*, 8251.

[3]   V. G. Malkin, O. L. Malkina, M. E. Casida, D. R. Salahub, *J. Am. Chem. Soc.* **1994**, *116*, 5898.

[4]   a) H.-U. Siehl, in: *Stable Carbocation Chemistry* (Eds.: G. K. S. Prakash, P. v. R. Schleyer), Wiley, New York, **1997**, p. 165; and references therein.

       b) H.-U. Siehl, Pure Appl. Chem. **1995**, *67*, 769.

[5]   B. Müller, Ph.D. thesis, Universität Tübingen, **1995**.

Matrix Isolation Studies of the **Reactions** of Silicon Atoms

Günther Maier*, Hans Peter Reisenauer, Heiko Egenolf
Institut für Organische Chemie
Justus-Liebig-Universität Gießen
Heinrich-Buff-Ring 58, D-35392 Gießen, Germany
Fax: Int. code + (641)99-34309

Keywords: Matrix Isolation / Photoisomerization / Cocondensation

Summary: Evaporation of silicon atoms and consecutive cocondensation with a suitable gas in an argon matrix was used to generate silylenes. Their isolation, matrix-spectroscopic investigation and photochemical isomerization demonstrates the potential of matrix isolation spectroscopy and leads further onto the relatively new field of silylene rearrangements. Structural assignments for the observed species are aided by calculated IR spectra and thus are a good example for the combination of quantum chemical calculations and experimental work. Apart from that, many of the described species as well as their carbon analogues are assumed to play an important role in interstellar chemistry.

Introduction

Evaporation of metal atoms followed by condensation, possibly after reaction with another substrate, is the key step of chemical vapor deposition (CVD) processes, which today are wide-spread. If the metal used is silicon, this method leads to the basis of the data processing industry.

On a small scale, cocondensation of silicon atoms with a gaseous reactant in an argon matrix provides a facile access to silylenes. The versatility of this method has already been demonstrated by the isolation of fluorosilylene (**1**), hydroxysilylene (**2**), and the parent silylene, SiH_2 (**3**), by Margrave and co-workers [1].

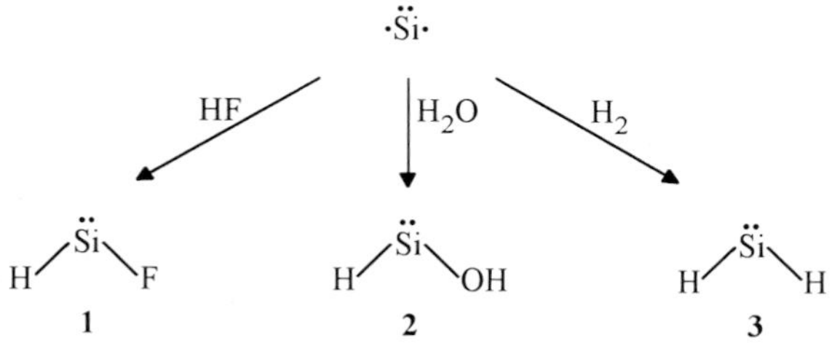

Scheme 1.

Among the cyclic silylenes **4-7**, SiC$_2$ (**4**) [2] and 1-silacyclopropenylidene (**5**) [3] have already been investigated by matrix spectroscopy. This series is now completed by silylene **6**, the formal hydrogenation product of **5**. We also report on the isolation of **7**, which can be derived from **5** by formal substitution of one CH unit by a nitrogen atom.

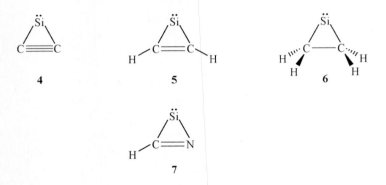

Fig. 1.

Silicon is vaporized from a boron nitride crucible at temperatures between 1450 and 1550°C. Resistance heating is applied by means of a tungsten wire wound around the crucible. The silicon atoms are cocondensed with a gaseous mixture of the substrate and argon of varying concentrations (the best results are achieved with ratios of 1:100-200) onto a CsI window at 10 K. When cocondensed with "pure" argon, the high reactivity of the silicon atoms can be seen in the formation of products such as SiH$_4$, Si$_2$H$_6$, SiNN, SiO, SiH$_2$ and presumably even Si$_2$H$_2$ (IR absorption at 1093 cm^{-1} in good agreement with ab initio calculations [4]). The appearance of the hydrogen containing products is most likely due to hydrogen impurities in the metal surfaces of the cryostat, whereas the presence of compounds containing nitrogen or oxygen may be caused by impurities contained in the argon used or by minor leaks in the apparatus.

Reactions of silicon atoms with acetylene, C$_2$H$_2$

In the past two years, we have been able to isolate four C$_2$H$_2$Si isomers in an argon matrix after pulsed flash pyrolysis of 2-ethynyl-1,1,1-trimethyldisilane [3]. As was proposed earlier, another access to the C$_2$H$_2$Si hypersurface consists of the reaction of silicon atoms with acetylene [5]. Based on this information about the C$_2$H$_2$Si isomers, it was obvious to take the Si/acetylene system to refine our silicon evaporation techniques.

As expected, the primary product of the cocondensation is 1-silacyclopropenylidene (**5**). The photochemical interconversion to the other known C$_2$H$_2$Si isomers could be reproduced; this is shown in the scheme below for completion.

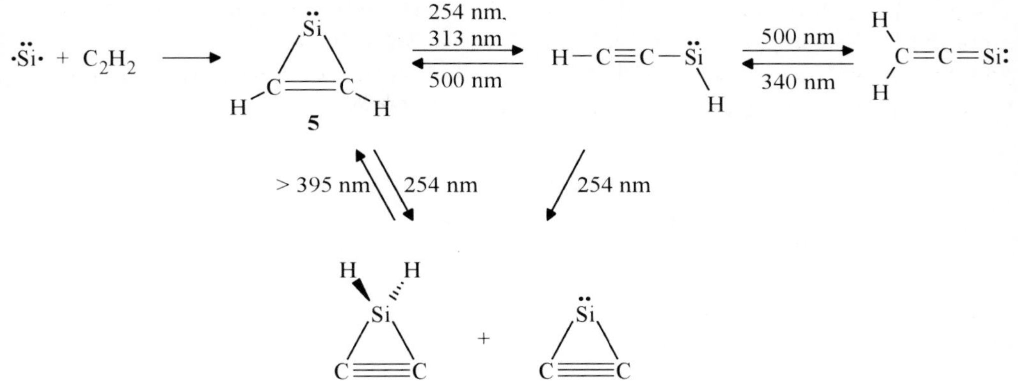

Scheme 2.

Reactions of silicon atoms with ethylene, C_2H_4

In analogy to the reaction of silicon atoms with acetylene, cocondensation of silicon atoms and ethylene leads to a cyclic silylene as the primary product, namely 1-silacyclopropanylidene (**6**). In this reaction, the π-system again serves as Lewis base. Compound **6** could be identified by comparison of its IR spectrum (no Si–H vibration observable, most intense band at 593 cm^{-1}) with that calculated by ab initio methods. Upon irradiation of **6** with light of wavelengths > 395 nm, the ring opens with cleavage of a Si–C bond. Simultaneous migration of a hydrogen atom from carbon to silicon leads to the formation of vinylsilylene (**8**), a previously unknown C_2H_4Si isomer (Si–H vibration at 1980 cm^{-1}). A comparison of experimental and calculated IR spectra and consideration of the most probable atomic migrations in the isomerization of **6** to **8** suggest that the conformation in **8** is *anti* with regard to the Si–C bond.

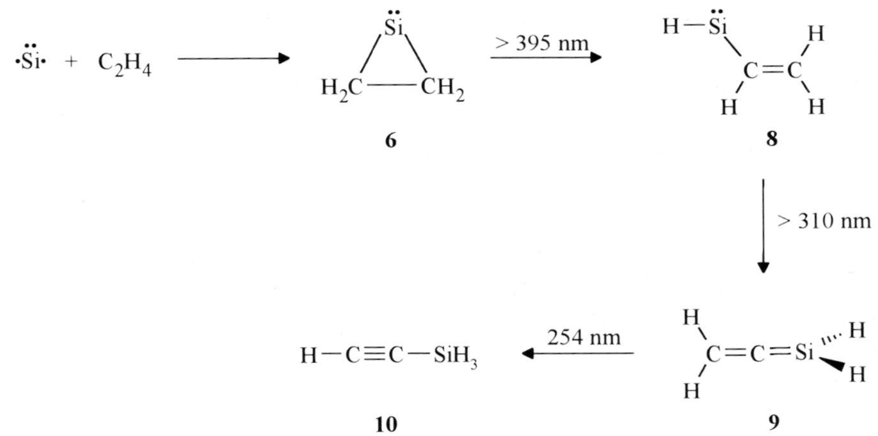

Scheme 3.

Another new and interesting C_2H_4Si isomer is formed upon irradiation of **8** with light of wavelengths > 310 nm: 1-silaallene (**9**, Si–H absorptions at 2166 and 2176 cm^{-1}). This assignment has to be made with some caution, since the conformity of observed and calculated IR spectra is not as good for compound **9** as for all other species mentioned in this report. On the other hand, this assumption may be confirmed by the isomerization of **9** to silylacetylene (**10**) by irradiation with light of wavelength 254 nm. This experimental fact can be taken as a hint that **9** is indeed a C_2H_4Si isomer. Moreover, the intermediacy of **9** would fit nicely into a series of molecules that show a successive migration of hydrogen atoms from carbon to silicon (the silicon atoms of **6**, **8**, **9**, and **10** bear 0, 1, 2, and 3 hydrogen atoms, respectively).

Reactions of silicon atoms with hydrogen cyanide, HCN

When silicon atoms are cocondensed with hydrogen cyanide, the main primary product is not the cyclic silylene **7**. The nitrogen lone pair serves as the Lewis base rather than the π-system, leading to the formation of 2-aza-1-silapropadiene-1,1-diyl (**11**). This behavior is analogous to the reaction of silicon atoms with nitrogen, producing the linear diazasilene molecule, SiNN [6]. The lowering of the frequency of the N–N stretching vibration in SiNN by about 600 cm^{-1} compared with the N_2 molecule indicates a relatively weak N–N bond. The same seems to hold for the C–N bond in **11**, as its stretching frequency (1538 cm^{-1}) is lowered by about 560 cm^{-1} compared with that of hydrogen cyanide (2097 cm^{-1}). Actually, it even lies below values observed for compounds with C–N double bonds (e. g., 1640 cm^{-1} for methanimine, $H_2C=NH$).

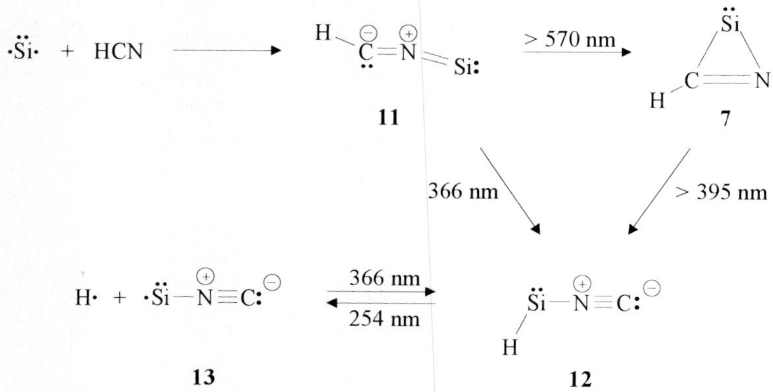

Scheme 4.

Compound **11** can be easily transformed into 2-aza-1-silacyclopropenylidene (**7**, weak IR absorption for the C–N stretching vibration at 1466 cm^{-1}), the cyclic analogue to **5** and **6** in the reactions described in the previous sections. The photoreaction requires only irradiation with visible light of wavelengths > 570 nm and is complete within a few minutes, indicative of a low activation barrier for this isomerization. In the course of irradiation, a small amount of another species is also formed. The IR absorptions of this second species increase upon irradiation with light of wavelengths > 395 nm, while those of **7** disappear. By comparison with calculated IR spectra, this species can be identified as isocyanosilylene (**12**, intense IR absorptions at 2018 cm^{-1} and 2040 cm^{-1} for the Si–H and C–N stretching vibrations, respectively).

Small amounts of **7** and **12** can even be observed immediately after the cocondensation and are also present when the matrix is protected from daylight, which can also initiate the reactions **11**→**7** and **7**→**12**. Presumably, the activation barriers for these reactions are so low that even the light emitted by the hot crucible can induce the isomerizations to some extent. Calculations to examine these observations are currently in progress.

The irradiation of **12** with light of wavelength 254 nm leads to a fourth species, which most likely is the SiNC radical, **13** (IR absorptions at 629 and 1945 cm^{-1}). The formation of **13** is reversible, irradiation with light of wavelength 366 nm leads back to isocyanosilylene (**12**).

References:

[1] a) Z. K. Ismail, L. Fredin, R. H. Hauge, J. L. Margrave, *J. Chem. Phys.* **1982**, *77*, 1626.
b) Z. K. Ismail, R. H. Hauge, L. Fredin, J. W. Kauffman, J. L. Margrave, *J. Chem. Phys.* **1982**, *77*, 1617.
c) L. Fredin, R. H. Hauge, Z. K. Ismail, J. L. Margrave, *J. Chem. Phys.* **1985**, *82*, 3542.
[2] R. A. Shepherd, W. R. M. Graham, *J. Chem. Phys.* **1988**, *88*, 3399, and references cited therein.
[3] G. Maier, H. P. Reisenauer, H. Pacl, *Angew. Chem.* **1994**, *106*, 1347; *Angew. Chem., Int. Ed. Engl.* **1994**, *33*, 1248.
[4] G. Maier, H. P. Reisenauer, A. Meudt, H. Egenolf, *Chem. Ber./Recueil* **1997**, *130*, 1043.
[5] M.-D. Su, R. D. Amos, N. C. Handy, *J. Am. Chem. Soc.* **1990**, *112*, 1499.
[6] R. R. Lembke, R. F. Ferrante, W. Weltner, jr., *J. Am. Chem. Soc.* **1977**, *99*, 416.

Cycloaddition Reactions of Dimethylaminomethyl-silylene with Dienes and Heterodienes

Susanne Meinel*, Joachim Heinicke

Institut für Anorganische Chemie
Ernst-Moritz-Arndt-Universität Greifswald
Soldtmannstr. 16, D-17487 Greifswald, Germany
Tel./Fax: Int. code + (3834)75459 (new (3834)864316 / 864337)

Keywords: Silylene / Thermolysis / Cycloadditions

Summary: The silylenes MeSiNMe$_2$ and MeSiNH(iPr) can be thermally generated from disilanes. In rapid copyrolytic gas phase reactions between disilanes and trapping reagents the intermediate silylenes and dienes or heterodienes furnish a variety of five-membered unsaturated silicon heterocycles in 35-65 % yield.

Much effort has been devoted to the chemistry of substituted silylenes since the end of the sixties. Great interest exists in π-stabilized compounds, especially in cyclodelocalized, stable 1,3,2λ^2-diazasiloles [1, 2].

Whereas hexamethyldisilane is decomposes homolytically at higher temperature, the hetero-substituted disilanes decomposed unsymmetrically [3]. Dimethylamino- and isopropylaminodisilanes undergo intramolecular α-elimination already at 420°C (residence time 50-70s) to corresponding aminosilylenes and aminosilanes [4]. A temperature above 550°C is necessary to generate silylenes from methylchlorodisilanes [5]. Quantum chemical investigations refer to a stabilization through (p-p)π-interactions in an aminosilylene (22.3 kcal/mol), which could contribute to the lower temperature of formation [6].

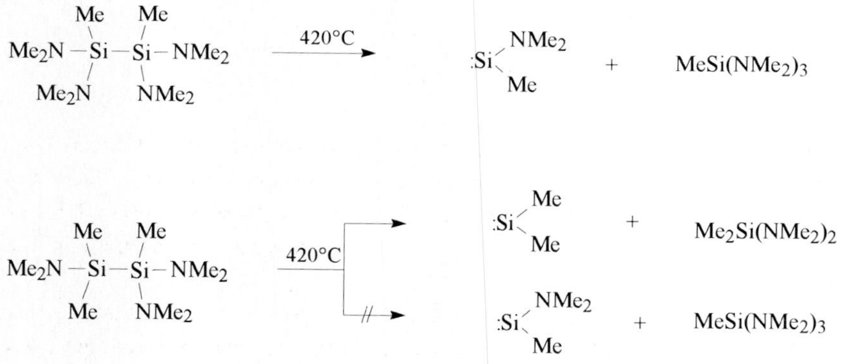

Scheme 1.

Reaction of technical 1,2-dimethyltetrachlorodisilane and 1,1,2-trimethyltrichlorodisilane with Me$_2$NH gave 1,2-dimethyl-1,1,2,2-tetrakis(dimethylamino)disilane and 1,1,2-trimethyl-1,2,2-tris(dimethylamino)disilane, respectively. These are favorable starting compounds to generate dimethylaminomethylsilylene (see Scheme 1). The isopropylaminomethylsilylene may be produced in a similar way.

The chemical characterization of the aminosilylenes was performed by classical trapping reactions with substituted 1,3-butadienes, which rapidly form characteristic and sufficiently thermally stable silacyclopentenes.

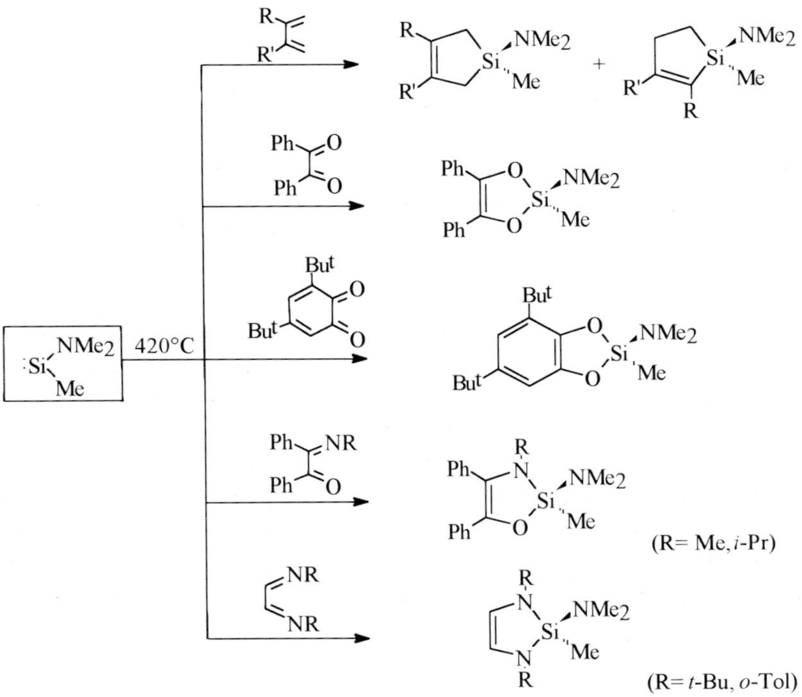

Scheme 2.

Then we investigated the usability of copyrolytic cycloadditions with α,β-unsaturated keto and imino compounds, respectively, to synthesize nitrogen and oxygen containing silicon-functionalized silaheteroterocycles (Scheme 2). The 1,4-diheterodienes and Me$_2$N(Me)Si: afford 1,3,2-diheterosilacyclopent-4-enes. Yields are reasonable to high for α-diimines or α-ketimines but low for α-diketones.

In copyrolytic reactions of the aminosilylenes with unsaturated ketones or imines (heterodienes) we mainly obtained isomeric mixtures. The chemo- and regioselectivity of main- and byproducts can be explained with multistep-cycloadditions. We assume a primary Lewis acid-base interaction between the lone electron pair of the heteroatom (oxygen or nitrogen) and the electron gap at silylene, which is followed by a [2+1]-cycloaddition and a radical ring-opening ring-closure reaction.

The selectivity is controlled by the first step (chemoselectivity) and not by substituents as observed in regioselective cycloadditions of substituted dienes (Lei-Gaspar mechanism) [7].

These trapping reactions, summarized in Schemes 2 and 3, open the possibility to form a number of unsaturated five-membered silaheterocycles with Si–N, Si–O or Si–N, and Si–O bonds in 35-65 % yield.

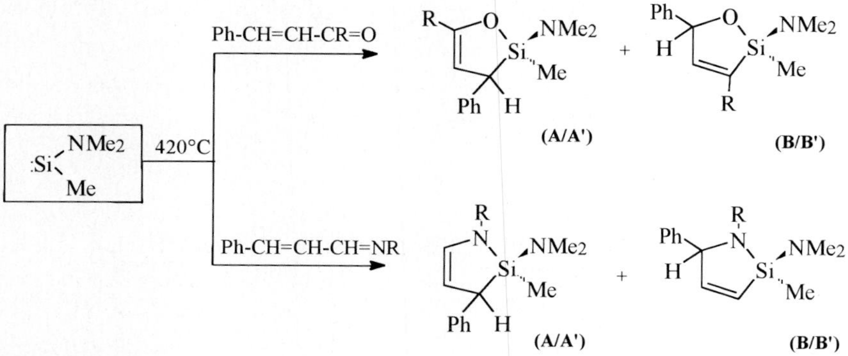

Scheme 3.

Complicated reactions occur in the absence of trapping agents. Many signals can be observed besides those of aminosilanes in the ^{29}Si spectrum, possibly oligosilanes by insertion of silylenes into Si–N bonds [8]. More investigations are necessary.

Furthermore, we found that trapping products with an NH-functionalized silylene, MeSiNH*i*Pr, can be obtained in reasonable amounts. No hints for thermal isomerization were found in cycloadditions with 2,3-dimethylbutadiene or oxadienes. Such reactions are at least slower than cycloadditions.

Acknowledgement: This work was supported by the *Deutsche Forschungsgemeinschaft* and the *Fonds der Chemischen Industrie*.

References:

[1]   M. Denk, R. Lennon, R. Hayashi, R. West, A. V. Belyakov, H. P. Verne, A. Haaland, M. Wagner, N. Metzler, *J. Am. Chem. Soc.* **1994**, *116*, 2691.

[2]   B. Gehrhus, M. F. Lappert, J. Heinicke, R. Boese, D. Bläser, *J. Chem. Soc., Chem. Commun.* **1995**, *19*, 1931.

[3]   W. H. Atwell, DD 1 921 833 (C1: C07d 103/02), **27.11.69**.

[4]   J. Heinicke, B. Gehrhus, S. Meinel, *J. Organomet. Chem.* **1994**, *474*, 71.

[5]   J. Heinicke, D. Vorwerk, G. Zimmermann, *J. Anal. Appl. Pyrol.* **1994**, *28*, 93.

[6]   B. T. Luke, J. A. Pople, M. B. Krogh-Jespersen, Y. Apeloig, J. Chandrasekar, P. v. R. Schleyer, *J. Am. Chem. Soc.* **1986**, *108*, 260.

[7]   D. Lei, P. P. Gaspar, *J. Chem. Soc., Chem. Commun.* **1985**, 1149.

[8]   G. Fritz, H. Amann, *Z. Anorg. Allg. Chem.* **1992**, *616*, 39.

Do Unsubstituted Silacyclobutadienes Exist?

Günther Maier*, Hans Peter Reisenauer, Jörg Jung, Andreas Meudt, Harald Pacl
Institut für Organische Chemie
Justus-Liebig-Universität Gießen
Heinrich-Buff-Ring 58, D-35392 Gießen, Germany
Fax: Int. code + (641)99-34309

Keywords: Matrix Isolation / *ab initio* Calculations / Silylene

Summary: A number of isomers of composition C_3H_4Si and $C_2H_4Si_2$ have been generated by pulsed flash pyrolysis of appropriate precursors and isolated in an argon matrix. Their photochemical interconversions were studied. Quantum chemical calculations have been performed at the BLYP/6-31G* level of theory. They play a decisive role in the identification of the highly reactive intermediates. Although silacyclobutadienes were shown to be minima on their respective energy hypersurfaces, no experimental evidence for the existence of such compounds was found. Instead, silylenes were detected, which undergo a variety of mutual interconversions.

Introduction

The urge to break existing rules or, more accurately, to fathom their limits is an impetus in science. In silicon chemistry one of the greatest challenges is the classical double bond rule, which tells us that a molecule with a third row element involved in a π system should not exist. It is also known that molecules with four cyclicly conjugated π electrons are strongly destabilized and so attract the chemists interest. Both characteristics, in combination with high values of ring strain energy, are exemplified in silacyclobutadienes, making them very rewarding synthetic targets.

In the present work we examined the chemistry of sila- and disilacyclobutadiene isomers. So far, no experimental data were available for $C_2H_4Si_2$ and few for C_3H_4Si species. In the latter case the formation of silacyclobutadiene (**4**) was reported [1] but its characterization was based solely on mass spectroscopic data. Furthermore a highly substituted analogue is known [2].

Scheme 1.

Our attempts to matrix-isolate silacyclobutadiene (**4**) in analogy to silabenzene (**2**) were unsuccessful. While the aromatic silabenzene (**2**) could be generated in a retro-ene reaction from silacyclohexadiene **1** [3], the analogous silacyclobutadiene precursor **3** did not react in the same way [4].

Consequently, we used a different strategy in our recent approach to the long-sought unsubstituted silacyclobutadienes. Instead of trying to generate the target compounds directly from an appropriate precursor, we examined approaches to energetically lower lying species on the respective hypersurface. The photochemistry of these compounds then was studied in the hope of finding the desired cyclobutadienes.

C_3H_4Si Isomers

Obviously silacyclobutenylidene **5** should be a suitable intermediate on the way to silacyclobutadiene (**4**), lying just 5.1 kcal/mol above the global minimum (silacyclopropenylidene **14**; see Fig. 1) and being separated from **4** only by a 1,2-hydrogen shift. This type of reaction is well-known in silylene chemistry. For example, methylsilylene (**6**) and silaethene (**7**) are readily interconvertible photochemically [5].

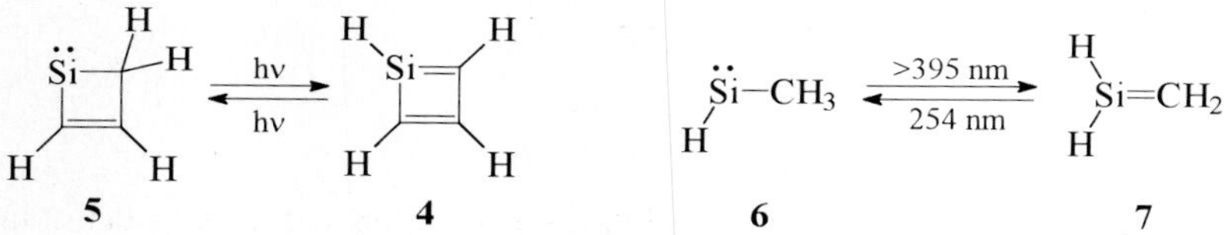

Scheme 2.

Analogous to the isolation of silacyclopropenylidene upon thermal trimethylsilane extrusion from 2-ethynyl-1,1,1-trimethyldisilane [6], pyrolysis of **8** could lead to silacyclobutenylidene **5**. Therefore, propargyldisilane **8** and all other stable open chain isomers of **8** were prepared and examined.

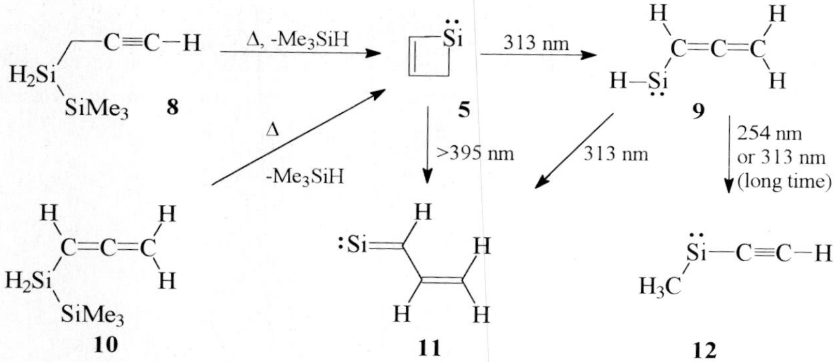

Scheme 3.

In the pyrolysis experiments propargyldisilane **8** and allenyldisilane **10** yielded the expected product **5**. However, isomerization to **4** did not occur during irradiation; ring-opening took place instead. Depending on the wavelength of the light, either allenylsilylene (**9**) or 1-silabutadienylidene **11** was generated. After a long irradiation period the formation of ethynylmethylsilylene (**12**) was observed.

The pyrolysis of propynyldisilane **13** took the same course as that of 1,1,1-trimethyl-2-ethynyldisilane. In the initial step, silacyclopropenylidene **14**, the global minimum on the C_3H_4Si hypersurface, was formed. Irradiation with light of wavelength $\lambda = 313$ nm yielded propynylsilylene (**15**). Further irradiation at 254 nm resulted chiefly in the generation of ethynylmethylsilylene (**12**). This isomer is an important link between the chemistry of precursors **8**, **10** and **13**. Moreover, silylene **12** is the primary product in the pyrolysis of ethynyldisilane **16**. The already known species **5** is observed after irradiation with 254 nm as well, but in this case only the most intense IR absorption was registered.

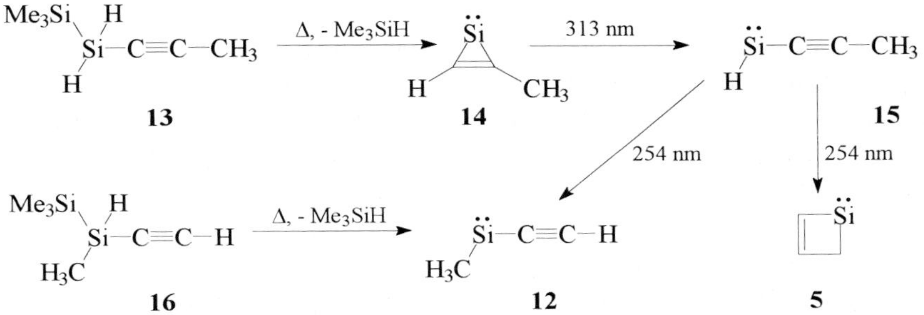

Scheme 4.

However, in none of these experiments could the desired cyclobutadiene **4** be detected.

$C_2H_4Si_2$ Isomers

The formal substitution of a saturated carbon atom in compounds **8**, **13** and **16** by silicon results in precursors **21**, **17** and **20**, which could all be synthesized. In contrast to the results above, pulsed flash pyrolysis of these oligosilanes gave rise to the formation of only one $C_2H_4Si_2$ isomer, namely **18**. Actually, irradiation of matrix-isolated 2-silylsilacyclopropenylidene (**18**) led to silylene **19** in analogy to reaction **14**→**15**.

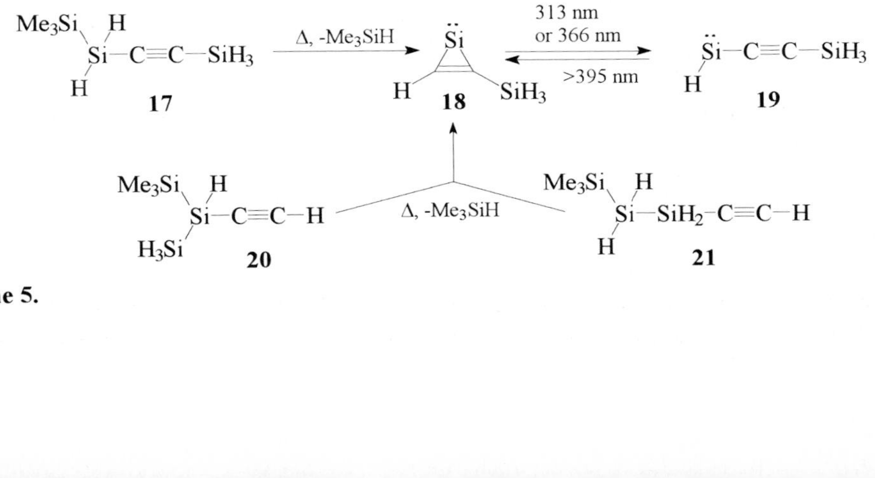

Scheme 5.

As in the experiments described above, a disilacyclobutadiene was not detected during photolysis.

Calculations

Since no ab initio frequency data for C_3H_4Si species except for silacyclobutadiene (**4**) [7] are given in the literature, calculations on the BLYP/6-31G* level of theory have been carried out. The relative energies of some selected minima on the C_3H_4Si hypersurface are shown in Fig. 1.

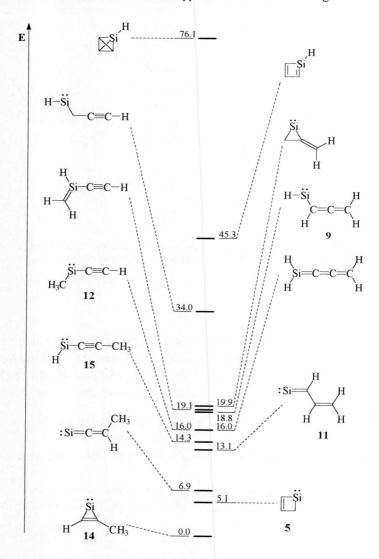

Fig. 1. Relative energies [kcal/mol] of selected C_3H_4Si species. Calculations were performed at the BLYP/6-31G* level of theory. Energy values are corrected by zero point vibrational energies.

In Fig. 2 the results of the same kind of calculations are depicted for $C_2H_4Si_2$ isomers. 1,2-Disilacyclobutadiene (**22**) deserves a special comment. In accordance with earlier calculations [8], one silicon-bonded hydrogen atom in this molecule is in a position perpendicular to a Si_2C plane. This fact and other unusual topological features of **22** indicate that this compound is not a real cyclobutadiene but rather a disilylene with some stabilizing interaction between the two silylene centers.

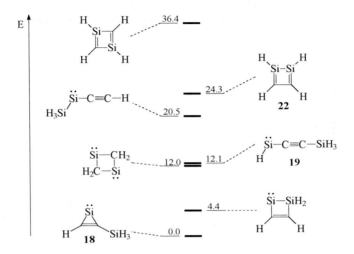

Fig. 2. Relative energies [kcal/mol] of selected $C_2H_4Si_2$ species. Calculations were performed at the BLYP/6-31G* level of theory. Energy values are corrected by zero point vibrational energies.

References:

[1] T. M. Gentle, E. L. Muetterties, *J. Am. Chem. Soc.* **1983**, *105*, 304.

[2] M. J. Fink, D. B. Puranik, M. P. Johnson, *J. Am. Chem. Soc.* **1988**, *110*, 1315.

[3] G. Maier, G. Mihm, H. P. Reisenauer, *Angew. Chem.* **1980**, *92*, 58; *Angew. Chem., Int. Ed. Engl.* **1980**, *19*, 52.

[4] P. Lingelbach, *Ph. D. Thesis*, Gießen, **1989.**

[5] H. P. Reisenauer, G. Mihm, G. Maier, *Angew. Chem.* **1982**, *94*, 864; *Angew. Chem., Int. Ed. Engl.* **1982**, *21*, 854.

[6] G. Maier, H. P. Reisenauer, H. Pacl, *Angew. Chem.* **1994**, *106*, 1347; *Angew. Chem., Int. Ed. Engl.* **1994**, *33*, 1248.

[7] H. E. Colvin, H. F. Schaefer III, *Faraday Symp. Chem. Soc.* **1984**, *19*, 39.

[8] T. A. Holme, M. S. Gordon, S. Yabushita, M. W. Schmidt, *Organometallics* **1984**, *3*, 583.

A Thermally Stable Silylene: Reactivity of the Bis(amino)silylene Si[{N(CH₂*t*Bu)}₂C₆H₄-1,2]

Barbara Gehrhus*, Peter B. Hitchcock, Michael F. Lappert
School of Chemistry and Molecular Sciences
University of Sussex
Brighton, BN1 9QJ, United Kingdom
Tel.: Int. code + (1273)678316 – Fax: Int. code + (1273)677196
E-mail: B.Gehrhus@susx.ac.uk

Keywords: Silylene / Bis(amino)silylene

Summary: The thermally stable silylene Si[{N(CH₂*t*Bu)}₂C₆H₄-1,2] **1** readily undergoes addition reactions with compounds containing CO and CN multiple bonds (e.g., ketones, imines, nitriles or isonitriles). The products, most of them characterized by their X-ray structures, contain novel skeletal frameworks: SiCOSi, SiOSiCCC, SiCNSi, CSi(CN), CSiSi(CN) and SiNCCC, respectively.

Introduction

Although transient silylenes have been known since the middle of the sixties, the first stable divalent and two-coordinate silicon compound, the bis(amino)silylene Si[N(*t*Bu)CHCHN*t*Bu], was reported only in 1994, including its gas phase structure by electron diffraction [1].

We have been using the dianions derived from *N,N'*-disubstituted *o*-phenylenediamines as chelating ligands to stabilize compounds of low valent Group 14 elements; a carbon(II) compound had been characterized as the electron-rich olefin [C{N(Me)}₂C₆H₄-1,2]₂ [2], and the stable stannylenes Sn[{N(R)}₂C₆H₄-1,2] (R = CH₂*t*Bu or SiMe₃) have been isolated and structurally authenticated [3]. In our quest for a thermally stable, crystalline silylene, the ligands [N(R')]₂C₆H₃-1,2-R-4 (R' = CH₂*t*Bu and R = H or Me) were chosen.

In a preliminary communication we reported

- (i) that reduction of the dichlorosilane Si[{N(CH₂*t*Bu)}₂C₆H₃-1,2-R-4]Cl₂ (R = H or Me) with potassium yielded the crystalline silylene Si[{N(CH₂*t*Bu)}₂C₆H₃-1,2-R-4] (R = H or Me);
- (ii) the first X-ray structure of a silylene, namely **1**; and
- (iii) that oxidative addition of MeI or EtOH to Si[{N(CH₂*t*Bu)}₂C₆H₄-1,2] gave Si[{N(CH₂*t*Bu)}₂C₆H₄-1,2](X)Y (X = Me and Y = I or X = OEt and Y = H) [4].

We also reported oxidative addition reactions of **1** with a chalcogen yielding [(Si{N(CH₂*t*Bu)}₂C₆H₄-1,2)(μ-E)]₂ (E = S, Se or Te) and provided single-crystal X-ray diffraction data for the resulting cyclodisilaselenane and -tellurane [5a]; PE spectra of **1** and molecular orbital calculations have been described [5b].

The chemistry of transient silylenes is well established [6] but very little is known about the reactivity of thermally stable silylenes. There are only a few reactions described for the stable silylene Si[N(*t*Bu)CHCHN*t*Bu] (with azides (e.g., Ph₃CN₃ or Me₃SiN₃ [7]) or with Ni(CO)₄ [8]).

Results and Discussion

Reactivity of 1 towards CO multiple bonds

The bis(amino)silylene **1** (abbreviated as "SiN₂") reacted readily at room temperature with benzophenone to give the 1,2,3-oxadisiletane **2** (Scheme 1; X-ray structure in Figure 1). These oxadisiletanes have only previously been observed by reaction of a disilene with a ketone [9, 10]. The formation of **2** is thought to proceed via an oxasilacyclopropane intermediate. The reaction of **1** with *t*Bu(Me)C=O yielded a similar disilaoxetane.

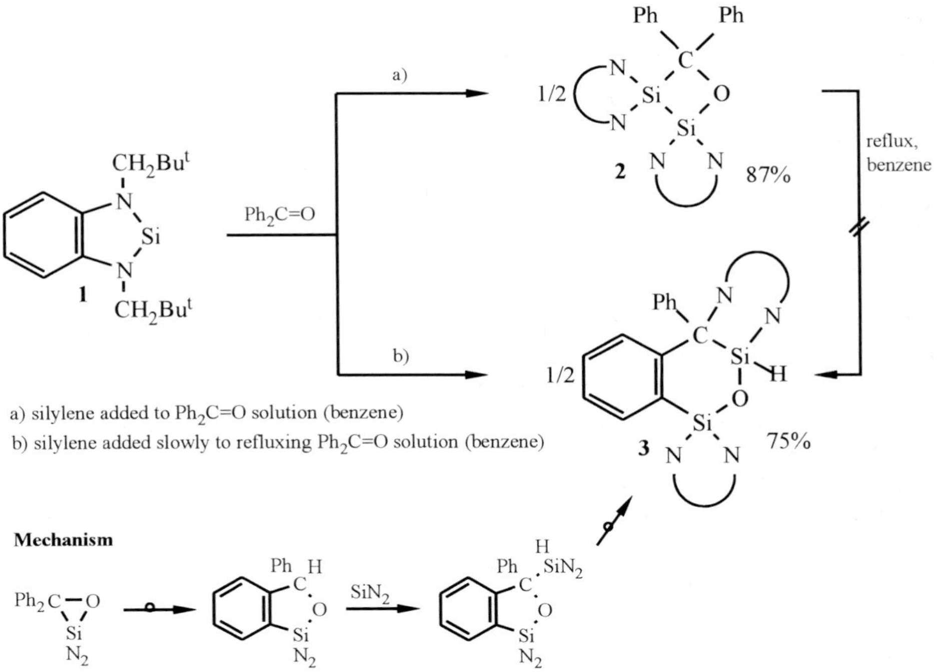

a) silylene added to Ph₂C=O solution (benzene)
b) silylene added slowly to refluxing Ph₂C=O solution (benzene)

Scheme 1.

However, by varying the reaction conditions and adding a benzene solution of **1** slowly to a refluxing Ph₂C=O solution in benzene, **3** was obtained (X-ray structure in Fig. 1). The proposed mechanism for the formation of **3** (Scheme 1) may be compared with that suggested for **5** in Scheme 3.

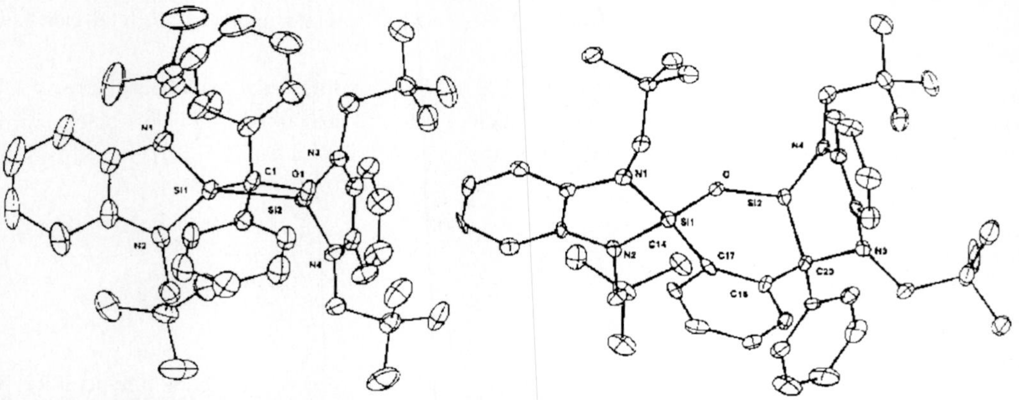

Fig. 1. A view of the X-ray structure of **2** and **3**. Selected bond lengths and (Å) and angles (°):
compound **2**: Si(1)–Si(2) 2.352(3), Si(1)–C(1) 1.963(6), Si(2)–O 1.649(4), O–C(1) 1.502(6), C(1)–Si(1)–Si(2) 72.9(2), C(1)–O–Si(2) 110.1(3), O–Si(2)–Si(1) 79.6(2), O–C(1)–Si(1) 97.2(3);
compound **3**: Si(1)–O 1.624(8), Si(2)–O 1.632(8), Si(1)–C(17) 1.854(12), Si(2)–C(23) 1.903(11), C(17)–C(18) 1.41(2), C(18)–C(23) 1.53(2), Si(2)–N(4) 1.711(9), C(23)–N(3) 1.535(13).

Reactivity of 1 towards CN multiple bonds

a) With Ph₂C=NSiMe₃

Treatment of **1** with $Ph_2C=NSiMe_3$ gave the isolable intermediate **4** (Scheme 2), which was obtained as a microcrystalline solid. It was stable at room temperature in benzene solution but on heating **4** yielded compound **5**.

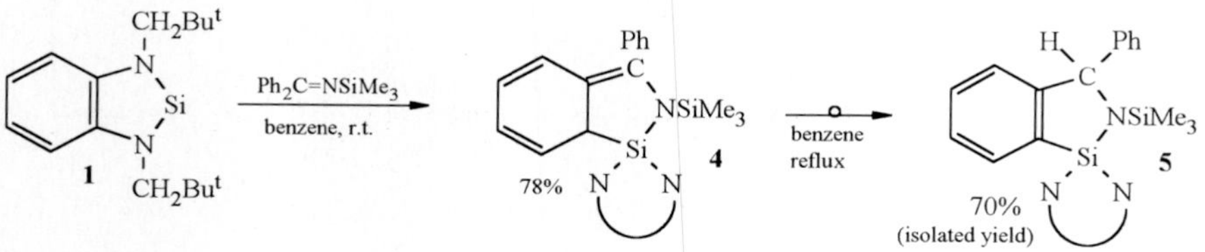

Scheme 2.

The structure of **4**, including its non-aromatic nature, was confirmed by ^1H NMR spectroscopy (Fig. 2).

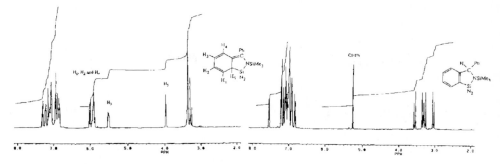

Fig. 2. Section of the ¹H NMR spectra of **4** and **5**.

The proposed mechanism is similar to that postulated for the gas phase reaction between Me₂Si: and Ph₂C=O at 700°C [11] (Scheme 3).

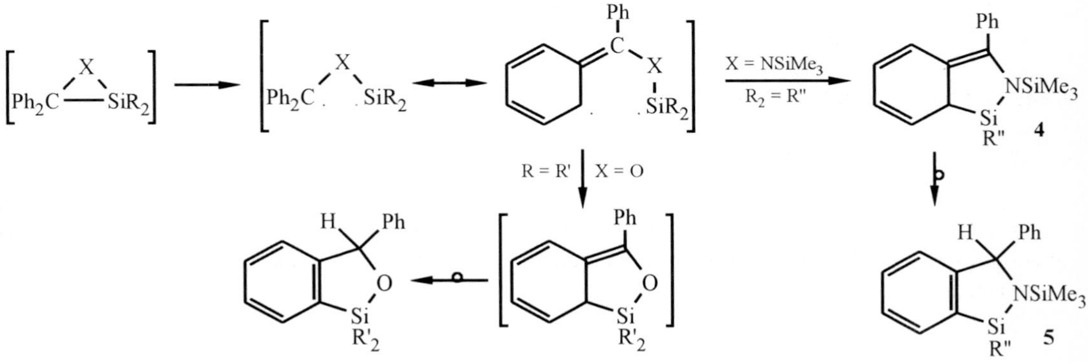

Scheme 3.   R' = Me [11]; R' = 2-(Me₂NCH₂)C₆H₄ [12], R" = 1,2-C₆H₄[N(CH₂*t*Bu)]₂ (this work)

b) with *t*BuCN

Treatment of **1** with *t*BuCN yielded **6** (Scheme 4; X-ray structure of **6** in Fig. 3). Such a disilaazetine has only previously been described by the reaction of a disilene with a nitrile [13]. The formation of **6** is thought to proceed via an azasilacyclopropene intermediate, that does not dimerize (*cf.*, Weidenbruch, et al. [13]) but undergoes insertion of another silylene molecule.

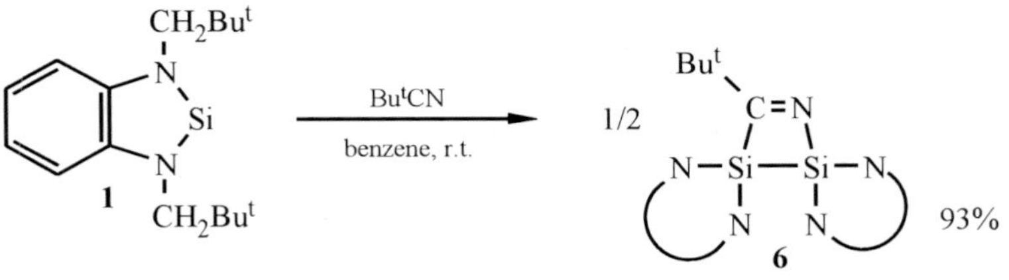

Scheme 4.

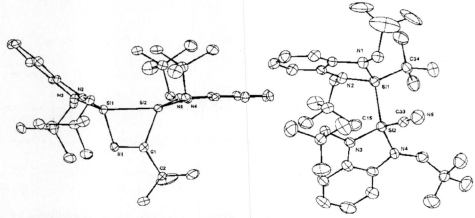

Fig. 3. A view of the X-ray structure of **6** and **8**. Selected bond lengths and (Å) and angles (°):
compound **6**: Si(1)–Si(2) 2.385(3), Si(2)–C(1) 1.925(6), N(1)–C(1) 1.275(7), Si(1)–N(1) 1.769(5), N(1)–Si(1)–Si(2) 77.1(2), C(1)–Si(2)–Si(1) 69.8(2), C(1)–N(1)–Si(1) 106.0(4), N(1)–C(1)–Si(2) 108.1(4).
compound **8**: Si(1)–Si(2) 2.423(3), Si(1)–C(34) 1.892(7), Si(2)–C(33) 1.862(8), C(33)–N(5) 1.148(7), Si(2)–C(33)–N(5) 175.2(7).

c) with *t*BuNC

Treatment of **1** with *t*BuNC gave two different products, **7** or **8**, depending on the reaction conditions (Scheme 5; X-ray structure of **8** in Fig. 3). A similar rearrangement of a silaketeneimine intermediate was observed by Weidenbruch, et al. only for very sterically demanding isonitriles [14].

Surprisingly a mixture of **7** and **1** did not yield **8**, despite the fact that formally **8** is the product of such an insertion reaction.

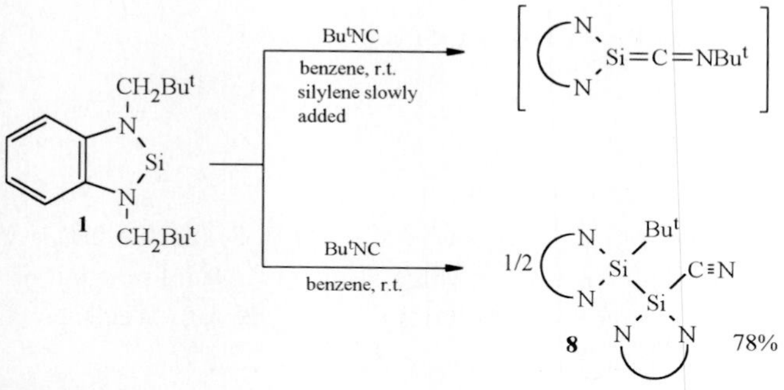

Scheme 5.

Acknowledgement: We thank the EPSRC for support.

References:

[1]   M. Denk, R. Lennon, R. Hayashi, R. West, A. V. Belyakov, H. P. Verne, A. Haaland, M. Wagner, N. Metzler, *J. Am. Chem. Soc.* **1994**, *116*, 2691.

[2]   E. Çetinkaya, P. B. Hitchcock, H. Küçükbay, M. F. Lappert, S. Al-Juaid, *J. Organomet. Chem.* **1994**, *481*, 89.

[3]   H. Braunschweig, B. Gehrhus, P. B. Hitchcock, M. F. Lappert, *Z. Anorg. Allg. Chem.* **1995**, *621*, 1922.

[4]   B. Gehrhus, M. F. Lappert, J. Heinicke, R. Boese, D. Bläser, *J. Chem. Soc., Chem. Commun.* **1995**, 1931.

[5]   a) B. Gehrhus, P. B. Hitchcock, M .F. Lappert, J. Heinicke, R. Boese, D. Bläser, *J. Organomet. Chem.* **1996**, *221*, 211; b) P. Blakeman, B. Gehrhus, J. C. Green, J. Heinicke, M. F. Lappert, M. Kindermann, J. Veszprémi, *J. Chem. Soc. Dalton Trans.* **1996**, 1475.

[6]   M. Weidenbruch, *Coord. Chem. Rev.* **1994**, *130*, 275; and refs. therein.

[7]   M. Denk, R. K. Hayashi, R. West, *J. Am. Chem. Soc.* **1994**, *116*, 10813.

[8]   M. Denk, R. K. Hayashi, R. West, *J. Chem. Soc., Chem. Commun.* **1994**, 33.

[9]   M. J. Fink, D. J. DeJoung, R. West, J. Michl, *J. Am. Chem. Soc.* **1983**, *105*, 1070.

[10] A. Schäfer, M. Weidenbruch, *J. Organomet. Chem.* **1985**, *282*, 305.

[11] W. Ando, M. Ikeno, A. Sekiguchi, *J. Am. Chem. Soc.* **1977**, *99*, 6448.

[12] J. Belzner, *J. Organomet. Chem.* **1992**, *430*, C51.

[13] M. Weidenbruch, A. Schäfer, *J. Organomet. Chem.* **1986**, *314*, 25; M. Weidenbruch, B. Flintjer, S. Pohl, W. Saak, *Angew. Chem., Int. Ed. Engl.* **1989**, *28*, 95.

[14] M. Weidenbruch, B. Brand-Roth, S. Pohl, W. Saak, *Polyhedron*, **1991**, *10*, 1147.

Pyrido[b]-1,3,2λ²-diazasilole:
The First Stable Unsymmetrical Silylene

Anca Oprea, Joachim Heinicke*
Institut für Anorganische Chemie
Ernst-Moritz-Arndt-Universität Greifswald
Soldtmannstr. 16, D-17487 Greifswald, Germany
Tel./Fax: Int. code + (3834)75459 (new (3834)864316 / 864337)

Keywords: Silylene / Dichlorosilanes

Summary: The synthesis and characteristic NMR data of 1,3-dineopentyl-pyrido[b]-1,3,2λ²-diazasilole are discussed.

Introduction

Although silylenes have been known for a long time as transient intermediates in thermal, photochemical or dehalogenation reactions, stable species have become available only recently. Si(II) compounds with coordination number larger than two have already been isolated in the late eighties, i.e., decamethylsilicocene prepared by Jutzi et al [1] and the tetracoordinated phosphine derivative obtained by Karsch et al [2].

Si(II) compounds with coordination number two are more effectively stabilized by bulky substituted amino groups at silicon. First representatives are the stable Arduengo-type silylene isolated by Denk, West et al [3], and the benzoannelated diaminosilylenes made by Gehrhus, Lappert et al [4]. They are obtained by reductive halogen elimination of the corresponding dihalogenated derivatives. The aim of this work is to extend the knowledge on stable diaminosilylenes by studying the effects of electron-poor or electron-rich annelation. Here we present several new bis(amino)dichlorosilanes as potential precursors and the synthesis of the first stable unsymmetrical silylene.

Results and Discussion

The bis (amino)dichlorosilanes were made by a reaction sequence similar to that described by Gehrhus, Lappert et al. [4]. The substituents R are neopentyl groups possessing bulky *t*-butyl units in the β-position, which are shifted away from the annellated cycle and should be quite favorable in shielding the silicon site.

Eq. 1.

Some characteristic NMR data of **3 a-c** are compiled in Table 1. Attempts to reduce these compounds to stable silylenes have met with limited success. So far, only the 1,3-dineopentyl-pyrido[b]-1,3,2λ²-diazasilole **4** has been obtained by stirring the 2,3-disubstituted pyridine derivative **3a** with potassium for 2d at room temperature and could be isolated by sublimation at 115°C/10⁻⁵ mbar. The yield was low.

Eq. 2.   **3a**   **4**

The formation of the silylene **4** is indicated by the strong low field ^{29}Si NMR shift (95.1 ppm), which is near to that found for dineopentyl-benzo-1,3,2λ²-diazasilole (96.9 ppm) [4]. Moreover, the NCH₂ protons of **4** are considerably shifted to low field ($\Delta\delta$ = 0.43 ppm) indicating some ring current effect in the π system of the planar 1,3,2λ²-diazasilole part (Table 1).

^{1}H and ^{13}C NMR data are in accordance with the proposed structure. The carbon chemical shifts of the pyridine are not much varied if Si(IV) is replaced by Si(II). This is surprising since the 1,3,2λ²-diazasilole ring is formally a π-excess system and due to the deshielding of silicon the charge density at the nitrogen and carbon atoms should be increased and influence also the pyridine ring.

Table 1. Selected NMR data of diazasiloles **3 a-c** and silylene **4**.

Compound	δ^1H [ppm]; N–CH$_2$	$\delta^{29}Si$ [ppm]
3a	3.70 / 3.00	-24.0
3b	3.34 / 3.41	-20.7
3c	3.25	-22.4
4	4.13 / 3.43	95.1

Acknowledgement: The authors thank the *Deutsche Forschungsgemeinschaft* for financial support.

References:

[1]   P. Jutzi, U. Holtmann, D. Kanne, C. Krüger, R. Blom, R. Gleiter, I. Hyla-Kryspin, *Chem. Ber.* **1989**, *122*, 1629.

[2]   H. H. Karsch, U. Keller, S. Gamper, G. Müller, *Angew. Chem., Int. Ed. Engl.* **1990**, *29*, 295.

[3]   M. Denk, R. Lennon, R. Hayashi, R. West, A. V. Belyakov, H. P. Verne, A. Haaland, M. Wagner, Metzler, *J. Am. Chem. Soc.* **1994**, *116*, 2691.

[4]   B. Gehrhus, M. F. Lappert, J. Heinicke, R. Boese, D. Bläser, *J. Chem. Soc., Chem. Commun.* **1995**, 1931.

A New Route to Silaheterocycles: Heterobutadiene Cycloaddition

Hans H. Karsch*, Peter Schlüter

Anorganisches-Chemisches Institut
Technische Universität München
Lichtenbergstraße 4, D- 85747 Garching, Germany
Tel./Fax: Int. code + (89)32093132
E-mail: Hans.H.Karsch@lrz.tu-muenchen.de

Keywords: Cycloaddition / Trichlorosilane / 1,4-Diaza-1,3-butadienes / Silacyclo-pentenes

Summary: The reaction of trichlorosilane with DABCO provides the anion $SiCl_3^-$, which can be used instead of $SiCl_2$ in cycloaddition reactions. In the presence of the 1,4-diaza-1,3-butadienes **1a-g**, the five membered rings **2a-g** are found. **2b** was characterized by an X-ray structure determination.

Product distribution and yield vary with the substituents on nitrogen. Trifluoroalkyl substituted 1,3-diaza-1,3-butadienes **3a,b** give the 1-trichlorosilylsubstituted 1,3-diaza-2-butenes **4a,b**. In the case of **4a**, the X-ray structure determination shows only a weak Si–N(2) contact. All compounds were identified by NMR and mass spectroscopy.

Introduction

Since $SiCl_2$ belongs to the most promising building blocks for silaheterocyclic synthesis, a preparative simple access to the carbene homologue is highly desirable.

It is well known that the reaction of trichlorosilane in the presence of *tert.*-amines, provides the anion $SiCl_3^-$ (Benkeser type system) [1].

$$HSiCl_3 + NR_3 \longrightarrow [HNR_3][SiCl_3]$$

Eq. 1.

This system in the presence of heterobutadienes also can be used to incorporate $SiCl_2$ into five-membered heterocycles.

Recently, we reported that reactive trifluoroalkyl substituted 1,3-diazadienes in a formal [4+1] cycloaddition form five-membered heterocycles, e.g., *t*BuN–C(Ph)=N–C(CF₃)₂–SiCl₂ [2]. New investigations have shown that this strategy can be extended, with some decisive modifications, to other heterobutadienes, e.g., 1,4-diaza-1,3-butadienes **1**.

Results and Discussion

The 1,4-diaza-1,3-butadienes **1** in dichloromethane react under mild conditions with an equimolar amount of $HSiCl_3$ and in the presence of DABCO to give the five-membered silaheterocycles **2** [3].

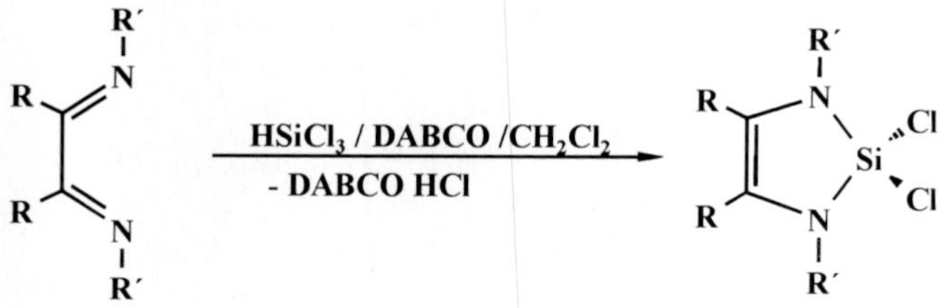

	R	R'	Yield [%]
a	H	$C(CH_3)_3$	29.5
b	C_6H_5	$C(CH_3)_3$	50.7
c	$C_{12}H_8$	$C(CH_3)_3$	–
d	H	C_6H_{11}	54.3
e	H	$2,6-(iC_3H_7)_2\ C_6H_3$	99.7
f	CH_3	$2,6-(iC_3H_7)_2\ C_6H_3$	–
g	H	$2,4,6-(CH_3)_3\ C_6H_2$	91.3

Scheme 1.   Synthesis of **2**.

Product distribution and yield, however, are strongly influenced by the nature of the substituents at nitrogen. The influence of the solvent is also important: indeed, there is almost no reaction in THF, pentane or toluene. Thus a number of five-membered heterocycles, especially with bulky substituents, is available. In the case of $tBuN–C(Ph)=C(Ph)–NtBu–SiCl_2$, **2b** the molecular structure has been determined by X-ray diffraction (Fig. 1).

Table 1. Characteristic bond lengths [Å] and angles [°]of **2b**.

Bond Lengths		Bond Angles	
Si–Cl(2)	2.053(1)	Cl(1)–Si–Cl(2)	103.5(3)
Si–Cl(1)	2.056(1)	N(2)–Si–N(1)	97.17(7)
Si–N(1)	1.704(1)	Si–N(1)–C(1)	106.4(1)
N(1)–C(1)	1.436(1)	N(1)–C(1)–C(2)	115.0(2)
N(2)–Si	1.697(2)	C(1)–C(2)–N(2)	114.4(2)
N(2)–C(2)	1.440(2)	C(2)–N(2)–Si	106.7(1)
C(1)–C(2)	1.384(2)		

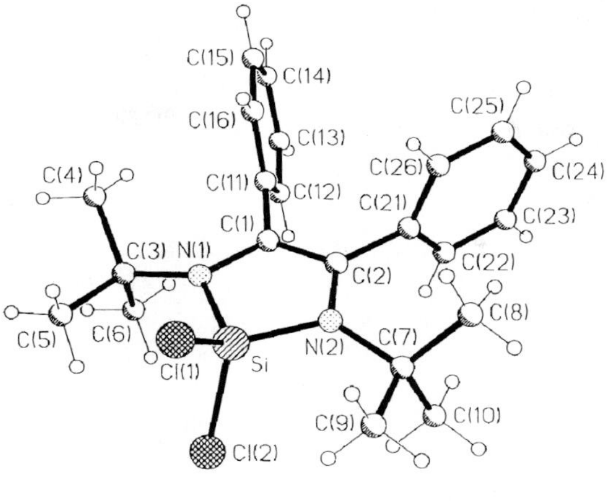

Fig. 1. Molecular structure of **2b**.

The silicon atom is surrounded tetrahedrally by two chlorine and two nitrogen atoms. The ring skeleton is nearly planar and the bond lengths of C(1)–C(2) agree with usual C=C-double bonds. All distances and angles correspond to values derived from analogous compounds [2, 4, 5].

The reaction sequence starts with a nucleophilic attack of nitrogen at the SiCl₃⁻ anion. In the transition state, probably with a Ψ-pentacoordinated silicon, the Si–Cl bonding is weakened by the influence of the lewis base DAD. The elimination of chloride is followed by a subsequent cyclocondensation (Scheme 2).

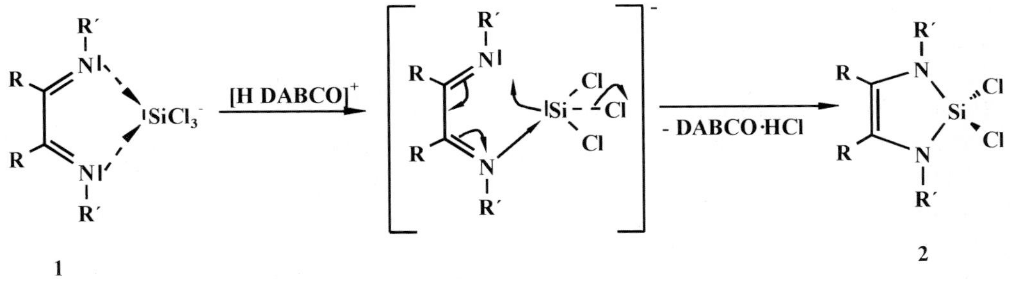

Scheme 2. Proposed pathway for the formation of heterocycles **2a-g**.

New results in the reactions of trifluoroalkyl substituted 1,3-diaza-1,3-butadienes with HSiCl₃/NR₃ suggest, that in the first step, the base acts catalytically and SiCl₃⁻ is added to the heterobutadiene forming an amidinate type structure (Scheme 3). The further reaction depends on the ligands on nitrogen and carbon, an HCl elimination and formation of the pentacycle only being observed in sterically favorable cases.

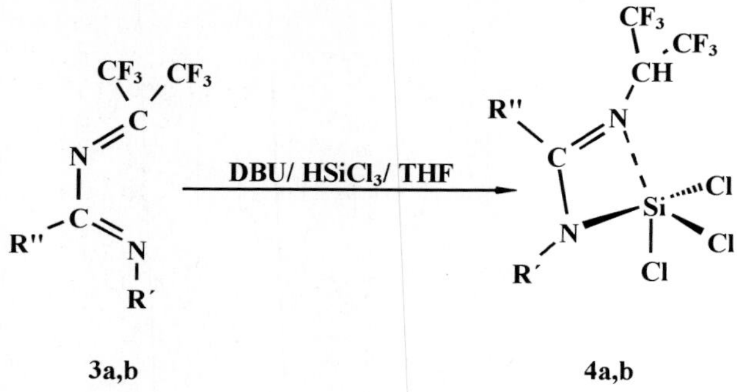

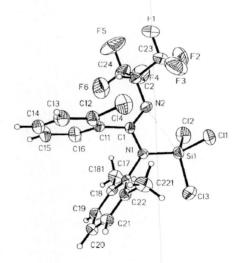

Scheme 3. Reaktion of **3a, b** with HSiCl₃/ DBU.

	R'	R"
a	(2,6-(CH₃)₂–C₆H₃)	*o*-Cl–C₆H₄
b	(2,4,6-(CH₃)₃–C₆H₂)	C₆H₅

For example, $(CF_3)_2CH-N=C(Cl-C_6H_4)N[2,6-Me_2C_6H_3]-SiCl_3$ **4a** is obtained from the reaction of the 1,3-diaza-1,3-butadiene with equimolar amounts of HSiCl₃ and DBU. As shown by X-ray structure determination, the amidinate acts as a semi-chelating ligand with a very weak Si–N(2) contact (Fig. 2).

Fig. 2. Molecular structure of **4a**.

Table 2. Characteristic bond lengths [Å] and angles [°] of **4a**.

Bond Lengths		Bond Angles	
Si–Cl(2)	2.013(2)	N(1)–C(1)–N(2)	114.0(3)
Si–Cl(3)	2.017(2)	Cl(3)–Si–N(1)	104.4(1)
Si–N(1)	1.735(3)	Cl(3)–Si–N(2)	160.1(1)
N(1)–C(1)	1.395(5)	N(2)–Si–Cl(1)	85.9(1)
N(2)–Si	2.684(4)	N(2)–Si–Cl(2)	83.7(1)
N(2)–C(2)	1.448(5)	Cl(3)–Si–Cl(2)	105.3(8)
C(1)–N(2)	1.273(5)	N(2)–Si–N(1)	56.0(1)
		Si–N(1)–C(1)	115.1(3)

Conclusions

The system $HSiCl_3$/ DABCO (DBU) may be used to generate $SiCl_3^-$ which formally acts as a silylene synthon. 1,4-Diazabuta-1,3-dienes predominately react to give the five membered heterocycles. Obviously this route can be applied both to nucleophilic and electrophilic heterodienes. Similar to the reactions of the 1,3- and 1,4-diazabutadienes, perfluoroalkyl-substituted vinylketones also give five-membered rings. However, in all cases the reaction conditions have to be optimized individually.

References:

[1]   R. A. Benkeser, *Acc. Chem. Res.* **1971**, *94*, 94.
[2]   H. H. Karsch, F. Bienlein, A. Sladek, M. Heckel, K. Burger, *J. Am. Chem. Soc.* **1995**, *117*, 5160.
[3]   H. H. Karsch, P. Schlüter, F. Bienlein, M. Herker, E. Witt, A. Sladek, M. Heckel, *Z. Anorg. Allg. Chem.* **1997**, in press.
[4]   M. Weidenbruch, H. Piel, *Organomet.* **1993**, *12*, 2882.
[5]   M. Weidenbruch, A. Lesch, *J. Organomet. Chem.* **1991**, *407*, 31.

Base Coordination: A Way to Nucleophilic Silylenes?

Johannes Belzner

Institut für Organische Chemie der

Georg-August-Universität Göttingen

Tammannstr. 2, D-37077 Göttingen, Germany

Fax: Int. code + (551)399475

E-mail: jbelzne@gwdg.de .

Keywords: Silylenes / Nucleophilicity

Summary: The reactivity of silylene **4** towards halogenated alkanes, diphenylacetylenes, chlorosilanes, and a stable, nucleophilic carbene was investigated. Product studies, kinetic measurements, and calculations provide good evidence for **4** being a nucleophilic species due to an intramolecular coordination of the dimethylamino groups to the silicon center.

It has been theoretically and experimentally well established that silylenes have a singlet ground state [1]. Such species posses a free electron pair in a σ-orbital and an empty orbital of π-symmetry; therefore, they are a priori ambiphilic compounds, which can react either as an electrophile or as a nucleophile towards appropriate substrates. However, most silylenes have revealed a distinctive "electrophilic character". Dimethylsilyene, e.g., adds to olefins and alkynes in the gas phase via a rate-controlling step that is accelerated by electron-donating substituents [2]; these experimental results are in good agreement with a theoretical study of the reaction of :SiH$_2$ with ethylene, which shows that this cycloaddition proceeds via an initial electrophilic phase in which the silylene LUMO interacts with the π-electron system of the double bond [3]. Up to now, only some stable silylenes, such as recently described **1** [4] or silicocene **2** [5] have shown nucleophilic reactivity.

Fig. 1.

In the course of our studies of the interaction of Lewis bases with silylenes we have found that easily available cyclotrisilane **3** is in equilibrium with silylene **4** at room temperature [6]. At this point, it is tempting to assume that the ease of formation of **4** from **3** is due to intramolecular coordination of the dimethylamino groups to the silylene center. This hypothesis is backed mainly by two results:

- An ab initio study of the reaction of $:SiH_2$ with ammonia showed that the coordinative interaction between the nitrogen and the silicon center is exothermic by 23±2 $kcal\,mol^{-1}$ [7].
- The isolation and structural characterization of pentacoordinated siliconium ions bearing the 2-(dimethylaminomethyl)phenyl substituent proves that the geometry of this substituent is well suited for a twofold coordination of a trigonal planar silicon center [8].

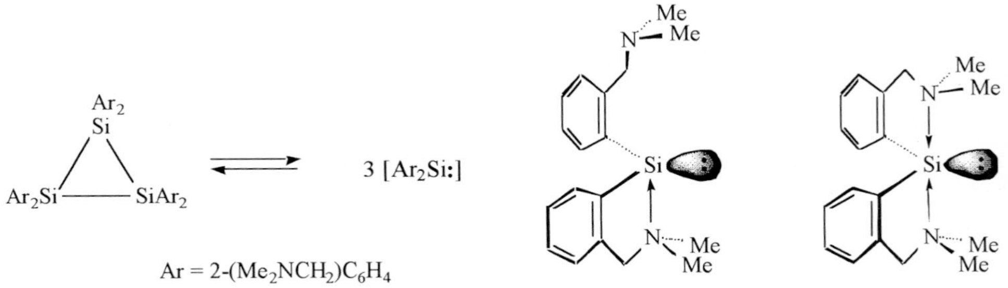
Ar2
Si
Ar2Si
SiAr2
3 [Ar2Si:]
Ar = 2-(Me2NCH2)C6H4
Me
N
Me
Si
N
Me
Me

3 **4** **4a** **4b**

Fig. 2.

Further confirmation of thermodynamic stabilization of **4** by intramolecular coordination is expected to arise from the investigation of its reactivity. The inherent electron demand of a silylene may be satisfied in a highly coordinated species such as **4a** or **4b** by the intramolecular coordination of one or two dimethylamino groups; accordingly, such species should show more nucleophilic character. Here we will present experimental and theoretical results that show that **4** indeed acts as a nucleophile towards appropriate substrates; these results corroborate the hypothesis of **4** being an intramolecularly coordinated silylene.

As a model system for elucidating the reactivity of silylene towards electrophiles the reaction of **3** with halogenoalkanes was chosen. Stirring a solution of **3** with three equivalents of iodomethane in toluene for seven days at room temperature afforded siliconium ion **5a** in 90% yield (Eq. 1). Analogously, **5b** and **c** were obtained.

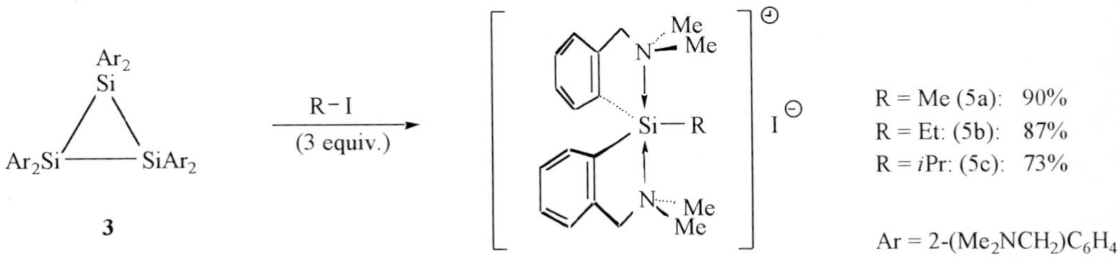
Ar2
Si
Ar2Si
SiAr2
3
R-I
(3 equiv.)
Me
N
Me
Si—R
N
Me
Me
I
R = Me (5a): 90%
R = Et: (5b): 87%
R = iPr: (5c): 73%
Ar = 2-(Me2NCH2)C6H4

5a–c

Eq. 1. Synthesis of siliconium ions **5a-c**.

The reaction of **3** with *t*-butyl chloride proceeded under clean formation of covalent **6** (Eq. 2). The satisfactory yields and selectivity of these transformations are in marked contrast to the reactions of photochemically generated trimethylsilylphenylsilylene **7** with chloroalkanes, which afforded, depending on the substrate, the silylene insertion product **8** and/or the hydrogen chloride abstraction product **9** in low yield (Eq. 3) [9].

$$
\begin{array}{ccc}
\underset{\substack{\\ Ar_2Si \text{———} SiAr_2}}{\overset{\substack{Ar_2 \\ Si \\ \triangle}}{}} & \xrightarrow[\text{(3 equiv.)}]{t\text{-BuCl}} & \underset{t\text{-Bu}}{\overset{Ar_2}{Si}}\diagdown Cl
\end{array}
$$

3   **6**

Eq. 2.   Reaction of **3** with *t*-butyl chloride (Ar = 2-(Me$_2$NCH$_2$)C$_6$H$_4$).

$$
\underset{Ph}{\overset{Me_3Si}{\diagdown}}Si: \quad \xrightarrow{\text{R}-\text{Cl}} \quad \underset{Ph}{\overset{Me_3Si}{\diagdown}}Si(Cl)R \quad + \quad \underset{Ph}{\overset{Me_3Si}{\diagdown}}Si(Cl)H \quad + \quad \text{olefin}
$$

7   **8**   **9**

	8	9	
R = *n*-Oct:	9%	0%	0%
R = *sec*-Bu:	7%	10%	12%
R = *tert*-Bu:	0%	27%	28%

Eq. 3.   Reaction of **7** with chloroalkanes.

In order to get more insight into the different reactivity of **4** and **7** a preliminary ab initio study of the reaction of :SiH$_2$ with chloromethane was performed. At the MP4/6-311G**//MP2(fu)/6-31G* level of theory a dative bond between the chloro substituent and :SiH$_2$ is formed. This step, in which :SiH$_2$ acts as an electrophile, is followed by nucleophilic attack of the weakly bound silylene on the carbon center. This results eventually, via a three-membered, cyclic transition state **B**, in the insertion of :SiH$_2$ into the carbon-chlorine bond (Fig. 3). The initial formation of complex **A** nicely rationalizes the observed competing reactions of **7** with different chloroalkanes: The coordinated silylene can undergo an intramolecular attack on the carbon center forming eventually the insertion product **8**, or, alternatively, may abstract a proton from a β-carbon; this reaction ultimately will result in the formation of chlorosilane **9**. The situation changes dramatically when ammonia enters the stage as a third partner (Fig. 4): Now :SiH$_2$ preferably coordinates to ammonia; the Si···N bond of the resulting complex **D** is by far more stable than the Si···Cl bond of **A**. Complex **D** does not undergo further coordination to chloromethane; all attempts to localize a corresponding minimum have failed so far. Moreover, no cyclic transition state analogous to **B** was found in this ternary system; instead, an approximately trigonal bipyramidal transition state **E** was localized. **E** is interpreted as the transition state structure of an S$_N$2 reaction, in which a silylene-ammonia complex plays the role of the nucleophile and chloride acts as leaving group. Thus, these theoretical results show clearly that the interaction of :NH$_3$ with :SiH$_2$ changes its "normal" electrophilic reactivity into the chemical behavior of a nucleophile. In addition, these computational results may explain the observation that no hydrogen halogenide abstraction was formed, when **3** or **4** was allowed to react with ethyl iodide, *i*-propyl iodide, or *t*-butyl chloride. In these cases the free electron pair of the approaching silylene **4a** or **4b** is involved in the process of Si–C bond formation and accordingly is not available for abstraction of a proton.

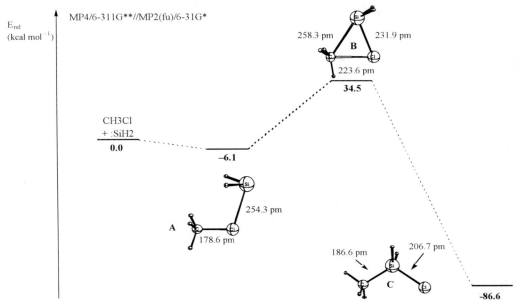

Fig. 3. Reaction of :SiH$_2$ with chloromethane (ab initio results).

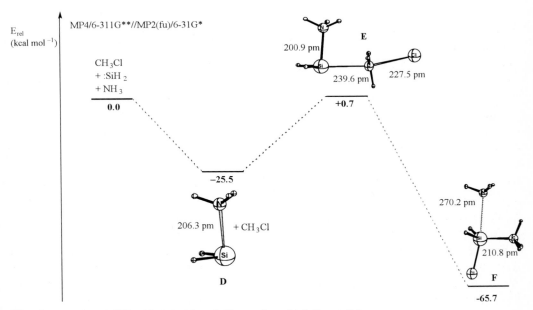

Fig. 4. Reaction of :SiH$_2$ with ammonia and chloromethane (ab initio results).

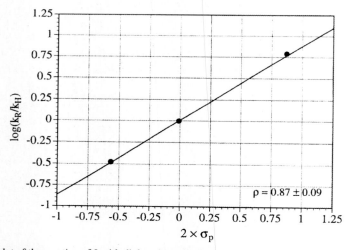

3   **10a-c**   **11**

Eq. 4.   Reaction of **3** with diphenylacetylenes (Ar = 2-(Me$_2$NCH$_2$)C$_6$H$_4$; R = Me, H, CO$_2$Me).

Another type of reaction allowed us to study the philicity of silylene **4**. As we have shown earlier, **4** adds smoothly to a variety of alkynes giving way to the silacyclopropene framework (Eq. 4) [10]. Varying the *para*-substituents of diphenylacetylene offers us the possibility to tune the electron density of the triple bond, and we now studied the rates of the reaction of **3** with diphenylacetylenes **10a-c**. The *absolute* reaction rates of these three first-order reactions are identical ($k_1 = 6.3 \pm 0.2 \times 10^{-4}\,s^{-1}$); this result is in accordance with a mechanism, in which the formation of silylene **4** from cyclotrisilane **3** is the rate determining step [6]. However, the *relative* reaction rates of the addition of **4** to the triple bond of **10a-c**, which were determined by competition experiments, turned out to differ appreciably from each other. Electron withdrawing substituents favor the addition of **4** to the alkyne, whereas electron donating substituents, such as a methyl group, slow down the reaction rate. As shown in Fig. 5, the relative reactions rates correlate well with the σ-values of the substituents R. Moreover, the ρ-value of 0.87 for this reaction compares well with the value of 1.05, which was found for the addition of nucleophilic tropylidene to substituted styrenes [11].

<!-- Figure 5 plot -->
y-axis: $\log(k_R/k_H)$
values: 1.25, 1, 0.75, 0.5, 0.25, 0, -0.25, -0.5, -0.75, -1
x-axis: $2 \times \sigma_p$
values: -1, -0.75, -0.5, -0.25, 0, 0.25, 0.5, 0.75, 1, 1.25

$\rho = 0.87 \pm 0.09$

Fig. 5.   Hammett plot of the reaction of **3** with diphenylacetylenes **10a-c**.

Thus, the reactivity of **4** towards alkynes contrasts that of dimethylsilylene, which adds in the gas phase to electron-rich triple bonds faster than to electron-poor [2], and is similar to the reactivity of typical nucleophilic carbenes. Accordingly, **4** must be characterized as a nucleophilic silylene.

Comparable results were obtained when **3** reacted with di- and trichlorosilanes (Eq. 5). The ρ-values (1.57 and 1.04, respectively) of these reactions, which allow the convenient synthesis of disilanes **13**, show once more that the nucleophilic character of **4** is of importance for the rate determining step of these conversions.

Eq. 5. Reaction of **3** with chlorosilanes (Ar = 2-$(Me_2NCH_2)C_6H_4$; R = NMe_2, Me, H, Cl; n = 2, 3).

Two concluding experiments highlight the nucleophilic properties of **4**:
- Reaction of **3** with three equivalents of the nucleophilic carbene **14** gave rise to the clean formation of **15** (Eq. 6). The formation of a silene as the product of a electrophilic reaction of **4** via its LUMO with the HOMO of **14** was not observed [12].

Eq. 6. Reaction of **3** with nucleophilic carbene **14** (Ar = 2-$(Me_2NCH_2)C_6H_4$).

- The reaction of **3** with *p*-toluenesulfonic acid yielded, besides an unidentified by-product, siliconium ion **16** (Eq. 7); having in mind the well known transformation of nucleophilic carbenes into carbenium ions by proton acids [13] this reaction may be interpreted as the protonation of a nucleophilic silylene **4**.

Eq. 7. Protonation of **3**.

Acknowledgement: The author is indebted to his co-workers U. Dehnert, D. Schär, and V. Ronneberger for performing the experiments and to the *Deutsche Forschungsgemeinschaft* and the *Fonds der Chemischen Industrie* for financial support of this work.

References:

[1]   Y. Apeloig, in: *The Chemistry of Organic Silicon Compounds* (Eds.: S. Patai, Z. Rappoport), Wiley, Chichester, **1989**, p. 55; and references cited therein.

[2]   J. E. Bagott, M. A. Blitz, H. M. Frey, P. D. Lightfoot, R. Walsh, *J. Chem. Soc., Faraday Trans.* **1988**, *84*, 515.

[3]   F. Anwari, M. S. Gordon, *Isr. J. Chem.* **1983**, *23*, 129.

[4]   a) M. Denk, R. Lennon, R. Hayashi, R. West, A. V. Belyakov, A. Haaland, M. Wagner, N. Metzler, *J. Am. Chem. Soc.* **1994**, *116*, 2691.

b) A. J. Arduengo, III., H. Bock, H. Chen, M. Denk, D. A. Dixon, J. C. Green, W. A. Herrmann, N. L. Jones, M. Wagner, R. West, *J. Am. Chem. Soc.* **1994**, *116*, 6641.

[5]   a) P. Jutzi, in: *Frontiers of Organosilicon Chemistry* (Eds.: A. R. Bassindale, P. P. Gaspar), Royal Society of Chemistry, Cambridge, **1991**, p. 305.

b) P. Jutzi, in: *Organosilicon Chemistry: From Molecules to Materials* (Eds.: N. Auner, J. Weis), VCH, Weinheim, **1994**, p. 87.

[6]   J. Belzner, H. Ihmels, B. O. Kneisel, R. O. Gould, R. Herbst-Irmer, *Organometallics* **1995**, *14*, 305.

[7]   R. T. Conlin, D. Laakso, P. Marshall, *Organometallics* **1994**, *13*, 838.

[8]   J. Belzner, D. Schär, B. O. Kneisel, R. Herbst-Irmer, *Organometallics* **1995**, *14*, 1840.

[9]   M. Ishikawa, K. Nakagawa, S. Katayama, M. Kumada, *J. Organomet. Chem.* **1981**, *216*, C48.

[10]   J. Belzner, H. Ihmels, *Tetrahedron Lett.* **1993**, *34*, 6541.

[11]   L. W. Christensen, E. E. Waali, W. M. Jones, *J. Am. Chem. Soc.* **1972**, *94*, 2118.

[12]   Weidenbruch has recently shown that a stannylene forms a donor-acceptor complex with a nucleophilic carbene: A. Schäfer, M. Weidenbruch, W. Saak, S. Pohl, *J. Chem. Soc., Chem. Commun.* **1995**, 1157.

[13]   See e.g.: W. Kirmse, K. Loosen, H. D. Sluma, *J. Am. Chem. Soc.* **1981**, *103*, 5935.

Isoelectronic Replacement of Si by P^+:
A Comparative Study of the Structures of the Spiro–cyclic E^{II} Compounds $E[C(PMe_2)_2(X)]_2$ (E = Si, Ge, Sn; X = PMe_2, $SiMe_3$) and a Novel Spirocyclic 10 e-Phosphorus Cation (P^{III}) $P[C(PPh_2)_2(SiMe_3)]_2^+$

Hans H. Karsch*, Eva Witt

Anorganisch-Chemisches Institut
Technische Universität München
Lichtenbergstr. 4, D-85747 Garching, Germany
Tel./Fax: Int. code + (89)32093132
E-mail: Hans.H. Karsch@lrz.tu-muenchen.de

Keywords: Phosphinomethanides / Spirocyclic E^{II} Compounds / Phosphorus Cation(P^{III})

Summary: The synthesis of stable group 14 compounds $E[(Me_2P)_2CX]_2$ (E = Si, Ge, Sn; X = PMe_2, $SiMe_3$) in the formal oxidation state + 2 is achieved by reacting Si_2Cl_6/$LiC_{10}H_8$ ($GeCl_2$·dioxane and $SnCl_2$) with $LiC(PMe_2)_2(SiMe_3)$, **1**, and $LiC(PMe_2)_3$ (**3**) respectively. An X-ray structure analysis of $Si[(Me_2P)_2C(SiMe_3)]_2$ (**2**) $Ge[(Me_2P)_3C]_2$ (**4b**) $Sn[(Me_2P)_2C(SiMe_3)]_2$ (**5a**) and $Sn[(Me_2P)_3C]_2$ (**5b**) shows, that in the crystal, **2**, **4b**, **5a,b**, have a distorted pseudo-trigonal-bipyramidal (Ψ-tbp) structure, in which E (E = Si, Ge, Sn) as a spiro center, is surrounded by four phosphorus atoms. A common feature of the four compounds is also the marked difference between axial and equatorial element-phosphorus bond lengths. An analogous P^+ compound $P[(Ph_2P)_2C(SiMe_3)]_2^+BPh_4^-$ was obtained by reaction of the triphosphete, **6**, and $(Cl)(Ph_2)P=C(PPh_2)(SiMe_3)$, **7**, with $NaBPh_4$. The cation adopts a similiar Ψ-tbp geometry as the E(II) compounds (E = Si, Ge, Sn), but differences are also important.

Introduction

Stable Group 14 compounds (Si, Ge, Sn) in the formal oxidation state +2 have attracted considerable interest in recent years. In all cases, intramolecular donor stabilization is the key for their successful isolation. Diphosphinomethanides **I** and Triphosphinomethanides **II**, respectively (Scheme 1) have been shown to be particularly useful for access to low oxidation states.

Decisive properties of the phosphinomethanides are their stability towards reduction, the separation of the negative charge of the anionic ligand (centered at the formal carbanion) from the electron rich E(II) (E = Si, Ge, Sn) center by the phosphane donor atoms, the thus generated additional electrostatic stabilization and the four-membered chelate ring formation [1].

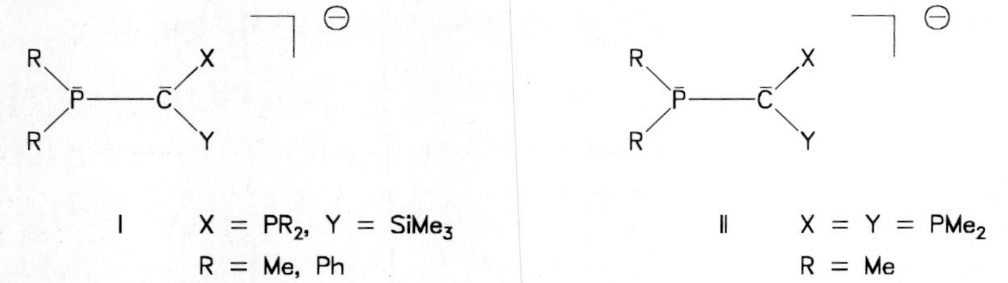

Scheme 1.

Results

Interaction of Si₂Cl₆ with LiC(PMe₂)₂(SiMe₃) (1) in the presence of LiC₁₀H₈

According to (Eq. 1), Si_2Cl_6 reacts with four equivalents of $LiC(PMe_2)_2(SiMe_3)$ (**1**) in the presence of $LiC_{10}H_8$ forming the first stable silicon(II) compound with σ-bonds to silicon [2]. An X-ray structure analysis (Fig. 1) shows that **2** posseses a (distorted) pseudo-trigonal-bipyramidal (Ψ-tbp) structure in which silicon, as a spiro center, is surrounded by four phosphorus atoms. The marked difference between axial and equatorial element phosphorus bond lenghts (Table 1) is noteworthy. One equatorial position within the tbp structure remains unoccupied and can formally be assigned to the lone pair at silicon.

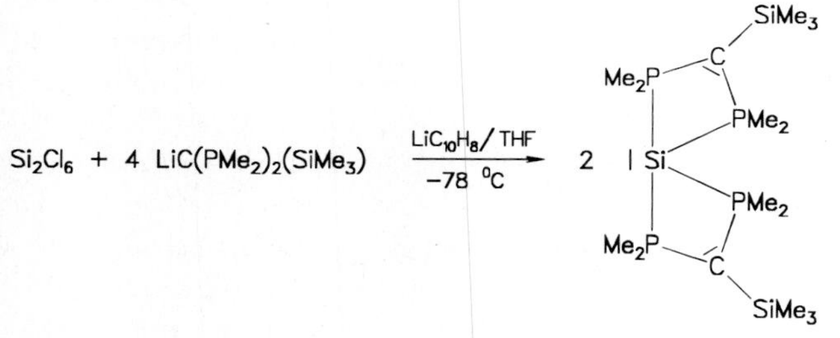

Eq. 1.   Synthesis of **2**.

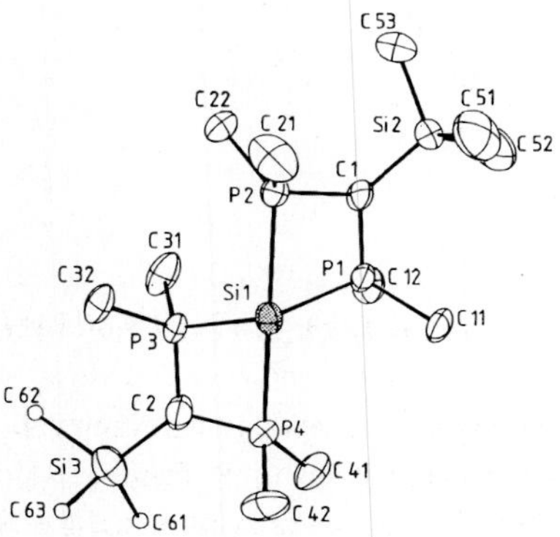

Fig. 1.   Molecular structure of **2**.

Interaction of GeCl₂·dioxan with LiC(PMe₂)₂(SiMe₃) (1) and LiC(PMe₂)₃ (3)

GeCl₂·dioxane reacts with two equivalents of LiC(PMe₂)₂(SiMe₃) (**1**) and LiC(PMe₂)₃ (**3**) respectively, (Eq. 2) yielding the germanium complexes **4a,b** [3]. The structure of the latter was determined by X-ray structure analysis (Fig. 2). Although generally the results obtained in the reactions of germanium halides with phosphinomethanides are analogues to those in the case of silicon, there are some important differences. Thus, in contrast to **2** and the analogous tin complexes Sn[(Me₂P)₂C(SiMe₃)]₂ (**5a**) [4] and Sn[(Me₂P)₃C]₂ (**5b**) [5], the Ge–P$_{ax}$ bond lengths in **4b** differ by as much as 0.38 Å. Therefore, germanium in **4b** appears to be in an intermediate state between tri- and tetra-coordinate (10 e). We attribute the different behavior of Si, Ge and Sn to the difference in bond polarity, i.e. the difference in electronegativity (ΔEN), the phosphorus-germanium bond being essentially nonpolar.

$$\text{GeCl}_2 \bullet \text{diox} + 2 \text{ LiC(PMe}_2)_2\text{X} \xrightarrow[\text{-78 °C}]{\text{THF}}$$

3

4a: X = SiMe₃
4b: X = PMe₂

Eq. 2. Synthesis of **4a,b**.

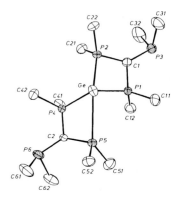

Fig. 2. Molecular structure of **4b**.

Interaction of the Triphosphete 6 and (Cl)(Ph₂)P=C(PPh₂)(SiMe₃) (7) with NaBPh₄

Likewise, for comparison, an analogous P⁺ compound was highly desirable. The synthesis of the first 10 e-phosphorus cation **8** with a homonuclear P₅-framework was achieved by reacting the triphosphete **6** with the chloroylide **7**, both being synthesized for the first time, in the presence of NaBPh₄ (Eq. 3).

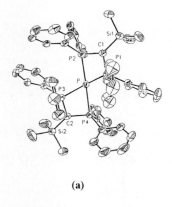

Eq. 3.   Synthesis of **8**.

Particularly indicative for the nature of the cation of **8** is the $^{31}P\{^{1}H\}$ NMR spectrum: a doublet at $\delta = 37.73$ and a quintet at $\delta = -76.26$ ppm $(J(P_AP_B) = 137.3$ Hz), not only demonstrate the presence of four magnetically equivalent phosphorus nuclei around a phosphorus atom (the first PP_4 arrangement) but also the hypervalent nature with a free electron pair at the central P atom due to its high field shift. Obviously, the cation suffers dynamic nonrigidity in solution (pseudorotation): on cooling (-107 °C), a Ψ-tbp ground state is revealed by the appearance of an A_2B_2M spin system. These observations are confirmed by an X-ray structure determination. The distorted Ψ-trigonal bipyramidal geometry of the cation (Fig. 3) is reminiscent to the structures described above, but there are also unexpected differences: a compilation of relevant data is given in Table 1. In the cation of **8**, both axial distances are not very different from each other. In the sense of the above argument, the P–P bonds in **8**$^+$ should be quite polar and indeed, this most likely is the case, as it is also shown by a representation of **8**$^+$ with formal charges. There are some other differences between **8**$^+$ and its isoelectronic counterparts: the four-membered chelate rings are not planar: C1/C2 deviates from the planes P1 P P2 and P3 P P4 by 0.480(2) and 0.467(3) Å, respectively. Furthermore, the angles P2 P P4 and P1 P P3 are much closer to those expected for an ideal tbp-structure at the expense of the orthogonality of the axial bonds to the equatorial plane. This, at least in part, may be ascribed to the steric demand of the P-Phenyl groups.

(a)	(b)

Fig. 3.   Molecular structure of **8**.

Table 1.   Characteristic bond lengths.

	X = SiMe₃	X = SiMe₃	X = PMe₂	X = PMe₂	X = SiMe₃
E–P$_{ax}$ (Å)	2.41/2.46	2.51/2.61	2.55/2.93	2.79/2.84	2.80/2.85
E–P$_{eq}$ (Å)	2.20/2.20	2.27/2.27	2.36/2.36	2.60/2.60	2.59/2.62
P$_{ax}$EP$_{ax}$ (°)	176	154	147	143	142
P$_{eq}$EP$_{eq}$ (°)	125	112	108	106	97
Δ EN	?	0.4	0.1	0.4	0.4

Conclusion

As revealed by the structural investigations, comparisons based on bond polarity considerations are only meaningful within a given group of elements. Different charges on elements (or respective molecules) lead to structural changes which obscure comparative studies. On the other hand, silicon(II) and [phosphorus(III)]$^+$ behave similarly enough to establish a high degree of analogy even in those unique, hypervalent structures as in **2** and **8**.

References:

[1]   H. H. Karsch, A. Appelt, B. Deubelly, K. Zellner, J. Riede, G. Müller, *Z. Naturforsch.* **1988**, *43b*, 1416.

[2]   H. H. Karsch, U. Keller, S. Gamper, G. Müller, *Angew. Chem.* **1990**, *102*, 297; *Angew. Chem., Int. Ed. Engl.* **1990**, *29*, 295.

[3]   H. H. Karsch, B. Deubelly, G. Hanika, J. Riede, G. Müller, *J. Organomet. Chem.* **1988**, *344*, 153.

[4]   H. H. Karsch, A. Appelt, G. Müller, *Organometallics* **1986**, *5*, 1664; G. Ferazin, *Dissertation*, Technische Universität München **1993**.

[5]   H. H. Karsch, A. Appelt, G. Müller, *Angew. Chem.* **1985**, *97*, 404; *Angew. Chem., Int. Ed. Engl.* **1985**, *24*, 402.

[6]   H. H. Karsch, E. Witt, F. E. Hahn. *Angew. Chem.* **1996**, *108*, 2380; *Angew. Chem. Int. Ed. Engl.* **1996**, *35*, 2242.

The Main Group Carbonyls RLi–CO and R₂Si–CO: An Ab Initio Study

Matthias Tacke
University College Dublin
Department of Chemistry
Belfield, Dublin 4, Ireland
E-mail: matthias.tacke@ucd.ie

Summary: Infrared spectroscopy in combination with liquid xenon (LXe) or liquid nitrogen (LN2), which are totally transparent in the frequency region used, was shown be able to detect complexes of CO and N_2 with permethylsilicocene. But the nature of the bonding of carbon monoxide and nitrogen to silicocene was unknown. In this paper it is shown by ab initio calculations that an equilibrium between species of η_5- and η_1-bonded cyclopentadienyl rings is the first step before CO or N_2 forms an adduct with the bis(monohapto)silicocene by releasing a few kcal/mole of adduct enthalpy:

In further investigations it was possible to show that *n*-butyl lithium is also able to complex carbon monoxide. On warming to -30°C the CO molecule inserts into the lithium-carbon bond in contrast to the silicocene case and even this pentoyl lithium intermediate decomposes at -20°C:

$$nBuLi + CO \xrightarrow[-100°C]{LXe} [nBuLi\text{-}CO] \xrightarrow[-30°C]{LXe} [nBu\text{-}C(O)Li] \xrightarrow[-20°C]{LXe} \text{ further reaction}$$

For the organyllithium case a better model for ab initio calculations is introduced, since tetrameric methyllithium serves as a substitute for $(LiH)_4$. With this progress a more accurate determination of geometries and energies of the reaction with carbon monoxide is available. From these new ab initio data a discussion about the bonding properties of RLi–CO and R₂Si–CO is deduced. Finally it is studied whether the intramolecular solvated $(HLiCO)_4$ and $(LiCOH)_4$ are able to dissociate to their corresponding dimers. Moreover, studies on a relative high level of theory demonstrate that both processes are highly endothermic and will not be detected in solution, when the CO to Li ratio rises to 1:1.

Introduction

Carbonyls of main group elements have recently moved into the focus of interest in inorganic chemistry [1]. Apart from experimental approaches this field of research is highly attractive for theoretical explorations of bonding and structure of such species.

In matrix experiments photochemically generated dimethylsilylene is found to react with trapped carbon monoxide and the resulting adduct was characterized by its IR and UV/VIS spectra [2, 3].

$$1/6 \ (Me_2Si)_6 \xrightarrow{\ h\nu\ } Me_2Si \xrightarrow{\ CO\ } Me_2Si{-}CO$$

Eq. 1.

In the IR spectrum the carbonyl absorption at 1965 cm^{-1} indicates a distinct back bonding of the CO molecule. The lone pair of the carbon donates electron density into a silicon p-orbital, while back bonding is achieved by an interaction of the lone pair at the silicon with the p*-orbital of the CO, which should result in a non-linear SiCO arrangement.

Results and Discussion

This bonding thesis is verified by an ab initio calculation since a carbon bonded Me₂SiCO with a pyramidal silicon center is the global energetic minimum. The SiCO substructure has a bonding angle of 159° and a long silicon-carbon bond distance of 202 pm. The sum of the bonding angles at the silicon atom is 288°, which shows the degree of pyramidalization of the species. The reaction enthalpy, which is corrected with the zero-point energy, is found to be slightly exothermic (-6.5 kcal/mole).

$$+ \ CO$$
$$\Delta H = -6.5 \ \text{kcal/mole}$$

d(Si–C): 202 (215) pm \ α(Si–C–O): 153 (166)°
ΔH_{0K} = -6.5 (-1.1) kcal/mole

Eq. 2.   Ab initio results on HF/STO-3G and HF/6-31G** level in parenthesis.

In our group we decided to investigate the above mentioned reaction with the stable and monomeric permethylsilicocene. We used LXe and LN2 as totally transparent solvents for IR spectroscopy and were therefore able to characterize the reversible adduct formation with carbon monoxide and nitrogen [4]. The coordinated nitrogen absorbs at 2053 cm^{-1}, which is value observed in common d-metal nitrogen complexes. For the silicon carbonyl (2065 cm^{-1}) again a back bonded CO is observed, but it is weaker than in Me₂SiCO.

$$Cp_2Si \ + \ CO \ \xrightleftharpoons{\text{LXe, 253K}} \ Cp_2SiCO$$

$$Cp_2Si \ + \ N_2 \ \xrightleftharpoons{\text{LN2, 88K}} \ Cp_2SiN_2$$

Eq. 3.

These spectroscopic results are in agreement with the quantum mechanical calculations. Cp_2Si is used as a model for permethylsilicocene in order to reduce the number of internal coordinates. It reacts as bis(monohapto)silicocene [5] with CO to a similiar adduct as found for the dimethylsilylene case. The SiCO angle is calculated to be 159° and the silicon-carbon bond to be 210 pm, which underlines the reduced back-bonding in comparison to Me_2SiCO. The reaction enthalpy is, with -6.4 kcal/mole, nearly identical.

The Cp_2SiN_2 is an extremely labile complex according to a reaction enthalpy of -0.9 kcal/mole. The geometric and energetic parameters are illustrated in Fig. 1.

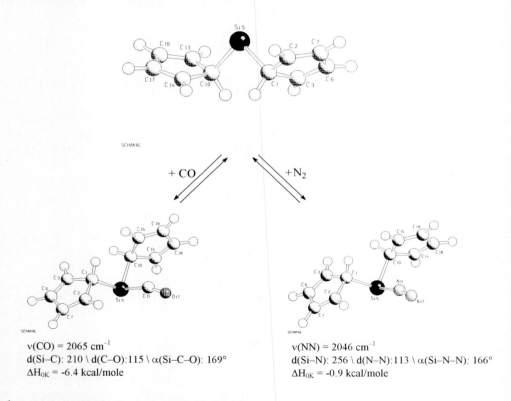

$\nu(CO) = 2065$ cm^{-1}
d(Si–C): 210 \ d(C–O):115 \ α(Si–C–O): 169°
ΔH_{0K} = -6.4 kcal/mole

$\nu(NN) = 2046$ cm^{-1}
d(Si–N): 256 \ d(N–N):113 \ α(Si–N–N): 166°
ΔH_{0K} = -0.9 kcal/mole

Fig. 1.   Ab initio results for the reaction of Cp_2Si with CO and N_2 (bond distances in pm).

Recently, we were able to show that lithium organyls are also able to interact with CO [6] in a newly developed LXe cell constructed from one piece of single crystal silicon [7]. In a first step carbon monoxide is complexed by back-bonding to nBuLi (n(CO): 2047 cm^{-1}) and inserts in a second step at higher temperature into the lithium-carbon bond (n(CO): 1635 cm^{-1}). Further warming to -20°C results in decomposition of the observed pentoyl lithium intermediate. Probably a lithiated oxycarbene is formed favoured by formation of a strong lithium-oxygen bond.

$$nBuLi + CO \xrightarrow[-100°C]{LXe} [nBuLi\text{-}CO] \xrightarrow[-30°C]{LXe} [nBu\text{-}C(O)Li] \xrightarrow[-20°C]{LXe} \text{further reaction}$$

Eq. 4.

With tetrameric lithium hydride as a model substance for ab initio calculations of oligomeric organolithium species it was possible to obtain first structural information [6] about the complexation and insertion of the CO molecule. But the insertion turned out to be a sligthly exothermic partial reaction, which is hard to believe.

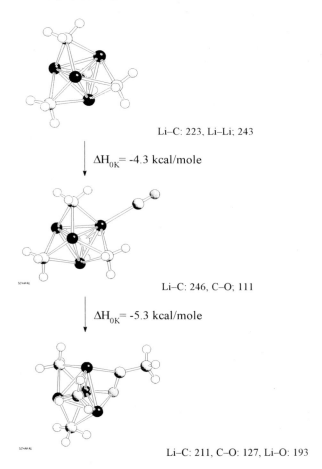

Li–C: 223, Li–Li; 243

$$\Delta H_{0K} = -4.3 \text{ kcal/mole}$$

Li–C: 246, C–O; 111

$$\Delta H_{0K} = -5.3 \text{ kcal/mole}$$

Li–C: 211, C–O: 127, Li–O: 193

Fig. 2. Molecular projections of Li_4Me_4, Li_4Me_4CO and $Li_4Me_3(COMe)$. Selected bond lengths in pm.

With tetrameric methyl lithium as a much better substitute we can now show that the wrong sign of the reaction enthalpy is caused by the inadequate model. The addition of CO to $(LiMe)_4$ releases - 4.3 kcal/mole with the formation of a linear LiCO substructure with a relative by long lithium-carbon distance of 246 pm. The insertion reaction of carbon monoxide into the lithium-carbon bond is now exothermic (-5.4 kcal/mole) and the resulting acetyl group coordinates with its carbon atom to one and with the oxygen to two lithium atoms of a lithium-tetrahedron plane. This coordination in a μ_3-fashion helps to find an exothermic reaction pathway for the insertion. Fig. 2 illustrates the results on the HF/6-31G** level of theory of the above mentioned reaction steps.

The Oligomerization of HLiCO and LiCOH

Since HLiCO and LiCOH are intramolecular solvated organolithium species – in contrast to LiH – it is a question of interest, whether oligomerization is still a strongly exothermic process. Therefore again ab initio calculations on HF/6-31G** level of theory are used to explore geometries and reaction enthalpies of dimers and tetramers, when the lithium to carbon monoxide ratio is 1:1.

Dimeric HLiCO is a species with D_{2h} symmetry. It is formed by bridging hydride ligands and CO coordinating the lithium with the carbon atom (Li–Li 236; Li–H 182; Li–C 235; C–O 111). The coordination number of lithium is thereby three. This intermediate is still able to form a tetrameric HLiCO in T_d symmetry (solvated heterocubane structure) in a strongly exothermic reaction. The main cause is the higher coordination number of lithium (7), since three neighbouring lithium and hydrogen atoms and one carbon from the ligand is found (Li–Li 256; Li–H 191; Li–C 237; C–O 111):

$$+ \, CO$$
$$\Delta H = -19.6 \text{ kcal/mole} \longrightarrow 0.5$$

Eq. 5

Thus, it can be concluded that coordination of CO does not significantly effect the oligomeric clusters. Solvation by carbon monoxide does not succeed in breaking up the larger clusters to small species like dimers.

$$+ \, CO$$
$$\Delta H = -14.8 \text{ kcal/mole} \longrightarrow 0.5$$

Eq. 6.

Similar behaviour is found for the lithium formyl species. Even the formyl group, which seems to be a good intramolecular donor, cannot stabilize smaller species than tetramers. The tetrameric LiCOH (Li–Li 279; Li–C 206; C–O 126; Li–O 197) in its unusual S_4 summetry breaks up to dimers (Li–Li 292; Li–C 214; C–O 124; Li–O 182), which are six-membered rings of C_{2h} symmetry, in an endothermic process. But the reaction enthalpy is lowered to -14.8 kcal/mole, since the formyl group reduces the subcoordination of the dimer due to its more flexible coordination mode. In the dimer it is bridging the Li$_2$ and in the tetramer the Li$_4$ core to reach a maximum of coordination numbers.

Acknowledgment: The author thanks the *Deutsche Forschungsgemeinschaft* and the *Fonds der Chemischen Industrie* for support.

References:

[1]   J. E. Ellis, W. Beck, *Angew. Chem.* **1995**, *107*, 2695.

[2]   C. A. Arrington, J. T. Petty, S. E. Payne, W. C. K. Haskins, *Amer. Chem. Soc.* **1988**, *110*, 6240.

[3]   M. Pearsall, R. West, *J. Amer. Chem. Soc.* **1988**, *110*, 7228.

[4]   M. Tacke, Ch. Klein, D. J. Stufkens, A. Oskam, P. Jutzi, E. A. Bunte, in: *Organosilicon Chemistry: From Molecules to Materials* (Eds.: N. Auner, J. Weis), VCH, Weinheim, **1994**, p. 93.

[5]   P. Jutzi, U. Holtmann, D. Kanne, C. Krüger, R. Blom, R. Gleiter, I. Hyla-Kryspin, *Chem. Ber.* **1989**, *122*, 1629.

[6]   M. Tacke, *Chem. Ber.* **1995**, *128*, 1051.

[7]   M. Tacke, P. Sparrer, R. Teuber, H.-J. Stadter, F. Schuster, in: *Organosilicon Chemistry II: From Molecules to Materials* (Eds.: N. Auner, J. Weis), VCH, Weinheim, **1996**, p. 837.

Unprecedented Multistep Reactions of Decamethyl-silicocene, $(Me_5C_5)_2Si:$, with CO_2, CS_2, COS, RNCS (R = Me, Ph), with CF_3CCCF_3, and with $HMn(CO)_5$

Peter Jutzi*, Dirk Eikenberg
Fakultät für Chemie
Universität Bielefeld
Universitätsstraße 25, D-33615 Bielefeld, Germany
Tel.: Int. code + (521)1066181 – Fax: Int. code + (521)106602 6

Keywords: Decamethylsilicocene / Silanone / Silathione / Silylene / Transition metal hydride

Summary: Decamethylsilicocene (**1**), the first Si(II) compound stable under ordinary conditions [1], is a hypercoordinated nucleophilic silylene, which reacts preferentially with electrophilic substrates [2]. In the reaction of **1** with the electrophilic heterocumulenes CO_2, COS and RNCS (R = Me, Ph), double bond species of the type $Cp*_2Si=X$ (X = O, S) are formed, which are stabilized via different routes to the silaheterocycles **I-IV** [3]. Multistep rearrangement processes are postulated to explain the formation of the dithiasiletane derivatives **V** and **VI** in the reaction of **1** with CS_2. A surprising polycyclic silaheterocycle **VII** is obtained in the reaction of **1** with hexafluorobutyne. With $HMn(CO)_5$ **1** reacts to the dimanganese-substituted silane **VIII**. In all reactions, the lone pair at silicon is involved, an η^5-η^1- rearrangement of the Cp* ligands take place, and the formal oxidation state at Si changes to +4 in the final products.

Introduction

Recent years have brought much progress in the attempts to synthesize stable monomeric compounds with divalent silicon. Species analogous to Arduengo's carbene have been prepared [4] as well as a higher coordinated Si(II) compound with two diphosphinomethanide ligands [5]. The first Si(II) compound stable under ordinary conditions was decamethylsilicocene (**1**), in which two pentamethylcyclopentadienyl ligands are bonded in a η^5 fashion to the silicon atom. The chemistry of **1** has already been investigated in some detail. As a result of these investigations, **1** is regarded as a hypercoordinated electron-rich nucleophilic silylene. Unprecedented reactions in silylene chemistry have been observed in the reaction of decamethylsilicocene (**1**) with the heterocumulenes carbon dioxide, carbon disulfide, carbon oxysulfide, methyl isothiocyanate and phenyl isothiocyanate, with hexafluorobutyne, and with hydridopentacarbonyl-manganese. In these reactions the compounds **I-VIII** are obtained as the final products (see Scheme 1). The products have been characterized by NMR (^1H, ^{13}C, ^{29}Si) spectroscopy, IR spectroscopy, mass spectrometry, microanalytical data and, with the exception of **III**, by X-ray crystallography. The formation of these interesting silicon containing compounds will be discussed.

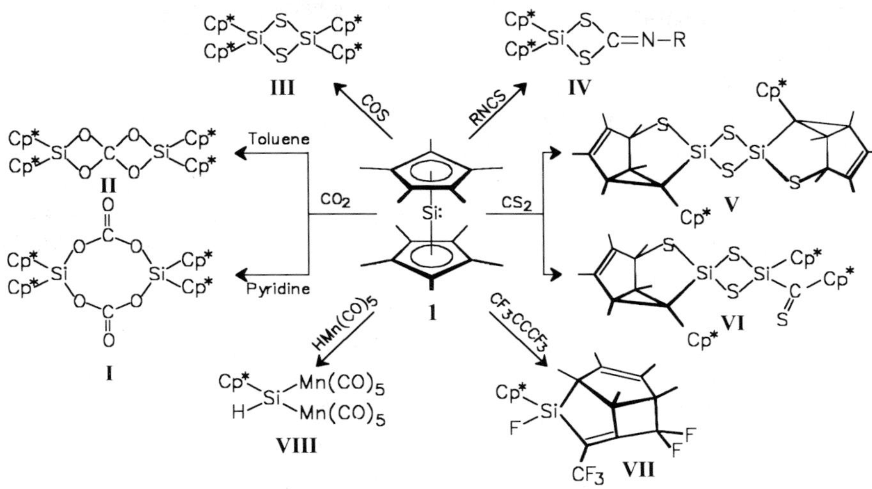

Scheme 1. Unprecedented multistep reactions of decamethylsilicocene (**1**).

Reaction of Decamethylsilicocene (1) with CO₂, COS, and RNCS (R = Me, Ph)

Decamethylsilicocene (**1**) reacts with carbon dioxide already under mild conditions. Surprisingly, the products obtained depend on the solvent used. Bubbling CO_2 at room temperature for about 3 through a solution of **1** in pyridine led to the eight-membered cyclic compound **I** in about 65 % yield, whereas in toluene as solvent the spiro heterocyclic compound **II** was formed in about 70 % yield. **1** reacts with carbon oxysulfide already under very mild conditions: A toluene solution of **1** was added to liquid COS at -78°C. The dithiadisiletane **III** was isolated in about 50 % yield after a reaction time of 2 at this temperature. Compound **III**, already known in the literature as the reaction product of **1** with sulfur [6], is poorly soluble in organic solvents.

In the reaction of decamethylsilicocene (**1**) with methyl isothiocyanate, the dithiasiletane **IV** was isolated in about 60 % yield after a reaction time of 16 h at room temperature (see Scheme 2). Slightly more drastic conditions (5 h, 65°C) were necessary for the reaction of **1** with phenyl isothiocyanate, which led to the corresponding dithiasiletane **IV** in 65 % yield. The formation of the **IV** was independent of the stoichiometry of the reactands.

The pathway of the reaction of **1** with the above heterocumulenes is discussed as follows (see Scheme 2): The first intermediate is a highly reactive [2+1] cycloaddition product or its ring-opened isomer, which easily loses CO or in the case of isothiocyanates the corresponding isonitrile RNC to give the silanone or silathione **2**. The formation of CO was proved in the reaction of **1** with CO_2 by reaction with the iron complex $(H_5C_5)Fe(SMe_2)_3^+$ BF_4^- [7]. In the reaction of **1** with isothiocyanates the formation of RNC was proved using NMR spectroscopy. The intermediates of the type $Cp*_2Si=X$ (X = O, S) are not stable and stabilized via different routes to the final products. In the reaction of **1** with CO_2 the double bond species $Cp*_2Si=O$ is transformed by CO_2 to the formal [2+2] cycloaddition product **3**, which is once more a reactive intermediate. In toluene as solvent, **3** reacts with the silanone to give the final product **II**. In pyridine as solvent, the intermediate silanone is deactivated [8]. As a result, **3** does not react with Cp*$_2$Si=O, but forms the dimerisation product **I** after ring opening at one of the Si–O bonds.

In the case of the reaction of **1** with COS the intermediate silathione dimerizes to the dithiadisiletane derivative **III**. In the reaction of **1** with RNCS the silathione does not dimerize, but it is trapped by the respective isothiocyanate in form of the formal [2+2] cycloaddition product **IV**. The intermediates Cp*$_2$Si=X (X = O, S) (**2**) could not be isolated but derivatized by trapping reactions. In the presence of *t*-butyl methyl ketone the silanone (Cp*$_2$Si=O) reacts in an ene-type reaction [9] to give the addition product **5**; the silathione (Cp*$_2$Si=S) is transformed in a first step to an addition product analogous to **5**, in a further step this species reacts with **1** in an oxidative addition process to form the final product **4** (see Scheme 2).

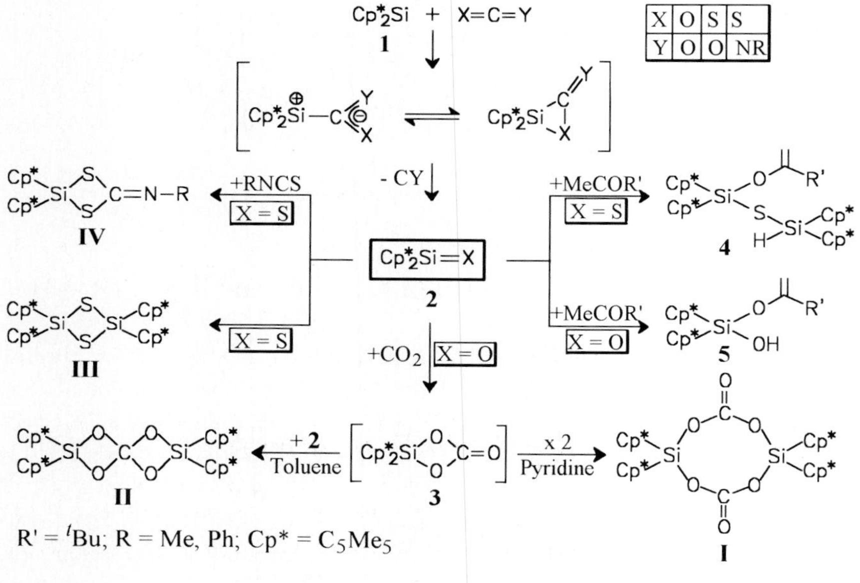

Scheme 2. Reaction of decamethylsilicocene with the heterocumulenes CO_2, COS, and RNCS (R = methyl, phenyl).

Reaction of Decamethylsilicocene (1) with Carbon Disulfide

In the reaction of **1** with CS_2, a highly surprising multistep reaction is observed. Excess CS_2 was added to a solution of **1** in benzene and after 16 h at room temperature the dithiadisiletane **V** was isolated in about 60 % yield. The dithiadisiletane **VI** can be isolated as byproduct in about 5 % yield.

The pathway of the underlying process is described as follows (see Scheme 3): A highly reactive formal [2+1] cycloaddition product or its ring-opened isomer is the first intermediate. In a subsequent multistep rearrangement, the ligands at carbon and at silicon have to be completely exchanged: after nucleophilic attack of **1** at the carbon atom of CS_2 a dyotropic rearrangement process [10] leads to the intermediate **6**. In a further dyotropic rearrangement process the sulfur atom migrates from carbon to silicon and the Cp* ligand from silicon to carbon under formation of a dipolar ring-opened intermediate. This species once more is highly reactive and gains stabilization by a rearrangement process transforming a classical carbenium ion into an allyl-type cation in form of the intermediate **7**. Comparable rearrangement processes are described in the literature [11]. In a further step **7** is transformed by carbon-sulfur bond formation to the silathione **8**, which reacts in a dimerazation reaction to the dithiadisiletane derivative **V**. This reaction pathway is confirmed through the formation of the dithiadisiletane derivative **VI** as by-product, which is the formal [2+2] cycloaddition product of the intermediates **6** and **8**.

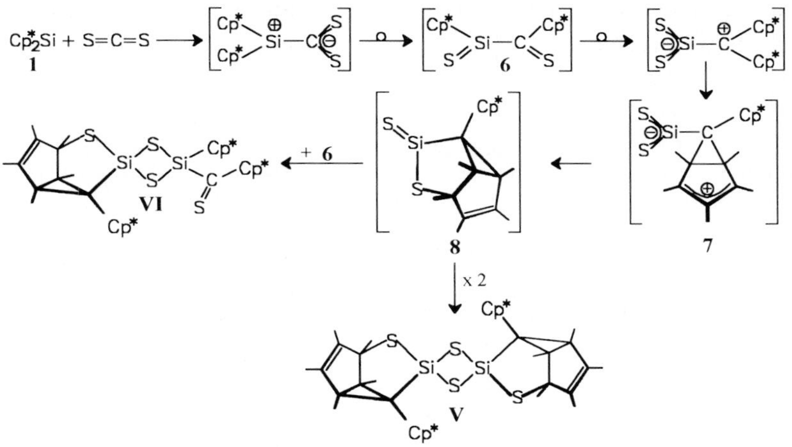

Scheme 3.   Reaction of decamethylsilicocene with CS_2

Reaction of Decamethylsilicocene (1) with Hexafluorobutyne

Decamethylsilicocene (**1**) reacts with alkynes having electron-withdrawing substituents ($MeO_2CC\equiv CCO_2Me$, $Me_3SiC\equiv CSO_2Ph$) under formation of silacyclopropene derivatives [12]. A surprising reaction is observed between **1** and hexafluorobutyne.

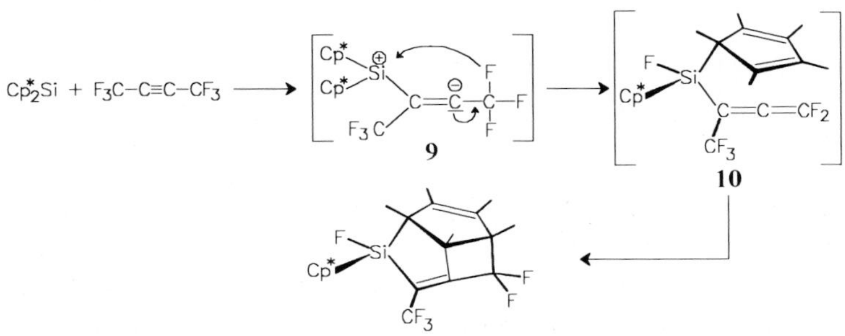

Scheme 4.   Reaction of decamethylsilicocene with $CF_3–C\equiv C–CF_3$.

Treatment of a solution of **1** in toluene with the butyne derivative at -30°C affords the polycyclic silaheterocycle **VII** in high yield. For the above reaction we propose the following pathway (see Scheme 4): First a nucleophilic attack of **1** to one of the alkyne carbon atoms takes place under formation of the dipolar intermediate **9**, which is stabilized by migration of a fluorine atom from a CF_3 group to the silicon atom. The resulting cumulene **10** is not stable under the reaction conditions and reacts instantly with one of the Cp* ligands in a [2+2] cycloaddition to form the final product **VII**.

Reaction of Decamethylsilicocene (1) with Hydridopentacarbonylmanganese

The insertion of silylenes into heteronuclear single bonds is a well-documented type of reaction. Insertion reactions of silylenes into transition-metal-hydrogen bonds are rarely investigated; only one attempt has been reported [13]. Decamethylsilicocene (**1**) reacts with $HMn(CO)_5$ under mild conditions (-30°C) to the dimetallo-substituted silane **VIII** in good yield. The reaction is independent of the stoichiometry of the substrate. The first step (see Scheme 5) is an oxidative addition reaction between **1** and $HMn(CO)_5$ to form the silane **11**, which reacts with a further molecule of $HMn(CO)_5$ to the final product **VIII** and to pentamethylcyclopentadiene. As in many other processes, the Cp* ligand acts as a leaving group [14].

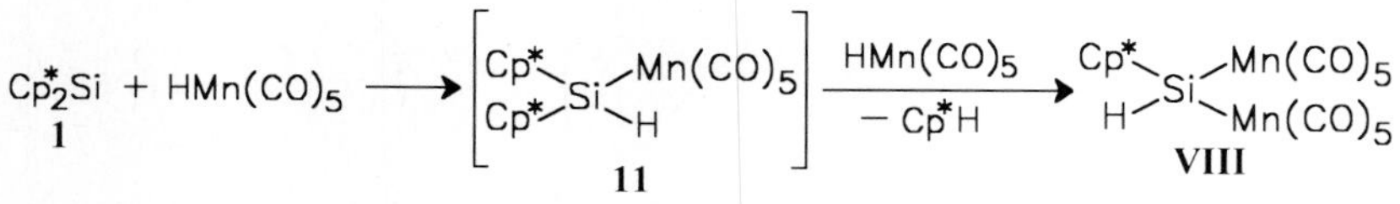

$$Cp_2^*Si + HMn(CO)_5 \longrightarrow \left[\begin{matrix} Cp^* \\ Cp^* \end{matrix} Si \begin{matrix} Mn(CO)_5 \\ H \end{matrix} \right] \xrightarrow[-Cp^*H]{HMn(CO)_5} \begin{matrix} Cp^* \\ H \end{matrix} Si \begin{matrix} Mn(CO)_5 \\ Mn(CO)_5 \end{matrix}$$

1   **11**   **VIII**

Scheme 5.   Reaction of decamethylsilicocene with $HMn(CO)_5$.

Acknowledgements: Sincere thanks are extended to the coworkers mentioned in the references. Generous support by the *Deutsche Forschungsgemeinschaft* is also gratefully acknowledged.

References:

[1] a) P. Jutzi, D. Kanne, C. Krüger, *Angew. Chem., Int. Ed. Engl.* **1986**, *25*, 164.
   b) P. Jutzi, U. Holtmann, D. Kanne, C. Krüger, R. Blom, R. Gleiter, I. Hyla-Kryspin, *Chem. Ber.* **1989**, *122*, 1629.

[2] For example, HX, BX_3, PX_3, AuCl, RX, R_2NH, RSH, $RCOCH_2COR$, R_2CO, RCN, RCCR, see Abstract *Sendai International Symposium on the Frontiers of Organosilicon Chemistry*, Japan, **1994**.

[3] P. Jutzi, D. Eikenberg, A. Möhrke, B. Neumann, H.-G. Stammler, *Organometallics* **1996**, *15*, 753.

[4] U. Denk, R. Lennon, R. West, A. V. Belgalov, H. P. Verne, A. Haaland, M. Wagner, H. Ketzler, *J. Am. Chem. Soc.* **1994**, *116*, 2691.

[5] H. H. Karsch, U. Keller, S. Gamper, G. Müller, *Angew. Chem., Int. Ed. Engl.* **1990**, *29*, 295.

[6] P. Jutzi, A. Möhrke, A. Müller, H. Bögge, *Angew. Chem., Int. Ed. Engl.* **1989**, *28*, 1518.

[7]   Reaction of CO with [(C$_5$H$_5$)Fe(Me$_2$S)$_3$]BF$_4$ leads to a carbonyl complex IR(CsI): $\nu_{(CO)}$ = 1980, 1730 cm^{-1}.

[8]   C. Chult, R. J. P. Corriu, C. Reye, J. C. Young , *Chem. Rev.* **1993**, *93*, 1371.

[9]   N. Wiberg, G. Preiner, G. Wagner, *Z. Naturforsch.* **1987**, *B 42*, 1062.

[10] M. T. Reetz, *Adv. Organomet. Chem.* **1977**, *16*, 33.

[11] a) R. F. Childs, M. Sakai, B. D. Parrington, S. Winstein, *J. Am. Chem. Soc.* **1974**, *96*, 6403.
      b) R. F. Childs, S. Winstein, *J. Am. Chem. Soc.* **1974**, *96*, 6409.

[12] P. Jutzi, A. Möhrke, S. Pohl, W. Saak, unpublished.

[13] D. H. Berry, Q. Jiang, *J. Am. Chem. Soc.* **1987**, *109*, 6210.

[14] a) P. Jutzi, D. Kanne, M. Hursthouse, A. J. Howes, *Chem. Ber.* **1988**, *121*, 1299.
      b) P. Jutzi, U. Holtmann, H. Bögge, A. Müller, *J. Chem. Soc., Chem Commun.* **1988**, 305.

Rearrangement of Bis(hypersilyl)silylene and Related Compounds: An Unusual Way to Three-Membered Rings

Karl Wilhelm Klinkhammer[*]

Institut für Anorganische Chemie
Universität Stuttgart
Pfaffenwaldring 55, D-70550 Stuttgart, Germany
Fax: Int. code + (711)6854241
E-mail: Karl.Klinkhammer@rus.uni-stuttgart.de

Keywords: Bis(hypersilyl)silylene / Bis(hypersilyl)germylene / Rearrangement / Three-Membered Rings

Summary: Solvent-free alkali metal tris(trimethylsilyl)silanides (hypersilanides) react with bis[bis(trimethylsilyl)amino]germylene or trichlorosilane yielding hexakis(tri-methylsilyl)germadisilirane and the related trisilirane. Plausible intermediates are bis(hypersilyl)germylene and silylene, respectively. Using tris(trimethylsilyl)germanides instead of hypersilanides, the formation of the analogous trigermirane and digermasilirane is observed.

Recently we reported the syntheses of bis(hypersilyl)stannylene and plumbylene, $[(Me_3Si)_3Si]_2Sn$ and $[(Me_3Si)_3Si]_2Pb$ [1], using solvent-free alkali metal derivatives of tris(trimethylsilyl)silane (hypersilane) [2]. Based on this success we tried to utilize the same method for the synthesis of the hitherto unknown corresponding germylene and silylene. Thus, bis[bis(trimethylsilyl)amino]-germylene and, as source of divalent silicon, trichlorosilane had been reacted with sodium or lithium hypersilanide in *n*-pentane at -30 to -60°C (Schemes 1 and 2).

Scheme 1.   E = Si, Ge; M = Na; Ts = SiMe$_3$.

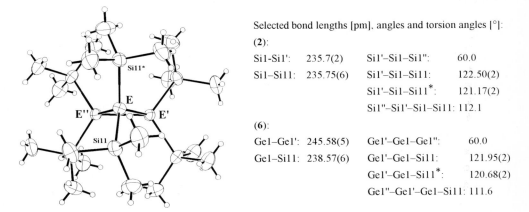

3/2

<Pentan>
-30°C

+

+ 3 MCl

+ Ts₃E-H

Scheme 2. E = Si, Ge; M = Li, Na; Ts = SiMe₃.

The only reaction products that could be isolated in high yield, identified by various spectroscopic methods, were hexakis(trimethylsilyl)disilagermirane (**1**) and hexakis(trimethylsilyl)trisilirane (**2**). Whereas the former had been already identified by Heine and Stalke as the main product of a related reaction of solvated lithium hypersilanide with GeCl₂ [3], the trisilirane **2**, to the best of our knowledge, had not been synthesized before. A mechanism for the formation of disilagermirane **1** via the reaction of bis(trimethylsilyl)germylene (**3**) and tetrakis(trimethylsilyl)disilene (**4**), as proposed by Stalke [3] seems not to be very likely because it is not at all obvious how these molecules could be formed. Instead, we suggest the intermediate formation of bis(hypersilyl)germylene (**5**) and its fast rearrangement, the driving force being the formation of one more strong single bond. Strong arguments for this proposal are the lack of side-products and the failure to intercept any reactive intermediate so far. Moreover, we could show that this reaction type is easily transferable to other similar species. Using tris(trimethylsilyl)germanides instead of hypersilanides under the same conditions leads to the clean formation of the – now expected – trigermirane **6** and siladigermirane **7** (Schemes 1 and 2). At this point it should be noted that similar rearrangements are already known or at least postulated in the case of some simple dialkylsilylenes. In these cases siliranes were formed instead [4].

Selected bond lengths [pm], angles and torsion angles [°]:

(2):

Si1-Si1':	235.7(2)	Si1'-Si1-Si1":	60.0
Si1-Si11:	235.75(6)	Si1'-Si1-Si11:	122.50(2)
		Si1'-Si1-Si11*:	121.17(2)
		Si1"-Si1'-Si1-Si11: 112.1	

(6):

Ge1-Ge1':	245.58(5)	Ge1'-Ge1-Ge1":	60.0
Ge1-Si11:	238.57(6)	Ge1'-Ge1-Si11:	121.95(2)
		Ge1'-Ge1-Si11*:	120.68(2)
		Ge1"-Ge1'-Ge1-Si11: 111.6	

Fig. 1. Molecular structure of (Me₃Si)₆Si₃ (**2**) (Me₃Si)₆Ge₃ (**6**) and (thermal ellipsoids at 50 % probability level).

The molecular structures of trisilirane **2** and trigermirane **6** have been derived from X-ray diffraction data on well-shaped trigonal colorless crystals. Despite the bulkiness of the Me_3Si ligands, these cycles comprise very short endocyclic Si–Si and Ge–Ge bonds, respectively (Fig. 1), within molecules of perfect D_3-symmetry. In fact, these are the shortest observed so far in such cyclic compounds [5]. Presumably the shortening is caused by the strongly electron donating substituents.

The good quality X-ray data allowed for the localization of maxima of electron density from difference-Fourier maps corresponding to the banana bonds typical for small ring systems, thus deriving "true" endocyclic bond angles not far away from an ideal tetrahedral angle (Fig. 2).

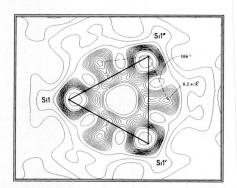 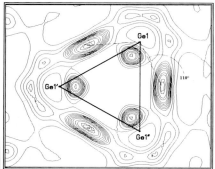

Fig. 2. Difference electron density plots from X-ray data showing density according to Si–Si and Ge–Ge bond bending in the cyclotrisilane **2** and cyclotrigermane **6**, respectively.

Outlook

We have to prove the proposed mechanism by theoretical calculations on suitable model systems. If the mechanism is true, we should be able to tranfer this synthesis to other three-membered ring systems even with a distinct pattern of substituents just by varying the alkali metal salt employed.

References:

[1]   K. W. Klinkhammer, W. Schwarz, _Angew. Chem._ **1995**, _107_, 1449.

[2]   K. W. Klinkhammer, W. Schwarz, _Z. Anorg. Allg. Chem._ **1993**, _619_, 1777; K. W. Klink-hammer, G. Becker, W. Schwarz, _"Alkali Metal Derivatives of Tris(trimethylsilyl)silane – Syntheses and Molecular Structures"_, in: _Organosilicon Chemistry II: From Molecules to Materials_ (Eds.: N. Auner, J. Weis), VCH, Weinheim, **1996**, p. 493; K. W. Klinkhammer, _Chem. Eur. J._ **1997**, _3_, 1418.

[3]   A. Heine, D. Stalke, _Angew. Chem._ **1994**, _106_, 121.

[4]   See for example: I. M. T. Davidson, G. H. Morgan, _Organomet._ **1993**, _12_, 289; and cited refs. therein.

[5]   Reviews: T. Tsumuraya, S. A. Batcheller, S. Masamune, _Angew. Chem._ **1991**, _103_, 916; M. Weidenbruch, _Chem. Rev._ **1995**, _95_, 1479.

[6] Crystal data of compounds **2** and **6**: $Mo_{K\alpha}$, -100°C, rhombohedral, space group $R\bar{3}c$, **2**: $a = 19.156(2)$ Å, $a = 31.98(2)°$. 1415 independent reflections ($4.5 < 2\Theta < 56°$), $R_1 = 0.023$. **6**: $a = 19.235(2)$ Å, $\alpha = 31.91(2)°$. 1301 independent reflections ($4.5 < 2\Theta < 54°$), $R_1 = 0.022$.

Oxidation of Silenes and Silylenes:
Matrix Isolation of Unusual Silicon Species

Wolfram Sander*, Martin Trommer, Andreas Patyk
Lehrstuhl für Organische Chemie II
Ruhr-Universität Bochum
D-44780 Bochum, Germany
Fax: Int. code + (234)7094353

Keywords: Matrix Isolation / Oxidation / Silene / Silylene

Summary: The oxidation of several silenes and silylenes was investigated using the matrix isolation technique. Depending on the substitution these unsaturated species react under cryogenic conditions with molecular oxygen either thermally or photochemically (irradiation into the long wavelength absorption of charge transfer complexes). The principal oxidation products of the silylenes are dioxasiliranes. Silanone O-oxides are – if formed at all – too short lived to be observed even under the conditions of matrix isolation. The oxidation of silenes results in the cleavage of the Si=C bond, presumably via dioxasiletanes, and formation of silanone-aldehyde complexes. These complexes rearrange via siloxycarbenes to the final products.

Silenes and silylenes are highly reactive molecules that rapidly react with molecular oxygen [1-7]. Despite the interest in the structure and synthetic potential of these species, the oxidation mechanisms have long been obscure. In several cases siloxanes as the polymers of silanones were observed [4, 5]. To elucidate the oxidation mechanisms we investigated a series of silenes and several silylenes using the matrix isolation technique [2, 3, 6, 7]. This technique allows to directly observe bimolecular thermal reactions in O_2-doped inert gas matrices [8, 9]. A necessary prerequisite is an activation barrier for the oxidation process of less than several kcal/mol. The advantage of this method is that the photochemical generation of the reactive species, e.g., a silene or silylene, and the subsequent thermal oxidation are completely separated. In the first step the reactive intermediate is prepared by photolysis of a suitable matrix isolated precursor at 10 K, or by gas phase thermolysis of a precursor with subsequent trapping of the products in solid argon, and characterized by IR and UV/vis spectroscopy. In the second step the thermal oxidation is induced by annealing the matrix at 30-80 K (the temperature depends on the inert gas, 30-45 K for argon and 50-80 K for xenon) to allow diffusion of the trapped oxygen molecules and other species. During the annealing process changes in the matrix are monitored spectroscopically. The oxidation products were identified and characterized by independent synthesis, by comparison of the IR spectra with ab initio calculated spectra (in most cases at the RHF and MP2/6-31G(d,p) level of theory), by [18]O isotopic labeling, and by characteristic photochemical behavior.

Oxidation of Silylenes

Silylenes **1** are highly reactive homologues of the carbenes, and we have been interested to compare the reactivity and primary products of the oxidation of these divalent species. In principal one can expect two different primary adducts of a silylene and molecular oxygen: the formal "end-on" adducts silanone *O*-oxide **2** or "side-on" adducts dioxasilirane **3** (Scheme 1). It was shown by numerous matrix studies [8-12], experiments in solution using time resolved spectroscopy [13-16], and a preparative scale synthesis in solution [17] that triplet as well as singlet carbenes yield carbonyl *O*-oxides as the primary oxidation products, while dioxiranes are products of secondary photolysis. Ando et al. reported on the synthesis of the silanone *O*-oxide **2e** by the reaction of dimesitylsilylene **1e** and O_2 in solid argon [1]. This is so far the only experimental evidence for a silanone *O*-oxide.

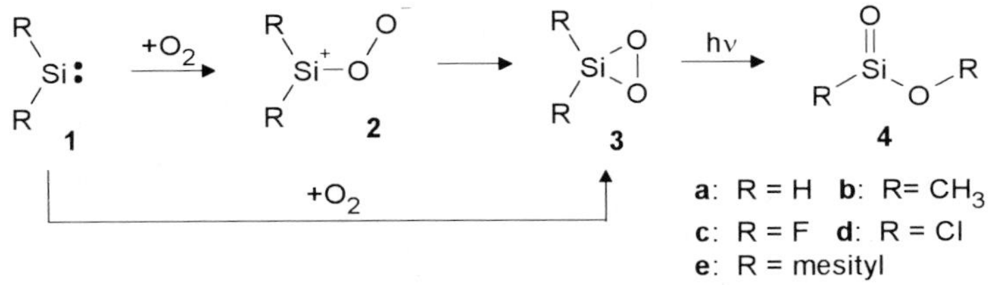

Scheme 1. Oxidation of silylenes **1**.

In our hands the thermal oxidation of dimethylsilylene **1b** could be directly observed by annealing O_2-doped argon matrices (0.1-0.5 % O_2) at 35-42 K [2]. The only oxidation product identified was dimethyldioxasilirane **3b**, silanone *O*-oxide **2b** was not observed. This is in line with ab initio calculations by Cremer et al., which demonstrate that the cyclization of **2a** is highly exothermic (63.8 kcal/mol) with an activation barrier of only 6.5 kcal/mol [18]. In contrast, the barrier for the rearrangement of formaldehyde *O*-oxide to dioxirane is 22.8 kcal/mol [8, 18]. This difference rationalizes the difficulties in isolating silanone oxide **2b**, while carbonyl oxides are frequently obtained in matrices. Siladioxirane **3b** was also obtained by the reaction of silylene **1b** and molecular oxygen in the gas phase with subsequent trapping of the products in an argon matrix. The photochemistry of **3b** is analogous to that of dioxiranes: irradiation at $\lambda = 436$ nm results in the rearrangement to silaester **4b**, presumably via cleavage of the O–O bond in **3b** with subsequent migration of a methyl group.

Halogen-substituted silylenes **1c** and **1d** are less reactive than alkyl-substituted silylenes. This is drastically shown by the matrix isolation of **1c** and **1d** in solid oxygen: up to 45 K, where O_2 rapidly evaporates, no reaction is observed. However, irradiation at 365 nm and 575 nm for **1c** and **1d**, respectively, rapidly yields dioxasiliranes **3c** and **3d** as the only products. Since there is no migratory aptitude of halogen from silicon to oxygen, these dioxasiliranes are photochemically stable even at 220 nm irradiation [3].

Oxidation of Silenes

The silenes **5b-d** were synthesized by irradiation of the corresponding silyldiazomethanes [7, 19, 20]. This method provides a clean, almost quantitative, source of **5**, which are completely stable in argon at 10 K. However, annealing at 35-45 K allows the diffusion of **5** and dimerization to 1,3-disiletanes. Annealing of the electron rich silene **5d** in 1-2% O_2-doped argon matrices results in a partial thermal oxidation (approximately 35-40 % within 40 min). The reaction can be completed by irradiation ($\lambda >$ 360 nm) at 35 K. As opposed to the oxidation of **5d**, silenes **5b** and **5c** require photochemical activation in combination with very high O_2 concentrations (irradiation at $\lambda > 575$ nm, $> 10\%$ O_2-doped argon matrices) to be oxidized. This differing behavior can be rationalized by the increasing ionization potential from **5a** to **5d** (Scheme 2), otherwise the oxidation mechanisms are similar for all three silenes.

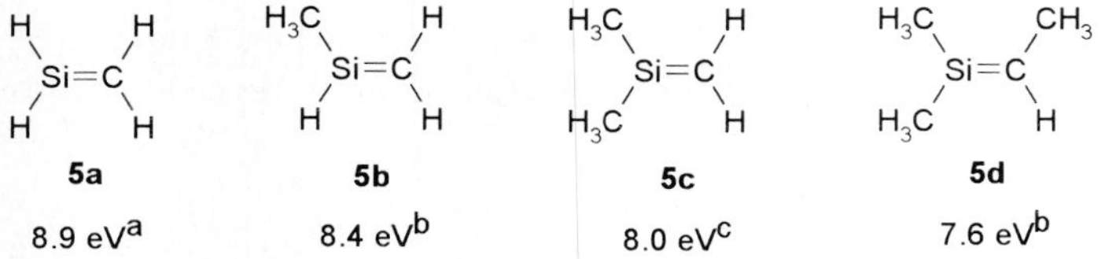

5a	**5b**	**5c**	**5d**
8.9 eV[a]	8.4 eV[b]	8.0 eV[c]	7.6 eV[b]

Scheme 2.   Ionization potentials of silene **5a** and some methyl substituted derivatives. [a] Ref. 21 – [b] Estimated by AM1 calculations scaled with the experimental values of silene **5a** and 1,1-dimethylsilene **5c** – [c] Ref. 22.

The first step in the oxidation is the formation of a triplet diradical T-**6**, which can interconvert to the singlet diradical S-**6** via intersystem crossing (ISC). Diradicals **6** are unstable even under the conditions of matrix isolation and cannot be observed by direct spectroscopic methods. However, the formation of hydroperoxide **7** – the formal product of an ene reaction – as one of the two major products of the thermal oxidation of **5d** is rationalized best by postulating peroxy diradical **6d** as an intermediate (Scheme 3).

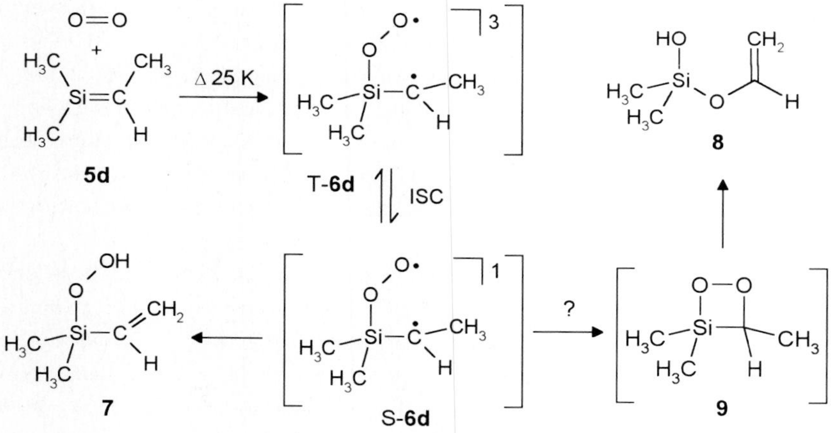

Scheme 3.   Oxidation of 1,1,2-trimethylsilene **5d**.

The second product of the oxidation is vinyloxysilane **8** in the *s-Z*-conformation, which could be clearly demonstrated by comparison of the IR spectrum of **8** with that of *s-Z*- and *s-E*-trimethyl(vinyloxy)silane [23, 24]. The formation of **8** requires the insertion of an oxygen atom into a

Si–C bond and the transfer of a hydrogen atom. A mechanism for this rearrangement, which also addresses the question whether dioxasiletane **9** is involved as an intermediate, will be given later.

The photooxidation of **5b** and **c** in O_2-doped argon matrices (up to 10% O_2) is induced by irradiation with red light (Scheme 4). Since neither **5** nor O_2 absorb at 575 nm, we assume that the photooxidation of these silenes is caused by excitation of a **5**···O_2 charge transfer complex, which could not be characterized spectroscopically. The product of the 575 nm-irradiation is adduct **10** with a highly characteristic IR spectrum which is in good accordance with ab initio calculations [7].

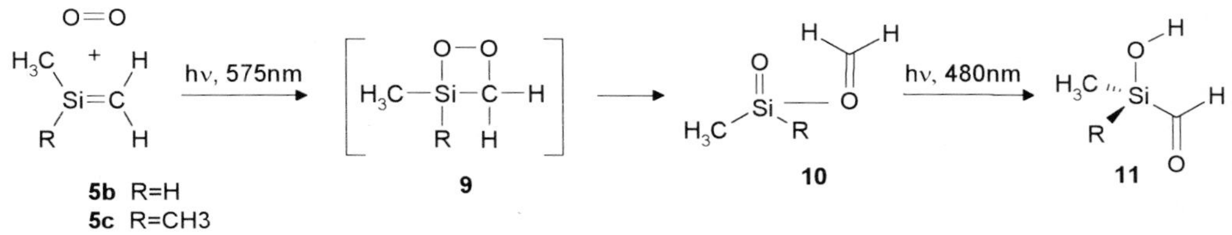

Scheme 4.   Oxidation of 1-methylsilene **5b** and 1,1-dimethylsilene **5c**.

The red-shift of the C=O and Si=O stretching vibration of 100 and 49 cm^{-1}, respectively, the calculated bond energy of complex **10** of more than 20 kcal/mol (RHF/6-31G(d,p)), and the calculated "non-bonding" Si–O distance of 1.98 Å, being only slightly longer than the normal Si–O bond distance, indicate the unusual electronic properties of **10**. Evidently dioxasiletane **9** is formed as a thermally or photochemically unstable intermediate which rapidly decomposes to the silanone-formaldehyde complex.

According to ab initio calculations the stabilization of complexes of type **10** depends on the substitution pattern of the methyl groups (Scheme 5). Methyl groups at the silanone moiety destabilize the complex by 1-2 kcal/mol, while acetaldehyde is more strongly bound than formaldehyde by ca. 3 kcal/mol.

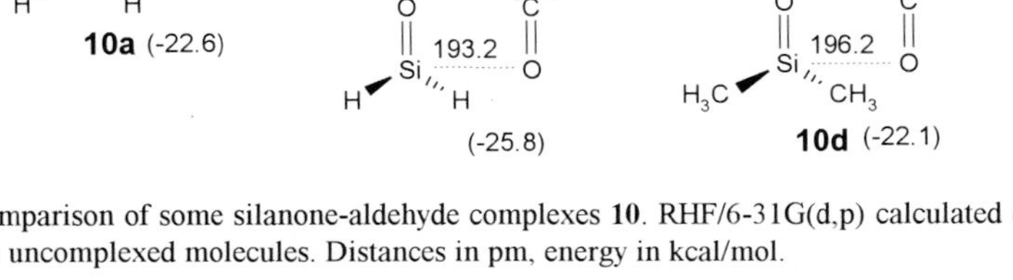

Scheme 5.   Comparison of some silanone-aldehyde complexes **10**. RHF/6-31G(d,p) calculated energies relative to the uncomplexed molecules. Distances in pm, energy in kcal/mol.

At shorter wavelength irradiation (> 480 nm) **10** rearranges to formylsilanol **11**. This rearrangement is remarkable, since the strong Si···O interaction in **10** obviously does not lead to a new Si–O bond. Instead, an Si–C bond is formed and a hydrogen atom transferred from the formaldehyde to the silanone moiety. The mechanism of the reaction between silanone and formaldehyde can be understood by comparison with the analogous reaction of silenes with formaldehyde (Scheme 6) [25]. Silenes **5b** and **5c** thermally react with formaldehyde even at a temperature as low as 30 K. A complex of **5** and formaldehyde is not observed, although calculated to be slightly stabilized (3-4 kcal/mol at the RHF level of theory). The major thermal product is the highly unstable siloxycarbene **13**, not oxasiletane **15**, the product of a formal [2+2] cycloaddition. The formation of **13** is rationalized by high level ab initio calculations (CCSD(T)//MP2/6-31G(d,p)) for the parent system (CH$_2$=SiH$_2$ + CH$_2$=O) [26]. Diradical **12** lies in a very shallow minimum 20.3 kcal/mol above **10**, with an activation barrier of only 0.3 kcal/mol for the hydrogen abstraction. Although the exothermicity of the formation of **15** is more than three times as large (67.2 kcal/mol), the barrier for the cyclization is 1.6 kcal/mol, and thus slightly higher than for the hydrogen abstraction.

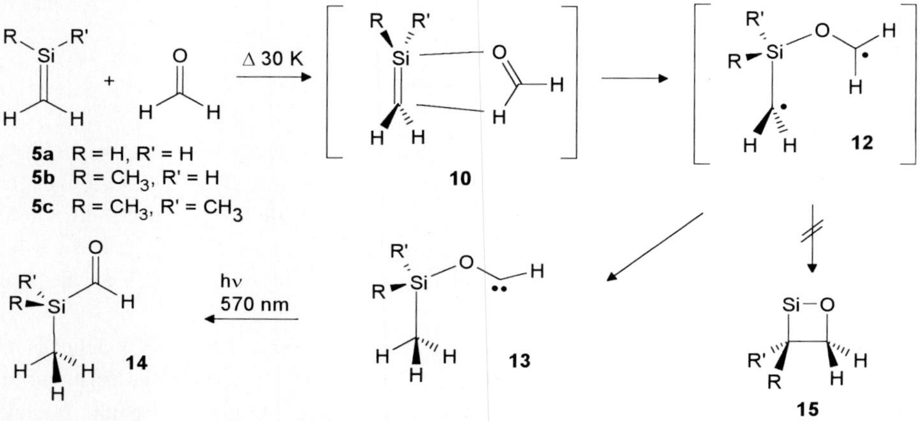

Scheme 6.   Reaction of silenes **5** with formaldehyde.

Carbene **13** is extremely photolabile, and 570 nm irradiation results in the rearrangement to acylsilanes **14**. The calculated activation barrier for this highly exothermic rearrangement is only 4.7 kcal/mol. If we assume that the analogous reaction of silanone and formaldehyde also leads to a labile siloxycarbene, the formation of formylsilanol **11** is easily explained.

Vice versa, siloxycarbenes have been generated as short lived intermediates by irradiation of acylsilanes [27-29]. Irradiation of matrix isolated acylsilanes **14c**, **e** and **f** resulted both in the α-cleavage to give radical pairs and in the rearrangement to siloxycarbenes **13** [29]. Since the formation of these intermediates is reversible, the radical pairs and carbenes **13** could only be identified by oxygen trapping.

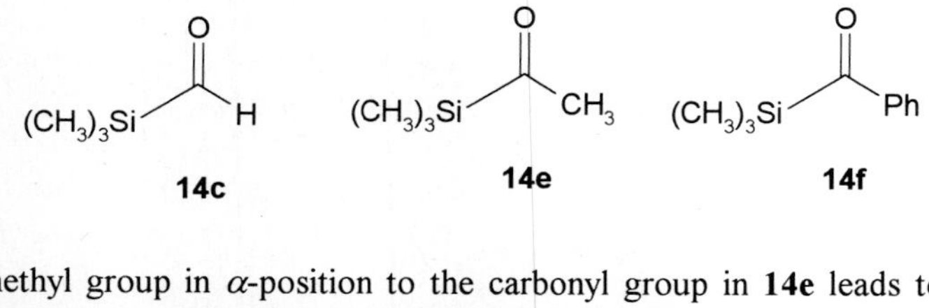

Fig. 1.

However, a methyl group in α-position to the carbonyl group in **14e** leads to methoxycarbene **13e**, where the hydrogen shift can compete with the silyl shift. Since the hydrogen shift is irreversible, vinyloxysilane **16** is formed as the major product.

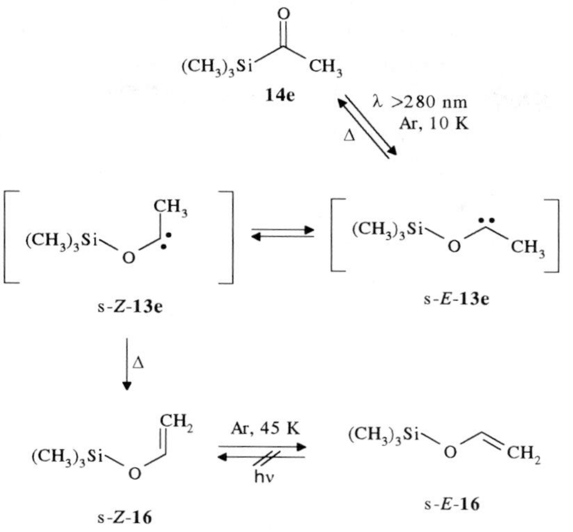

Scheme 7. Photochemistry of acetyl(trimethyl)silane **14e**.

Comparison with published spectra of **16** [24, 23] clearly reveals that only the *s-Z* conformer is formed. Annealing the matrix at 45 K slowly leads to the thermodynamically more stable conformer *s-E-16*, which is photochemically stable and especially does not rearrange back to *s-Z-16*.

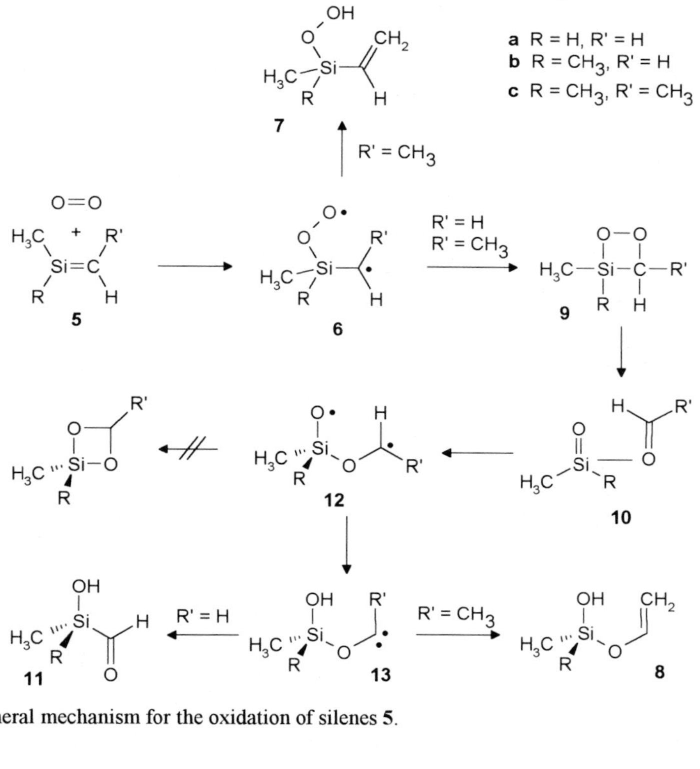

Scheme 8. General mechanism for the oxidation of silenes **5**.

The most reasonable explanation for this is that the photochemical [1,2]silyl shift in **14e** yields carbene **13e** in its *s-E* conformation. This carbene either rearranges back to **14e** or produces the conformational isomer *s-Z*-**13e** (Scheme 7). Since the migrating trimethylsilyl group is located *syn* to the methyl group, the silyl shift in *s-Z*-**13e** is expected to be less feasible than in the *s-E* conformer, and thus the [1,2]H shift to *s-Z*-**16** can compete. Since the vinyloxysilanes **16** and **8** are both formed in the *s-Z* conformation it is tempting to assume that a similar mechanism is operating.

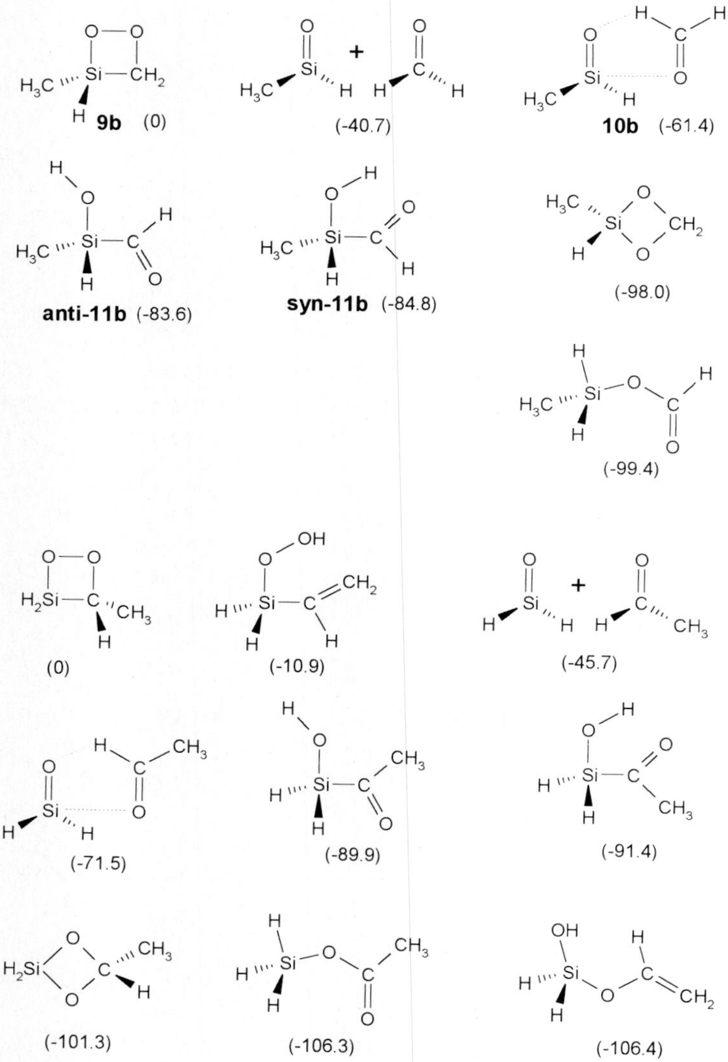

Scheme 9.   Energies of some oxidation products (only closed shell molecules) of 1-methylsilene **5b** and 2-methylsilene, calculated at the RHF/6-31G(d,p) level of theory. Energies in kcal/mol.

All these observations can now be summarized in a general mechanism for the oxidation of the silenes **5** (Scheme 8):

1. The first step of the oxidation of silenes **5** is the formation of charge transfer complexes, which in the case of the electron-rich silene **5d** thermally produce diradicals **6**, while **5b** and **c** require photochemical activation.

2. Diradicals **6** are too labile to be observed even under the conditions of matrix isolation and ring-closure to dioxasiletanes **9** is the major route. Only in **6d** hydrogen abstraction to the stable hydroperoxide **7** can compete with the ring-closure.

3. Dioxasiletanes are also short-lived and are cleaved to a silanone and an aldehyde, which form a complex **10** with a very short "nonbonding" Si–O distance. The cleavage and formation of **10** is exothermic by 60-70 kcal/mol (Scheme 9).

4. Thermal or photochemical activation of the complexes results in the formation of siloxycarbenes **13** via diradicals **12**. Carbenes **13** lie in a very shallow minimum and are stabilized by a [1,2]H shift, if an α-hydrogen atom is available, or by [1,2]silyl migration. Vinyloxysilanol **8** or formylsilanol **11** are the final oxydation products in argon matrices.

Acknowledgement: This work was financially supported by the *Deutsche Forschungsgemeinschaft* and the *Fonds der Chemischen Industrie*. We thank Prof. Dr. D. Cremer and C.-H. Ottosson for numerous discussions and assistance of our ab initio calculations.

References:

[1] T. Akasaka, S. Nagase, A. Yabe, W. Ando, *J. Am. Chem. Soc.* **1988**, *110*, 6270.

[2] A. Patyk, W. Sander, J. Gauss, D. Cremer, *Angew. Chem.* **1989**, *101*, 920; *Angew. Chem., Int. Ed. Engl.* **1989**, *28*, 898.

[3] A. Patyk, W. Sander, J. Gauss, D. Cremer, *Chem. Ber.* **1990**, *123*, 89.

[4] I. M. T. Davidson, C. E. Dean, F. T. Lawrence, *J. Chem. Soc., Chem. Commun.* **1981**, 52.

[5] A. G. Brook, S. C. Nyburg, F. Abdesaken, B. Gutekunst, G. Gutekunst, R. K. M. R. Kallury, Y. C. Poon, Y.-M. Chang, W. Wong-Ng, *J. Am. Chem. Soc.* **1982**, *104*, 5667.

[6] W. Sander, M. Trommer, *Chem. Ber.* **1992**, *125*, 2813.

[7] M. Trommer, W. Sander, A. Patyk, *J. Am. Chem. Soc.* **1993**, *115*, 11775.

[8] W. Sander, *Angew. Chem.* **1990**, *102*, 362-72; *Angew. Chem., Int. Ed. Engl.* **1990**, *29*, 344.

[9] W. Sander, G. Bucher, S. Wierlacher, *Chem. Rev.* **1993**, *93*, 1583.

[10] G. A. Bell, I. R. Dunkin, *J. Chem. Soc., Chem. Commun.* **1983**, 1213.

[11] W. Sander, *Angew. Chem.* **1986**, *98*, 255-6; *Angew. Chem., Int. Ed. Engl.* *25*, 255.

[12] G. A. Ganzer, R. S. Sheridan, M. T. H. Liu, *J. Am. Chem. Soc.* **1986**, *108*, 1517.

[13] T. Sugawara, H. Iwamura, H. Hayashi, A. Sekiguchi, W. Ando, M. T. H. Liu, *Chem. Lett.* **1983**, 1261.

[14] Y. Fujiwara, Y. Tanimoto, M. Itoh, K. Hirai, H. Tomioka, *J. Am. Chem. Soc.* **1987**, *109*, 1942.

[15] H. L. Casal, S. E. Sugamori, J. C. Scaiano, *J. Am. Chem. Soc.* **1984**, *106*, 7623.

[16] N. H. Werstiuk, H. L. Casal, J. C. Scaiano, *Can. J. Chem.* **1984**, *62*, 2391.

[17] A. Kirschfeld, S. Muthusamy, W. Sander, *Angew. Chem.* **1994**, *106*, 2261; *Angew. Chem., Int. Ed. Engl.* **1994**, *33*, 2212.

[18] D. Cremer, T. Schmidt, J. Gauss, T. P. Radhakrishnan, *Angew. Chem.* **1988**, *100*, 431; *Angew. Chem., Int. Ed. Engl.* **1988**, *27*, 427.

[19] O. L. Chapman, C. C. Chang, J. Kolc, M. E. Jung, J. A. Lowe, T. J. Barton, M. L. Tumey, *J. Am. Chem. Soc.* **1976**, *98*, 7844.

[20] M. R. Chedekel, M. Skoglund, R. L. Kreeger, H. Shechter, *J. Am. Chem. Soc.* **1976**, *98*, 7846.

[21] P. Rosmus, H. Bock, B. Solouki, G. Maier, G. Mihm, *Angew. Chem.* **1981**, *93*, 616; *Angew. Chem., Int. Ed. Engl.* **1981**, *20*.

[22] J. M. Dyke, G. D. Josland, R. A. Lewis, A. Morris, *J. Phys. Chem.* **1982**, *86*, 2913.

[23] J. Dedier, A. Marchand, M. T. Forel, E. Frainnet, *J. Organomet. Chem.* **1974**, *81*, 161.

[24] J. Dédier, A. Marchand, *Spectrochim. Acta, Part A* **1982**, *38*, 339.

[25] M. Trommer, W. Sander, C.-H. Ottosson, D. Cremer, *Angew. Chem.* **1995**, *107*, 999.

[26] R. Tian, J. C. Facelli, J. Michl, *J. Phys. Chem.* **1988**, *92*, 4073.

[27] J. M. Duff, A. G. Brook, *Can. J. Chem.* **1973**, *51*, 2869.

[28] A. G. Brook, *J. Organometal. Chem.* **1986**, *300*, 21.

[29] M. Trommer, W. Sander, *Organometallics* **1996**, *15*, 189.

New Silaheterocycles: Formation and Properties

Edwin Kroke, Manfred Weidenbruch*
Fachbereich Chemie
Universität Oldenburg
Carl-von-Ossietzky-Straße 9, D-26111 Oldenburg, Germany
Tel.: Int. code + (441)7983655 – Fax: Int. code + (441)7983329

Keywords: Disilenes / Silylenes / Silaheterocycles / Thermolysis / Isonitriles

Summary: Tetra-*t*-butyldisilene **3** was generated together with di-*t*-butylsilylene **2** by photolysis of hexa-*t*-butylcyclotrisilane **1**. In the presence of alkenes and 1,3-dienes [1+2]-cycloadditions of **2** furnished a series of siliranes and vinylsiliranes. Irradiation of **1** in the presence of cyclopentadiene and furan yielded among other products the [2+4]-cycloadducts **34** and **37,** which were completely characterized. The analogous reaction with thiophene and selenophene led to the unexpected chalcogen abstraction products, namely the disilathiirane **38** and the diselenadisiletane **41**. Thermolysis of the siliranes **21-23** furnished various products in almost quantitative yields. Finally, isonitrile insertions into the Si–C bonds of siliranes provided an access to the novel silacyclobutanimine class of compounds, for example **51**. The structures of **34**, **37**, **38**, **41**, and **51** were determined by X-ray crystallography.

Although silylenes and disilenes bearing bulky substituents have been known for a considerable time, their synthetic potential has by no means been fully exploited. In particular, only very few routes are available for the preparation of alkyl-substituted representatives [1]. Hexa-*t*-butylcyclotrisilane is an ideal starting material for the generation of di-*t*-butylsilylene **2** and tetra-*t*-butyldisilene **3**.

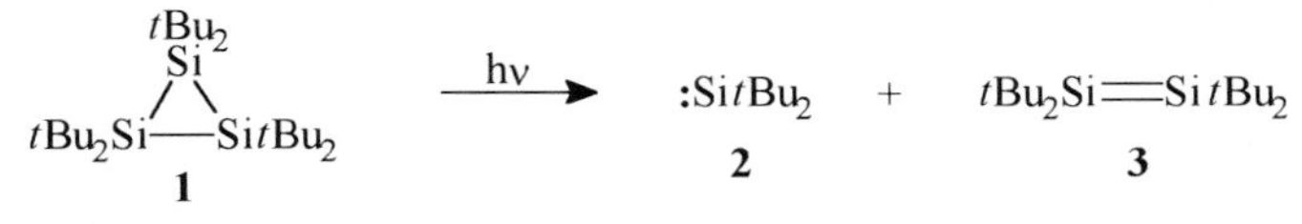

Scheme 1.

The chemistry of **2** and **3** has been investigated over the past ten years. Thus, for example, reactions of these species with nitriles, isonitriles, ketones, 1,4-dihetero-1,3-dienes, and many other multiple bond systems have been realized [2]. However, the important groups of the alkenes and 1,3-dienes were missing from this series. The objective of the present work was to rectify this omission.

On photolysis of **1** in the presence of alkenes or open-chain and cyclic 1,3-dienes, the generated di-*t*-butylsilylene participates in a [1+2] cycloaddition to furnish siliranes. In the cases of cyclopentene **4**, cyclohexene **5** and the monosubstituted ethylenes **6** and **7**, the respective siliranes **8-11** were obtained as air-sensitive, colorless oils in good to very good yields [3, 4]. This behavior is typical for most silylenes [1a]. Compounds **8** and **9** have previously been prepared by reduction of di-*t*-butyldichlorosilane in the presence of the corresponding alkene [5].

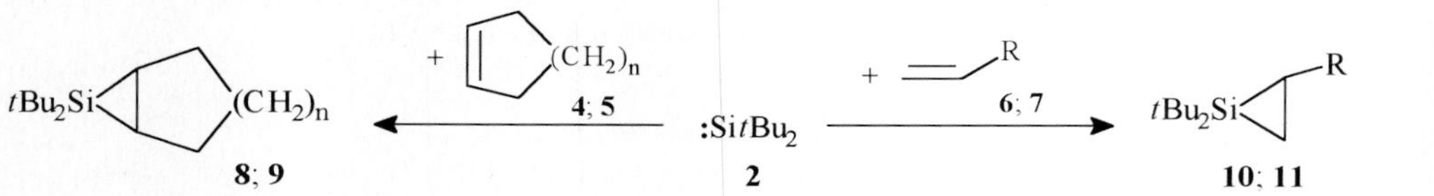

4; 8: n = 1   5; 9: n = 2

6; **10**: R = 2-MePh   7; **11**: R = CH$_2$Ph

Scheme 2.

Open-chain 1,3-dienes such as **12-14** reacted with the silylene to furnish the vinylsiliranes **15-17**. These products are also air-sensitive, colorless oils which, in the pure state, are only stable at room temperature for short times [3, 4].

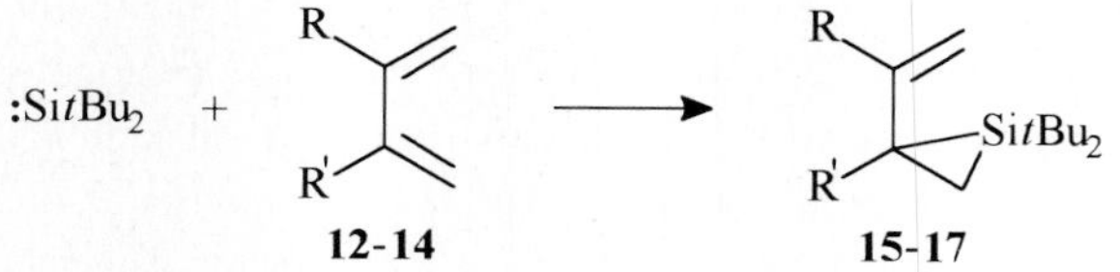

12; **15**: R = R' = Me
13; **16**: R = R' = OMe
14; **17**: R = Me, R' = H

Scheme 3.

The fact that the vinylsiliranes can be isolated is worthy of note since the previously reported reactions of practically all silylenes including those bearing very voluminous substituents had given rise to [1+4] cycloaddition products exclusively. Only with dimesitylsilylene, generated by photolysis of 2,2-dimesitylhexamethyltrisilane, was it possible to characterize mixtures of [1+2] and [1+4] cycloadducts, but no pure vinylsilirane was isolated [6].

The [1+2] cycloaddion products **21-23** were obtained in yields of 77-100 % from the reactions of the silylene **2** with the carbocyclic 1,3-dienes **18** and **19** or with norbornadiene **20** [3-5].

18; **21**: n = 1
19; **22**: n = 2

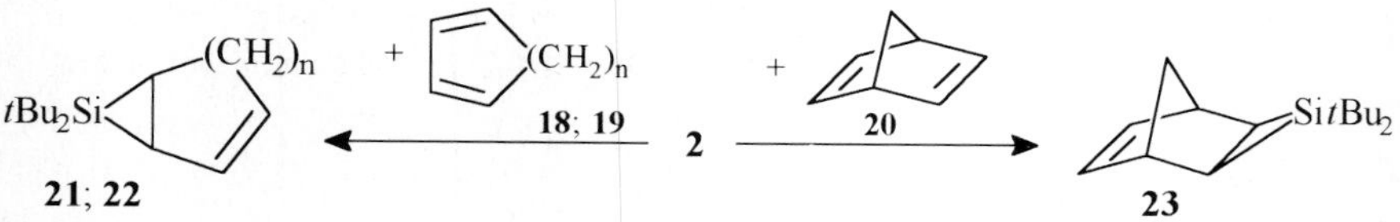

Scheme 4.

Solely the photolysis of **1** with the heterocyclic, five-membered ring compounds furan, thiophene and selenophene furnished other products instead of the respective siliranes (see below). The reactions of tetra-*t*-butyldisilene **3** with alkenes gave rise to a more complex palette of products. In the cases of cyclopentene and cyclohexene, no reactions with the alkene were observed. Instead, compounds **24**, **25**, **27**, and **28** were formed; these results can be rationalized in terms of a dimerization of the disilene with β-elimination of isobutene **26** and subsequent reactions of the disilene and the silylene with **26** [7, 8].

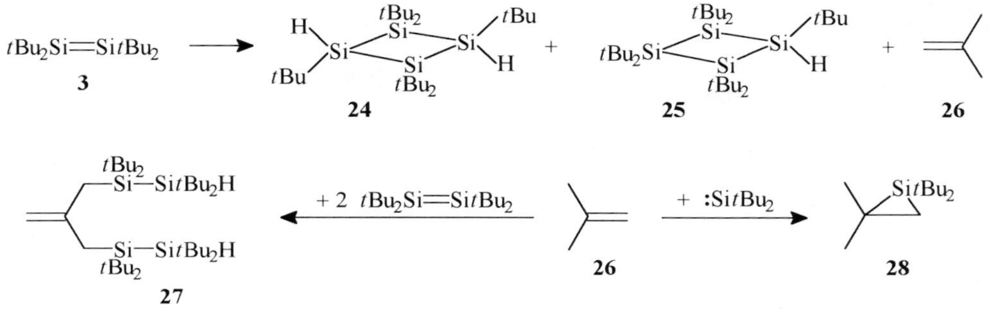

Scheme 5.

Differing disilene reaction products were obtained from the reactions with *o*-methylstyrene and allylbenzene. The reaction of *o*-methylstyrene gave rise to the [2+2] cycloaddition product [3]. On the other hand, the reaction of allylbenzene furnished the ene reaction product [8]. Further information on the cycloaddition behavior of the disilene **3** has been obtained from photolysis of **1** in the presence of the carbocyclic dienes **18-20**. It was initially assumed that the fixation of the *syn*-geometry of the conjugated dienes would favor the Diels-Alder reaction. However, cyclohexadiene **19** and norbornadiene **20** reacted to form the [2+2]-adducts **29** and **30** and no indications for the formation of Diels-Alder adducts were seen.

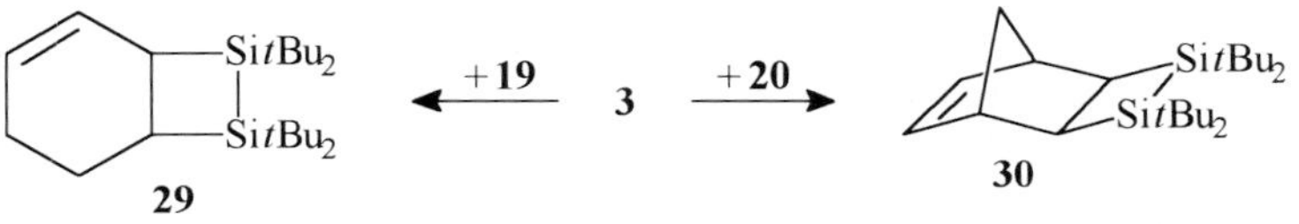

Scheme 6.

It is well known in organic chemistry that the 1,4-separation of conjugated dienes can decisively influence their readiness to participate in Diels-Alder reactions [9]. In a first approach to examine this relationship, hexa-*t*-butylcyclotrisilane **1** was allowed to react with cyclopentadiene **18** as well as the heterocyclic 1,3-dienes furan **31**, thiophene **32**, and selenophene **33** and the following different products were isolated [8, 10]: Like cyclopentadiene, furan yielded initially a Diels-Alder product, which subsequently reacted with **2** to furnish the strained tricyclic system **37**. Compounds **35** and **36** are most likely formed through prototropic 1,2-shifts of the ene reaction product. Silatropic rearrangements are also possible, but would appear to be improbable on account of the spatial requirements of the tetra-*t*-butyldisilanyl group. The bulkiness of this group is also held to be responsible for the fact that the disilanyl group undergoes bonding exclusively to the olefinic carbon atoms of the cyclopentadienyl ring [11]. The disilanorbornene structure of the cycloaddition product **34** was confirmed by an X-ray crystal structure analysis. The Si–Si bond in the molecule has a length of 248.5 pm and is thus similar to the Si–Si bond length in **37** (247 pm), highly stretched. On the other hand, the acute C–Si–C angle in the silirane part of **37** corresponds well with the respective angles at the silicon atoms in other siliranes [12]. The very reactive silylene **2** surprisingly does not seem to react with furan. Only after the less reactive disilene **3** has formed a [2+4] adduct does a

[1+2]-cycloaddition of **2** with the double bond formed in the Diels-Alder reaction take place to furnish **37**.

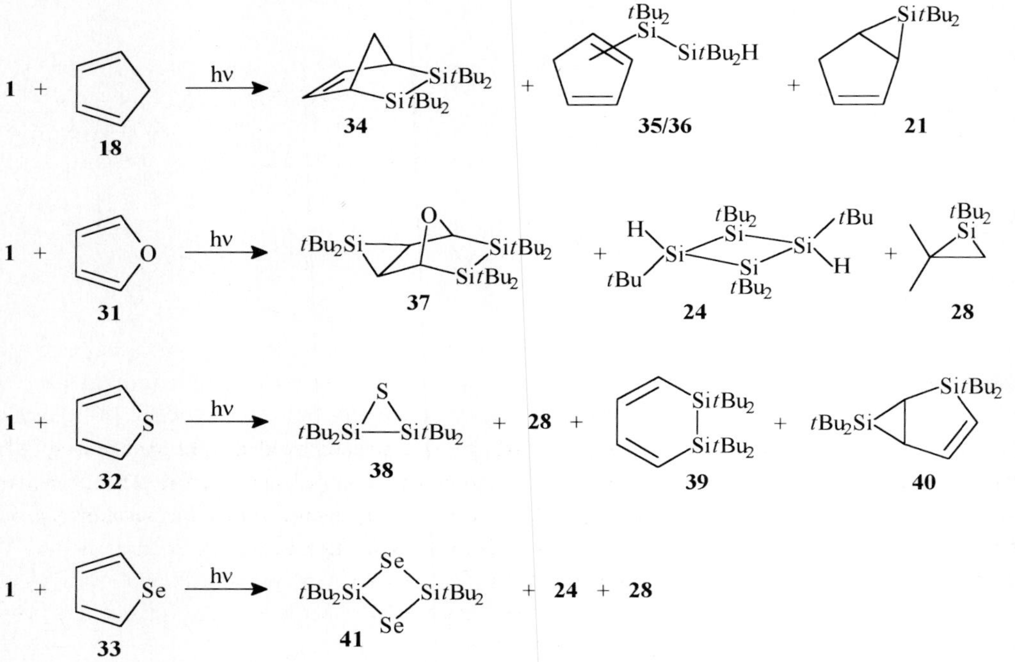

Scheme 7.

In the reaction with thiophene **32**, the disilathiirane **38** was unexpectedly formed in about 50 % yield by way of sulfur extraction. The X-ray structure analysis of the disilathiirane **38** revealed a very short Si–Si bond length of 230.5 pm and an almost planar environment at the silicon atoms (angular sum C–Si–C + C–Si–Si + Si–Si–C = 358.7°) [10]. These features are typical for related three-membered rings of this type and were also observed for other disilathiiranes [13]. As illustrated by the isolation of the 1,2-disilacyclohexadiene **39**, sulfur abstraction from **32** seems to be initiated by [2+4]-cycloaddition of disilene **3**. The bicyclic compound **40** is most likely formed by photoisomerization of **39** [14].

Similarly, selenophene – which, like thiophene, possesses a higher aromaticity than furan – undergoes fragmentation by extraction of the heteroatom on reaction with **2** and **3**. Compound **41** is the first 1,3-diselena-2,4-disiletane for which the structure has been determined [15]. In comparison with the four-membered ring units in SiSe$_2$ [16], the Si–Se bond lengths, the Si/Si separation, and the endocyclic angles at the selenium atoms are enlarged whereas the endocyclic angles at the silicon atoms are smaller than those in silicon diselenide. This may be attributed to the bulkiness of the *t*-butyl groups.

In the second part of the present work we have investigated the thermal and photochemical stabilities of the siliranes synthesized in the first part of the study. Photolysis experiments with the siliranes demonstrate that all of these molecules are photolabile. In general, the product mixtures were so complex that mostly only those compounds could be identified that were, in part, formed quantitatively in the thermolysis experiments and for which spectra accordingly were available for comparison. On thermolysis, the three siliranes **21-23** gave the products shown in the following scheme in almost quantitative yields and without any side reactions [8].

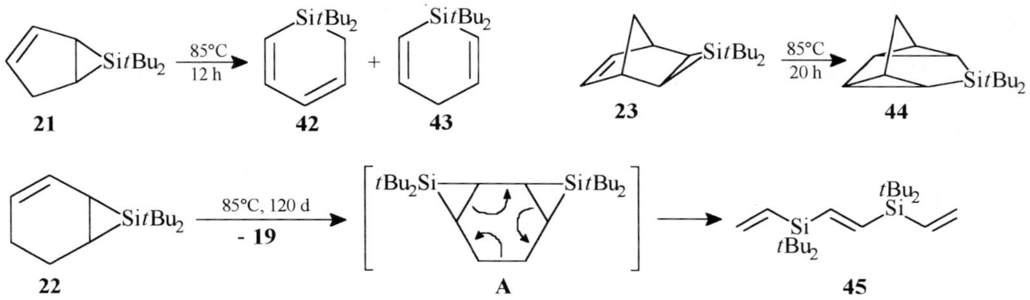

Scheme 8.

The compounds **21** and **23** underwent rearrangement to the products **42** and **43** or **44**, respectively, in reactions that obey first-order kinetics. In contrast, compound **22** underwent dissociation, presumably with formation of the silylene **2**, which reacted with a further molecule of the substrate to furnish the final product **45** by way of the tricyclic intermediate **A**.

Finally, siliranes have been allowed to react for the first time with isocyanides. Phenyl isonitrile **47** and *p*-nitrophenyl isonitrile **48** underwent insertion into the Si–C bond at room temperature with formation of the silacyclobutanimines **49-51** [17].

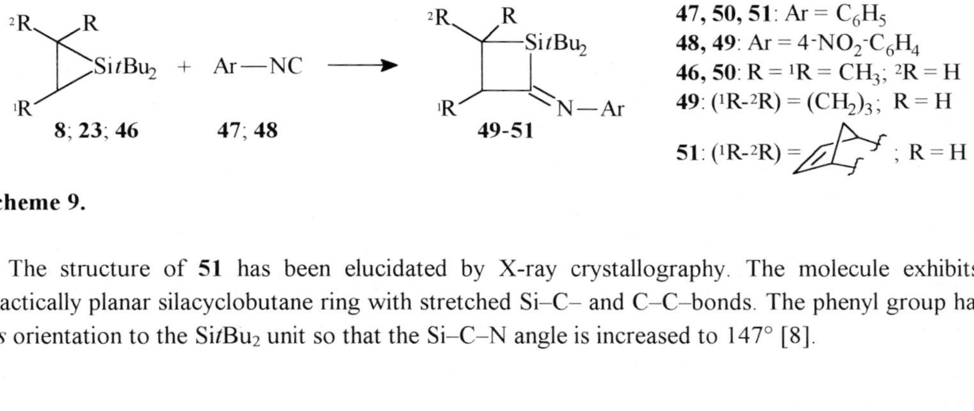

Scheme 9.

The structure of **51** has been elucidated by X-ray crystallography. The molecule exhibits a practically planar silacyclobutane ring with stretched Si–C– and C–C–bonds. The phenyl group has a *cis* orientation to the Si*t*Bu$_2$ unit so that the Si–C–N angle is increased to 147° [8].

Acknowledgements: Financial support of our work by the *Deutsche Forschungsgemeinschaft*, the *Volkswagen Stiftung*, and the *Fonds der Chemischen Industrie* is gratefully acknowledged.

References:

[1] a) For silylenes see e.g.: P. P. Gaspar, in: *Reactive Intermediates, Vol. 3* (Eds.: M. Jones, R. A. Moss), Wiley, New York, **1985**, p. 333.
b) For disilenes: G. Raabe, J. Michl, in: *The Chemistry of Organic Silicon Compounds* (Eds.: S. Patai, Z. Rappoport), Wiley, New York, **1989**, p. 1013.

[2] a) M. Weidenbruch, in: *Frontiers of Organosilicon Chemistry* (Eds.: A. R. Bassindale, P. P. Gaspar), Royal Society of Chemistry, Cambridge, **1991**, p. 122.
b) M. Weidenbruch, *Coord. Chem. Rev.* **1994**, *130*, 275.

[3] M. Weidenbruch, E. Kroke, H. Marsmann, S. Pohl, W. Saak, *J. Chem. Soc., Chem. Commun.* **1994**, 1233.

[4] E. Kroke, P. Will, M. Weidenbruch, in: *Organosilicon Chemistry II: From Molecules to Materials* (Eds.: N. Auner, J. Weis), VCH, Weinheim, **1996**, p. 309.

[5] P. Boudjouk, U. Samaraweera, R. Sooriyakumaran, J. Chrisciel, K. R. Anderson, *Angew. Chem.* **1988**, *100*, 1406; *Angew. Chem., Int. Ed. Engl.* **1988**, *27*, 1355.

[6] S. Zhang, R. T. Conlin, *J. Am. Chem. Soc.* **1991**, *113*, 4272.

[7] M. Weidenbruch, E. Kroke, S. Pohl, W. Saak, H. Marsmann, *J. Organomet. Chem.* **1995**, *499*, 229.

[8] E. Kroke, *Dissertation*, Universität Oldenburg, **1995**: *Silylen- und Disilen-Additionen an Alkene und Diene sowie Folgereaktionen der Cycloaddukte*, Shaker Verlag, Aachen, **1996**.

[9] a) R. Sustmann, M. Böhm, J. Sauer, *Chem. Ber.* **1979**, *112*, 883.
b) H.-D. Scharf, H. Plum, J. Fleischhauer, W. Schleker, *Chem. Ber.* **1979**, *112*, 862.

[10] E. Kroke, M. Weidenbruch, W. Saak, S. Pohl, H. Marsmann, *Organometallics* **1995**, *14*, 5695.

[11] P. Jutzi, *Chem. Rev.* **1986**, *86*, 983.

[12] See e.g.: a) G. L. Delker, Y. Wang, G. D. Stucky, R. L. Lambert, Jr., C. K. Haas, D. Seyferth, *J. Am. Chem. Soc.* **1976**, *98*, 1779.
b) D. H. Pae, M. Xiao, M. Y. Chiang, P. P. Gaspar, *J. Am. Chem. Soc.* **1991**, *113*, 1281.

[13] a) R. West, D. J. DeYoung, K. J. Haller, *J. Am. Chem. Soc.* **1985**, *107*, 4942.
b) J. E. Mangette, D. R. Powell, R. West, *Organometallics* **1995**, *14*, 3551.

[14] Y. Nakadaira, S. Kanouchi, H. Sakurai, *J. Am. Chem. Soc.* **1974**, *96*, 5623.

[15] Three spectroscopically characterized disiladiseletanes are known: see P. Jutzi, A. Möhrke, A. Müller, H. Bögge, *Angew. Chem.* **1989**, *101*, 1527; *Angew. Chem., Int. Ed. Engl.* **1989**, *28*, 1518; and literature cited therein.

[16] J. Peters, B. Krebs, *Acta. Crystallogr.* **1982**, *B38*, 1270.

[17] E. Kroke, S. Willms, M. Weidenbruch, W. Saak, S. Pohl, H. Marsmann, *Tetrahedron Lett.* **1996**, *37*, 3675.

Cycloaddition Reactions of
1,1-Dichloro-2-neopentylsilene with Monoterpenes

Claus-Rüdiger Heikenwälder, Norbert Auner*

Fachinstitut für Anorganische und Allgemeine Chemie
Humboldt-Universität zu Berlin
Hessische Str. 1-2, D-10115 Berlin, Germany
Tel.: Int. Code + (30)20936965 – Fax: Int. Code + (30)20936966
E-mail: Norbert=Auner@chemie.hu-berlin.de

Keywords: Cycloaddition Reactions / Monoterpenes / Silacyclobutanes / Silene / Silaheterocycles

Summary: The reactions of monoterpenes with 2-neopentyl-1-silenes $R_2Si=CHCH_2tBu$ [R = Cl (**1**), Me (**2**)], which are *in situ* generated from chlorovinylsilanes and *t*-butyllithium, are described. Reactions of **1** with α-terpinene lead to the formation of the stereoisomeric *E/Z* [2+2] cycloadducts **3** and the Diels-Alder products *exo/endo*-**4**, whereas the ene product **5** and the [2+2] adducts *E/Z*-**6** are formed using S-limonene as trapping agent. The silacylobutanes *E/Z*-**7** and *E/Z*-**8** are obtained from **1** and myrcene, whereas the *E/Z* isomeric [2+2] compounds **9** and **10**, together with **11–14**, result from the reaction of silene **2** with myrcene. From reactions of **5/6** and **7/8** with methyllithium the stereo- and regioisomeric differently substituted derivatives **9/10** and **15/16** could be isolated.

The cycloaddition behavior of 1,1-dichloro-2-neopentyl-1-silene, $Cl_2Si=CHCH_2tBu$ (**1**), formed by treating trichlorovinylsilane with *t*-butyllithium in nonpolar solvents, has been studied extensively [1]. Compared with diorganosubstituted neopentylsilenes, such as $Me_2Si=CHCH_2tBu$ (**2**) [2], **1** shows a quite different cycloaddition potential; its tendency to preferentially form [2+2] and [4+2] cycloadducts in the presence of dienes has aroused interest concerning its electronic structure.

The [4+2] cycloadditions with anthracene [3] and 1,3-cyclopentadiene [3] proceed as expected; surprisingly, dienes of lower activity, such as naphthalene [4] or furans [5], form Diels-Alder products likewise. The reactions of **1** with pentafulvenes [6] and 1,3-cyclohexadiene [7] give stereoisomeric *E/Z* [2+2] cycloadducts and [4+2] products, whereas with butadienes [8] only [2+2] compounds are obtained.

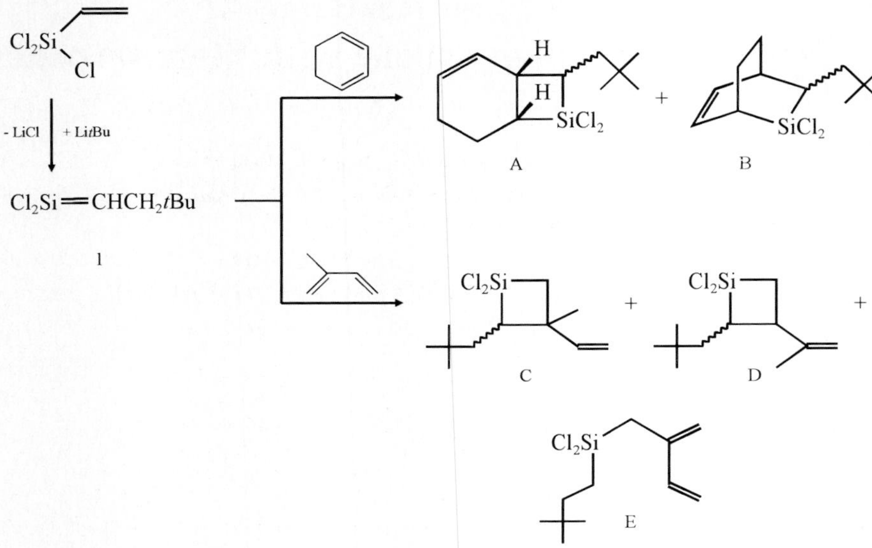

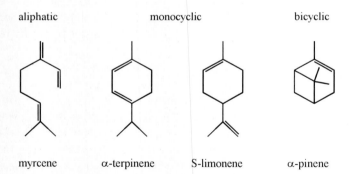

Scheme 1.   Cycloadditions of silene **1** with 1,3-cyclohexadiene [7] and 2-methyl-1,3-butadiene (isoprene) [8].

In this contribution, we describe the investigations on the cycloaddition behavior of **1** in the presence of several monoterpenes.

aliphatic	monocyclic		bicyclic
myrcene	α-terpinene	S-limonene	α-pinene

Fig. 1.

Terpenes are most familiar, at least by odor, as compounds of the so-called essential oils obtained by steam distillation or ether extraction of various plants. Thousands of different terpenes are known. According to the isoprene rule proposed by L. Ruzicka in 1921, they can be considered to arise from head-to-tail joining of simple five-carbon isoprene (2-methyl-1,3-butadiene) units. Terpenes are subdivided into groups depending on the number of isoprene units [9]. For example, monoterpenes are 10-carbon substances biosynthesized from two isoprene units, which can be divided into aliphatic, monocyclic, or bicyclic species. Some typical exponents of each monoterpene subgroup are shown in Fig. 1.

Although the monocyclic terpenes α-terpinene and S-limonene have the same basic carbon skeleton [10], they reveal quite different cycloaddition behavior:

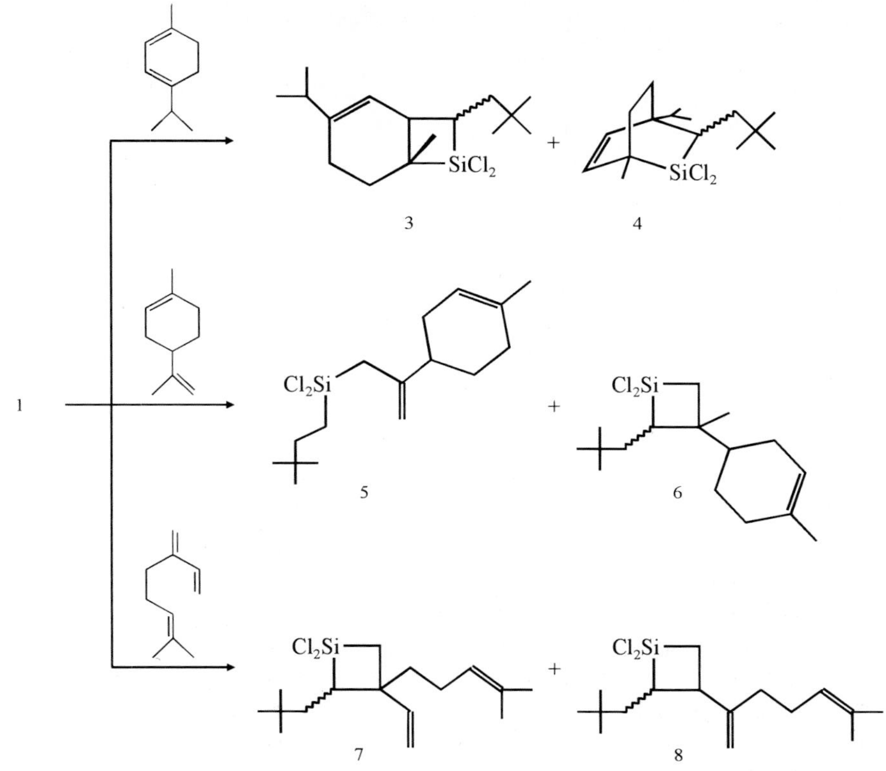

Scheme 2.   Cycloadditions of silene **1** with α-terpinene, S-limonene, and myrcene.

As shown in Scheme 2, the reaction of **1** with α-terpinene leads to the stereoisomeric *E/Z* [2+2] cycloadducts **3** in competition with the formation of the Diels-Alder products *exo/endo*-**4** (ratio: *E/Z*-**3**:*exo/endo*-**4** = 39:41:8:12). This result conforms with investigations on the cycloadditon between **1** and 1,3-cyclohexadiene [7]. An analogous rearrangement of the [2+2] isomer **A** to the more stable bicyclic 2-silaoctene **B** (Scheme 1) via a dipolar intermediate could not be observed for the homologous terpinene cycloadducts.

With S-limonene **1** yields **5** and the silacyclobutanes *E/Z*-**6** in a ratio of 73:12:15. Compound **5** is formed by an ene reaction, which is well known for reactions of **1** with alkines [11], strained ring systems, like norbornene or norbornadiene [12], and butadienes [8]. In contrast to the reaction of **1** with isoprene (ratio: *E/Z*-**C**: *E/Z*-**D**:**E** = 38:38:14:8:2) the formation of ene product **5** is favored. 2D-NMR investigations prove these results [13].

Using an aliphatic monoterpene like myrcene as reaction partner for **1** the formation of silacylobutanes *E/Z*-**7** and *E/Z*-**8** in a ratio of 34:30:22:14 is observed; an ene product could not be detected. In this case myrcene behaves like an isoprene derivative, which forms only [2+2] cycloadducts with **1** [8]. The reaction with **2**, however, yields a product mixture containing the silacyclobutanes *E/Z*-**9** and *E/Z*-**10**, the Diels-Alder compounds **11** and **12**, the ene product **13**, and the silene dimer *E/Z*-**14** [14, 15] (Fig. 2).

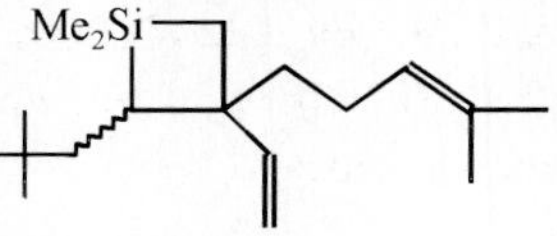

9

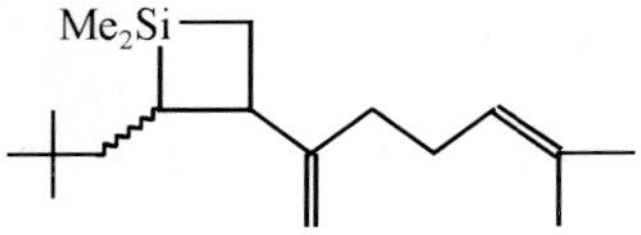

10

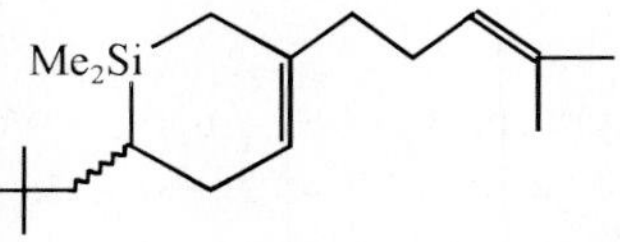

11

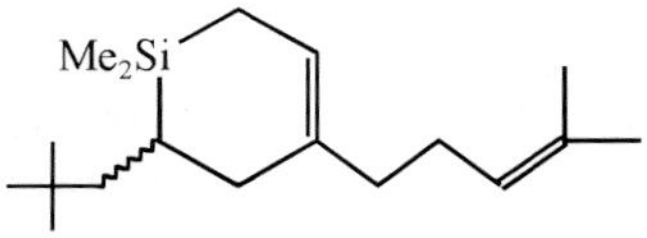

12

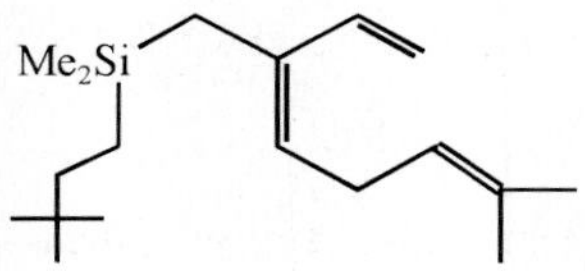

13

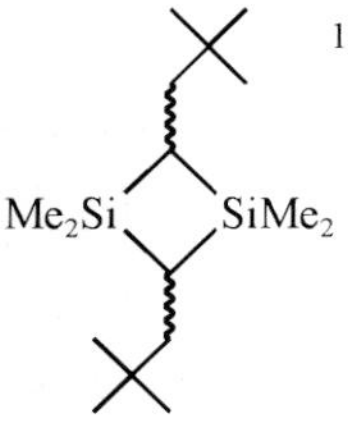

14

Fig. 2.

This result is in agreement with former investigations on cycloadditions of **2** with butadienes [2]; the [4+2] and the ene reaction are favored, wheareas the [2+2] adduct and the dimer are observerd in a minor amount.

A useful method to prove the formation of silacyclobutanes and ene products is their derivatization by organo groups [1]. Thus, the reactions of **5/6** and **7/8** with two equivalents of methyllithium yield compounds **9/10** and **15/16** (Fig. 3).

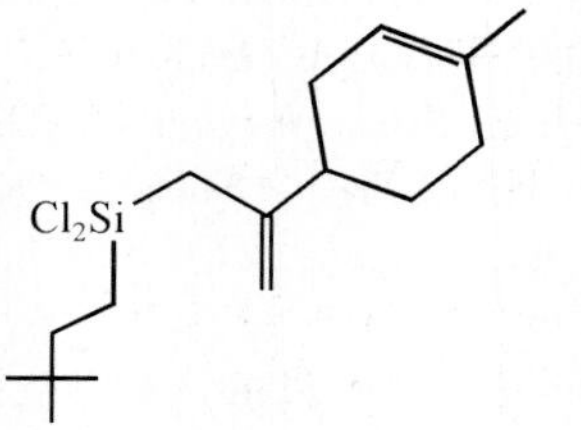

15

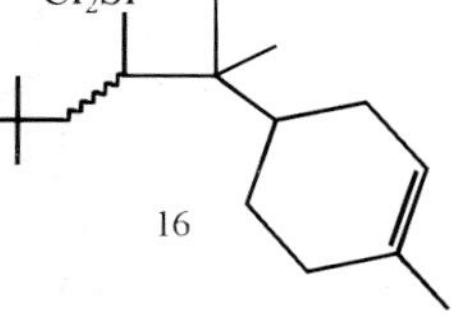

16

Fig. 3.

All described compounds are characterized by standard analytical and NMR spectroscopic methods [13]. The reaction of **1** with the bicyclic monoterpene α-pinene does not reveal efficient results.

Acknowledgment: We gratefully thank *Wacker-Chemie GmbH* (Burghausen, Germany) for chlorosilanes, *Chemetall GmbH* (Frankfurt/Main, Germany) for lithium alkyls, and *Dow Corning Corp.* (Midland, USA) and *Dow Corning Ltd.* (Barry, Wales) for financial support.

References:

[1] For recent reviews on 1,1-dichloro-2-neopentylsilene see: N. Auner, *"Neopentylsilenes – Laboratory Curiosities or Useful Building Blocks for the Synthesis of Silaheterocycles"*, in: *Organosilicon Chemistry: From Molecules to Materials* (Eds.: N. Auner, J. Weis), VCH, Weinheim, **1994**, p. 103; N. Auner, *J. Prakt. Chem.* **1995**, *337*, 79.

[2] P. R. Jones, T. F. O. Lim, *J. Am. Chem. Soc* **1977**, *99*, 2013; P. R. Jones, T. F. O. Lim, *J. Am. Chem. Soc.* **1977**, *99*, 8447; P. R. Jones, T. F. O. Lim, R. A. Pierce, *J. Am. Chem. Soc.* **1980**, *102*, 4970.

[3] N. Auner, *J. Organomet. Chem.* **1988**, *353*, 275.

[4] N. Auner, C. Seidenschwarz, N. Sewald, E. Herdtweck, *Angew. Chem.* **1991**, *103*, 425; *Angew. Chem., Int. Ed. Engl.* **1991**, *30*, 444.

[5] N. Auner, A. Wolff, *Chem. Ber.* **1993**, *126*, 575.

[6] C.-R. Heikenwälder, N. Auner, *"Cycloaddition Reactions of 1,1-Dichloro-2-neopentylsilene with Pentafulvenes"*, in: *Organosilicon Chemistry II: From Molecules to Materials* (Eds.: N. Auner, J. Weis), VCH, Weinheim, **1996**, p. 399; N. Auner, C.-R. Heikenwälder, *Z. Naturforsch.* **1997**, *54b*, 500.

[7] N. Auner, C. Seidenschwarz, N. Sewald, *Organometallics* **1992**, *11*, 1137.

[8] N. Sewald, W. Ziche, A. Wolff, N. Auner, *Organometallics* **1993**, *12*, 4123.

[9] For further information on terpenes see: J. D. Connolly, R. A. Hill, *Dictionary of Terpenoides, Vol 1–3*, Chapman & Hall, London, **1991**; E. Breitmaier, G. Jung, *Organische Chemie II*, Thieme Verlag, Stuttgart, **1995**.

[10] Monoterpenes derive from *trans-p*-menthane (4-isopropyl-1-methyl-cyclohexane), respectively from the unsaturated menthene, $C_{10}H_{18}$, or menthadiene, $C_{10}H_{16}$; α-terpinene and S-limonene are isomeric forms of menthadiene; see: H. Beyer, W. Walter, *Lehrbuch für Organische Chemie*, Hirzel Verlag, Stuttgart, **1984**.

[11] N. Auner, C. Seidenschwarz, E. Herdtweck, *Angew. Chem.* **1991**, *103*, 1172; *Angew. Chem., Int. Ed. Engl.* **1991**, *30*, 1151; N. Auner, C.-R. Heikenwälder, C. Wagner, *Organometallics* **1993**, *12*, 4135.

[12] A. Wolff, *Ph.D. thesis*, Technische Universität München, **1991**.

[13] C.-R. Heikenwälder, *Synthese und Reaktivität neuartiger Silaheterocyclen*, Cuvillier, Göttingen, **1997**; N. Auner, C.-R. Heikenwälder, B. Herrschaft, *Organometallics*, in press.

[14] For an X-ray structure of *E/Z*-1,1,3,3-Tetramethyl-2,4-dineopentyl-1,3-disilacyclobutane (**14**) see: C. Seidenschwarz, *Ph.D. thesis*, Technische Universität München, **1991**.

[15] Ratio of product mixture: *E/Z*-**9**:*E/Z*-**10**:**11**:**12**:**13**:*E/Z*-**14** = 10:10:4:2:26:20:18:6:7.

Silaspirocycles as Precursors for a 2-Silaallene

Brigitte Goetze, Bernhard Herrschaft, Norbert Auner*
Fachinstitut für Anorganische und Allgemeine Chemie
Humboldt-Universität zu Berlin
Hessische Straße 1-2, D-10115 Berlin, Germany
Tel.: Int. code + (030)20936965 – Fax: Int. code + (030)20936966
E-mail: Norbert=Auner@chemie.hu-berlin.de

Keywords: Cycloadditions / Dichlorodivinylsilane / 2-Silaallene / Silaspirocycles

Summary: Reactions of dichlorodivinylsilane and Li*t*Bu in a molar ratio 1:1 and 1:2 lead to highly reactive intermediates, which can be trapped by suitable trapping agents. From that mono and double addition products are formed, which provide experimental hints for the intermediate formation of the neopentylsilene $H_2C=CH(Cl)Si=CHCH_2tBu$ **3** and the 2-silaallene *t*BuCH$_2$CH=Si=CHCH$_2$*t*Bu **4**. In particular, the formation of double cycloadducts from **2** with Li*t*Bu is a preparatively facile route for the synthesis of silaspirocycles such as **12**, **14**, and **15,** which could be characterized by single crystal X-ray structure analysis.

The reaction of chlorovinylsilanes $R^1R^2Si(Cl)CH=CH_2$ with Li*t*Bu leads to lithiated intermediates $R^1R^2Si(Cl)CH(Li)CH_2tBu$ and via 1,2-LiCl elimination to highly reactive neopentylsilenes $R^1R^2Si=CHCH_2tBu$. The only stable derivative was isolated for $R^1 = R^2 = $ Mes [1]. Our experimental investigations during the last few years indicate that dichloroneopentylsilene ($R^1 = R^2 = $ Cl) **1** in particular is characterized by an extraordinarily high reactivity [2]. Although the reaction of chlorovinylsilanes with Li*t*Bu and various trapping agents may be explained by alternative reaction pathways (e.g., from the lithiated species $Cl_3SiCH(Li)CH_2tBu$) we use this general principle of silene generation for the preparation of a 2-silaallene building block.

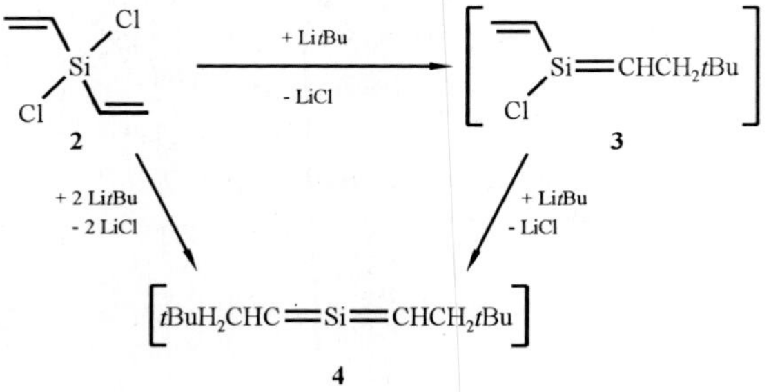

Scheme 1. Reaction of dichlorodivinylsilane with one or two equivalents of Li*t*Bu.

It is well known in the literature that silaspirocycles are suitable precursors for the pyrolytic generation of 2-silaallenes [3]. Thus, the double (cyclo)adducts of **4** are possible educts for the retrosynthetic generation of 2-silaallenes, allowing trapping experiments and spectroscopic analysis in low temperature matrices or in the gas phase [4, 5].

For our investigations we needed dichlorodivinylsilane **2** in high purity and on a preparative scale. Therefore, we developed a new, highly selective method to generate **2**.

$$Ph_2SiCl_2 + 2\ ViMgCl \longrightarrow \underset{\mathbf{5}}{Ph_2SiVi_2} + 2 MgCl_2$$

$$\underset{\mathbf{5}}{Ph_2SiVi_2} + 2\ F_3CSO_3H \longrightarrow (F_3CSO_3)_2SiVi_2 + 2\ PhH$$

$$(F_3CSO_3)_2SiVi_2 + 2\ NEt_3HCl \longrightarrow \underset{\mathbf{2}}{Cl_2SiVi_2} + 2\ NEt_3H(SO_3CF_3)$$

Scheme 2. Synthesis of dichlorodivinylsilane starting from dichlorodiphenylsilane.

Depending on the degree of Si=C polarity and, thus, on the influence of substituents on the Si=C moiety, neopentylsilenes react with carbon-carbon unsaturated compounds to yield cycloaddition products ([2+2] and/or [4+2] addition) [2, 6]. In the absence of trapping agents neopentylsilenes dimerize to yield 2,4-dineopentyl-1,3-disilacyclobutanes as stable products [7]. Not surprisingly, the reaction of **2** with *one* equivalent of Li*t*Bu and Me₃SiOMe gives the silene addition products **9** and **10** in 40 and 20 % yield, respectively. From the reaction of **2** with *two* equivalents of Li*t*Bu and Me₃SiOMe only the double addition product **10** was obtained (yield: 65 %). The latter result gives experimental hints on the formation of silaallene **4** during the course of reaction as no trace of **9** can be detected.

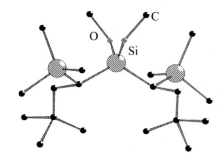

Fig. 1. Structure of **10**.

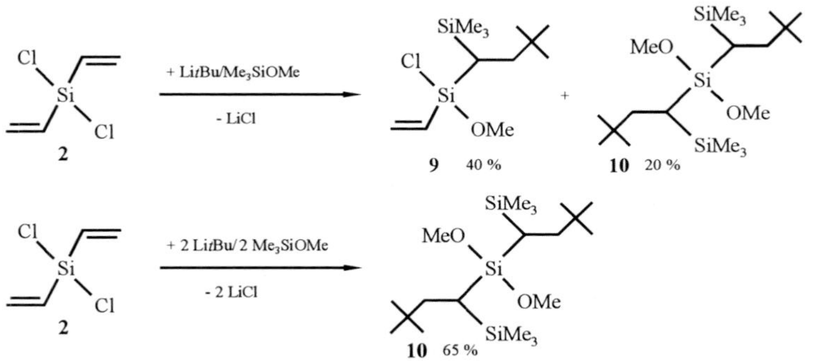

Scheme 3. Trapping reaction of the mixture dichlorodivinylsilane/Li*t*Bu with Me₃SiOMe.

Compound **10** is thermally very stable: in the gas phase it takes a temperature of 980°C for elimination of *two* equivalents of Me$_3$SiOMe. Thus, **10** might be a potential precursor for the retrosynthetic formation of silaallene **4**.

Anthracene is reported to be a very efficient dienophile for neopentylsilenes [8]. From the mixture of **2**/Li*t*Bu/anthracene the [4+2] cycloadduct **11** was isolated [9]. Further reaction of **11** with Li*t*Bu and anthracene yields the crystalline double cycloadduct **12**. Alternatively, **12** can be synthesized in a one-step reaction, from **2** with 2 equivalents of Li*t*Bu and anthracene; the results of the single crystal X-ray analysis of **12** is shown in Fig. 2.

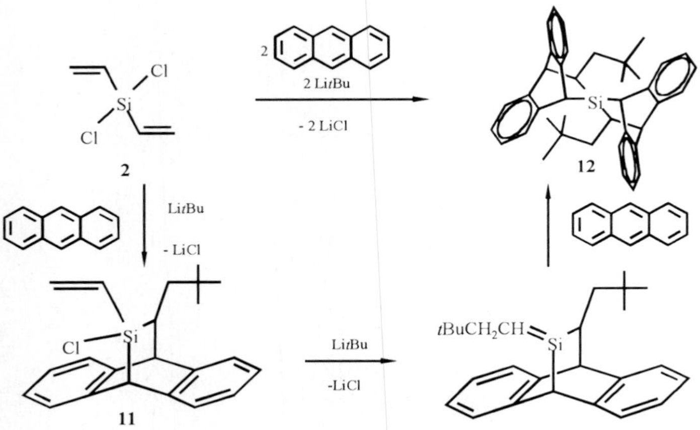

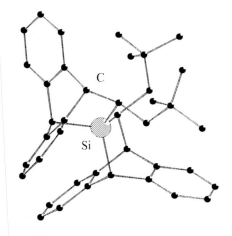

Scheme 4.   Reactions of dichlorodivinylsilane/Li*t*Bu and anthracene.

Both reactions yield the crystalline product **12**. Pyrolysis experiments show that **12** eliminates *two* equivalents of anthracene at about 950°C giving an identical product as isolated in matrix from the pyrolysis of **10**; the nature of this compound is not yet known and is currently under investigation [4, 5].

Trapping reactions of equimolar amounts of **2**/Li*t*Bu and norbornadiene [10-12] clearly prove the formation of **3**, which is trapped to give **13** in a [2+2+2] cycloaddition. Very similar to **11**, **13** may serve as a silene precursor: It is able to react slowly with a second equivalent of Li*t*Bu and norbornadiene to give the double [2+2+2] adduct **14**, which is isolated alternatively from the reaction mixture of **2** with two equivalents of Li*t*Bu and norbornadiene in a one step procedure. In strong contrast to the stepwise synthesis, the latter reaction is very fast.

Fig. 2.   Structure of **12**.

When the mono adduct **13** is only treated with Li*t*Bu the 1,3-disilacyclobutane **15** is obtained in a very slow reaction and at room temperature.

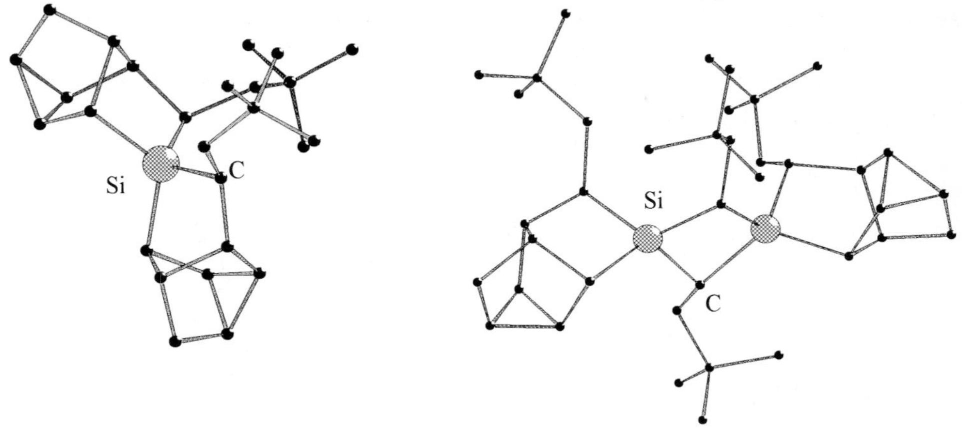

Fig. 3. Structure of **14** and **15**.

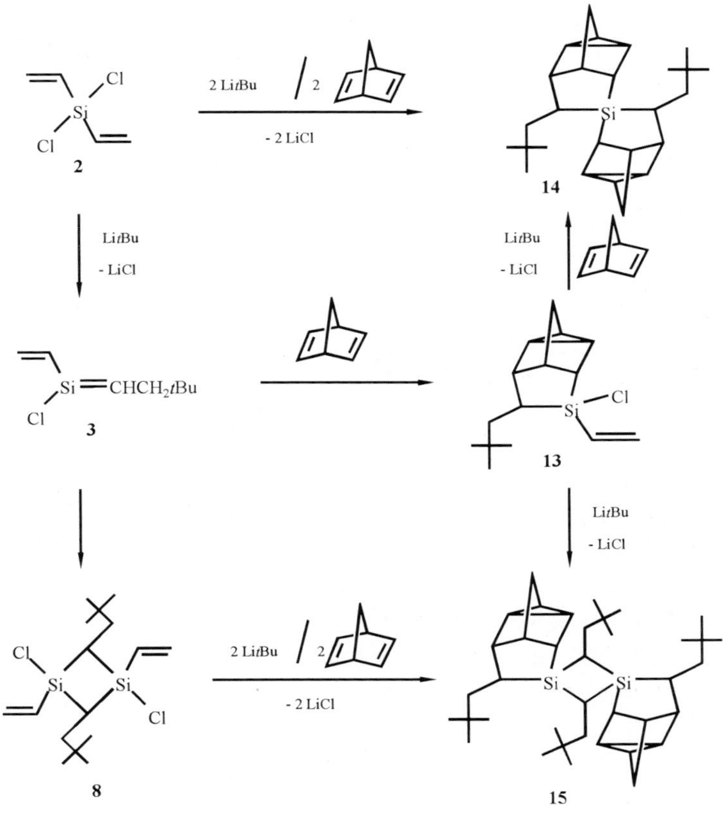

Scheme 5. Reaction of dichlorodivinylsilane with Li*t*Bu and norbornadiene.

These findings were proven by the reaction of 1,3-disilacyclobutane **8** with two equivalents of Li*t*Bu and in the presence of execess norbornadiene. Again, **15** was isolated after distillation (Scheme 5).

The synthesis of a *mixed* double cycloadduct **16** is possible by either allowing **11** to react with Li*t*Bu in the presence of norbornadiene or by reaction of **13** with Li*t*Bu and anthracene (Scheme 6).

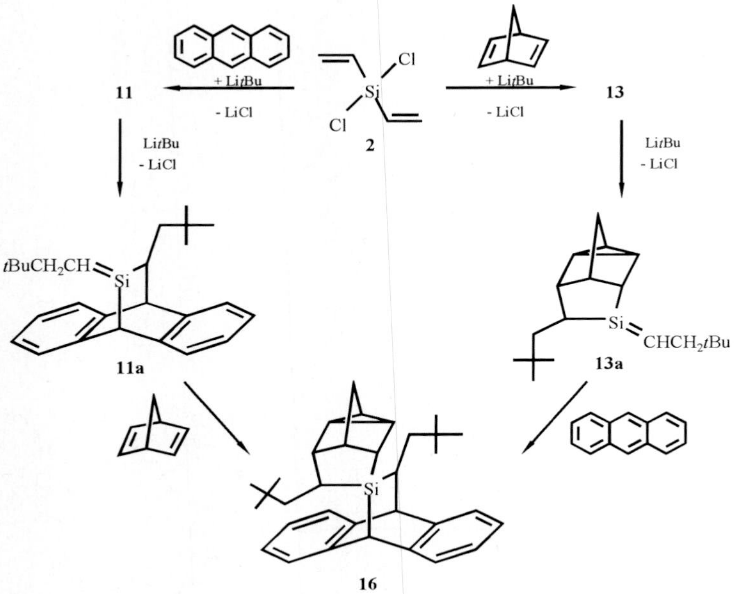

Scheme 6. Reaction of dichlorodivinylsilane with Li*t*Bu, anthracene and norbornadiene.

In the latter reaction the formation of the disilacyclobutane **15** competes with the trapping reaction. Both products could be clearly identified but not separated from each other. Thus, starting from **11** is by far the more efficient route to **16**; a dimerization product resulting from corresponding precursor **11a** could not be observed [9].

While dichloroneopentylsilene **1** is trapped by quadricyclane forming a stereoisomeric pair of silacyclobutanes, **3** – as known from diorgano substituted neopentylsilenes – does not react with this reagent. The products formed are identical to those from the reaction of **3** with Li*t*Bu in the absence of trapping agents. The product formation strongly depends on the conditions: The faster Li*t*Bu is added to the solution, the higher is the yield of silicon *t*-butyl substituted 1,3-disilacyclobutanes **17** and **18** (in addition to **8**). Evidently, the substitution reaction Si–chlorine to Si–*t*-butyl is favored over silene formation when the local concentration of Li*t*Bu is comparably high. Reaction of the chlorovinylsilane/quadricyclane mixture with two equivalents of Li*t*Bu leads to oligomers containing disilacyclobutane spirochains **19**.

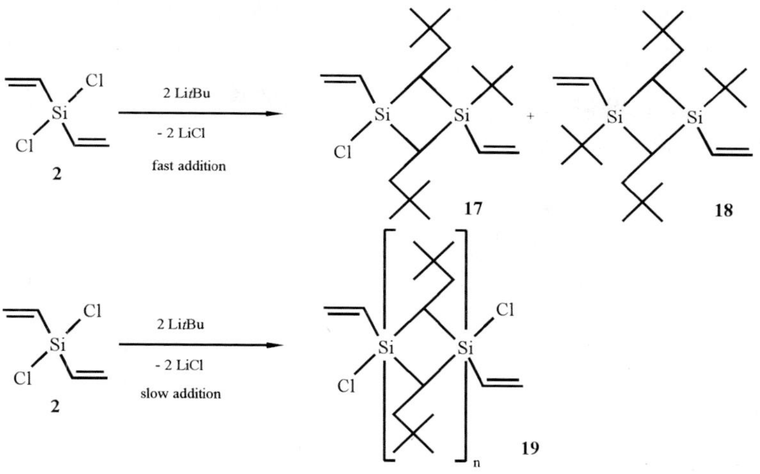

Scheme 7. Reaction of dichlorodivinylsilane with two equivalents of Li*t*Bu.

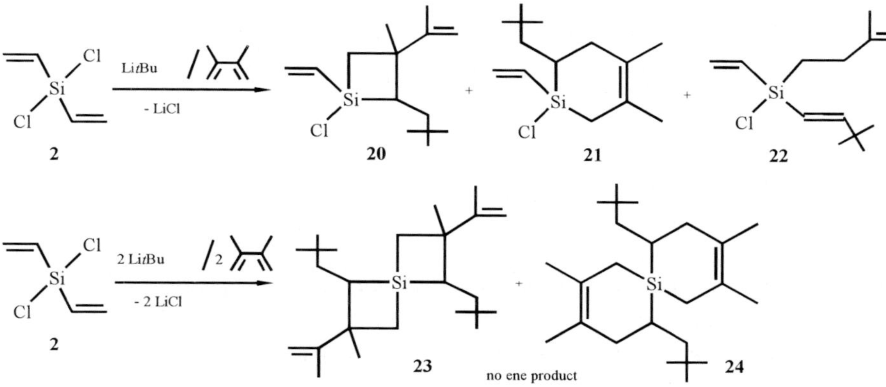

Scheme 8. Reaction of dichlorodivinylsilane with Li*t*Bu and 2,3-dimethyl-1,3-butadiene.

While from **1** and butadienes silacyclobutanes are easily available, competitive [4+2]/[2+2] additions and the ene reaction are reported ($R^1 = R^2 = Me$, Ph) [13-15] from diorganosubstituted silenes $R^1R^2Si=CHCH_2t$Bu. Not surprisingly, the reaction of **2**/Li*t*Bu and 2,3-dimethyl-1,3-butadiene leads to an inseparable mixture of each, the silacyclobutanes as well as the Diels-Alder adducts and the products of the ene reaction; the two main products are the cycloadducts **20** or **21** and the ene product **22** (Scheme 8). Starting from a mixture of **2**/2Li*t*Bu and the diene the results become even worse. More than 20 products are formed and only some of them could be identified. GCMS analysis proved the formation of the double addition products **23** and **24** and from the ^{13}C NMR spectra it is concluded that no ene product has been formed during the reaction.

Reaction of **1** with diphenylacetylene is described to be a high yield synthesis of a dichlorosilacyclobutene whereas diorganosubstituted neopentylsilenes do not react [16, 17]. From the reaction of **3** and tolane the *E*/*Z* isomeric silacyclobutenes **25** are obtained in about 10 % yield. When **2** i reacts with two equivalents of the Li*t*Bu and tolane only traces of **26** were identified by

NMR and MS spectroscopic methods (yield < 5 %). The main product is an oligomeric white powder **27** [18].

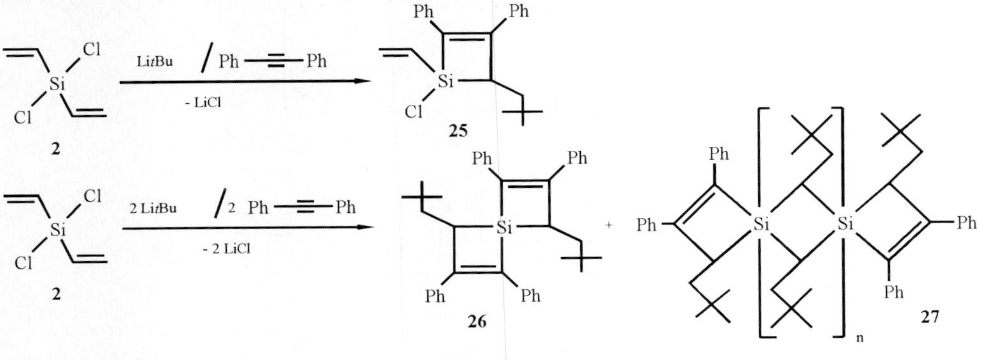

Scheme 9.   Reaction of dichlorodivinylsilane with LitBu and tolane.

Acknowledgement: We gratefully thank *Dow Corning Ltd.* for financial support, *Wacker-Chemie GmbH* for chlorosilanes, and *Chemetall GmbH* for lithium alkyls.

References:

[1]   G. Delpon-Lacaze, C. Couret, *J. Organomet. Chem.* **1994**, *480*, C14.

[2]   N. Auner, *J. prakt. Chem.* **1995**, *337*, 79.

[3]   G. Bertrand, G. Manuel, P. Mazerolles, *Tetrahedron* **1978**, *34*, 1951.

[4]   IR spectroscopic studies on the pyrolysis products at 10 K are currently under investigation: W. Sander, Ruhr Universität Bochum, **1997**.

[5]   PE spectroscopic studies of the pyrolysis products in the gas phase are currently under investigation: H. Bock, B. Solouki, Johann Wolfgang von Goethe Universität, Frankfurt am Main, **1997**.

[6]   N. Auner, *"Neopentylsilenes: Laboratory Curiosities or Useful Building Blocks for the Synthesis of Silaheterocycles"* in: *Organosilicon Chemistry: From Molecules to Materials* (Eds.: N. Auner, J. Weis), VCH, Weinheim, **1994**, p. 103.

[7]   N. Auner, R. Gleixner, *J. Organomet. Chem.* **1990**, *393*, 33.

[8]   N. Auner, *J. Organomet. Chem.* **1988**, *353*, 275.

[9]   N. Auner, *J. Organomet. Chem.* **1989**, *377*, 175.

[10]   N. Auner, C. Wagner, W. Ziche, *Z. Naturforsch.* **1994**, *49b*, 831.

[11]   N. Auner, C.-R. Heikenwälder, W. Ziche, *Chem. Ber.* **1993**, *126*, 2177.

[12]   W. Ziche, N. Auner, J. Behm, *Organometallics* **1992**, *11*, 2494.

[13]   N. Sewald, W. Ziche, A. Wolff, N. Auner, *Organometallics* **1993**, *12*, 4123.

[14]   P. R. Jones, F. F. O. Lim, R. A. Pierce, *J. Am. Chem. Soc.* **1980**, *102*, 4970.

[15]   P. R. Jones, M. E. Lee, L. T. Lin, *Organometallics* **1983**, *2*, 1039.

[16]   N. Auner, C. Seidenschwarz, E. Herdtweck, *Angew. Chem.* **1991**, *103*, 1172.

[17]   N. Auner, C.-R. Heikenwälder, C. Wagner, *Organometallics* **1993**, *12*, 4135.

[18]   B. Goetze, B. Herrschaft, N. Auner, *Chem. Eur. J.* **1997**, *3*, 948.

Catalytic Carbon-Carbon Hydrogenation of Silicon-Functionalized Olefins

Hans-Uwe Steinberger, Norbert Auner*
Institut für Anorganische und Allgemeine Chemie
Humboldt-Universität zu Berlin
Hessische Straße 1-2, D-10115 Berlin, Germany
Tel.: Int. code + (030)20936965 – Fax: Int. code + (030)20936966
E-mail: Norbert=Auner@chemie.hu-berlin.de

Keywords: Hydrogenation / Silanorbornenes / Silanorbornanes / Silicon-Functionalized Olefins

Summary: From catalytic hydrogenation of olefinic silicon-functionalized compounds (chloro-Si, alkoxy-Si, alkyl-Si, and aryl-Si) the saturated products are available in good yield. In general, the chloro- and alkoxy substituents are unaffected and for silaheterocyclic compounds the cyclic or the bicyclic moieties, respectively, remain intact. Thus, the silanorbornenes **1**, **2**, and **3**, compounds containing cyclopentenyl groups **13** and **15**, and various carbon vinyl substituted silacyclobutanes **7**, **8**, and **11** were hydrogenated in a simple apparatus. The reactions were performed in ether and THF as solvents; the hydrogenation catalyst Pd/C can be used several times.

The silanorbornenes **1**, **2**, and **3** can be synthesized according to the literature [1, 2]. Hydrogen chloride cleavage of **3** leads to the silicon chlorinated precursor of **13**, which is obtained by a Grignard reaction with naphthyl magnesium bromide in high yield (86 %). Similarly, **15** is available from HCl cleavage of **1** and reaction of the latter with fluorenyl lithium. The silicon-chlorinated silacyclobutanes **7**, **8**, and **11** are synthesized by [2+2]-cycloaddition of dichloroneopentylsilene with isoprene and dimethylbutadiene, respectively [3].

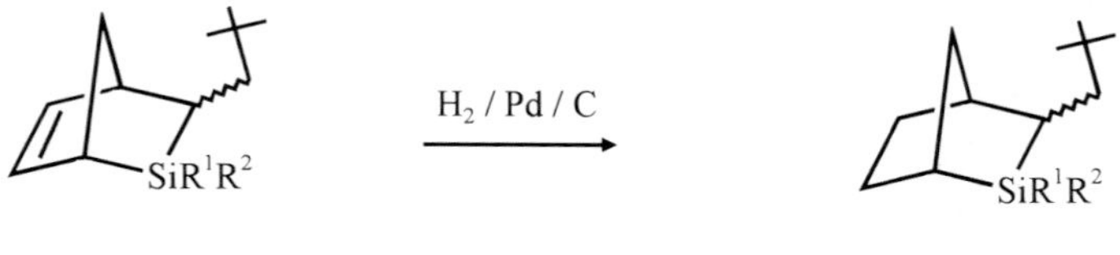

$R^1 = R^2 = Me$	:	**1**
$R^1 = R^2 = Cl$	:	**2**
$R^1 = R^2 = OMe$	:	**3**

$R^1 = R^2 = Me$	:	**4**
$R^1 = R^2 = Cl$	:	**5**
$R^1 = R^2 = OMe$	:	**6**

Scheme 1.

The silanorbornenes **1**, **2**, and **3** were hydrogenated overnight in an apparatus described in the experimental part. After work up of the reaction mixture the silanorbornanes are obtained in high yield: **4** (92 %), **5** (78.8 %), and **6** (82 %) (Scheme 1). Hydrogenation of the silicon hydrido substituted silanorbornene was not successful: instead hydrosilylation took place and a polymer was formed.

Analogous to the hydrogenation of sp^2 carbons within a cyclic system, unsaturated carbons, located in the *exo*cyclic position in the silacyclobutane subunits are transformed into their saturated derivatives. This is exemplified by the silacyclobutanes **7**, **8**, and **11** (Scheme 2) as model compounds. From those the products **9**, **10**, and **12** can be easily obtained in good yield, leaving the four membered ring moiety unchanged.

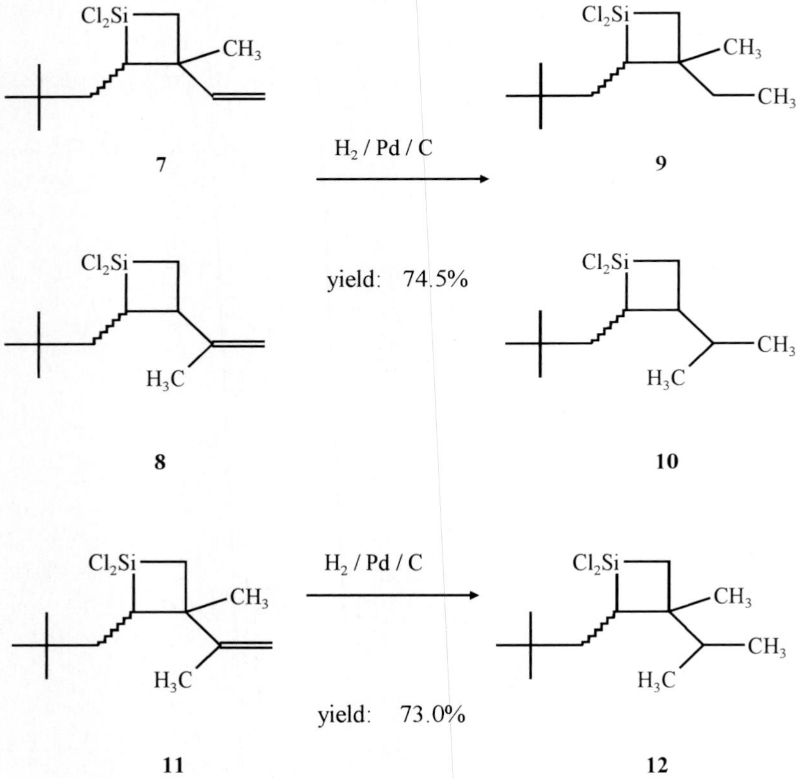

Scheme 2.

Not surprisingly, from the cyclopentenyl substituted compounds **13** and **15** (model compounds for cyclic sp^2 carbon atoms with *exo*cyclic silicon) the cyclopentane analogous derivatives **14** and **16** are available in more than 80 % yield (Scheme 3).

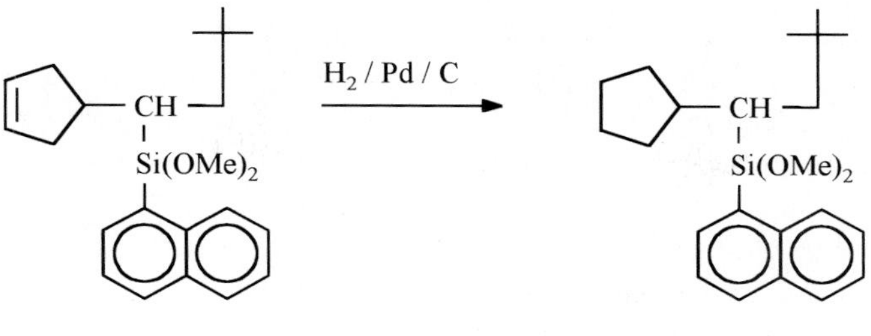

13 → H₂ / Pd / C → **14**

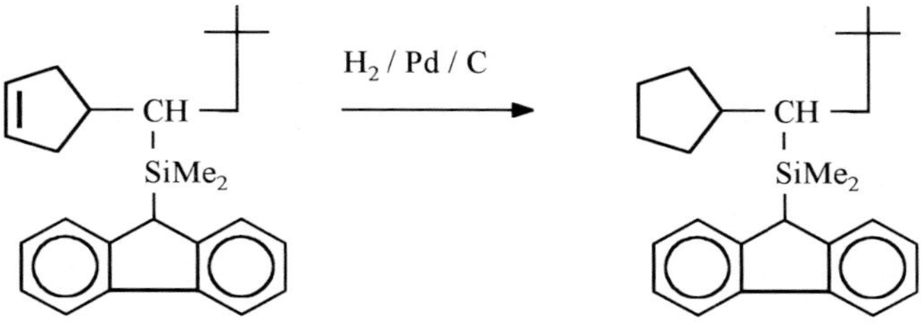

15 → H₂ / Pd / C → **16**

Scheme 3.

Experimental Part

Two dropping funnels are connected by a flexible tube and silicone oil is used as sealing fluid. The oil within the first dropping funnel is displaced by hydrogen. The volume is marked and then the tap of the reaction flask is opened. During the hydrogenation the amount of hydrogen consumed is monitored as change in volume.

All compounds described in this work have been completely characterized by NMR, MS, and elemental analysis [4]. As a representative characterization the δ ^{29}Si chemical shifts are listed in Table 1.

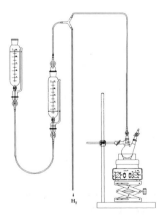

Fig. 1.   Description of the hydrogenation apparatus.

Table 1. ^{29}Si NMR data (ppm) of stereoisomeric pairs of educts and products.

No.	δ(^{29}Si) [ppm]		No.	δ(^{29}Si) [ppm]	
1	8.33 (*exo*)	8.11 (*endo*)	9	18.07 (*exo*)	18.83 (*endo*)
2	27.60 (*exo*)	25.90 (*endo*)	11	19.40 (*exo*)	19.60 (*endo*)
3	-3.37 (*exo*)	-4.10 (*endo*)	12	19.08 (*exo*)	19.62 (*endo*)
4	16.23 (*exo*)	10.06 (*endo*)	13	-18.62	
5	40.60 (*exo*)	37.57 (*endo*)	14	-21.71	
6	4.55 (*exo*)	0.34 (*endo*)	15	8.27	
7	20.00 (*exo*)	18.50 (*endo*)	16	8.43	

Acknowledgement: We thank *Wacker-Chemie GmbH* and *Chemetall GmbH* for gifts of chlorosilanes and lithium alkyls and *Dow Corning Ltd./Corp.* for financial support.

References:

[1]   P. R. Jones, T. F. O. Lim, R. A. Pierce, *J. Am. Chem. Soc.* **1980**, *102*, 4970.

[2]   a) N. Auner, *J. Organomet. Chem.* **1988**, *353*, 275.
        b) H.-U. Steinberger, N. Auner, *Z. Naturforsch.* **1994**, *49b*, 1743.

[3]   N. Sewald, W. Ziche, A. Wolff, N. Auner, *Organometallics* **1993**, *12*, 4123.

[4]   H.-U. Steinberger, B. Herrschaft, N. Auner, *Organometallics*, in preparation.

Dieno- and Enophilicity of Sila-, Germa-, and Stannaethenes

N. Wiberg*, S. Wagner

Institut für Anorganische Chemie
Ludwig-Maximilians-Universität München
Meiserstrasse 1, D-80333 München, Germany
Tel.: Int. code + (89)5902232 – Fax: Int. code + (89)5902578
E-mail: niw@anorg.chemie.uni-muenchen.de

Keywords: Silaethenes / Germaethenes / Stannaethenes / [4+2] Cycloadditions / [2+2] Cycloadditions / Ene Reactions

Summary: The reactivity of unsaturated compounds $Me_2E=C(SiR_3)_2$ (E = Si, Ge, Sn; SiR_3 = $SiMe_3$, $SiMePh_2$) with a variety of organic dienes is studied. The relative yields of products found for [4+2], [2+2], and ene reactions are explained by the electronegativity of E (Si ≈ Sn < Ge), the radius of E (Si ≈ Ge < Sn) and the energy of the E–C bond formed (Si–C > Ge–C > Sn–C). As a result, we found that the trend towards ene reaction and [2+2] cycloaddition decreases and the trend towards [4+2] cycloaddition increases in the direction sila-, germa-, stannaethene.

Sila-, germa-, and stannaethenes $Me_2E=C(SiR_3)_2$ (E = Si, Ge, Sn / SiR_3 = $SiMe_3$, $SiMePh_2$) undergo Diels-Alder and ene reactions as well as [2+2] cycloadditions with methyl derivatives of organic dienes.

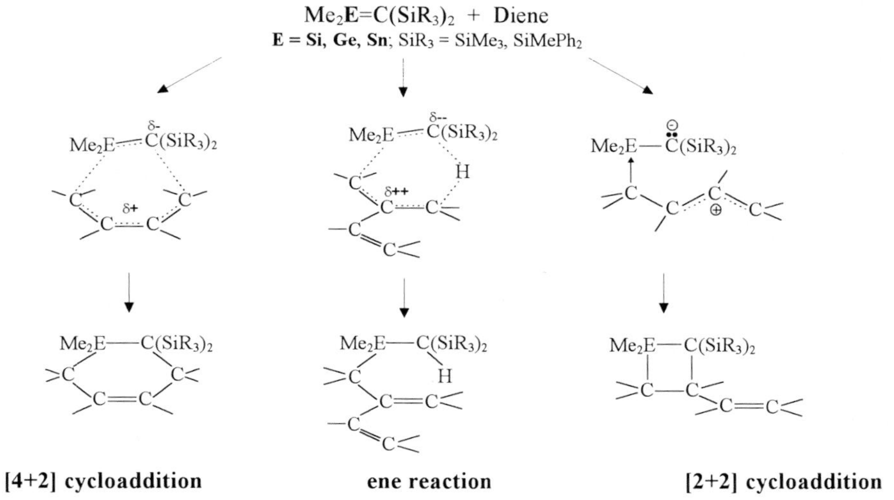

Scheme 1.

Mechanistic studies show that both Diels-Alder [1-3] and ene reactions [3, 4] of sila- and germaethenes proceed in a concerted way. On the other hand [2+2] cycloadditions proceed via a two-step mechanism [5]. Scheme 1 illustrates the increasing asymmetry of the transition state in the direction [4+2] → ene → [2+2] reaction.

All effects that result in a high symmetry of the transition state, or – in other words –that cause a transition state with relatively weak E-C bonding relation, will favor the Diels-Alder reaction. Important effects are the electronegativity of E, the radius of E and the energy of the new E-C bond formed. These are summarized in Table 1.

Table 1. Some relevant characteristics of silicon, germanium, and tin.

E	Radius [Å]	Electronegativity	Energy of E–C bond [kJ/mol]
Si	1.17	1.74	364
Ge	1.22	2.02	339
Sn	1.40	1.72	289

Sila-, germa-, and stannaethenes were allowed to react with a large number of 1,3-dienes, such as isoprene, dimethylbutadiene, *trans*- or *cis*-pentadiene, and *trans,trans*- or *cis,trans*-hexadiene [1-5]. Using dimethylbutadiene or *cis*-pentadiene, for instance, [4+2], [2+2], and ene products are formed in yields shown in Table 2.

Table 2. Relative yields of [4+2], [2+2], and ene products.

+	[4+2] cycloadduct	ene product	[2+2] cycloadduct
>Si=C<	76	24	0
>Ge=C<	85	15	0
>Sn=C<	93	7	0

+	[4+2] cycloadduct	ene product	[2+2] cycloadduct
>Si=C<	0	0	100
>Ge=C<	44	39	17
>Sn=C<	21	79	0

The yields can be explained in the following way (see Table 1):

- Going from sila- to germaethenes there is only a slight change in the radius of the metal atom, but an evident decrease in the polarity of the E=C double bond., i. e., a more symmetrical transition state is favored. As a result, germaethenes react with dienes to give less ene reaction products or [2+2] cycloadducts than silaethenes.

- Going from sila- to stannaethenes the polarity of the double bond remains nearly constant. But the formation of an asymmetrical transition state with stannaethenes is disadvantaged for two reasons: the tin atomic radius is much greater than the silicon or germanium radius and Sn–C single bonds (which have to be newly formed) are much weaker than Si–C and Ge–C bonds, respectively. Therefore, stannaethenes are poorer enophiles and [2+2] cycloadduct partners in comparison to sila- or germaethenes.

Acknowledgement: The authors thank the *Deutsche Forschungsgemeinschaft* for financial support.

References:

[1]   N. Wiberg, G. Fischer, S.Wagner, *Chem. Ber.* **1991**, *124*, 769.
[2]   N. Wiberg, S. Wagner, G. Fischer, *Chem. Ber.* **1991**, *124*, 1981.
[3]   N. Wiberg, S. Wagner, *Z. Naturforsch.* **1996**, *51b*, 838.
[4]   N. Wiberg, S. Wagner, *Z. Naturforsch.* **1996**, *51b*, 629.
[5]   N. Wiberg, G. Fischer, K. Schurz, *Chem. Ber.* **1987**, *120*, 1605.

Iminosilanes and Silaamidides:
Synthesis and Reactions

Jörg Niesmann, Andrea Frenzel, Uwe Klingebiel*
Institut für Anorganische Chemie
Georg-August-Universität Göttingen
Tammannstr. 4, D-37077 Göttingen, Germany

Keywords: Iminosilanes / Silaamidides / Addition Reactions / Crystal structure

Summary: One of the remarkable features of lithiated aminofluorosilanes and dilithiated diaminofluorosilanes with bulky substituents is that they do not undergo the expected substitution on treatment with Me_3SiCl on the nitrogen atom but instead on the silicon atom to give lithium derivatives of aminochlorosilanes. These salts are far less stable than the analogous fluorine compounds. LiCl elimination leads to the formation of iminosilanes and monomeric and dimeric silaamidides.

Addition and cycloaddition reactions of unsaturated Si=N compounds with organic and inorganic molecules are reported and their crystal structures are presented.

Introduction

In 1986, five years after the synthesis of the first silaethene and disilene, two independent routes for the preparation of uncoordinated iminosilanes were published by Wiberg et al. and our group [1-3]. We developed a synthesis based on fluorine-chlorine exchange of lithiated fluorosilylamines.

Synthesis of Iminosilanes

Aminofluorosilanes have NH- and SiF-functional groups. Lithiation of the NH-function makes either intramolecular elimination of lithium fluoride or fluorine-chlorine-exchange by Me_3SiCl possible.

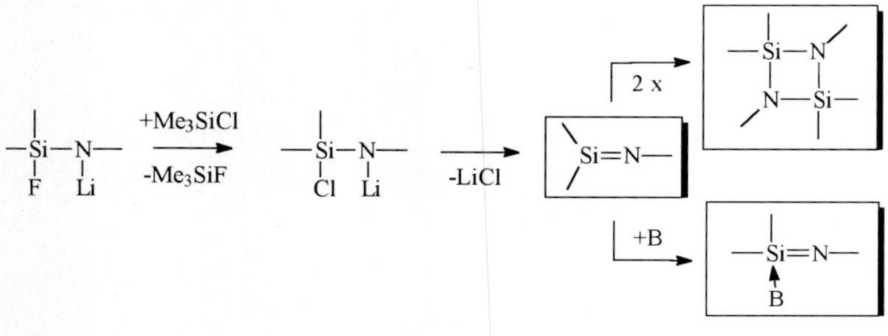

Scheme 1.

The last variant requires thermally less drastic conditions for the elimination of lithium chloride, and monomeric or dimeric iminosilanes are formed.

Altogether, three free iminosilanes, one THF-adduct of an iminosilane and all intermediate products have been synthesized and characterized by NMR and some by X-ray analysis [3]. Our monomeric iminosilanes **1** and **2** are remarkably stable. They could be seperated from LiCl by distillation, **1** (*bp* 65°C/0.01 mbar) without loss of THF and **2** (*bp* 108°C/0.01 mbar) as the free iminosilane.

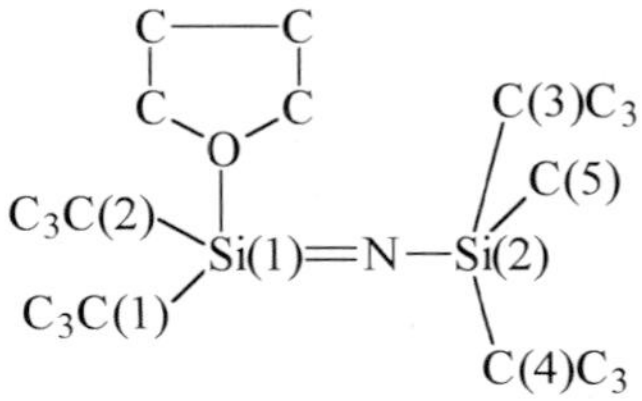

Fig. 1.　Crystal structure of **1** [4].
Si(1)–N:　　　159.6 pm
Si(2)–N:　　　166.1 pm
Si(1)–O:　　　190.2 pm
Si(1)–N–Si(2): 174°

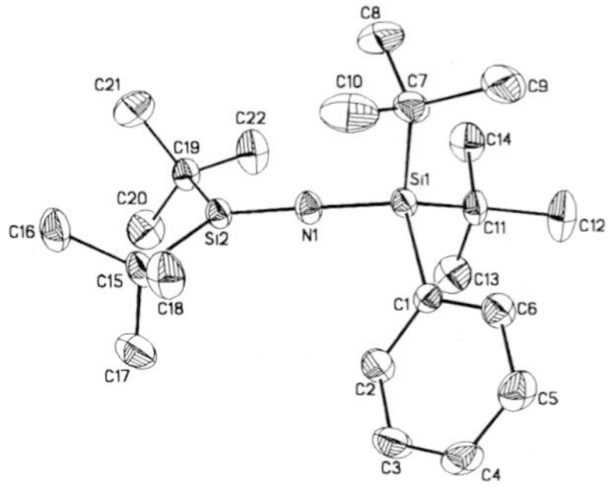

Fig 2.　Crystal structure of **2** [5].
Si(2)–N(1):　　157.3 pm
Si(1)–N(1):　　169.3 pm
Si(1)–N(1)–Si(2): 171°

Intra- and intermolecularly stabilized iminosilanes are obtained by using heteroaromatic substituted aminofluorosilanes. For example, the bicyclic system **3** is obtained in an insertion reaction of the iminosilane into a polar C–H bond of the pyrrole substituent. Using smaller substituents the [2+2] cycloaddition product **4** is obtained [6].

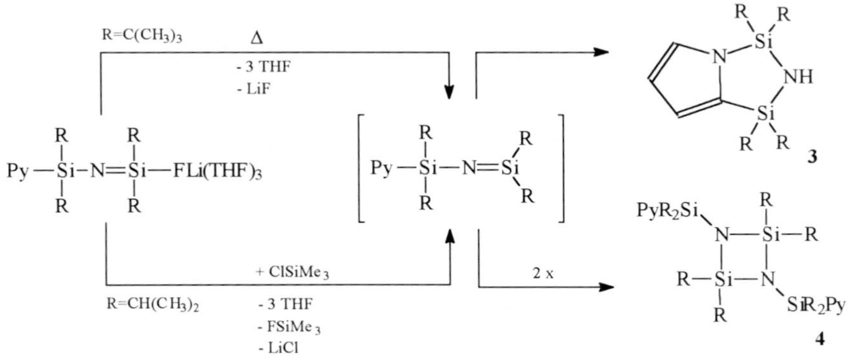

Scheme 2.

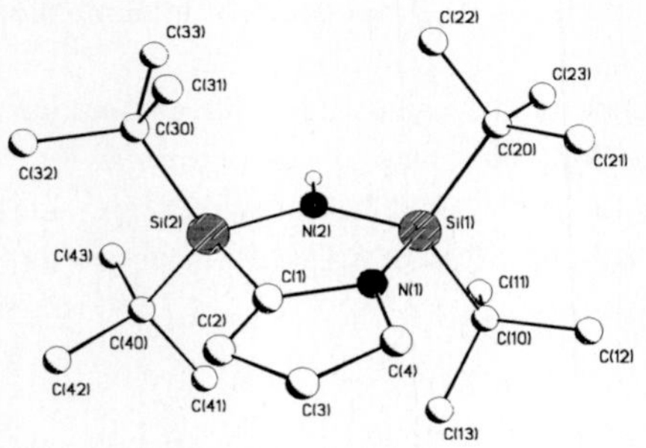

Fig. 3.   Crystal structure of **3** [6].

Si(1)–N(2):   172.7 pm
N(2)–Si(2):   173.1 pm
Si(1)–N(2)–Si(2):   118.4°

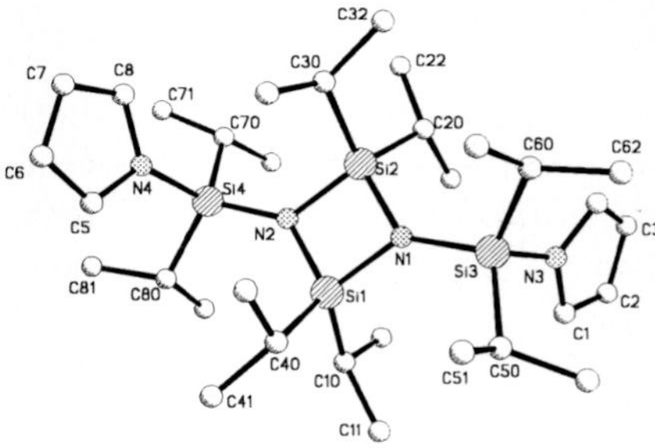

Fig. 4.   Crystal structure of **4** [6].

Si(1)–N(1):   176.8 pm
Si(1)–N(2):   177.0 pm
N(1)–Si(3):   173.3 pm
N(1)–Si(1)–N(2):   89.3°

Reactions of Iminosilanes [3, 7, 8]

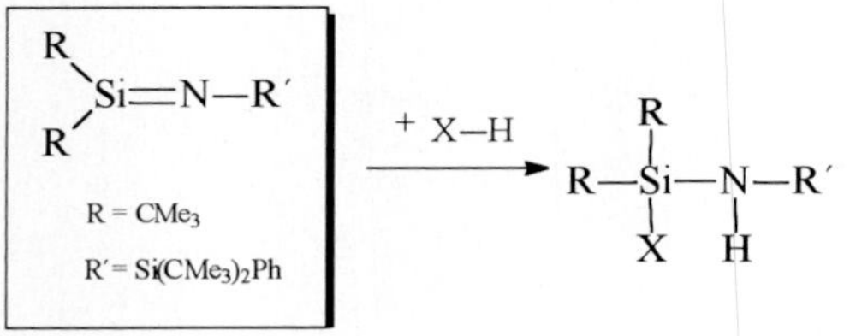

	X
5	OH
6	OSi(CMe$_3$)$_2$NHSi(CMe$_3$)$_2$Ph
7	Cl
8	OEt

Scheme 3.   Insertion reactions [7]

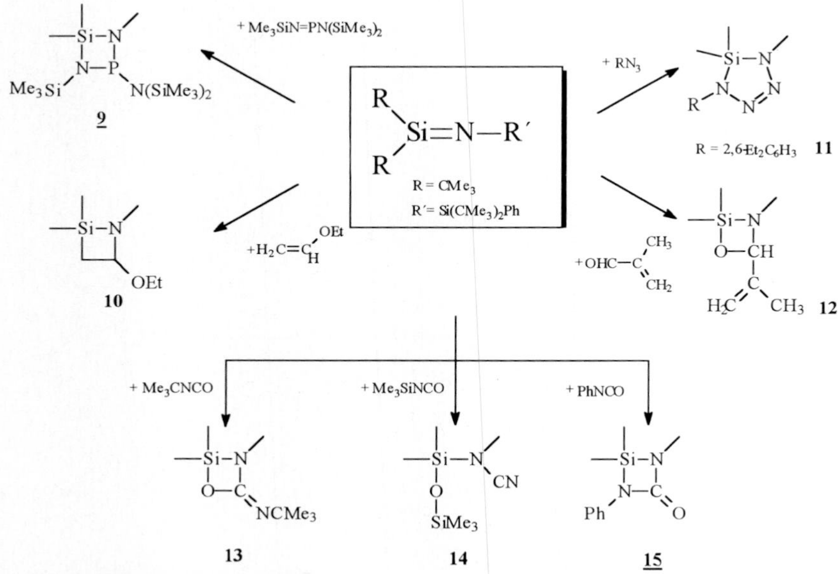

Scheme 4.   Cycloaddition reactions [3, 8].

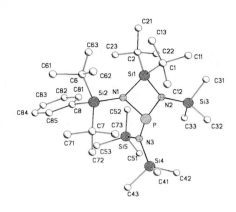

Fig. 5. Crystal structure of **9**

Si(1)–N(2):	175.5 pm
Si(1)–N(1):	179.4 pm
N(1)–P:	178.98 pm
N(1)–Si(1)–N(2):	85.49°
N(1)–P–N(2):	85.83°

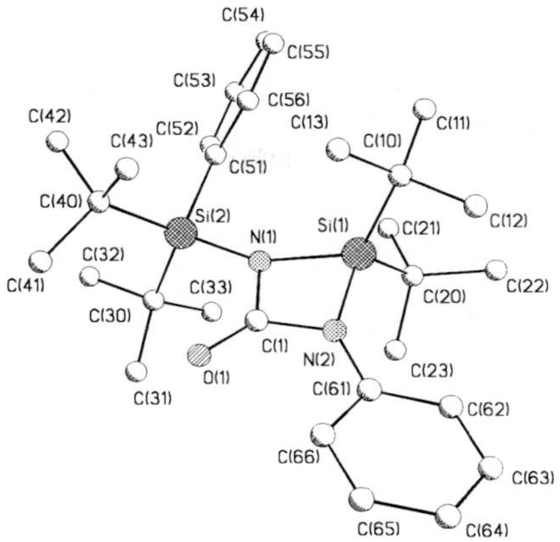

Fig. 6. Crystal structure of **15**

Si(1)–N(1):	177.94 pm
Si(1)–N(2):	176.07 pm
C(1)–N(1):	140.9 pm
N(1)–Si(1)–N(2):	76.24°
Si(1)–N(1)–Si(2):	148.17°

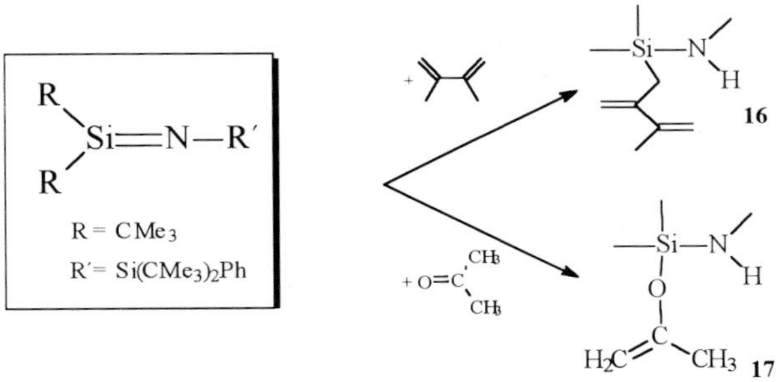

Scheme 5. Ene reactions [7].

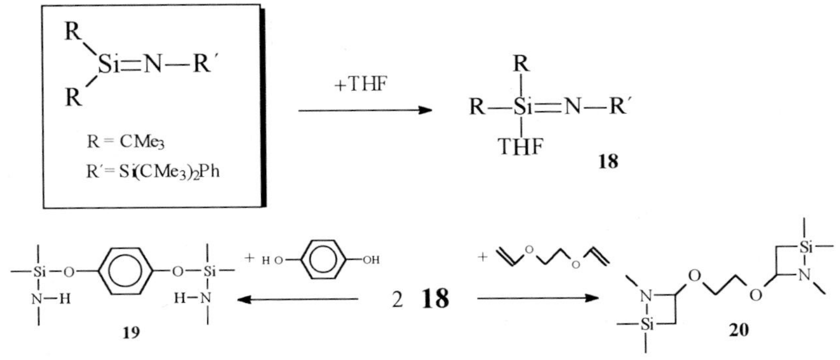

Scheme 6. THF-adduct formation; insertion reactions [7].

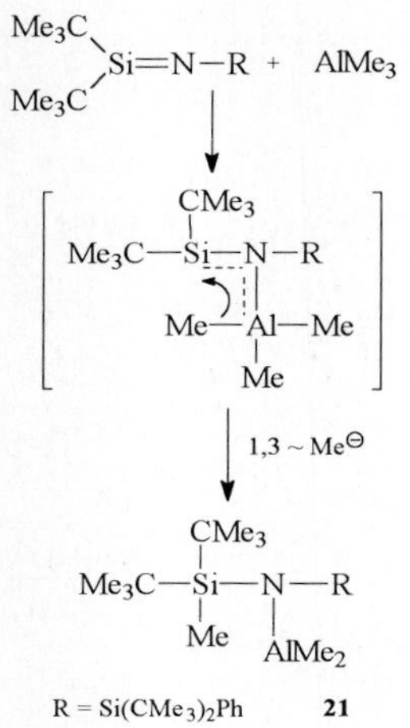

Scheme 7. Reaction with Me₃Al; a monomeric Me₂AlNSi₂-unit [5, 9].

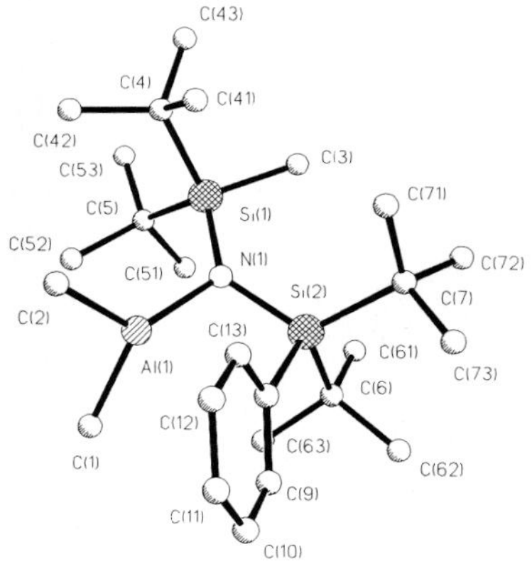

Fig. 7. Crystal structure of **21**
Al(1)–N(1):   186.9 pm
Si(1)–N(1):   175.2 pm
Si(2)–N(1):   175.0 pm
Σ < N(1) = 360°
Σ < Al(1) = 358°

Silaamidides – A facile Synthesis of Cyclodisilazanes in cis-Conformation

Treatment of dilithiated diaminofluorosilanes with Me₃SiCl leads to the formation of silaamidides. At higher temperature a [2+2] cycloaddition occurs and dimeric silaamidides are obtained. Hydrolysis of these silaamidides is the only facile syntheseis of cyclodisilazanes in cis-conformation [10, 11].

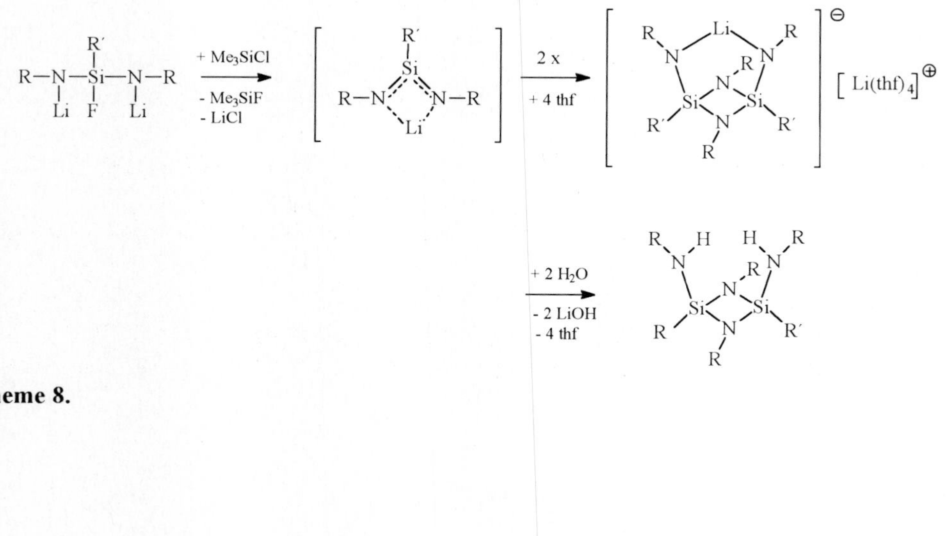

Scheme 8.

e.g., crystal structures of

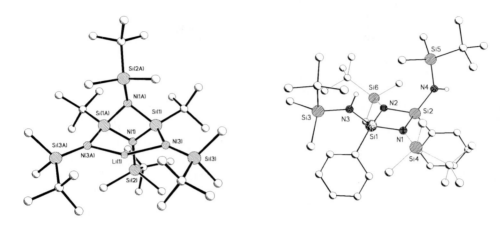

Fig. 8.   **Fig. 9.**

Acknowledgement: We thank the *Deutsche Forschungsgemeinschaft* and *the Fonds der Chemischen Industrie* for financial support.

References:

[1] G. Raabe, J. Michl, *"Multiple bonds to silicon"*, in: *The Chemistry of Organic Silicon Compounds Part 2* (Eds.: S. Patai, Z. Rappoport), Wiley, Chichester, **1989**, p. 1015.

[2] N. Wiberg, H.-W. Lerner, *"Silaneimines"*, in: *Organosilicon Chemistry II: From Molecules to Materials* (Eds.: N. Auner, J. Weis), VCH, Weinheim, **1996**, p. 405.

[3] I. Hemme, U. Klingebiel, *Adv. Organomet. Chem.* **1996**, *39*, 159.

[4] S. Walter, U. Klingebiel, *Coord. Chem. Rev.* **1994**, *130*, 481.

[5] J. Niesmann, U. Klingebiel, M. Noltemeyer, R. Boese, *J. Chem. Soc., Chem. Comm.* **1997**, 365.

[6] A. Frenzel, U. Klingebiel, R. Herbst-Irmer, S. Rudolph, *J. Organomet. Chem.* **1996**, *524*, 203.

[7] J. Niesmann, U. Klingebiel, M. Noltemeyer, *J. Organomet. Chem.*, **1996**, *521*, 191.

[8] J. Niesmann, U. Klingebiel, S. Rudolph, R. Herbst-Irmer, M. Noltemeyer, *J. Organomet. Chem.*, **1996**, *515*, 43.

[9] J. Niesmann, U. Klingebiel, M. Noltemeyer, to be published.

[10] I. Hemme, M. Schäfer, R. Herbst-Irmer, U. Klingebiel, *J. Organomet. Chem.* **1995**, *493*, 223.

[11] U. Klingebiel, B. Tecklenburg, R. Herbst-Irmer, T. Müller, M. Noltemeyer, D. Schmidt-Bäse, to be published.

Metastable Compounds Containing Silicon-Phosphorus and Silicon-Arsenic Multiple Bonds: Syntheses, Structures, and Reactivity

Matthias Driess*, Stefan Rell, Uwe Winkler
Lehrstuhl für Anorganische Chemie I
Fakultät für Chemie
Ruhr-Universität Bochum
Universitätsstraße 150, D-44801 Bochum, Germany
Fax: Int. code + (234)7094378
E-mail: driess@ibm.anch.ruhr-uni-bochum.de

Hans Pritzkow
Anorganisch-chemisches Institut
Ruprecht-Karls-Universität Heidelberg
Im Neuenheimer Feld 270, D-69120 Heidelberg, Germany

Keywords: Silicon-Phosphorus Compounds / Silicon-Arsenic Compounds / Silylidene-phosphanes and -arsanes / Silaheterocycles / Allyl Analogues

Summary: The monomeric lithium(fluorosilyl)phosphanides $Is_2Si(F)$-$P[LiL_n]R$ **1**, (Is = 2,4,6-triisopropylphenyl; L = THF/DME, R = organosilyl and germyl groups) and the related lithium(fluorosilyl)arsanides $Is_2Si(F)$-$As[Li(THF)_2]R$, **3**, (R = organosilyl groups), eliminate LiF and THF/DME on heating to give the corresponding silylidene-phosphanes and -arsanes **2** and **4**, respectively. The reactivity of the Si=E bonds (E = P, As) towards P_4, elemental sulfur, and tellurium is very similar and leads to the corresponding 1,2,3-triphospha-4-silabicyclo[1.1.0]butane **12** but to two isomers of arsa-diphospha-silabicyclo[1.1.0]butanes, namely the 1-triisopropylsilyl-4,4-diisityl-1-arsa-2,3-diphospha-4-silabicyclo[1.1.0]butane **13**, and the 3-triisopropylsilyl-4,4-diisityl-1-arsa-2,3-diphospha-4-silabicyclo[1.1.0]butane **14**, respectively. The reaction of S_8 and Te with **2** and the related phospha- and arsa-silenes $(t$Bu)IsSi=E(SiiPr$_3$) **6** (E = P) and **7** (E = As), respectively, gives rise to the corresponding thia- and tellura-cyclopropanes **15**, **16**, and **17a-d**, respectively, which have unexpectedly low inversion barriers at phosphorus and arsenic. The reaction of **2** with phenylacetylene yields $Is_2Si(CCPh)$-$PH(SiiPr_3)$ **18**, and benzonitrile reacts with the Si=E bond to give [2+2]-cycloadducts.

Only the Si=As bond reacts with *t*-butylphosphaacetylene to give the [2+2] cycloadduct **21** with an As–P and additional Si–C bond. Compound **2** reacts with cyclopentadiene to provide the expected hetero Diels-Alder product **22**. The reaction of **2** with benzophenone leads to the corresponding 2,1,3-silaphosphaoxetane **23** and treatment of **2** with 1,2-diphenyl-1,2-diketone unexpectedly provides the [2+2] cycloadduct **27** as the thermodynamic product. The 2,4-di-*t*-butyl-*o*-quinone solely yields with the Si=P bond in **2** and **6** the [2+4] cycloadducts **28** and **28a**, respectively. The cycloaddition of **2** and **4** with diphenyldiazomethane, mesitylazide, and mesitylisocyanide were also investigated and the respective [2+*n*]-cycloadducts ($n = 1, 2, 3$) were characterized by means of NMR spectroscopy. For example, the Si=As bond in **4** reacts with diphenyldiazomethane to give the unusual [2+1]-cycloadduct **29**. Furthermore, we discuss the synthesis, structure and reactivity of a potentially conjugated phosphasilene, namely the 2-phospha-1,3-disilaallylfluoride **37**. The latter possesses isolated Si=P and Si–P bonds in the solid-state and in solution at room temperature, but shows fast fluctuation of the fluorine atom on the NMR time scale at 40°C. Compound **37** was transformed to the disilaphosphacyclopropanes **41** and **42**, probably via the 2-phospha-1,3-disilaallylide anion in **40**, simply by reduction with two mole equivalents of Li metal and subsequent elimination of LiF. The Si_2P-heterocycle **42** is the first disilaphosphacyclopropane, the structure of which has been established by X-ray crystallography.

Introduction

Silicon compounds bearing low coordinated silicon and Si=E bonds (E = main group element), which are stable at room temperature, are convenient and valuable synthetic building blocks in silicon-heteroatom chemistry. This has been widely demonstrated by the use of disilenes (Si=Si) [1], silenes (Si=C) [2] and silanimines (Si=N) [3] as starting materials for the syntheses of several new classes of silaheterocycles. Although stable phosphasilenes (Si=P) with sterically demanding organyl groups at silicon and phosphorus (type **I**, Fig. 1) have been known since 1984 [4], their isolation, and therefore the study of reactivity, has proven to be very difficult until recently.

In 1991 we reported on the synthesis of *P*-silyl and -phosphanyl substituted silylidenephosphanes (phosphasilenes) **2**, i.e., compounds of type **IIA** and **III**, which are accessible by thermally induced elimination of LiF from corresponding *P*-lithium-(fluorosilyl)phosphanides **IV** [5, 6] and possess a remarkable thermal stability (up to 110°C). In 1993 the first crystalline phosphasilene **5** [7], a compound of type **IV**, had been prepared and its structure was established by X-ray diffraction [8], whereas the Si=P compounds of type **V** were merely characterized by means of NMR spectroscopy [9]. Compound **5** possesses a relatively long Si=P bond length and a non-planar geometry around the λ^4,σ^3-Si atom, which can be explained in terms of steric hindrance or a second-order Jahn-Teller distortion [10, 11]. In comparison, the Si=P bond in **6** [12] is significantly shorter and the Si atom is trigonal planar coordinated.

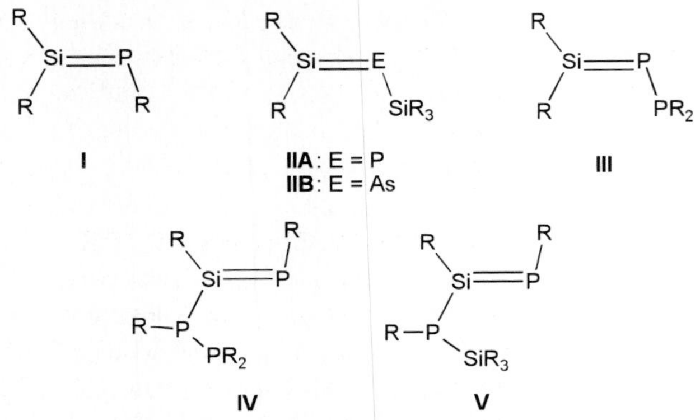

Fig. 1. Types of compounds containing isolated Si=E bonds (E = P, As).

It has been shown by calculations that the Si–P-π-bond in $H_2Si=P(SiH_3)$ is strengthened by hyperconjugation due to the silyl substituent at phosphorus [13]. Furthermore, the stabilizing influence of silyl groups provides that even the Si=As bond in silylidenearsanes (arsasilenes), i.e., compounds of type **IIB**, are surprisingly easy to access. They are similarly formed by thermally initiated elimination of LiF from *As*-lithium-(fluorosilyl)arsanides **3** [13, 14]. Recently, we performed the structural characterization of compound **7** [12]. Despite the progress in structural investigations of Si=P and Si=As compounds, knowledge of their reactivity is still limited. The thermolability of compounds of type **I**, **IV**, and **V** only allowed for reactions with water, methanol, elemental tellurium, and lithiumalkanides; these reactions proceed with addition onto the Si=P bond [4]. Further insight into the reactivity of the Si=P bond was provided by trapping experiments of a transient phosphasilene [15]. Because of the convenient handling of the similar reactive phospha- and arsasilene derivatives **2**, **4**, **6**, and **7** (Scheme 1), we have carried out reactions with P_4 [5], elemental Te [12, 14], diphenyldiazomethane [16], and 1,6-diisocyanohexane [17], respectively.

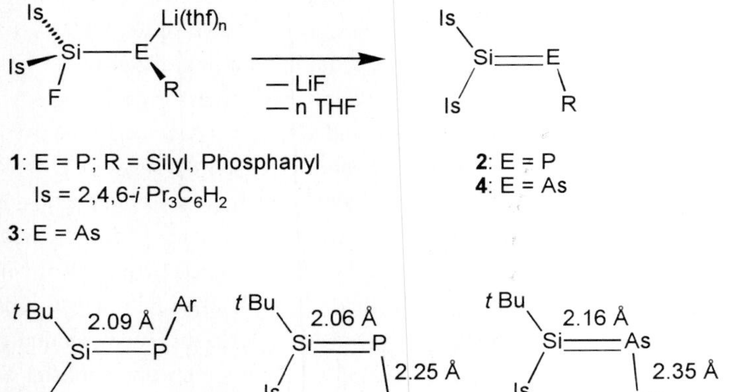

Scheme 1. Synthesis of silylidene-phosphanes and -arsanes **2**, **4**, starting from **1** and **3**, and Si–E distances (E = P, As; Ar = 2,4,6-tri-*tert*-butylphenyl) in **5**, **6**, and **7**.

The most recent developments in the synthesis and reactivity of isolated Si=E bonds (X = P, As) in **2**, **4**, and **6** will be the main focus in this paper. Furthermore, we discuss the synthesis and reactivity of the first 1,3-disila-2-phosphaallylfluoride derivative, that is, a silylidenephosphane bearing a potentially conjugated allylic Si$_2$P-π system [18].

Synthesis of Silylidenephosphanes and -arsanes with Isolated Si=E-Bonds (E = P, As)

The synthesis of the phosphasilenes **2a-h** and of the arsasilenes **4a-c** was achieved via the corresponding lithium-(fluorosilyl)pnictides **1a-h** and **11a-c**, respectively (Scheme 2). It has been shown that perfect steric protection of the highly reactive Si=E bonds (E = P, As) in **2** and **4** is provided by the 2,4,6-triisopropylphenyl substituent (Is = isityl) attached to the low-coordinated silicon center. The appropriate precursors **1** and **3** were prepared in a multiple-step procedure, starting from **8** and **10**, respectively.

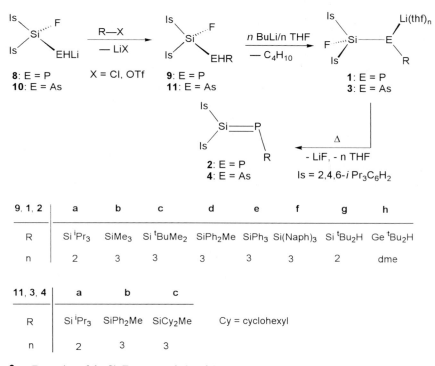

Scheme 2. Formation of the Si=E compounds **2** and **4**.

9, 1, 2	a	b	c	d	e	f	g	h
R	Si ^{i}Pr$_3$	SiMe$_3$	Si ^{t}BuMe$_2$	SiPh$_2$Me	SiPh$_3$	Si(Naph)$_3$	Si ^{t}Bu$_2$H	Ge ^{t}Bu$_2$H
n	2	3	3	3	3	3	2	dme

11, 3, 4	a	b	c		
R	Si ^{i}Pr$_3$	SiPh$_2$Me	SiCy$_2$Me	Cy = cyclohexyl	
n	2	3	3		

The compounds **8** and **10** are formed by the reaction of Is$_2$SiF$_2$ with two equiv [LiEH$_2$(DME)] (E = P, As; DME = 1,2-dimethoxyethane); they show no tendency to eliminate LiF in THF solution, surely because of the inherent strength of the Si–F bond. Silylation/germylation of **8** furnished the compounds **9** and **11**, respectively, which were subsequently lithiated on phosphorus and arsenic to yield **1** and **3**, respectively. In the first instance, the nature of the substituent at phosphorus determines the structures of the precursors **1** and **3**, respectively. This has been demonstrated by a study of a structure-reactivity relationship for several compounds **1** [6]. The latter investigation revealed that lithiumphosphanides **1** are monomeric in the solid state and in solution, whereby the number x of the donor solvent molecules attached to the lithium center (x = 2, 3) is strongly dependent on the nature of the substituent at phosphorus. It was further observed that phenyl substituted silyl groups attached to phosphorus in **1i** (R = Si(H)Ph$_2$) cause a P(n)/π^*-hyperconjugation which essentially seems to stabilize the trigonal-planar geometry around phosphorus and leads to remarkably short Si–P distances (2.16, 2.18 Å).

An X-ray structure elucidation of the related phosphanide **1f**, bearing a tris-α–naphthylsilyl group at phosphorus, revealed that the P atom is *not* trigonal-planar surrounded (sum of bond angles at phosphorus 350.4°) [19]. Other interesting structural features were observed by an X-ray crystal structure determination of the di-*t*-butylgermyl substituted lithiumphosphanide **1h** (Fig. 2) [19]. In this case the lithium center is chelated by one dimethoxyethane (DME) molecule and at the same time attached to the phosphorus and fluorine atoms, as observed in **1j** (R = Si(NMe$_2$)$_2$Me) and **1k** (R = P(N*i* Pr$_2$)$_2$) [6]. This causes a large pyramidalization on phosphorus (sum of bond angles 268.4°). The Li–F distance is 1.983(6) Å and thus significantly shorter than that observed in **1k** (2.060(8) Å). The Li–F bond seems to be preferred in **1h** for mainly two reasons: the DME ligand is unable to electronically saturate the lithium center and the spatial requirements of the *t*-butyl groups at germanium force a different conformation of the (F)Si–P(Li) moiety than that observed in **1a, b**, and **1i**. In the case of **1k** (R = P(N*i*Pr$_2$)$_2$), however, the Li–F bond is solely caused by a chelate effect [6]. One methyl hydrogen atom of the *t*-butyl groups at germanium in **1h** is close by and also contributes to the electronic saturation of the lithium center. This is supported by a relatively short Li–H distance of 2.37 Å. Therefore, the Li center can be regarded as five-fold coordinated. The structural features of the arsanides **3** are, as expected, identical with those of the phosphorus analogues **1** [14].

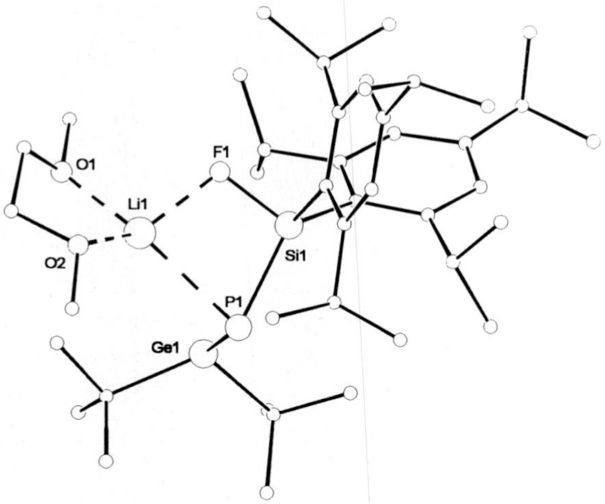

Fig. 2.   Solid-state structure of the lithiumdisilylphosphanide **1h**.

The phosphasilenes **2a-h** and the arsasilenes **4a-c** were formed by heating solutions of the corresponding derivatives **1a-h** and **3a-c**, respectively (see Scheme 2). With exception of **2b** (R = SiMe$_3$) and **2e** (R = SiPh$_3$), the Si=P compounds **2** were isolated in the form of yellow or orange-red oils. Compounds **2a-h** exhibit characteristic ^{31}P and ^{29}Si NMR spectroscopic data [19]. Noteworthy is the relatively large shielding for the ^{31}P nucleus in the ^{31}P NMR spectra of **2** (δ = 28.1

to -7.8), which is quite unusual for compounds bearing two-coordinate phosphorus [20]. Evidently, this is caused by the strong σ-donor ability of the silyl and germyl group, which is also reflected in the comparison of calculated ^{31}P chemical shifts for the parent compounds $H_2Si=PH$ ($\delta = 85$) and $H_2Si=P(SiH_3)$ ($\delta = 48$) [12, 13]. In the case of **2g**, the shielding of the phosphorus represents the largest value in this series ($\delta = -7.8$) observed to date. An even more strongly shielded ^{31}P nucleus in phosphasilenes was detected for **6** ($\delta = -29.9$) [12]. This unusual finding for **2g** and **6** clearly indicates that steric effects seem to have a considerable influence on the ^{31}P chemical shifts. In line with this, the phosphorus bond angle in **6** (112.7°) significantly exeeds the value calculated for $H_2Si=P(SiH_3)$ (100°) [13]. In the ^{29}Si NMR spectra the low-coordinate Si atoms reveal doublets at very low field ($\delta = 167.8$-181.7). The $^1J_{Si-P}$ coupling constants (149-160 Hz) are diagnostic for Si–P π-bonds, whereas much smaller values were obtained for saturated silylphosphanes [20]. The resonance signals of the [29]Si nuclei of the SiR_3 groups lie in the expected range for saturated silicon compounds, however, the $^1J_{Si-P}$ coupling constants are found to be significantly larger (70-78 Hz) due to the low coordination of the phosphorus. The arsasilenes **4a-c** are as thermally resistant as **1**.

The low-coordinate Si atom exhibits a singlet at very low field in the ^{29}Si NMR spectrum. The largest deshielding for low-coordinated silicon in arsasilenes known to date has been detected for **7** ($\delta = 228.8$) [12]. Remarkably, the Lewis-donor ability of the solvent employed does not significantly affect the [29]Si chemical shifts; the same is true for Si=P compounds. This is in contrast to the behavior of silanimines (Si=N) [3], which easily furnish donor solvent adducts. The relatively strong deshielding of the [29]Si nuclei in **2** and **4** in comparison to that observed for silanimines is, in the first instance, caused by the smaller HOMO-LUMO gap (π–$\pi*$) and probably by a stronger contribution of the polar resonance structure **A** in the electronic ground state (Scheme 3).

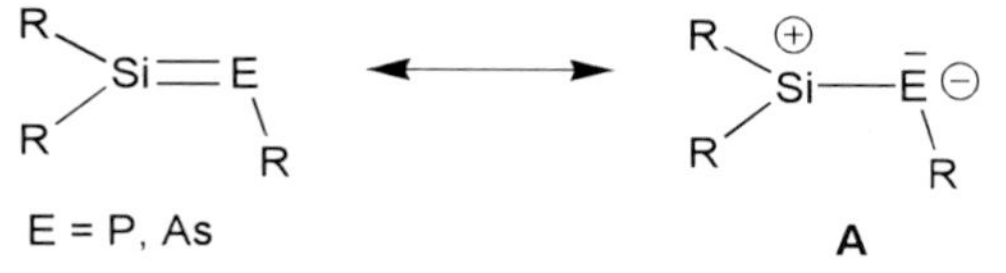

Fig. 3.   Resonance structure **A**.

(E)/(Z)-Isomerization has not yet been observed for both types, phospha- and arsa-silenes, and ab initio calculations predict Si=E π-bond energies of 34 (E = P) and 30 kcal mol^{-1} (E = As), respectively [13]. We assume that the rotation barrier for the silylated phospha- and arsa-silenes **2** and **4** is much lower, since no isomerization was observed on the NMR time scale within the temperature range -80 to 100°C. The unusually low barrier could be explained by the presence of sterically demanding substituents. A very similar electronic situation of Si=As and Si=P bonds is further reflected by their UV/visible spectra [19].

Reactivity of the Phosphasilenes 2a, 6, and the Arsasilene 4a

The reactivity of the representative derivatives **2a** and **4a** towards P_4, elemental sulfur, tellurium, phenylacetylene, benzonitrile, *t*-butylphosphaacetylene, and cyclopentadiene (see Scheme 3), and furthermore towards benzophenone, 1,2-diphenyl-1,2-diketone, 3,5-di-*t*-butyl-*o*-quinone, diphenyldiazomethane, mesitylazide, and mesitylisocyanide was studied (see Scheme 6) [19].

Compounds **2a** and **4a** react with P$_4$ in the molar ratio of 2:1 to furnish the SiP$_3$ butterfly-like compound **12** and the SiAsP$_2$ analogous isomers **13** and **14**, respectively. These products are analogous to those observed by the degradation reactions of P$_4$ and As$_4$ with disilenes, which lead to Si$_2$P$_2$- and Si$_2$As$_2$-bicyclo[1.1.0]butanes, respectively [21, 22]. The ^{31}P and ^{29}Si NMR spectra indicate that **12** prefers the *exo*-configuration at 25°C. At higher temperature (> 38°C) the inversion of configuration of the peripheral phosphorus atom is observed in the ^{31}P NMR spectrum and **12** rearranges into its *endo* isomer [19]. For this process a inversion barrier of 23 kcal mol^{-1} has been estimated. Following results from MO calculations of P$_2$Si$_2$-bicyclo[1.1.0]butanes [23], we assume that the SiP$_3$ system has a very high barrier for ring inversion. The analogous reaction of **4a** with P$_4$ in the molar ratio 2:1 occurs readily at 40°C, in which a complicated product mixture was formed [19]. Upon heating of this mixture at 100°C, the heterocycles **13** and **14** were observed. The approximate ratio of **13** and **14** is 3:2, and they cannot be transformed into each other.

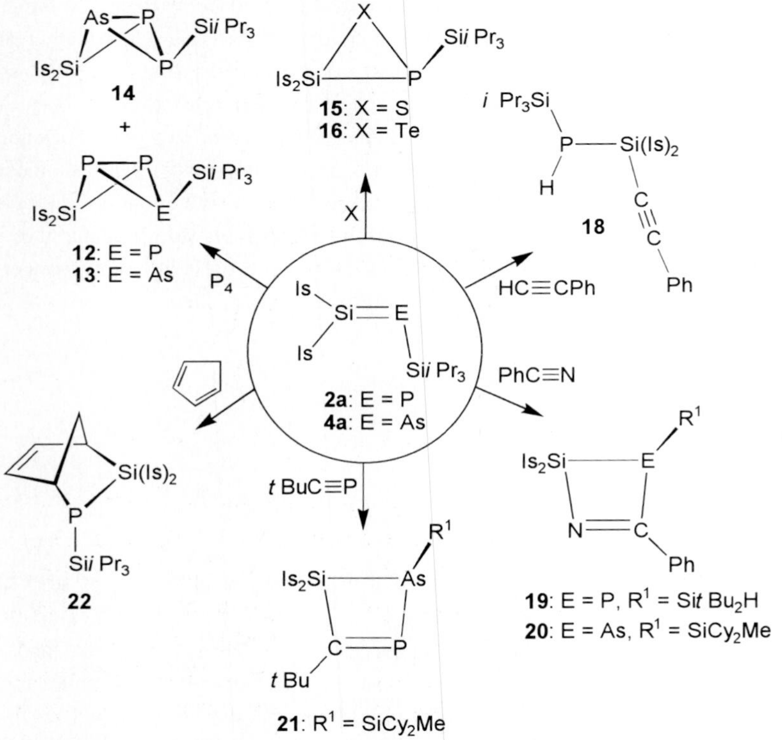

Scheme 3.   Reactivity of **2a** and **4a**.

The oxidation of **2a** with elemental sulfur and tellurium quantitatively furnish the three-membered heterocycles **15** and **16**, respectively. The latter do not yield 1,3-dichalcogen-2,4-phosphasila-cyclobutanes on treatment with excess elemental sulfur or tellurium. The reaction of **6** and **7** with elemental tellurium furnished a 1:1 mixture of the diasteromeres **17a,b** and **17c,d** (*t*-butyl/Si(*i*Pr)$_3$ *cis*-and *trans*-forms), respectively, which were characterized by means of NMR spectroscopy ([29]P, [29]Si, [125]Te) [12, 19].

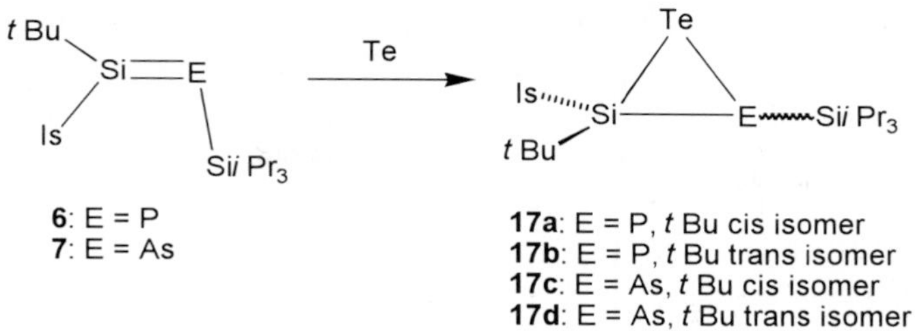

Scheme 4. Synthesis of **17a,b** and **17c,d**.

Interestingly, only one diasteriomeric form (racemic mixture *R,R* and *S,S* configuration) was observed in the crystal in both cases [12, 24]. Therefore, we were able to characterize **17a** and **17c** by single crystal X-ray diffraction, as shown for **17a** in Fig. 4.

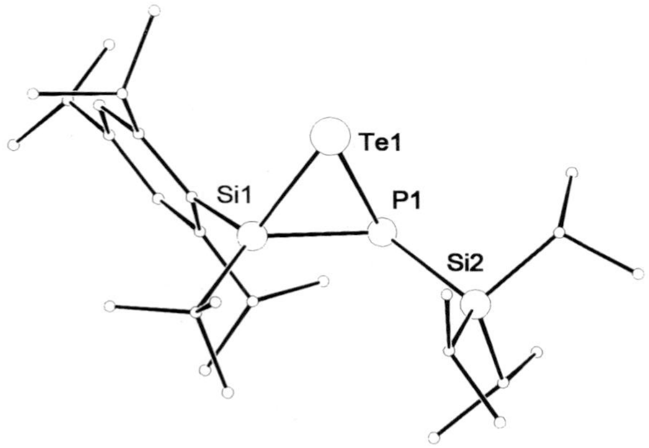

Fig. 4. Solid-state structure of the *cis* configured phosphasilatelluracyclopropane **17a**.

The Si–Te (2.496(2) Å), Si–P (2.235(2) Å), and P–Te-distances (2.524(2) Å) in **17a**, as well as the endocyclic angles, are in very good agreement with ab initio calculated values of the parent compound **17*** (basis set: B3LYP/3-21G*) [25]. Heating of a solution of a 1:1 mixture of **17a** and **17b** in toluene initiates a temperature-dependent equilibration, whereby coalescence occurs at 59°C in the ^{31}P NMR spectrum. The free activation enthalpy (Gibbs energy) for this interconversion process was estimated to be 15.8 kcal mol^{-1}. It seems rather unusual that this process is probably *not* due to phosphorus-inversion *within the intact* SiPTe three-membered skeleton, since ab initio calculation of the phosphorus-inversion of **17*** reveals a much larger value (32.06 kcal mol^{-1}) [25].

The latter fact and the energetically disfavored process via heterolytic bond cleavage of the SiPTe skeleton may rather imply a nonclassical *conversion process*, in which both bonds to the Te atom (Si–Te, P–Te) were simultaneously cleaved and reformed after phosphorus inversion. Remarkably, the dissociation energy for the latter process takes only about 20 kcal mol^{-1} [25]. This observation strongly supports the description of such small ring compounds as π-complexes in terms of the Dewar-Chatt-Duncanson model (see Scheme 5).

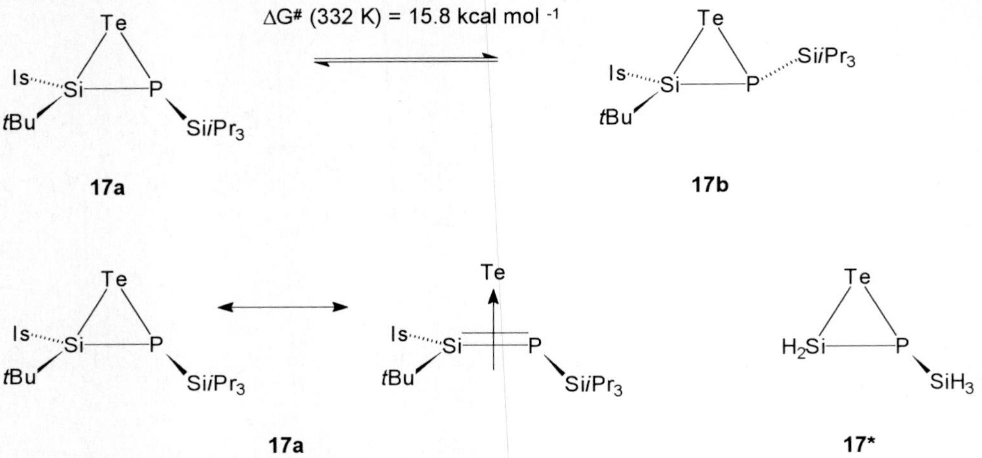

ΔG# (332 K) = 15.8 kcal mol⁻¹

17a

17b

17a

17*

Scheme 5.   Interconversion process of **17a** and **17b** by phosphorus-inversion.

The Si=E bond in **2a** and **4a**, respectively, turned out to be chemically inert toward dialkyl-, diaryl-, or disilyl-substituted alkynes and cyclooctyne. However, **2a** reacts with phenylacetylene to furnish the C–H insertion product **18** [19]. The acetylene moiety has added to the Si atom and the phosphorus was protonated, according to the bond polarity of the Si=P bond (Si^+–P^-). It is interesting to note that the reaction of tetramesityldisilene (Si=Si) with phenylacetylene exclusively yielded the [2+2] cycloaddition product 1,1,2,2-tetramesityl-3-phenyl-disilacyclobut-3-ene [26]. Experiments to convert **18** into the corresponding cyclic silaphosphaheterocyclobut-3-ene, using AIBN (azodiisobutyronitrile) as a radical starter for hydrophosphinations, have failed. The Si=E bond in **2a** and **4a** reacts readily with the strongly polarized C≡N triple bond in mesitylcyanide at 25°C to furnish the [2+2]-cycloaddition products **19** and **20**, respectively. The molecular structure of **20** was established by a single crystal X-ray diffraction analysis (Fig. 5) [19].

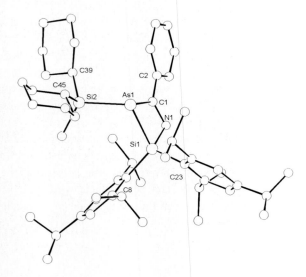

Fig. 5.   Solid-state structure of **20**.

The four-membered SiNAsC framework is slightly puckered and the As–Si distances are slightly longer than the respective values observed in the four-membered SiOAsC skeleton in a 1,2-silaarsa-oxetane derivative [14]. However, the As–C bond length (1.988(6) Å) and the endocyclic arsenic bond angle (65.1(2)°) are significantly smaller than the respective values in the latter compound. Evidently, the short C1–N1 distance (1.292(7) Å) in **20** forces a relatively small endocyclic bond angle at silicon. The different bond lengths in the four-membered ring are responsible for distinctly smaller inner angles in comparison to the ideal value of 120° at carbon and nitrogen (112.2(4) and 101.3(4)°), respectively. Analogous adducts, which were generated by the reaction of organo cyanides and disilenes are also known [27]. Surprisingly, the even more reactive C≡P triple bond in *t*-butylphosphaacetylene does not react with the Si=P bond but rather with the Si=As bond in **4c** to furnish the novel heterocycle **21**. According to the reverse polarity of the C≡P triple bond compared with the C≡N triple bond, the silicon ring atom in **21** is bound to the carbon atom of the C=P moiety. The first [2+4]-cycloaddition the Si=P bond was verified by the reaction of **2a** with cyclopentadiene, which leads to **22**. Interestingly, a Brønsted-acid- (C–H) base (Si=P) reaction, as in case of the reaction of **2a** with phenylacetylene, was not observed. The Si=P bond in **2a** readily reacts like **4a** [14] with benzophenone at -80°C to furnish the heterocyclobutane **23** (Scheme 6). The molecular structure of **23** was established by an X-ray structure analysis [19]. The compound is isotypic with its arsenic homologue [14]. A remarkably long P–C distance (1.935(5) Å) is observed, which is caused by steric hindrance.

Compound **23** decomposes on thermolysis at 160°C in a sila-Wittig-type reaction to furnish the transient silanone Is₂Si=O, which immediately dimerizes to the 1,3-disila-2,4-dioxetane **24**, and the phosphaalkenes **25** and **25A** (Scheme 6). 1,3-Disila-2,4-dioxetanes are well known from the work of West et al. [28]. Because of the drastic reaction conditions, the expected phosphaalkene **25** partially isomerizes under phenyl and silyl group migration to **25A**. However, such an isomerization process was previously unknown for phosphaalkenes.

If **2a** was allowed to react with PhC(O)–C(O)Ph at 25°C, surprisingly, the [2+4] cycloadduct **26** was formed in only 10 % yield (see Scheme 6). Interestingly, the unexpected [2+2] cycloadduct **27** has been isolated in diastereomerically pure form as the major product. The silyl group at phosphorus in **27** has the *trans*-position relative to the orientation of the phenyl group attached to the chiral carbon center. This has been proven by means of ¹H NMR NOESY experiments. On heating, **26** completely rearranges to **27**; that is, **27** is the thermodynamic product. This enone-ketone isomerization is unexpected since a relatively strong P–O bond is broken. However, the decrease in ring strain and the formation of the C–O π- and P–C σ-bonds in **27** may provide an plausible explanation. The formation of the four-membered CPSiO-framework as in **27** can be prevented if **2a** is treated with 4,6-di-*t*-butyl-*o*-quinone to furnish the thermally resistant benzo-condensed heterocycle **28** (see Scheme 6). Its ³¹P chemical shift in the ³¹P NMR spectrum (δ = 114.3) is almost identical with the value observed for **26** (δ = 116.3). The final molecular structure of the analogous derivative **28a**, prepared by reaction of **6** with the *o*-quinone (see Eq. 1), was established by X-ray crystallography, as shown in Fig 6.

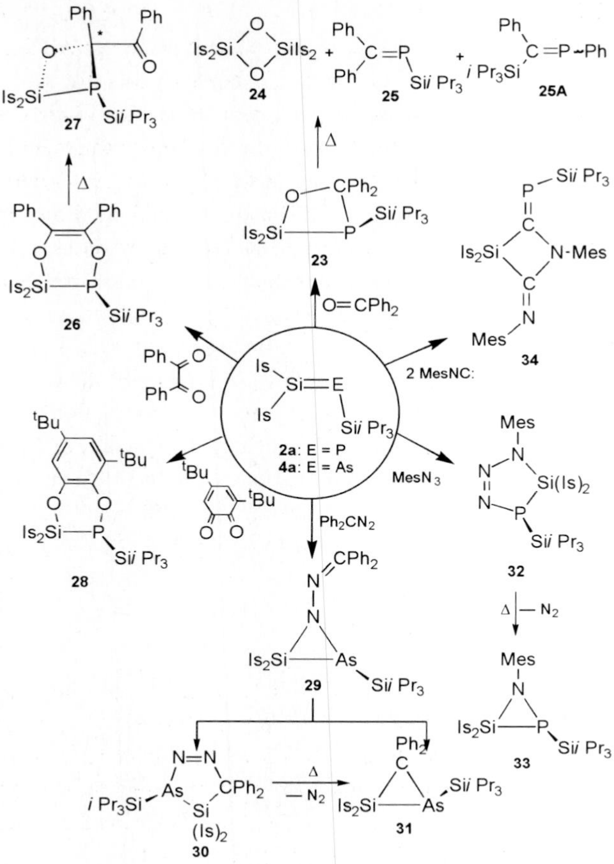

Scheme 6.   Cycloaddition reactions of **2a** and **4a**.

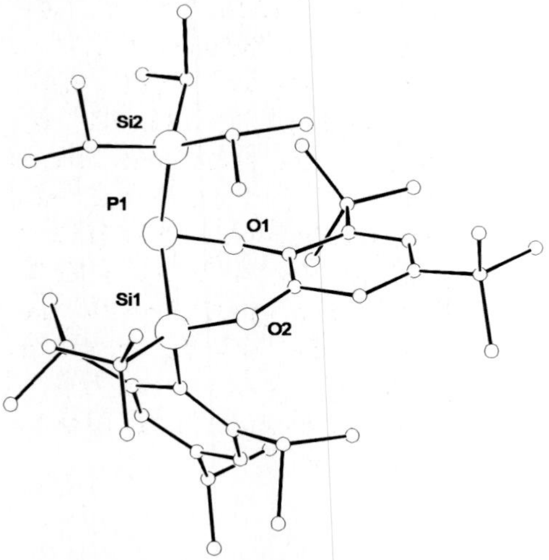

Fig. 6.   Solid state structure of **28a**. Selected distances [Å] and angles [°] (averaged values): P1–O1 1.73(1), P1–Si1 2.26(1), P1–Si2 2.30(1), Si1–O2 1.68(1), O1–C1 1.33(2), O2–C2 1.31(2); P1–Si1–O2 115.4(3), Si1–P1–O1 89.5(5), C1–O1–P1 126.5(10), C2–O2–Si1 124(1).

A [2+1] cycloaddition between the Si=As bond in **4a** and diphenyldiazomethane has been observed. From this reaction mixture the yellow, crystalline [2+1] cycloadduct **29** was isolated and its molecular structure was established by an X-ray structure analysis [19].

Compound **29** is isotypic with the analogous phosphorus derivative [16]. The endocyclic angle at arsenic is 45.6(2)° and, therefore, significantly smaller than the corresponding value at phosphorus observed in the phosphorus analogue (49.2°). Compound **29** rearrranges in toluene if heated at 90°C for 2 days to furnish a mixture of the isomeric [2+3] cycloadduct **30** and the arsasilirane **31**. Further thermolysis leads to tetraphenylethene and as yet unidentified products, whereas the analogous phosphorus compound of **31** cleanly rearranges to an unusual benzo-bicyclic product [16].

Eq. 1. Synthesis of **28a**.

In contrast to the latter reaction, the Si=P bond in **2a** readily reacts with mesitylazide at -80°C in toluene to furnish the [2+3] cycloadduct **32** as the primary product and **33** (see Scheme 6) [19]. Compound **32** is transformed completely into **33** on heating at 40°C with loss of N_2. The five-membered $SiPN_3$ skeleton in **32** is unambiguously proven by a relatively strong deshielding of the ring Si atom in the ^{29}Si NMR spectrum ($\delta = -16.7$) compared with the values of **29** ($\delta = -81.6$) disilaaziridines (δ -50 to -54.6) [29] and **33** ($\delta = -64.7$), respectively.

The intriguing reactivity of the Si=E bonds (E = P, As) is also shown by the behavior of **2a** toward mesitylisocyanide. During this reaction, two equivalents of isocyanide were consumed, even if an equivalent molar ratio was used, which leads to the unusual heterocycle **34** bearing an exocyclic imino and phosphaalkenylidene group (see Scheme 6). Analogous results were obtained by the reaction of **2a** and **4a** with 1,6-diisocyanohexane [17] and by the analogous reaction of the Si=As bond in **4a** with cyclohexylisocyanide [30]. These results further indicate that Si=E bonds (E = P, As) react amazingly similar to silenes (Si=C) [31]. Interestingly, in the reaction of silenes with isocyanides, the isolation of intermediates has been successful. The Si=Si bond in disilenes, however, reacts with two equivalents of isocyanide in a stepwise process to furnish silacyclopropanimines [32] and, subsequently, 1,3-disilacyclobutan-2,4-diimines [27]. The structure of **34** was proven by NMR and IR spectroscopy, showing particularly diagnostic $\delta(^{13}C)$ values for the carbon atoms of the C=N-(169.4) and P=C-moiety (207.2), respectively, as well as their J_{C-P} coupling constants. The $^1J_{C-P}$ value lies in the upper range of values hitherto obtained for phosphaalkenes [20, 33].

Conjugated Si=P Bonds of the Allylic Type

The synthesis and experimental study of metastable compounds bearing *conjugated* Si=E bonds (E = main group element) is still very limited. In fact, compounds having conjugated Si=Si bonds are hitherto unknown. The first metastable compounds, **35a** and **35b**, containing conjugated Si=E bonds (E = P, N), were synthesized by Niecke et al. [7, 8] and West et al. [34].

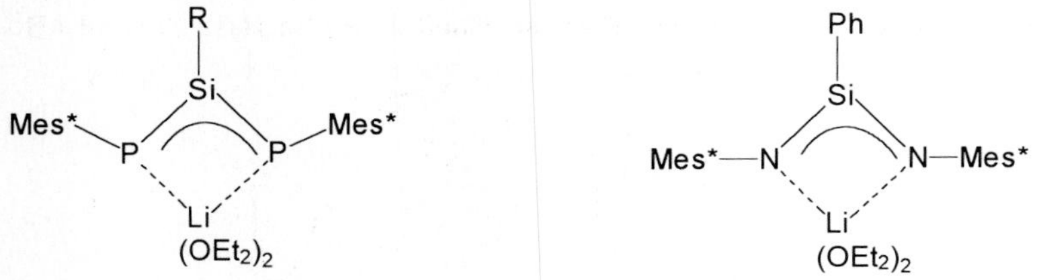

35a: Mes* = 2,4,6-*t*Bu$_3$C$_6$H$_2$ **35b**: Mes* = 2,4,6-*t*Bu$_3$C$_6$H$_2$

Fig. 7.   The first metastable conjugated Si=E compounds (E = P, N).

Recently, we have expanded our method to synthesize metastable compounds with isolated Si=P bonds (phosphasilenes) for the preparation of potentially conjugated Si–P π-sytems (e.g., allyl-, butadiene-, and benzene-analogues). Thus, we synthesized the first 2-phospha-1,3-disilaallylfluoride **37**, which is accessible by thermally-induced elimination of LiF from the corresponding lithium phosphanide **36** [18]. The latter was prepared by lithiation of the phosphane **38**. Compound **36** shows remarkable structural features and crystallizes in an enatiomerically pure form in the space group $P2_12_12_1$. Interestingly, the fourfold coordinated lithium center is only bonded to one fluorine atom (Fig. 8), whereas in an analogous derivative [35] it is chelated by the two fluorine atoms but not coordinated to the phosphorus center.

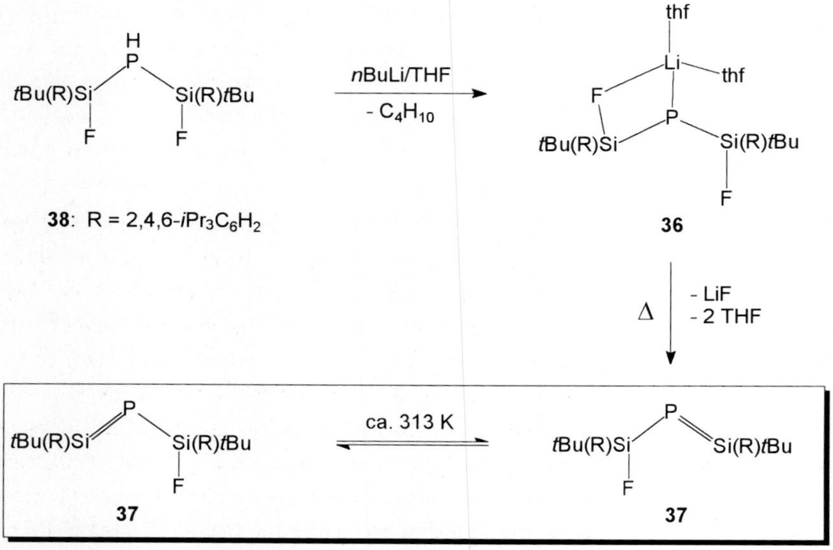

Scheme 7.   Synthesis of **36** and **37**.

Compound **37** was isolated in the form of pale-yellow crystals, which crystallize even in hot toluene. The ^{31}P NMR spectrum of **37** shows a doublet signal slightly at higher field (δ = -33.0, J_{P-F} = 44 Hz) compared with the value observed for the phosphasilene **6** (δ = -29.9). The structure of the parent compound $H_2Si=P(SiH_2F)$ **B** was determined by ab initio calculations (MP2/6-31G*). The calculated Si–P-distances (Si=P 2.070 Å, Si–P 2.235 Å), however, prove that this compound prefers isolated Si=P and Si–P bonds, respectively. This was confirmed by an X-ray crystal structure analysis of **37**, which has further proven that there is no intra- and intermolecular Si–F–Si contact (Fig. 9) [18]. The Si–P-distances in **37** are similar to the calculated values for $H_2Si=P(SiH_2F)$ **B**.

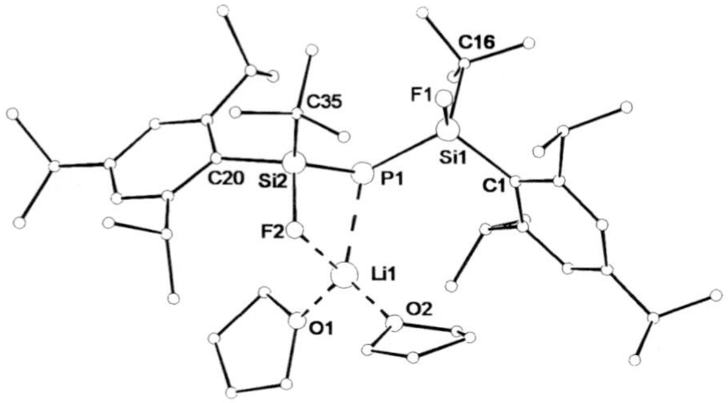

Fig. 7. Solid-state structure of **36**.

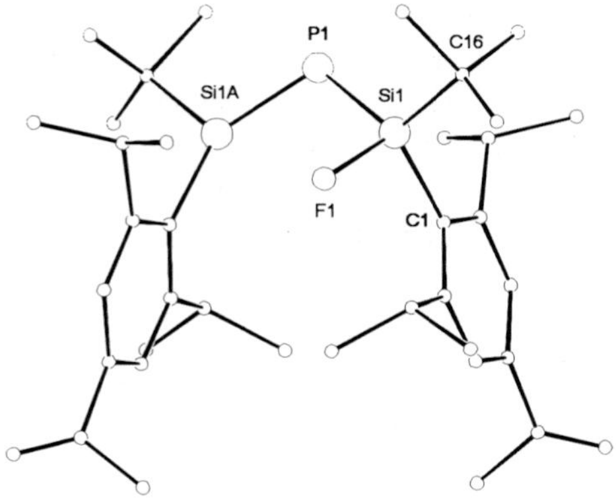

Fig. 8. Solid-state structure of **37**.

The ^{1}H NMR spectrum of **37** in toluene at 25°C shows two sets of *t*-butyl and aryl groups, indicating that a isolated Si=P bond is present even in solution. However, on heating a sample at 40°C, coalescence of the *t*-butyl groups is observed, which implies a suprafacial [1,3]-sigmatropic shift of the fluorine atom. The Gibbs activation energy for this process was estimated to be 16 kcal mol[-1]. A related fluctuation process was hitherto only known for the silaethene $Me_2Si=C(SiMe_3)_2$ [36]. The activation energy for **37** is in very good agreement with the ab initio calculated value for the parent compound $H_2Si=P(SiH_2F)$ **B** (14.7 kcal mol[-1]) [18]. Interestingly, the transition state structure for the fluorine exchange process is described by the unsymmetrical form **C**, bearing two very different Si–F interactions, instead of the symmetrical form **D** (Scheme 8). The latter turns out to be metastable on the potential hypersurface, that is, it should be possible to synthesize a metastable derivative by providing suitable steric protection.

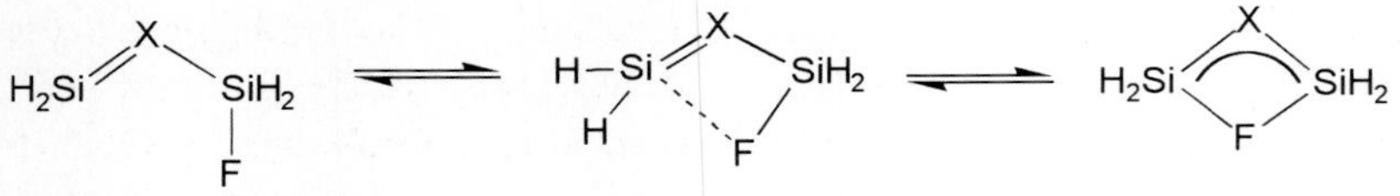

Relative Free Gibbs-Energy (298 K)ΔG (kcal mol^{-1})

B (C$_s$, Min.)	C (C$_1$, TS)	D (C$_s$, Min.)
0.0	14.7	12.5

Scheme 8.   Fluorine exchange in **B** via the transition structure **C**.

Another task was to synthesize a metastable 2-phospha-1,3-disilaallyl anion, which may be prepared by reduction of **37** with two mole equivalents elemental lithium and subsequent elimination of LiF. However, the reduction with Li metal, via the lithiumsilanidyl-lithium phosphanide **39** as reactive intermediate, does not furnish the desired allyl anion **40**. Instead, the disilaphosphacyclopropanes **41**, a valence isomer of **40**, and **42**, the protonated product, were formed but only **42** was isolated and structurally characterized by X-ray crystallography (Fig. 10) [18].

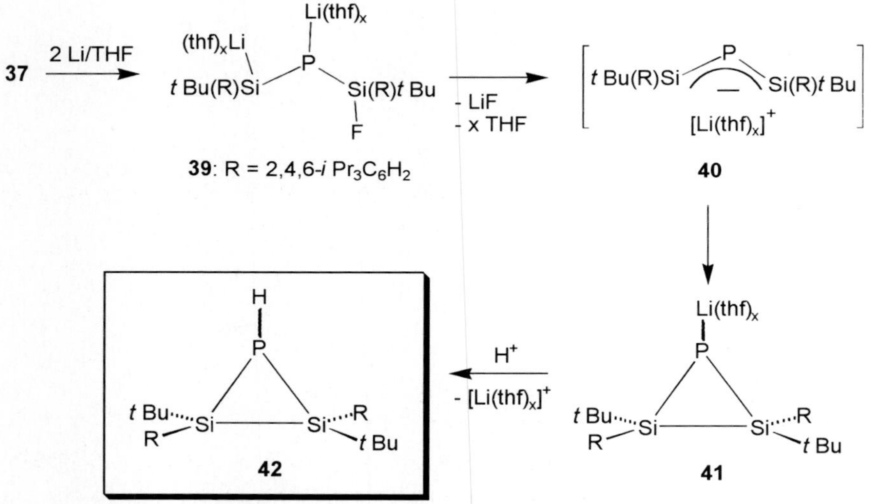

Scheme 9.   Formation of 39, 41, and42.

Although no experimental evidence for the formation of **40** is presently available, we believe that **40** is an intermediate. The latter is supported by theoretical studies [25]. It is tempting to assume that the driving force for the rearrangement of **40** to **41** derives from the unfavorable electronic situation in the anionic skeleton of **40**, where the negative charge is mostly localized in the 1,3-position at the silicon centers (HOMO of the allylic anion). This leads to the unusual situation that the electropositive silicon atoms bear more electron density than the phosphorus center. Upon conrotatory ring closure and formation of the Si–Si bond in **41**, the negative charge is transferred to the phosphorus center. According to the electronic reasons for this rearrangement, it seems rather clear that the analogous 1,3-diphospha- and 1,3-diaza-2-silaallyl anions in **35** and **36** do not isomerize to the corresponding diphospha- and diazasilacyclopropanes, respectively. The relatively fragile Si–Si bond in **41** and **42**, respectively, predestinates these compounds to be useful potential precursors for the novel 2-phospha-1,3-disilaallyl ligand (four or three π electron donor) in the coordination chemistry of transition metals. Investigations whether **41** acts as a four-or three-electron donor towards transition metal complex fragments are in progress.

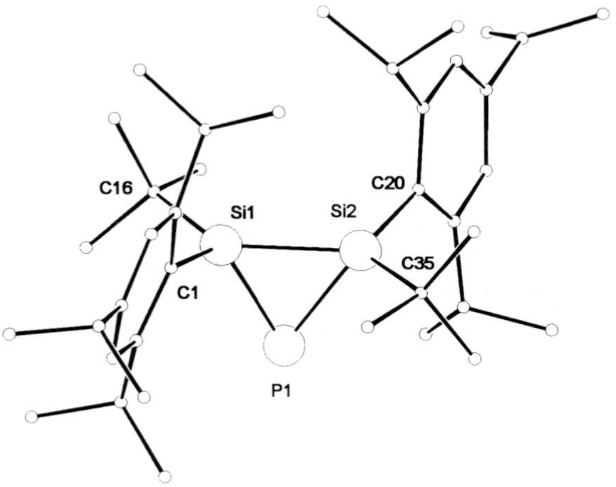

Fig. 10.   Solid-state structure of **42**.

Acknowledgements: This work was generously supported by the *Deutsche Forschungs-gemeinschaft* and the *Fonds der Chemischen Industrie.*

References:

[1]   a) R. West, *Angew. Chem., Int. Ed. Engl.* **1987**, *26*, 1201.
      b) R. West, *Pure Appl. Chem.* **1984**, *56*, 163.
      c) S. Masamune, S. A. Batcheller, T. Tsumuraya, *Angew. Chem., Int. Ed. Engl.* **1991**, *30*, 902.
[2]   a) A. G. Brook, K. M. Baines, *Adv. Organomet. Chem.* **1986**, *25*, 1.
      b) G. Raabe, J. Michl, *"Multiple Bonds to Silicon"*, in: *The Chemistry of Organic Silicon Compounds, Part 2* (Eds.: S. Patai, Z. Rappoport), Wiley, New York, **1989**, p. 1044.
[3]   a) N. Wiberg, K. Schurz, G. Fischer, *Angew. Chem., Int. Ed. Engl.* **1985**, *24*, 1053.
      b) M. Hesse, U. Klingebiel, *Angew. Chem., Int. Ed. Engl.* **1986**, *25*, 649.
      c) N. Wiberg, K. Schurz, G. Müller, J. Riede, *Angew. Chem., Int. Ed. Engl.* **1988**, *27*, 935.
[4]   a) C. N. Smit, F. M. Lock, F. Bickelhaupt, *Tetrahedron Lett.* **1984**, *25*, 3011.
      b) C. N. Smit, F. Bickelhaupt, *Organometallics* **1987**, *6*, 1156.
      c) Y. van den Winkel, H. M. M. Bastiaans, F. Bickelhaupt, *J. Organomet. Chem.* **1991**, *405*, 183.

[5]   M. Driess, *Angew. Chem., Int. Ed. Engl.* **1991**, *30*, 1022.

[6]   M. Driess, U. Winkler, W. Imhof, L. Zsolnai, G. Huttner, *Chem. Ber.* **1994**, *127*, 1031.

[7]   E. Niecke, E. Klein, M. Nieger, *Angew. Chem., Int. Ed. Engl.* **1989**, *28*, 751.

[8]   E. Niecke, H. R. G. Bender, M. Nieger, *J. Am. Chem. Soc.* **1993**, *115*, 3314.

[9]   H. R. G. Bender, *Ph. D. thesis*, Universität Bonn, **1993**.

[10] a) J. K. Dykema, P. N. Troung, M. S. Gordon, *J. Am. Chem. Soc.* **1985**, *107*, 4535.
      b) P. v. R. Schleyer, D. Kost, *J. Am. Chem. Soc.* **1988**, *110*, 2105.

[11] T. A. Albright, J. K. Burdett, M.-H. Whangbo, in: *Orbital Interactions in Chemistry*, Wiley, New York, **1985**, p. 95.

[12] M. Driess, S. Rell, H. Pritzkow, *J. Chem. Soc., Chem. Commun.* **1995**, 253.

[13] M. Driess, R. Janoschek, *J. Mol. Struct. (Theochem.)* **1994**, *313*, 129.

[14] M. Driess, H. Pritzkow, *Angew. Chem., Int. Ed. Engl.* **1992**, *31*, 316.

[15] N. Wiberg, H. Schuster, *Chem. Ber.* **1991**, *124*, 93.

[16] a) M. Driess, H. Pritzkow, *Angew. Chem., Int. Ed. Engl.* **1992**, *31*, 751.
      b) M. Driess, H. Pritzkow, *Phosphorus, Sulfur, Silicon Rel. Elem.* **1993**, *76*, 57.

[17] M. Driess, H. Pritzkow, *J. Chem. Soc., Chem. Commun.* **1993**, 1585.

[18] M. Driess, S. Rell, H. Pritzkow, R. Janoschek, *Angew. Chem., Int. Ed. Engl.* **1997**, *36*, 1326.

[19] a) M. Driess, H. Pritzkow, S. Rell, U. Winkler, *Organometallics*, **1996**, *15*, 1845.
      b) M. Driess, *Coord. Chem. Rev.* **1995**, *145*, 1.
      c) M. Driess, *Adv. Organomet. Chem.* **1996**, *39*, 193.

[20] K. Karaghiosoff, *Survey of 31 P NMR Data* in *Multiple Bonds and Low Coordination in Phosphorus Chemistry* (Eds.: M. Regitz, O. J. Scherer), Thieme, Stuttgart, **1990**, p. 463.

[21] a) M. Driess, A. D. Fanta, D. R. Powell, R. West, *Angew. Chem., Int. Ed. Engl.* **1989**, *28*, 1038.
      b) M. Driess, H. Pritzkow, M. Reisgys, *Chem. Ber.* **1991**, *124*, 1923.
      c) A. D. Fanta, R. P. Tan, N. M. Comerlato, M. Driess, D. R. Powell, R. West, *Inorg. Chim. Acta* **1992**, *198-200*, 733.

[22] R. P. Tan, N. M. Comerlato, D. R. Powell, R. West, *Angew. Chem., Int. Ed. Engl.* **1992**, *31*, 1217.

[23] M. Driess, R. Janoschek, *Angew. Chem., Int. Ed. Engl.* **1992**, *31*, 460.

[24] M. Driess, H. Grützmacher, *Angew. Chem., Int. Ed. Engl.* **1996**, *35*, 851.

[25] R. Janoschek, M. Driess, unpublished results.

[26] J. D. De Young, R. West, *Chem. Lett.* **1986**, *6*, 883.

[27] a) M. Weidenbruch, J. Hamann, H. Diel, D. Lentz, H. G. v. Schnering, K. Peters, *J. Organomet. Chem.* **1992**, *426*, 35.
      b) M. Weidenbruch, *"Novel Ring Systems from Cyclotrisilanes and Cyclotristannanes"*, in: *The Chemistry of Inorganic Ring Systems, Studies in Inorganic Chemistry, Vol. 14* (Ed.: R. Steudel), Elsevier, Amsterdam, **1992**, p. 51.

[28] K. L. McKillop, G. R. Gillette, D. R. Powell, R. West, *J. Am. Chem. Soc.* **1992**, *114*, 5203; See also ref. [1].

[29] G. R. Gillette, R. West, *J. Organomet. Chem.* **1990**, *394*, 45.

[30] M. Driess, H. Pritzkow, M. Sander, *Angew. Chem., Int. Ed. Engl.* **1992**, *32*, 283.

[31] a) A. D. Brook, Y. Kun Kong, A. K. Saxena, J. F. Sawyer, *Organometallics* **1988**, *7*, 2245.
      (b) A. D. Brook, A. K. Saxena, J. F. Sawyer, *Organometallics* **1989**, *8*, 850.

[32] H. B. Yokelson, A. J. Millevolte, K. J. Haller, R. West, *J. Chem. Soc., Chem. Commun.* **1987**, 1605.

[33] E. Fluck, G. Heckmann, *"Chemical Shift Interpretations in Phosphorus - 31 P NMR Spectroscopy"*, in: *Stereochemical Analysis* (Eds.: J. G. Verkade, L. D. Quin), VCH, Weinheim, **1987**, p. 76.

[34] G. E. Underiner, R. P. Tan, D. R. Powell, R. West, *J. Am. Chem. Soc.* **1991**, *113*, 8437.

[35] U. Klingebiel, M. Meyer, U. Pieper, D. Stalke, *J. Organomet. Chem.* **1991**, *408*, 19.

[36] N. Wiberg, H. Köpf, *Chem. Ber.* **1987**, *120*, 653.

Silole and Germole Dianions and their Dilithium Derivatives: Are they Aromatic?

Thomas Müller*
Fachinstitut für Anorganische und Allgemeine Chemie
Humboldt-Universität zu Berlin
Hessische Str. 1-2, D-10115 Berlin, Germany
Tel.: Int. code + (30)20937369 – Fax: Int. code + (30)20936966
E-mail : h0443afs@joker.rz.hu-berlin.de

Yitzhak Apeloig
Department of Chemistry and Lise Meitner Minerva Center for Computational Quantum Chemistry
Technion-Israel Institute of Technology
32000 Haifa, Israel

Honglae Sohn, Robert West
Department of Chemistry
University of Wisconsin
Madison, U.S.A.

Keywords: *Ab initio* calculations / Aromaticity / Germole Dianion / Silaaromatics / Germaromatics

Summary: According to structural, thermodynamic, electronic, and magnetic criteria silole and germole dianions show indications of cyclic delocalization. The degree of aromaticity in the silole- **1**(Si) and germole dianions **1**(Ge) is of the same magnitude as that in the cyclopentadienyl anion.

Introduction

Recent years have witnessed renewed interest in the problems of chemical bonding in Group IVa metallole mono- and dianions. Several metallole anions have been synthesized by reductive dehalogenation of dihalometalloles as reported by the groups of Joo [1], Boudjouk [2], Tilley [3] and West [4].

In this paper we report on our studies of the bonding in the free silole and germole dianions and their lithium derivatives using *ab initio* quantum mechanical methods [5]. Several criteria for aromaticity, suggested recently by Schleyer et al. [6], were evaluated to test the degree of delocalization in Group IVa metallole dianions.

Cyclic delocalization in metallole dianions

The silole- **1**(Si) and the germole dianions **1**(Ge) are calculated to be planar (inner-cyclic dihedral angle $\vartheta = 0°$, see Table 1). The structural data, summarized in Table 1, clearly indicate the occurence of cyclic delocalization in the dianions. Thus, the C–C bond distances in the butadiene part of the molecule are nearly equal, so that both for **1**(Si) and for **1**(Ge) the Julg parameter A [6a] is 1 or close to 1. The bond length equalization in the butadiene-part is accompanied by a shortening of the element carbon bond length. Formal deprotonation of dimethylsilane yielding the dimethylsilyldianion results in a Si–C bond length *elongation* of 0.181 Å (at MP2(fc)/6-31+G*). In contrast, the Si–C bond in the silole dianion **1**(Si) is by 0.007 Å *shorter* than in silole **2**. The short E–C bond and the nearly equalized C–C bonds in the butadiene moiety especially when taken together are a strong indication of cyclic conjugation in **1**(Si) and **1**(Ge).

Table 1. Optimized geometries of metallole dianions and their lithio derivatives (at MP2(fc)/6-31+G*).

E	cpd.	r_1	r_2	r_3	a	b	c	d	ϑ	A
Si[a]	1	1.869	1.421	1.420					0	1.000
	3	1.892	1.438	1.434	2.489	2.200	2.149		0	1.000
	4	1.855	1.420	1.426	2.495	2.194	2.151	2.440	6.5°	0.999
	5	1.884	1.429	1.425	2.464	2.154	2.125		4.3°	1.000
Ge[a]	1	1.935	1.415	1.420					0	0.999
	3	1.963	1.433	1.434	2.539	2.218	2.152		0	1.000
	4	1.912	1.417	1.425	2.532	2.225	2.156	2.438	4.8	0.998
	5	1.947	1.426	1.424	2.523	2.166	2.132		5.3	1.000

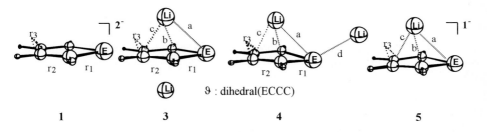

ϑ : dihedral(ECCC)

     1   **3**   **4**   **5**

The electron distribution according to a NBO analysis [7] of the MP2(fc)/6-31+G* wave function supports our structural arguments. Thus, from 6 electrons in π-type orbitals of the metallole dianions (4 electrons from the butadiene-part and 2 electrons from the formal negative charge) nearly 5e$^-$ (4.81e$^-$ and 4.62e$^-$ in **1**(Si) and in **1**(Ge), respectively) are delocalized on the butadiene part of the molecule, indicating a strong π-charge transfer from Si or Ge to the ring C-atoms. Consequently, the

NPA charge distribution places nearly two negative atomic charges at the butadiene part of the metallole dianions (-1.96e$^-$ and -1.70e$^-$ in **1**(Si) and in **1**(Ge), respectively), while the charges at the heteroatom are only -0.04e$^-$ and -0.30e$^-$ at Si and Ge, respectively.

Eq. 1.

Cyclic conjugation is accompanied by thermodynamic stabilization. The aromatic stabilization energy (ASE), which is defined as the reaction enthalpy of the isodesmic equation (1) is a gauge for the stabilization of the dianion by resonance between the conjugated double bonds and the lone pairs at the negatively charged Si or Ge. The ASE is calculated to be 17.9 kcal/mol (at MP2(fc)/6-31+G*) for **1**(Si) and 13.0 kcal/mol for **1**(Ge) (at B3LYP/6-311+G*). This is significantly smaller than the ASE calculated for the cyclopentadienyl anion or for thiophen (28.8 kcal/mol and 22.4 kcal/mol at MP2(fc)/6-31G*, respectively), but nevertheless quite substantial.

Another important manifestation of cyclic delocalization is the ring current in aromatic molecules, which results in special magnetic properties. Thus, all aromatic molecules are characterized by a high anisotropy of the magnetic susceptibility $\Delta\chi$. For the cyclopentadienyl anion, a typical aromatic molecule, we calculate a high $\Delta\chi$ of 56.6 ppm cgs (see Table 2). The calculated $\Delta\chi$ for the metallole dianions **1**(Si) and **1**(Ge) are even larger: 75.9 ppm cgs and 92.4 ppm cgs, respectively. This again supports a significant aromatic character in the dianions.

Table 2.   Calculated magnetic properties of metallole dianions and their lithio derivatives.

	C$_5$H$_5^-$	1(Si)	3(Si)	4(Si)	5(Si)	1 (Ge)	3(Ge)	4(Ge)	5(Ge)
$\Delta\chi$[a, b, c]	56.6	75.9	55.3	57.8		92.4	59.7	59.6	
NICS[d, e]	-14.3	-12.8	-17.6	-18.4	-17.1	-11.2	-18.4	-21.4	-17.7

[a] in ppm cgs – [b] $\Delta\chi = \chi_x - (\chi_y - \chi_z)/2$ – [c] Calculated at IGAIM/B3LYP/6-311+G(2df,p)//MP2(fc)/-6-31+G* – [d] at GIAO/HF/6-31G*//MP2/(fc)/6-31+G* – [e] in ppm.

Protons placed outside an aromatic ring but in the plane of the skeleton are deshielded. Sondheimer et al. have demonstrated that protons that are placed inside a cyclic delocalized π-system are strongly shielded [8]. Both effects are due to the magnetic properties of the ring current. Schleyer et al. [9] suggested the calculated shielding of a "ghost atom" in the center of the ring as an efficient probe for the aromaticity of cyclic systems. Like a proton that is placed inside an aromatic ring, the "ghost atom" should exhibit a strong shielding. We have calculate these nucleus independent chemical shifts (NICS) for **1**(Si) and **1**(Ge) (see Table 2). Both NICS values indicate strong shielding in the center of the ring of these molecules. The NICS values for **1**(Si) and for **1**(Ge) are comparable to that of the cyclopentadienyl anion (see Table 2).

The calculations support unequivocally the notion that dianions of germoles and siloles are highly delocalized and that the degree of aromaticity in these molecules is nearly as large as in the "classic" aromatic cyclopentadienyl anion.[10]

Lithio Complexes of Group IVa Metallole Dianions

To mimic possible metal complexes of metallole dianions we did calculated the structures of the dilithio complexes of **1**(Si) and **1**(Ge). Two different types of complexes were found to be minima at the correlated MP2(fc)/6-31+G* level. One of these structures, **3**, is of C_{2v} symmetry with both lithiums η^5 coordinated to the ring, one above and one below the cycle forming an inverse sandwich structure. In the other structure, **4**, one lithium is bonded to silicon (or to germanium) in a η^1-fashion and the other is located above the plane of the molecule in close contact with all five ring atoms. The Madison group was able to crystallize the η^1,η^5 isomer of the dilithio tetraphenylsilole dianion [2b] and both isomers of the dilithio tetraphenyl germole dianion as solvates with THF or dioxane, respectively [4c]. All the lithio complexes investigated show planar metallole rings with equalized C–C bond lengths. The general good agreement between experiment and theory can be demonstrated by comparing the crystal structure of the solvated η^5,η^5 dilithio complex of the tetraphenyl germole dianion with the calculated structure of the hydrated **3** (Ge), shown in Fig. 1. In particular, note the good agreement for the Ge–Li and C–Li bond distances.

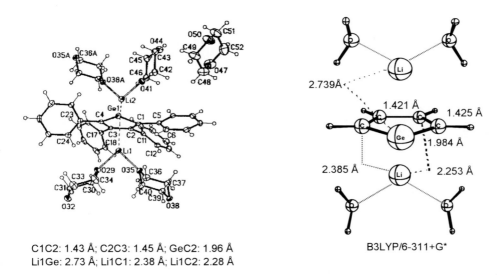

C1C2: 1.43 Å; C2C3: 1.45 Å; GeC2: 1.96 Å
Li1Ge: 2.73 Å; Li1C1: 2.38 Å; Li1C2: 2.28 Å

B3LYP/6-311+G*

Fig. 1. Molecular structure of $Li_2([PhC]_4Ge\cdot5\ C_4H_8O_2$ and the calculated structure of the tetrawater complex of **3**(Ge) [4c].

The energy difference between **3** and **4** is nearly the same for silole- and germole dianions (21.0 and 20.9 kcal/mol, respectively at MP2(fc)/6-31+G*) with **3** being in both cases the more stable isomer. The calculations predict that the η^1, η^5 isomer **4** exists only in a flat potential well (see Fig. 2). For example varying the angle α in **4**(Ge) from 30° to 160° requires only 2.8 kcal/mol. The interaction between the Si or Ge and the η^1-Li in **4** is mostly electrostatic and therefore not directional. Thus, the actual position of the η^1 bonded Li is strongly influenced by the complexing molecules and/or by lattice effects. Our computational results suggest that the very different positions of the η^1-Li atoms in the X-ray structures of **4**(Si) and **4**(Ge) (see Fig. 3) are *not* due to different bonding situations but are merely due to lattice effects in the crystal.

These arguments are supported by NMR measurements for the dilithio tetraphenylsilole dianion. The isotropic ^{29}Si NMR chemical shift for *solid* dilithio tetraphenylsilole dianion, 87.3 ppm, in excellent agreement with the calculated ^{29}Si NMR chemical shift of the η^5-monolithio silole dianion **5**(Si) (83.7 ppm at GIAO/MP2/tz2p(Si),tzp(C,Li),dz(H)//MP2(fc)/6-31+G*). However, in solution the ^{29}Si NMR signal for the dilithio tetraphenylsilole dianion is distinctly highfield shifted, 68.5 ppm. This value is still much higher than predicted by the calculations for a free silole dianion **1**(Si) (18.5 ppm at GIAO/MP2/tz2p(Si), tzp (C,Li),dz(H)//MP2(fc)/6-31+G*) but close to the computed ^{29}Si NMR chemical shift for **3**(Si) (55.7 ppm at GIAO/MP2/tz2p(Si),tzp(C,Li),dz(H)// MP2(fc)/6-31+G*). Thus, our calculations suggest that in solution a η^5,η^5 dilithio complex similar to **3**(Si) predominates while the solid state structure of the dilithio tetraphenylsilole and germole are better interpreted as formed from a solvated lithium cation and a η^5-monolithio dianions similar to **5**, weakly interacting with a solvated lithium cation.

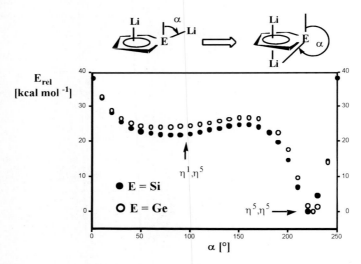

Fig. 2. Relative energy of η^1,η^5 isomer **4** compared to the η^5,η^5 species **3** as a function of a (RHF/3-21+G*).

Cyclic Delocalization in Lithio Complexes of Group IVa Metallole Dianions

The lithio complexes **3**, **4**, and **5** of the metallole dianion also show indications of significant cyclic delocalization. The equalized C–C bond lengths, together with the short E–C bond distances indicate strong resonance interaction between the negatively charged Si or Ge and the butadiene part of the molecule. Consequently, all calculated Julg parameters A for the lithio complexes **3-5** are 1 or very close to 1 (see Table 1). The calculated magnetic properties of **3-5** also point to cyclic conjugation within the metallole cycle. The anisotropy of the magnetic susceptibility $\Delta\chi$ is smaller than in the free dianions **1**, but $\Delta\chi$ is as large as for the cyclopentadienyl anion (see Table 2). The calculated NICS values for **3-5** are 5-9 ppm more negative than for the isolated dianions **1**. Whether this remarkably negative NICS value is a sign of an exeptional high aromaticity or is caused by diamagnetic contributions from bonds between Li and the ring atoms is a point for further investigations.

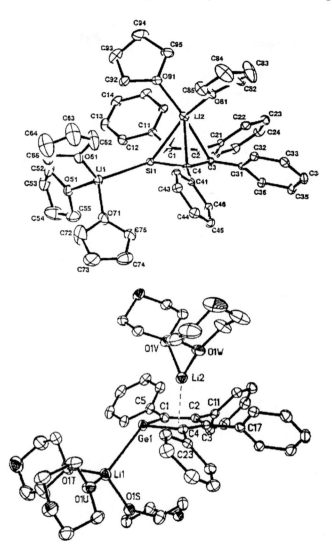

Fig. 3. Molecular structures of the η^1,η^5 dilithio complexes of the tetraphenylgermole and tetraphenylsilole dianions [4b, 4c].

Conclusion

According to structural, thermodynamic, electronic, and magnetic criteria the metallole dianions (and also their dilithio complexes) show indications of cyclic delocalization. The degree of "aromaticity" in silole and germole dianions is of the same magnitude as in the classic aromatic cyclopentadienyl anion.

Acknowledgement: This work was supported by the *National Science Foundation*, the *Israel-US Binational Foundation*, the *German-Israeli Foundation* (GIF), the *German Federal Ministry of Science, Research, Technology, and Education* and the *Minerva Foundation*. Thomas Müller thanks the *Fonds der Chemischen Industrie* for a Liebig Scholarship.

References:

[1]   W.-C. Joo, J.-H. Hong, S.-B. Choi, H.-E. Son, *J. Organomet. Chem.* **1990**, *391*, 27.

[2]   a) J.-H. Hong, P. Boudjouk, *J. Am. Chem. Soc.* **1993**, *115*, 5883.
b) J.-H. Hong, P. Boudjouk, S. Castellino, *Organometallics* **1994**, *13*, 3387.
c) J.-H. Hong, P. Boudjouk, I. Stoenescu, *Organometallics* **1996**, *15*, 2179.
d) J.-H. Hong, Y. Pan, P. Boudjouk, *Angew. Chem.* **1996**, *108*, 213.

[3]   a) W. P. Freeman, T. D. Tilley, A. L. Rheingold, *J. Am. Chem. Soc.* **1994**, *116*, 8428.
b) W. P. Freeman, T. D. Tilley, F. P. Arnold, A. L. Rheingold, P. K. Gantzel, *Angew. Chem.* **1995**, *107*, 2029.
c) W. P. Freeman, T. D. Tilley, G. P. A. Yap, A. L. Rheingold, *Angew. Chem.* **1996**, *108*, 960.

[4]   a) U. Bankwitz, H. Sohn, D. R. Powell, R. West, *J. Organomet. Chem.* 1995, *499*, C7.
b) R. West, H. Sohn, U. Bankwitz, J. Calabrese, Y. Apeloig, T. Müller, *J. Am. Chem. Soc.* **1995**, *117*, 11608.
c) R. West, H. Sohn, D. R. Powell, T. Müller, Y. Apeloig, *Angew. Chem.* **1996**, *108*, 1095.

[5]   The Gaussian 94 programme was used: Gaussian 94: M. J. Frisch, G. W. Trucks, H. B. Schlegel, P. M. W. Gill, B. G. Johnson, M. A. Robb, J. R. Cheeseman, T. A. Keith, G. A. Petersson, J. A. Montgomery, K. Raghavachari, M. A. Al-Laham, V. G. Zakrzewski, J. V. Otiz, J. B. Foresman, J. Cioslowski, B. Stefanov, A. Nanayakkara, M. Challacombe, C. Y. Peng, P. Y. Ayala, W. Chen, M. W. Wong, J. L. Andres, E. S. Repogle, R. Gomperts, R. L. Martin, D. J. Fox, J. S. Binkley, D. J. Defrees, J. Baker, J. P. Stewart, M. Head-Gordon, C. Gonzalez, J. A. Pople, Gaussian, Inc. Pittsburgh PA, 1995.

[6]   a) P. v. R. Schleyer, P. K. Freeman, H. Jiao, B. Goldfuss, *Angew. Chem.* **1995**, *107*, 332.
b) P. v. R. Schleyer, B. Goldfuss, *Pure Appl. Chem.* **1996**, *28*, 209.

[7]   A. E. Reed, L. A. Curtiss, F. Weinhold, *Chem. Rev.* **1988**, *88*, 899.

[8]   For a review see: F. Sondheimer, *Acc. Chem. Res.* **1972**, *5*, 81.

[9]　P. v. R. Schleyer, C. Maerker, A. Dransfeld, H. Jiao, N. J. R. v. E. Hommes, *J. Am. Chem. Soc.* **1996**, *118*, 6317.

[10]　During the editorial process of this paper several computational studies reporting similar results has been published: a) B. Goldfuss, P. v. R. Schleyer, F. Hampel, *Organometallics* **1996**, *15*, 1755.

b) B. Goldfuss, P. v. R. Schleyer, *Organometallics* **1997**, *16*, 1543.

Supersilylmetal Compounds

N. Wiberg*, K. Amelunxen, H. Nöth, A. Appel, M. Schmidt, K. Polborn
Institut für Anorganische Chemie
Ludwig-Maximilians-Universität München
Meiserstrasse 1, D-80333 München, Germany
Tel.: Int. code + (89)5902230 – Fax: Int. code + (89)5902578
E-mail: niw@anorg.chemie.uni-muenchen.de

Keywords: Supersilyl / Group 1 metals / Group 12 metals / Group 13 metals

Summary: We discuss the structures and properties of supersilylmetal compounds $(t\mathrm{Bu_3Si})_m\mathrm{M}_n\mathrm{X}_p$ that were synthesized in the following way: (i) $t\mathrm{Bu_3SiNa}$ from $t\mathrm{Bu_3SiBr}$ and Na, (ii) $(t\mathrm{Bu_3Si})_2\mathrm{M}$ (M = Zn, Cd, Hg), $t\mathrm{Bu_3SiZnBr}$, $t\mathrm{Bu_3SiAlCl_2 \cdot THF}$, $t\mathrm{Bu_3SiGaCl_2 \cdot Me_2NEt}$, $(t\mathrm{Bu_3Si})_2\mathrm{MHal}$ (M = Al, Ga), $(t\mathrm{Bu_3SiAl})_4$ and $(t\mathrm{Bu_3Si})_2\mathrm{M}$–$\mathrm{M}(\mathrm{Si}t\mathrm{Bu_3})_2$ (M = In, Tl) from their metal halogenides and $t\mathrm{Bu_3SiNa}$, and $t\mathrm{Bu_3SiBBr_2}$ from $\mathrm{BBr_3}$ and $(t\mathrm{Bu_3Si})_2\mathrm{Zn}$.

Supersilyl Compounds of Group 1 Metals

Reaction of supersilylbromide $t\mathrm{Bu_3SiBr}$ ($t\mathrm{Bu_3Si}$ = supersilyl) with sodium in heptane at 100°C leads to supersilylsodium (Eq. 1).

$$t\mathrm{Bu_3SiBr} + 2\,\mathrm{Na} \xrightarrow[100°C]{\text{heptane}} t\mathrm{Bu_3SiNa} + \mathrm{NaBr}$$

Eq. 1.

On the other hand, adducts $t\mathrm{Bu_3SiNa(Do)_2}$ are formed (Do = THF, $n\mathrm{Bu_2O}$), when $t\mathrm{Bu_3SiBr}$ and Na are refluxed in THF or $n\mathrm{Bu_2O}$ [1].

Solvent-free $t\mathrm{Bu_3SiNa}$ can be crystallized from heptane in well-shaped orange octahedrons (decomp. 112°C). Crystal structure determination shows the compound to be dimeric in the solid state (Fig. 1).

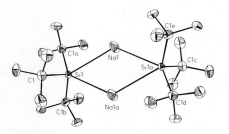

Crystal system: orthorhombic, *Pa*-3
$R_1 = 0.054$, $wR2 = 0.123$

Important bond length [Å] and angles [°]:
Si–Na (av.):	3.067
Si–C (av.):	1.978
Na1–Si1–Na1A:	47.41(11)
Si1–Na1–Si1A:	124.56(12)
C–Si–C (av.):	107.3

Fig. 1.   Crystal structure of $t\mathrm{Bu_3SiNa}$.

To obtain donor adducts of *t*Bu₃SiNa, either the solvent-free compound or *t*Bu₃SiNa(THF)₂ is treated with the corresponding donor (e.g., PMDTA = pentamethyldiethylenetriamine; Eq. 2) in hydrocarbon solution.

$$\textit{t}Bu_3SiNa(THF)_2 \; + \; PMDTA \; \xrightarrow{\text{pentane}} \; \textit{t}Bu_3SiNa(PMDTA) \; + \; 2\, THF$$

Eq. 2.

Crystals of *t*Bu₃SiNa·PMDTA precipitate from a pentane solution as pale yellow rhomboids that decompose at 87°C. The X-ray structure analysis shows a monomeric molecule with one PMDTA coordinated to the sodium through its N-atoms (Fig. 2).

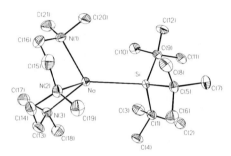

Crystal system: monoclinic, $P2(1)/c$
$R_1 = 0.058$, $wR2 = 0.144$

Important bond lengths [Å] and angles [°]:

Si–Na:	2.9677(14)	Na–Si–C (av.):	112.3
Si–C (av.):	1.993	C–Si–C (av.):	106.5
Na–N1:	2.502(3)	N1–Na–N2:	73.50(10)
Na–N2:	2.485(3)	N2–Na–N3:	74.41(9)
Na–N3:	2.533(3)	N1–Na–N3:	104.03(9)

Fig. 2.   Crystal structure of *t*Bu₃SiNa PMDTA.

Supersilyl Compounds Group 12 Metals

Reaction of *t*Bu₃SiNa with ZnBr₂, CdI₂, and HgCl₂ in THF at room temperature leads to colorless to pale yellow *t*Bu₃SiMHal and (*t*Bu₃Si)₂M (M = Zn, Cd, Hg), respectively.

$$MHal_2 \; \xrightarrow[- NaHal]{+ \textit{t}Bu_3SiNa} \; \textit{t}Bu_3SiMHal \; \xrightarrow[- NaHal]{+ \textit{t}Bu_3SiNa} \; (\textit{t}Bu_3Si)_2M$$

Eq. 3.

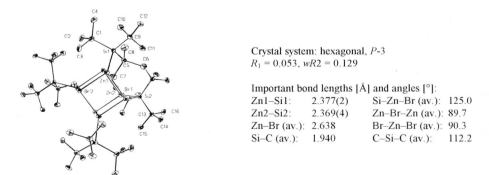

Crystal system: hexagonal, $P\text{-}3$
$R_1 = 0.053$, $wR2 = 0.129$

Important bond lengths [Å] and angles [°]:

Zn1–Si1:	2.377(2)	Si–Zn–Br (av.):	125.0
Zn2–Si2:	2.369(4)	Zn–Br–Zn (av.):	89.7
Zn–Br (av.):	2.638	Br–Zn–Br (av.):	90.3
Si–C (av.):	1.940	C–Si–C (av.):	112.2

Fig. 3.   Crystal structure of *t*Bu₃SiZnBr.

Crystal structure determination shows *t*Bu₃SiZnBr to be tetrameric of heterocubane type with alternating M and X at the edges of the cube (Fig. 3) and (*t*Bu₃Si)₂M to be monomeric with the two *t*Bu₃Si-substituents bonded in a linear fashion to the metal (Fig. 4).

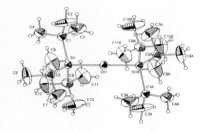

Crystal system: triclinic, P-1

Important bond length [Å] and angles [°]:

	M = Zn	**M = Cd**	**M = Hg**
R₁:	0.051	0.036	0.033
wR2:	0.139	0.091	0.085
M–Si (av.):	2.384	2.524	2.495
Si–M–Si:	180	180	180
C–Si–C (av.):	111.8	111.9	112.2

Fig. 4. Crystal structure of (*t*Bu₃Si)₂M.

Photolysis of alkane solutions of (*t*Bu₃Si)₂M (M = Zn, Cd, Hg) leads to the disilane *t*Bu₃Si–Si*t*Bu₃ and the corresponding metal.

Stirring (*t*Bu₃Si)₂Zn with BBr₃, GeCl₄, or PCl₃ in hydrocarbon solution at room temperature gives the monoslilylated element halogenides (Eq.4).

$$(t\text{Bu}_3\text{Si})_2\text{Zn} + \text{EX}_n \longrightarrow t\text{Bu}_3\text{SiZnHal} + t\text{Bu}_3\text{SiEX}_{n-1}$$

Eq. 4.

Supersilyl Compounds of Group 13 Metals

As shown in Eq. 4, *t*Bu₃SiBBr₂ can be prepared by reaction of (*t*Bu₃Si)₂Zn with BBr₃ in heptane at room temperature. When pyridine is added to the reaction mixture, colorless needles of *t*Bu₃SiBBr₃·py precipitate at -20°C. Crystal structure determination shows a boron atom tetracoordinated by one supersilyl group, two bromine atoms and the nitrogen atom of pyridine (Fig. 5).

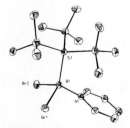

Crystal system: monoclinic, P2(1)/c
$R_1 = 0.046$, $wR2 = 0.096$

Important bond lengths [Å] and angles [°]:
B–Si: 117(5)
B–Br (av.):   2.052
B–N: 1.617(6)
C–Si–C (av.): 109.5

Fig. 5. Crystal structure of *t*Bu₃SiBBr₂ py.

Reaction of *t*Bu₃SiNa with AlCl₃·THF and GaCl₃·Me₂NEt at -78°C in hydrocarbon solvents leads to colorless *t*Bu₃SiAlCl₂·THF and *t*Bu₃SiGaCl₂·Me₂NEt, respectively (Eq. 5).

$$t\text{Bu}_3\text{SiNa} + \text{MCl}_3(\text{Do}) \xrightarrow[-\text{NaCl}]{\text{alkane, -78°C}} t\text{Bu}_3\text{SiMCl}_2(\text{Do})$$

Eq. 5.

In contrary, reaction of $t\text{Bu}_3\text{SiNa}$ with AlBr_3 and GaCl_3 at room temperature in alkane solution without donor always leads to the double silylated products $(t\text{Bu}_3\text{Si})_2\text{AlBr}$ and $(t\text{Bu}_3\text{Si})_2\text{GaCl}$ (Eq. 6).

$$2\ t\text{Bu}_3\text{SiNa} + \text{MHal}_3 \xrightarrow[-\ 2\ \text{NaHal}]{\text{alkane}} (t\text{Bu}_3\text{Si})_2\text{MHal}$$

Eq. 6.

Reaction of $t\text{Bu}_3\text{SiNa}$ with AlCl in a mixture of toluene, dibutylether and diethylether gives, among other products, deep violet $(t\text{Bu}_3\text{SiAl})_4$ (Eq. 7), which can be isolated by sublimation at 180°C under high vacuum [2].

$$4\ t\text{Bu}_3\text{SiNa} + 4\ \text{AlCl} \xrightarrow[-\ 4\ \text{NaCl}]{\text{toluene, Et}_2\text{O, Bu}_2\text{O}} (t\text{Bu}_3\text{SiAl})_4$$

Eq. 7.

Reaction of InBr or TlBr in THF at -78°C leads to the dimetal compounds $(t\text{Bu}_3\text{Si})_2\text{M-}$ $\text{M(Si}t\text{Bu}_3)_2$ (M = In, Tl) (Eq. 8) [2].

$$4\ t\text{Bu}_3\text{SiNa} + 4\ \text{MBr} \xrightarrow[-\ 4\ \text{NaBr}]{\text{THF, -78°C}} (t\text{Bu}_3\text{Si})_2\text{M-M(Si}t\text{Bu}_3)_2 + 2\ \text{M}$$

Eq. 8.

Crystal structure determination shows the two compounds to crystallize isomorphically (Fig. 6). The central structural element is the M–M grouping with a very long distance, compared with similar compounds. The M atoms are each coordinated in a trigonal planar fashion by two Si atoms and one M atom, the MMSi_2 planes are orthogonal.

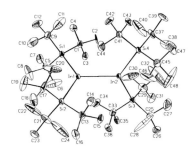

Fig. 6. Crystal structure of $(t\text{Bu}_3\text{Si})_2\text{In-In(Si}t\text{Bu}_3)_2$.

Crystal system: orthorhombic, *Pa*-3
In: $R_1 = 0.086$, wR2 = 0.229
Tl: $R_1 = 0.135$, wR2 = 0.417

Important bond lengths [Å] and angles [°]:

	M = In	**M = Tl**
M–M:	2.9217(11)	2.966(2)
Si–M–Si (av.):	129.3	130.2
M–M–Si (av.):	115.3	114.9
C–Si–C (av.):	109.7	

References:

[1]   N. Wiberg, G. Fischer, P. Karampatses, *Angew. Chem., Int. Ed. Engl.* **1984**, *23*, 59.

[2]   N. Wiberg, K. Amelunxen, H. Nöth, M. Schmidt, H. Schwenk, *Angew. Chem., Int. Ed. Engl.* **1996**, *35*, 65.

Trialkylsilyl-Substituted Homobimetallic Phosphanides of the Alkaline Earth Metals as well as Zinc

Matthias Westerhausen*, Bernd Rademacher, Manfred Hartmann,
Michael Wieneke, Matthias Digeser
Institut für Anorganische Chemie
Ludwig-Maximilians-Universität München
Meiserstraße 1, D-80333 München, Germany
Tel.: Int. code + (89)5902638 – Fax: Int. code + (89)5902578
E-mail: maw@ anorg.chemie.uni-muenchen.de

Keywords: Alkaline Earth Metal / Phosphanide / Trialkylsilyl / Zinc

Summary: Trialkylsilyl-substituted dimeric alkaline earth metal bis(phosphanides) show unexpected structures in the solid state as well as in solution. Depending on the steric demand of the trialkylsilyl group at the phosphorus atom, monocyclic or bicyclic structures of the types R–M(μ-R)$_2$M–R and R–M(μ-R)$_3$M, respectively, can be isolated. In this context the tri-*i*-propylsilylphosphandiide ligand proves to stabilize a variety of bimetallic molecules such as Bis(alkylzinc)tri-*i*-propysilylphosphane (RZn)$_2$PSi*i*Pr$_3$, the first structurally characterized geminal organozinc-substituted derivative.

The metallation of trialkylsilyl-substituted phosphanes with the easily accessible alkaline earth metal bis[bis(trimethylsilyl)amides] in ether such as tetrahydrofuran or 1,2-dimethoxyethane yields quantitatively the alkaline earth metal phosphanides according to Eq. 1 [1]. The ether content depends on the size of the alkaline earth metal.

$$M[N(SiMe_3)_2]_2 \; + \; 2\,HPR_2 \; \rightarrow \; M(PR_2)_2 \; + \; 2\,HN(SiMe_3)_2$$

Eq. 1.   M = Ca, Sr, Ba; R = SiMe$_2$CH$_2$, SiMe$_3$, SiMe$_2$*i*Pr

   In ether solution these compounds are monomeric, however, in toluene solution an equilibrium between the monomeric and the dimeric species is observable in the NMR spectra. Fig. 1 shows the temperature dependency of the ^{31}P{^{1}H} NMR spectra in [D$_8$]toluene of (THF)$_3$Sr[μ-P(SiMe$_3$)$_2$]$_3$Sr–P(SiMe$_3$)$_2$ [2]. At low temperature (253 K) only the monomeric tetrakis(tetrahydrofuran) complex can be detected, but at elevated temperature a second dimeric molecule dominates the spectra.

   Whereas at 253 K only the monomeric molecule is detectable, at 283 K a small signal arises from the dimeric species (Fig. 2). At 313 K coalescence is observed between these two species, furthermore, an intramolecular exchange reaction of the terminal and the bridging phosphanide ligands of the dimer broadens the singlets. The latter process is fast on the NMR time scale. The spectroscopic data are compared in Table 1. For the barium derivative shown in Fig. 3 the bicyclic structure has been proven by the small $^2J_{\text{P–P}}$ coupling of 6.7 Hz at 213 K which leads to a quartet for the terminal ligand and a doublet for the three bridging phosphanide substituents in the ^{31}P{^{1}H} NMR spectrum [3]. Eq. 2 displays the temperature-dependent equilibrium in toluene solution, whereas the intramolecular exchange reaction leads to a broadening of the signals under loss of the coupling pattern before the intermolecular dynamic monomer-dimer process can be detected.

The size of the trialkylsilyl groups determines the appearance of the dimeric molecule. Thus, the dimethyl-*i*-propylsilyl substituent enforces a monocyclic dimeric barium bis(phosphanide) with two terminal and two bridging phosphanide ligands (Fig. 4) [4].

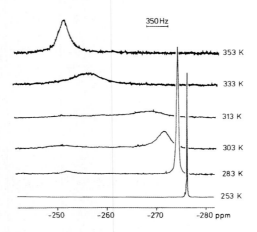

Fig. 1.   Dynamic ^{31}P{^{1}H} NMR spectra of a [D$_8$]toluene solution of tetrakis(tetrahydrofuran-*O*)strontium bis[bis(trimethylsilyl)phosphanide], 81.015 MHz.

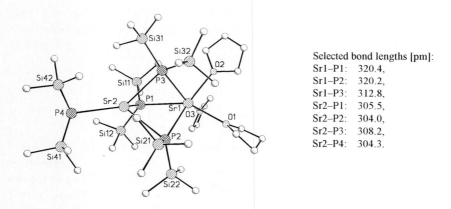

Selected bond lengths [pm]:
Sr1–P1:   320.4,
Sr1–P2:   320.2,
Sr1–P3:   312.8,
Sr2–P1:   305.5,
Sr2–P2:   304.0,
Sr2–P3:   308.2,
Sr2–P4:   304.3.

Fig. 2.   Molecular structure of (Me$_3$Si)$_2$P–Sr[μ-P(SiMe$_3$)$_2$]$_3$Sr(THF)$_3$.

Due to *ab initio* SCF calculations the favored bicyclic structure of the alkaline earth metal bis(phosphanides) can be explained by small, but structure dominating d orbital participation. Whereas the monocyclic magnesium bis(phosphanide) H$_2$P–Mg(μ-PH$_2$)$_2$Mg–PH$_2$ with a C_{2h} symmetry is favored by 27.9 kJ/mol against the bicyclic C$_1$ symmetric bicycle, the bicyclic compounds M(μ-PH$_2$)$_3$M–PH$_2$ of calcium and strontium are lower in energy [5]. In the case of the dimeric barium bis(phosphanide), the tricyclic derivative Ba(μ-PH$_2$)$_4$Ba is even favored by additional 8.0 kJ/mol compared to the bicyclic structure, which again is 52.0 kJ/mol lower in energy compared to the monocyclic C_{2h} symmetric structure. In these calculations neither the steric hindrance nor the bonded solvent molecules THF or DME are taken into account.

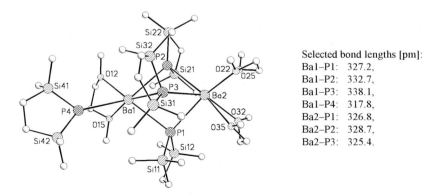

Eq. 2.

Table 1. NMR data of (THF)$_4$Sr[P(SiMe$_3$)$_2$]$_2$ in [D$_8$]toluene solution (81.015 MHz).

	(THF)$_4$Sr[P(SiMe$_3$)$_2$]$_2$	(THF)$_3$Sr[μ-P(SiMe$_3$)$_2$]$_3$Sr–P(SiMe$_3$)$_2$
Temperature [K]	233.0	353.0
δ(^{31}P{^{1}H}) [ppm]	-276.8	-256.2
δ(^{29}Si{^{1}H}) [ppm]	1.6	1.1
$^1J_{Si-P}$ [Hz]	38.4	34.1
δ(^{13}C{^{1}H}) [ppm]	7.8	8.0
$^2J_{C-P}$ [Hz]	9.4	10.2
δ(^{1}H) [ppm]	0.5	0.4

Selected bond lengths [pm]:
Ba1–P1: 327.2,
Ba1–P2: 332.7,
Ba1–P3: 338.1,
Ba1–P4: 317.8,
Ba2–P1: 326.8,
Ba2–P2: 328.7,
Ba2–P3: 325.4.

Fig 3. Molecular structure of tris(dme)dibarium tetrakis(2,2,5,5-tetramethyl-2,5-disilaphospholanide).

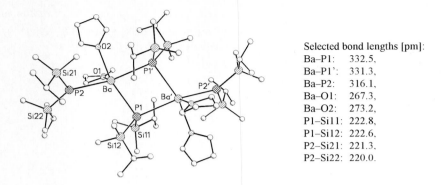

Selected bond lengths [pm]:
Ba–P1:　332.5,
Ba–P1':　331.3,
Ba–P2:　316.1,
Ba–O1:　267.3,
Ba–O2:　273.2,
P1–Si11:　222.8,
P1–Si12:　222.6,
P2–Si21:　221.3,
P2–Si22:　220.0.

Fig. 4.　Molecular structure of the monocyclic $(Me_2iPrSi)_2P–Ba(THF)_2[\mu-P(SiMe_2iPr)_2]_2Ba(THF)_2–P(SiMe_2iPr)_2$.

A comparison of these compounds with the corresponding zinc derivatives should clarify the influence of the empty d orbitals involved in the bonding situation of the alkaline earth metal bis(phosphanides). Whereas zinc bis[bis(trimethylsilyl)amide] is monomeric due to the steric demand of the bulky amide ligand [6], the trimethylsilyl substituted phosphanide leads to oligomers such as dimers or trimers [7]. The influence of the pnicogen atom is small, thus the phosphorus and arsenic derivatives (Fig. 5.) look very similar [8] or even crystallize isotypically. In contrast to the d^0 metals calcium, strontium and barium, zinc derivatives solely build up monocyclic ring systems.

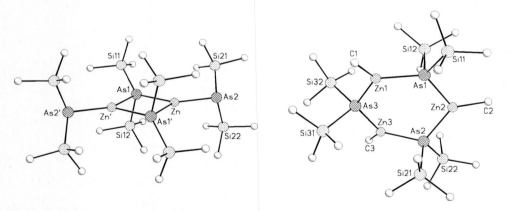

Fig. 5.　Molecular structures of dimeric zinc bis[bis(trimethylsilyl)arsenide] (left) and trimeric methyl zinc bis(trimethylsilyl)arsenide (right).

Since the ring size depends solely on the steric demand of the substituents at the zinc atom a monomeric derivative has been stabilized by the tris(trimethylsilyl)methyl group [7]. This molecule $(Me_3Si)_3C–Zn–P(SiMe_3)_2$ displays a chemical $^{31}P\{^1H\}$ shift of -245 ppm and a $^1J_{P–Si}$ coupling constant of 33.3 Hz. Melting of this compound at 75°C leads to decomposition and the loss of $P(SiMe_3)_3$. The enlargement of the trialkylsilyl substituent allows the isolation of the bis(alkylzinc) trialkylsilylphosphane $[(Me_3Si)_3C–Zn]_2P–SiiPr_3$ (Fig. 6) [9].

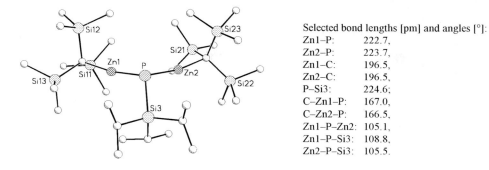

Selected bond lengths [pm] and angles [°]:

Zn1–P: 222.7,
Zn2–P: 223.7,
Zn1–C: 196.5,
Zn2–C: 196.5,
P–Si3:   224.6;
C–Zn1–P: 167.0,
C–Zn2–P: 166.5,
Zn1–P–Zn2: 105.1,
Zn1–P–Si3: 108.8,
Zn2–P–Si3: 105.5.

Fig. 6.   Molecular structure of bis[tris(trimethylsilyl)methylzinc] tri-*i*-propylphosphane.

This molecule shows a remarkable $^2J_{C-Zn-P}$ coupling of 6.9 Hz, a chemical $^{31}P\{^1H\}$ shift of -297 ppm with a $^1J_{P-Si}$ coupling constant of 48.2 Hz. Due to increasing interest in the zinc phosphide Zn_3P_2 as a ceramic material [10] this molecule gains interest, furthermore, only very few geminal diorganozinc-substituted molecules such as $RHC(ZnX)_2$ [11], $H_2C(ZnX)_2$ [12], $RN(ZnEt)_2$ [13] are known so far, and here the first structural characterized derivative is described.

Acknowledgements: We thank the *Deutsche Forschungsgemeinschaft* and the *Fonds der Chemischen Industrie* for the generous financial support.

References:

[1]   M. Westerhausen, W. Schwarz, *J. Organomet. Chem.* **1993**, *463*, 51.

[2]   M. Westerhausen, *J. Organomet. Chem.* **1994**, *479*, 141.

[3]   M. Westerhausen, M. Hartmann, W. Schwarz, *Inorg. Chem.* **1996**, *35*, 2421.

[4]   M. Westerhausen, G. Lang, W. Schwarz, *Chem. Ber.* **1996**, *129*, 1035.

[5]   M. Westerhausen, R. Löw, W. Schwarz, *J. Organomet. Chem.* **1996**, *513*, 213.

[6]   A. Haaland, K. Hedberg, P. P. Power, *Inorg. Chem.* **1984**, *23*, 1972.

[7]   a) B. Rademacher, W. Schwarz, M. Westerhausen, *Z. Anorg. Allg. Chem.* **1995**, *621*, 287.

        b) S. C. Goel, M. Y. Chiang, W. E. Buhro, *J. Am. Chem. Soc.* **1990**, *112*, 5636.

[8]   B. Rademacher, W. Schwarz, M. Westerhausen, *Z. Anorg. Allg. Chem.* **1995**, *621*, 1439.

[9]   M. Westerhausen, M. Wieneke, K. Doderer, W. Schwarz, *Z. Naturforsch.* **1996**, *51b*, 1439.

[10] a) W. E. Buhro, *Polyhedron* **1994**, *13*, 1131.

        b) S. C. Goel, W. E. Buhro, N. L. Adolphi, M. S. Conradi, *J. Organomet. Chem.* **1993**, *449*, 9.

[11] B. Martel, M. Varache, *J. Organomet. Chem.* **1972**, *40*, C53.

[12] a) S. Miyano, T. Ohtake, H. Tokumasu, H. Hashimoto, *Nippon Kagaku Kaishi* **1973**, 381.

        b) J. J. Eisch, A. Piotrowski, *Tetrahedron. Lett.* **1983**, *24*, 2043.

        c) K. Takai, T. Kakiuchi, Y. Kataoka, K. Utimoto, *J. Org. Chem.* **1994**, *59*, 2668.

[13] a) H. Tani, N. Oguni, *J. Polym. Sci., Part B* **1969**, *7*, 769.

        b) N. Oguni, T. Fujita, H. Tani, *Macromolecules* **1973**, *6*, 325.

The Tris(trimethylsilyl)silyl Substituent: An Old Hat With A New Feather

Andreas Heine, Lutz Lameyer, Dietmar Stalke*
Institut für Anorganische Chemie
Universität Würzburg
Am Hubland, D-97074 Würzburg, Germany
Fax: Int. code + (931)8884619
E-mail: dstalke@xiris.chemie.uni-wuerzburg.de

Keywords: Reaction Intermediates / Tris(trimethylsilyl)silyllithium / Lithium Silyl Cuprates / Disilagermirane

Summary: The results presented in this article prove that the tris(trimethylsilyl)silyl substituent is a very efficient ligand that is able to stabilize reaction intermediates or highly reactive compounds both kinetically and electronically. The steric requirement and the electron-releasing properties and the good availability of the lithiated species $(SiMe_3)_3SiLi(THF)_3$ facilitate the syntheses of compounds which are too unstable to be isolated otherwise. Cryo-crystal structure analysis and the handling of crystals at low temperature give access to freeze out intermediates at different positions along a reaction pathway and to determine their solid state structures. The cocrystallization product $[Si(SiMe_3)_4][(Me_3Si)_3SiLi(THF)_3]_2$ (**1**) and the pure $(Me_3Si)_3SiLi(THF)_3$ (**2**) were structurally characterized. The straightforward reaction of tris(trimethylsilyl)silyllithium with $GeCl_2$ yields the disilagermirane $\{(Me_3Si)_2Si\}_2Ge(SiMe_3)_2$ (**3**). In the reaction with metal halides, e.g., $AlCl_3$, $[Li(THF)_4][AlCl_3Si(SiMe_3)_3]$ (**4**), was isolated. At low temperature no LiCl elimination occurs. The anion is a Lewis base adduct of tris(trimethyl-silyl)silyl and $AlCl_3$ – possibly an early intermediate on the reaction pathway towards $AlSi(SiMe_3)_3Cl_2$. The reactions with CuCl and CuBr reveal the complexes of $[Li(THF)_4][Cu_5Cl_4\{Si(SiMe_3)_3\}_2]$ (**5**) and $[Cu_2\{Si(SiMe_3)_3\}_2BrLi(THF)_3]$ (**6**), only stable at ca. -30 °C. In both structures the central silicon atom is pentacoordinated due to μ_2 coordination to a remarkably short Cu–Cu distance.

Introduction

Tris(trimethylsilyl)silyllithium is a versatile reagent for the syntheses of a great variety of polysilyl derivatives. When coordinated with THF (tetrahydrofuran) first it dissolves well even in nonpolar solvents. Ever since the tris(trimethylsilyl)silyl ligand has been developed by Gilman and Smith [1,2] it was widely used both in transition metal chemistry [3-5] and main group chemistry [5-8]. The great steric demand as well as the electron releasing properties along with the usually good solubility of the products make it a very useful ligand. It allows investigation of highly reactive intermediates due to electronic and kinetic stabilization.

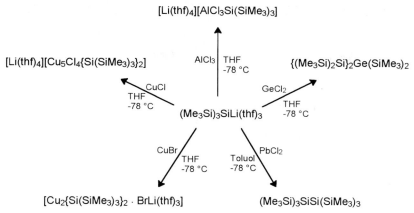

Scheme 1.

The straightforward synthesis of the disilagermirane $\{(Me_3Si)_2Si\}_2Ge(SiMe_3)_2$ was achieved without using sodium naphtalide or similar reductive agents to initialize the coupling reaction [9].

The reaction of tris(trimethylsilyl)silyllithium with $AlCl_3$ at -78°C yields the aluminate $[Li(THF)_4][AlCl_3Si(SiMe_3)_3]$ [10]. This Lewis base adduct can be interpreted as an LiCl-containing intermediate of the reaction towards $[(Me_3Si)_3SiAlCl_2]$. As a building block this intermediate makes soluble alumosilicates feasible that can be used in homogeneous catalysis.

The thermally unstable products of the reactions of tris(trimethylsilyl)silyllithium with copper(I) halides again prove the great potential of this ligand. The reaction with CuCl, for example, yields $[Cu_5Cl_4\{Si(SiMe_3)_3\}_2][Li(THF)_4]$ [10], which can be considered a reaction intermediate on the pathway towards a silylcuprate of the type $\{(Me_3Si)_3Si\}_2CuLi(L)_n$. This is particularly interesting since cuprates are among the most widely used organometallic reagents in organic chemistry though very little is known about the structural properties of these compounds [11]. Also in technical processes like the Rochow process silicon copper compounds are likely to play an important role in catalytic cycles.

The application of tris(trimethylsilyl)silyllithium to stabilize reactive intermediates and to form small rings of group 14 elements are summarized in Scheme 1.

Preparation and Structure of Tris(tetrahydrofuran)tris(trimethylsilyl)silyllithium

Following the original 1968 recipe of Gilman and Smith [2] a white crystalline product (**1**) is obtained in 69% yield. The crystal structure of this compound [12] (Fig. 1) revealed it to consist of 2 equiv. of the desired tris(trimethylsilyl)silyllithium which cocrystallized with 1 equiv. of the starting material tetrakis(trimethylsilyl)silane.

Obviously, lithiation was not complete since 1 out of 3 equiv. of the starting material is still present. The minor modification of stirring the reaction mixture for 4 d instead of 24 h yielded the completely metallated product **2** as shown in Fig. 2. Crystallization is not necessarily providing pure products and the use of this cocrystallized material **1** is one of the reasons for not very clear reactions which have been reported with $(Me_3Si)_3SiLi(THF)_3$.

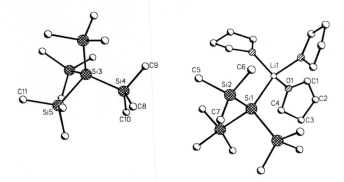

Fig. 1.   Structure of [Si(SiMe₃)₄][(Me₃Si)₃SiLi(THF)₃]₂ (**1**) in the solid state (1 equiv. tris(trimethylsilyl)silyllithium omitted).

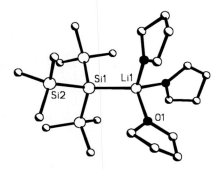

Fig. 2.   Structure of (Me₃Si)₃SiLi(THF)₃ (**2**) in the solid state.

As expected the structure of the lithiated component in **1** is almost identical to the structure of the pure tris(trimethylsilyl)silyllithium **2**. Interestingly, the lengths of the Si–C and Si–Si bonds are almost unaffected by the lithiation compared to those of the starting material [12]. The average Si–Li bond length of 266 pm is consistent with other Si–Li bond lengths previously reported [13].

The parameter most affected by the replacement of a SiMe₃ group with Li(THF)₃ is the Si–Si–Si angle. It is reduced by 7.4° from the ideal tetrahedral value of 109.5° to a mean value of 102.1°. The steric bulk of the Li(THF)₃ group alone cannot account for this reduction since in the even more sterically crowded disilane (Me₃Si)₃SiSi(SiMe₃)₃ the average Me₃Si–Si–SiMe₃ angle is reduced only by 4.7° to 104.8° [12]. The nature of the Si-metal bond might have an influence on the Si–Si–Si angle of the ligand. The difference in electronegativity between silicon and lithium is about 0.9, whereas it is, for example, 0.3 for silicon and aluminum and 0.0 for silicon and copper. That means that the bond between silicon and lithium is much more ionic than Si–Al (Si–Si–Si 107.5° in (Me₃Si)₃SiAlPh₂(THF) [14] or Si–Cu (Si–Si–Si 110.0° in [Li(THF)₄][Cu₅Cl₄{Si(SiMe₃)₃}₂] [10]. If one considers the extreme resonance description (Me₃Si)₃Si⁻ ⁺Li(THF)₄, the lone pair of the 'anion' is stereochemically active. The more ionic the Si-metal bond, the greater the influence of this 'lone pair' on the reduction of the Si–Si–Si angle. This thesis is supported by the ²⁹Si NMR chemical shift of the central silicon atom. In **2** the shift is -189.40 ppm (Fig. 3), in the disilane (Me₃Si)₃SiSi(SiMe₃)₃ it is -130.04 ppm, and in the (Me₃Si)₃SiAlCl₃ anion of **4** it is −115.66 ppm, indicating that the electron density at silicon decreases as the Si–Si–Si angle increases.

Although the Si–Si–Si angle indicates the strong ionic character of the Si–Li bond, there must be a significant covalent contribution because a $^1J_{(Si-Li)}$ NMR coupling in both the ^{29}Si and the ^{7}Li NMR spectra can be observed as shown in Fig. 3.

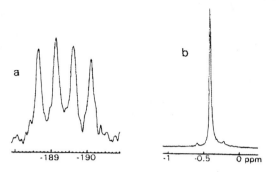

Fig. 3. (a) ^{29}Si NMR spectrum and (b) ^{7}Li NMR spectrum at -50°C of [(Me$_3$Si)$_3$SiLi(THF)$_3$ (**2**).

The $^1J_{(Si-Li)}$ coupling constant is 38.6 Hz in the ^{29}Si NMR spectrum at ambient temperature and 36.7 Hz in the ^{7}Li NMR spectrum at -50 °C. These values are of similar magnitude to previously observed $^1J_{(Si-Li)}$ coupling constants [15, 16].

The question whether organolithium compounds are essentially ionic, covalent, or intermediate has been vigorously debated [17] and similar considerations clearly apply to silyllithium derivatives.

Synthesis and Crystal Structure of the Disilagermirane {(Me$_3$Si)$_2$Si}$_2$Ge(SiMe$_3$)$_2$ (3)

Three-membered ring systems of heavier group 14 elements are known to be accessible by reductive cyclization of diorganodihalo derivatives of silicon and germanium [18, 19].These compounds were considered unobtainable until the beginning of the 1980s. The use of bulky organic ligands like 2,6 dimethylphenyl [19a] and *t*-butyl [19b], however, kinetically stabilizes the small rings.

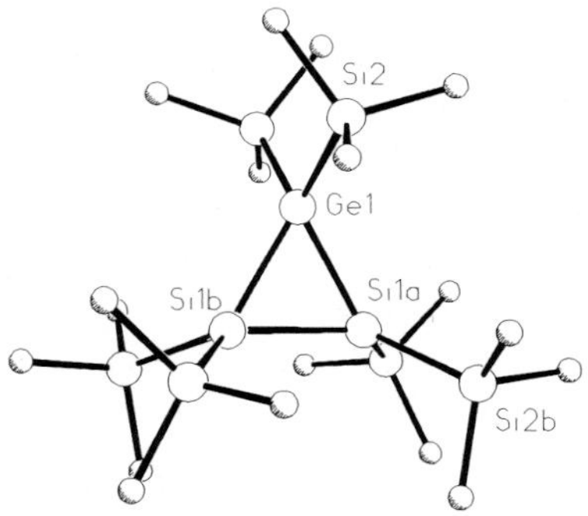

Fig. 4. Structure of {(Me$_3$Si)$_2$Si}$_2$Ge(SiMe$_3$)$_2$ (**3**) in the solid state.

Unfortunately, in all of the known cases reductive cyclization with, e.g., lithium naphtalide was necessary to obtain the products. With this route the synthesis of three-membered rings containing various group 14 metals is limited and yields are poor. In contrast, reaction of 2 equiv. of tris(trimethylsilyl)silyllithium with $GeCl_2$ at -78 C yields $\{(Me_3Si)_2Si\}_2Ge(SiMe_3)_2$ **(3)** in a straightforward way [9] (Fig. 4).

The reaction mechanism for the formation of the disilagermirane has not been confirmed unequivocally. It seems possible, however, that initially a germylene is formed intermediately by simultaneous LiCl elimination and migration of two Me_3Si groups. The $(Me_3Si)_2Ge|$ then reacts with the intermediate disilene $(Me_3Si)_2Si=Si(SiMe_3)_2$ to give the product **3**. This reaction might be a general approach towards three-membered rings containing varions group 14 metals simply by reacting $(Me_3Si)_3ELi(THF)_n$ (E=Si, Ge, Sn) with group 14 dihalides.

Isolation of the Kinetically Stabilized Reaction Intermediate $[Li(THF)_4][AlCl_3Si(SiMe_3)_3]$ (4)

As a product of the reaction between tris(trimethylsilyl)silyllithium and $AlCl_3$ at low temperature the aluminate **4** was isolated [10] (see Fig. 5).

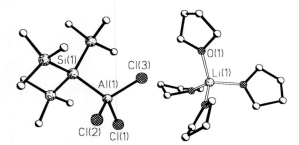

Fig. 5.   Structure of the aluminate $[Li(THF)_4][AlCl_3Si(SiMe_3)_3]$ **(4)** in the solid state.

This aluminate can be seen as an intermediate of the reaction

$(Me_3Si)_3SiLi(thf)_3$ + $AlCl_3$ $\longrightarrow$ $(Me_3Si)_3SiAlCl_2$ +   LiCl

Eq. 1.

which yields tris(trimethylsilyl)silylaluminum dichloride as the final product. An analogous reaction with AlR_3 should yield the silylaluminate $[(Me_3Si)_3SiAlR_3]^-$. Moreover, as the trihalogenated parent compound can react with either silanols or lithium silanolates in arbitrary molar ratios, molecular alumosiloxanes of spherical structure like, for example, alumosesquioxanes or zeolites (Linde type A), should be accessible [20].

Today, silicates and zeolites are interesting carrier materials for heterogeneous catalysts. With $[(Me_3Si)_3SiAlCl_3]^-$ as a starting material the synthesis of modified alumosiloxanes seems feasible to yield molecular soluble carriers, hence catalysis in homogeneous phase seems possible.

Reaction of Tris(trimethylsilyl)silyllithium with Cu^{I}Cl and Cu^{I}Br

The importance of silyl groups in organic synthesis is undoubted although very few structural details are known of the intermediates involved in the lithium cuprate reactions [21]. These species have commonly been employed as *in situ* reagents in both organic [11] and inorganic synthesis. The use of silyl cuprates [22] and more recently even of stannyl cuprates [23] in organic synthesis was pioneered by Fleming. The growing interest in organo copper compounds, in particular consideration of the theoretical aspects [24], together with the report of Eaborn and coworkers in 1983 on the preparation and structure of the lithiumcuprate [Li(THF)$_4$][Cu{C(SiMe$_3$)$_3$}$_2$] [25], encouraged us to react tris(trimethylsilyl)silyllithium with copper(I)halides. (Me$_3$Si)$_3$SiLi(THF)$_3$ [26] was reacted with Cu^{I}Cl and Cu^{I}Br at -78°C. The different reactivity of Cu^{I}Cl compared to that of Cu^{I}Br (e.g., the latter shows a lower tendency to disproportionate) should give rise to different unprecedented intermediates in the lithium silyl cuprate reaction.

(Me$_3$Si)$_3$SiLi(THF)$_3$ was reacted with Cu^{I}Cl in the ratio of 1:2 and with Cu^{I}Br in 1:1 ratio in THF [10, 27]. Pure (Me$_3$Si)$_3$SiLi(THF)$_3$ is extremely pyrophoric and should be handled with care [12]. All manipulations should be performed under an inert atmosphere of dry argon. The first reaction mixture was transferred to a deep freeze at -30°C immediately, while the second was allowed to warm up to room temperature first.

$$5\ CuCl\ +\ 2\ (Me_3Si)_3SiLi(THF)_3 \xrightarrow[-LiCl]{THF} [Li(THF)_4][Cu_5Cl_4\{Si(SiMe_3)_3\}_2]$$

$$2\ CuBr\ +\ 2\ (Me_3Si)_3SiLi(THF)_3 \xrightarrow[-LiBr]{THF} [Cu_2\{Si(SiMe_3)_3\}_2BrLi(THF)_3]$$

Eq. 2.

Under these conditions the first two lithium silylcuprates were isolated and structurally characterized. The complexes were found to be extremely air-and moisture-sensitive. During crystal selection and mounting the flask and the flushing argon were cooled down to ca. -50°C. At a temperature slightly above -30°C both compounds decompose instantaneously.

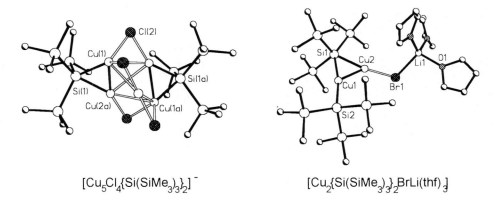

$$[Cu_5Cl_4\{Si(SiMe_3)_3\}_2]^-$$

$$[Cu_2\{Si(SiMe_3)_3\}_2BrLi(thf)_3]$$

Fig. 6. The structures of the lithiumsilylcuprates [Cu$_5$Cl$_4$\{Si(SiMe$_3$)$_3$\}$_2$]$^-$ and [Cu$_2$\{Si(SiMe$_3$)$_3$\}$_2$BrLi(THF)$_3$] **6.**

The structure analysis of the first complex reveals the formula [Li(THF)$_4$][Cu$_5$Cl$_4${Si(SiMe$_3$)$_3$}$_2$] [10], (Fig. 6, left). This not predictable formula can be rationalized a posteriori as a [Li(THF)$_4$][Cu{Si(SiMe$_3$)$_3$}$_2$] containing an excess of 4 CuCl units. At the end of the reaction this cuprate, similar to that of Eaborn et al., might be the main product, although we have not been able to prove that ultimately.

Fig. 6 shows the cuprate anion. The Li(THF)$_4$ cation has been omitted for clarity. To the best of our knowledge it is the first structurally characterized lithium silylcuprate [28]. Besides the fact that there is only one other Si–Cu bond determined by X-ray analysis (in Ph$_3$SiCu(PMe$_3$)$_3$) [29], the silicon atom is in quite an unusual bonding situation. The central Si(1) is bound to the three trimethylsilyl groups and to two copper atoms, leading to pentacoordinated silicon. μ_2-Bridging for silyl groups is very rare [30], although well documented for alkyl and aryl bridging carbons. The two Si–Cu bonds are equally long (234.8(3) and 233.4(3) pm) and of the same length as in Ph$_3$SiCu(PMe$_3$)$_3$ (234.0(2) pm). The bridged Cu1–Cu2a bond with 240.3(2) pm is by far the shortest of the Cu–Cu distances and 25 pm shorter than the next (Cu1–Cu2 265.4(2); Cu1–Cu3 273.6(2); Cu2–Cu2a 295.2(3) pm). Although slightly shorter Cu–Cu distances are known where the Cu atoms are forced in close contact by chelating ligands (234.8(1) and 236.1(1) pm) [31] this distance is the shortest in Cu$_5$-complexes [32] to the best of our knowledge. So the almost equilateral triangle SiCu$_2$ can be described in terms of a three-center two-electron bond. The two Cu1–Cu2 and Cu1–Cu3 distances are μ_2-bridged by chlorine atoms. The long Cu2···Cu2a vector remains unbridged.

In contrast to **5**, [Cu$_2${Si(SiMe$_3$)$_3$}$_2$BrLi(THF)$_3$], **6**, is a contact ion pair. It contains two differently bound (Me$_3$Si)$_3$Si groups and an additional LiBr equivalent. The tris(trimethylsilyl)silyl group around Si1 is coordinated *side on* to a very short Cu1–Cu2 bond, again leading to a pentacoordinated central Si1 atom. The Cu1–Cu2 bond is 236.9(1) pm and more than 3 pm shorter than in the last complex. In this example the three-center two-electron bond does not form an equilateral triangle. While the Si1–Cu2 bond is 228.3(2) pm long, the Si1–Cu1 bond is more than 12 pm longer (240.6(2) pm). The shortest Si–Cu bond is that of the terminally-bound tris(trimethylsilyl)silyl group around Si2 and Cu(1) (226.6(2) pm). In view of the two other known CuSi distances (in Ph$_3$SiCu(PMe$_3$)$_3$: 234.0(2) [29] and in [Li(THF)$_4$][Cu$_5$Cl$_4${Si(SiMe$_3$)$_3$}$_2$]: av. 234.1(3) pm), this bond seems to be very short. Cu2 is coordinated by a (THF)$_3$LiBr group. The Cu2–Br1 distance of 226.0(5) pm is notably longer than in CuBr (217.3 pm) [33], but comparable to the published value in [CuBrCH(SiMe$_3$)$_2$]$^-$ (226.7(2) pm) [34]. The Li–Br bond length is 250(1) pm and comparable to the distances found in molecular lithium bromide (250.8 pm in coordination to PMDTA (PMDTA = MeN(CH$_2$CH$_2$NMe$_2$)$_2$) and 251.5 and 255.1 pm, respectively, as dimeric compound with acetone) [35].

Therefore, the (THF)$_3$LiBr group could be considered a leaving group of the intermediate. When reacted completely Cu$_2${Si(SiMe$_3$)$_3$}$_2$ might be the product. On the other hand CuBr might leave the complex to give [Li(THF)$_4$][Cu{Si(SiMe$_3$)$_2$}], which would presumably be isostructural with [Li(THF)$_4$][Cu{C(SiMe$_3$)$_3$}] [25]. Further experiments to isolate other intermediates in the reactions of copper(I)halides, as well as to elucidate the final products, are underway.

Conclusion

The results presented here clearly demonstrate the great synthetic capacity of tris(trimethylsilyl)silyllithium. Three synthetic routes to target materials have been developed and the solid state structures of these species were characterized by cryo-crystal structure analyses. The title compound gives access to three membered rings containing varions group 14 metals, alumosiloxanes and lithium silyl cuprates, provided the reaction temperatures are low. Particularly the structural feature of the pentacoordinated silicon atom inducing the short Cu–Cu distances gleaned from the lithium silyl cuprates have to be investigated by MO calculations on model systems.

References:

[1]   H. Gilman, C. L. Smith, *J. Organomet. Chem.* **1967**, *8*, 245.

[2]   H. Gilman, C. L. Smith, *J. Organomet. Chem.* **1968**, *14*, 91.

[3]   J. Arnold, T. D. Tilley, A. L. Rheingold, S. Geib, *J. Inorg. Chem.* **1987**, *26*, 2106.

[4]   J. Meyer, J. Willnecker, U. Schubert, *Chem. Ber.* **1989**, *122*, 223.

[5]   H. Piana, H. Wagner, U. Schubert, *Chem. Ber.* **1991**, *124*, 63.

[6]   A. M. Arif, A. H. Cowley, T. M. Eklins, *J. Organomet. Chem.* **1987**, *325*, C11.

[7]   S. P. Mallela, R. A. Geanagel, *Inorg. Chem.* **1990**, *29*, 3525.

[8]   A. M. Arif, A. H. Cowley, T. M. Elkins, R. A. Jones, *J. Chem. Soc., Chem. Commun.* **1986**, 1776.

[9]   A. Heine, D. Stalke, *Angew. Chem.* **1994**, *106*, 121; *Angew. Chem., Int. Ed. Engl.* **1994**, *33*, 113.

[10] A. Heine, D. Stalke, *Angew. Chem.* **1993**, *105*, 90; *Angew. Chem., Int. Ed. Engl.* **1993**, *32* 121.

[11] B. H. Lipshutz, S. Sengupta, *Org. React.* **1992**, *41*, 135.

[12] A. Heine, R. Herbst-Irmer, G. M. Sheldrick, D. Stalke, *Inorg. Chem.* **1993**, *32*, 2694.

[13] a) T. F. Schaaf, W. Butler, M. D. Glick, J. P. Oliver, *J. Am. Chem. Soc.* **1974**, *96*, 7593.
      b) W. H. Ilsley, T. F. Schaaf, M. D. Glick, J. P. Oliver, *J. Am. Chem. Soc.* **1980**, *102*, 3769.
      c) B. Teclé, W. H. Ilsley, J. P. Oliver, *Organometallics* **1982**, *1*, 875.
      d) G. Becker, H.-M. Hartmann, A. Münch, H. Riffel, *Z. Anorg. Allg. Chem.* **1985**, *530*, 29.
      e) G. Becker, H.-M. Hartmann, E. Hengge, F. Schrank, *Z. Anorg. Allg. Chem.* **1989**, *572*, 63.
      f) H. V. R. Dias, M. M. Olmstead, K. Ruhlandt-Senge, P. P. Power, *J. Organomet. Chem.* **1993**, *462*, 1.
      g) J. Belzner, U. Dehnert, D. Stalke, *Angew. Chem.* **1994**, *106*, 2580; *Angew. Chem., Int. Ed. Engl.* **1994**, *33*, 2450.

[14] M. L. Sierra, V. S. J. de Mel, J. P. Olive, *Organometallics* **1989**, *8*, 2312.

[15] U. Edlund, T. Lejon, T. K. Venkatachalam, E. Buncel, *J. Am. Chem. Soc.* **1985**, *107*, 6408.

[16] U. Edlund, T. Lejon, P. Pyykkö, T. K. Venkatachalam, E. Buncel, *J. Am. Chem. Soc.* **1987**, *109*, 5982.

[17] a) A. Sreitwieser, jr., J. E. Williams, jr., S. Alexandratos, J. M. McKelvey, *J. Am. Chem. Soc.* **1976**, *98*, 4778.

b) J. D. Dill, P. v. R. Schleyer, J. S. Binkley, J. A. Pople, *J. Am. Chem. Soc.* **1977**, *99*, 6159.

c) G. D. Graham, D. S. Marynik, W. N. Lipscomb, *J. Am. Chem. Soc.* **1980**, *102*, 4572.

d) T. Clark, C. Rohde, P. v. R. Schleyer, *Organometallics* **1983**, *2*, 1344.

e) S. M. Bachrach, A. Steitwieser, jr., *J. Am. Chem. Soc.* **1984**, *106*, 2283.

[18] Review: T. Tsumuraya, S. A. Batcheller, S. Masamune, *Angew. Chem.* **1991**, *103*, 916; *Angew. Chem., Int. Ed. Engl* **1991**, *30*, 902; and references therein.

[19] a) S. Masamune, Y. Hanzawa, S. Murakami, T. Bally, J. J. Blount, *J. Am. Chem. Soc.* **1982**, *104*, 1150.

b) A. Schäfer, M. Weidenbruch, K. Peters, H. G. von Schnering, *Angew. Chem.* **1984**, *96*, 311; *Angew. Chem., Int. Ed. Engl.* **1984**, *23*, 302.

c) K. H. Baines, J. A. Cooke, *Organometallics* **1991**, *10*, 3419.

d) K. H. Baines, J. A. Cooke, N. C. Payne, J. J. Vittal, *Organometallics* **1992**, *11*, 1408.

[20] a) N. Winkhofer, A. Voigt, H. Dorn, H. W. Roesky, A. Steiner, D. Stalke, A. Relle, *Angew. Chem.* **1994**, *106*, 1414; *Angew. Chem., Int. Ed. Engl.* **1994**, *33*, 1352.

b) U. Ritter, N. Winkhofer, H.-G. Schmidt, H. W. Roesky, *Angew. Chem.* **1996**, *108*, 591; *Angew. Chem., Int. Ed. Engl.* **1996**, *35*, 524; and references therein.

[21] a) H. O. House, *Acc. Chem. Res.* **1976**, *9*, 59.

b) M. M. Olmstead, P. P. Power, *Organometallics* **1990**, *9*, 1720.

[22] I. Fleming, H. M. Hill, D. Parker, D. Waterson, *J. Chem. Soc., Chem. Commun.* **1985**, 318.

[23] A. Barbero, P. Cuadrado, I. Fleming, A. M. Gonzáles, F. J. Pulido, *J. Chem., Soc. Chem. Commun.* **1992**, 351.

[24] d^{10}-d^{10} interactions:

a) P. K. Mehrotra, R. Hoffmann, *Inorg. Chem.* **1978**, *17*, 2187.

b) K. M. Merz, R. Hoffmann, *Inorg Chem.* **1988**, *27*, 2120.

computational approach:

c) R. Ahlrichs, M. Bär, M. Häser, H. Horn, C. Kölmel, *Chem. Phys. Lett.* **1989**, *162*, 165.

d) F. Haase, R. Ahlrichs, *J. Comp. Chem.* **1993**, *14*, 907.

recent chemical examples of Cu:

e) M. Håkansson, M. Örtendahl, S. Jagner, M. P. Sigalas, O. Eisenstein, *Inorg. Chem.* **1993**, *32*, 2018.

f) S. Dehnen, A. Schäfer, D. Fenske; R. Ahlrichs, *Angew. Chem.* **1994**, *106*, 786; *Angew. Chem., Int. Ed. Engl.* **1994**, *33*, 746.

[25] C. Eaborn, P. B. Hitchcock, J. D. Smith, A. C. Sullivan, *J. Organomet. Chem.* **1984**, *263*, C23.

[26] G. Gutekunst, A. G. Brook, *J. Organomet. Chem.* **1982**, *225*, 1.

[27] A. Heine, R. Herbst-Irmer, D. Stalke, *J. Chem. Soc., Chem. Commun.* **1993**, 1729.

[28] Cambridge Crystallographic Database, Version **1995**.

[29] A. H. Cowley, T. M. Elkins, R. A. Jones, C. M. Nunn, *Angew. Chem.* **1988**, 100, 1396; *Angew. Chem., Int. Ed. Engl.* **1988**, 27, 1349.

[30] J. C. Calabrese, L. F. Dahl, *J. Am. Chem. Soc.* **1971**, 6042.

[31] a) J. Beck, J. Strähle, *Angew. Chem.* **1985**, 97, 419; *Angew. Chem. Int. Ed. Engl.* **1985**, *27*, 409.

b) R. Schmid, J. Strähle, *Z. Naturforsch.* **1988**, 44b, 105.

c) G. van Koten, *J. Organomet. Chem.* **1990**, *400*, 283; and literature quoted there.

[32] P. G. Edwards, R. W. Gellert, M. W. Marks, R. Bau, *J. Am. Chem. Soc.* **1982**, *104*, 2072.

[33] A. F. Wells, *Structural Inorganic Chemistry*, Clarendon Press, Oxford, 5th. edn., **1984**, p. 444.

[34] H. Hope, M. M. Olmstead, P. P. Power, J. Sandell, X. Xu, J. *Am. Chem. Soc.* **1985**, *107*, 4337.

[35] a) S. R. Hall, C. L. Raston, B. W. Skelton, A. H. White, *Inorg. Chem.* **1983**, *22*, 4070.
      b) R. Amstutz, J. Dunitz, T. Laube, W. B. Schweizer, D. Seebach, *Chem. Ber.* **1986**, *119*, 434.

Functionalized Trisilylmethanes and Trisilylsilanes as Precursors of a New Class of Tripodal Amido Ligands

Martin Schubart, Bernd Findeis, Harald Memmler, Lutz H. Gade*

Institut für Anorganische Chemie
Universität Würzburg
Am Hubland, D-97074 Würzburg, Germany
E-mail: Lutz.Gade@mail.uni-wuerzburg.de

Keywords: Trisilylmethanes / Trisilylsilanes / Tripodal Amido Ligands

Summary: A series of amino-functionalized trisilylmethanes $HC(SiMe_2NHR)_3$ and trisilylsilanes $MeSi(SiMe_2NHR)_3$ (R = alkyl, aryl) has been synthesized and studied structurally both in solution and in the solid state. The corresponding trilithium amides were obtained readily by reaction with *n*-butyllithium and display adamantoid cage structures. They were found to be the ideal ligand transfer reagents for the coordination of the novel tripodal amido ligands to early transition metals.

There has recently been a burgeoning interest in the development of formally highly charged polydentate ligands due to their ability to stabilize large sectors in the coordination sphere of high valent early transition metal complexes. The reactivity of a transition metal complex is thereby limited to a remaining reactive site occupied by ligands which may be substituted readily or removed completely.

The chemistry of functionalized trisilylmethanes and trisilylsilanes has thus far received little attention outside the context of basic research in silane chemistry [1, 2]. The only previous attempt to prepare tripodal amido complexes based on an amino-functionalized trisilylmethane was reported by Bürger and coworkers, who prepared $HC(SiMe_2NHMe)_3$ but were not successful in coordinating this ligand precursor as a tripod ligand to a transition metal [3].

Synthesis of the Trisilylmethane and Trisilylsilane Backbones

There are two principal synthetic routes reported in the literature for the synthesis of trisilylmethanes. Merker and Scott developed a Grignard-related method based on the reductive coupling of chlorosilanes with bromoform (or other hydrocarbons containing a CBr_3 group) using magnesium metal ("Merker-Scott Coupling") [1]. The alternative route is the coupling of chloroform with chlorosilanes in the presence of lithium, a strategy based on early work by Gilman and coworkers. The latter is also the method of choice for the synthesis of trisilylsilanes replacing chroroform by a trichlorosilane ("Gilman Coupling") (Scheme 1) [2].

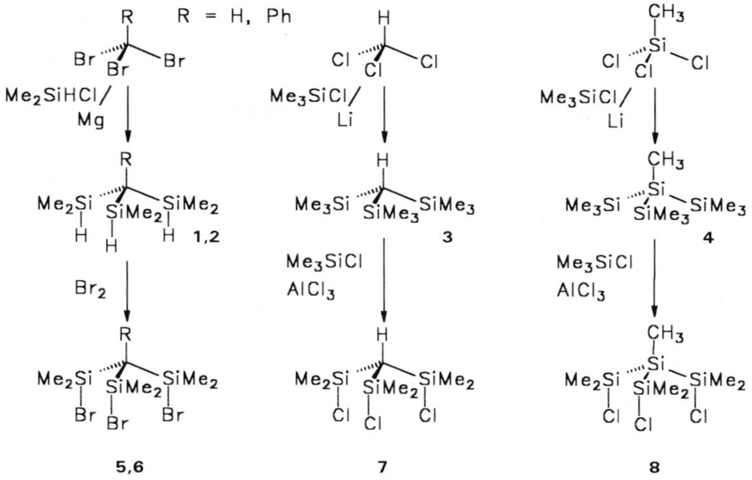

Scheme 1. Synthesis of the trisilylmethane and trisilylsilane backbones and conversion to their silylhalide derivatives.

The key intermediates in the ligand backbone synthesis are the compounds $RC(SiMe_2H)_3$ ($R = H$: **1**, Ph: **2**) or $HC(SiMe_3)_3$ **3** and the trisilylsilane $MeSi(SiMe_3)_3$ **4**. While **1** and **2** are converted to the respective bromo-functionalized derivatives $RC(SiMe_2Br)_3$ ($R = H$: **5**, Ph: **6**) by direct bromination, the chlorosilanes $HC(SiMe_2Cl)_3$ **7** and $MeSi(SiMe_2Cl)_3$ **8** are obtained by $AlCl_3$-catalyzed Me/Cl-exchange by the method first reported by Ishikawa et al. for related systems [4, 5].

Synthesis of the Tris(aminosilyl)methanes and -silanes

Both compounds **5** and **8** readily react with primary amines giving the corresponding tris(amino-silyl)methanes and -silanes (Scheme 2).

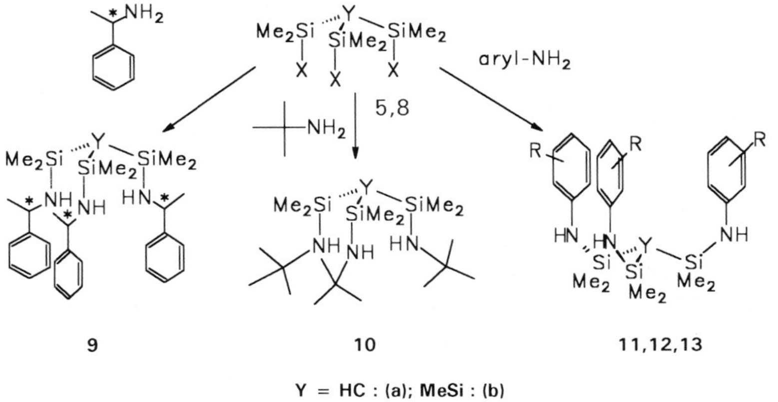

$Y = HC : (a); MeSi : (b)$

Scheme 2. Synthesis of the tris(aminosilyl)methanes **9a-13a** ($Y = HC$) and -silanes **9b-13b** ($Y = MeSi$); $R = 4\text{-}CH_3C_6H_4$ **11**, $4\text{-}CH_3OC_6H_4$ **12**, $2\text{-}F\text{-}C_6H_4$ **13**; auxiliary base: Triethylamine.

The reactions proceed smoothly provided the primary amine is sufficiently nucleophilic. In general, the trisilylsilane **8** was found to be significantly more reactive than the corresponding trisilylmethane **7**. Electron-withdrawing substituents in arylamines may inhibit the desired condensation with the silylhalide to an extent that aminolysis occurs only partially or not at all. 2-Fluoroaniline, for example, reacts rapidly and cleanly with the reactive bromosilane **5**, but requires several days to be converted to **13a** using the chloro derivative **7**. Pentafluoroaniline is completely inert towards **11-13**, which is certainly not a consequence of steric effects in view of the high reactivity of, e.g., *t*-butylamine [6, 7].

Structures of the Tris(aminosilyl)methanes

The trisilylmethane derivatives described in the section above display two conformational extremes already indicated in Scheme 2. While the alkyl-substituted derivatives **9a** and **10a** have an adamantoidal structure which is probably determined by weak intramolecular N–H···N hydrogen bonding, the aryl-substituted species **11a**, **12a**, and **13a** adopt an "inverted" conformation both in solution (established by NOESY experiments at variable temperature) and in the solid state (X-ray structure analysis) (Fig. 1) [7].

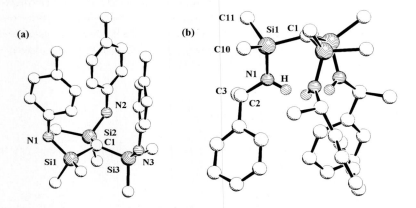

Fig. 1. Molecular structures of **11** (a) and **9** (b) in the solid as determined by X-ray structure analysis.

The proposed weak N–H···N hydrogen bonding in the structures of the alkyl-substituted compounds is backed up by a significant shift of the ν(N–H) bands in the IR spectra of these compounds to lower wavenumbers. The interaction itself may be a consequence of the increased basicity at the N-atoms in comparison to **11a-13a**. An equally pronounced preference of a certain structural arrangement is not observed for the trisilylsilane derivatives **9b-13b**, probably due to the steric demand of the apical methyl group, which prevents the occupation of the completely inverted conformation as shown in Fig. 1a.

Structures of the Lithiated Tris(aminosilyl)methanes and -silanes

Lithiation of the tripodal triamines **9**, **10**, and **11** with *n*BuLi yields the corresponding trilithium triamides (Fig. 2). The metallation itself is a highly cooperative process in which the trimetallated species is already generated upon addition of only a fraction of the required amount of *n*BuLi. This is thought to be due to the kinetic and thermodynamic stability of the adamantoid triamide, which is able to adopt a more compact structure than the partially metallated species [6].

R	
(S)-CHMePh	14a
tBu	15a
4-CH₃C₆H₄	16a

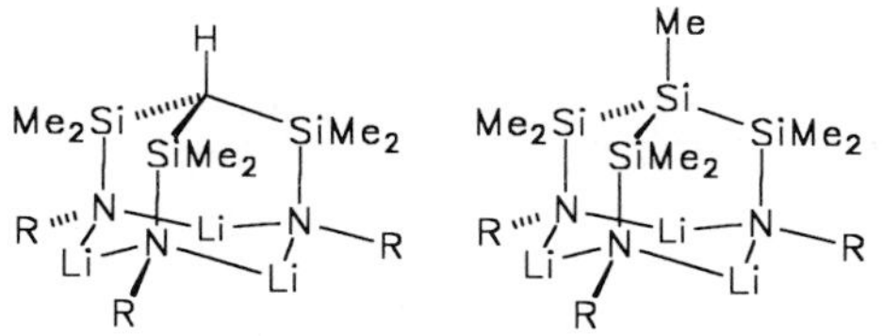

R	
(S)-CHMePh	14b
tBu	15b
4-CH₃C₆H₄	16b

Fig. 2. Tripodal lithium triamides with an adamantane-related structure.

X-Ray structure analysis of **14a**, **15a**, and **15b** has established the central adamantane cage which is characteristic of the unsolvated triamides [6, 8]. Whereas there is little appreciable interaction between the periphery of the molecules in the structures of the *t*-butyl substituted species **15a** and **15b**, the two-coordinate Li-atoms are "internally solvated" by the phenyl groups in the chiral triamide **14a** (Fig. 3). The orientation of the aryl rings towards the puckered six-membered (LiN)₃-ring and the short contact between Li and C7 indicates a direct interaction between the metal atoms and the periphery.

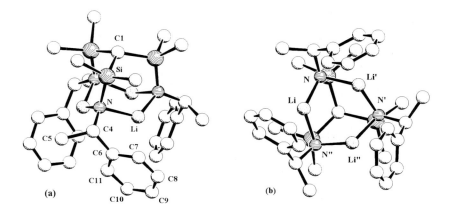

Fig. 3. a) Molecular structure of the chiral lithium amide **14a** in the solid.
b) View along the threefold molecular axis.

While the puckered (LiN)₃ ring is the central structural unit in these unsolvated amides, a ring-ladder interconversion may be observed upon addition of coligand to the lithium amides. This structural rearrangement follows the same pattern as observed in the chemistry of monofunctional amides [9]. The bis-THF-adduct of **15a**, HC{SiMe₂N(Li)tBu}₃(THF)₂ **17a**, is an example of such a solvated Li-amide which has been characterized by X-ray crystallography [6].

Transition Metal Complexes Containing Tripodal Amido Ligands with Trisilylmethane and Trisilylsilane Frameworks

The lithium triamides discussed in the previous section were found to be the ideal ligand transfer reagents in the synthesis of transition metal amido complexes. The tripodal ligands effectively shield a large sector in the coordination sphere of the metal thus focussing the reactivity of the complex upon a single coordination site. Application of this strategy to the tetravalent metals of the Ti-triad [5, 7] provided valuable building blocks for the synthesis of early-late dinuclear complexes (Fig. 4) [10, 11].

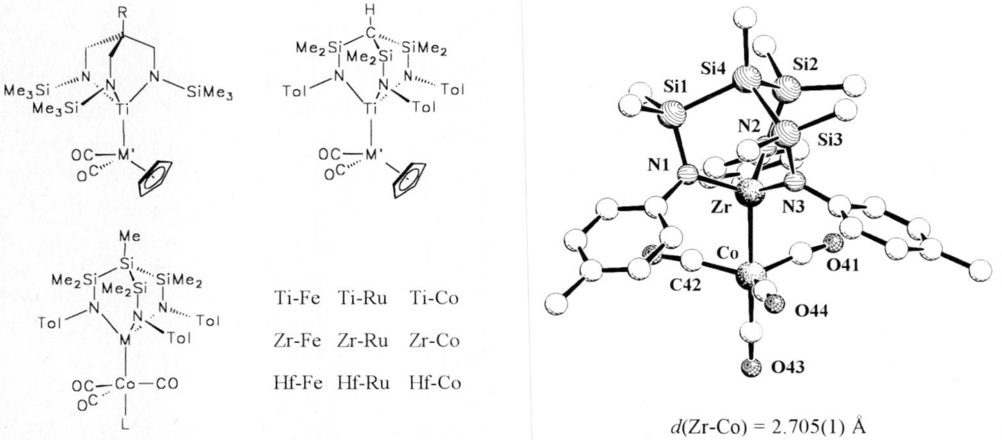

Ti-Fe	Ti-Ru	Ti-Co
Zr-Fe	Zr-Ru	Zr-Co
Hf-Fe	Hf-Ru	Hf-Co

d(Zr-Co) = 2.705(1) Å

Fig. 4.   Early-late heterobimetallic complexes with unsupported metal-metal bonds stabilized by tripodal amido ligands.

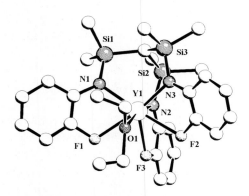

Fig 5.   Molecular structure of HC{SiMe₂N(2-FC₆H₄)}₃Y(OEt₂).

The concept may be extended by additional coordination of weakly binding peripheral donor functions to the metal centre, as exemplified by the yttrium complex displayed in Fig. 5. The 2-fluorophenyl groups provide an "active" ligand periphery which stabilizes the highly acidic Y-center [12].

Acknowledgements: The X-ray structure determinations referred to in this paper were carried out by Prof. M. McPartlin, Dr. I. J. Scowen, and Dr. W. S. Li, *University of North London* (UK) as well as Prof. J. W. Lauher, *State University of New York at Stony Brook* (USA). This work has been generously supported by the *Deutsche Forschungsgemeinschaft*. We also thank *Wacker AG* for providing valuable basic chemicals.

References:

[1]   R. L. Merker, M. J. Scott, *J. Org. Chem.* **1964**, *29*, 953; R. L. Merker, M. J. Scott, *J. Organomet. Chem.* **1965**, *4*, 98; H. Sakurai, K. Tominaga, T. Watanabe, M. Kumada, *Tetrahedron Lett.* **1966**, 5493; C. Eaborn, P. B. Hitchcock, P. D. Lickiss, *J. Organomet. Chem.* **1984**, *269*, 235; R. Hager, O. Steigelmann, G. Müller, H. Schmidbaur, *Chem. Ber.* **1989**, *122*, 2115 and refs. cited therein; C. Eaborn, P. B. Hitchcock, P. D. Lickiss, *J. Organomet. Chem.* **1983**, *252*, 281; M. Baier, P. Bissinger, H. Schmidbaur, *Z. Naturforsch.* **1993**, *48b*, 1672.

[2]   H. Gilman, J. M. Holmes, C. L. Smith, *Chem. Ind. (London)* **1965**, 849; H. Gilman, C. L. Smith, *J. Organomet. Chem.* **1968**, *14*, 91; H. C. Marsmann, W. Raml, E. Hengge, *Z. Naturforsch.* **1980**, *35b*, 1541; J. B. Lambert, J. L. Pflug, C. L. Stern, *Angew. Chem.* **1995**, *107*, 106; *Angew. Chem., Int. Ed. Engl.* **1995**, *34*, 98.

[3]   H. Bürger, R. Mellies, K. Wiegel, *J. Organomet. Chem.* **1977**, *142*, 55.

[4]   M. Ishikawa, M. Kumada, H. Sakurai, *J. Organomet. Chem.* **1970**, *23*, 63; K. Hassler, *Monatsh. Chem.* **1986**, *117*, 613.

[5]   M. Schubart, B. Findeis, L. H. Gade, W; S. Li, M. McPartlin, *Chem. Ber.* **1995**, *128*, 329; B. Findeis, M. Schubart, L. H. Gade, *Inorg. Synth.* **1997**, *32*, in press.

[6]   L. H. Gade, C. Becker, J. W. Lauher, *Inorg. Chem.* **1993**, *32*, 2308.

[7]   H. Memmler, L. H. Gade, J. W. Lauher, *Inorg. Chem.* **1994**, *33*, 3064.

[8]   H. Memmler, L. H. Gade, J. W. Lauher, submitted.

[9]   D. Barr, W. Clegg, R. E. Mulvey, R. Snaith, *J. Chem. Soc., Chem. Commun.* **1984**, 285; D. R. Armstrong, R. E. Mulvey, G. T. Walker, D. Barr, R. Snaith, W. Clegg, D. Reed, *J. Chem. Soc., Dalton Trans.* **1988**, 617.

[10] S. Friedrich, H. Memmler, L. H. Gade, W. S. Li, M. McPartlin, *Angew. Chem.* **1994**, *106*, 705; *Angew. Chem., Int. Ed. Engl.* **1994**, *33*, 676; S. Friedrich, H. Memmler, L. H. Gade, W. S. Li, I. J. Scowen, M. McPartlin, C. E. Housecroft, *Inorg. Chem.* **1996**, *35*, 2433.

[11] B. Findeis, M. Schubart, C. Platzek, L. H. Gade, I. J. Scowen, M. McPartlin, *Chem. Commun.* **1996**, 219.

[12] H. Memmler, K. Walsh, L. H. Gade, J. W. Lauher, *Inorg. Chem.* **1995**, *34*, 4062.

Methoxy-bis[tris(trimethylsilyl)silyl]methane: The First Geminal Di(hypersilyl) Compound

E. Jeschke, T. Gross, H. Reinke, H. Oehme*
Fachbereich Chemie, Abteilung Anorganische Chemie
Universität Rostock
Buchbinderstr. 9, D-18051 Rostock, Germany
Telefax: Int. code + (381)4981763

Keywords: Hypersilyl compounds / Tris(trimethylsilyl)silyl derivatives / Polysilanes

Summary: Methoxy-bis[tris(trimethylsilyl)silyl]methane (**4**), the first geminal di(hypersilyl) compound with a central carbon atom, was prepared by the reaction of tris(trimethylsilyl)silyl lithium with dichloromethyl methyl ether. The structure of **4**, which is characterized by considerable distortions due to the spatial demand of the two $(Me_3Si)_3Si$ groups, is discussed on the basis of an X-ray crystal structure analysis.

The tris(trimethylsilyl)silyl group, the hypersilyl group, proved to be a function with unusual steric and electronic properties as well as a broad synthetic utility. The three trimethylsilyl substituents and the central silicon atom form an extended hemispherical shield providing the center, to which the $(Me_3Si)_3Si$ group is fixed, with exceptional steric protection. This was discussed in terms of the cone angles of the substituent (which was calculated to be 199° for the $C–Si(SiMe_3)_3$ group) [1] and was also demonstrated by structure analyses of, e.g., hexakis(trimethylsilyl)disilane $(Me_3Si)_3SiSi(Si Me_3)_3$ [2a, 2b, 3] and 1,4-bis[tris(trimethylsilyl)silyl]benzene $(Me_3Si)_3Si–C_6H_4–Si(SiMe_3)_3$ [2a, 2c].

Of special interest – particularly from a structural point of view – are geminal di(hypersilyl) compounds, since the spatial demand of the two extended hemispherical $(Me_3Si)_3Si$ groups is expected to cause tremendous steric distortions of the molecules. Some geminal di(hypersilyl) derivatives, mainly of transition metals and of heavier main group elements, are known. Confined to the elements of group 14 only very few compounds were described. For example $(Me_3Si)_3Si–SnCl_2–Si(SiMe_3)_3$ was made by the reaction of $SnCl_4$ with $(Me_3Si)_3SiLi$ and the Si–Sn–Si angle was found to be 142.5 °[4]. $(Me_3Si)_3Si–SiMe_2–Si(SiMe_3)_3$ was obtained by aluminum chloride-catalyzed rearrangement of permethylated polysilanes, but was not structurally characterized [5]. There are reports in the literature about unsuccessful attempts of the synthesis of bis[tris(trimethylsilyl)silyl] derivatives of germanium [4, 6] and silicon [4, 7]. Geminal di(hypersilyl) compounds with a central carbon atom are unknown so far.

Methoxy-bis[tris(trimethylsilyl)silyl]methane (**4**) – the first compound bearing two hypersilyl groups at a carbon atom – was synthesized by the reaction of tris(trimethylsilyl)silyllithium (**1**) with dichloromethyl methyl ether in a yield of 35 %. In view of the extreme bulkiness of the two hypersilyl substituents, the ease of the formation of **4** is really surprising. But the reaction pathway is easily understood as a consecutive replacement of the two chlorine atoms of the dichloromethyl methyl ether by the silanide **1** (Eq. 1).

$$\underset{\textbf{1}}{\overset{\displaystyle Me_3Si}{\underset{\displaystyle Me_3Si}{Me_3Si-Si-Li}}} \;+\; Cl_2CH-OCH_3 \quad\xrightarrow[-\,LiCl]{}\quad \underset{\textbf{2}}{\overset{\displaystyle Me_3Si}{\underset{\displaystyle Me_3Si\;\;Cl}{Me_3Si-Si-CH-OCH_3}}}$$

$$-Cl^- \Big\downarrow$$

$$\left[\;\underset{\textbf{3}}{\overset{\displaystyle Me_3Si\;\;\;OCH_3}{\underset{\displaystyle Me_3Si\;\;\;H}{Me_3Si-Si-C^{\oplus}}}}\;\right]$$

$$+\,\mathbf{1}\;\Big|\;-Li^+$$

$$\underset{\textbf{4}}{\overset{\displaystyle Me_3Si\;\;\;SiMe_3}{\underset{\displaystyle Me_3Si\;\;\;SiMe_3}{\overset{\displaystyle Me_3Si-Si}{\underset{\displaystyle Me_3Si-Si}{\diagdown CH-OCH_3\diagup}}}}}$$

Eq. 1.   Synthesis of methoxy-bis[tris(trimethylsilyl)silyl]methane (**4**).

Obviously, after replacement of the first chlorine atom of the dichloromethyl methyl ether by a hypersilyl group, the approach of the second lithium silanide **1** to the central sp^3 carbon atom of the intermediate **2** under conditions of a bimolecular nucleophilic substitution, i.e., with inversion of the configuration at the reaction center, is hardly conceivable due to the extreme steric shielding by the $(Me_3Si)_3Si$ group. Therefore, we suppose that the reaction proceeds via a carbenium ion transition state **3**, which is easily formed because it is stabilized by the methoxy group as well as the polysilanyl substituent. Thus, the silicon nucleophile **1** can attack the electrophilic carbon atom from the "front side", and steric problems caused by already fixed substituents become less significant.

The structure of **4** was elucidated on the basis of NMR and MS data and is in full agreement with the results of an X-ray crystal structure analysis. The crystals are considerably disordered. Thus, there are two possible configurations at C-1, which obviously occur approximately in a 1:1 ratio. As can be seen in Fig. 1, the two hemispherical shells of the $(Me_3Si)_3Si$ groups, separated by one sp^3 carbon atom as a spacer, are squeezed together at one side and open a narrow slit, where the methoxy group and the hydrogen atom are placed. The spatial demand of the two $(Me_3Si)_3Si$ groups forces an extreme widening of the Si1–C1–Si2 angle to a value of 132.7° and additionally also the central carbon-silicon bonds are significantly elongated (C1–Si1 1.94 Å; C1–Si2 1.95 Å). Consequently, the Si–Si–Si angles in the $(Me_3Si)_3Si$ proups are pressed together, the average angle being 106.0°. On the other hand, the average bond angles C1–Si1–Si and C1–Si2–Si of more than 112.5 ° are rather high, but the values of the individual angles differ remarkably. At the positions, where the two hemispheres contact each other, the angles are widened (C1–Si1–Si3 120.3°, C1–Si1–Si5 112.0°, C1–Si2–Si7 111.6°, C1–Si2–Si6 120.0°) and the whole shell is pushed aside with deformation of the originally tetrahedral configuration at the central silicon atom pressing the C1–Si1–Si4 angle (105.4°) and the C1–Si2–Si8 angle (105.8°) together. The methoxy group is fully enclosed by the (Me₃Si)₃Si hemispheres, filling the gap between these two shells. Thus, the intramolecular distances between the methoxy carbon atom and the next neighboring Me₃Si carbon atoms are extremely short. The distances C6–C20 (3.25 Å) and C17–C20 (3.30 Å) are approximately 19 % or 17.5 %, respectively, smaller than the sum of van der Waals radii of two methyl groups amounting to about 4.00 Å [8].

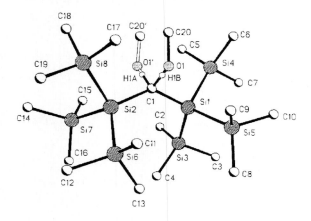

Fig. 1.   Molecular structure of **4** in the crystal, hydrogen atoms omitted (except H1A and H1B), selected bond lengths [Å] and angles [°]: C1–Si1 1.942(7), C1–Si2 1.954(7), C6–C20 3.25(3), C17–C20 3.30(3), Si1–C1–Si2 132.69(37), C1–Si1–Si3 120.31(25), C1–Si2–Si4 105.45(23), C1–Si1–Si5 111.98(25), C1–Si2–Si6 120.02(24), C1–Si2–Si7 111.64(26), C1–Si2–Si8 105.83(23), Si3–Si1–Si4 103.68(10), Si3–Si1–Si5 108.02(12), Si4–Si1–Si5 106.17(11), Si6–Si2–Si7 108.25(11), Si6–Si2–Si8 103.86(10), Si7–Si2–Si8 106.06(11), Si1–C1–O1 101.59(68), Si2–C1–O1 108.56(69), O1–C1–O1' 98.86(113).

The ¹H-NMR spectrum of **4** shows only three signals with the correct intensities of 54:3:1. Also the ¹³C-NMR spectrum consists of only three, the ²⁹Si NMR spectrum of two signals. Dynamic NMR studies up to -80°C did not change the signal patterns. The magnetic equivalence of all six trimethylsilyl groups is really surprising and may suggest the picture of a cogwheel like gearing of the two (Me₃Si)₃Si hemispheres.

Unfortunately, the extreme steric congestion gives rise to a remarkable chemical inertness of **4**. Intended transformations of **4** into carbanionic or carbenium ion transition states, which were expected to undergo further rearrangements, failed. Thus, after reaction with strong acids (e.g., HCl) or strong bases (e.g., organolithium compounds), **4** was always recovered unchanged.

Acknowledgement: We gratefully acknowledge the support of our research by the *Fonds der Chemischen Industrie*.

References:

[1]   M. Aggarwal, M. A. Ghuman, R. A. Geanangel, *Main Group Met. Chem.* **1991**, *14*, 263.
[2]   a) H. Bock, J. Meuret, K. Ruppert, *Angew. Chem.* **1993**, *105*, 413; *Angew. Chem., Int. Ed. Engl.* **1993**, *32*, 414.
       b) H. Bock, J. Meuret, K. Ruppert, *J. Organomet. Chem.* **1993**, *445*, 19.
       c) H. Bock, J. Meuret, R. Baur, K. Ruppert, *J. Organomet. Chem.* **1993**, *446*, 113.
[3]   A. Heine, R. Herbst-Irmer, G. M. Sheldrick, D. Stalke, *Inorg. Chem.* **1993**, *32*, 2694.
[4]   S. P. Mallela, R. A. Geanangel, *Inorg. Chem.* **1990**, *29*, 3525.
[5]   M. Ishikawa, J. Iyoda, H. Ikeda, K. Kotake, T. Hashimoto, M. Kumada, *J. Am. Chem. Soc.* **1981**, *103*, 4845.
[6]   S. P. Mallela, R. A. Geanangel, *Inorg. Chem.* **1991**, *30*, 1480; *Inorg. Chem.* **1994**, *33*, 1115.
[7]   Y. Derouiche, P. D. Lickiss, *J. Organomet. Chem.* **1991**, *407*, 41.
[8]   H. Bock, K. Ruppert, C. Näther, Z. Havlas, H.-F. Herrmann, C. Arad, I. Göbel, A. John, J. Meuret, S. Nick, A. Rauschenbach, W. Seitz, T. Vaupel, B. Solouki, *Angew. Chem.* **1992**, *104*, 564; *Angew. Chem., Int. Ed. Engl.* **1992**, *31*, 550.

The Use of the Tris(trimethylsilyl)silyl Group in Stabilization of Low Valent Gallium Compounds

Wolfgang Köstler, Gerald Linti*

Institut für Anorganische Chemie
Universität (TH) Karlsruhe
Engesserstr. Geb. 30.45, D-76128 Karlsruhe, Germany
Tel.: Int. code + (721)6082822 – Fax: Int. code + (721)6084854
E-mail: linti@achpc9.chemie.uni-karlsruhe.de

Keywords: Gallium Silyls / Digallanes / Gallium(I) compounds

Summary: The synthesis and structural characterization of hypersilyl-(=A) substituted digallanes A_4Ga_2 and $[A_2Ga_2Cl_2]_2$ and tetrahedral $[AGa]_4$ is described. In addition, ab initio calculations are used to understand stabilities of gallium(I)clusters.

Introduction

Well characterized examples of gallium species in oxidation state one (except halides [1]) are confined to organyl derivates [2]. Examples include monomeric, volatile GaCp* [3, 4], $(RGa)_3Na_2$ (R = 2,6-$Mes_2C_6H_3$) with a trigonal bipyramidal Ga_3Na_2 core [5] and $[GaC(SiMe_3)_3]_4$ with a tetrahedral core of gallium atoms [6]. This gallium cluster dissociates into monomers in solution and in the gas phase [7].

Compounds with two valent gallium and gallium gallium bonds have been known for several years as dioxane adducts of the digallium tetrahalides [8] and hexahalodigallates [9]. More recently, tetraorganyl-and tetra(amino) substituted digallanes $R_2Ga-GaR_2$ (R = $CH(SiMe_3)_2$ [10], 2,4,6-triisopropylphenyl [11], 2,4-bis(trimethylsilyl)-2,4-dicarba-*nido*-hexaborate(2⁻) [12], tmp [13]) have been investigated.

We will present the use of the tris(trimethylsilyl)silyl group (hypersilyl) [14] to stabilize low valent gallium species.

Reactions

In analogy to the preparation of $[GaC(SiMe_3)_3]_4$ [6], three equivalents of $(Me_3Si)_3SiLi(THF)_3 \cdot 0.5(Me_3Si)_4Si$ were combined with $Ga_2Cl_4 \cdot$ 2dioxane (Eq. 1), leading to an intensively violet colored solution. Colorless crystals of $[(Me_3Si)_3Si]_2GaCl_2Li(THF)_2$ [15] and excess $(Me_3Si)_4Si$ crystallized from this solution. From the concentrated mother liquor deeply violet plates of the gallium(I) silyl **1** precipitated.

$$Ga_2Cl_4 \cdot 2Dioxan \xrightarrow[\substack{-2\ LiCl \\ -\ A_2GaCl_2Li(thf)_2}]{+\ 3\ Li(THF)_3A} 0.25\ A—Ga \overset{\overset{\displaystyle A}{|}\underset{\displaystyle Ga}{|}}{\underset{\underset{\displaystyle A}{|}}{Ga}}Ga—A \quad (1)$$

1

$A = Si(SiMe_3)_3$

Eq. 1.

If only two equivalents of lithium tris(trimethylsilyl) silanide·3THF are combined with $Ga_2Cl_4 \cdot 2$dioxane (Eq. 2) colorless crystals of the dimeric bis(hypersilyl)dichloro digallane **2** are isolated. The latter **2** is also formed – in a non uniform reaction – if $GaGaCl_4$ is used instead of applying a 1:1 molar ratio.

$$2\ Ga_2Cl_4 \cdot 2dioxane + 4\ Li(THF)_3Si(SiMe_3)_3 \xrightarrow[\substack{-\ dioxane \\ -\ THF}]{2\ LiCl} \quad (2)$$

2

Eq. 2.

If two or more mole equivalents lithium tris(trimethylsilyl) silanide·3THF are used in the reaction with $GaGaCl_4$ (Eq. 3) **3** is isolated. Compound **3** is also formed as a byproduct in the synthesis of **1**.

$$Ga[GaCl_4] + 4\ Li(THF)_3Si(SiMe_3)_3 \xrightarrow[-\ THF]{-\ 4\ LiCl} \quad \underset{(Me_3Si)_3Si}{\overset{(Me_3Si)_3Si}{}}Ga—Ga\overset{Si(SiMe_3)_3}{\underset{Si(SiMe_3)_3}{}} \quad (3)$$

3

Eq. 3.

X-Ray Structure

The compound crystallizes in the tetragonal space group *P*4/*ncc*. Compound **1** consists of nearly ideal gallium tetrahedra, possessing crystallographic $\overline{4}$ symmetry (Fig. 1). Compound **1** crystallizes together with $(Me_3Si)_4Si$ molecules (1:1), which are highly symmetry disordered; a crystallographic fourfold symmetry is imposed on the T_d symmetric molecules. The gallium gallium distances in **1** average 258.4(2) pm and are nearly equal. This bond length is 6 – 10 pm longer than those known for tetra(organyl)digallanes [10]. This is understandable in terms of the ring strain in molecules of **1** and because there are missing four electrons necessary for a description of this cluster with 2e2c bonds. The gallium(I) organyl $\{GaC(SiMe_3)_3\}_4$ [6] also tetrahedral exhibits gallium-gallium bonds that are 10 pm longer [$d_{Ga–Ga} = 258.4(2)$ pm]. At first glance this is explained with steric arguments. The $(Me_3Si)_3C$ substituent is much more sterically demanding than its silicon analogue, the hypersilyl group. A gallium-carbon bond is approx. 40 pm shorter than a gallium-silicon bond. On the other hand the exchange of a carbon for a silicon atom will affect the electronic properties of these substituents.

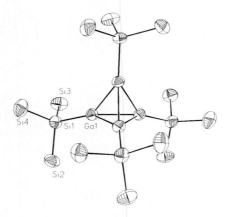

Fig. 1. View of a molecule of **1**, methyl groups have been omitted for clarity. Selected bond lengths [pm] and angles [°]: Ga(1)–Si(1) 240.6(2), Ga(1)–Ga(1)a 256.7(2), Ga(1)–Ga(1)b 258.7(2), Ga(1)–Ga(1)c 259.2(2); Si(2)Si(1)–Si(3) 110.3(2), Si(2)Si(1)Si(4) 110.6(2), Si(3)Si(1)Si(4) 111.2(2), Si(2)Si(1)Ga(1) 109.0(2), Si(3)Si(1)Ga(1) 107.5(1), Si(4)Si(1)Ga(1) 108.1(2).

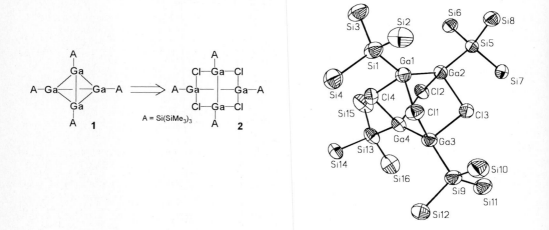

Fig. 2. View of a molecule of **2**, methyl groups have been omitted for clarity. Selected bond lengths [pm] and angles [°]: Ga(1)–Ga(2) 250.14(10), Ga(3)–Ga(4) 250.93(12), Ga–Cl 238.9(2) - 240.1(2), Ga–Si 239.0(2) - 240.2(2); Si–Ga–Ga 145.02(7) - 146.76(7), Ga–Cl–Ga 99.85(6) - 100.54(7), Cl–Ga–Cl 91.02(7) - 92.28(7), Cl–Ga–Ga 100.31(6) - 102.33(6), Cl–Ga–Si 100.79(9) - 103.71(7)

The analysis of the crystal structure (space group $Pca2_1$) of **2** (Fig. 2) shows a cage consisting of four gallium and four chlorine atoms, in which the four gallium atoms reside on the corners of a bisphenoid. Pairs of gallium atoms are bonded to one another [d_{Ga-Ga} = 250.5 pm (average)]. The other four edges of the bisphenoid are so long [d_{Ga-Ga} = 365 bis 369 pm], that no gallium-gallium interactions are found here. Just these four edges are bridged by four chlorine atoms, forming a square [d_{Cl-Cl} = 345 pm]. The nearly symmetrical GaClGa bridges [d_{Ga-Cl} = 240.1 pm (average)] are of normal length. This resembles the structure of regular As_4S_4. Caused by the planar arrangement of the Ga_2Si_2 units (torsional angle Si(1)Ga(1)–Ga(2)Si(5): 1.3°) this allows maximum release of steric strain between the hypersilyl groups. **2** can be deduced from tetrahedrally built **1** as product of a partial oxidation, during which the preformed structure of the gallium frame is mainly preserved.

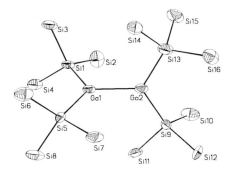

Fig. 3. View of a molecule of **3**, methyl groups have been omitted for clarity. Selected bond lengths [pm] and angles [°]: Ga(1)–Ga(2) 259.9(4); Si–Ga–Si 122.1.

Compound **3** (Fig. 3) has a gallium-gallium bond that is 9 pm longer [d_{Ga-Ga} = 259.9(4) pm], together with a nearly orthogonal arrangement of the Si_2Ga planes (interplanar angle: 80°). This is explained by the space filling hypersilyl groups.

Quantum Chemical Studies at Gallium Clusters Ga_4R_4

On the ab initio (SCF) level (Table 1) all compounds Ga_4R_4 under investigation are stable towards dissociation into monomers, if the tetramerisation energy (E_{tetra}), defined according to Eq. 4, is used as a scale for cluster stability,

$$4 \ GaR \ \rightarrow \ Ga_4R_4 \qquad E_{tetra} = E_{SCF}(Ga_4R_4) - 4 \bullet E_{SCF}(GaR)$$

Eq. 4.

Nevertheless there are remarkable differences. Hydrogen and organyl-substituted gallium(I) compounds have a tetramerisation energy that is approx. 100 kJ/mol smaller than that of silylsubstituted gallanes. It should be added that MP2 calculations making allowance for electron correlation predict an additional stabilization of the clusters towards dissociation of 300 kJ/mol (230 kJ in the case of NH_2) (Table 2). This is also true for gallium halide clusters [16]. With similar calculations on aluminium compounds Al_4R_4 (R = Cp, H, SiR_3, F, Cl) these additional cluster stabilization was found to be 200 kJ/mol [17].

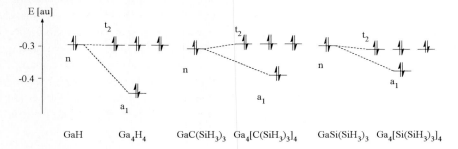

Fig. 4.   Frontier orbitals from the MO-scheme of Ga_4R_4.

Table 1.   Results of ab initio(SCF) calculations on $[RGa]_n$ (base set: DZP, bond lengths in [pm]).

R	n	Symmetry	d_{Ga-R}	d_{Ga-Ga}	E_{SCF} [au]	E_{tetra} [kJ/mol]
H	1	$C1$	167.2	-	-1923.743545	-
H	4	T_d	158.5	259.7	-7695.064181	-236
Me	1	$C3v$	205.0	-	-1962.787408	-
Me	4	T_d	200.4	261.0	-7851.234817	-224
CMe_3	1	$C3$	208.5	-	-2079.878160	-
CMe_3	4	T	204.2	261.7	-8319.599772	-229
SiH_3	1	$C3v$	262.6	-	-2213.829459	-
SiH_3	4	T_d	247.0	259.7	-8855.428542	-291
$SiMe_3$	1	$C3$	262.2	-	-2330.956776	-
$SiMe_3$	4	T	248.1	260.0	-9323.958800	-346
$C(SiH_3)_3$	1	$C3$	213.6	-	-2833.044232	-
$C(SiH_3)_3$	4	T	203.3	263.2	-11332.25152	-196
$Si(SiH_3)_3$	1	$C3$	262.7	-	-3084.086944	-
$Si(SiH_3)_3$	4	T	247.4	260.8	-12336.45971	-294
NH_2	1	$C2v$	186.3	-	-1978.814626	-
NH_2	4	$D2d$	184.6	266.6	-7915.285629	-71
				262.0		
NH_2	4	$D4h$	183.9	253.0	-7915.205789	+138

The trimethylsilyl group apparently is best able to stabilize clusters of all the silyl groups tested. Its tetramerisation energy is 50 kJ/mol larger than that one of SiH_3 and $Si(SiH_3)_3$ substituted compounds. On the other hand, the $C(SiH_3)_3$ group appears to cause the lowest tetramerisation energy of all organyl substituents in our series. Thus, our quantum chemical results can take account of the different stabilities of hypersilyl and tris(trimethylsilyl) substituted gallium clusters.

The calculated gallium-gallium distances in Ga_4R_4 are between 259.8 and 263.2 pm, whereby the silyl derivatives have the shortest bond lengths. The $Si(SiH_3)_3$ reproduces the experimental value for the hypersilyl derivative quite well. The longest gallium-gallium bonds are found in $[GaC(SiH_3)_3]_4$, but they are still 6 pm shorter than determined for $\{GaC(SiMe_3)_3\}_4$.

Table 2. Results of MP2-calculations on Ga_4R_4-clusters (bond lengths [pm]).

	d_{Ga-Ga}	d_{Ga-R}	E_{tetra}^{MP2} **[kJ/mol]**
Ga_4H_4	252.9	159.6	-556
Ga_4Me_4	255.5	198.4	-528
$Ga_4(SiH_3)_4$	254.5	242.3	-596
$Ga_4(NH_2)_4$	-	-	-300

Conclusion

We could show that the hypersilyl substituent is a very useful group in the synthesis of low valent gallium compounds. Not only the steric demand of this group but also its electronic properties, as theoretical studies have confirmed, contribute to its cluster stabilizing ability [18].

Acknowledgement: We thank the *Deutsche Forschungsgemeinschaft* for support.

References:

[1]   D. Loos, H. Schnöckel, D. Fenske, *Angew. Chem.* **1993**, *105*, 1124; *Angew. Chem., Int. Ed. Engl.* **1993**, *32*, 1059.

[2]   For a review article see: H. Schnöckel, C. Dohmeier, *Angew. Chem.* **1996**, *108*, 141; *Angew. Chem., Int. Ed. Engl.* **1996**, *35*, 129.

[3]   D. Loos, H. Schnöckel, *J. Organomet. Chem.* **1993**, *463*, 37.

[4]   A. Haaland, K.-G. Martinsen, H. V. Volden, D. Loos, H. Schnöckel, *Acta Chem. Scand.* **1994**, *48*, 172.

[5]   X. W. Li, W. T. Pennington, G. H. Robinson, *J. Am. Chem. Soc.* **1995**, *117*, 7578.

[6]   W. Uhl, W. Hiller, M. Layh, W. Schwarz, *Angew. Chem.* **1992**, *104*, 1378; *Angew. Chem., Int. Ed. Engl.* **1992**, *31*, 1364.

[7]   A. Haaland, K.-G. Martinsen, H. V. Volden, W. Kaim, E. Waldhör, W. Uhl, U. Schütz, *Organometallics* **1996**, *15*, 1146.

[8]   J. C. Beamish, R. W. H. Small, I. J. Worrall, *Inorg. Chem.* **1979**, *18*, 220.

[9]   K. L. Brown, D. Hall, *J. Chem. Soc. Dalton* **1973**, 1843.

[10]   W. Uhl, *Angew. Chem.* **1993**, *105*, 1449; *Angew. Chem., Int. Ed. Engl.* **1993**, *32*, 1386.

[11] X. He, R. A. Bartlett, M. M. Olmstead, K. Ruhlandt-Senge, B. E. Sturgeon, P. P. Power, *Angew. Chem.* **1993**, *105*, 761; *Angew. Chem., Int. Ed. Engl.* **1993**, *32*, 717.

[12] A. K. Saxena, H. Zhang, J. A. Maguire, N. S. Hosmane, A. H. Cowley, *Angew. Chem.* **1995**, *107*, 378; *Angew. Chem., Int. Ed. Engl.* **1995**, *34*, 332.

[13] G. Linti, R. Frey, M. Schmidt, *Z. Naturforsch.* **1994**, *49b*, 958.

[14] H. Gilman, C. L. Smith, *J. Organomet. Chem.* **1968**, *14*, 91.

[15] A. M. Arif, A. H. Cowley, T. M. Elkins, R. A. Jones, *J. Chem. Soc., Chem. Commun.* **1986**, 1776.

[16] U. Schneider, Dissertation, Universität Karlsruhe, **1994**.

[17] R. Ahlrichs, M. Ehrig, H. Horn, *Chem. Phys. Lett.* **1991**, *183*, 227; and Lit. [16].

[18] for a more detailed information, see: a) G. Linti, *J. Organomet. Chem.* **1996**, *520*, 107; b) G. Linti, W. Köstler, *Angew. Chem.* **1996**, *108*, 593; *Angew. Chem. Int. Ed. Engl.* **1996**, *35*, 550.

Synthesis, Structure, and Reactions of Tris(trimethylsilyl)silyl Gallanes and Gallates

Horst Urban, Ronald Frey, Gerald Linti*
Institut für Anorganische Chemie
Universität (TH) Karlsruhe
Engesserstr. Geb. 30.45, D-76128 Karlsruhe, Germany
Tel.: Int. code + (721)6082822 – Fax: Int. code + (721)6084854
E-mail: linti@achpc9.chemie.uni-karlsruhe.de

Keywords: Hypersilyl Gallanes / Hypersilyl Gallates

Summary: The synthesis and some reactions of $(Me_3Si)_3SiGaCl_2 \cdot THF$ are described. Cleavage of gallium-nitrogen bonds in $tmp_2GaSi(SiMe_3)_3$ affords the dimeric diethoxyhypersilylgallane. The gallium-silicon bond length is apparently very sensitive to steric and electronic influences.

Introduction

Organogallium compounds have been objects of intensive investigations. In contrast, the chemistry of compounds with bonds between gallium and silicon has not gathered much attention. *Rösch* et al. [1] were the first to synthesize tris(trimethylsilyl)gallane **1**, a compound with a gallium-silicon bond. Compounds **2** and **3** [2, 3] are the only structurally characterized compounds with gallium-silicon bonds so far. This work describes the synthesis, structure and reactions of new silyl gallanes and gallates.

Reactions

The reaction of lithium tris(trimethylsilyl)silanide·3THF [6] with an excess of gallium(III)chloride affords **4** in good yields (Eq. 1).

$$2\ GaCl_3\ +\ Li(thf)_3Si(SiMe_3)_3\ \xrightarrow[\substack{-\ LiGaCl_4 \\ -\ 2\ THF}]{Et_2O}\ \begin{array}{c} Cl \\ \diagdown \\ Ga \\ \diagup \\ Cl \end{array}\!\!\!\!\overset{\substack{thf \\ |}}{\underset{}{}}\!\!\!\!Si\!\!\begin{array}{c} SiMe_3 \\ -SiMe_3 \\ SiMe_3 \end{array}$$
4

Eq. 1.

If only a 1:1 or a 1:2 stoichiometry is applied the main product is **2**. Compound **4** is also formed as the only isolated product by addition of lithium tris(trimethylsilyl)silanide·3THF to a solution of gallium(I, III) chloride in toluene (Eq. 2).

$$GaGaCl_4\ +\ Li(thf)_3Si(SiMe_3)_3\ \xrightarrow[\substack{-\ Ga,\ -\ thf \\ -\ LiX\ (X\ =\ Cl,\ GaCl_4)}]{toluene}\ \mathbf{4}$$

Eq. 2.

With humid 2,6-dimethylpyridine it forms the hydrolysis product **5** (Eq. 3), and in a similar pathway it adds 2,6-diisopropylanilinium hydrochloride to yield **6** (Eq. 4).

$$\mathbf{4}\ +\ NC_6H_3(2,6\text{-}Me)_2\ +\ H_2O\ \xrightarrow[(Me_3Si)_3Si]{pentane}\ 0.5\ \begin{array}{c} Cl \\ \diagdown \\ Ga \end{array}\!\!\!\overset{\substack{H \\ | \\ O}}{\underset{\substack{O \\ | \\ H}}{}}\!\!\!\begin{array}{c} Si(SiMe_3)_3 \\ \diagup \\ Ga \\ \diagdown \\ Cl \end{array}$$
$$\cdot\ 2\ NC_6H_3(2,6\text{-}Me)_2\quad \mathbf{5}$$

Eq. 3.

$$\mathbf{4}\ +\ [H_3NC_6H_3(2,6\text{-}iPr)_2]^+Cl^-\ \longrightarrow\ \begin{array}{c}[(Me_3Si)_3SiGaCl_3]^- \\ [H_3NC_6H_3(2,6\text{-}iPr)_2]^+ \\ \mathbf{6}\end{array}$$

Eq. 4.

Cleavage of the gallium-nitrogen bonds with hydrogen chloride affords a similar trichlorohypersilylgallate [tmpH$_2$][Cl$_3$GaSi(SiMe$_3$)$_3$]. With ethanol the dimeric diethoxygallane **7** (Eq. 5) is formed. In both cases no cleavage of the gallium-silicon bond was observed.

$$2\ \mathbf{3}\ +\ 4\ EtOH\ \xrightarrow[-\ 4\ tmpH]{pentane}\ \begin{array}{c} EtO \\ \diagdown \\ (Me_3Si)_3Si \end{array}\!\!\!\overset{\substack{Et \\ | \\ O}}{\underset{\substack{O \\ | \\ Et}}{}}\!\!\!\begin{array}{c} Si(SiMe_3)_3 \\ \diagup \\ OEt \end{array}$$
7

Eq. 5.

Crystal Structure Analysis

Compound **4** crystallizes orthorhombic, space group *Pccn*. This compound has a distorted tetrahedrally coordinated gallium atom (Fig. 1). The gallium-oxygen distance [$d_{\text{Ga-O}} = 201.0(3)$ pm] in **7** is 10 pm shorter than in an analogous compound with two methyl groups instead of chlorine. The gallium-silicon bond length [$d_{\text{Ga-Si}} = 236.2(1)$ pm] is shortened by 5 pm compared to that in the above mentioned compound and is 10 pm shorter than in **3** [3]. The gallium-chlorine bonds are comparable to terminal ones in organogalliumhalides.

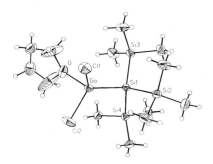

Fig. 1.   ORTEP plot of **4**; hydrogen atoms omitted for clarity; $d_{\text{Ga-Si}} = 236.2(1)$ pm.

Compound **5** crystallizes as a centrosymmetric dimer with bridging hydroxy groups in the triclinic system, space group $P\overline{1}$. The diamond-shaped Ga_2O_2 ring has inner ring angles of approx. 80° at the gallium atoms and approx. 100° at the oxygen atoms (Fig. 2). The gallium-silicon bond length ($d_{\text{Ga-Si}} = 238.4(3)$ pm) is comparable to that in **4**. The silicon-silicon bond lengths in both compounds are nearly equal. The gallium-chlorine bond is longer than usual for terminal gallium-chlorine bonds in dimeric gallium compounds [4-6].

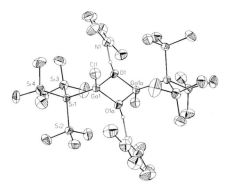

Fig. 2.   ORTEP plot of **5**; hydrogen atoms (except hydroxylic ones) omitted for clarity; $d_{\text{Ga-Si}} = 238.4(3)$ pm.

Compound **6** crystallizes in the monoclinic space group $P2_1/n$. This structure consists of hydrogen bond linked quartupels of two gallate and two ammonium ions (Fig. 3). The gallium-silicon bond [$d_{\text{Ga-Si}} = 235.8(1)$ pm] is comparable to that in **4** and **5**.

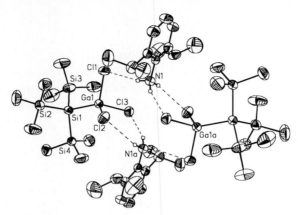

Fig. 3. ORTEP plot of **6**; hydrogen atoms (except NH) omitted for clarity; d_{Ga-Si} = 235.8(1) pm.

Compound **7**, crystallizing in the monoclinic crystal system, space group $P2_1/c$, is a centrosymmetric dimer. One of the ethoxy groups at the gallium atom occupies a bridging position (Fig. 4). The planar Ga_2O_2 ring has acute angles at the gallium atoms and wide ones at the oxygen atoms. The gallium-oxygen distances in the Ga_2O_2 four-membered ring are in the typical range [d_{Ga-O} = 193.9 pm (average)] found for other dimeric gallium ethoxides [5, 7]. The gallium-silicon bond in **7** is shorter than in **3**, but comparable to those in the compounds described above. This is understandable in terms of contraction of the effective covalent radius of gallium by electronegative bonding partners (chlorine, oxygen).

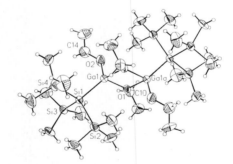

Fig. 4. ORTEP plot of **7**; hydrogen atoms omitted for clarity; d_{Ga-Si} = 238.8(2) pm.

Quantum-Chemical Results

X-Ray crystallographic studies indicate a range of 10 pm for the gallium-silicon bonds between 236.2 and 247.8 pm. Table 1 summarizes the results of ab initio calculations on the SCF level for model compounds $(H_2N)_2GaSiH_3$, $R_2GaSi(SiH_3)_3$ and $R_2Ga(OH_2)Si(SiH_3)_3$ (R = NH_2, Me, Cl). For molecules with a tricoordinated gallium atom the gallium-silicon bond lengths vary from 239.7 to 245.7 pm. Here more polar gallium-R bonds are parallel to shorter gallium-silicon bonds. A shorter gallium-silicon bond for **4** compared with the analogous dimethyl compound was predicted by calculations, but the absolute calculated values are larger than the experimental ones. Compound **3** has the longest observed gallium-silicon bond, in contrast to $(H_2N)_2GaSi(SiH_3)_3$, for which a gallium-silicon bond length shorter than in the methyl derivative was predicted. This reflects the influence of steric factors on the gallium-silicon bond length. The same conclusion may be drawn from simple MM+ force-field calculations. Thus, for **3** a gallium-silicon bond length of 246 pm was calculated, for $(H_2N)_2GaSi(SiMe_3)_3$ only 243 pm. By similar calculations on **4** and its dimethyl analogue, gallium-silicon distances of 237 and 238 pm were found. These values are in good accordance with the X-ray data. The steric demand can influence the gallium-silicon bond to a great extent, because this bond seems to be "elastic". The change in energy for $(H_2N)_2GaSi(SiH_3)_3$, as derived from single-point ab initio calculations, is only 2 kJ/mol if the gallium-silicon bond is elongated between 236 and 247 pm. This is the known range for gallium-silicon bonds. Obviously, very small changes in the bulkiness or electronegativity of the substituents results in large changes in the gallium-silicon distances.

Table 1. Bond lengths [pm] as results of ab initio (SCF) calculations on silyl gallium compounds.

Compound	d_{Ga-Si}	d_{Ga-X}	d_{Ga-O}	d_{Si-Si}
$(H_2N)Ga–SiH_3$	242.9	179.5		
$(H_2N)Ga–Si(SiH_3)_3$	242.2	179.5		235.1
$Cl_2Ga–Si(SiH_3)_3$	239.7	220.0		235.3
$Cl_2Ga(OH_2)–Si(SiH_3)_3$	239.9	223.8	202.6	235.0
$Me_2Ga–Si(SiH_3)_3$	245.7	198.2		235.0
$Me_2Ga(OH_2)–Si(SiH_3)_3$	248.1	199.0	212.6	235.0

Conclusions

From X-ray crystallographic data and quantum-chemical calculations it was concluded that the gallium-silicon bond is sensitive to electronic and steric influences. Compound **4** will be a versatile starting material for the synthesis of new compounds containing gallium-silicon bonds.

Acknowledgements: We thank Mrs. *A. Appel* (Univ. of Munich) and Mrs. *E. Möllhausen* (Univ. of Karlsruhe) for their assistance in X-ray structure determination. Financial support by the *Deutsche Forschungsgemeinschaft*, the *Fonds der Chemischen Industrie* and *Chemetall GmbH* is gratefully acknowledged.

References:

[1]   L. Rösch, H. Neumann, *Angew. Chem.* **1980**, *92*, 62; *Angew. Chem., Int. Ed. Engl.* **1980**, *19*, 55.

[2]   A. M. Arif, A. H. Cowley, T. M. Elkins, R. A. Jones, *J. Chem. Soc., Chem. Commun.* **1986**, 1776.

[3]   R. Frey, G. Linti, K. Polborn, *Chem. Ber.* **1994**, *127*, 101.

[4]   M. A. Petrie, P. P. Power, H. V. R. Dias, K. Ruhlandt-Senge, K. M. Waggoner, R. J. Wehmschulte, *Organometallics* **1993**, *12*, 1086.

[5]   D. A. Atwood, A. H. Cowley, R. A. Jones, M. A. Mardones, J. L. Atwood, S. G. Bott, *J. Coord. Chem.* **1992**, *26*, 285.

[6]   D. A. Atwood, A. H. Cowley, R. A. Jones, M. Mardones, J. L. Atwood, S. G. Bott, *J. Coord. Chem.* **1992**, *25*, 233.

[7]   A. H. Cowley, S. K. Mehratra, J. L. Atwood, W. Hunter, *Organometallics* **1985**, *4*, 1115.

Novel Pathways in the Reactions of Vinylsilanes with Lithium Metal [1]

Adalbert Maercker*, Kerstin Reider, Ulrich Girreser
Institut für Organische Chemie
Universität Siegen
D-57068 Siegen, Germany
Tel.: Int. code + 271/7404356

Keywords: Dilithiovinylsilanes / Lithium Hydride Elimination / Reductive Metalation / Solvent Effect / Vinyllithium Dimerization

Summary: Vinylsilanes are known to react with lithium metal either to 1,2-dilithioethanes by reduction or to 1,4-dilithiobutanes by reductive dimerization. The reaction of substituted vinylsilanes with lithium metal is employed in the approach to vicinal and geminal dilithiated vinylsilanes by two consecutive additions of lithium metal and subsequent eliminations of lithium hydride. A mechanistic investigation in the reactivity of α- and β-substituted vinylsilanes towards lithium metal discloses several new reaction pathways, whereby the choice of solvent plays an important role; in apolar solvents like toluene vinyllithium compounds are obtained. Compound **14**, R = Ph, which is not stable under the reaction conditions, finally affords the 1,4-dilithium compound **27**. Compound **18**, R = SiMe$_3$, on the other hand either adds to the starting vinylsilane (forming the monolithium compound **39**) or shows an unusual dimerization to **47**, which is studied in detail.

Introduction

Vinylsilanes as mono- and geminal disilyl-substituted C=C-double bonds like **1** [2] or **3** [3] afford, when brought to reaction with lithium metal in THF, the products of reductive dimerization, i.e., the 1,4-dilithiobutanes **2** and **4**. This type of reaction is known as Schlenk dimerization [4]. Symmetrically tetrasilyl-substituted C=C-double bonds as in **5** on the other hand add lithium metal with formation of 1,2-dilithioethanes (Scheme 1); as stable intermediates in these reactions radical anions, like **7**, can be observed, which are then reduced once again, here to the dianion **6** [5]. These two types of reaction are analogous to the reductions of the corresponding styrene derivatives 1,1-diphenylethylene [6] and stilbene [7]. Obviously, silyl groups have the same ability to stabilize negative charges in these polyanions as have aromatic substituents. The comparison holds even true for the structures observed for either 1,2-dilithio-1,1,2,2-tetrakis(trimethylsilyl)ethane **6** and 1,2-dilithio-1,2-diphenylethane **8**, both show a double lithium bridge in *trans* configuration, interacting with the solvent (Fig. 1) [8, 9].

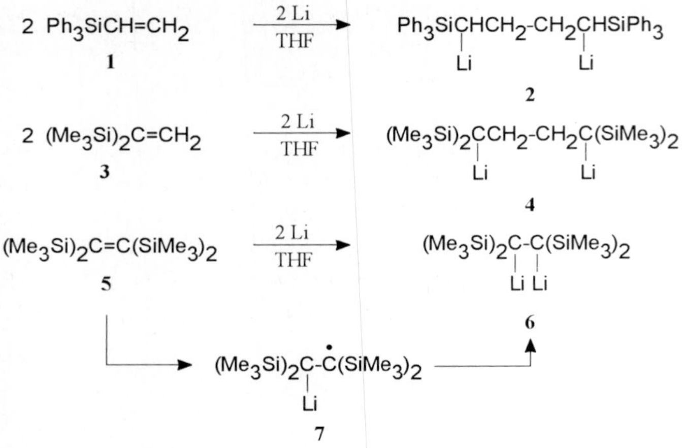

$$2 \ Ph_3SiCH=CH_2 \xrightarrow[\text{THF}]{2 \ Li} \underset{\underset{Li}{|}}{Ph_3SiCHCH_2}-\underset{\underset{Li}{|}}{CH_2CHSiPh_3}$$

1   **2**

$$2 \ (Me_3Si)_2C=CH_2 \xrightarrow[\text{THF}]{2 \ Li} \underset{\underset{Li}{|}}{(Me_3Si)_2CCH_2}-\underset{\underset{Li}{|}}{CH_2C(SiMe_3)_2}$$

3   **4**

$$(Me_3Si)_2C=C(SiMe_3)_2 \xrightarrow[\text{THF}]{2 \ Li} \underset{\underset{Li \ Li}{|\ \ |}}{(Me_3Si)_2C-C(SiMe_3)_2}$$

5   **6**

$$\longrightarrow \underset{\underset{Li}{|}}{(Me_3Si)_2C-\overset{\bullet}{C}(SiMe_3)_2} \longrightarrow$$

7

Scheme 1.   Reported reductions of vinylsilanes.

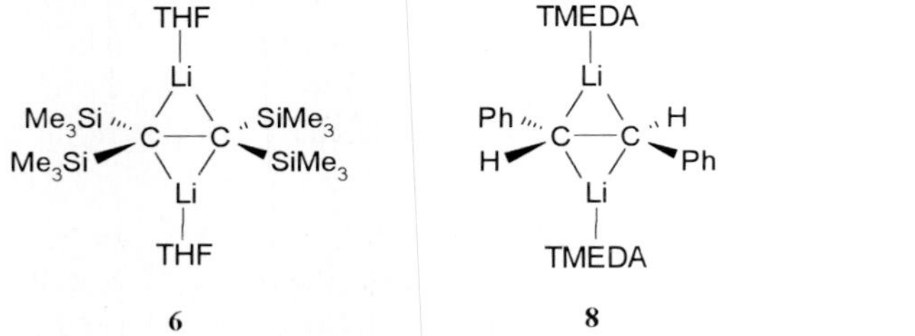

Fig. 1.   Structures of **6** and **8** in the solid state (TMEDA = N,N,N',N'-tetramethylethylene diamine).

Recently, Khotimskii and coworkers reported a similar Schlenk dimerization of trimethylvinylsilane with lithium metal in THF [10]. Interestingly, when employing hexane as the solvent a 1:1 mixture of the corresponding vinyllithium compound **11** and the acetylide **12** were formed (Scheme 2), besides lithium hydride. For this reaction no mechanistic explanation was given, however, the result resembles the reactivity of ethylene [11-13] and alkyl-substituted alkenes [14] towards lithium metal, the reaction pathway is mainly determined by the solvent employed: ethylene forms quantitatively vinyllithium when using special catalysts as shown by Rautenstrauch [11] and Bogdanovic [12], when using dioxane or DME (1,2-dimethoxyethane) as the solvent dilithioacetylide (lithium carbide) is formed as the final product [13].

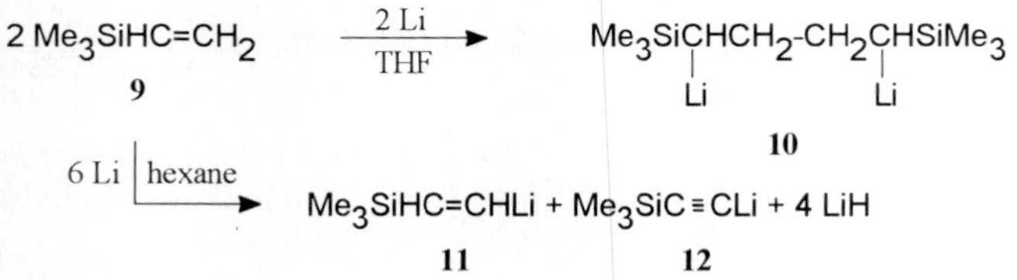

$$2 \ Me_3SiHC=CH_2 \xrightarrow[\text{THF}]{2 \ Li} \underset{\underset{Li}{|}}{Me_3SiCHCH_2}-\underset{\underset{Li}{|}}{CH_2CHSiMe_3}$$

9   **10**

$$\xrightarrow[\text{hexane}]{6 \ Li} Me_3SiHC=CHLi + Me_3SiC\equiv CLi + 4 \ LiH$$

11   **12**

Scheme 2.   Khotimskii´s results in the reaction of trimethlvinylsilane **9** with lithium metal.

In all these reactions polylithiumorganic intermediates have to be postulated. We have shown for 1,1-dilithioethanes, as well as *cis-* and *trans-*dilithioethylene, the ease of lithium hydride elimination through independent synthesis of these reactive intermediates [15, 16]; 1,2-dilithioethane can be trapped in a small amount at -120 °C [17].

So we propose a general mechanism (Scheme 3) for the reaction of vinylsilanes with lithium metal, which should also allow a general access to vicinal and geminal dilithiovinylsilanes by repetitive addition of lithium metal to the C=C-double bond and subsequent elimination of lithium hydride. In order to explore this synthetic approach the reduction of a series of either α- or β-substituted vinylsilanes with lithium was examined, here the substituent R ($\neq$ H) in **15** and **19** is introduced to prevent the last lithium hydride elimination.

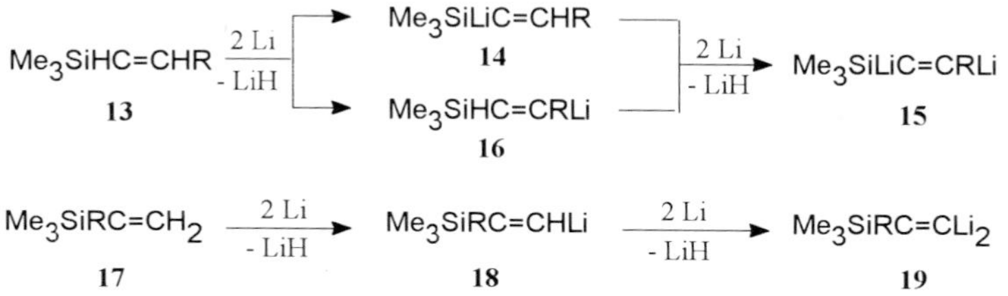

Scheme 3.   Dilithiovinylsilanes **15** and **19** from vinylsilanes by repetitive lithium addition and lithium hydride elimination.

It has to be mentioned that 2,2-diphenylvinyllithium, which corresponds to structure **18**, does not afford 1,1-dilithio-2,2-diphenylethene upon treatment with lithium [18]. Furthermore, two additional results are of interest when discussing the introduction of substituents into the β-position of the vinylsilane (Scheme 4): Seyferth et al. reported that (Z)-propenyltrimethylsilane, (Z)-**13**, R = Me, was isomerized quantitatively to the corresponding (E)-isomer through the intermediate radical anion **20** when brought into contact with a catalytic amount of lithium metal in THF [19], the alkyl substituent deactivates the double bond, no further reduction was observed. Eisch and Gupta, on the other hand, showed that (E)-styryltrimethylsilane, (E)-**13**, R = Ph, afforded the expected products of reduction and of reductive dimerization, **21** and **22** (*meso* compound), respectively [20].

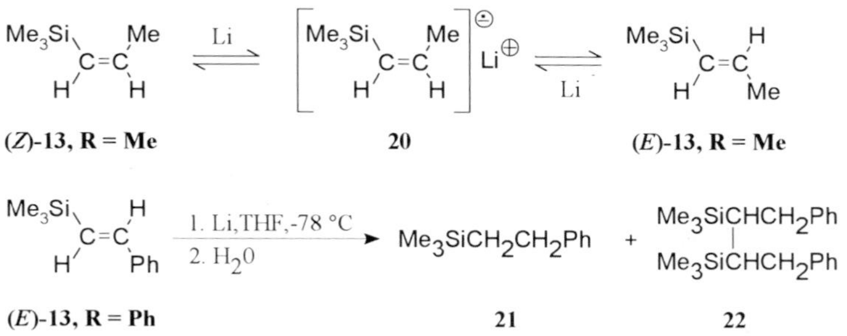

Scheme 4.   Known reactions of β-substituted vinylsilanes with lithium metal.

Results and Discussion

Reaction of β-Substituted Vinylsilanes with Lithium Metal

We started our investigation with the reduction of (Z)-styryltrimethylsilane, (Z)-**13**, R = Ph [21]. When brought to reaction with lithium metal in diethyl ether the usual product of reduction **23** is found, trapped as the dimethyl derivative **24** after work-up with dimethyl sulfate. **24** is isolated in 87 % yield as a 1:3 mixture of *erythro* and *threo* compounds. On the other hand, upon heating (Z)-**13**, R = Ph, for 10 hours in toluene **28** is obtained derived from the 1,4-dilithium intermediate **27** (Scheme 5).

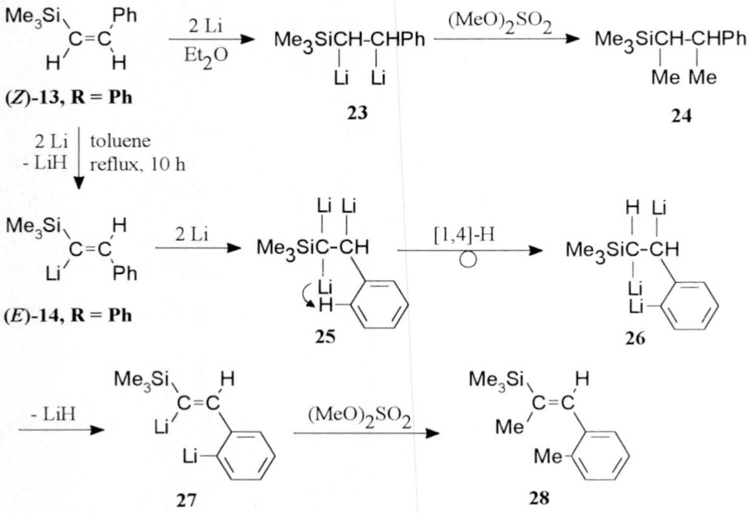

Scheme 5.   Reaction of (Z)-**13**, R = Ph, with lithium metal in either diethyl ether or toluene.

Thus, besides two additions of lithium and two consecutive lithium hydride eliminations a [1,4]-proton shift has occurred. In order to prove the proposed reaction mechanism all three intermediates, i.e., the vinyllithium compound (E)-**14**, R = Ph, and the trilithium compounds **25** and **26**, were synthesized independently.

The formation of **27**, by the way, as the final product in this reaction cascade is only surprising at first sight. *o*-Lithiostyryllithium is easily obtained by a lithium-tellurium exchange reaction (transmetalation) from 1-benzotellurole [22]. We have synthesized a number of substituted *o*-lithiostyryllithium derivatives (Fig. 2) alternatively by cleavage of methylenecyclopropanes and -butanes, which rearrange via an intermolecular proton shift [23, 24]. The *ortho* position of the aromatic ring is activated by agostic interactions with the lithium in the vinylic position, as driving force the stable doubly bridged structure has to be anticipated, which was shown by calculations of Schleyer et al. for the monomeric species [25]. In solution a dimeric species is present, according to NMR spectroscopic investigations of Günther and coworkers [26].

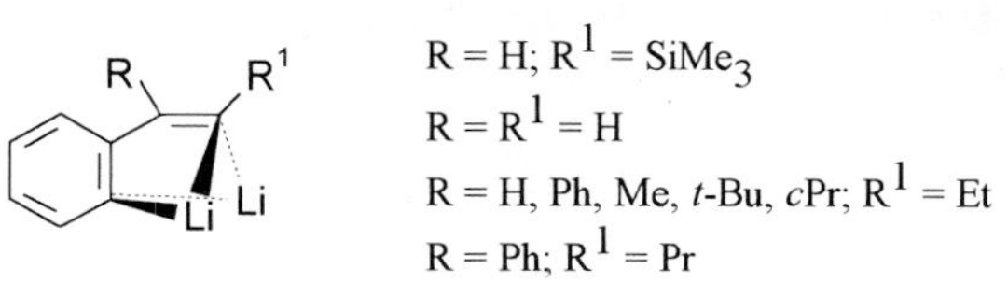

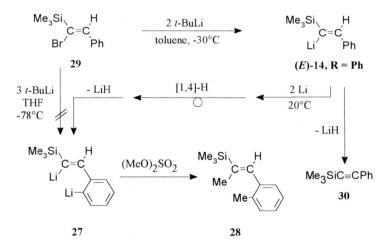

Fig. 2. Structure of **27** and other 1,4-dilithiostyrenes as monomeric and dimeric species.

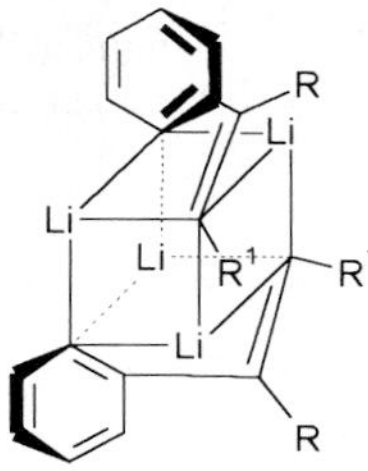

The synthesis of the vinyllithium derivative (*E*)-**14**, R = Ph is straightforward: bromine-lithium exchange starting from the vinyl bromide **29** [27] with *t*-butyllithium in toluene according to Seyferth et al. [28] affords (*E*)-**14**, R = Ph which, upon treatment with lithium metal at room temperature and subsequent work-up with dimethyl sulfate affords a 63 % yield of **28**. In the less polar solvent hexane the same reaction can only be brought about by adding 5 % of THF (86 % of **28**). Interestingly, the vinyllithium compound itself is prone to lithium hydride elimination, thus 17 % of trimethyl-1-propynylsilane (**30**) (Scheme 6) is found when the reaction is performed in toluene, none is formed in the above mentioned hexane/THF mixture. A second metalation of (*E*)-**14**, R = Ph to **27** with an excess of *t*-butyllithium in THF at -78 °C is not possible, although *n*-butyllithium in diethyl ether works well at room temperature [29].

Scheme 6. Formation of *o*-lithiostyryllithium **27** starting from the vinyllithium derivative (*E*)-**14**, R = Ph.

When starting from (*E*)-1-*o*-bromophenyl-2-trimethylsilylethene **31** [30] the postulated trilithium compound **26** is obtained by addition of lithium metal to **32**; in this case diethoxymethane (DEM) [31] is the most suitable solvent. Compound **26**, however, eliminates lithium hydride very easily, so only a small amount (7 %) of the trimethyl derivatives (*erythro* and *threo* **33**, 1:4) is obtained. As the main product **28** (66 % yield) is again observed, additionally 9 % of (*E*)-1-*o*-tolyl-2-trimethylsilylethene originating from **32** is found (Scheme 7).

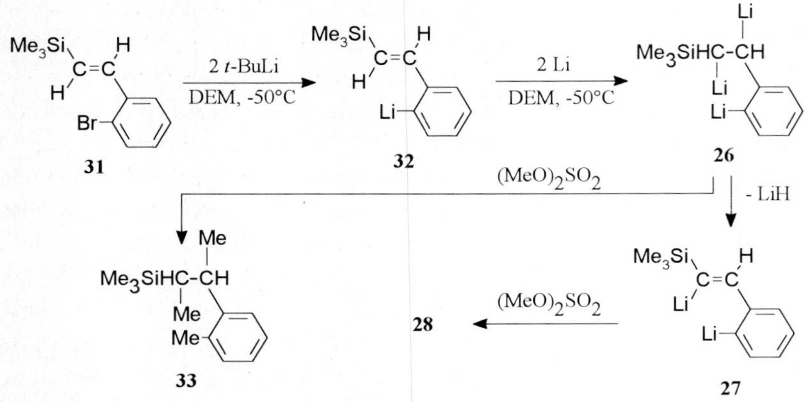

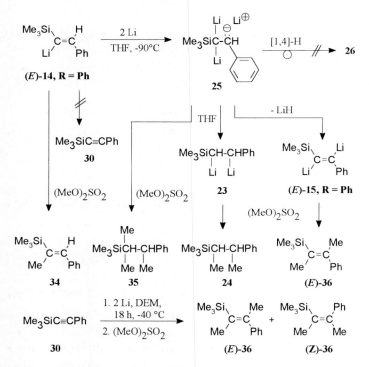

Scheme 7.   Approach to the trilithium intermediate **26** by addition of lithium metal to **32**.

Finally the intermediate **25** is synthesized by adding lithium metal to (*E*)-**14**, R = Ph, in the more polar solvent THF, here not only the [1,4]-proton shift is supressed, neither does **14** eliminate lithium hydride under the reaction conditions, probably a solvent separated ion pair has to be anticipated for the lithium in the benzylic position of **25** as the cause for this stability.

Scheme 8.   Synthesis of the trilithium intermediate **25** by addition of lithium metal to (*E*)-**14**, R = Ph, in THF.

When derivatizing the reaction mixture obtained by treatment of (*E*)-**14**, R = Ph, with lithium metal with dimethyl sulfate a mixture consisting of 22 % of **34**, 63 % of **24**, and only 3 % of the trimethyl derivative **35** is found, obviously the trilithium intermediate **25** reacts with the solvent with formation of **23**. When performing this reaction in perdeuterated THF, which is known to be less acidic and therefore more stable towards cleavage [32], the yield of **35** increases (22 %) while the yield of **34** (14 %) and **24** (44 %) decreases. Additionally in perdeuterated THF a small amount (1 %) of (*E*)-1,2-dimethyl-1-trimethylsilyl-2-phenylethene (*E*)-**36** is found, formed by lithium hydride elimination from **25**. That **15**, R = Ph, is indeed a stable intermediate can be shown by addition of lithium to trimethyl-1-propynylsilane (**30**), very slow addition of the alkyne to the lithium metal affords up to 20 % of (*E*)- and (*Z*)-**36** in a ratio of 1:4, besides mainly dimeric products, which are exclusively formed in THF or diethyl ether. This result is again in accordance with the corresponding addition of lithium metal to diphenylacetylene in diethyl ether [33]. The *Z*-isomer is the major product in this reduction, a doubly bridged structure is discussed as the stabilizing feature of the *cis*-dilithioalkene by calculations of Schleyer et al. [34]. In this case, however, isomerization during the work-up cannot be ruled out, considering the known inversion of α-silyl and α-phenylvinyllithium compounds [35, 36]. Vinyllithium itself [37] and 1,2-dilithioethylene [16] as well as 1,2-dilithioalkenes with aliphatic substituents [38] on the other hand are configurationally stable.

Reaction of α-Substituted Vinylsilanes with Lithium Metal

When introducing a phenyl substituent into the α-position of the vinylsilane, as in **17**, R = Ph, the reaction with lithium metal affords the usual product of Schlenk-dimerization **37**, which yields a mixture of *d*, *l*, and *meso* dimethyl derivatives **38** (92%). Even when heating α-trimethylsilylstyrene in either hexane or toluene no reaction to **18**, R = Ph, can be enforced.

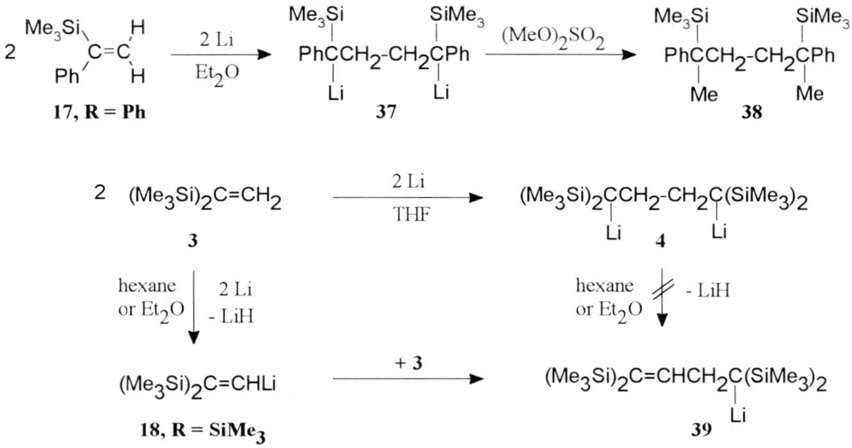

Scheme 9. Formation of the dimeric monolithium compound **39** upon reaction of **3** with lithium metal.

This is, however, possible, when interchanging the phenyl moiety with a second trimethylsilyl group. As stated above the reaction with lithium in THF is known [3]; again, switching to the less polar solvent diethyl ether or hexane proves successful for the synthesis of **18**,

R = SiMe₃.Interestingly, the vinyllithium compound is not stable under these conditions but adds to the starting material with formation of **39**, a dimeric monolithium derivative (Scheme 9). The formation of **39** by lithium hydride elimination from **4** could be excluded experimentally.

Introduction of an additional substituent in β-position would prevent **18** from dimerization. Therefore **40** [39] is brought to reaction with lithium metal whereupon reduction of the double bond to **41** occurs. No lithium hydride elimination takes place even in the apolar solvent mixture hexane with 3 % of diethyl ether. The vicinal dilithioalkane **41** affords the corresponding dimethyl derivative in 91 % yield when quenching with dimethyl sulfate. But this additional substituent in β-position can itself act as a leaving group, which is one of the modes of reaction when **42** [40] is treated with lithium in THF (less polar solvents do not work in this case). A second pathway is the rearrangement of a phenyl group, the so-called Grovenstein-Zimmerman rearrangement [41], thereby a 1,3-dilithium compound is formed, which affords the usual product upon hydrolysis (**45**, H instead of Li). When the reaction mixture is quenched with dimethyl sulfate, **45** is not trapped, but decomposes to **40** and the cumyllithium **48**, which demonstrates the instability of **46**, hydrolysis is obviously much faster in this case as is the derivatization with dimethyl sulfate. Very surprising in this reaction is the faith of 2,2-bis(trimethylsilyl)vinyllithium, **18**, R = SiMe₃, which does not react further with lithium, but dimerizes to **47**.

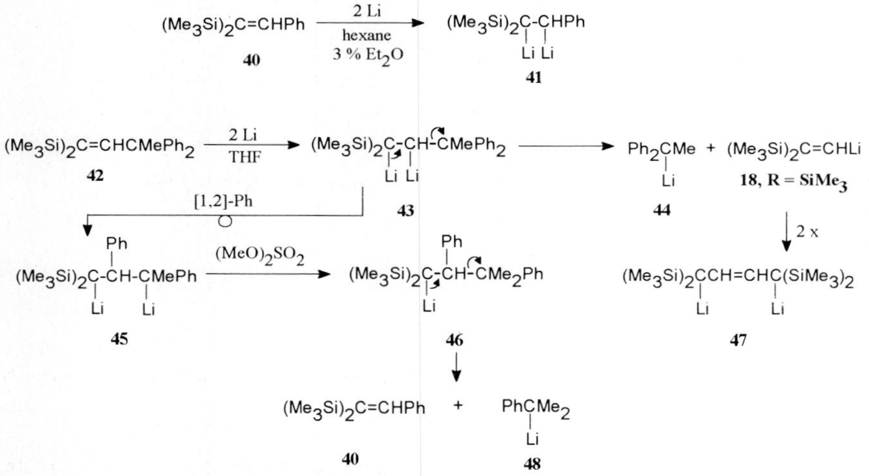

Scheme 10.   Reaction of 1,1-bis(trimethylsilyl)-3,3-diphenyl-1-butene **42** with lithium metal in THF.

In order to study this unusual dimerization in detail **49** [40] was synthesized and subjected to the same reaction conditions. The cumyl group inhibits the Grovenstein-Zimmerman rearrangement as in this case the formation of an anion at a tertiary center would be encountered, so only cleavage to the vinyllithium compound **18** is possible. Indeed, after reacting **49** with lithium metal for 4 h at room temperature and subsequent work-up with dimethyl sulfate the dimer **51** is isolated in 82 % yield, with *t*-butylbenzene (**53**) as the other product in 84 % yield (Scheme 11).

When quenching the reaction mixture already after 1.5 hours, 20 % of the starting vinylsilane **49** is still found, besides 11 % of **52** derived from the intermediate vinyllithium compound **18**, R = SiMe₃, and only 64 % of **51** and 61 % of **53**, respectively, are observed. This result clearly proves the proposed mechanism.

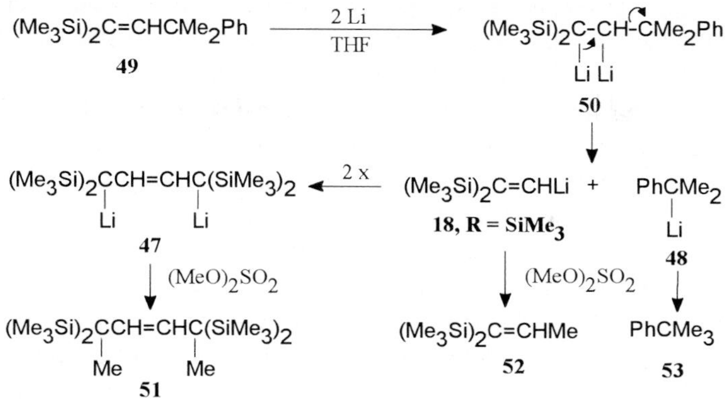

$$(Me_3Si)_2C=CHCMe_2Ph \xrightarrow[\text{THF}]{2\ Li} (Me_3Si)_2C-CH-CMe_2Ph$$

49

$$(Me_3Si)_2C-CH-CMe_2Ph$$
$$\underset{Li}{|}\ \underset{Li}{|}$$

50

$$(Me_3Si)_2CCH=CHC(SiMe_3)_2 \xleftarrow{2\ x} (Me_3Si)_2C=CHLi\ +\ PhCMe_2$$
$$\underset{Li}{|}\qquad \underset{Li}{|}$$

47

18, R = SiMe₃

$$PhCMe_2$$
$$\underset{Li}{|}$$

48

↓ (MeO)₂SO₂ ↓ (MeO)₂SO₂

$$(Me_3Si)_2CCH=CHC(SiMe_3)_2 \qquad (Me_3Si)_2C=CHMe \qquad PhCMe_3$$
$$\underset{Me}{|}\qquad \underset{Me}{|}$$

51 **52** **53**

Scheme 11. Reaction of 1,1-bis(trimethylsilyl)-3-methyl-3-phenyl-1-butene (**49**) with lithium metal in THF.

This interesting dimerization is without precedent in the literature. Recent results in our laboratory have shown that 2,2-diphenylvinyllithium dimerizes as well. This reaction must be anticipated as a true dimerization (no lithium is consumed) of a vinyllithium compound (Scheme 12), catalytic amounts of the metal are sufficient, the reaction is catalyzed by electron transfer [18].

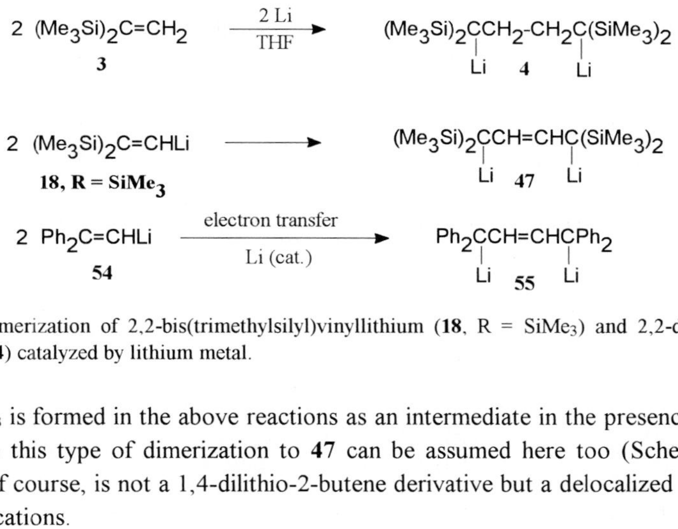

$$2\ (Me_3Si)_2C=CH_2 \xrightarrow[\text{THF}]{2\ Li} (Me_3Si)_2CCH_2-CH_2C(SiMe_3)_2$$
3 $$\underset{Li}{|}\quad \underset{4}{}\quad \underset{Li}{|}$$

$$2\ (Me_3Si)_2C=CHLi \longrightarrow (Me_3Si)_2CCH=CHC(SiMe_3)_2$$
18, R = SiMe₃ $$\underset{Li}{|}\ \underset{47}{}\ \underset{Li}{|}$$

$$2\ Ph_2C=CHLi \xrightarrow[\text{Li (cat.)}]{\text{electron transfer}} Ph_2CCH=CHCPh_2$$
54 $$\underset{Li}{|}\ \underset{55}{}\ \underset{Li}{|}$$

Scheme 12. Dimerization of 2,2-bis(trimethylsilyl)vinyllithium (**18**, R = SiMe₃) and 2,2-diphenylvinyllithium (**54**) catalyzed by lithium metal.

18, R = SiMe₃ is formed in the above reactions as an intermediate in the presence of an excess of lithium metal, so this type of dimerization to **47** can be assumed here too (Schemes 10 and 11). Compound **47**, of course, is not a 1,4-dilithio-2-butene derivative but a delocalized dianion with two solvated lithium cations.

Conclusions

The addition of lithium metal to vinylsilanes is a suitable approach to vinyllithium compounds as the addition of lithium metal is usually followed by lithium hydride elimination, especially when performing the reaction in less polar solvents like toluene. So far, the formation of vicinal or geminal dilithiovinylsilanes by two consecutive lithium addition and lithium hydride elimination sequences is observed only to a small amount, which is caused by two effects: On the one hand, the vinyllithium compounds obtained after one addition-elimination sequence are very reactive under the reaction conditions (addition to the vinylsilane employed as the starting material or dimerization in the presence of catalytic amounts of lithium metal). On the other hand, lithium hydride elimination in either the first or second addition-elimination sequence does not occur, other reaction pathways (Grovenstein-Zimmerman rearrangement or [1,4]-proton shift) are preferred.

Acknowledgement: We thank the *Volkswagen-Stiftung* and the *Fonds der Chemischen Industrie* for the support of this study. K. R. would like to thank the *Arbeitsgemeinschaft zur Förderung wissenschaftlicher Projekte an der Universität-GH Siegen (AFP)* for the support given.

References:

[1]   Preliminary communication: A. Maercker, K. Reider, in: *Organosilicon Chemistry: From Molecules to Materials* (Eds.: N. Auner, J. Weis), VCH, Weinheim, **1994**, p. 123; see also: K. Reider, *Zur Reaktionsweise von Vinylsilanen mit elementarem Lithium*, Shaker, Aachen, **1994**.

[2]   J. J. Eisch, R .J. Beuhler, *J. Org. Chem.* **1963**, *28*, 2876.

[3]   M. Kira, T. Hino, Y. Kubota, N. Matsuyama, H. Sakurai, *Tetrahedron Lett.* **1988**, *29*, 6939.

[4]   W. Schlenk, E. Bergmann, *Justus Liebigs Ann. Chem.* **1928**, *463*, 1; *Justus Liebigs Ann. Chem.* **1930**, *479*, 58, 78.

[5]   H. Sakurai, Y. Nakadaira, H. Tobita, *Chem. Lett.* **1982**, 771.

[6]   D. R. Weyenberg, L. H. Toporcer, A. E. Bey, *J. Org. Chem.* **1965**, *30*, 4096.

[7]   W. Schlenk, E. Bergmann, *Justus Liebigs Ann. Chem.* **1928**, *463*, 106.

[8]   A. Sekiguchi, T. Nakanishi, C. Kabuto, H. Sakurai, *J. Am. Chem. Soc.* **1989**, *111*, 3748.

[9]   M. Walczak, G. Stucky, *J. Am. Chem. Soc.* **1976**, *98*, 5531.

[10]   V. S. Khotimskii, I. S. Bryantseva, S. G. Durgar´yan, P. V. Petrovskii, *Izv. Akad. Nauk SSSR, Ser. Khim.* **1984**, 479; *Chem. Abstr.* **1984**, *100*, 209971.

[11]   V. Rautenstrauch, *Angew. Chem.* **1975**, *87*, 254; *Angew. Chem., Int. Ed. Engl.* **1975**, *14*, 259.

[12]   a) B. Bogdanovic, B. Wermeckes, *Angew. Chem.* **1981**, *93*, 691; *Angew. Chem., Int. Ed. Engl.* **1981**, *20*, 684.
       b) B. Bogdanovic, *Angew. Chem.* **1985**, *97*, 253; *Angew. Chem., Int. Ed. Engl.* **1985**, *25*, 262.

[13]   A. Maercker, B. Grebe, *J. Organomet. Chem.* **1987**, *334*, C21.

[14]   D. L. Skinner, D. J. Peterson, T. J. Logan, *J. Org. Chem.* **1967**, *32*, 105.

[15]   A. Maercker, M. Theis, A. J. Kos, P. v. R. Schleyer, *Angew. Chem.* **1983**, *95*, 755; *Angew. Chem., Int. Ed. Engl.* **1983**, *22*, 733.

[16]   A. Maercker, T. Graule, W. Demuth, *Angew. Chem.* **1987**, *99*, 1075; *Angew. Chem., Int. Ed. Engl.* **1987**, *26*, 1032.

[17]   N. J .R. van Eikemma Hommes, F. Bickelhaupt, G. W. Klumpp, *Angew. Chem.* **1988**, *100*, 1100; *Angew. Chem., Int. Ed. Engl.* **1988**, *27*, 1083.

[18]   a) A. Maercker, M. T. Hajgholipour, unpublished results.
       b) M. T. Hajgholipour, *Versuche zur Darstellung polylithiumorganischer Verbindungen durch Addition von Lithium an Vinyllithium-Verbindungen*, Shaker, Aachen, **1993**.

[19] D. Seyferth, R. Suzuki, L. G. Vaughan, *J. Am. Chem. Soc.* **1966**, *88*, 286.

[20] J. J. Eisch, G. Gupta, *J. Organomet. Chem.* **1979**, *168*, 139.

[21] R. B. Miller, G. McGarvey, *J. Org. Chem.* **1978**, *43*, 2739.

[22] A. Maercker, H. Bodenstedt, L. Brandsma, *Angew. Chem.* **1992**, *104*, 1387; *Angew. Chem., Int. Ed. Engl.* **1992**, *31*, 1339.

[23] A. Maercker, V. E. E. Daub, *Tetrahedron* **1994**, *50*, 2439.

[24] A. Maercker, K. D. Klein, *J. Organomet. Chem.* **1991**, *401*, C1.

[25] W. Bauer, U. Feigel, G. Müller, P. v. R. Schleyer, *J. Am. Chem. Soc.* **1988**, *110*, 6033.

[26] a) O. Eppers, H. Günther, K.-D. Klein, A. Maercker, *Magn. Reson. Chem.* **1991**, *29*, 1065.
b) O. Eppers, T. Fox, H. Günther, *Helv. Chim. Acta* **1992**, *75*, 883.
c) H. Günther, O. Eppers, H. Hausmann, D. Hüls, H.-E. Mons, K.-D. Klein, A. Maercker, *Helv. Chim. Acta* **1995**, *78*, 1913.

[27] Prepared according to: G. Zweifel, W. Lewis, *J. Org. Chem.* **1978**, *43*, 4424.

[28] D. Seyferth, J. L. Lefferts, R. L. Lambert, Jr., *Organomet. Chem.* **1977**, *142*, 39.

[29] J. Kurita, M. Ishii, S. Yasuike, T. Tsuchiya, *Chem. Pharm. Bull.* **1994**, *42*, 1437.

[30] Prepared according to: K. Karabelas, A. Hallberg, *J. Org. Chem.* **1986**, *51*, 5286.

[31] B. Venepalli, *Spec. Chem.* **1995**, *15*, 108.

[32] A. Maercker, W. Theysohn, *Justus Liebigs Ann. Chem.* **1971**, *747*, 70.

[33] G. Levin, J. Jagur-Grodzinski, M. Szwarc, *J. Am. Chem. Soc.* **1970**, *92*, 2268.

[34] Y. Apeloig, T. Clark, A. J. Kos, E. D. Jemmis, P. v. R. Schleyer, *Israel J. Chem.* **1980**, *20*, 43.

[35] R. Knorr, T. v. Roman, *Angew. Chem.* **1984**, *96*, 349; *Angew. Chem., Int. Ed. Engl.* **1984**, *23*, 366.

[36] R. Knorr, E. Lattke, *Tetrahedron Lett.* **1977**, 3969.

[37] D. Seyferth, L. C. Vaughan, *J. Am. Chem. Soc.* **1964**, *86*, 883.

[38] A. Maercker, U. Girreser, *Tetrahedron* **1994**, *50*, 8019.

[39] B.-T. Gröbel, D. Seebach, *Chem. Ber.* **1977**, *110*, 852.

[40] Prepared by Peterson olefination from tris(trimethylsilyl)methyllithium and the corresponding aldehyde, see ref. [39].

[41] a) E. Grovenstein, Jr., *Adv. Organomet. Chem.* **1977**, *16*, 167.
b) E. Grovenstein, Jr., *Angew. Chem.* **1978**, *90*, 317; *Angew. Chem. Int., Ed. Engl.* **1978**, *17*, 313.
c) A. Raja, L. M. Tolbert, *J. Am. Chem. Soc.* **1987**, *109*, 1782.

New Organosilicon Reagents: Synthesis, Structure, and Reactivity of (Lithiomethyl)(aminomethyl)silanes

B. C. Abele, C. Strohmann*

Institut für Anorganische Chemie
Universität des Saarlandes
Postfach 1150, D-66041 Saarbrücken, Germany
Tel.: Int. code + (0681)3022465
E-mail: c.strohmann@rz.uni-sb.de

Keywords: (Lithiomethyl)silanes / (Aminomethyl)silanes

Summary: (Aminomethyl)silanes have been prepared by new synthetic routes starting from lithiosilanes or (lithiomethyl)amines. (Lithiomethyl)(aminomethyl)silanes were formed by metallation with lithioalkyls. A sulfur-substituted (lithiomethyl)(aminomethyl)silane and a THF adduct are characterized by single crystal X-ray diffraction. In reactions of (lithiomethyl)(aminomethyl)silanes with aldehydes or ketones no clear addition/deprotonation profile was observed.

Introduction

(Lithiomethyl)silanes are important reagents in organic synthesis, for example, Peterson olefination [1]. Addition and deprotonation reactions with ketones and aldehydes have been reported. Klumpp et al. have studied the reactivity of (lithiomethyl)(aminomethyl)silanes (**A**) with respect to their aggregation in solution [2]. Addition should be observed with a higher degree of aggregation and deprotonation by a low degree of aggregation. Lithiated allylsilanes of the type **B** show high regioselectivity in addition reactions [3]. Chiral (lithiomethyl)silanes of the type **C** induce high diastereoselectivity in addition reactions [4]. Despite these interesting and important reactivity patterns no solid state structure of these lithiated (aminomethyl)silanes is available.

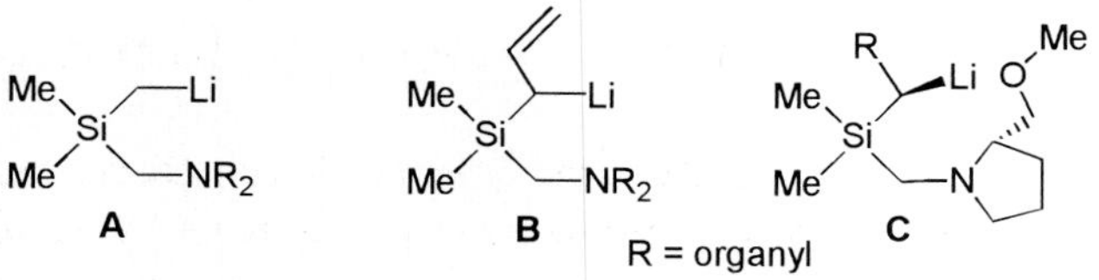

Scheme 1.

Unaware of the above mentioned work we explored, almost at the same time, the similar theme and encountered some very interesting and novel findings about the synthesis and structure of (lithiomethyl)(aminomethyl)silanes. We are presenting new synthetic routes to (aminomethyl)silanes and (lithiomethyl)(aminomethyl)silanes and we want to give experimental information on two questions concerning (lithiomethyl)silanes: (i) aggregation of (lithiomethyl)(aminomethyl)silanes in the solid state; (ii) stabilizing effects of silicon in (lithiomethyl)silanes.

Synthesis of (Aminomethyl)silanes and (Lithiomethyl)(aminomethyl)silanes

We have prepared (aminomethyl)silanes (**F**) by the known reaction (i) and three new (ii-iv) synthetic routes (Scheme 1):

(i)   Reaction of amines with (chloromethyl)silanes (**D**) [5];
(ii)   Reaction of lithiosilanes with iminium salts (**G**);
(iii) and (iv)   Reaction of (lithiomethyl)amines (**I**) with chlorosilanes (**H**) [6].

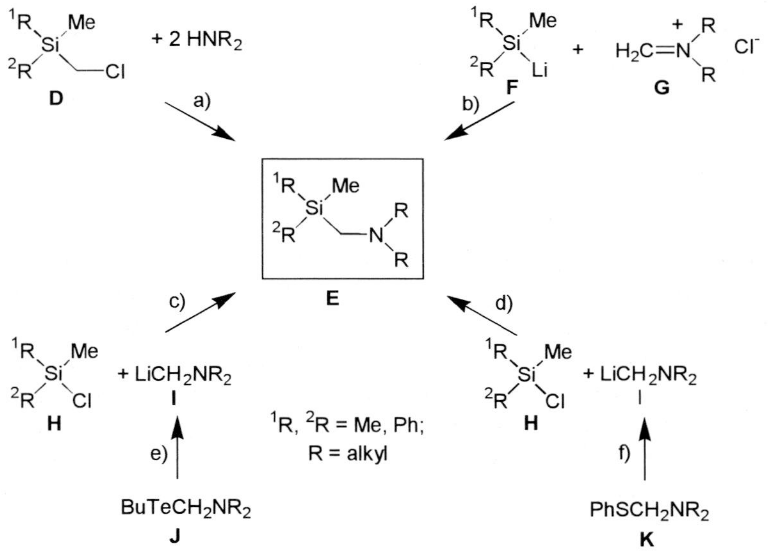

Scheme 2.   a) - $H_2NR_2^+ Cl^-$; b) - LiCl; c) - LiCl; d) - LiCl; e) + nBuLi/- TeBu$_2$; f) + 2 LiC$_{10}$H$_8$/- LiSPh.

(Aminomethyl)silanes (**E**) can be metallated by lithiumalkyls and characterized by trapping with disulfides (Scheme 3). The (phenylthiomethyl)(aminomethyl)silanes **G** can be metallated again in toluene or THF. In reactions of (lithiomethyl)(aminomethyl)diphenylsilanes with aldehydes or ketones we observed no clear addition/deprotonation profile. The (lithiomethyl)silanes **H** add to ketones with *de* up to 70 %.

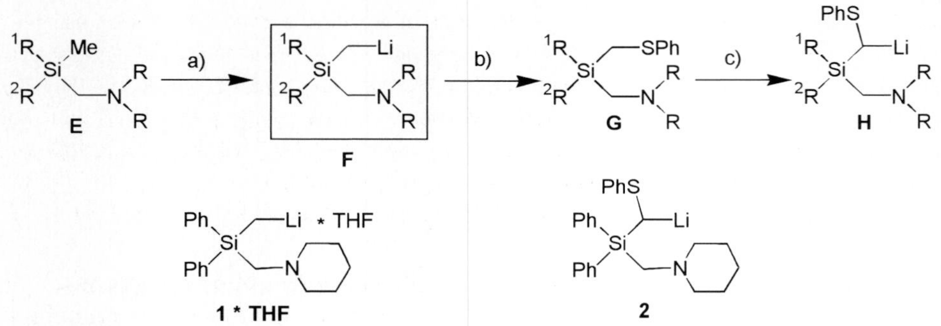

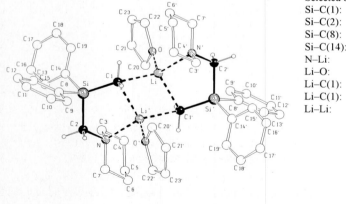

Scheme 3.   a) + *n*BuLi/- *n*BuH; b) + PhSSPh/- LiSPh; c) + BuLi/- BuH.

We were able to characterize the (lithiomethyl)silanes **1·THF** and **2** by single crystal X-ray diffraction.

Crystal Structures of (Lithiomethyl)(aminomethyl)silanes

1·THF crystallizes from hexane as a dimer. The central four membered ring is formed by the two lithium atoms and two metallated carbon atoms. The carbon atoms show two lithium contacts. Similarly, the lithium atoms have two carbon contacts. An intramolecular coordination of the piperidino nitrogen and the oxygen of the THF molecule complete the coordination sphere of lithium. A ball and stick drawing of the molecular structure of **1·THF** and selected bond distances and angles are given in Fig. 1.

Selected bond distances (pm) and angles (°):

Si–C(1):	180.7(4)	C(1)–Si–C(2):	110.2(2),
Si–C(2):	188.9(4)	C(1)–Si–C(14):	112.9(2),
Si–C(8):	189.6(4)	C(2)–Si–C(14):	111.0(2),
Si–C(14):	189.4(4)	C(1)–Si–C(8):	116.7(2),
N–Li:	217.8(6)	C(2)–Si–C(8):	101.1(2),
Li–O:	202.3(6)	C(14)–Si–C(8):	104.1(2),
Li–C(1):	224.4(6)	C(2)–N–Li:	98.3(2),
Li–C(1):	227.0(6)	O–Li–N:	102.1(2),
Li–Li:	253.0(10)	N–Li–C(1):	99.0(2),
		O–Li–C(1):	108.0(3)
		O–Li–C(1):	110.4(3)
		N–Li–C(1):	124.0(3)
		C(1)–Li–C(1):	111.8(2)
		O–Li–N:	103.9(2)
		Si–C(1)–Li:	97.7(2)
		Si–C(1)–Li:	143.4(2)
		Li–C(1)–Li:	68.2(2)
		N–C(2)–Si:	116.6(2).

Fig. 1. View of the molecular structure of **1·THF**.

Compound **2** crystallizes from toluene/hexane similar to **1·THF** as a dimer with half a molecule of toluene, which shows no interaction with **2**. Two different views of the molecular structure of **2** and selected bond distances and angles are given in Fig. 2.

In the first view (Fig. 2, left) a four membered ring is formed by the two lithium atoms and two metallated carbon atoms. The carbon atoms have two lithium contacts [224.3(5) pm and 238.3(5) pm]. The lithium atoms have two carbon contacts. An intramolecular lithium-nitrogen and an intramolecular lithium-sulfur contact with a small Li–C(1)–S angle of 69.2(2)° complete the coordination sphere of lithium.

In the second view (Fig. 2, right) two monomeric molecules (with a C–Li distance of 224.3(5) pm) are connected in such a way that the lithium atoms complete their coordination sphere by a nearly symmetric bridged C(1)–S bond with Li–C (238.3(5) pm) and Li–S (241.2(5) pm) contacts.

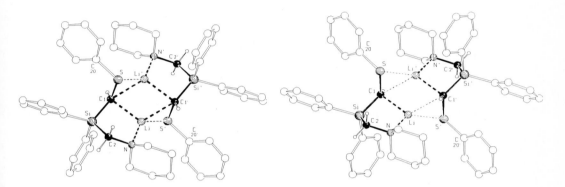

Fig. 2.   Two views of the molecular structure of **2**. Selected bond distances (pm) and angles (°): Si–C(1) 183.4(3), Si–C(14) 188.2(3), Si–C(8) 189.5(2), Si–C(2) 189.9(3), S–C(1) 177.4(3), S–C(20) 179.9(3), S–Li1 241.2(5), N–C(2) 149.9(3), N–Li 206.7(5), Li–C(1) 224.3(5), Li–C(1) 238.3(5), C(1)–Si–C(2) 105.76(13), C(1)–S–Li 67.4(2), C(20)–S–Li 119.42(14), C(2)–N–Li 102.6(2), N–Li–C(1) 102.7(2), N–Li–C(1) 138.3(2), C(1)–Li–C(1) 109.3(2), N–Li–S 134.4(2), C(1)–Li–S' 118.8(2), C(1)–Li–S 43.41(10), S–C(1)–Si 115.93(14), S–C(1)–Li 102.3(2), Si–C(1)–Li 90.7(2), S–C(1)–Li 69.2(2), Si–C(1)–Li 161.4(2), Li–C(1)–Li 70.7(2), N–C(2)–Si 114.4(2).

The silicon-carbon bond length (180.4(4) pm for **1·THF** and 183.4(3) pm for **2**) to the metallated carbon atoms of are shorter than the "normal" silicon-carbon bond length of 186-189 pm reported in the literature [7].This effect is known for metallated carbon atoms bonded to silicon [8]. However, there is only one crystal structure described for a (lithiomethyl)silane which has no substituent on the metallated carbon atom [9]. For (lithiomethyl)trimethylsilane (hexamer) [9] a Si–CH$_2$Li bond length of 176.8 pm was found. No significantly longer unmetallated Si–C bonds were observed (these bond lengths are 183.5-186.5 pm). Crystal structures with additional substituents on the metallated carbon atoms, e.g., **2**, can give no clear information of silicon stabilizing effects. Therefore the question arises: Is silicon stabilizing the metallated carbon atoms by polarization or negative hyperconjugation? Longer silicon-carbon bond lengths antiperiplanar to the C–Li bond are expected in the case of negative hyperconjugation.

We observed slightly longer Si–C bonds of the unmetallated carbon atoms of **1·THF** (Si–C(2) 188.9(4) pm, Si–C(8) 189.6(4) pm, Si–C(14) 189.4(4) pm) and **2** (Si–C(2) 189.9(3) pm, Si–C(8) 189.5(2) pm, Si–C(14) 188.2(3) pm). No conformation indicating negative hyperconjugation was found for **1·THF** and no significant lengthening of one or two Si–C-bond was observed. The crystal structure of **1·THF** is in agreement with stabilizing metallated carbon atoms bonded to silicon through polarization effects of silicon. Theoretical studies by Lambert and Schleyer support the stabilization by polarization effects of silicon [8]. For metallated carbon atoms bonded to phosphorous or sulfur stabilization by negative hyperconjugation was discussed [8]. For **2** the C(20)–S–C(1)–Li torsion angle is 177.9°. Thus, the aryl-S bond (179.9(3) pm) of each monomeric fragment is antiperiplanar to the C–Li bond indicating, sulfur stabilization of the metallated carbon atom of **2** by negative hyperconjugation.

Further related work on the synthesis, structure, and reactivity of (lithiomethyl)(aminomethyl)silanes is currently under way to understand more about the behavior of these systems in solution and in the solid state.

Acknowledgements: We are grateful to the *Fonds der Chemischen Industrie* (FCI) for a PhD grant to B. C. A. and to the *Deutschen Forschungsgemeinschaft* (DFG) for financial support. We would like to thank Prof. M. Veith (*Universität Saarbrücken*) for supporting this work.

References:

[1]   D. J. Ager, *Org. Reactions* **1990**, *38*, 1.
[2]   a) H. Luitjes, F. J. J. de Kanter, M. Schakel, R. F. Schakel, R. F. Schmitz, G. W. Klumpp, *J. Am. Chem. Soc.* **1995**, *117*, 4179.
       b) H. Luitjes, M. Schakel, R. F. Schmitz, G. W. Klumpp, *Angew. Chem.* **1995**, *107*, 2324.
[3]   a) R. F. Horvath, T. H. Chan, *J. Org. Chem.* **1989**, *54*, 317.
       b) T. H. Chan, D. Wang, *Chem. Rev.* **1992**, *92*, 995.
[4]   T. H. Chan, P. Pellon, *J. Am. Chem. Soc.* **1989**, *111*, 8737.
[5]   a) D. Labrecque, K. T. Nwe, T. H. Chan, *Organometallics* **1994**, *13*, 332.
       b) G. Lambrecht, G. Gmelin, K. Rafeiner, C. Strohmann, R. Tacke, E. Mutschler, *Eur. J. Pharmacol.* **1988**, *151*, 155.
[6]   C. Strohmann, B. C. Abele, *Angew. Chem.* **1996**, *108*, 2514.
[7]   W. S. Sheldrick, in: *The Chemistry of Organic Silicon Compounds, Part 1* (Eds.: S. Patai, Z. Rappoport), Wiley, Chichester, **1989**, p.227.
[8]   C. Lambert, P. v. R. Schleyer in: *Methoden Org. Chem. (Houben-Weyl) 4th ed.* Vol. E19d, **1993**, p. 1.
[9]   B. Tecle, A. F. M. M. Rahman, J. P. Oliver, *J. Organomet. Chem.* **1986**, *317*, 267.

(Phenylthiomethyl)silanes as New Bifunctional Assembling Ligands for the Construction of Heterometallic Complexes

M. Knorr*, S. Kneifel, C. Strohmann*

Institut für Anorganische Chemie
Universität des Saarlandes
Postfach 1150, D-66041 Saarbrücken, Germany
E-mail: m.knorr@sbusol.rz.uni-sb.de; c.strohmann@rz.uni-sb.de

Keywords: Silyl Complexes / Platinum / Rhenium / Chelate / Heterobimetallics

Summary: Functionalized hydridosilanes such as $(PhSCH_2)_2Si(Me)H$ (**1a**) or $(PhSCH_2)_3SiH$ (**1b**) add oxidatively to the platinum(0) complex $[Pt(PPh_3)_2(CH_2=CH_2)]$ to afford hydrido silyl complexes like the structural characterized complex *cis*-$[Pt(H)\{Si(Me)(CH_2SPh)_2\}(PPh_3)_2]$ (**2a**). The donor ability of the thioether groups of the silyl ligand can be used to attach a second metal center to construct heterobimetallic complexes. This concept has been successfully applied for the synthesis of the heterobimetallic platinum-rhenium complex $[Pt(H)(PPh_3)_2\{Si(Me)(CH_2SPh)_2\}(Re(CO)_3Br)]$ (**5**).

Introduction

The pioneering work of Eaborn on platinum silyl complexes has shown that a wide range of commercially available hydridosilanes can be oxidatively added to $[Pt(PPh_3)_2(CH_2=CH_2)]$ to afford hydrido silyl complexes of the type *cis*-$[Pt(H)(SiR_3)(PPh_3)_2]$ [1]. Our aim was to extend this route using functionalized hydridosilanes bearing donor groups in order to achieve (i) an intramolecular coordination in monometallic complexes (**A**), (ii) to synthesize and stabilize bimetallic complexes of type (**B**) displaying a bridging bonding mode, and (iii) to attach a second metal center by formation of inorganic ring systems of type (**C**).

Since thioethers R–S–R, which are incorporated in potentially chelating systems, are known to form stable dative bonds with transition metals [2], we set out to prepare functionalized hydridosilanes possessing two or three thioether groups to coordinate a second metal complex, as shown in (**C**).

Synthesis of the [(Phenylthio)methyl]hydridosilanes 1

The [(phenylthio)methyl]hydridosilanes **1a/1b** used as bifunctional assembling ligands were prepared by the reaction of [(phenylthio)methyl]lithium with the corresponding chlorosilanes, similar to the synthesis of dimethyl[(phenylthio)methyl]silane [3a] and diphenylbis[(phenylthio)methyl]silane [3b].

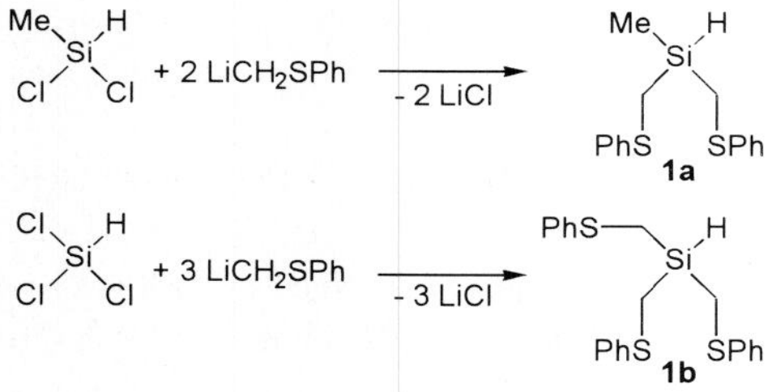

Scheme 1.

Synthesis of the Platinum Hydrido Silyl Complexes 2

The stable yellow platinum complexes **2** were prepared in high yield by reaction of [Pt(PPh$_3$)$_2$(CH$_2$=CH$_2$)] with an excess of **1** under mild conditions according to Scheme 2.

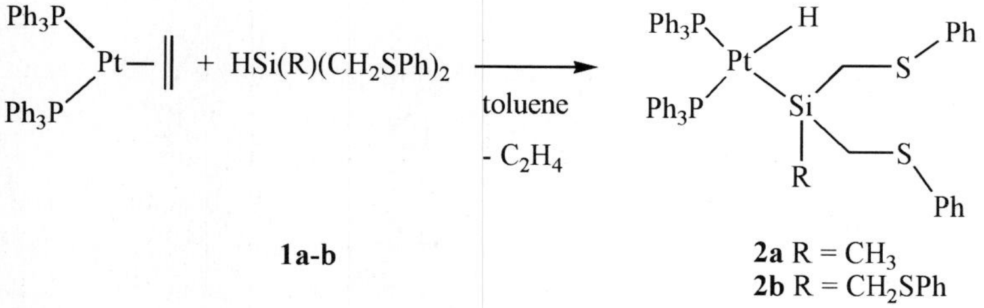

Scheme 2.

These fluxional complexes have been fully characterized in solution by multinuclear NMR techniques. The data are consistent with a *cis*-arrangement of the hydride and the silyl ligand. In the ^{1}H NMR spectrum of **2a** (in CDCl$_3$ at 243 K) the hydride resonance is found at $\delta = -2.66$ ppm and consists of a doublet of doublets ($^2J_{P-H} = 19.6, 152.7$ Hz), which is flanked by platinum satellites ($^1J_{Pt-H} = 972.3$ Hz). A single crystal X-ray determination performed on **2a** revealed that intramolecular short contacts between a thioether group and the platinum center are absent in the solid state.

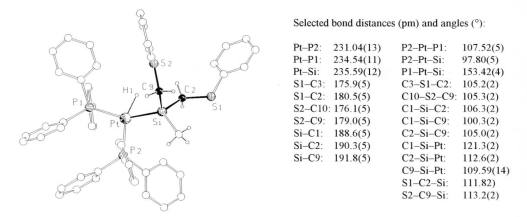

Selected bond distances (pm) and angles (°):

Pt–P2:	231.04(13)	P2–Pt–P1:	107.52(5)
Pt–P1:	234.54(11)	P2–Pt–Si:	97.80(5)
Pt–Si:	235.59(12)	P1–Pt–Si:	153.42(4)
S1–C3:	175.9(5)	C3–S1–C2:	105.2(2)
S1–C2:	180.5(5)	C10–S2–C9:	105.3(2)
S2–C10:	176.1(5)	C1–Si–C2:	106.3(2)
S2–C9:	179.0(5)	C1–Si–C9:	100.3(2)
Si–C1:	188.6(5)	C2–Si–C9:	105.0(2)
Si–C2:	190.3(5)	C1–Si–Pt:	121.3(2)
Si–C9:	191.8(5)	C2–Si–Pt:	112.6(2)
		C9–Si–Pt:	109.59(14)
		S1–C2–Si:	111.82)
		S2–C9–Si:	113.2(2)

Fig. 1. View of the molecular structure of **2a**; the position of the hydride ligand was not located with certainty.

Synthesis of Platinum and Rhenium Complexes with a Chelating (Phenylthiomethyl)silane Ligand

In order to study the ability of the (phenylthiomethyl)silanes **1** to chelate metal complexes and the associated stereochemical problems we first prepared simple model compounds using diphenylbis[(phenylthio)methyl]silane **1c**. On reaction of **1c** with [PtCl₂(PhCN)₂] the very stable square planar chelate complex **3** was isolated. NMR investigations at variable temperature show that *meso-* and DL-isomers coexist in a temperature-dependent equilibrium due to a facile inversion process at the sulfur atoms [2] in solution. At higher temperature (325 K) the pyramidal inversion is sufficiently fast that only one "averaged planar" conformation is observed in the ¹H NMR spectrum.

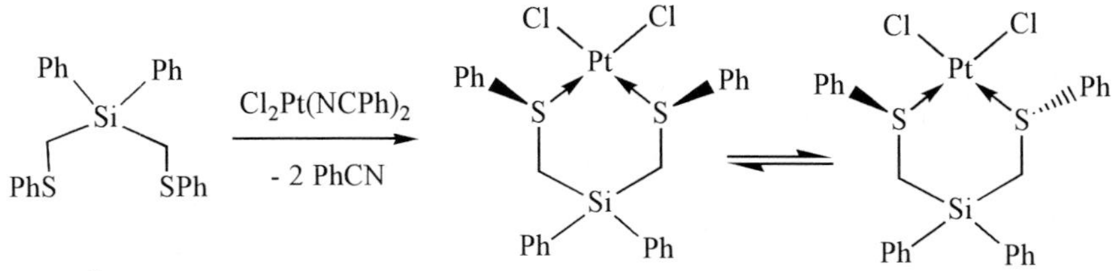

Scheme 3.

A single crystal X-ray diffraction study shows that the *meso* conformation is preferred in the solid state for this platinum(II) chelate complex **3**.

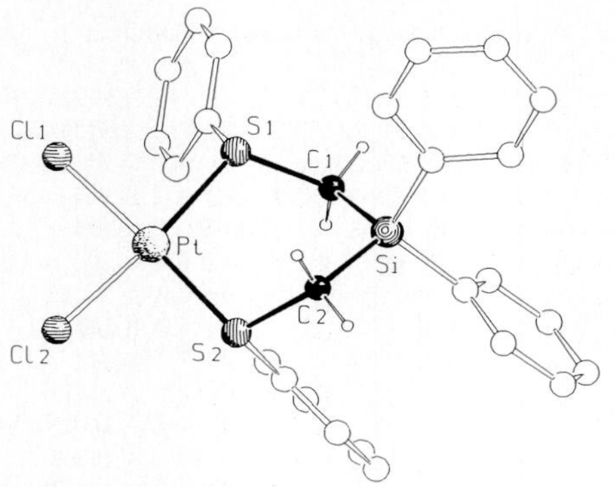

Selected bond distances (pm) and angles (°):

Pt–S2:	227.1(2)	S2–Pt–S1:	100.25(6)
Pt–S1:	228.1(2)	S2–Pt–Cl1:	173.02(6)
Pt–Cl1:	231.7(2)	S1–Pt–Cl1:	84.49(7)
Pt–Cl2:	232.5(2)	S2–Pt–Cl2:	85.20(7)
S1–C1:	182.2(6)	S1–Pt–Cl2:	174.49(6)
S2–C2:	182.2(6)	Cl1–Pt–Cl2:	90.02(8)
Si–C2:	188.3(7)	C1–S1–Pt:	112.0(2)
Si–C1:	189.5(6)	C2–S2–Pt:	108.4(2)
		S1–C1–Si:	115.9(3)
		S2–C2–Si:	118.4(4)

Fig. 2. View of the molecular structure of **3**.

A stable chelate complex is also formed when **1c** reacts with [Re(μ-Br)(CO)₃THF]₂. ¹H, ¹³C, and ²⁹Si NMR spectra of *fac*-[Re(Br)(CO)₃{(PhSCH₂)₂SiPh₂}] (**4a**) all indicate that, in contrast to the case of **3** at ambient temperature, only one of the several conceivable isomers exists in solution. At present we cannot conclude whether the solution structure of **4a** corresponds to the ligand arrangement found in the solid state structure of **4a**.

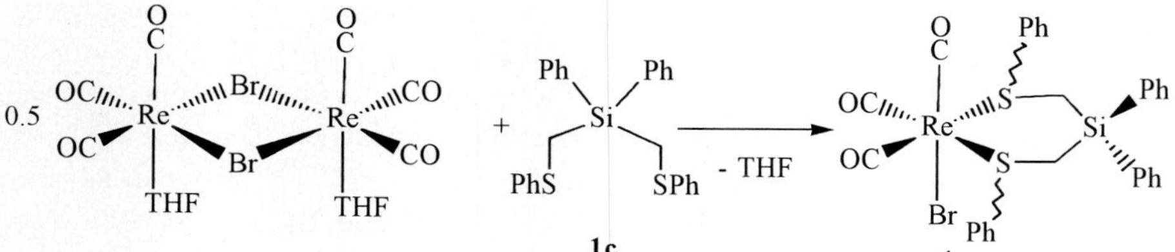

Scheme 4.

The latter determination performed on a single crystal shows a rhenium(I) center in an octahedral environment with a facial arrangement of the three carbonyl ligands. The diphenylbis[(phenylthio)methyl]silane **1c** forms a six-membered cycle with the rhenium center, in which the phenyl rings of the thioether groups are both orientated in the same direction as the bromide ligand.

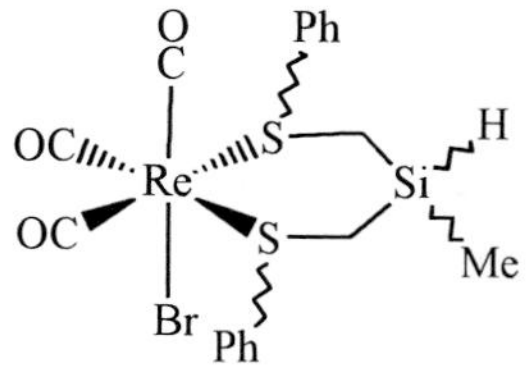

When [Re(μ-Br)(CO)₃THF]₂ reacts with **1a**, formation of the chelate complex **4b** is preferred instead of oxidative addition. Several isomers were obtained in this case.

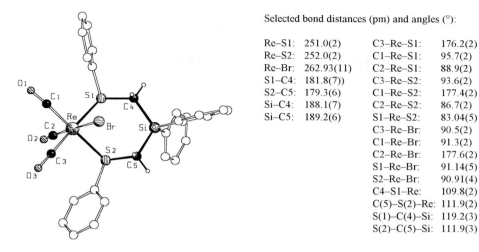

Selected bond distances (pm) and angles (°):

Re–S1:	251.0(2)	C3–Re–S1:	176.2(2)
Re–S2:	252.0(2)	C1–Re–S1:	95.7(2)
Re–Br:	262.93(11)	C2–Re–S1:	88.9(2)
S1–C4:	181.8(7))	C3–Re–S2:	93.6(2)
S2–C5:	179.3(6)	C1–Re–S2:	177.4(2)
Si–C4:	188.1(7)	C2–Re–S2:	86.7(2)
Si–C5:	189.2(6)	S1–Re–S2:	83.04(5)
		C3–Re–Br:	90.5(2)
		C1–Re–Br:	91.3(2)
		C2–Re–Br:	177.6(2)
		S1–Re–Br:	91.14(5)
		S2–Re–Br:	90.91(4)
		C4–S1–Re:	109.8(2)
		C(5)–S(2)–Re:	111.9(2)
		S(1)–C(4)–Si:	119.2(3)
		S(2)–C(5)–Si:	111.9(3)

Fig. 3. View of the molecular structure of **4a**.

Synthesis of a Heterobimetallic Platinum-Rhenium Complex

The fluxional bimetallic complex **5** has been prepared by route (a) or alternativley by route (b). Both strategies combine the ability of the functionalized silanes **1a/b** to coordinate metal centers by oxidative addition (reactivity of the Si–H function) or to form dative bonds using the donor properties of the thioether groups.

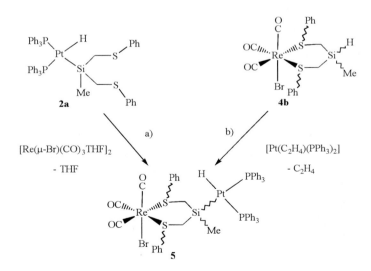

Scheme 5.

Acknowledgement: We are grateful to the *Deutschen Forschungsgemeinschaft* for financial support. We also thank Prof. M. Veith (Universität Saarbrücken) for supporting this work.

References:

[1]   a) C. Eaborn, B. Ratcliff, A. Pidcock, *J. Organomet. Chem.* **1974**, *65*, 181.
b) H. Azizian, K. R. Dixon, C. Eaborn, A. Pidcock, N. M. Shuaib, J. Vinaixa, *J. Chem. Soc., Chem. Commun.* **1982**, 1020; L. A. Latif, C. Eaborn, A. Pidcock, *J. Organomet. Chem.* **1994**, *474*, 217.

[2]   a) F. Hartley, S. G. Murray, W. Levason, H. E. Soutter, C. A. McAulife, *Inorg. Chim. Acta* **1979**, *220*, 129.
b) E. W. Abel, S. K. Bhargava, K. G. Orrell, *Prog. Inorg. Chem.* **1984**, *32*, 1.

[3]   a) J. Yoshida, H. Tsujishima, K. Nakano, S. Isoe, *Inorg. Chim. Acta* **1994**, *220*, 129.
b) C. Strohmann, S. Lüdtke, E. Wack, *Chem. Ber.* **1996**, *129*, 799.

Synthesis of 1,3-Disilacyclobutanes, 1,3-Digermacyclobutanes, and 1-Germa-3-silacyclobutanes with New 1,3-Dimetallated Organoelement Building Blocks

Carsten Strohmann*, Eric Wack
Institut für Anorganische Chemie
Universität des Saarlandes
Postfach 1150, D-66041 Saarbrücken, Germany
Tel.: Int. code + (681)3022465
E-mail: c.strohmann@rz.uni-sb.de

Keywords: 1,3-Disilacyclobutanes / 1-Germa-3-silacyclobutanes / 1,3-Digermacyclobutanes / Bis(lithiomethyl)silanes

Summary: Bis(lithiomethyl)silanes and -germanes, prepared by reductive cleavage of C–S bonds in THF with lithium naphthalenide ($LiC_{10}H_8$) have been allowed to react with dichlorosilanes and -germanes to give 1,3-disilacyclobutanes, 1,3-digermacyclobutanes, and 1-germa-3-silacyclobutanes. In addition, some one ring-carbon-atom substituted 1,3-disilacyclobutanes have been synthesized. Crystal structures of 1,3-disilacyclobutanes and 1,3-digermacyclobutanes show that the nonplanar nature of 1,3-disilacyclobutanes is not only due to the effects of substituents, but a significant contribution is caused by the packing of molecules.

Introduction

As a part of our systematic studies [1] on the structural unit "–CR$_2$–El–CR$_2$–" (El = element of group 14-16, R = alkyl, aryl) we have been investigating bis(lithiomethyl)silanes and -germanes [2]. Bis(lithiomethyl)silanes and -germanes were prepared by the reductive cleavage of C–S bonds in THF with lithium naphthalenide ($LiC_{10}H_8$) (see Scheme 1) [2]. With these new 1,3-dimetallated compounds it was possible to synthesize symmetrical and as asymmetrical sila- and germacycles. In this section we report on the use of bis(lithiomethyl)silanes and -germanes as building blocks for the synthesis of 1,3-disilacyclobutanes, 1-germa-3-silacyclobutanes, and 1,3-digermacyclobutanes and structural features of the synthesized cyclic systems.

Most of the known 1,3-disilacyclobutanes were synthesized by [2+2] cycloaddition of the corresponding silicon-carbon double bonds or the "Kriner Reaction" of chloro(chloromethyl)silanes with magnesium or lithium [3]. The 1,3-disilacyclobutanes prepared by these reactions are normally symmetrically substituted (Scheme 1).

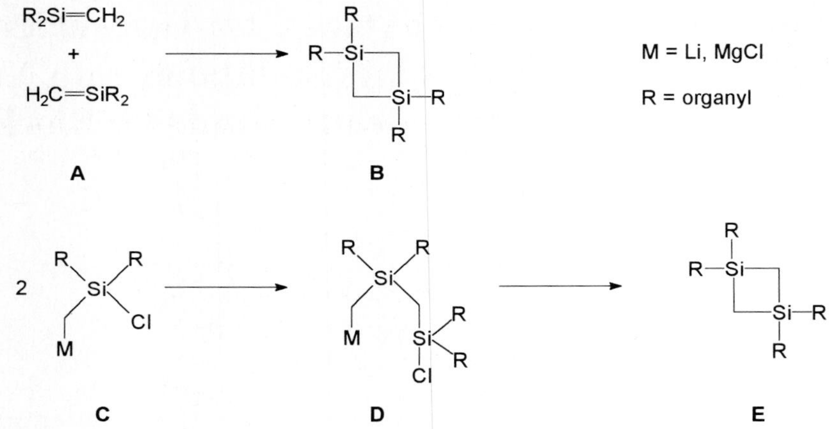

Scheme 1.

There is no synthetic route available for selective synthesis of unsymmetrically substituted 1,3-disilacyclobutanes, 1-germa-3-silacyclobutanes, or 1,3-digermacyclobutanes.

Synthesis of 1,3-Disila-, 1-Germa-3-sila-, and 1,3-Digermacyclobutanes

The 1,3-disilacyclobutanes **4a-4f** were formed by the addition of the appropriate dichlorosilane at -50 °C to a freshly prepared solution of bis(lithiomethyl)diphenylsilane (**3a**) (Scheme 2) or (lithiomethyl)[lithio(trimethylsilyl)methyl]diphenylsilane (**8**) (Scheme 3). After aqueous work-up, the 1,3-disilacyclobutanes **4a-4f** were isolated by Kugelrohr distillation or cystallization in 34-62 % yield. Polymeric materials or higher ring systems were formed as byproducts, reducing the yield. The synthesis of **4a**, **4d**, and **4f** by another synthetic route has been described previonsly [4].

	El	El'	R^1	R^2
4a	Si	Si	Me	Me
4b	Si	Si	Me	Et
4c	Si	Si	Me	Vi
4d	Si	Si	Me	Ph
4e	Si	Si	Et	Ph
4f	Si	Si	Ph	Ph
5a	Si	Ge	Me	Me
5b	Si	Ge	Et	Et
5c	Si	Ge	Ph	Ph
6	Ge	Ge	Ph	Ph

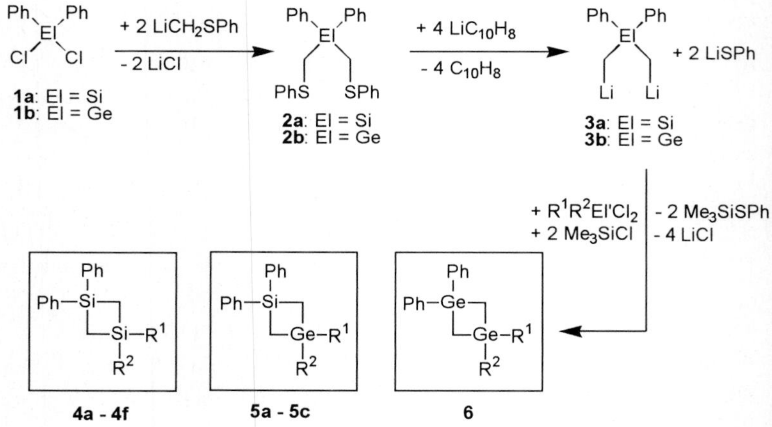

Scheme 2.

The advantage of the synthetic pathway described here over the known synthetic routes to 1,3-disilacyclobutanes is the selective preparation of unsymmetrical 1,3-disilacyclobutanes. The only known 1,3-disilacyclobutanes with one substituent on one ring carbon atom were prepared by Seyferth by metallation of 1,1,3,3-tetramethyl-1,3-disilacyclobutane [6] and trapping with electrophiles. 1-Germa-3-silacyclobutanes and 1,3-digermacyclobutanes were prepared analogously using bis(lithiomethyl)silanes and -germanes as the starting material (Scheme 2).

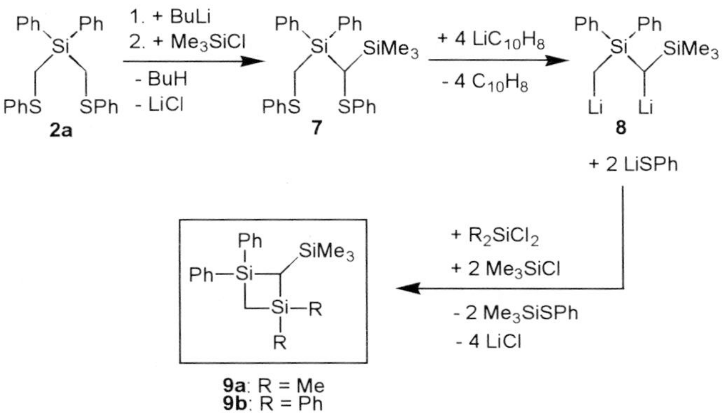

Scheme 3.

Crystal Structures of 1,3-Disilacyclobutanes and 1,3-Digermacyclobutanes

Two features of the reported solid state structures of 1,3-disilacyclobutanes (i) the distance between the two silicon atoms in 1,3-position and (ii) the folding of the central cyclobutane ring are discussed in the literature [6].

Theoretical studies indicate that relative short distances of the silicon atoms in 1,3-disilacyclobutanes are a result of the geometry of the four-membered rings [6a]. There is no bonding between these two silicon atoms. Nevertheless, it is not clear, which effects cause the folding or non-folding of the cyclobutane rings. Steric effects of the substituents on the silicon or carbon atoms of 1,3-disilacyclobutanes have been discussed as a cause for the structures with different folding angles in the solid state [6b].

We present two crystal structures of 1,3-disilacyclobutanes and the structure of a 1-germa-3-silacyclobutane. The first examples of a homologous pair of 1,3-disila- and 1,3-digermacyclobutanes are shown in Figures 1 and 2.

The geometry of **4f** and **6** can be explained with the help of the Electron Localisation Function (ELF). The geometry is a result of tetrahedral "pseudo bond angles" on the silicon, germanium, and carbon atoms and connecting the ring fragments together with the formation of bent bonds [7]. The smaller angles on germanium compared with silicon are the geometric effects of the longer germanium-carbon bonds.

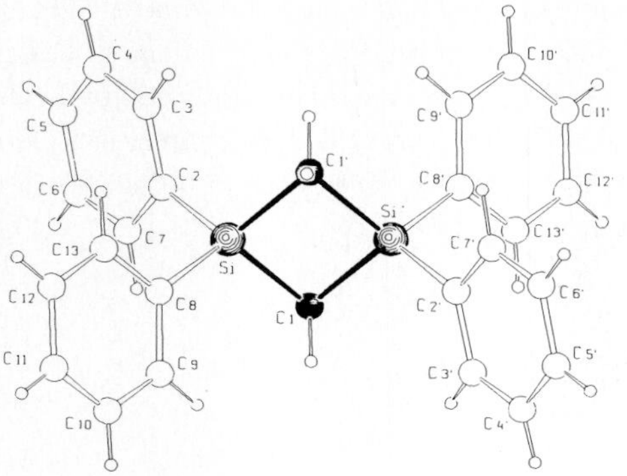

Fig. 1. View of the molecular structure of **4f**.

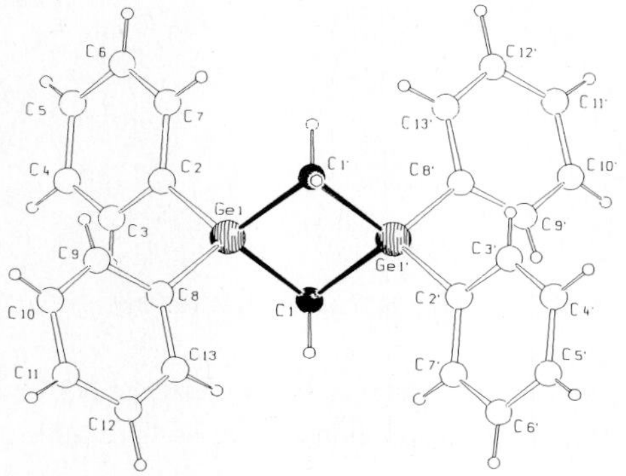

Selected bond distances (pm) and angles(°)
Si–C(8):   187.3(2)   C(1)–Si–C(1)´: 92.22(10)
Si–C(2):   187.4(2)   Si–C(1)–Si: 87.78(10)
Si–C(1):   188.3(2)   Si ··· Si' (pm): 261.4(2), planar
Si–C(1)´: 188.8(2);)   cyclobutane ring

Fig. 2. View of the molecular structure of **6**.

Selected bond distances (pm) and angles(°):
Ge(1)–C(2) :   193.2(4)   C(1)–Ge(1)–C(1) :   90.9(2)
Ge(1)–C(8) :   193.9(4)   Ge(1)–C(1)–Ge(1) :   89.1(2)
Ge(1)–C(1)´ : 195.6(4)   Ge(1) ··· Ge(1) (pm): 274.5(2),
Ge(1)–C(1) :   195.8(4)   planar cyclobutane ring.

For the substituted 1,3-disilacylobutane **9b** two different conformers were found in the asymmetric unit of the crystal structure. A view of the molecular structure of one molecule with selected bond lengths and angles is given in Fig. 3. A side view of the two molecules is shown in Fig. 4. The main difference in the two conformers is the folding of the central four membered ring (Folding: 2.9° (molecule A), 14.3° (molecule B)). The two molecules of **9b** were found in the same crystal and the different geometries observed are the result of packing effects.

Our results show that the nonplanar nature of 1,3-disilacyclobutanes is not only due to the effects of substituents, but is significantly contributed to by the packing of molecules.

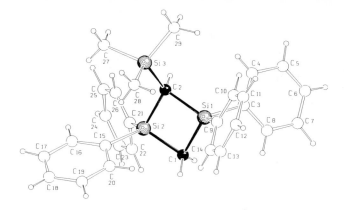

Fig. 3. View of the molecular structure of **9b** (molecule A). Selected bond distances (pm) and angles (°): Si(1)–C(3) 185.8(4), Si(1)–C(9) 185.8(4), Si(1)–C(1) 186.3(5), Si(1)–C(2) 188.0(3), Si(2)–C(21) 185.7(4), Si(2)–C(15) 186.0(4), Si(2)–C(1) 186.5(3), Si(2)–C(2) 188.3(5), Si(3)–C(2) 185.1(4); C(3)–Si(1)–C(9) 107.6(2), C(3)–Si(1)–C(1) 111.2(2), C(9)–Si(1)–C(1) 115.5(2), C(3)–Si(1)–C(2) 114.66(14), C(9)–Si(1)–C(2) 114.80(14), C(1)–Si(1)–C(2) 92.7(2), C(1)–Si(2)–C(2) 92.5(2), Si(1)–C(1)–Si(2) 87.9(2), Si(3)–C(2)–Si(1) 125.2(2), Si(3)–C(2)–Si(2) 122.7(2), Si(1)–C(2)–Si(2) 86.9(2); Si(1) ⋯ Si(2) (pm): 258.8(6).

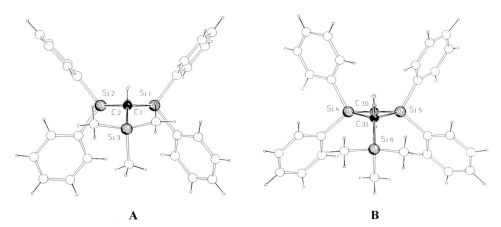

A	**B**

Fig. 4. Side view on the molecular structure of the two conformers of **9b** in the crystal. Folding: 2.9° (**A**), 14.3° (**B**).

Acknowledgements: We are grateful to the *Fonds der Chemischen Industrie* (FCI) and to the *Deutschen Forschungsgemeinschaft* (DFG) for financial support. We would like to thank Prof. M. Veith (*Universität Saarbrücken*) for supporting this work.

References:

[1]   a) C. Strohmann, *Chem. Ber.* **1995**, *128*, 167.
b) C. Strohmann, *Angew. Chem.* **1996**, *108*, 600.

[2]   a) C. Strohmann, S. Lüdtke, E. Wack, *Chem. Ber.* **1996**, *129*, 799.
b) C. Strohmann, S. Lüdtke, *Mono-, Bis-, Tris- and Tetrakis(lithiomethyl)silanes: "New Building Blocks for Organosilicon Compounds"* in: *Organosilicon Chemistry II: From Molecules to Materials* (Eds.: N.Auner, J.Weis), VCH, Weinheim, **1996**, p. 499.

[3]   N. Auner, J. Grobe, *J. Organomet. Chem.* **1980**, *188*, 151.

[4]   a) P. Jutzi, P. Langer, *J. Organomet. Chem.* **1980**, *222*, 401.
b) A. M. Devine, P. A. Griffin, R. N. Haszeldine, M. J. Newlands, A. E. Tipping, *J. Chem. Soc. Dalton. Trans.* **1975**, 1822.

[5]   J. L. Robinson, W. M. Davis, D. Seyferth, *Organometallics* **1991**, *10*, 3385.

[6]   a) A. Savin, H. J. Flad, J. Flad, H. Preuß, H. G. von Schnering, *Angew. Chem.* **1992**, *104*, 185.
b) N. Auner, W. Ziche, E. Herdtweck, *J. Organomet. Chem.* **1992**, *426*, 1.

[7]   C. Strohmann, F. Wagner, unpublished results.

Trialkylsilyldiazomethane Derivatives: Wonderful Chemical Building Blocks

Guy Bertrand

Laboratoire de Chimie de Coordination du CNRS
205 route de Narbonne, 31077 Toulouse Cédex, France
Tel.: Int. code + 561333124 – Fax: Int. code + 561553003
E-mail: gbertran@lcctoul.lcc-toulouse.fr

Keywords: Diazomethane / Nitrilimines / Carbodiimides / Carbenes / Antiaromaticity

Summary: Trimethyl- and triisopropyl-silyl diazomethane derivatives have been used for the synthesis of new types of diazo derivatives including those featuring a heavier main-group element or a transition metal directly bound to the diazo-carbon. They are also precursors to a variety of stable nitrilimines and carbodiimides, and stable pseudo-diazoalkenes and carbenes. (Phosphino)(silyl)carbenes have been used in the synthesis of new types of heterocycle including some stable four-π-electron four-membered derivatives.

Introduction

Diazomethane is explosive [1]! In contrast, no explosions or toxicity have been noted with the silyldiazomethane derivatives [$(R_3SiC(N_2)H$: **1**, R=Me [2]; **2**, R=*i*Pr [3]), which can be purified by distillation (**1**, 75°C/760 mm Hg; **2**, 39°C/10^{-1} mm Hg). Compound **1** is commercially available, while **2** is easily prepared in 45 % isolated yield by the sequential addition of an ethereal solution of diazomethane followed by triisopropylsilyl triflate at -20°C to neat diisopropylethylamine [3].

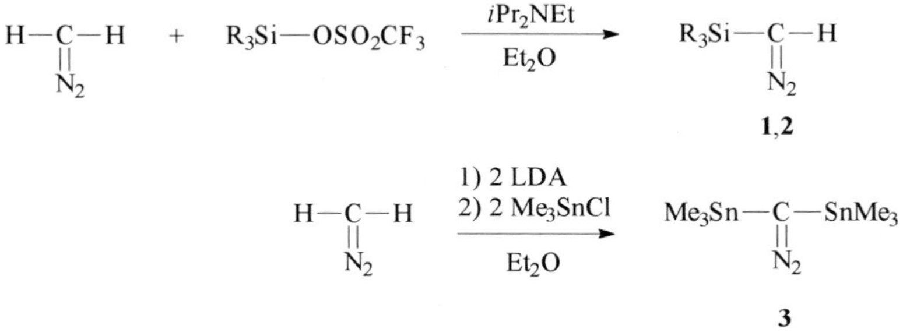

1: R = Me (74 % yield)
2: R = *i*Pr (45 % yield)
3: (81 % yield)

Scheme 1. Synthesis of silyl and stannyl diazomethane derivatives **1-3**.

These two compounds **1** and **2** cannot always be considered as synthetic equivalents of diazomethane; for example, they do not react with carboxylic acids to give esters. In fact, they play a very different role in synthesis, and it is demonstrated in this review that they are powerful synthetic building blocks. Since tin chemistry is often compared to that of silicon, a few results involving bis(trimethylstannyl)diazomethane **3** [4] will also be presented (Scheme 1). Juggling the relative steric bulk of the triisopropylsilyl versus the trimethylsilyl group, and the enhanced reactivity of tin-carbon compared to silicon-carbon bonds, the three diazomethane derivatives **1-3** are complementary synthons. Not only are they attractive precursors for the synthesis of stable compounds hitherto believed to be only transient intermediates, but also of new heterocycles or even of polymer crosslinking agents.

Transfer of the CN$_2$ Fragment Leading to Diazo Compounds or Nitrilimines

Both the C–H and C–Si bonds of **1** and **2** are not sufficiently labile to interact with electrophiles, and therefore the corresponding lithium salts **4** and **5** have to be prepared, which is easily done using butyllithium or LDA. Both **4** and **5** react with a variety of electrophiles affording the corresponding diazo compounds in good yields. Interestingly, this route allows access to novel types of diazo derivatives, which include the first α-diazophosphine **6** (95 % yield) [5], a very rare example of a *C*-substituted diazoalkane transition metal compound **7** (45 % yield) [6], and even to date the only known organometallic complex **8** bearing a diazomethane-functionalized ligand (75 % yield) [7] (Scheme 2).

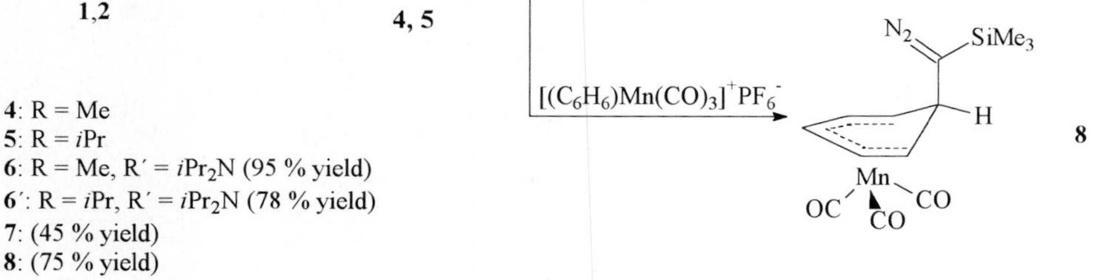

4: R = Me
5: R = *i*Pr
6: R = Me, R′ = *i*Pr$_2$N (95 % yield)
6′: R = *i*Pr, R′ = *i*Pr$_2$N (78 % yield)
7: (45 % yield)
8: (75 % yield)

Scheme 2. Synthesis of novel types of diazo derivatives.

Although, it was generally admitted that the reaction of an electrophile with a diazo lithium salt always led to the corresponding substituted diazo derivative [8], we found that in some cases their thermodynamically less favoured nitrilimine isomers were obtained as stable compounds [9]. Note that until recently [10], nitrilimines, like several other 1,3-dipoles, were considered reactive intermediates, and were only spectroscopically observed in organic glasses below 85 K [11a-c] and in the gas phase [11d]. In our hands, the reactions of trimethyl- and triphenylchlorosilanes with the diazo lithium salts **4** and **5** were not clean [12]. However, addition of triisopropylchlorosilane to diazo lithium salt **5**, gave rise to the bis(silylated)nitrilimine **9** in 80 % yield. Surprisingly, **9** is stable enough to be purified by distillation at 90-100°C under vacuum (5 10^{-2} mm Hg) [3] (Scheme 3).

Interestingly, nitrilimine **9** can be obtained in 90 % yield by treating the bis(stannyl)diazo derivative **3** with two equivalents of triisopropylchlorosilane at room temperature in the absence of solvent [13]. The same reaction performed in solution (THF, CH_2Cl_2, or toluene) leads to the formation of the (triisopropylsilyl)(trimethylstannyl)diazomethane **10**, which can be purified by distillation (55 % yield). The dramatic difference observed in the reaction of **3** with triisopropylchlorosilane, with and without solvent, is of primary interest. We have verified that (silyl)(stannyl)diazomethane **10** does not react with triisopropylchlorosilane, which proves that **10** is not the intermediate leading to nitrilimine **9**. This implies that under both experimental conditions, the first step is certainly an *N*-silylation giving rise to the *C*-stannyl *N*-silyl nitrilimine **11**. Then, in the absence of solvent, **11** undergoes a substitution reaction with another molecule of chlorosilane affording bis(silyl)nitrilimine **9**, while in the presence of a solvent, nitrilimine **11** has enough time to rearrange into the isomeric diazo compound **10** [13a] (Scheme 3).

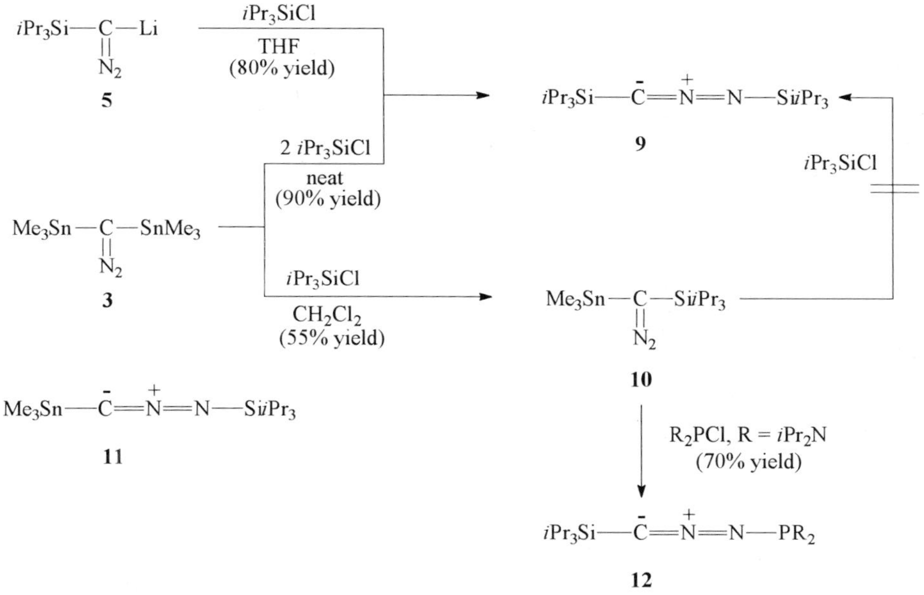

Scheme 3.   Synthesis of nitrilimines.

Addition of chlorophosphane to the mixed (silyl)(stannyl)diazo derivative **10** affords the nitrilimine **12** (70 % yield), which was unobtainable using the silylated lithium salt **5**, demonstrating the superiority of stannyl diazo compounds over diazolithium salts for the synthesis of stable nitrilimines [13, 14] (Scheme 3).

Lastly, it should be mentioned that using **3** the synthesis of the only known purely organic nitrilimine **13** (90 % yield) was possible [13]. To illustrate the importance of this type of 1,3-dipole in heterocyclic synthesis, note that nitrilimine **13** undergoes regio-, stereo-, and even diastereo-selective [2+3]-cycloadditions [13, 15] with methyl acrylate, methylpropionate, dimethyl fumarate, dimethyl maleate and (R)-α-acryloxy-β,β-dimethyl-γ-butyrolactone, affording the corresponding five-membered heterocycles in high yields (Scheme 4).

Transfer of the CN$_2$ Fragment Leading to Carbodiimides

We have seen that diazo compounds **1-3** are suitable starting materials for the synthesis of other diazo derivatives or nitrilimines. We found that **1-3** are also direct or indirect precursors for carbodiimides.

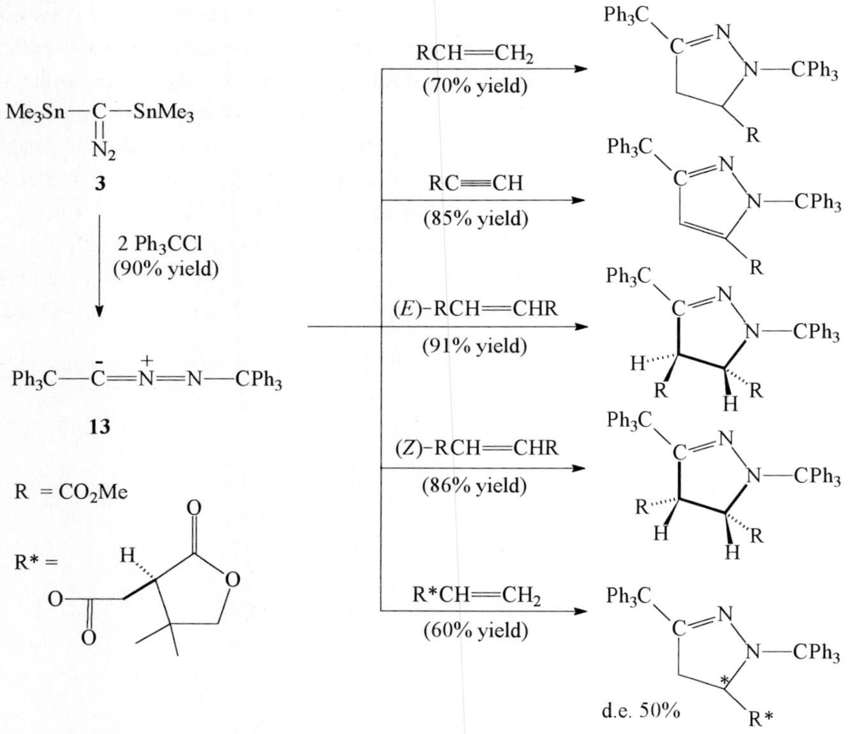

Scheme 4.   [2+3]-cycloaddition reactions involving nitrilimine **13**.

Indeed, under photolytic conditions, bis(silyl)nitrilimine **9** isomerizes into bis(silyl)carbodiimide **14** in quantitative yield [3b]. More surprisingly, when a THF solution of diazo derivative **3** is stirred overnight at room temperature in the presence of a catalytic amount of tetrakis(triphenylphosphine) palladium, bis(stannyl)carbodiimide **15** is obtained in 90 % yield [16a] (Scheme 5).

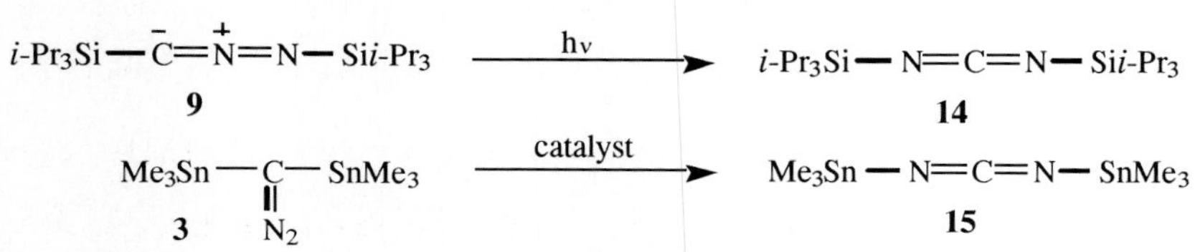

Scheme 5. Rearrangements of nitrilimines and diazo derivatives into carbodiimides.

The isomerization of **3** into **15** implies the migration of the two substituents from the carbon atom to the two nitrogen atoms, and also a rearrangement of the CNN skeleton. Since under these experimental conditions diazo derivatives usually loose dinitrogen, this complicated process is very surprising. One can easily anticipate that the migrating aptitude of the stannyl substituents is a key point of the process.

While the well known bis(trimethylsilyl)carbodiimide [17] is rather inert, the two nitrogen-heteroatom bonds of the bis(trimethylstannyl)carbodiimide **15** are reactive towards electrophiles. For example, room temperature addition of two equivalents of trimethylchlorosilane or even tributylchlorostannane to carbodiimide **15** affords bis(trimethylsilyl)- and bis(tributylstannyl)carbodiimide in 85 and 90 % yields, respectively, along with trimethylchlorostannane. However, using bulky electrophiles such as triisopropylchlorosilane, the reaction can be controlled and stopped at the first substitution giving the unsymmetrical carbodiimide **16**, which was isolated by sublimation in 85 % yield. This new mixed substituted carbodiimide **16** is still reactive towards electrophiles affording a route to other carbodiimides. Thus, addition of triisopropylchlorosilane and bis(diisopropylamino)-chlorophosphine at room temperature affords bis(triisopropylsilyl)carbodiimide **17** (90 % yield) and [bis(diisopropylamino)phosphino](triisopropylsilyl)carbodiimide **18** (78 % yield). More interestingly, carbodiimide **16** reacts in THF solution with half an equivalent of diethylaminodichlorophosphine to give bis(carbodiimide)phosphine **19** which was isolated after sulfuration as the thioxophosphoranyl derivative **20** (90 % yield) [16] (Scheme 6). The polymer cross-linker properties of this bis(carbodiimide) are under active investigation.

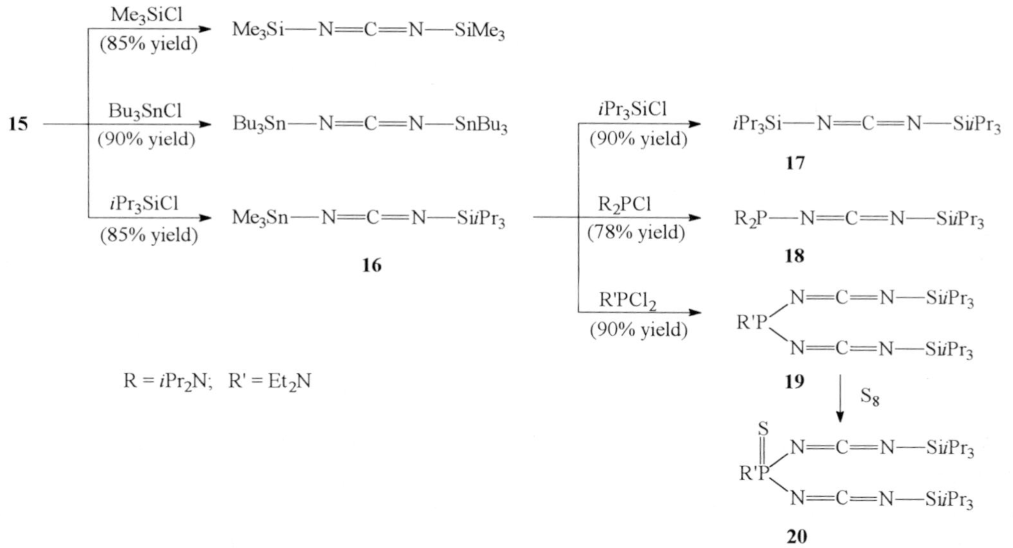

Scheme 6. Synthesis of carbodiimides

Precursors of Pseudo-Diazoalkenes

Diazoalkenes have attracted considerable interest in the last few years as potential sources of unsaturated carbenes [18], but it was only in 1990 that they were spectroscopically characterized in a matrix at 11 K [19]. Prior to this in 1987, we reported that addition of carbon tetrachloride (or tetrabromide) to the *C*-silylated α-diazophosphine **6** at room temperature led to stable diazomethylene-phosphoranes **21** and **21'**, respectively [20] (Scheme 7). The stability of these species has been explained by the ylidic character of the P=C bond which decreases the strain at the diazo carbon atom.

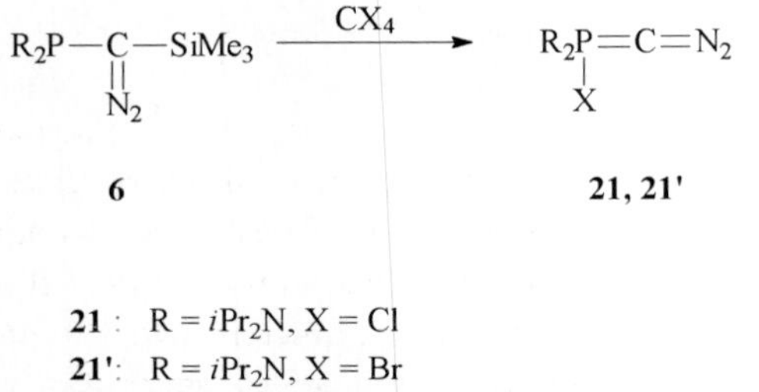

$$R_2P-\underset{\underset{N_2}{\|}}{C}-SiMe_3 \xrightarrow{\quad CX_4 \quad} R_2P{=}C{=}N_2 \atop \underset{X}{\big|}$$

6   **21, 21'**

21 : R = *i*Pr₂N, X = Cl
21': R = *i*Pr₂N, X = Br

Scheme 7.   Synthesis of diazomethylenephosphoranes

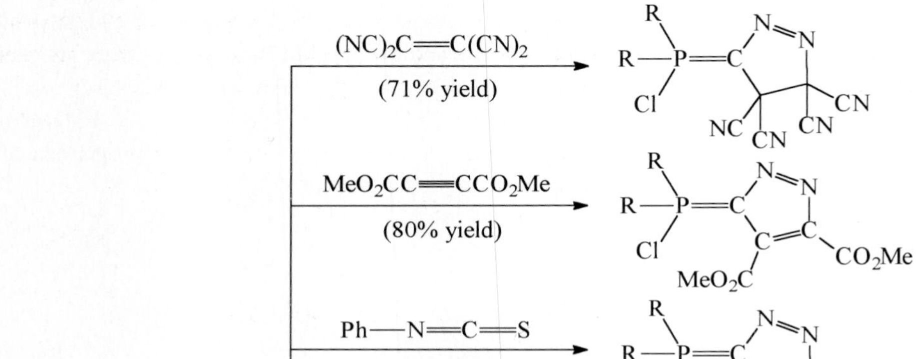

Scheme 8.   Diazomethylenephosphorane **21** as a powerful building block in heterocyclic chemistry

Besides their fundamental interest as pseudo-unsaturated diazo derivatives, diazomethylene-phosphoranes **21** and **21'** are highly functionalized molecules: they possess a diazo group which is a well-known 1,3-dipole, and a Wittig ylide moiety which is itself among the most useful building blocks in organic and inorganic chemistry. Scheme 8 gives some examples of the reactivity of **21** and clearly demonstrates the synthetic potential of these cumulenic diazo compounds in heterocyclic chemistry [20, 21]. Note that although some heterocycles result from classical [3+2] cycloaddition reactions of the diazo group with the dipolarophiles, others involve both the CN_2 group and the PC bond.

Diazomethylenephosphorane **21** can also be used as a synthetic equivalent of "naked carbon". Indeed, when a toluene solution of **21** was heated (70°C) in the presence of excess elemental sulfur, the chlorophosphine was obtained along with CS_2 (88 % yield). Carbon disulfide formally results from the trapping of "naked carbon" by sulfur. However, it is quite clear that the first step of the reaction is the nucleophilic attack of the ylidic carbon at sulfur, as observed in the reaction of elemental sulfur with Wittig ylides. The transient adduct **22** could then eliminate the chlorophosphine, giving the thioxodiazomethane **23** which finally reacts with sulfur with loss of dinitrogen (Scheme 9).

The use of **21** as "naked carbon-like" reagents merits further development since the generation of "C" usually requires very high temperatures in conjunction with extremely unselective trapping reactions [22].

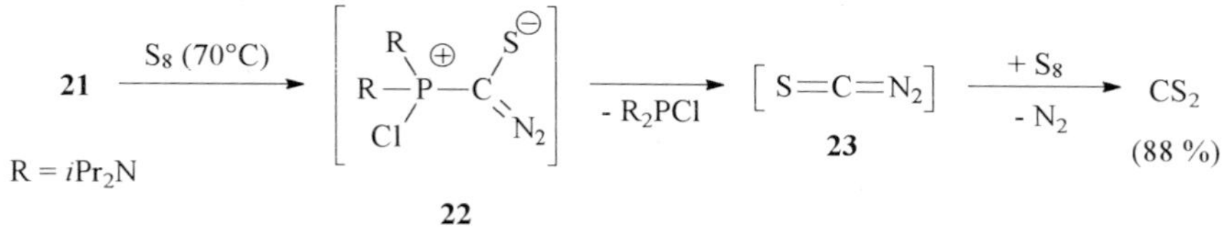

Scheme 9. Diazomethylenephosphorane **21** as a synthetic equivalent to naked carbon.

Carbene Precursors

Diazo compounds are classical precursors for carbenes. Silyl diazo compounds have been the subject of numerous papers, especially as silene precursors [23]. A few years ago, we demonstrated that diazo derivative **6** could be used in the synthesis of the stable (phosphino)(silyl)carbene **24** [5, 24]. The influence of the silyl group is apparent when comparing the stability of **24** with that of the hydrogeno **25** or tin **26** analogues. Indeed, whereas the silyl carbene **24** is stable enough to be purified by flash distillation at 75-80°C under 10^{-2} mm Hg, **25** [5b] and **26** [25] are unobservable spectroscopically even at -78°C. A theoretical study, carried out in collaboration with Arduengo's group [26] (who have discovered the only other class of stable carbene [27]), shows that the presence of the silyl group, compared to hydrogen, results in a larger bond angle at carbon; moreover, the carbon-silicon bond is somewhat shorter than a usual C–Si single bond. In other words, in the singlet state (which is about 6 kcal mol^{-1} lower in energy than the triplet), **24** has an ylidic structure **24a** (Scheme 10).

Whether carbene, germylene, and silylene are justifiable terms for stabilized versions of the reactive species is a very debatable question [28]. However, for synthetic chemists, the most important issue is to know whether these stable species feature the reactivity of transient carbenes. Formation of azaphospholidines upon thermolysis (intramolecular carbene insertion into a carbon-hydrogen bond), cyclopropanation reactions, and [1+1]-addition to isocyanides giving keteneimines are nice examples of the carbenic behaviour of phosphinocarbene **24** (Scheme 11). Since most of these results have been the subject of a recent review [29], this present paper will focus on other new topics.

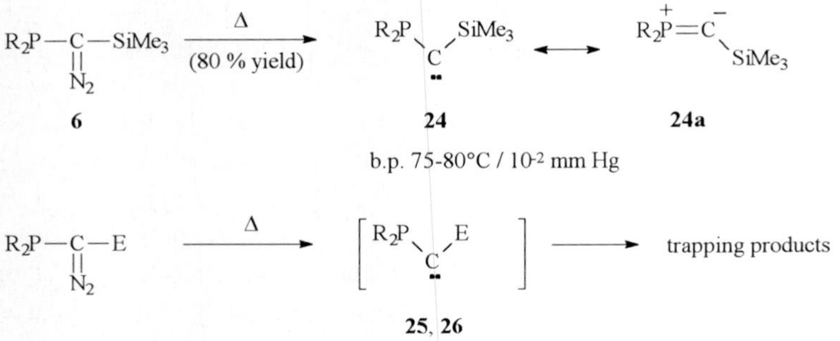

$$R_2P-\underset{\underset{N_2}{\|}}{C}-SiMe_3 \xrightarrow[\text{(80 \% yield)}]{\Delta} \underset{\mathbf{24}}{R_2P\diagdown\underset{\cdot\cdot}{C}\diagup SiMe_3} \longleftrightarrow \underset{\mathbf{24a}}{\overset{+}{R_2P}{=}\overset{-}{C}\diagdown SiMe_3}$$

6

b.p. 75-80°C / 10⁻² mm Hg

$$R_2P-\underset{\underset{N_2}{\|}}{C}-E \xrightarrow{\Delta} \left[R_2P\diagdown\underset{\cdot\cdot}{C}\diagup E \right] \longrightarrow \text{trapping products}$$

25, 26

24 : R = iPr$_2$N
25 : R = iPr$_2$N, E = H
26 : R = iPr$_2$N, E = SnMe$_3$

Scheme 10.   Synthesis of phosphino carbenes

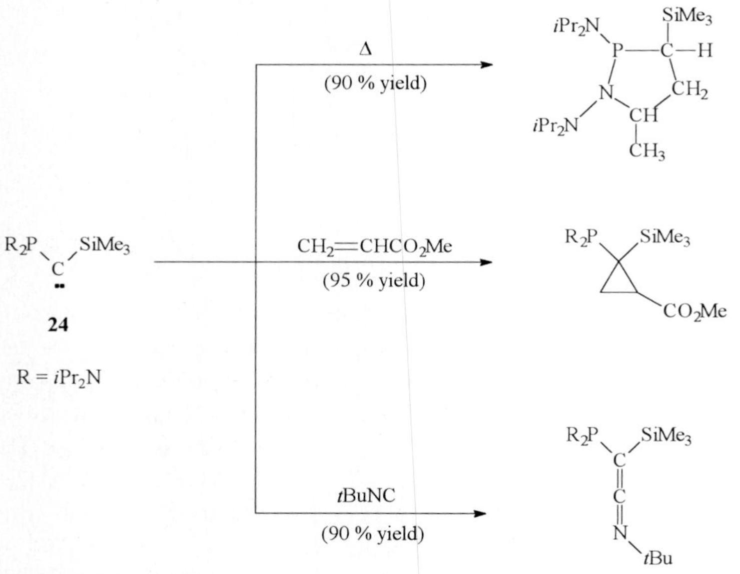

Scheme 11.   Classical carbene reactivity of **24**

Having used well known reactions of transient species to demonstrate the existence of their stabilized versions, we are now in a position to use the stabilized versions to discover new reactions applicable to the transient species. The first example is the unprecedented synthesis of a 2*H*-azirine by a [1+2]-cycloaddition. Indeed, carbene **24** reacts with a large excess of benzonitrile in toluene, at room temperature, affording azirine **27** in 85 % isolated yield [30] (Scheme 12). Interestingly, addition at room temperature of a catalytic amount of (*p*-cymene)ruthenium (II) chloride to a dichloromethane solution of **27**, leads to the $1,2\lambda^5$-azaphosphete **28** in 95 % yield [30]. The four-π-electron four-membered heterocycle **28** is indefinitely stable at room temperature (Scheme 12)!

In a very similar way, carbene **24** reacts with *t*-butyl-λ^3-phosphaalkyne affording the corresponding 2*H*-phosphirene **29** in 72 % yield [31]. Again, heating the three-membered ring **29** at 35°C gives rise to the $1\lambda^5,2\lambda^3$-diphosphete **30** in 90 % yield! Note, that the photolysis of the corresponding diazo compound **6**, in the presence of *t*-BuCP, directly led to the four-membered heterocycle **30** (90 % yield), the initial adduct **29** not being observable [32]. This is a nice example of the importance of possessing a stable carbene (Scheme 12).

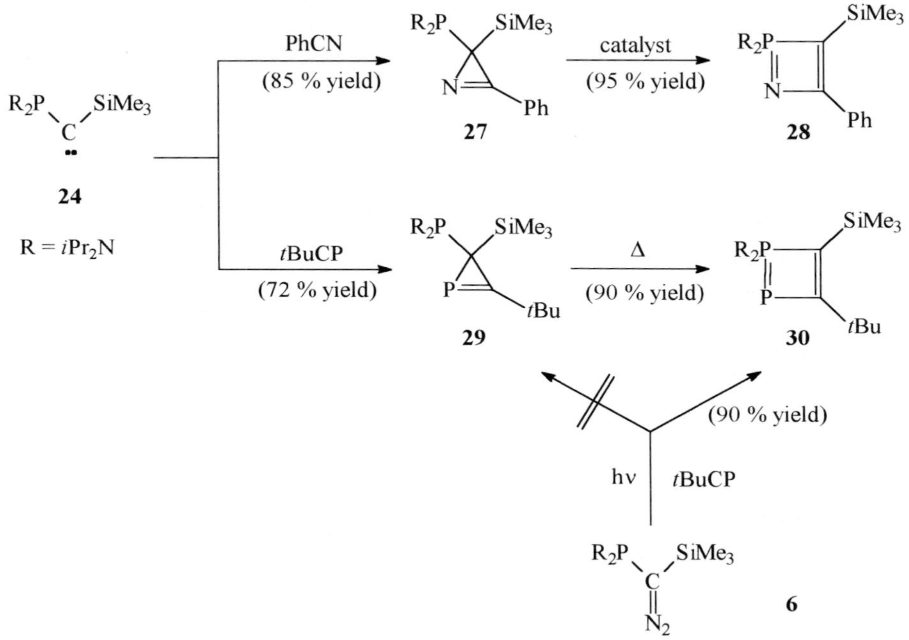

Scheme 12. Synthesis of stable four-π-electron four-membered heterocycles from **24**.

A few years ago, we found that addition of trimethylsilyl triflate to carbene **24** led to a stable methylene phosphonium salt **31** [33]. This new class of phosphorus cation is of special interest as they are valence isoelectronic with olefins. Since that time, it has been shown that the silyl groups play a key role in the stability of these types of salt, by stabilizing the positive charge in the β-position. In the absence of one of the two silyl groups, the phosphorus cation is extremely electrophilic, a property that has been used for the synthesis of another four-π-electron four-membered heterocycle **34** [34]. Indeed, when the second methylene substituent is an aryl group intramolecular ring closure occurs leading to the dihydrophosphetium salt **32**. All attempts to deprotonate **32** failed; however, in the presence of pyridine, an irreversible and quantitative isomerization into **33** occurs. Again, deprotonation of this silylated heterocycle **33** was not possible, but after cleavage of the C–Si bond by hydrolysis, addition of sodium bis(trimethylsilyl)amide afforded the benzo-λ^5-phosphete **34** in 76 % yield (Scheme 13).

The stability and thus the structure of heterocycles **28**, **30** and **34** deserves some comment. They are annulenes possessing (4n)-π-electrons and are therefore anti-aromatic according to the so-called Hückel rules. The aromatic and antiaromatic character is perturbed when one or more second-row heteroatoms are present in the ring [35]. With a third-row element (or heavier) possessing available p-orbitals to build the π-system, the comparison is even more striking: for example, neither silabenzene [36] nor germabenzene [37] are stable. When the heteroatom has no p-orbital available for the π-system, it has been shown that the six-membered ring is no longer a Hückel aromatic system [38] and, of course, the four-membered ring should not be antiaromatic. Starting from this hypothesis, six-π-electron six-membered and four-π-electron four-membered rings featuring a third row element should not be particularly stable and unstable, respectively. However, looking at the history of these species, it is amazing to realize that the first six-membered heterocycle of this type, namely cyclotriphosphazene **35** [39], was discovered in 1834, while the first vinylogue, the cyclodiphosphazene **36** (Scheme 14) was reported by our group only in 1984 [40]!

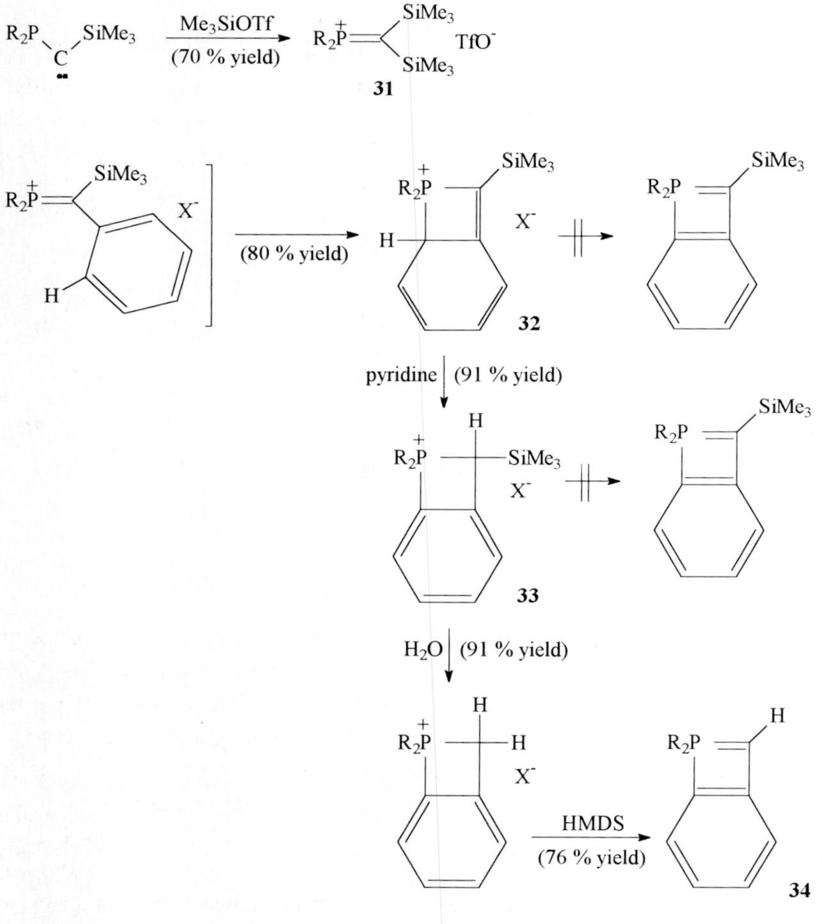

Scheme 13. Synthesis of methylephosphonium salts and subsequent formation of benzophosphete

In fact, there are very few other four-π-electron four-membered ring systems featuring a third-row (or heavier) main group element. Indeed, to the best of our knowledge, there are no stable examples involving group 14 (although silacyclobutadiene has been observed in a matrix [41]) and 16 elements, and only a couple of very recent examples involving group 13 elements [42]. According to X-ray data and ab initio calculations, heterocycles **28**, **30**, and **34** (as well as the related compounds that we have recently prepared [43]) have a rhomboidal zwitterionic structure of type **37** (Scheme 14). The positive charge is located at phosphorus, while the negative charge is delocalized on an allylic fragment.

35 **36** **37**

Scheme 14.

Just to illustrate the real synthetic interest of these heterocyclobutadienes, it appeared that trimethylsilyl isocyanate inserts into the P–N bond of **28** leading to **38**, which after hydrolysis afforded **39** in 50 % yield. Interestingly, **39** exists as a hydrogen-bonded dimer in the solid state and has a structure very similar to that of cytosine, the C(NH$_2$) being replaced by a P(NiPr$_2$)$_2$ group; it could be considered as a labelled cytosine (Scheme 15).

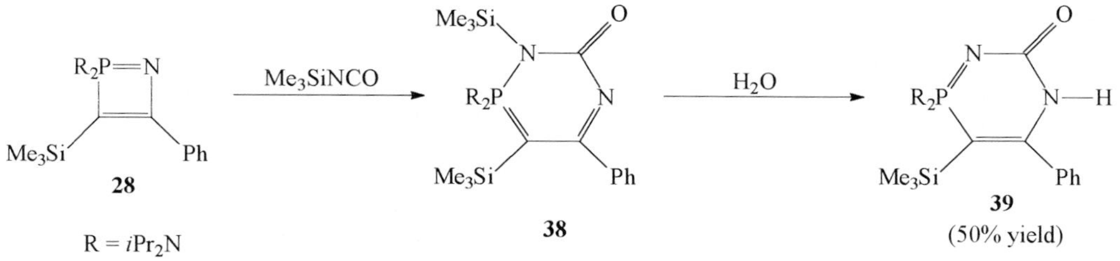

28

R = iPr$_2$N

38

39

(50% yield)

Scheme 15. Synthesis of a labeled cytosine from azaphosphete **28**

Conclusion

Trialkylsilyldiazomethane derivatives can be used to prepare new types of diazo derivatives not readily available *via* other routes. They are direct or indirect precursors of stable nitrilimines which undergo regio, stereo and diastereoselective [3+2]-dipolar cycloaddition reactions leading to a variety of five-membered heterocycles. From the silylated nitrilimines or the bis(stannyl)diazo-methane, several symmetrical and unsymmetrical carbodiimides can be synthesized including some interesting phosphorus biscarbodiimides, which are polymer cross-linking agents.

Starting from (silyl)(phosphino)diazomethane derivatives, stable pseudo-diazoalkenes and carbenes have been obtained. The diazomethylenephosphoranes are very reactive dipolarophiles but are also synthetic equivalents of naked carbon "C". The stable carbenes react in a similar manner to their transient analogues, but also exhibit new reactions; the synthesis of unsaturated three-membered heterocycles is particularly noteworthy. Ring expansion reactions involving the latter are of special interest since they afford novel non-antiaromatic four-π-electron four-membered heterocycles.

Last but not least, it should be mentioned that although this review has focused on our own results, other very spectacular findings using trimethylsilyldiazomethane have been reported in the literature. As an illustration, I have selected the synthesis of a pentacoordinate carbon cation [44] by Professor Hubert Schmidbaur, the recipient of the Wacker Prize 1996 (Scheme 16).

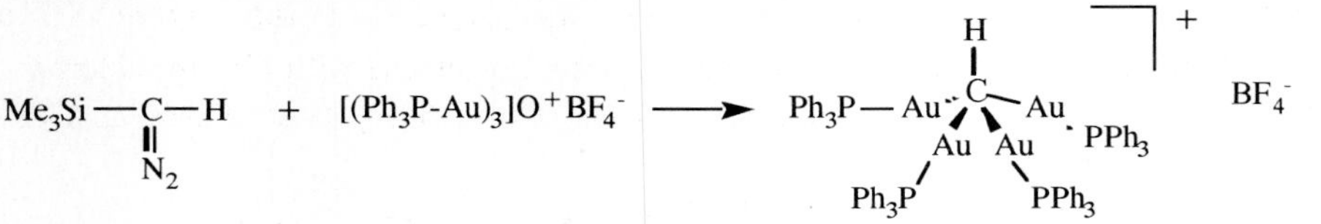

$$Me_3Si - \overset{\underset{\parallel}{\text{C}}}{\underset{N_2}{\text{C}}} - H \quad + \quad [(Ph_3P\text{-}Au)_3]O^+BF_4^- \quad \longrightarrow \quad$$

Scheme 16. Synthesis of hypercoordinate carbon

Acknowledgment: I would like to thank all my co-workers for their dedication, intellectual contribution and hard work, and the *CNRS* for generous support of this work.

References:

[1]   T. J. De Boer, H. J. Baker, *Organic Syntheses, Collect. Vol 4*, Wiley, New York, **1963**, p. 250.

[2]   a) M. Martin, *Synth. Commun.* **1983**, *13*, 809
       b) D. Seyferth, H. Menzel, A. Dow, T. Flood, *J. Organomet. Chem.* **1972**, *44*, 279.
       c) T. J. Barton, S. K. Hoekman, *Synth. React. Inorg. Met.-Org. Chem.* **1979**, *9*, 297.

[3]   a) F. Castan, A. Baceiredo, G. Bertrand, *Angew. Chem., Int. Ed. Engl.* **1989**, *28*, 1250.
       b) F. Castan, A. Baceiredo, D. Bigg, G. Bertrand, *J. Org. Chem.* **1991**, *56*, 1801.
       c) F. Castan, G. Bertrand, *Synthetic Methods of Organometallic and Inorganic Chemistry*, *Vol. 1* (Eds.: W. A. Herrmann, N. Auner, U. Klingebiel), Thieme, Stuttgart, **1996**, p. 70.

[4]   a) M. F. Lappert, J. Lorberth, J. S. Poland, *J. Chem. Soc. A* **1970**, 2954.
       b) R. Réau, G. Bertrand, *Synthetic Methods of Organometallic and Inorganic Chemistry*, *Vol. 2*, Eds.: W. A. Herrmann, A. Salzer), Thieme, Stuttgart, **1996**, p. 286.

[5]   a) A. Baceiredo, G. Bertrand, G. Sicard, *J. Am. Chem. Soc.* **1985**, *107*, 3945.
       b) A. Igau, H. Grützmacher, A. Baceiredo, G. Bertrand, *J. Am. Chem. Soc.* **1988**, *110*, 6463.

[6]   M. J. Menu, P. Desrosiers, M. Dartiguenave, Y. Dartiguenave, G. Bertrand, *Organometallics* **1987**, *6*, 1822.

[7]   R. Réau, R. W. Reed, F. Dahan, G. Bertrand, *Organometallics* **1993**, *12*, 1501.

[8]   a) S. Patai, *The chemistry of diazonium and diazo groups*, Wiley, New York, **1978**.
       b) M. Regitz, *Diazoalkanes*, Georg Thieme Verlag, Stuttgart, **1977**.
       c) M. Regitz, G. Maas, *Diazo Compounds, Properties and Synthesis*, Academic Press Inc., Orlando, **1986**.

[9]   For a review on stable nitrilimines see: G. Bertrand, C. Wentrup, *Angew. Chem., Int. Ed. Engl.* **1994**, *33*, 527.

[10]   a) G. Sicard, A. Baceiredo, G. Bertrand, *J. Am. Chem. Soc.* **1988**, *110*, 2663.
       b) M. Granier, A. Baceiredo, Y. Dartiguenave, M. Dartiguenave, M. J. Menu, G. Bertrand, *J Am. Chem. Soc.* **1990**, *112*, 6277.

[11]   a) H. Meier, W. Heinzelmann, H. Heimgartner, *Chimia* **1980**, *34*, 504; *Chimia* **1980**, *34*, 506.
       b) N. H. Toubro, A. Holm, *J. Am. Chem. Soc.* **1980**, *102*, 2093.

c) C. Wentrup, S. Fischer, A. Maquestiau, R. Flammang, *Angew. Chem., Int. Ed. Engl.* **1985**, *24*, 56.

d) H. Bock, R. Dammel, S. Fischer, C. Wentrup, *Tetrahedron Lett.* **1987**, *28*, 617.

[12] The reaction of the lithium salt of (trimethylsilyl)diazomethane with trimethylchlorosilane was reported to give bis(trimethylsilyl)diazomethane: D. Seyferth, T. C. Flood, *J. Organomet. Chem.* **1971**, *29*, C25.

[13] a) R. Réau, G. Vénéziani, G. Bertrand, *J. Am. Chem. Soc.* **1992**, *114*, 6059.

b) R. Réau, G. Vénéziani, F. Dahan, G. Bertrand, *Angew. Chem., Int. Ed. Engl.* **1992**, *31*, 439.

[14] C. Leue, R. Réau, B. Neumann, H. G. Stammler, P. Jutzi, G. Bertrand, *Organometallics* **1994**, *13*, 436.

[15] J. L. Fauré, R. Réau, M. W. Wong, R. Koch, C. Wentrup, G. Bertrand, *J. Am. Chem. Soc.* **1997**, *119*, 2819.

[16] a) G. Vénéziani, R. Réau, G. Bertrand, *Organometallics* **1993**, *12*, 4289.

b) G. Vénéziani, P. Dyer, R. Réau, G. Bertrand, *Inorg. Chem.* **1994**, *33*, 5639.

[17] M. V. Vovk, L. I. Samarai, *Russ. Chem. Rev.* **1992**, *61*, 548.

[18] For reviews on alkylidene carbenes see: a) P. J. Stang, *Chem. Rev.* **1978**, *78*, 383.

b) P. J. Stang, *Acc. Chem. Res.* **1978**, *11*, 107.

c) P. J. Stang, *Acc. Chem. Res.* **1982**, *15*, 348.

d) D. Bourissou, G. Bertrand, *C. R. Acad. Sci. Paris,* **1996**, *322*, 489.

[19] J. C. Brahms, W. P. Dailey, *J. Am. Chem. Soc.* **1990**, *112*, 4046.

[20] J. M. Sotiropoulos, A. Baceiredo, G. Bertrand, *J. Am. Chem. Soc.* **1987**, *109*, 4711.

[21] a) J. M. Sotiropoulos, A. Baceiredo, K. Horchler von Locquenghien, F. Dahan, G. Bertrand, *Angew. Chem., Int. Ed. Engl.* **1991**, *30*, 1154.

b) J. M. Sotiropoulos, A. Baceiredo, G. Bertrand, *Bull. Soc. Chim. Fr.* **1992**, *129*, 367.

c) A. Baceiredo, M. Nieger, E. Niecke, G. Bertrand, *Bull. Soc. Chim. Fr.* **1993**, *130*, 757.

d) A. Baceiredo, R. Réau, G. Bertrand, *Bull. Soc. Chim. Belg.* **1994**, *103*, 531.

e) N. Dubau-Assibat, A. Baceiredo, F. Dahan, G. Bertrand, *Bull. Soc. Chim. Fr.* **1995**, *132*, 1139.

[22] a) M. L. Mckee, M. L. Paul, P. B. Shevlin, *J. Am. Chem. Soc.* **1990**, *112*, 3374.

b) P. B. Shevlin, *Reactive Intermediates* (Ed. R. A. Abramovitch), Plenum, New York, **1990**, p. 1.

[23] a) N. Wiberg, *J. Organomet. Chem.* **1984**, 273, 141.

b) A. G. Brook, K. M. Kaines, *Adv. Organomet. Chem.* **1974**, *66*, 29.

c) M. Trommer, W. Sander, A. Patyk, *J. Am. Chem. Soc.* **1993**, *115*, 11775.

d) M. Trommer, W. Sander, C. H. Ottosson, D. Cremer, *Angew. Chem., Int. Ed. Engl.* **1995**, *34*, 929.

[24] a) A. Igau, A. Baceiredo, G. Trinquier, G. Bertrand, *Angew. Chem., Int. Ed. Engl.* **1989**, *28*, 621.

b) see also: G. Gillette, A. Baceiredo, G. Bertrand, *Angew. Chem., Int. Ed. Engl.* **1990**, *29*, 1429.

[25] N. Emig, J. Tejeda, R. Réau, G. Bertrand, *Tetrahedron Lett.* **1995**, *36*, 4231.

[26] D. A. Dixon, K. D. Dobbs, A. J. Arduengo, G. Bertrand, *J. Am. Chem. Soc.* **1991**, *113*, 8782.

[27] a) A. J. Arduengo, R. L. Harlow, M. Kline, *J. Am. Chem. Soc.* **1991**, *113*, 361.

b) A. J. Arduengo, H. V. R. Dias, R. L. Harlow, M. Kline, *J. Am. Chem. Soc.* **1992**, *114*, 4430.

c) A. J. Arduengo, H. V. R. Dias, D. A. Dixon, R. L. Harlow, W. T. Klooster, T. F. Koetzle, *J. Am. Chem. Soc.* **1994**, *116*, 6812.

d) A. J. Arduengo, D. A. Dixon, K. K. Kumashiro, C. Lee, W. P. Power, K. W. Zilm, *J. Am. Chem. Soc.* **1994**, *116*, 6361.

e) A. J. Arduengo, H. Bock, H. Chen, M. Denk, D. A. Dixon, J. C. Green, W. A. Herrmann, N. L. Jones, M. Wagner, R. West, *J. Am. Chem. Soc.* **1994**, *116*, 6641.

[28] a) R. Dagani, *Chem. Eng. News* **1994**, *72(18)*, 20

b) R. Dagani, *Chem. Eng. News* **1991**, *69(4)*, 19.

c) M. Regitz, *Angew. Chem., Int. Ed. Engl.* **1991**, *30*, 674.

[29] G. Bertrand, R. Reed, *Coordination Chem. Rev.* **1994**, *137*, 323.

[30] G. Alcaraz, U. Wecker, A. Baceiredo, F. Dahan, G. Bertrand, *Angew. Chem., Int. Ed. Engl.* **1995**, *34*, 1246.

[31] M. Sanchez, R. Réau, G. Bertrand, unpublished results.

[32] R. Armbrust, M. Sanchez, R. Réau, U. Bergstrasser, M. Regitz, G. Bertrand, *J. Am. Chem. Soc.* **1995**, *117*, 10785.

[33] A. Igau, A. Baceiredo, H. Grützmacher, H. Pritzkow, G. Bertrand, *J. Am. Chem. Soc.* **1989**, *111*, 6853.

[34] U. Heim, H. Pritzkow, U. Fleischer, H. Grützmacher, M. Sanchez, R. Réau, G. Bertrand, *Eur. J. Chem.* **1996**, *2*, 68.

[35] W. W. Schoeller, T. Busch, *Angew. Chem., Int. Ed. Engl.* **1993**, *32*, 617.

[36] a) B. Solouki, P. Rosmus, H. Bock, G. Maier, *Angew. Chem., Int. Ed. Engl.* **1980**, *19*, 51.

b) G. Maier, G. Mihn, P. Reisenauer, *Angew. Chem., Int. Ed. Engl.* **1980**, *19*, 523.

[37] G. Märkl, D. Rudnick, R. Schulz, A. Schweig, *Angew. Chem., Int. Ed. Engl.* **1982**, *21*, 221.

[38] a) D. H. R. Barton, M. B. Hall, Z. Lin, S. I. Parekh, *J. Am. Chem. Soc.* **1993**, *115*, 955.

b) M. J. S. Dewar, E. A. C. Lucken, M. A. Whitehead, *J. Chem. Soc.* **1960**, 2423.

[39] a) J. Liebig, *Ann. Chem.* **1834**, *11*, 139.

b) H. Rose, *Ann. Chem.* **1834**, *11*, 131.

[40] a) A. Baceiredo, G. Bertrand, J. P. Majoral, G. Sicard, J. Jaud, J. Galy, *J. Am. Chem. Soc.* **1984**, *106*, 6088.

b) A. Baceiredo, G. Bertrand, J. P. Majoral, F. El Anba, G. Manuel, *J. Am. Chem. Soc.* **1985**, *107*, 3945.

[41] J. R. Gee, W. A. Howard, G. L. McPherson, M. J. Fink, *J. Am. Chem. Soc.* **1991**, *113*, 5461.

[42] S. Schulz, L. Häming, R. Herbst-Irmer, H. W. Roesky, G. M. Sheldrick, *Angew. Chem., Int. Ed. Engl.* **1994**, *33*, 969.

[43] a) J. Tejeda, R. Réau, F. Dahan, G. Bertrand, *J. Am. Chem. Soc.* **1993**, *115*, 7880.

b) G. Alcaraz, A. Baceiredo, M. Nieger, G. Bertrand, *J. Am. Chem. Soc.* **1994**, *116*, 2159.

c) K. Bieger, J. Tejeda, R. Réau, F. Dahan, G. Bertrand, *J. Am. Chem. Soc.* **1994**, *116*, 8087.

d) G. Alcaraz, V. Piquet, A. Baceiredo, F. Dahan, W. W. Schoeller, G. Bertrand, *J. Am. Chem. Soc.* **1996**, *118*, 1060.

e) K. Bieger, G. Bouhadir, R. Réau, F. Dahan, G. Bertrand, *J. Am. Chem. Soc.* **1996**, *118*, 1038.

f) G. Alcaraz, A. Baceiredo, M. Nieger, W. W. Schoeller, G. Bertrand, *Inorg. Chem.* **1996**, *35*, 2458.

[44] H. Schmidbaur, F. P. Gabbai, A. Schier, J. Riede, *Organometallics*, **1995**, *14*, 4969.

A New Route to Silaheterocycles:
Nucleophilic Aminomethylation

Hans H. Karsch*, Kai A. Schreiber

Anorganisch-Chemisches Institut
Technische Universität München
Lichtenbergstr. 4, D-85747 Garching, Germany
Tel./Fax: Int. code + (89)32093132
E-mail: Hans.H.Karsch@lrz.tu-muenchen.de

Keywords: Diazasilaheterocycles / Aminomethylation / Lithium methylamines

Summary: By metalation of the respective aminals with t-BuLi, two different double lithiated methylmethylenediamines, namely $LiCH_2N(CH_3)CH_2N(CH_3)CH_2Li$ and $LiCH_2N(C_6H_{11})CH_2N(C_6H_{11})CH_2Li$, were obtained and isolated. The reaction of both compounds with difunctional chlorosilanes yield six-membered heterocycles including $(CH_3)_2SiCH_2N(CH_3)CH_2N(CH_3)CH_2$ by nucleophilic amino methylation. A spiro heterocycle $Si[CH_2N(CH_3)CH_2N(CH_3)CH_2]_2$ can be synthsized in the same way by using two equivalents of lithiated tetramethylmethylenediamine and $SiCl_4$. The compounds have been characterized by mass, ^1H, ^{13}C, and ^{29}Si NMR spectroscopy and elemental analysis.

Introduction

Aminomethylation reactions are of major interest in elementorganic and organic chemistry. Normally these reactions are restricted to electrophilic aminomethylation reactions, for example, the well known Mannich reaction. This is due to the fact that postively charged carbon centers in the α-position to nitrogen atoms are highly stabilized through iminium ion formation. Metalation of tertiary amines needs activated species in most cases [1]. Lithiation often successfully is performed via the stannyl intermediate [2] or via telluride-lithium-exchange [3], but the overall yield is modest, the procedure is time consuming and the separation from the byproducts LiCl and SnBu$_4$, which are discarded, sometimes is difficult. Metalation reactions of simple, unsubstituted aminomethyl functions with alkyllithium bases normally are unsuccessful or proceed with low yields. These difficulties can be avoided by using methyl-substituted 1,3-diaza-compounds. A series of silicon containing compounds has now been synthsized (Eq. 1) by using the double lithiated tetramethylmethylenediamine (TMMDA) **1a** or dicyclohexyldimethylmethylenediamine (CMMDA) **1b**, which precipitate after treatment of TMMDA or CMMDA with t-BuLi in pentane as highly pyrophoric substances.

Stabilization of **1a,b** probably is achieved by the formation of inter- or intramolecular Li-N adducts.

Eq. 1.   Synthesis of **1a,b**.

Results

Compounds **1a,b** have been identified by elemental analysis and by formation of derivatives, which confirm that metalation takes place in the described manner. Thus, **1a,b** are silylated by monochlorsilanes $ClSiR_3$ (R = CH$_3$, Ph) (Eq. 2). The formation of open chain compounds **2a,b** and **3a** was performed in toluene.

Eq. 2.   Synthesis of **2a,b**, and **3a**.

These compounds were obtained as colorless or pale yellow liquids and were identified by spectroscopic methods. The $^{13}C\{^1H\}$ NMR spectra of **2a,b** and **3a** show a $^{13}C^{29}Si$-coupling of the methylene carbon atoms in the α-position to the silicon centers. Thus, silylation of the former lithiated methylene groups has been achieved. These reactions demonstrate that nucleophilic aminomethylation with double lithiated aminals TMMDA or CMMDA at silicon centers is feasible and open the way for the synthesis of 1,3-diaza-5-silacyclohexanes if dichlorosilanes are used.

Substitution reactions of **1a,b** with dichlordimethylsilane led to the formation of *N*-methyl or *N*-cyclohexyl substituted 1,3-diaza-5-silacyclohexanes **4a,b** (Eq. 3).

Eq. 3.   Synthesis of **4a,b**.

Again, using toluene as solvent gives the best results. Cyclic ethers, polyethers or diethyl ether are not recommended because reactions with **1a,b** lead to decomposition of these solvents below room temperature. The methyl-substituted compound **4a** is purified by distillation and identified by elemental analysis, mass and NMR spectroscopy. ^{1}H NMR investigations on **4a** at -100°C lead to the conclusion that ring inversion is fast on the NMR time scale. NMR studies and conformational analysis on hexahydropyrimidines [4] are consistent with slowly inverting rings at low temperature (-70°C). In our case ring inversion is fast even at low temperature probably because of the elongated Si–C bond. The tetrahedral bond angles at silicon should be more deformable than those at carbon and this may lead to some relief of internal strain when a carbon atom is replaced by silicon.

The pure compound **4a** is used for further reactions. In order to obtain solid or crystalline products, the 1,3-diaza-compound can be transferred into the mono-quarternary ammonium- salt **5a** [5]. (Eq. 4) by using MeI as methylating agent.

Eq. 4.   Synthesis of **5a**.

The bis-quarternary ammonium salt has not yet been synthesized, even with a polar solvent (CH$_3$CN). **5a** is obtained as colorless crystals. Elemental analysis and NMR spectroscopy confirm the considered structure.

Nucleophilic aminomethylation is even possible with sterical demanding dichlorosilanes. Eq. 5 shows the synthesis of *N*-methyl substituted 1,3-diaza-silacyclohexanes with other substituents than methyl connected to the silicon atom [6].

6a : R$_1$ = Ph, R$_2$ = CH$_3$
7a : R$_1$ = ^{t}Bu, R$_2$ = Ph
8a : R$_1$ = Ph, R$_2$ = Ph

Eq. 5.   Synthesis of **6a**, **7a**, and **8a**.

Compounds **6a**, **7a**, and **8a** were isolated as viscous, colorless liquids. These compounds can be identified with NMR and GC-MS spectroscopy. In some cases unwanted and yield reducing side reactions like oligomerization or monoalkylation of the dichlorosilanes occur. Side products are not detected by NMR specroscopy, but GC-MS studies definitely show peaks at higher masses. In the ^{1}H NMR spectra of **6a** and **7a** well-separated signals can be observed for the diastereotopic methylene protons at room temperature. Obviously the ring inversion process is slowed down if bulky substituents are used.

Symmetrical silicon-containing spiranes have already been prepared directly from alkyldilithium reagents and silicon tetrachloride [7]. Using dilithiated 1,3-diamines as alkylating agents a new way of synthesizing spirocycles with nitrogen atoms in the β-positions to the silicon is achieved (Eq. 6).

Eq. 6.   Synthesis of **9a**.

Compound **9a** is obtained as a colorless liquid and purified by fractional distillation. The structure is confirmed by NMR and mass spectroscopy.

Conclusion

Although there are no informations available concerning the structure of the dilithiated 1,3-diamines in solid phase or in solution, subsequent reactions of **1a,b** with di- and trichlorosilanes support the existence of these dianions. **1a,b** are important building blocks in organoelement synthesis, not only restricted to the field of silicon chemistry. Remarkable results have been achieved in reactions with *N*- or *C*-substituted dichlorophosphanes. An extension to germanium and tin is possible, but side reactions or lower yields are often observed. Nevertheless, the facile double metalation of TMMDA and CMMDA is unique for tertiary amines (e.g., the phosphorus homologue $(CH_3)_2PCH_2P(CH_3)_2$ is only metalated at the methylene carbon atom [8]).

References:

[1]   a) P. Beak, D. B. Reitz, *Chem. Rev.* **1978**, *78*, 275.
       b) D. Seebach, J. J. Lohmann, M. A. Syfrig, M. Yoshifuji, *Tetrahedron* **1983**, *39*, 1938.

[2]   D. J. Peterson, H. R. Hays, *J. Org. Chem.* **1965**, *30*, 1939.

[3]   N. Petragnani, J. V. Comasseto, *Synthesis* **1991**, *10*, 897.

[4]   a) F. G. Riddell, *J. Chem. Soc. (B)* **1967**, 560.
       b) R. F. Farmer, J. Hamer, *Tetrahedron* **1967**, *24*, 829.

[5]   H. Böhme, M. Dähne, W. Lehners, E. Ritter, *Liebigs Ann. Chem.* **1969**, *723*, 34.

[6]   H. H. Karsch, K. A. Schreiber, *Chem. Ber.*, submitted.

[7]   R. West, *J. Am. Chem. Soc.* **1954**, *76*, 6012.

[8]   H. H. Karsch, H. Schmidbaur, *Z. Naturforsch.* **1977**, *32b*, 762.

Infrared and Raman Spectra, ab initio Calculations, and Rotational Isomerism of Methylated Disilanes

Karla Schenzel, Anke Jähn
Institut für Analytik und Umweltchemie
Martin-Luther-Universität Halle
Kurt-Mothes-Str. 2, D-06120 Halle, Germany

Margot Ernst
Institut für Theoretische Chemie
Karl Franzens Universität
Mozartgasse 14, A-8010 Graz, Austria

Karl Hassler
Institut für Anorganische Chemie
Technische Universität Graz
Stremayrgasse 16, A-8010 Graz, Austria
E-mail: hassler@fscm1.tu-graz.ac.at

Keywords: Methylhalodisilanes / IR Spectroscopy / Raman Spectroscopy / ab initio
Calculations

Summary: Methylated halodisilanes MeX_2SiSiX_2Me and $Me_2XSiSiXMe_2$ with X = F, Cl, Br and I exist as mixtures of *anti* and *gauche* rotamers, as can be deduced from the temperature dependence of their dipole moments [1,2]. We have investigated the Raman spectra of $CH_3Cl_2SiSiCl_2CH_3$ and $CD_3Cl_2SiSiCl_2CD_3$ at various temperatures. Energy differences $H_{anti} - H_{gauche}$ of -0.9 $\pm$ 0.2 and -1.2 $\pm$ 0.2 kJmol^{-1} have been determined for the two isotopomers using the Van´t Hoff equation. The vibrational spectra of the *anti* and *gauche* rotamers of both isotopomers have been assigned with the help of ab initio harmonic force fields. The calculations were performed using standard RHF theory, core potentials (ECP´s) for the heavy atoms and a double zeta basis with polarization functions on all atoms.
Further studies of the Raman spectra of MeX_2SiSiX_2Me and $Me_2XSiSiXMe_2$ revealed that all these disilanes exist as mixtures of *anti* and *gauche* rotamers.

By analogy with 1,2-disubstituted ethanes it is reasonable to assume that 1,2-disubstituted disilanes will also exist as mixtures of two conformers, *anti* and *gauche*. If the barrier separating the two conformers is considerably larger than 2.5 kJmol^{-1} (= kT at room temperature), they will exist as well-defined species on the time scale which is typical for infrared and Raman spectroscopy. Energy differences ΔH between rotational isomers can then be determined with temperature dependent infrared and Raman spectroscopy by applying the Van't Hoff equation: $\ln I_a/I_g = -\Delta H/RT + \text{const.}$

I_a and I_g are the intensities of vibrational bands belonging to rotamer *anti* (*a*) or *gauche* (*g*). By plotting $\ln I_a/I_g$ against T^{-1}, $\Delta H/R$ can be determined. For methylated halodisilanes, useful informations of molecular conformations can be obtained from skeletal vibrations. The SiSi and SiX stretching vibrations of chloro, bromo, and iododisilanes that are most sensitive to molecular conformations fall into the range between 550 and 250 cm^{-1}. Skeletal deformations that are also influenced by molecular conformations lie below 250 cm^{-1}. Fig. 1 presents the Raman spectra of $CH_3Cl_2SiSiCl_2CH_3$ and $CD_3Cl_2SiSiCl_2CD_3$ in the liquid (300 K) and crystalline (173 K) state.

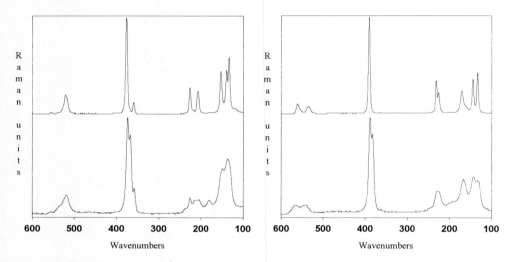

Fig. 1.   Raman spectra < 600 cm^{-1} of MeCl$_2$SiSiCl$_2$Me (left) and the D$_6$-isotopomer (right) in the liquid and crystalline state (upper spectra).

As can be clearly observed from Fig. 1, the stretching vibration ν SiSi around 385 cm^{-1} (H$_6$-isotopomer) and 375 cm^{-1} (D$_6$-isotopomer) is split into a doublet by rotational isomerism. Their intensity ratio is temperature-dependent and can be determined by deconvolution, as presented in Fig. 2. Van't Hoff plots of the intensity ratios against inverse temperature are depicted in Fig. 3.

The resulting values for ΔH ($= H_{anti} - H_{gauche}$) are -0.9 ± 0.2 and -1.2 ± 0.2 kJmol^{-1} for the H$_6$ and the D$_6$-isotopomer, respectively. The values agree satisfactorily within the estimated error limits of ± 0.2 kJmol^{-1}, the *anti* conformation being more stable than the *gauche* conformation. From empirical force field calculations, a ΔH value of -5.15 kJmol^{-1} (-1.23 kcalmol^{-1} [3]) was deduced, which is significantly larger than the value obtained by Raman spectroscopy.

For the determination of the sign of ΔH it was necessary to assign the fundamental vibrations of both *anti* and *gauche* 1,2-dimethyltetrachlorodisilane. We have therefore carried out ab initio calculations of structures, harmonic frequencies, and harmonic force fields. As this work is part of a series of similar investigations of methylated halodisilanes Me$_n$Si$_2$X$_{6-n}$ with X= F, Cl, Br, and I, the core electrons of the heavy atoms were represented by a non relativistic core potential (ECP) described in [4] in order to ensure comparability of the results within the halogen series and as a compromise between accuracy and computational costs. A symmetry constrained geometry optimization for both *anti* and *gauche* 1,2-dimethyltetrachlorodisilane was performed using standard restricted Hartree Fock (RHF) theory employing a double zeta basis with polarization functions on all atoms. Harmonic frequency analysis was performed by numerical differentiation at the equilibrium geometries. As Hartree Fock frequencies are known to be too high by a factor of approximately 0.90, a single scaling factor of 0.92 was chosen for best agreement between calculated and experimental data. The program package GAMESS was used to perform these calculations.

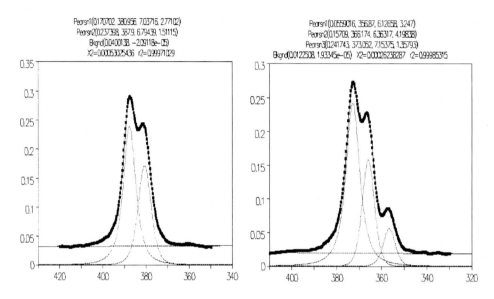

Fig. 2. Deconvolution of ν_s SiCl$_2$ of CH$_3$Cl$_2$SiSiCl$_2$CH$_3$ (left) and CD$_3$Cl$_2$SiSiCl$_2$CD$_3$ (right).

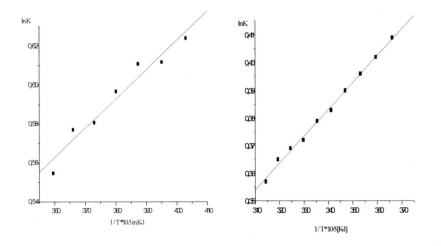

Fig. 3. Van't Hoff plots of the intensity ratio of ν_s SiCl$_2$ for CD$_3$Cl$_2$SiSiCl$_2$CD$_3$ (373, 366 cm^{-1}; left) and CH$_3$Cl$_2$SiSiCl$_2$CH$_3$ (388, 382 cm^{-1}; right).

Table 1 summarizes the infrared and Raman spectra of $CH_3Cl_2SiSiCl_2CH_3$ and $CD_3Cl_2SiSiCl_2CD_3$. As we were not interested in ν_s, ν_{as}, δ_s, and δ_{as} CH_3/CD_3 vibrations, they have been omitted from Table 1.

Table 1. Infrared and Raman spectra [cm^{-1}] of $CH_3Cl_2SiSiCl_2CH_3$ and $D_3Cl_2SiSiCl_2CD_3$ in the region < 900 cm^{-1}.

CH₃Cl₂SiSiCl₂CH₃			CD₃Cl₂SiSiCl₂CD₃		
Ir (l)	Ra (l)	Ra (s)	Ir (l)	Ra (l)	Ra(s)
848 w					
789 vs		781 vw	705 sh	708 m	711 s
765 vs					
		750 sh			
731 vs	743 m, b	745 m	676 vs	665 w, b	668 ms
			642 vs		628 w
			610 s		
	680 vvw	680 vvw	602 s		605 w
	565 m	561 m	550 sh	552 vw	552 w
550 vvs, b	545 m		532 vs	536 m	
	535 m	535 m		520 ms	520 vs
477 vs			480 sh		
469 vs			462 vs		
	388 vs, p	390 vs, p		373 vs, p	375 vs
380 ms	382 vs, p		366 mw	366 vs, p	
	363 vw	363 vw	355 w	357 m	359 m
243 vs	245 vvw		234 s	236 vw	236 vw
		231 vs	224 s	225 m	225 vs
228 sh	226 m, b	225 s	215 sh	216 m	216 w
195 sh	190 mw, b			203 m	206 vs
181 s	165 s	170 s	177 s	178 m	
		165 sh			
145 w	141 s	144 s		150 vs	151 vs
	133 s	133 s		135 vs	138 vs
			128 mw		133 vs
94 w					120 m
	85 w	85 vw		84 w, b	90 w

The ab initio calculations predict CSiSiC torsion angles of 179.23° and 69.34° for the *anti* and *gauche* conformation, respectively. *Anti* $CH_3Cl_2SiSiCl_2CH_3$ therefore belongs to a good approximation to point group C_{2h}, the *gauche* conformer to point group C_2. The distribution of 36 normal modes between the symmetry species as well as Infrared and Raman activities will be

$$C_{2h}: \Gamma_{vib} = 11\ Ag(Ra,p) + 7\ Bg(Ra) + 8\ Au(\ IR) + 10\ Bu(IR)$$
$$C_2: \Gamma_{vib} = 19\ A(Ra,p,IR) + 17\ B(Ra,IR)$$

The spectra of the *anti* conformer will follow the rule of mutual exclusion, as the molecules possess an inversion centre. The selection rules predict five Raman active skeleton deformations for the *anti* conformer. For the crystalline state (Fig. 1), five deformations are observed, leaving no doubt that the *anti* conformer is energetically favored. Table 2 summarizes the calculated and observed skeleton stretching vibrations of both isotopomers. According to the correlation tables, A_g and A_u vibrations of the *anti* conformer (point group C_{2h}) combine into A vibrations of point group C_2, B_g, and B_u vibrations into vibrations belonging to symmetry species B.

Table 2. Calculated and observed heavy atom stretching vibrations (cm^{-1}) of H_6 and D_6 1,2-dimethyl-tetrachlorodisilane.

anti $CH_3Cl_2SiSiCl_2CH_3$					*gauche* $CH_3Cl_2SiSiCl_2CH_3$			
species	ab initio	scaled	exp.	vibration	species	ab initio	scaled	exp.
A_g	404.89	372	388	ν SiSi	A	397.41	366	382
	596.05	548	565	$ν_s$ $SiCl_2$		609.84	561	565
	778.72	716	745	ν SiC		768.69	707	743
A_u	587.23	540	550	$ν_{as}$ $SiCl_2$		609.84	561	565
B_g	574.81	529	535	$ν_{as}$ $SiCl_2$	B	589.19	542	545
B_u	500.96	461	466	$ν_s$ $SiCl_2$		501.40	461	477
	766.45	705	731	ν SiC		764.56	703	743

anti $CD_3Cl_2SiSiCl_2CD_3$					*gauche* $CD_3Cl_2SiSiCl_2CD_3$			
species	ab initio	scaled	exp.	vibration	species	ab initio	scaled	exp.
A_g	389.36	358	366	ν SiSi	A	381.94	351	373
	556.93	512	520	$ν_s$ SiCl2		550.81	507	520
	708.73	652	665	ν SiC		696.46	641	642
A_u	566.81	521	532	$ν_{as}$ $SiCl_2$		572.31	527	552
B_g	556.75	512	520	$ν_{as}$ $SiCl_2$	B	567.80	522	536
B_u	490.83	452	462	$ν_s$ $SiCl_2$		490.75	451	462
	709.10	652	642	ν SiC		704.36	648	642

The calculated harmonic force field given in cartesian coordinates was transformed into a force field defined by symmetry coordinates. A normal coordinate analysis was performed using the geometry and the symmetry force constants from ab initio calculations, resulting in a description of the normal modes by their potential energy distribution (PED). Table 3 presents the potential energy distribution for ν SiSi and ν_s SiCl$_2$ for both rotamers, and Table 4 summarizes the SiSi, SiCl, and SiC stretching force constants.

Table 3. Potential energy distribution of ν SiSi and ν_s SiCl$_2$ for *anti* and *gauche* 1,2-dimethyltetrachlorosilane.

anti CH$_3$Cl$_2$SiSiCl$_2$CH$_3$	*gauche* CH$_3$Cl$_2$SiSiCl$_2$CH$_3$
388: 32 (ν SiSi), 38 (ν_s SiCl$_2$)	382: 35 (ν SiSi), 38 (ν_s SiCl$_2$)
565: 33 (ν_s SiCl$_2$), 35 (ν SiSi)	520: 42 (ν_s SiCl$_2$), 28 (ν SiSi)
anti CD$_3$Cl$_2$SiSiCl$_2$CD$_3$	*gauche* CD$_3$Cl$_2$SiSiCl$_2$CD$_3$
366: 33 (ν SiSi), 34 (ν_s SiCl$_2$)	373: 37 (ν SiSi), 33 (ν_s SiCl$_2$)
561: 48 (ν_s SiCl$_2$), 22 (ν SiSi)	520: 31 (ν_s SiCl$_2$), 13 (ν SiSi)

The potential energy distributions given in Table 3 illustrate the strong vibrational coupling between SiSi and SiCl stretching modes, which is common to all halogenated disilanes. Because the mass of an Si atom is quite similar to that of an Cl atom, the vibrational coupling is at its maximum for chlorodisilanes and gets successively weaker with increasing as well as with decreasing mass of the halogen atom. The SiSi force constants calculated by ab initio methods are considerably smaller than one would infer from normal coordinate calculations (calculated for Si$_2$Cl$_6$: 240Ncm^{-1} [5]). Results for other methylated disilanes Me$_n$Si$_2$X$_{6-n}$ will be reported in the near future.

Table 4. SiSi, SiCl and SiC valence force constants [Ncm^{-1}] for *anti* and *gauche* 1,2-di-methyltetrachlorosilane.

force constants	*anti*	*gauche*	force constants	*anti*	*gauche*
f (SiSi)	1.69	1.68	f (SiSi/SiCl)	0.035	0.030
f (SiCl)	2.56	2.65/2.55	f (SiCl/SiCl)	0.144	0.144
f (SiC)	2.80	2.79	f (SiSi/SiC)	0.025	0.025
f (SiC/SiC)	0.00	0.00	f (SiC/SiCl)	0.08	0.08

Acknowledgment: Two of the authors (K.S. and A.J.) thank the *Land Sachsen-Anhalt* for financial support of project no. 60501-01-301.

References:

[1]   J. Nagy, S. Ferenczi-Gresz, E. Hengge, S. Waldhör, *J. Organomet. Chem.* **1975**, *96*, 199.
[2]   J. Nagy, G. Zsombok, E. Hengge, W. Veigl, *J. Organomet. Chem.* **1982**, *232*, 1.
[3]   R. Stolevik, P. Bakken, *J. Mol. Struct.*, **1989**, *197*, 131.
[4]   W. J. Stevens, H. Basch, M. Krauss, *J. Chem. Phys.* **1984**, *81*, 6026.
[5]   F. Höfler, W. Sawodny, E. Hengge, *Spectrochim. Acta* **1970**, *26A*, 819.

1,2-Di-*tert*-butyltetrafluorodisilane:
A Highly Fluxional Molecule

Robert Zink, Karl Hassler
Institut für Anorganische Chemie
Technische Universität Graz
Stremayrgasse 16, A-8010 Graz, Austria
E-mail: hassler@fscm1.tu-graz.ac.at

Norbert W. Mitzel, Bruce A. Smart, David W. H. Rankin
Department of Chemistry
University of Edinburgh
West Mains Road, Edinburgh, EH9 3JJ, UK
E-mail: dwhr01@ed.ac.uk

Keywords: Di-*tert*-butyldisilane / IR and Raman Spectroscopy / ab initio / Electron Diffraction

Summary: Recent investigations have shown that the Raman spectra of disilanes of the general formula $tBuSiX_2SiX_2tBu$ (X = H, F, Cl, Br, I) do not change with temperature. This is unexpected as rotational isomerism of related disilanes such as $RMe_2SiSiMe_2R$ (R = $SiMe_3$ [1], Cl [2]) can be observed by Raman spectroscopy. This work reports a combined ab initio, vibrational spectroscopic (IR, Ra) and gas-phase electron diffraction (GED) study of $tBuSiF_2SiF_2tBu$. Ab initio calculations predict that $tBuSiF_2SiF_2tBu$ is a highly fluxional molecule.

Synthesis of *t*BuSiF₂SiF₂*t*Bu

$tBuSiF_2SiF_2tBu$ was prepared by reaction of $tBuSiBr_2SiBr_2tBu$ with ZnF_2 in Et_2O. $tBuSiBr_2SiBr_2tBu$ was synthesized from $tBuSiPh_2SiPh_2tBu$ by reaction with gaseous HBr in the presence of $AlBr_3$.

Ab initio Calculations

Calculations on both *anti* (C_{2h} symmetry, CSiSiC dihedral angle 180°) and *gauche* (C_2 symmetry, CSiSiC dihedral angle expected to be 60°) conformers of $tBuSiF_2SiF_2tBu$ were performed along with the transition state (also C_2 symmetry) connecting them. Calculated bond lengths and bond angles were never found to differ by more than 0.4 pm or 0.4° between the two isomers. The optimized 6-31G*/SCF geometric parameters of the *anti* conformer are presented in Table 1 and the relative energies of the two conformers and the transition state are shown in Table 2. It is of some note that the C–Si–Si–C dihedral angle of the '*gauche*' structure is far from the idealised value of 60° (138.3°

at the 6-31G*/SCF level of theory) and appears to be a result of interactions between the hydrogen atoms of the *t*-butyl groups and fluorine. Furthermore, this interaction appears to stabilize the *gauche* conformer. Calculations predict that the two isomers are very close in energy, the *anti* isomer being more stable by just 0.64 kJmol^{-1} (6-31G*/MP2//6-31G*/SCF). In addition, the isomers are separated by a barrier of only 0.48 kJmol^{-1} (6-31G*/MP2//6-31G*/SCF). Since inclusion of the effects of zero point energy (ZPE) at the 3-21G*/SCF level was found to reduce this barrier by more than 1 kJmol^{-1}, it is possible that the barrier to interconversion lies below the ground state vibrational level. Rotation about the Si–Si bond over a range of 100° is predicted to lead to a change in the total molecular energy that does not exceed 1 kJmol^{-1}. This molecule is therefore highly fluxional.

Table 1. Calculated and experimental geometric parameters of *t*BuSiF$_2$SiF$_2$*t*Bu[a]

Geometric parameters	6-31G*/SCF calculations	Electron diffraction (GED)
r(Si–Si)	234.9	236.6 (6)
r(Si–C)	188.0	187.2 (3)
r(Si–F)	159.9	160.1 (2)
r(C–C)	154.2[b]	154.6 (4)
r(C–H)	108.6[b]	109.6 (2)
∠Si–Si–C	117.6	115.8 (3)
∠Si–Si–F	107.6	108.8 (2)
∠F–Si–F	105.2	103.2 (7)
∠C–C–C	109.6[b]	110.8 (2)
∠H–C–C	111.2[b]	110.9 (10)
tilt, (CH$_3$)$_3$C–Si	3.7	2.2 (5)
C–Si–Si–C twist, *anti*	180	185.1 (26)
C–Si–Si–C twist, *gauche*	138.3	139.8 (9)
CH$_3$ twist	180	200.3 (20)
(CH$_3$)$_3$C twist	180	190.4 (4)

[a] All distances in pm, all angles in degrees. Calculated bond lengths and bond angles refer to the *anti* conformer. – [b] Weighted average of all values.

The presence of either near degenerate isomers or a very low barrier to interconversion should lead to vibrational spectra that change very little with temperature. Indeed, virtually no change in the experimental spectra is observed in the range of -25°C to 200°C.

Table 2. Relative energies (kJ mol⁻¹) of the *anti* and *gauche* conformers of *t*BuSiF₂SiF₂*t*Bu and the barrier between them.[a], [b]

	Anti	*Gauche*	**Barrier**
3-21G*/SCF	0 (0)	-0.88 (0.23)	2.44 (1.27)
6-31G*/SCF	0	1.20	0.03
6-31G*/MP2	0	0.64	0.48

[a] Values in parentheses are corrected for ZPE.
[b] The barrier is calculated relative to the gauche conformer.

Molecular Model and Structure Determination from GED Data.

The large number of geometric parameters necessary for the definition of the structure of *t*BuSiF₂SiF₂*t*Bu required the adoption of a number of assumptions to simplify the problem. Local C_{3v} symmetry was assumed for the CH₃ groups while the C(CH₃)₃ groups were restricted to local C_3 symmetry. C_s symmetry was adopted for the Si₂CF₂ fragments. Since *ab initio* calculations predicted that both the *anti* and *gauche* conformers represent local minima on the potential energy surface and are nearly degenerate in energy, both conformers were included in the refinement, each reflecting overall C_2 symmetry. As the *ab initio* calculations predicted that differences between individual bond lengths and bond angles in the two conformers were small, they were set to zero during the refinements. Table 1 contains the geometric parameters defining the molecular structure and presents their refined values. The tilt angle makes allowance for a deviation of the C_3 axes of the *t*-butyl groups from the Si–C bonds.

The molar composition of the gauche conformer was determined to be 66 (10) % according to a Hamilton test at the 95 % confidence level [3]. The GED data thus confirm the predictions of *ab initio* theory that the two conformers are nearly equal in energy. Fig. 1 shows the experimental and difference (experimental minus theoretical) radial distribution curves.

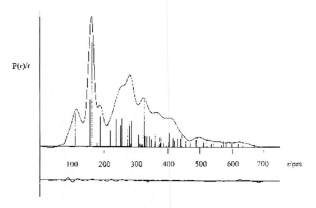

Fig. 1. Experimental and difference (experimental minus theoretical) radial-distribution curves for *t*BuSiF₂SiF₂*t*Bu.

The fit between experimental and calculated intensities is very good (R_G = 4.0 %), even though only two conformers were used to describe the large amplitude torsional motion of the molecule about the Si–Si bond. In principle, this could be modeled by a series of instantaneous conformers, but we do not believe this would lead to a significantly improved fit.

Vibrational Spectra and their Assignment

Fig. 2 depicts Raman spectra of *t*BuSiF$_2$SiF$_2$*t*Bu at two temperatures. In accordance with the *ab initio* predictions no temperature-dependent rotational isomerism can be observed. Subsequently, the vibrational analysis of *t*BuSiF$_2$SiF$_2$*t*Bu was carried out for the point group C_{2h}, since the vibrational spectra obey the rule of mutual exclusion (C_{2h} contains inversion as a symmetry operation).

Table 3. Experimental and calculated IR frequencies (<1250 cm^{-1}) and their assignment.

B_u	$\nu_{exp.}$[cm^{-1}]	$\nu_{calc.}$ [cm^{-1}]	A_u	$\nu_{exp.}$[cm^{-1}]	$\nu_{calc.}$ [cm^{-1}]
ρCH$_3$	1225	1260	ρCH$_3$	1186	1225
ρCH$_3$	1186	1225	ρCH$_3$	1005	1037
ρCH$_3$	1005	1037	ρCH$_3$	–	974
ν_{as}CC$_3$	942	947	ν_{as}CC$_3$	942	947
ν_sSiF$_2$	846	843	ν_{as}SiF$_2$	900	897
ν_sCC$_3$	804	798	δ_{as}CC$_3$	–	391
νSiC	606	590	ρCC$_3$	–	292
ρCC$_3$	422	420	τSiF$_2$	–	138
δ_sCC$_3$	362	358	ρSiF$_2$	–	100
δ_{as}CC$_3$	347	337			
δSiF$_2$	–	248			
γSiF$_2$	–	184			
δSiSiC	–	59			

The distribution of the 90 normal modes of *t*BuSiF$_2$SiF$_2$*t*Bu between the symmetry species of the point group C_{2h} can then be described as Γ_{vib}=25A$_g$(Ra)+20B$_g$(Ra)+21A$_u$(IR)+24B$_u$(IR). After removal of the torsional vibrations and the high frequency modes (ν_sCH$_3$, ν_{as}CH$_3$, δ_sCH$_3$, δ_{as}CH$_3$) one obtains Γ_{vib}=14A$_g$(Ra)+9B$_g$(Ra)+9A$_u$(IR)+13B$_u$(IR).

It is well known that the SiCC$_3$ fragment shows many absorptions in the range between 300 and 450 cm^{-1} and Si–Si stretching vibrations also lie in that region. Therefore, it was necessary to perform a normal coordinate analysis (NCA) to assign the large number of vibrations correctly. Both the NCA and the vibrational potential energy distribution analysis (PED) were performed with unscaled *ab initio* symmetry force constants derived from the Cartesian Hessian matrix (vibrational frequency calculation at the 6-31G*/SCF level of theory). Tables 3 and 4 show the experimental IR and Raman frequencies below 1250 cm^{-1}, their assignment to symmetry coordinates and the calculated *ab initio* frequencies (6-31G*/SCF, scaled by 0.92).

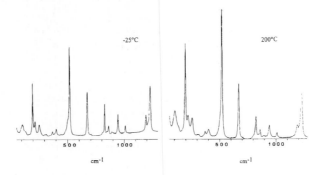

Fig. 2.   Raman spectra of *t*BuSiF$_2$SiF$_2$*t*Bu.

Table 4.   Experimental and calculated Ra frequencies (<1250 cm^{-1}) and their assignment.

A$_g$	$\nu_{exp.}$[cm^{-1}]	$\nu_{calc.}$[cm^{-1}]	B$_g$	$\nu_{exp.}$[cm^{-1}]	$\nu_{calc.}$[cm^{-1}]
ρCH$_3$	1230	1260	ρCH$_3$	1190	1225
ρCH$_3$	1190	1225	ρCH$_3$	1010	1037
ρCH$_3$	1010	1037	ρCH$_3$	–	974
ν_{as}CC$_3$	943	947	ν_{as}CC$_3$	943	947
ν_sSiF$_2$	861	854	ν_{as}SiF$_2$	890	888
ν_sCC$_3$	824	820	δ_{as}CC$_3$	402	397
νSiC	671	662	ρSiF$_2$	315	310
νSiSi	517	511	τSiF$_2$	197	196
δ_sCC$_3$	402	397	ρCC$_3$	133	138
δ_{as}CC$_3$	367	361			
δSiF$_2$	253	248			
ρCC$_3$	219	213			
γSiF$_2$	197	196			
δSiSiC	105	100			

It should be noted that the PED did not allow an unambiguous assignment of some low frequency modes (ρCC$_3$, ρSiF$_2$, δSiF$_2$, τSiF$_2$). These symmetry coordinates are highly coupled. NCA and PED probably suffer from the fact that the *ab initio* Si–Si force constant is small (f(Si/Si) = 2.12 Ncm^{-1}, unscaled) and that some off diagonal symmetry force constants are larger than one would expect from NCAs with spectroscopic force field parameters [4].

In the near future we will report spectroscopic and theoretical studies of the related compounds $SiMe_3SiX_2SiX_2SiMe_3$ and $SiMe_3SiX_2SiX_2tBu$ (X=H,F,Cl,Br,I).

Acknowledgement: Two of the authors (R.Z. and K.H.) thank the *Austrian Science Foundation (FWF)* for financial support of project no. 10283-CHE. We thank the *U.K. Engineering and Physical Science Research Council* for financial support.

References:

[1]   C. A. Ernst, A. L. Alfred, M. A. Ratner, *J. Organomet. Chem.* **1979**, *178*, 119.
[2]   A. Jähn, K. Schenzel, K. Hassler, to be published.
[3]   W. C. Hamilton, *Acta Cryst.* **1965**, *18*, 502.
[4]   B. Reiter, K. Hassler, *J. Organomet. Chem.* **1994**, *467*, 21.

Amino-Substituted Disilanes by Reductive Coupling

Steffen Mantey, Joachim Heinicke*
Institut für Anorganische Chemie
Ernst-Moritz-Arndt-Universität
Soldtmannstr. 16, D-17487 Greifswald, Germany
Tel./Fax: Int code + (3834)75459 (new: Int. code + (3834)864316 / 864337)

Keywords: Aminodisilanes

Summary: The reductive coupling of aminochlorosilanes and aminohydridochlorosilanes with lithium is discussed, first cross-coupling products with trimethylchlorosilane are presented.

Introduction

Functionally-substituted disilanes are useful starting materials for silylenes, silicon containing heterocycles or for polymeric silicon-containing materials. The classical method to prepare functionalized disilanes is the cleavage of aryl groups by HCl/AlCl₃ [1] or by triflic acid [2]. An easier method is the introduction of chloro- or alkoxygroups by replacement of amino substituents.

As shown by Tamao et al [3] and Matsumoto et al. [4], the reduction of alkylaminochlorosilanes with lithium leads to the corresponding symmetrical amino-substituted disilanes. However, this procedure has not yet been successfully extended to the preparation of unsymmetrical amino-substituted disilanes.

This prompted us to find out conditions for the efficient cross-coupling of aminochlorosilanes with alkyl- or arylchlorosilanes to prepare unsymmetrical amino-substituted disilanes as well as amino-substituted hydridodisilanes. We also wanted to find criteria to estimate the selectivity for symmetrical or unsymmetrical coupling products.

Results and Discussion

We studied the co-reduction of chlorotriaminosilanes and, for practical purposes, of chloro-trimethylsilane with lithium in tetrahydrofuran. In a slow reaction unsymmetrical, distillable 1,1,1-triamino-2,2,2- trimethyldisilanes are formed.

$$(R_2N)_3\,Si\,Cl \;+\; Cl\,Si\,Me_3 \quad \xrightarrow[\text{room temp.}]{2Li\;/\;THF} \quad (R_2N)_3\,Si\,Si\,Me_3$$

Eq. 1.

It is supposed that the aminochlorosilanes are primarily reduced owing to the calculated lower LUMO energy (σ^*_{Si-Cl}). The resulting intermediate aminosilylanions prefer the cross-coupling with chlorotrimethylsilane. The variation of the amino groups with varied space demand allows the assumption that the cross-coupling and the long reaction times are mainly due to steric effects. In Table 1 the ^{29}Si-NMR data of the aminochlorosilanes and the coupling products are listed. Yields depend on the nature of the amino groups.

Aminochlorosilicon hydrides are also suitable for the preparation of aminohydridodisilanes by reductive coupling. The hydrido substituent is not markedly attacked in this procedure.

$$(R_2N)_2 \; Si \; H \; Cl \quad \xrightarrow[\text{room temp.}]{2Li \; / \; THF} \quad ((R_2N)_2Si \, H)_2$$

Eq. 2.

The ^{29}Si-NMR chemical shift of the bis(diethylamino)chlorosilane is -27.76 ppm and that of the tetrakis(diethylamino)disilane -23.05 ppm. In mixtures of $(R_2N)_2SiHCl$ and chlorotrimethylsilane with R = Me or Et, the symmetrical coupling is clearly favored compared to the cross-coupling. Studies allowing estimation of the dependence of the selectivity for symmetrical or cross-coupling products on the relative bulkiness of the substituents at both components are in progress.

Table 1. ^{29}Si NMR data for aminochlorosilanes and their coupling products

R	$(R_2N)_3SiCl$ [ppm]	$(R_2N)_3SiSiMe_3$ [ppm]
NEt$_2$	- 30.15	- 16.89 / - 25.87
piperidine	- 34.49	- 20.72 / - 24.51
morpholine	- 35.22	- 20.25 / - 24.91

The hydridodisilanes can be used for generation of aminosilylenes; the advantage is the easy migration of amino groups and cleavage of $HSi(NR_2)_3$. As the polarity of the silicon-hydrogen bond is low compared with that of the silicon-nitrogen bond, a modification of disilanes by stepwise substitution of amino groups and of hydride is supposed to be possible. Finally unsymmetrical amino-substituted disilanes may be interesting for the preparation of new silicon oligomers, polymers or materials.

Acknowledgement: This work was supported by the *Deutsche Forschungsgemeinschaft* and the *Fonds der Chemischen Industrie*.

References:

[1]   a) M. Kumada, M. Ishikawa, S. Maeda, *J. Organomet. Chem.* **1964**, *2*, 478.
       b) K. Tamao, M. Kumada, *J. Organomet. Chem.* **1971**, *30*, 329; *J. Organomet. Chem.* **1971**, *30*, 339.
[2]   a) W. Uhlig, A. Tschach, *J. Organomet. Chem.* **1989**, *378*, C1
       b) W .Uhlig, *J. Organomet. Chem.* **1991**, *421*, 189.
[3]   K. Tamao, A. Kawachi, Y. Ito, *Organometallics* **1993**, *12*, 580.
[4]   M. Unno, M. Saito, H. Matsumoto, *J. Organomet. Chem.* **1995**, *499*, 221.

Multifunctional Disilane Derivatives

Harald Stüger*, Paul Lassacher, Edwin Hengge
Institut für Anorganische Chemie
Technische Universität Graz
Stremayrgasse 16, A-8010 Graz, Austria
Tel.: Int. code + (316)8738202 – Fax: Int. code + (316)8738701
E-mail: Stueger@anorg.tu-graz.ac.at

Keywords: Disilanes / Aminosilanes / Hydrochlorodisilanes

Summary: Multifunctional disilane derivatives containing bonds of silicon to carbon, chlorine, hydrogen, nitrogen, and iron simultaneously were synthesized from bis(trimethylsilyl)aminopentachlorodisilane **(1)** or 1,2-bis[bis(trimethylsilyl)amino]-tetrachlorodisilane **(2)**. **1** and **2** are easily accessible from the reaction of Si_2Cl_6 with one or two equivalents of $LiN(SiMe_3)_2$ under mild reaction conditions. Partially hydrogenated chlorodisilanes were obtained from the corresponding aminodisilanes and dry HCl.

Introduction

The ease with which Si–N bonds can be made and broken [1] and the remarkable chemical stability of aminosilanes towards various nucleophiles [2] provides an attractive basis for the synthesis of multi-functional polysilane derivatives. $(Me_3Si)_2N$- groups attached to oligosilane backbones proved to be particularly inert thus allowing the performance of chemical reactions at the Si-center without affecting the Si–N bonds. By utilizing this approach we succeeded in elaborating a systematic access to multifunctional disilanes bearing several reactive substituents at the same time. Compounds of this type are important for the study of substituent-substituent interactions in polysilanes and as potential precursors to extended functionalized polysilane frameworks.

Results and Discussion

When Si_2Cl_6 reacts with one or two equivalents of $LiN(SiMe_3)_2$ under mild reaction conditions (-40°C, heptane solution), the corresponding mono- or diaminochlorodisilanes bis(trimethyl-silyl)aminopentachlorodisilane **(1)** or 1,2-bis[bis(trimethylsilyl)amino]tetrachlorodisilane **(2)**, are obtained nearly exclusively.

Attempts to attach more than two $(Me_3Si)_2N$- groups to the disilane moiety by the application of more severe reaction conditions (excess $LiN(SiMe_3)_2$, refluxing THF or DME) have not been successful. Even after reaction times of up to 48 h only the starting materials could be recovered.

$$\text{Cl}_3\text{SiSiCl}_3 \xrightarrow[\text{heptane, -40°C}]{+ \text{LiN(SiMe}_3)_2} \text{Cl}_3\text{SiSiCl}_2 - \text{N(SiMe}_3)_2$$

(1)

$$\text{Cl}_3\text{SiSiCl}_3 \xrightarrow[\text{heptane, -40°C}]{+ 2\ \text{LiN(SiMe}_3)_2} (\text{Me}_3\text{Si})_2\text{N} - \text{SiCl}_2\text{SiCl}_2 - \text{N(SiMe}_3)_2$$

(2)

Scheme 1.

The stability of the Si–N bonds in **1** towards nucleophilic attack subsequently allows the introduction of further functional groups, which leads to the formation of multifunctional disilane derivatives simultaneously bearing three different kinds of substituents. Thus, for instance, **1** can react with *t*-butyllithium or the transition metal anion [Fe(CO)$_2$Cp]$^-$ without any detectable scission of the Si–N linkages:

$$\xrightarrow{+ \text{Na[Fe(CO)}_2\text{Cp]}} [\text{Fe(CO)}_2\text{Cp}] - \text{SiCl}_2\text{SiCl}_2 - \text{N(SiMe}_3)_2$$

(3)

$$\text{Cl}_3\text{SiSiCl}_2 - \text{N(SiMe}_3)_2$$

$$\xrightarrow{+ (\text{CH}_3)_3\text{CLi}} (\text{CH}_3)_3\text{C} - \text{SiCl}_2\text{SiCl}_2 - \text{N(SiMe}_3)_2$$

(4)

Scheme 2.

Hydrogenation of the remaining Si–Cl bonds in **1-4** using LiAlH$_4$ in diethyl ether at 0°C additionally affords the corresponding disilanyl hydrides **5-8** in excellent yields. Neither Si–Si, Si–Fe or Si–N bond cleavage is observed:

(1) $\xrightarrow[\text{ether, 0°C}]{+ \text{LiAlH}_4}$ $(\text{Me}_3\text{Si})_2\text{N} - \text{SiH}_2\text{SiH}_3$

(5)

(2) $\xrightarrow[\text{ether, 0°C}]{+ \text{LiAlH}_4}$ $(\text{Me}_3\text{Si})_2\text{N} - \text{SiH}_2\text{SiH}_2 - \text{N(SiMe}_3)_2$

(6)

(3) $\xrightarrow[\text{ether, 0°C}]{+ \text{LiAlH}_4}$ $[\text{Fe(CO)}_2\text{Cp}] - \text{SiH}_2\text{SiH}_2 - \text{N(SiMe}_3)_2$

(7)

(4) $\xrightarrow[\text{ether, 0°C}]{+ \text{LiAlH}_4}$ $(\text{CH}_3)_3\text{C} - \text{SiH}_2\text{SiH}_2 - \text{N(SiMe}_3)_2$

(8)

Scheme 3.

The sensitivity of Si–N bonds towards acidic conditions subsequently can be utilized to remove the amino groups and to restore the original Si–Cl functionalities, which provides an easy access to partially hydrogenated chlorodisilanes, which otherwise are rather troublesome to obtain. The attempted synthesis of 1,2-dihalodisilanes X–(SiH$_2$)$_2$–X with X = Cl or Br from 1,2-diphenyldisilane and HX, for instance, usually affords azeotropic mixtures of the corresponding dihalodisilane and benzene, which cannot be separated by distillation [3, 4]. We found, that **6** and **8** are converted to 1,2-dichlorodisilane (**9**) and 1-*t*-butyl-2-chlorodisilane (**10**), respectively, simply by treatment with dry HCl gas in pentane for 30 minutes at 0°C (Scheme 4).

$$(\textbf{6}) \quad \xrightarrow[\text{pentane, 0°C}]{\text{+ dry HCl}} \quad \text{Cl-SiH}_2\text{SiH}_2\text{-Cl}$$

$$(\textbf{9})$$

$$(\textbf{8}) \quad \xrightarrow[\text{pentane, 0°C}]{\text{+ dry HCl}} \quad (\text{CH}_3)_3\text{C-SiH}_2\text{SiH}_2\text{-Cl}$$

$$(\textbf{10})$$

Scheme 4.

After filtration and distillation pure **10** is obtained in 80 % yield, whereas **9** unfortunately cannot be separated from Me$_3$SiCl, which is also formed in the reaction of **5** and HCl. Pure **9**, however, can be synthesized from 1,2-bis[bis(phenyldimethylsilyl)amino]disilane (**11**) and HCl, because in this case the second reaction product PhMe$_2$Cl exhibits a much higher boiling point than **9** and therefore can be easily removed by distillation:

$$(\text{PhMe}_2\text{Si})_2\text{N-SiH}_2\text{SiH}_2\text{-N(SiMe}_2\text{Ph})_2 \quad \xrightarrow[\text{decaline, 0°C}]{\text{+ dry HCl}} \quad (\textbf{9}) \ + \ \text{PhMe}_2\text{SiCl}$$

$$(\textbf{11})$$

Scheme 5.

All substances were characterized by common spectroscopic techniques like MS, IR, ^{1}H-, and ^{29}Si-NMR and by elemental analysis. As expected, ^{29}Si-NMR-spectroscopy turned out to be particularly useful to prove the proposed structures of the disilanes **1-10**. ^{29}Si NMR chemical shifts are depicted in Table 1.

The coupled ^{29}Si NMR-spectra of the disilanyl hydrides **5-10** exhibit rather complex splitting patterns due to extensive ^{29}Si^{1}H coupling. Selected coupling constants can be derived from Table 2. The signals for Si(1) and Si(2) are split into triplets (except compound **5**, where the SiH$_3$ group gives rise to a quartet) by couplings to the directly bonded hydrogens. The resulting resonance lines show further hyperfine structure caused by long range ^{29}Si^{1}H couplings to the hydrogen atoms at the adjacent silicon atoms and occasionally to the protons of the Me$_3$C- groups, which allows an unequivocal structure assignment in each case.

Table 1. ^{29}Si NMR chemical shifts δ[ppm] of multifunctional disilane derivatives **1-10** (C_6D_6 solution vs. ext. TMS).

		δSi(1)	δSi(2)	δSi(3)
$Cl_3\overset{2}{Si}\overset{1}{Si}Cl_2\overset{3}{N}(SiMe_3)_2$	**1**	-19.41	-0.11	9.50
$(Me_3\overset{3}{Si})_2N\overset{1}{Si}Cl_2SiCl_2N(SiMe_3)_2$	**2**	-14.55	–	7.04
$[Fe(CO)_2Cp]\overset{2}{Si}Cl_2\overset{1}{Si}Cl_2\overset{3}{N}(SiMe_3)_2$	**3**	-12.89	85.04	8.21
$(CH_3)_3C\overset{2}{Si}Cl_2\overset{1}{Si}Cl_2\overset{3}{N}(SiMe_3)_2$	**4**	-13.79	17.68	8.86
$(Me_3\overset{3}{Si})_2N\overset{1}{Si}H_2\overset{2}{Si}H_3$	**5**	-49.52	-96.55	6.27
$(Me_3\overset{3}{Si})_2N\overset{1}{Si}H_2SiH_2N(SiMe_3)_2$	**6**	-45.61	–	5.56
$[Fe(CO)_2Cp]\overset{2}{Si}H_2\overset{1}{Si}H_2\overset{3}{N}(SiMe_3)_2$	**7**	-41.89	-44.39	5.48
$(CH_3)_3C\overset{2}{Si}H_2\overset{1}{Si}H_2\overset{3}{N}(SiMe_3)_2$	**8**	-49.98	-38.16	6.59
$Cl\overset{1}{Si}H_2SiH_2Cl$	**9**	-30.53[a]	–	–
$(CH_3)_3C\overset{2}{Si}H_2\overset{1}{Si}H_2Cl$	**10**	-24.35	-44.49	

[a] data taken from reference 4.

Table 2. Selected ^{29}Si^{1}H coupling constants J[Hz] of multifunctional disilanyl hydrides (t = triplet; q = quartet; m = multiplet).

	Si(1)		Si(2)	
	$^1J_{Si-H}$	$^2J_{Si-H}$	$^1J_{Si-H}$	$^2J_{Si-H}$
$(Me_3\overset{3}{Si})_2N\overset{1}{Si}H_2\overset{2}{Si}H_3$	201.6 (t)	6.4 (q)	188.7 (q)	14.8 (t)
$(Me_3\overset{3}{Si})_2N\overset{1}{Si}H_2SiH_2N(SiMe_3)_2$	192.5 (t)	14.6 (t)	–	–
$[Fe(CO)_2Cp]\overset{2}{Si}H_2\overset{1}{Si}H_2\overset{3}{N}(SiMe_3)_2$	190.3 (t)	7.2 (t)	172.5 (t)	15.7 (t)
$(CH_3)_3C\overset{2}{Si}H_2\overset{1}{Si}H_2\overset{3}{N}(SiMe_3)_2$	194.4 (t)	7.2 (t)	179.5 (t)	m
$(CH_3)_3C\overset{2}{Si}H_2\overset{1}{Si}H_2Cl$	205.7 (t)	8.7 (t)	183.1 (t)	m

Acknowledgment: This work was supported by the *Fonds zur Förderung der wissenschaftlichen Forschung* (Wien). The authors are grateful to *Wacker-Chemie GmbH* for the supply of organochlorosilane starting materials.

References:

[1]   D. A. ("Fred") Armitage, *"Organosilicon Nitrogen Compounds"* in *The Silicon Heteroatom Bond* (Eds.: S. Patai, Z. Rappoport, Wiley, Chichester **1991**, pp. 367ff.

[2]   W. Uhlig, C. Tretner, *J. Organomet. Chem.* **1994**, *467*, 31.

[3]   A. Haas, R. Süllentrup, C. Krüger, *Z. Anorg. Allg. Chem.* **1993**, *619*, 819.

[4]   H. Söllradl, E. Hengge, *J. Organomet. Chem.* **1983**, *243*, 257.

New Transition Metal Substituted Oligosilanes

Wolfram Palitzsch, Uwe Böhme*, Gerhard Roewer

Institut für Anorganische Chemie
Technische Universität Bergakademie Freiberg
Leipziger Straße 29, D-09596 Freiberg, Germany
Tel.: Int. code + (3731)392108 – Fax: Int. code (3731)394058
E-mail: boehme@silicium.aoch.tu-freiberg.de

Keywords: Iron / Molybdenum / Oligosilanes / Transition Metal Silicon Compounds

Summary: The transition metal substituted disilanes (η^5-C$_5$Me$_4$Et)Fe(CO)$_2$SiMe$_2$SiMe$_2$Cl and (η^5-C$_5$Me$_4$Et)Mo(CO)$_3$SiMe$_2$SiMe$_2$Br were prepared by means of salt elimination. Both compounds are well crystallized and their structures determined by X-ray diffraction analysis. Surprisingly, the reaction of Na[(η^5-C$_5$Me$_4$Et)Fe(CO)$_2$] with Br(SiMe$_2$)$_6$Br gave exclusively the diiron hexasilane complex (η^5-C$_5$Me$_4$Et)(CO)$_2$Fe(SiMe$_2$)$_6$Fe(CO)$_2$(η^5-C$_5$Me$_4$Et).

Introduction

Organosilicon polymers in which silicon atoms are connected to transition metal atoms represent an area of considerable current interest [1]. We are presently studying various transition metal substituted oligosilanes. The series of (η^5-C$_5$H$_5$)Fe(CO)$_2$-silyl complexes has been spectroscopically investigated in detail. However, only a few structural studies have been published [2, 3]. This disadvantage probably arises from the use of the cyclopentadienyl ligand (C$_5$H$_5$).

The 1-ethyl-2,3,4,5-tetramethylcyclopentadienyl ligand (η^5-C$_5$Me$_4$Et) at the transition metal should give compounds suitable for X-ray diffraction analysis. Furthermore, the use of this sterally pretentious ligand might produce transition metal substituted oligosilanes with improved stability. A conceptually simple way in which a M–Si bond may be formed is the reaction between a metallate complex and a silicon halide with elimination of one equivalent of a salt [4]. Reactions of chlorine- and bromine-substituted disilanes and α,ω-dibromdodecamethylhexasilane with metallate complexes of molybdenum and iron is presented here.

New Disilane Derivatives of Iron and Molybdenum

Synthesis and Characterization of (η^5-C$_5$Me$_4$Et)Fe(CO)$_2$SiMe$_2$SiMe$_2$Cl (1)

The reaction of one equivalent of Na[(η^5-C$_5$Me$_4$Et)Fe(CO)$_2$] with ClMe$_2$SiSiMe$_2$Cl in THF/pentane at ambient temperature in formation of (η^5-C$_5$Me$_4$Et)Fe(CO)$_2$SiMe$_2$SiMe$_2$Cl (**1**, Eq. 1).

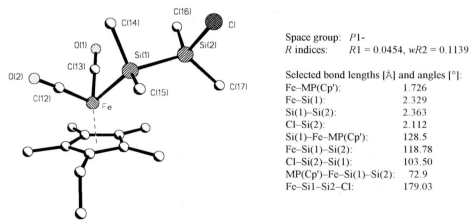

Eq. 1.

Salts were filtered off and the filtrate was concentrated. Recrystallization from pentane at -20°C gave analytically pure orange-yellow crystals of **1** suitable for an X-ray crystallographic study (Figure 1). The infrared spectrum of **1** shows two carbonyl stretching modes at 1908 and 1963 cm^{-1}, as expected for compounds of the type $(\eta^5\text{-}C_5Me_4Et)Fe(CO)_2L$. The NMR data of **1** are in agreement with the proposed structure (Table 1) and the observations are in accordance with other experimental results [4].

Molecular Structure of $(\eta^5\text{-}C_5Me_4Et)Fe(CO)_2SiMe_2SiMe_2Cl$ (1)

The molecular structure of **1** reveals a staggered conformation of the Fe–SiMe$_2$–SiMe$_2$–Cl unit with the iron and the chlorine atoms trans to each other (torsion angle nearly 180°). The Fe–Si bond length of 2.329 Å is in good agreement with published data for $(\eta^5\text{-}C_5H_5)Fe(CO)_2$-silyl compounds (Fe–Si bond length 2.348-2.365 Å) [3]. The C$_5$Me$_4$Et ligand occupies a definite conformation, no disorder was observed.

Space group: *P*1-
R indices: $R1 = 0.0454$, $wR2 = 0.1139$

Selected bond lengths [Å] and angles [°]:
Fe–MP(Cp'):	1.726
Fe–Si(1):	2.329
Si(1)–Si(2):	2.363
Cl–Si(2):	2.112
Si(1)–Fe–MP(Cp'):	128.5
Fe–Si(1)–Si(2):	118.78
Cl–Si(2)–Si(1):	103.50
MP(Cp')–Fe–Si(1)–Si(2):	72.9
Fe–Si1–Si2–Cl:	179.03

Fig. 1. Molecular structure of $(\eta^5\text{-}C_5Me_4Et)Fe(CO)_2SiMe_2SiMe_2Cl$ (**1**); hydrogen atoms omitted; measurement at –150 °C.

Synthesis and Characterization of $(\eta^5\text{-}C_5Me_4Et)Mo(CO)_3SiMe_2SiMe_2Br$ (2)

Treatment of a solution of BrMe$_2$SiSiMe$_2$Br in pentane with a solution of Na[η^5-C$_5$Me$_4$Et)Mo(CO)$_3$] in THF yields $(\eta^5\text{-}C_5Me_4Et)Mo(CO)_3SiMe_2SiMe_2Br$ (**2**, Eq. 2).

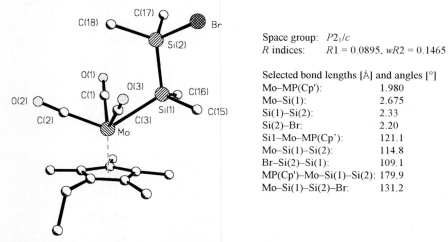

Eq. 2.

Recrystallization from pentane gave attractive yellow crystals. The structure of **2** could be determined by X-ray diffraction analysis (Figure 2.). The NMR spectroscopic results were consistent with the assigned structure (Table 1). The IR spectrum shows three ν(CO) absorptions at 1965, 1991 and 2044 cm^{-1}. This behavior is typical for metal carbonylates of the type *cis*-L$_2$M(CO)$_3$. Compound **2** is more stable than the previosly described (η^5-C$_5$H$_5$)M(CO)$_3$SiMe$_2$SiMe$_2$Br (M=W, Mo) [5]. Complex **2** decomposes thermally below 150°C and is stable against light. Decomposition under the influence of donor solvents was not observed.

Molecular Structure of (η^5-C$_5$Me$_4$Et)Mo(CO)3SiMe2SiMe$_2$Br (2)

The molecular structure of **2** confirms the *cis*-L$_2$M(CO)$_3$ geometry around the molybdenum atom as a distorted trigonal bipyramid. The torsion angle MP(Cp')–Mo–Si(1)–Si(2) is nearly 180°. This is in contrast to the structure of the iron compound **1**, in which the angle MP(Cp')–Fe–Si(1)–Si(2) was determined to be 72.9°. This effect is probably due to the coordination geometry around the molybdenum atom involving three CO groups. Therefore, the Si(1)Me$_2$–Si(2)Me$_2$–Cl ligand has a conformation in which only one methyl group of the second silicon atom Si(2) is located in the cone of the CO groups. Bond lengths and angles are in the normal range for Cp'Mo(CO)$_3$ compounds. The relatively high *R*-value results from the measurement at room temperature.

Space group:	*P*2$_1$/*c*	
R indices:	R1 = 0.0895, wR2 = 0.1465	

Selected bond lengths [Å] and angles [°]	
Mo–MP(Cp'):	1.980
Mo–Si(1):	2.675
Si(1)–Si(2):	2.33
Si(2)–Br:	2.20
Si1–Mo–MP(Cp'):	121.1
Mo–Si(1)–Si(2):	114.8
Br–Si(2)–Si(1):	109.1
MP(Cp')–Mo–Si(1)–Si(2):	179.9
Mo–Si(1)–Si(2)–Br:	131.2

Fig. 2.   Molecular structure of (η^5-C$_5$Me$_4$Et)Mo(CO)3SiMe2SiMe$_2$Br (**2**); 50 %probability; hydrogen atoms omitted; measurement at room temperature.

Synthesis and Characterization of [(η⁵-C₅Me₄Et)Fe(CO)₂(SiMe₂)₃]₂ (3)

Following the successful X-ray structure analysis of **1** and **2** the same synthetic route was applied to the synthesis of a compound with a longer silicon chain. A solution of $Br(SiMe_2)_6Br$ in pentane was added dropwise to a THF solution of $Na[(\eta^5\text{-}C_5Me_4Et)Fe(CO)_2]$ at -78°C. The reaction mixture was stirred at this temperature for 15 min and then brought to room temperature and stirred for 1 h. The solvent was removed, the orange residue was extracted with pentane, and the mixture was filtered. Recrystallization from pentane gave compound **3** (Eq. 3).

Eq. 3.

The compound **3** was identified by ist ^1H, ^{13}C, and ^{29}Si NMR, and IR spectra (Table 1). The IR spectrum of $(\eta^5\text{-}C_5Me_4Et)Fe(CO)_2(SiMe_2)_6Fe(CO)_2(\eta^5\text{-}C_5Me_4Et)$ exhibits two bands of ν(CO) at 1924 and 1979 cm^{-1} according to A' and A'' modes for local C_s symmetry at iron. The formation of the diiron complex **3** is highly favored in relation to that one of $(\eta^5\text{-}C_5Me_4Et)Fe(CO)_2(SiMe_2)_6Br$.

Table 1. NMR spectral properties of the new oligosilane derivatives of iron and molybdenum (data in ppm relative to TMS at 0.0 ppm, recorded in CDCl₃).

[a]	1[a]	2[a]	3[a]
δ ^{13}C=O	216.8	213.8, 216.8	217.7
δ ^{13}CH₃–Si	2.0, 3.9	1.3, 3.9	-3.4, -2.7, 4.3
δ ^{13}CH₃–CH₂	18.4	18.5	18.6
δ ^{13}CH₂–CH₃	14.5	14.5	14.6
δ ^{13}C–CH₃	94.6, 95.5	94.7, 95.5	94.2, 95.1
δ ^{13}C–CH₂–	100.5	100.5	100.1
δ ^{13}CH₃–C	9.8, 10.1	9.9, 10.1	9.9, 10.1
δ ^{29}Si–M	30.8	28.6	20.8
δ ^{29}Si–Si–Si–			-36.4
δ ^{29}Si–Si–M	15.1	4.1	-31.7
δ ^{1}H (Si–Me)	0.43 (s), 0.49 (s)	0.621 (s), 0.622 (s)	0.18 (s), 0.20 (s), 0.43 (s)
δ ^{1}H (C–Me)	1.86 (s,s)	2.0 (s,s)	1.82 (s), 1.83 (s)
δ ^{1}H (C–Et)	1.02 (t), 2.30 (q)	1.02 (t), 2.41 (q)	1.00(t), 2.25 (q)

[a] **1**: (η⁵-C₅Me₄Et)Fe(CO)₂(SiMe₂)₂Cl; **2**: (η⁵-C₅Me₄Et)Mo(CO)₃(SiMe₂)₂Br;
3: [(η⁵-C₅Me₄Et)Fe(CO)₂(SiMe₂)₃]₂.

Conclusion

The use of 1-ethyl-2,3,4,5-tetramethylcyclopentadienyl as protecting ligand at the transition metal center allows the isolation of well crystallizing and thermally stable transition metal substituted oligosilanes. The remaining halide atoms in compounds **1** and **2** can act as functional groups in subsequent reactions. Such silicon complexes represent useful synthons for metal-containing polysilanes. For this purpose reactive halide groups in the prepared products are highly recommended.

A series of silicon complexes with a variety of transition metals is being prepared at present and their chemistry will be explored further.

Acknowledgement: Financial support of this work by the *Deutsche Forschungsgemeinschaft* and the *Fonds der Chemischen Industrie* is gratefully acknowledged.

References:

[1]   I. Manners, J. M. Nelson, A. J. Lough, *Angew. Chem.* **1994**, *9*, 106, 1019.
[2]   H. K. Sharma, K. H. Pannell, *Chem. Rev.* **1995**, *95*, 1351; K. H. Pannell, S.-H. Lin, R. N. Kapoor, F. Cervantes-Lee, M. Pinon, L. Parkanyi, *Organometallics* **1990**, *9*, 2454; K. H. Pannell, J. Cervantes, L. Parkanyi, F. Cervantes-Lee, *Organometallics* **1990**, *9*, 859.
[3]   R. West, E. K. Pham, *J. Organomet. Chem.* **1991**, *402*, 215; L. Parkanyi, K. H. Pannell, C. Hernandez, *J. Organomet. Chem.* **1983**, *252*, 127; T. J. Drahnak, R. West, J. C. Calabrese, *J. Organomet. Chem.* **1980**, 198, 55.
[4]   E. Hengge, M. Eibl, B. Stadelmann, *Monats. Chem.* **1993**, *124*, 523.
[5]   W. Malisch, *J. Organomet. Chem.* **1974**, *82*, 185.

Regiospecific Chlorination and Oxygenation of Pentahydridodisilanyl Complexes of Iron and Ruthenium [1]

Stephan Möller, Heinrich Jehle, Wolfgang Malisch*, Wolfgang Seelbach

Institut für Anorganische Chemie
Universität Würzburg
Am Hubland, D-97074 Würzburg, Germany
Tel.: Int. code + (931)8885277 – Fax: Int. code + (931)8884618
E-mail: anor142@rzbox.uni-wuerzburg.de

Keywords: Metallo-Disilanes / Regiospecific Oxofunctionalization / Regiospecific Chlorination

Summary: Metalation of Si_2Cl_6 with the alkali-metalates $M'[M(CO)_2C_5R_5]$ (**1a-d**) (M = Fe, Ru; M' = Na, K; R = H, Me) yields the metallo-pentachlorodisilanes $C_5R_5(OC)_2M–Si_2Cl_5$ [M = Fe, R = H (**2a**), Me (**2b**); M = Ru, R = H (**2c**), Me (**2d**)], which can be converted to the bis(metallo)-tetrachlorodisilanes $[C_5R_5(OC)_2M–SiCl_2]_2$ [M = Fe, R = H (**3a**), Me (**3b**); M = Ru, R = H (**3c**), Me (**3d**)] by further metalation. Formation of the metallo-pentahydridodisilanes $C_5R_5(OC)_2M–Si_2H_5$ [M = Fe, R = H (**4a**), Me (**4b**); M = Ru, R = H (**4c**), Me (**4d**)] is achieved by Cl/H-exchange on **2a-d** with excess $LiAlH_4$. Activation of the α-Si–H units by the metal fragment is used to transform **3a-c** to the ferrio-dihydroxydisilanes $C_5R_5(OC)_2Fe–Si(OH)_2SiH_3$ (R = H (**5a**), Me (**5b**)) with dimethyldioxirane, and to the ferrio- and ruthenio-dichlorodisilanes $C_5R_5(OC)_2Fe–SiCl_2SiH_3$ [M = Fe, R = H (**6a**), Me (**6b**); M = Ru, R = H (**6c**)] with CCl_4.

Synthesis of the Metallo-Pentachloro and -Pentahydridodisilanes $C_5R_5(OC)_2M–Si_2X_5$ (M = Fe, Ru; R = H, Me; X = H, Cl)

Since the preparation of the $Cp(OC)_2Fe-$ [2], $Cp(OC)_3W-$ [3a], and $Cp(OC)_2(Me_3P)Mo/W-$ substituted [3b, 4] pentachlorodisilanes and their pentahydrido-derivatives, our interest is focussed on extending this series to the $C_5Me_5(OC)_2Fe-$ and $C_5R_5(OC)_2Ru-$substituted systems (R = H, Me). These investigations are motivated by the study concerning the electronic effect of transition metal centers on the chemical properties and reactivity of the α- and β-silicon in disilanyl-complexes.

The metallo-pentachlorodisilanes $C_5R_5(OC)_2M–Si_2Cl_5$ (**2a-d**) are obtained by heterogenous metalation of Si_2Cl_6 with the corresponding alkali-metalates **1a-d**. Compounds **2a-d** are isolated in good yields as yellow solids, which proved to be stable on thermal treatment or exposure to air. Further metalation yields the bis(metallo)-disilanes **3a-d** (Scheme 1).

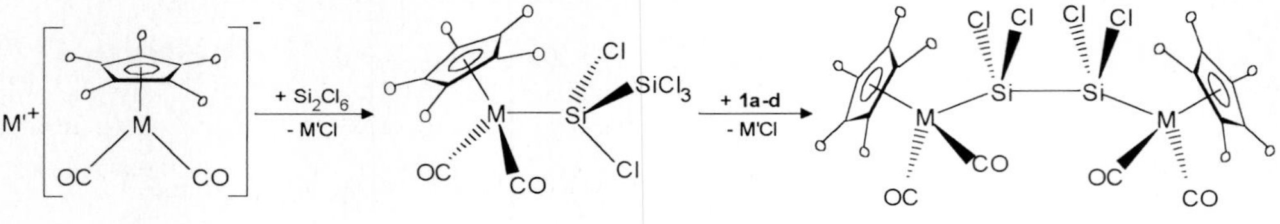

1a-d **2a-d** **3a-d**

	a	b	c	d
M'	Na	Na	Na	K
M	Fe	Fe	Ru	Ru
–o	H	Me	H	Me

Scheme 1. Synthesis of **2a-d** and **3a-d**.

As originally demonstrated by us, Cl/H-exchange at **2a-d** can be achieved with $LiAlH_4$ to give the metallo-pentahydridodisilanes **4a-d** as pale yellow oils (**4a** [5],**c**) or solids (**4b,d**), respectively (Scheme 2).

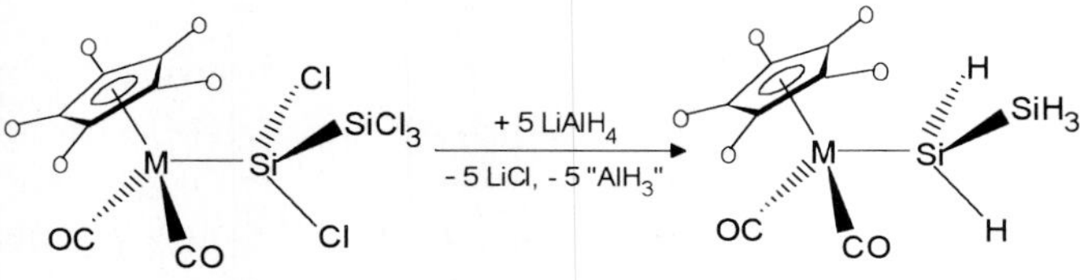

2a-d **4a-d**

	a	b	c	d
M	Fe	Fe	Ru	Ru
–o	H	Me	H	Me

Scheme 2. Synthesis of **4a-d**.

4a-d are soluble in aliphatic solvents and thermally stable but light-sensitive. The IR, ^1H, and, ^{29}Si NMR spectra reveal two kinds of Si–H units, characterizing the SiH_2-moiety as extremly electron-rich while the electronic character of the SiH_3-moiety is very similiar to that of free silanes.

Regiospecific Functionalization of 3a-c with Dimethyldioxirane and CCl₄

As shown in earlier works, the Si–H bonds in ferrio- and ruthenio-silanes $Cp(OC)_2M–SiR_{3-n}H_n$ (M = Fe, Ru; R = alkyl, aryl; n = 1-3) are extremely sensitive towards insertion of oxygen using dimethyldioxirane. This route has been utilized to generate diverse ferrio- and ruthenio-silanols, e.g., $Cp(OC)_2M–SiR_2OH$ (M = Fe, R = *t*Bu; M = Ru; R = *o*Tol) [6].

Application of this oxygenation procedure to the ferrio-disilanes **4a,b** reveals the following result: treatment of **4a,b** with 2 equivalents of dimethyldioxirane regiospecifically leads to insertion of oxygen into the α-Si–H bonds, yielding the metallo-(α-dihydroxy)disilanes **5a,b** (Scheme 3).

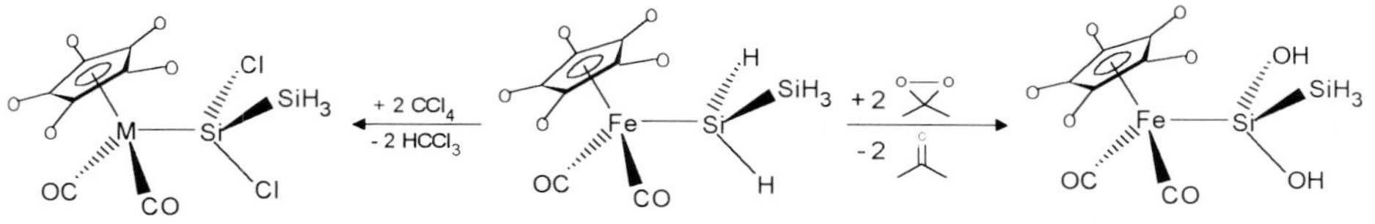

6a-c					5a,b	
	a	**b**	**c**		**a**	**b**
M	Fe	Fe	Ru			
–o	H	Me	H		**–o**	H	Me

Scheme 3.   Synthesis of **5a,b** and **6a-c**.

These products suggest a strong specific activation of the α-Si–H bond by the adjacent metal fragment. In addition, this effect determines the reaction of **4a-c** with two moles of CCl₄ leading to regiospecific chlorination of the α-silicon to give the ferrio- and ruthenio-dichlorodisilanes **6a-c** (Scheme 3).

Further investigations, including condensation reactions of **5a,b** with chlorosilanes and oxofunctionalization of **4c,d** are in progress.

Acknowledgement: This work has been generously supported by the *Deutsche Forschungsgemeinschaft* (SFB 347: *"Selektive Reaktionen Metall-aktivierter Moleküle"*).

References:

[1]   Part 16 in the series *"Metallo-Silanols and Metallo-Siloxanes"*. In addition, Part 42 of the series *"Synthesis and Reactivity of Silicon Transition Metal Complexes"*. Part 15/41, see W. Malisch, M. Neumayer, K. Perneker, N. Gunzelmann, K. Roschmann, in: *Organosilicon Chemistry III: From Molecules to Materials* (Eds.: N. Auner, J. Weis), VCH, Weinheim, **1997**, p. 407.

[2]   a) W. Malisch, M. Kuhn, *Chem. Ber.* **1974**, *107*, 979.
      b) B. Stadelmann, P. Lassacher, H. Stueger, E. Hengge, *J. Organomet. Chem.* **1994**, *482*, 201.

[3] a) W. Malisch, R. Lankat, W. Seelbach, J. Reising, M. Noltemeyer, R. Pikl, U. Posset, W. Kiefer, *Chem. Ber.* **1995**, *128*, 1109.

b) A. Zechmann, E. Hengge, *J. Organomet. Chem.* **1996**, *508*, 227.

[4] W. Malisch, S. Möller, R. Lankat, J. Reising, S. Schmitzer, O. Fey, in: *Organosilicon Chemistry II: From Molecules to Materials* (Eds.: N. Auner, J. Weis), VCH, Weinheim, **1996**, p. 575.

[5] The synthesis of the compounds **2a,b**, **3a**, and **4a,b** was originally described by: G. Thum, *Ph. D. thesis*, Universität Würzburg, **1984**.

[6] a) W. Adam, U. Azzena, F. Prechtl, K. Hindahl, W. Malisch, *Chem. Ber.* **1992**, *125*, 1409.

b) S. Möller, W. Malisch, W. Seelbach, O. Fey, *J. Organomet. Chem.* **1996**, *507*, 239.

Inter- and Intramolecular Oxidative Addition of Si–H Bonds

R. Karch, H. Gilges, U. Schubert*
Institut für Anorganische Chemie
Technische Universität Wien
Getreidemarkt 9, A-1060 Wien, Austria
Tel.: Int. code + (1)588014633 – Fax: Int. code + (1)5816668
E-mail: uschuber@fbch.tuwien.ac.at

Keywords: Disilanes / Oxidative Addition / Phosphinoalkylsilanes / Chelate Assistance / Hydrido-Silyl Complexes

Summary: The photochemical reaction of $Fe(CO)_5$ with $HMe_2Si–SiMe_2H$ yields $(PPh_3)_2(CO)_2Fe(H)_2$, and the thermal reaction of *mer*-$(CO)_3(PPh_3)Fe(H)(SiMe_3)$ with disilanes $HR_2Si–SiR_2H$ gives *mer*-$(CO)_3(PPh_3)Fe(H)(SiR_2–SiR_2H)$ and $[$*mer*-$(CO)_3(PPh_3)(H)Fe–SiR_2]_2$, depending on the substituents at silicon. By the reaction of the phosphinoalkylsilanes $Ph_2PCH_2CH_2SiR_2H$ with $W(CO)_6$ or $W(CO)_5(solv)$ the intramolecular addition products $W(CO)_4(PPh_2CH_2CH_2SiR_2)(H)$ are obtained, in which the W,H,Si three centre two-electron bond is stabilized by the chelating ligand.

Synthesis of the Hydrido-Disilanyl Complexes *mer*-$(CO)_3(PPh_3)Fe(H)(SiR_2–SiR_3)$

Hydrido-disilanyl complexes $L_nM(H)(SiR_2–SiR_2H)$ are interesting because H_2 elimination could support a possible β-Si–H elimination and result in disilene complexes or products derived thereof.

Analogous to the synthesis of the well-known hydrido-silyl complexes *mer*-$(PR_3)(CO)_3Fe(H)(SiR'_3)$ [1] the photochemical reaction of $(CO)_4(PPh_3)Fe$ with $HMe_2Si–SiMe_2H$ in an inert solvent was investigated. Formation of the expected hydrido-disilanyl complex *mer*-$(CO)_3(PPh_3)Fe(H)(SiMe_2–SiMe_2H)$ was not observed but instead a di(hydrido)-bis(phosphino) complex was isolated.

$$(CO)_4(PPh_3)Fe \ + \ HMe_2Si-SiMe_2H \ \xrightarrow{\ h\nu\ } $$

Eq. 1.

However, the synthesis of the desired hydrido-disilanyl complexes was achieved by the thermal reaction of *mer*-$(CO)_3(PPh_3)Fe(H)(SiMe_3)$ [1] with the disilanes $HR_2Si–SiR_3$. It is known that the trimethylsilyl complex easily eliminates $HSiMe_3$ at slightly elevated temperature and thus empties a coordination site for the oxidative addition of the disilanes.

Eq. 2.

The compounds were spectroscopically and analytically identified. The β-Si–H ^{1}H NMR signals of the hydrido-disilanyl complexes *mer*-(CO)$_3$(PPh$_3$)Fe(H)(SiR$_2$–SiR$_2$H) are shifted to lower field relative to the corresponding disilane. This shows the electron withdrawing properties of the group *mer*-(CO)$_3$(PPh$_3$)Fe(H) which activates the β-Si–H group towards a second addition reaction.

This activation can be observed in the reaction of HMe$_2$Si–SiMe$_2$H with *mer*-(CO)$_3$(PPh$_3$)Fe(H)(SiMe$_3$). In the first step of the reaction the hydrido-disilanyl complex *mer*-(CO)$_3$(PPh$_3$)Fe(H)(SiMe$_2$–SiMe$_2$H) is obtained. With increasing reaction time the ^{31}P NMR signal of this complex decreases and a new signalcc of the bridging compound [*mer*-(CO)$_3$(PPh$_3$)(H)Fe–SiMe$_2$]$_2$ arises. This reaction is even observed when *mer*-(CO)$_3$(PPh$_3$)Fe(H)(SiMe$_2$–SiMe$_2$H) is stored in frozen C$_6$D$_6$ at -30°C.

Eq. 3.

Similar results were obtained for hydrido-disilanyl complexes of other metals (e.g. Cp(CO)$_2$Re(H)(SiR$_2$–SiR$_2$H) or (C$_5$R$_5$)(CO)$_2$Mn(H)(SiR$_2$–SiR$_2$H)). They can be summarized as follows [2]:

1. The photochemical synthesis of such compounds is unfavorable. A better way is the thermal reaction of disilanes HR$_2$Si–SiR$_2$H with complexes with a labile ligand.
2. β-H-substituted hydrido-disilanyl complexes L$_n$M(H)(SiR$_2$–SiR$_2$H) are activated towards a second oxidative addition reaction and tend to form bridged compounds [L$_n$(H)M–SiR$_2$]$_2$.

3. The tendency to form bridged complexes is reduced with an increasing steric bulk of the substituents at the disilanyl ligand.

Chelate-Assisted Oxidative Addition of Si–H Bonds

Chelate assistance and the chelate effect are well documented phenomena for stabilizing complexes. The goal of our investigation was to get an estimate of the magnitude of stabilization brought about by the chelate effect in comparison to other factors promoting oxidative addition of Si–H bonds, e.g., small and electron-donating coligands at the metal and electron-withdrawing groups at silicon.

We used the bidentate ligands $Ph_2PCH_2CH_2SiRR'H$, first described by Stobart et al. [3] and reacted them with $W(CO)_6$ (Scheme 1) [4].

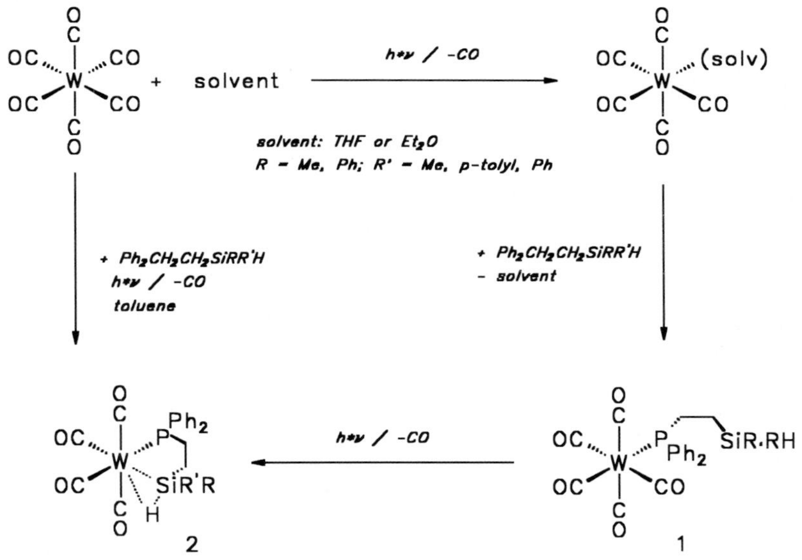

Scheme 1.

IR data prove the expected *cis* arrangement of P and Si enforced by the chelating ligand for the complexes **2**, while complexes **1** show the typical pattern of octahedral mono-phosphine complexes. Some characteristic NMR data are summarized in Table 1

Table 1.   (All spectra in C_6D_6).

Compound[a]	^{1}H (Si**H**) [ppm]	$^3J_{HC-SiH}$ [Hz]	^{31}P [ppm]	$^1J_{W-P}$ [Hz]	
1a	3.97 (m)	–	16.1	239.2	
1c	5.03 (t)	3.10	16.4	239.4	
	^{1}H (M**H**) [ppm]	J_{Si-H} [Hz]	J_{W-H} [Hz]	^{31}P [ppm]	$^1J_{W-P}$ [Hz]
2a[b]	-8.18 (s,br)	95.2	36.4	40.0	256.7
2b[b]	-7.58 (s,br)	96.8	37.8	40.3	255.6
2c	-6.90 (d)	98.1	37.0	42.8	254.0

[a] **a**: R, R' = Me; **b**: R = Me, R'= *p*tol; **c**: R, R' = Ph – [b] Spectroscopic evidence; prone to decomposition on work up.

The J_{Si-H} coupling constant is of special interest, because it is a versatile parameter to monitor the degree of oxidative addition. For uncoordinated silanes $^1J_{Si-H}$ is about 200 Hz, as observed for **1c** with 196.5 Hz. For hydrido silyl complexes without any Si,H interaction values below 20 Hz are observed. The coupling constants of **2a-c** of 95-98 Hz are typical for three-center two-electron bonds. They are among the highest values found up to now, confirming that the oxidative addition is "frozen" in an early state.

When nonchelating ligands are employed in complexes of the type $W(CO)_{5-n}(PR_3)_n(H)(SiR_3)$ (n = 0-2), the complex $W(CO)_3$(dppe)(H)(SiTolCl$_2$) (J_{Si-H} = 34.8 Hz). marks the borderline of stability towards reductive elimination of the silane. Complexes with less electronic stabilization, for example, the mono-phosphine complexes $W(CO)_4(PR_3)(H)(SiR_3)$ or complexes with trialkyl- or triarylsilyl ligands instantly decompose by reductive elimination of the silane. Bearing this in mind, the magnitude of stabilization brought about by incorporation of the SiH bond into the chelating ligand can be quantified as slightly larger than the electronic effect of an additional PR$_3$ ligand or the replacement of SiR$_3$ (R= alkyl, aryl) against SiCl$_3$.

References:

[1]   M. Knorr, U. Schubert, *Trans. Met. Chem.* **1986**, *11*, 268.

[2]   U. Schubert, R. Karch, *Inorg. Chim. Acta* **1997**, *259*, 151.

[3]   R. D. Holmes-Smith, R. D. Osei, S. R. Stobart, *J. Chem. Soc., Perkin Trans.* **1983**, *1*, 861.

[4]   H. Gilges, U. Schubert, *Organometallics* **1996**, *15*, 2373.

Novel Synthetic Approach to Molybdenum-Silicon Compounds: Structures and Reactivities

Peter Jutzi*, Stefan H. A. Petri
Fakultät für Chemie
Universität Bielefeld
Universitätsstr. 25, D-33615 Bielefeld, Germany
Tel.: Int. code + (521)1066181 – Fax.: Int. code + (521)1066026

Keywords: Exchange Reactions / MOCVD / Molybdenum Compounds / Silicon Compounds / X-Ray Crystal Structures

Summary: For several reasons, complexes with a transition metal-silicon bond are of special interest. A new and very convenient synthesis of molybdenum-silicon compounds is presented. The complexes $Cp_2Mo(H)SiBr_3$ (**1**) and $Cp_2Mo(H)Si_2Cl_5$ (**2**) are formed by simply stirring a solution of Cp_2MoH_2 in toluene in the presence of the corresponding halogenosilane. The X-ray crystal structure analyses reveal the shortest ever observed Mo–Si bond distances (**1**: 2.459(3)Å and **2**: 2.4636(8)Å). Several other new $Cp_2Mo(H)SiR_3$ complexes are synthesized by the well-known reductive elimination/oxidative addition reaction starting from Cp_2MoH_2 and the corresponding hydridosilane. Some typical reactions of these compounds are presented.

Introduction

In the last years transition metal-silyl complexes have received special attention for several reasons [1, 2]. On the one hand, they are assumed to be important intermediates in catalytic processes [2] (transition metal-catalyzed hydrosilylation reaction, dehydrogenative coupling of silanes to polysilanes, etc.), on the other metal-substituted silanes show special properties, which can be tuned systematically by judicious choice of the metal and its ligands [3]. Furthermore, silylenes (silanediyls) are stabilized by unsaturated transition metal fragments leading to metal-silicon double-bonds [4]. In the light of a possible application in MOCVD processes some of these complexes are of interest as potential single-source precursors for the manufacture of thin silicide films [5].

Synthesis

Common syntheses of these compounds are the alkaline salt elimination reaction, the insertion of unsaturated metal fragments into Si–H bonds, or metal-silicon bond formation by elimination of small molecules [1, 2].

The synthesis of molybdocene compounds containing a Mo–Si bond commonly proceeds via the oxidative addition of Si–H bonds to the 16 VE molybdocene fragment (vide infra). We present a new and very convenient synthetic approach to complexes of this type [6].

The complexes $Cp_2Mo(H)SiBr_3$ (1) and $Cp_2Mo(H)Si_2Cl_5$ (2) are formed via elimination of hydrogen halide starting from molybdocene dihydride and the corresponding halogenosilane.

1 ($XSiR_3$ = $BrSiBr_3$)
2 ($XSiR_3$ = $ClSi_2Cl_5$)

Eq. 1.   Synthesis of **1** and **2**.

This type of elimination reaction seems to be limited to halogenosilanes containing a silicon atom exhibiting a strong Lewis acidity. The analogous reaction of Cp_2MoH_2 with several other chlorosilanes only yields the monochloro compound $Cp_2Mo(H)Cl$.

Several other new $Cp_2Mo(H)SiR_3$ complexes are synthesized by the photolysis of Cp_2MoH_2 in the presence of the corresponding hydridosilane [6].

SiR_3	$SiEt_3$	$SiCl_3$	$Si(OEt)_3$	SiH_2Cp*	$SiH_2(C_5Me_4H)$	$SiH(C_5Me_4H)_2$
	3	4	5	6	7	8

Eq. 2.   Synthesis of **3-8**.

The reaction proceeds via photolytic induced extrusion of hydrogen from Cp_2MoH_2. The resulting unsaturated molybdocene fragment inserts into the Si–H bond of the hydridosilane. This type of reductive elimination/oxidative addition reaction is a well-known strategy in the synthesis of silyl hydride complexes [7, 8]. Compounds **3-8** are all yellow to orange, air- and moisture-sensitive solids showing spectroscopic data typical for $Cp_2Mo(H)SiR_3$ complexes.

Structures

Compounds **1**, **2**, **6**, **7**, and **8** are structurally characterized by single-crystal X-ray diffraction analysis. ORTEP drawings, showing 50 % probablity thermal ellipsoids, are shown in Figs. 1-5. Characteristic bonding parameters are listed in Table 1. These compounds show the typical geometry of bent-sandwich complexes: The molybdenum possesses pseudotetrahedral surroundings with the molybdenum and silicon lying in the plane bisecting the angle between the two η^5-Cp ligands. The methylated Cp groups in **6-8** are σ-bound to the silicon. Compound **8** crystallizes with two slightly different, but statistically equivalent, molecules in the asymmetric unit. Only one of these molecules is shown in Fig. 5.

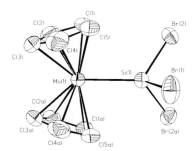

Fig. 1. Molecular structure of **1**.

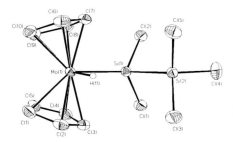

Fig. 2. Molecular structure of **2**.

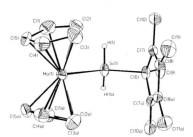

Fig. 3. Molecular structure of **6**

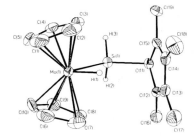

Fig. 4. Molecular structure of **7**.

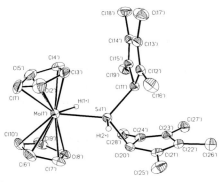

Fig. 5. Molecular structure of **8**.

Table 1. Selected bond lengths (Å) and bond angles (deg).

	Mo–Si	Mo–H	Si–E$_{eq}$	Cp1–Mo–Cp2[a]	Mo–Si–E$_{eq}$[b]	H–Mo–Si
1	2.459(3)	[c]	2.271(3)	143.5	116.81(10)	[c]
2	2.4636(8)	1.60(3)	2.3394(10)	144.0	115.00(3)	77.0(10)
6	2.5211(12)	[c]	1.970(3)	148.0	124.30(10)	[c]
7	2.516(2)	1.58(6)	1.956(4)	146.5	121.72(14)	76.0(2)
8[d]	2.5683(12)	1.65(5)	1.35(4)	144.3	107(2)	76.0(2)
	2.5637(11)	1.83(4)	1.51(4)	144.0	108(2)	75.9(14)

[a] Cp1 and Cp2 are the centroids of the η^5-Cp – [b] E$_{eq}$ = atom lying in the H–Mo–Si plane – [c] HMo not located – [d] Two molecules in the unit cell.

The Mo–Si bond lengths in compounds **6-8** are between 2.516(2)Å and 2.5683(12)Å, which correlates very well with M–Si (M = Mo, W) bond lengths already observed by Berry and coworkers [8] in similar metallocene complexes. On the other hand, compounds **1** and **2** possess significantly shorter Mo–Si bonds. In fact, to the best of our knowledge the SiBr$_3$ derivative **1** has the shortest ever observed Mo–Si bond. Its value of 2.459(3)Å is in the range of the distance calculated for the silylene (silanediyl) complex (CO)$_5$Mo=SiH$_2$ (D$_{Mo=Si}$ = 2.46Å) [9].

In solution compound **6** undergoes fast sigmatropic rearrangements [10] within the Si–Cp* fragment indicated by just one broad signal for the methyl protons in the ^{1}H NMR spectrum. Compounds **7** and **8** only show signals attributed to the isomers having the Si atom in allylic position of the C$_5$Me$_4$H group.

Reactivity

A typical reaction of metal hydride complexes is the conversion to the corresponding halogeno compounds. This is exemplarily illustrated by the clean H/Cl exchange with CCl$_4$ on **5** to Cp$_2$Mo(Cl)Si(OEt)$_3$ **9**. The reverse conversion can be done by hydrogenation with LiAlH$_4$ (Eq. 3).

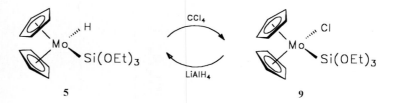

Eq. 3. Conversion of **5** to **9** and vice versa.

Many metal silyl complexes, especially those with early transition metals, have weak metal-silicon bonds. The molybdocene silyl hydride complexes seem to have molybdenum-silicon bonds of sufficient strength to withstand nucleophilic attack by strong hydride reagents (Eq. 4).

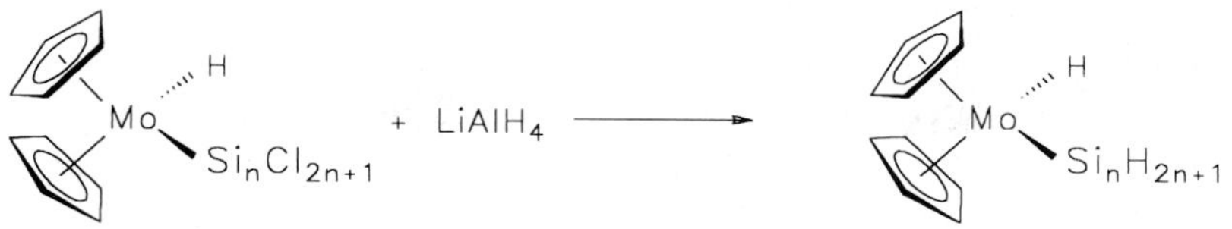

$n = 1$: **4**
   2: **2**

10
11

Eq. 4. Cl/H exchange on **4** and **2**.

On the other hand, the attempted conversion of the $SiCl_3$ derivative **4** to $Cp_2Mo(H)Si(OEt)_3$ **5** in the presence of ethanol and pyridine is not successful.

Perspectives

Some of the compounds mentioned, especially **10** and **11**, possibly have the potential to be used in MOCVD processes for the generation of thin molybdenum silicide films (Eq. 5). Deposition experiments are planned.

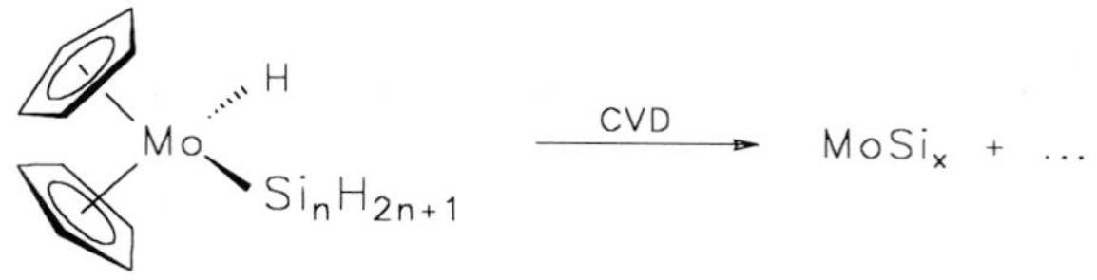

$n = 1$: **10**
   2: **11**

Eq. 5. Potential application of **10** and **11** in CVD processes.

Furthermore, the synthesis of silylene (silanediyl) complexes by appropriate elimination or abstraction reactions is under present investigation (Eq. 6).

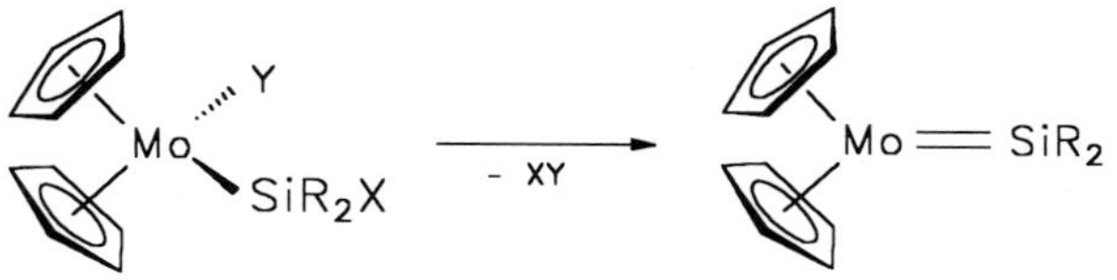

Eq. 6. Strategy for the synthesis of silylene (silanediyl) complexes.

Acknowledgement: This research was generously supported by a grant from the *Deutsche Forschungsgemeinschaft*. *Wacker-Chemie* (Burghausen) is acknowledged for a generous gift of Si_2Cl_6.

References:

[1]   B. J. Aylett, *Adv. Inorg. Chem. Radiochem.* **1982**, *25*, 1; U. Schubert, *Transition Met. Chem.* **1991**, *16*, 136; H. K. Sharma, K. H. Pannell, *Chem. Rev.* **1995**, *95*, 1351; P. Braunstein, M. Knorr, *J. Organomet. Chem.* **1995**, *500*, 21.

[2]   T. D. Tilley, *"Transition-Metal Silyl Derivatives"*, in: *The Chemistry of Organosilicon Compounds* (Eds.: S. Patai, Z. Rappoport), Wiley, New York, **1989**, p. 1415; and refs. cited therein; T. D. Tilley, Appendix to *"Transition-Metal Silyl Derivatives"*, in: *The Silicon-Heteroatom Bond* (Eds.: S. Patai, Z. Rappoport), Wiley, New York, **1991**, p. 309; and refs. cited therein.

[3]   S. Schmitzer, U. Weis, H. Käb, W. Buchner, W. Malisch, T. Polzer, U. Posset, W. Kiefer, *Inorg. Chem.* **1993**, *32*, 303; W. Malisch, S. Schmitzer, G. Kaupp, K. Hindahl, H. Käb, U. Wachtler, *"Organometal Fragment Substituted Silanols, Siloxanes, and Silylamines"*, in: *Organosilicon Chemistry: From Molecules to Materials* (Eds.: N. Auner, J. Weis), VCH, Weinheim, **1994**, p. 185; W. Malisch, S. Möller, R. Lankat, J. Reising, S. Schmitzer, O. Fey, *"Novel Metallo-Silanols, -Silandiols, and -Silantriols of the Iron and Chromium Group: Generation, Structural Characterization, and Transformation to Metallo-Siloxanes"*, in: *Organosilicon Chemistry II: From Molecules to Materials* (Eds. N. Auner, J. Weis), VCH, Weinheim, **1996**, p. 575; W. Malisch, R. Lankat, S. Schmitzer, J. Reising, *Inorg. Chem.* **1995**, *34*, 5701; W. Malisch, R. Lankat, W. Seelbach, J. Reising, M. Noltemeyer, R. Pikl, U. Posset, W. Kiefer, *Chem. Ber.* **1995**, *128*, 1109; W. Malisch, R. Lankat, S. Schmitzer, R. Pikl, U. Posset, W. Kiefer, *Organometallics* **1995**, *14*, 5622.

[4]   C. Zybill, *Top. Curr. Chem.* **1991**, *160*, 1; P. D. Lickiss, *Chem. Soc. Rev.* **1992**, *21*, 271, and refs. cited therein; H. Kobayashi, K. Ueno, H. Ogino, *Organometallics* **1995**, *14*, 5490.

[5]   T. T. Kodas, M. J. Hampden-Smith, in: *The Chemistry of Metal CVD*, VCH, Weinheim, **1994**.

[6]   P. Jutzi, S. H. A. Petri, B. Neumann, H.-G. Stammler, in preparation.

[7]   C. Aitken, J.-P. Barry, F. Gauvin, J. F. Harrod, A. Malek, D. Rousseau, *Organometallics* **1989**, *8*, 1732; S. Seebald, B. Mayer, U. Schubert, *J. Organomet. Chem.* **1993**, *462*, 225.

[8]   T. S. Koloski, D. C. Pestana, P. J. Carrol, D. H. Berry, *Organometallics* **1994**, *13*, 489.

[9]   A. Márquez, J. F. Sanz, *J. Am. Chem. Soc.* 1992, *114*, 2903.

[10]   P. Jutzi, *Chem. Rev.* **1986**, *86*, 983; P. Jutzi, *Comments Inorg. Chem.* **1987**, *6*, 123; P. Jutzi, *J. Organomet. Chem.* **1990**, *400*, 1.

Unexpected Reactivity of Bis-1,2-[(bromodiphenyl)methyl]-1,1,2,2-tetramethyldisilane

Frederik Pillong, Oliver Schütt, Carsten Strohmann*
Institut für Anorganische Chemie
Universität des Saarlandes
Postfach 1150, D-66041 Saarbrücken, Germany
Tel.: Int. code + (0681)3022465
E-mail: c.strohmann@rz.uni-sb.de

Keywords: Carbocations / Disilanes / Silacycles / (Bromomethyl)silanes

Summary: Bis-1,2-[(bromodiphenyl)methyl]-1,1,2,2-tetramethyldisilane (**3**) was prepared by bromination of the corresponding (diphenylmethyl)disilane with NBS in CCl_4. A solution of **3** in $CHCl_3$ loses HBr and forms a silacycle. 3,3,5,5-Tetraphenyl-1-oxa-2,4-disila-cyclopentane (**10**) was found as the reaction product of **3** with H_2O. The formation of the silacycles can be explained by an intramolecular rearrangement of the disilane **3** with a carbocation intermediate. Single crystal X-ray diffraction analysis has been performed on two silacycles.

Introduction

3-Silaoxetanes were prepared by the reaction of bis(halogenmethyl)silanes with H_2O [1].

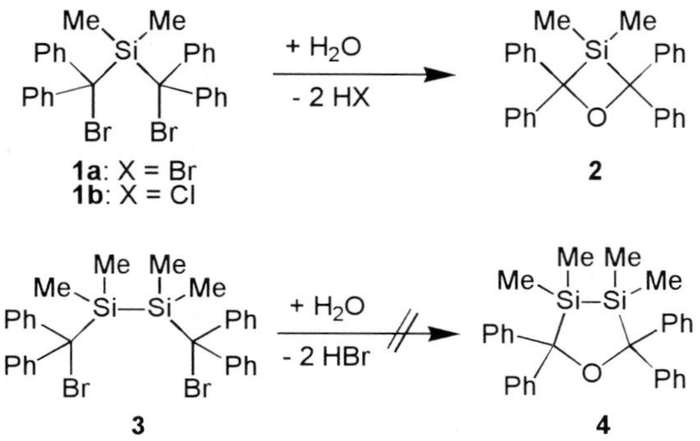

Scheme 1.

As a part of our systematic studies on the structural unit "–CR$_2$–El–CR$_2$–" (El = element of group 14-16, partly with substituents; R = H, alkyl, aryl) [2], we were interested in the preparation of 1-oxa-3,4-disilacyclopentanes in an analogous reaction sequence (Scheme 1) and also in the reactivity of these disilacycles. Unexpected reactivity of the starting material bis-1,2-[(bromodiphenyl)methyl]-1,1,2,2-tetramethyldisilane (**3**) was observed.

Synthesis and Reactivity of Bis-1,2-[(bromodiphenyl)methyl]-1,1,2,2-tetramethyl-disilane

The disilane **6** was prepared by reaction of **5** with diphenylmethyllithium. Bromination with NBS in CCl$_4$ gave the (bromomethyl)disilane **3** (see Scheme 2). A solution of **3** in CHCl$_3$ loses HBr and the silacycle **7** was formed in high yield. Reaction of **3** with H$_2$O/isopropyl alcohol (1:9) analogous to the synthesis of 3-silaoxetanes [1] gave the 3,3,5,5-tetraphenyl-1-oxa-2,4-disila-cyclopentane (**10**).

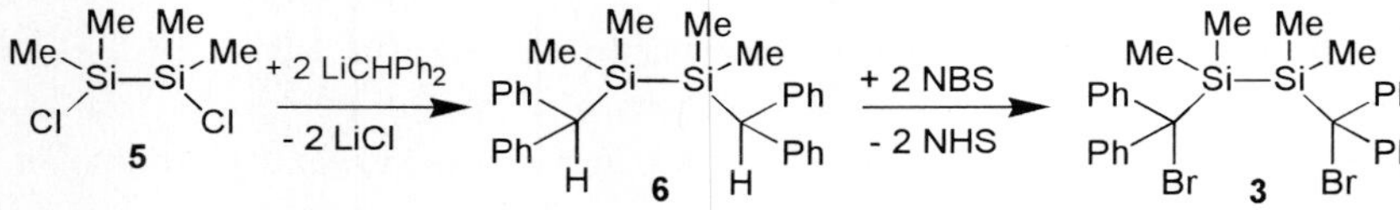

Scheme 2.

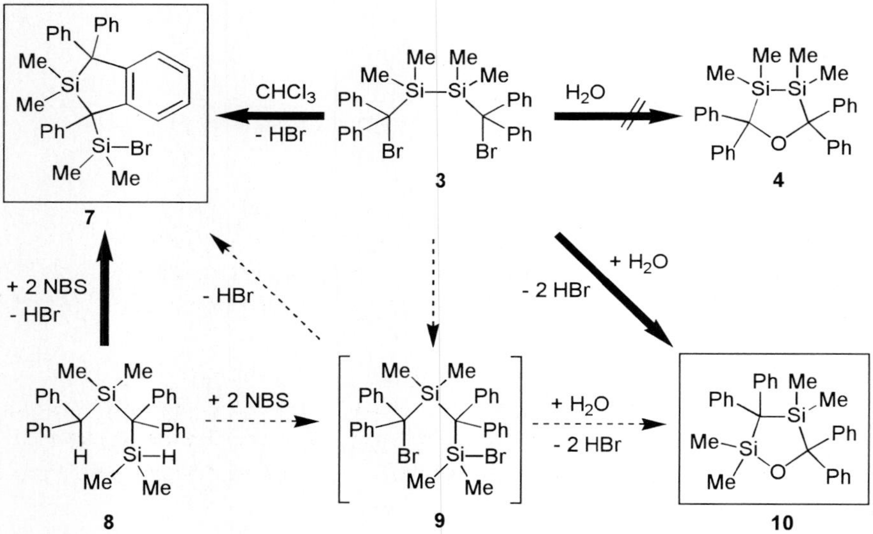

Scheme 3.

Both reactions can be explained by the formation of a carbocation and rearrangement to the bromosilane **9**. The silacycle **7** was formed from **9** through a second carbocation and C–C bond formation with HBr elimination. The reaction of **9** with H$_2$O gives the silacycle **10**. The existence of **9** was proved by ist synthesis by another synthetic route. Bromination of **8** with NBS produces the silacycle **7**.

A similar rearrangement of **1a** to the correspondimg silacycle could be initiated by Lewis acids. The formation of **9** from **3** is the first example of such a rearrangement under mild conditions, however, few examples are known for (chloromethyl)disilanes with bases or Lewis acids [3].

Crystal Structures

The molecular structure of **10** in the crystal shows a folded ring in envelope form. Selected bond lengths and angles are shown in Fig. 1.

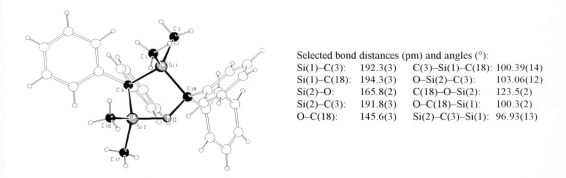

Selected bond distances (pm) and angles (°):

Si(1)–C(3):	192.3(3)	C(3)–Si(1)–C(18):	100.39(14)
Si(1)–C(18):	194.3(3)	O–Si(2)–C(3):	103.06(12)
Si(2)–O:	165.8(2)	C(18)–O–Si(2):	123.5(2)
Si(2)–C(3):	191.8(3)	O–C(18)–Si(1):	100.3(2)
O–C(18):	145.6(3)	Si(2)–C(3)–Si(1):	96.93(13)

Fig. 1. View of the molecular structure of **10**.

The silacycle **10** is the first example of these ring systems, that has been characterized by single crystal X-ray diffraction [4]. The bond lengths in the cyclopentane ring are longer than the normal Si–C, Si–O, and C–O bond lengths [5], indicating a sterically crowded molecule. In Fig. 2 a view of the molecular structure of the silacycle **7** is given.

Fig. 3 shows a side view of the silacycle **7** and for comparison, the silacycle **11**, which was formed from **1b** similar to the formation of **7** from **9**. Characteristic for **7** is a folded silacyclopentane ring. A lightly folded cyclopentane ring was observed for **11**. We propose the packing effects as the cause for the two different conformations. Influence of packing effects on ring folding is also found for 1,3-disilacyclobutanes [6] and 1,3,4-trisilacylopentanes [7].

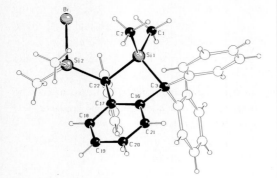

Selected bond distances (pm) and angles(°)

Br–Si(2):	224.1(2)	C(22)–Si(1)–C(3):	94.2(2)
Si(1)–C(22):	192.1(4)	C(22)–Si(2)–Br:	115.4(2)
Si(1)–C(3):	192.9(4)	C(16)–C(3)–Si(1):	96.1(2)
Si(2)–C(22):	190.7(4)	C(17)–C(16)–C(3):	116.7(3)
C(3)–C(16):	154.0(5)	C(16)–C(17)–C(22):	117.6(3)
C(16)–C(17):	140.1(5)	C(17)–C(22)–Si(2):	109.5(3)
C(17)–C(22):	153.3(5)	C(17)–C(22)–Si(1):	98.5(2)
		Si(2)–C(22)–Si(1):	117.9(2)

Fig. 2.   View of the molecular structure of **7**.

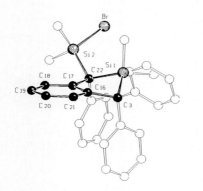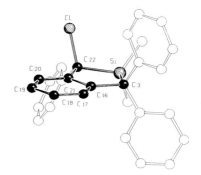

Fig. 3.   Side view on the molecular structure of **7** and **11** in the crystal.

Acknowledgements: We are grateful to the *Fonds der Chemischen Industrie* (FCI) and to the *Deutschen Forschungsgemeinschaft* (DFG) for financial support. We would like to thank Prof. M. Veith (Universität Saarbrücken) for supporting this work.

References:

[1]   C. Strohmann, *Chem. Ber.* **1995**, *128*, 167.

[2]   a) C. Strohmann, *Angew. Chem.* **1996**, *108*, 600.
       b) C. Strohmann, S. Lüdtke, E. Wack, *Chem. Ber.* **1996**, *129*, 799.
       c) C. Strohmann, S. Lüdtke, *"Mono-, Bis-, Tris-, and Tetrakis(lithiomethyl)silanes: New Building Blocks for Organosilicon Compounds"* in: *Organosilicon Chemistry II: From Molecules to Materials* (Eds.: N. Auner, J. Weis), VCH, Weinheim, **1996**, p. 499.

[3]   A. W. P. Jarvie, A. Holt, J. Thompson, *J. Chem. Soc.* (B) **1969**, 852.

[4]   J. Ohshia, H. Hasebe, Y. Masaoka, M. Ishikawa, *Organometallics* **1994**, *13*, 1064.

[5]   W. S. Sheldrick, in: *The Chemistry of Organic Silicon Compounds, Part 1* (Eds.: S.Patai, Z. Rappoport), Wiley, Chichester, **1989**, p.227.

[6]   C. Strohmann, E. Wack, *"Synthesis of 1,3-Disilacyclobutanes, 1,3-Digermacyclobutanes and 1-Germa-3-silacyclobutanes with New 1,3-Dimetallated Organoelement Building Blocks"* in: *Organosilicon Chemistry III: From Molecules to Materials* (Eds.: N. Auner, J. Weis), Wiley-VCH, Weinheim, **1997**, this volume p. 217.

[7]   C. Strohmann, F. Pillong, unpublished results.

Trichlorosilane/Triethylamine – An Alternative to Hexachlorodisilane in Reductive Trichlorosilylation Reactions ?

L.-P. Müller, A. Zanin, J. Jeske, P. G. Jones, W.-W. du Mont*
Institut für Anorganische und Analytische Chemie
Technische Universität Braunschweig
Hagenring 30, D-38106 Braunschweig, Germany
Tel.: Int. code + (0531)3915302 – Fax: Int. code + (0531)3915387
E-mail: dumont@mac1.anchem.nat.tu-bs.de

Keywords: Organophosphorus halides / Reductive Silylation / Silylphosphanes / Diphosphene

Summary: Alkyl- and dialkylaminodichlorophosphanes $RPCl_2$ **1a-f** (**1a**: R = *t*Bu, **1b**: R = *i*Pr$_2$N, **1c**: R = 1-adamantyl, **1d**: R = (Me$_3$Si)$_2$CH, **1e**: R = *i*Pr, **1f**: R = Et$_2$N) and chlorophosphanes RR'PCl **3a - e** (**3a**: R = *i*Pr, R' = *i*Pr$_2$N; **3b**: R = *i*Pr, R' = Ph$_2$N; **3c**: R = *t*Bu, R' = Et$_2$N; **3d**: R = *t*Bu, R' = Me$_3$SiCH$_2$; **3e**: R = *i*Pr, R' = Et$_2$N;) react with hexachlorodisilane or with trichlorosilane/triethylamine to provide trichlorosilyl-phosphanes $RP(SiCl_3)_2$ **2a-f** and RR'PSiCl$_3$ **4a-e**. Depending on the nature of R, bis(trichlorosilylations) leading to compound **2** are accompanied by cyclophosphane formation and other decomposition reactions. The trichlorosilane/triethylamine method allows further product formation and milder reaction conditions. Trichlorosilylation of *P*-halogenophosphaalkenes (Me$_3$Si)$_2$C=P–X **6a-c** (**6a**: X = Cl, **6b**: X = Br, **6c**: X = I) with hexachlorodisilane provided an unusual diphosphene [(Me$_3$Si)$_2$(Cl$_3$Si)CP]$_2$ **7**. With trichlorosilane/triethylamine, however, the reduction of **6a** is accompanied by hydrosilylation of the P=C double bond leading to **2d** as main product.

Recently the reductive trichlorosilylation of chlorophosphanes with hexachlorodisilane provided a novel, mild access to molecules with silicon-phosphorus bonds [1-4]. Dialkylchlorophosphanes and dialkylamino(alkyl)chlorophosphanes were easily converted into trichlorosilylphosphanes.

$$RR'PCl\ +\ Si_2Cl_6\ \longrightarrow\ RR'PSiCl_3\ +\ SiCl_4$$

$$\text{3a-d} \qquad\qquad\qquad\qquad\qquad \text{4a-d}$$

Eq.1. **a**: R = *i*Pr, R' = *i*Pr$_2$N; **b**: R = *i*Pr, R' = Ph$_2$N; **c**: R = *t*Bu, R' = Et$_2$N; **d**: R = *t*Bu, R = Me$_3$SiCH$_2$.

The bis(silylation) of alkyldichlorophosphanes and dialkylamino(dichloro)phosphanes, however, with two equivalents of hexachlorodisilane, leading to bis(trichlorosilyl)phosphanes, requires rather bulky substituents at phosphorus to prevent the undesired formation of cyclophosphanes or other decomposition products.

$$RPCl_2 + 2 Si_2Cl_6 \longrightarrow RP(SiCl_3)_2 + 2 SiCl_4$$
$$\textbf{1a-d} \hspace{6cm} \textbf{2a-d}$$

Eq. 2. **a**: R = *t*Bu; **b**: R = *i*Pr$_2$N; **c**: R = Ad; **d**: R = (Me$_3$Si)$_2$CH.

Various reactions of hexachlorodisilane in the presence of nucleophiles appear to be associated with latent trichlorosilyl anions. These anions are generated from one SiCl$_3$ group of Si$_2$Cl$_6$ when the other silicon atom is attacked by a nucleophile [5-7]. Recently, an amino(chloro)phosphane hexachlorodisilane adduct was detected by ^{31}P NMR and ^{29}Si NMR [4].

3a   **5a**   **4a**

Eq. 3. **5a**: δ(^{31}P) = 57.1 (s); δ([29]Si) = -69.6 (d) PSiCl$_3$SiCl$_3$; $^1J_{P-Si}$ = 160.7 Hz 9.8 (d) PSiCl$_3$SiCl$_3$; $^2J_{P-Si}$ = 24.8 Hz.

Trichlorosilyl anions are also key intermediates in numerous reactions involving trichlorosilane/triethylamine systems [8-10]. The reagent is known to reduce chlorophosphanes RPCl$_2$ and R$_2$PCl to the phosphanes RPH$_2$ and R$_2$PH. Previously amino(trichlorosilyl)phosphanes (containing *one* P–Si bond) had been generated by (cationic) trichlorosilylation of the P–H compounds with silicon tetrachloride. Avoiding strictly hydrolytic P–Si bond cleavage, a quite selective transformation of several dialkyl- and dialkylamino(alkyl)chlorophosphanes with trichlorosilane/triethylamine to the corresponding trichlorosilylphosphanes is possible [4].

$$RR'PCl + HSiCl_3 \xrightarrow{\text{+ NEt}_3/\text{- HNEt}_3Cl} RR'PSiCl_3$$
$$\textbf{3a-e} \hspace{6cm} \textbf{4a-e}$$

Eq. 4. **a**: R = *i*Pr, R' = *i*Pr$_2$N; **b**: R = *i*Pr, R' = Ph$_2$N; **c**: R = *t*Bu, R' = Et$_2$N; **d**: R = *t*Bu, R' = Me$_3$SiCH$_2$; **e**: R = *i*Pr, R' = NEt$_2$.

Similarly, alkyl- and dialkylaminodichlorophosphanes are converted into the bis(trichlorosilyl)-phosphanes. With less bulky substituents (**2e**, **2f**) the bis(trichlorosilyl)phosphanes are not isolated in the pure state. On attempts at distillation cyclophosphanes are formed (**2e**) or the product decomposes to form an insoluble precipitate (**2f**).

$$RPCl_2 + 2 HSiCl_3 \xrightarrow{\text{+ 2 NEt}_3/\text{- 2 HNEt}_3Cl} RP(SiCl_3)_2$$
$$\textbf{1a-f} \hspace{6cm} \textbf{2a-f}$$

Eq. 5. **a**: R = *t*Bu; **b**: R = *i*Pr$_2$N; **c**: R = Ad; **d**: R = (Me$_3$Si)$_2$CH; **e**: R = *i*Pr; **f**: R = Et$_2$N.

The large magnitude of $^1J_{P-Si}$ in the trichlorosilylphosphanes **2a-f** and **4a-e** (Tables 1 and 2) reflects the electron-withdrawing effect of the chlorine atoms attached to silicon and the steric requirements of the other substituents at phosphorus. There is some evidence that this large coupling coincides with strong P–Si bonds: trichlorosilylphosphanes (R_2PSiCl_3) are not cleaved by Me_3GeCl or Me_3SnCl; tri*methyl*silylphosphanes, however, give straightforward exchange reactions with these reagents, providing trimethylgermyl- and stannylphosphanes [11, 12]. It is also known that the P–Si bond of H_2PSiF_3 (220.7 pm) [13] is significantly shorter than the P–Si bond of H_2PSiH_3 (224.9 pm) [14]. Therefore, it was important to us to carry out a X-ray crystal structure determination on **2b** (Fig. 1).

Table 1. ^{31}P and ^{29}Si-NMR shifts and coupling constants $^1J_{P-Si}$ of the bis(trichloro-silyl)phosphanes.

	$\delta(^{31}P)$ [ppm]	$\delta(^{29}Si)$ [ppm]	$^1J_{P-Si}$ [Hz]
2a: $tBuP(SiCl_3)_2$	-55.3	7.0	77.3
2b: $iPr_2NP(SiCl_3)_2$	-9.3	1.8	70.9
2c: $1\text{-}AdaP(SiCl_3)_2$	-56.1	7.2	79.8
2d: $(Me_3Si)_2CHP(SiCl_3)_2$	-82.8	6.2	85.1
2e: $i\text{-}PrP(SiCl_3)_2$	-78.0	8.0	72.7
2f: $Et_2NP(SiCl_3)_2$	11.7	1.8	75.0

Table 2. ^{31}P and ^{29}Si-NMR shifts and coupling constants $^1J_{P-Si}$ of the trichloro-silylphosphanes.

	$\delta(^{31}P)$ [ppm]	$\delta(^{29}Si)$ [ppm]	$^1J_{P-Si}$ [Hz]
4a: $iPr(iPr_2N)PSiCl_3$	35.8	10.1	123.8
4b: $iPr(Ph_2N)PSiCl_3$	42.6	9.1	125.3
4c: $tBu(Et_2N)PSiCl_3$	74.0	6.6	125.7
4d: $tBu(Me_3SiCH_2)PSiCl_3$	-33.6	13.6	116.2
4e: $iPr(Et_2N)PSiCl_3$	55.3	10.9	131.8

The coordination geometry at phosphorus in **2b** (Fig. 1) is essentially pseudo-tetrahedral (all angles at phosphorus are between 102 and 107°); the nitrogen atoms in both molecules are planar. The P–Si bond lengths are "normal" (224 pm), closely similar to that in H_2PSiH_3; silicon-clorine distances (203 pm) are also as expected for $RSiCl_3$ species. In summary, the rather large magnitude of the NMR coupling $^1J_{P-Si}$ does not coincide with an unusually short P–Si bond length.

The reactions of *P*-halogeno-bis(trimethylsilyl)methylenphosphanes **6a-c** with hexachlorodisilane lead to a diphosphene **7** (instead of the expected $(Me_3Si)_2C=P\text{-}SiCl_3$) with a new trichlorosilyl-functionalized alkyl group $-C(Me_3Si)_2(SiCl_3)$ [1].

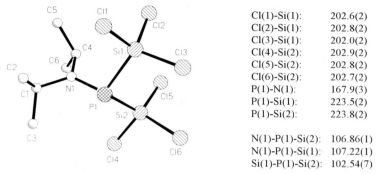

Cl(1)-Si(1):	202.6(2)
Cl(2)-Si(1):	202.8(2)
Cl(3)-Si(1):	202.0(2)
Cl(4)-Si(2):	202.9(2)
Cl(5)-Si(2):	202.8(2)
Cl(6)-Si(2):	202.7(2)
P(1)-N(1):	167.9(3)
P(1)-Si(1):	223.5(2)
P(1)-Si(2):	223.8(2)

N(1)-P(1)-Si(2):	106.86(1)
N(1)-P(1)-Si(1):	107.22(1)
Si(1)-P(1)-Si(2):	102.54(7)

Fig. 1. Molecular structure and selected geometrical parameters [pm, °] of iPr$_2$NP(SiCl$_3$)$_2$ **2b**.

$$2\ (Me_3Si)_2C{=}P{-}X\ +\ 2\ Si_2Cl_6\ \longrightarrow\ TSi^*{-}P{=}P{-}TSi^*\ +\ 2\ SiCl_3X$$

6a-c   **7**

Eq. 6. **6a**: X = Cl; **6b**: X = Br; **6c**: X = I; TSi* = C(SiCl$_3$)(SiMe$_3$)$_2$.

However, the similar silylation of the *P*-chlorophosphaalkene **6a** with two equivalents of the trichlorosilane/triethylamine reagent does not proceed with formation of a P=P double bond. The main product of this reaction is the stable bis(trimethylsilyl)methylbis(trichlorosilyl)phosphane **2d**, which indicates a hydrosilylation of the P=C double bond by trichlorosilane, like the known Benkeser additions at unsaturated carbon-carbon bond systems [8].

$$(Me_3Si)_2C{=}P{-}Cl\ +\ 2\ HSiCl_3\ \xrightarrow{\ NEt_3\ }\ (Me_3Si)_2CH{-}P(SiCl_3)_2\ +\ NHEt_3Cl$$

6a   **2d**

Eq. 7.

Acknowledgement: We thank the *Deutsche Forschungsgemeinschaft*, and the *Fonds der Chemischen Industrie* for financial support.

References:

[1]   Part IV: A. Zanin, M. Karnop, J. Jeske, P. G. Jones, W.-W. du Mont, *J. Organomet. Chem.* **1994**, *475*, 95

[2]   R. Martens, W.-W. du Mont, L. Lange, *Z. Naturforsch.* **1991**, *46b*, 1609

[3]   R. Martens, W.-W. du Mont, *Chem. Ber.* **1992**, *125*, 657.

[4]   L.-P. Müller, W.-W. du Mont, J. Jeske, P. G. Jones, *Chem. Ber.* **1995**, *128*, 651.

[5]   K. Naumann, G. Zon, K. Mislow, *J. Am. Chem. Soc.* **1969**, *91*, 7012.

[6]   G. Zon, K. E. de Bruin, K. Naumann, K. Mislow, *J. Am. Chem. Soc.* **1969**, *91*, 7023.

[7]   R. Martens, W.-W. du Mont, *Chem. Ber.* **1993**, *126*, 1015.

[8]   R. A. Benkeser, *Acc. Chem. Res.* **1971**, *4*, 94.

[9]   R. A. Benkeser, K. M. Foley, J. B. Grutzner, W. E. Smith, *J. Am. Chem. Soc.* **1970**, *92*, 697.

[10]   S. C. Bernstein, *J. Am. Chem. Soc.* **1970**, *92*, 699.

[11]   H. Schumann, L. Rösch, *Chem. Ber.* **1974**, *107*, 854.

[12]   R. Martens, W.-W. du Mont, J. Jeske, P. G. Jones, W. Saak, S. Pohl *J. Organomet. Chem.* **1995**, *501*, 251.

[13]   R. Demuth, H. Oberhammer, *Z. Naturforsch.* **1973**, *28a*, 1862.

[14]   C. Glidewell, P. M. Pinder, A. G. Robiette, G. M. Sheldrick, *J. Chem. Soc., Dalton Trans.* **1972**, 1402.

Disproportionation of Chloromethyldisilanes using Lewis Base Heterogeneous Catalysts – A Way to Influence the Polymer Structure

Thomas Lange*, Norbert Schulze, Gerhard Roewer, Robin Richter
Institut für Anorganische Chemie
Technische Universität Bergakademie Freiberg
Leipziger Str. 29, D–09596 Freiberg, Germany
Tel.: Int. code + (3731)394302
E-mail: thomasl@server.compch.tu-freiberg.de

Keywords: Disilanes / Lewis Base / Lewis Acid / Heterogeneously Catalyzed Disproportionation / Polysilane

Summary: The heterogeneous disproportionation of 1,1,2,2-tetrachlorodimethyldisilane to chloromethylsilanes and oligo(chloromethylsilane)s is catalyzed by Lewis bases like bis(dimethylamid)phosphoryl compounds and *N*-heterocycles. The oligosilanes undergo branching and crosslinking reactions controlled by reaction temperature and time schedule forming poly(chloromethylsilane)s that show a 3D polysilyne-type polymer skeleton.
Lewis acids, such as triphenylboron, dissolved in the starting disilane, prevent branching of the polymer backbone during the reaction course.

Introduction

During the last years several synthetic pathways to various polysilane backbones have been extensively studied. One interesting polysilane synthesis has been developed based on the disproportionation of chloromethyldisilanes, which are byproducts of the industrial chloromethylsilane production (Müller-Rochow Synthesis) [1].

The disproportionation of 1,1,2,2-tetrachlorodimethyldisilane (**2**) leads to trichloromethylsilane (**1**) and oligo(methylchloro)silanes catalyzed by Lewis bases (Eq. 1).

R= Cl, SiClCH₃R

Eq. 1

Formation of Oligo(chloromethylsilane)s Derived From 1,1,2,2-Tetrachlorodimethyldisilane

The heterogeneous catalytic disproportionation offers access to a poly(chloromethylsilane) free of catalyst, due to a perfect phase separation between the catalyst, the starting chloromethyldisilane and the reaction products [1]. It is thus possible to avoid subsequent uncontrollable cross-linking reactions.

The catalytically active entities, such as bis(dimethylamido)phosphoryl compounds and *N*-heterocycles were grafted onto the surface of a silica carrier via siloxane bonds, as shown in the following simplified scheme (Fig. 1).

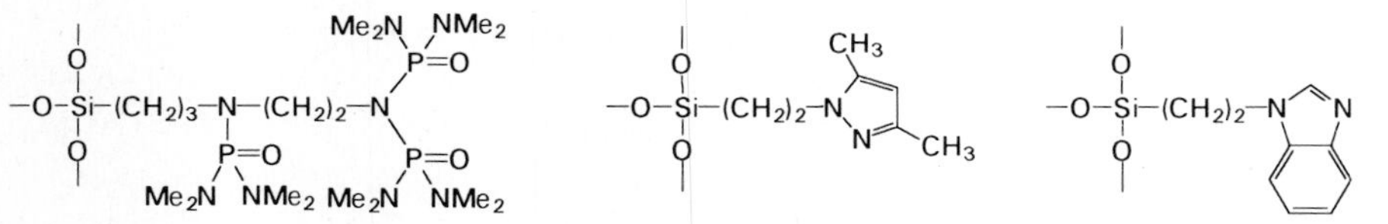

8   **9**   **10**

Fig. 1. Catalytic entities: bis(dimethylamid)phosphoryl compound (**8**), dimethylpyrazole groups (**9**) and benzimidazole groups (**10**) grafted on the surface of a silica carrier.

1,1,2,2-Tetrachlorodimethyldisilane is evaporated (*bp* 155°C) and then brought into contact with the catalyst stored in a fixed-bed reactor. The disilane disproportionates into a mixture of trisilane **3**, tetrasilane **4** and trichloromethylsilane MeSiCl$_3$ on the catalyst surface. The monosilane is distilled off due to its low boiling point (66°C). The oligomer mixture obtained at 175°C in the reaction pot contains, beside oligomers **3** and **4**, higher branched oligomers **5-7** (Fig. 2).

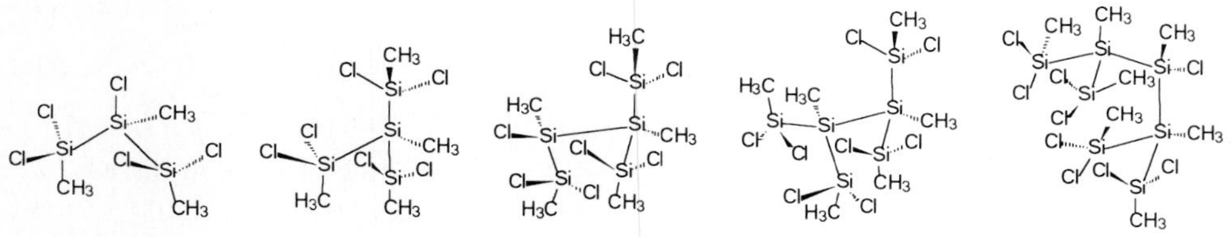

3   **4**   **5**   **6**   **7**

Fig. 2   Oligo(chloromethylsilane)s formed during the first period of the disproportionation reaction - trisilane, tetrasilane, pentasilane, hexasilane and heptasilane (from left to right).

The composition of the oligomer mixture depends on the catalyst. Between 155°C and 180°C grafted *N*-heterocycles (**9** and **10**) generate oligomer mixtures rich in **3** with no hexa- and heptasilane (**6** and **7**) in contrast to grafted bis(dimethylamido)phosphoryl groups (**8**) [2]. We suppose that the basicity of the electron pair donors is not the decisive criterion for the catalytic efficacy. The one-electron donor capability is correlated with the value of the first ionization potential of the Lewis bases. The lower the ionization potential the higher the electron donor capability expected. Our investigations have shown that Lewis bases with first ionization potentials smaller than 8.5 eV are suitable catalysts for Si-Si bond cleavage in $Si_2Cl_4Me_2$.

The Si-Si bond cleavage generates donor-stabilized silylene species (:SiClCH_3) that insert into Si-Cl bonds. It is suggested that **3** and **4** are formed in such a way [3]. Due to the functionality of **2**, the formed oligo(chloromethylsilane)s show a branched structure. The reactions that lead to higher oligo(chloromethylsilane)s **5-7** in the reaction pot are less understood so far. Their formation is probably caused by condensation involving **3** or **4** with formation of $MeSiCl_3$.

The Formation of Poly(chloromethylsilane)s

If the pot temperature is slightly increased up to 220°C, the oligosilanes undergo cross-linking reactions into highly branched poly(chloromethylsilane)s. Using ^{13}C and ^{29}Si NMR spectral editing techniques, an average composition of $MeSiCl_{0.73}$ is obtained, which is in rather good agreement with the mass balance analysis ($MeSiCl_{0.62}$).

The polymer skeleton is constituted with $MeCl_2Si-$, $MeClSi<$, and $MeSi(Si)_3$ groups, as shown in ^{13}C and ^{29}Si CP-MAS NMR spectra (Fig. 3) [3, 4].

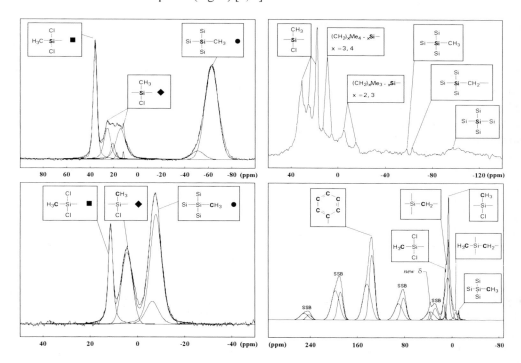

Fig. 3   Influence of the Lewis base (**8**) on the polymer structure: ^{29}Si CP-MAS NMR spectrum (left above), ^{13}C CP-MAS NMR spectrum (left below); Influence of the Lewis acid (BPh_3) on the polymer structure: ^{29}Si CP-MAS NMR spectrum (right above), ^{13}C CP-MAS-NMR spectrum (right below); (SSB: spinning side bands; CP: cross polarization).

A suggestion of the polysilane structure based on NMR and mass balance data is depicted in Fig. 4.

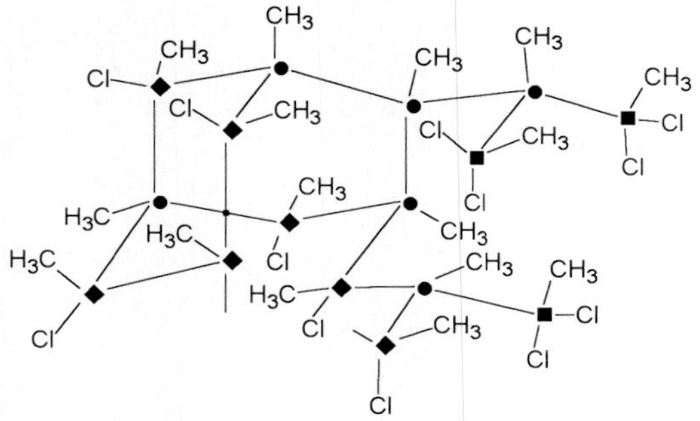

MeSiCl $_{0.73}$

CD: 2.2

Fig. 4   Structure of a poly(chloromethylsilane); the symbols ■, ◆ and ● (see also Fig. 3 left-hand side) are assigned to the Si atoms with different substituents (CD: cross-linking degree).

Gel permeation chromatography indicates a polymodal polydispersity with a broad molecular weight distribution. Currently it is not possible to specify the definite values of the average molecular weights due to a lack of comparable standards.

The addition of a weak Lewis acid like triphenylborane (BPh₃) to the oligo(chloromethylsilane)s has a considerable effect on the polymer building procedure, above all on the average molecular weight and on the structural groups of the resulting polymer. These polymers are characterized by higher molecular weights and a small polydispersity, which can be described as almost mono-modal. Compared to the polysilane represented in Figure 4 the characteristic end groups MeCl₂Si- and branched points MeSi(Si)₃ are missing or only found in very low amount (see Fig. 3 right-hand side). On the contrary, the linear units MeClSi< are the dominant sites. The presence of different carbosilane units, especially (-CH₂)ₓSiR₄₋ₓ (R = Me or Cl with *x* > 2), seems typical for triphenylborane modified polymers.

The oligomer formation (3-7) remains unchanged, when BPh₃ is already added to the starting disilane. However, the composition of the volatile compounds becomes different. It consists not only of trichloromethylsilane but also of dichlorodimethylsilane, benzene, trichloroborane, trimethylborane, and other diverse substituted silanes containing different combinations of chloro-, hydrido-, and methyl groups.

The reaction mechanism, including the possible incorporation of the triphenylborane, has not yet been understood in detail very well. This complex reaction may compete with the branching process. We think that the donor-acceptor interaction between BPh₃ and the MeCl₂Si- groups of the oligomers induces first a carbosilane formation, which is then followed by a conversion into MeClSi< groups. Therefore the modified polymer has more linear structural units resulting in lower branching.

Conclusion

The heterogeneously catalyzed disproportionation of 1,1,2,2-tetrachlorodimethyldisilane leads via oligo(chloromethylsilane)s to highly branched poly(chloromethylsilane)s. The disilane derived oligomer formation can be controlled by the nature of the Lewis base catalyst.

Modified polymers synthesized by the addition of the Lewis acid BPh_3 permit access to lower branched polymer skeleton.

Acknowledgment: The authors are grateful to the *Deutsche Forschungsgemeinschaft* and the *Fonds der Chemischen Industrie* for financial support. We gratefully acknowledge a *NATO* grant for travel between Freiberg and Paris. We particularly thank Dr. Florence Babonneau (*CNRS, Université Pierre et Marie Curie, Paris*) for CP/IRCP MAS-NMR measurements.

References:

[1]   U. Herzog, R. Richter, E. Brendler, G. Roewer, *J. Organomet. Chem.* **1996**, *507*, 221.

[2]   R. Richter, G. Roewer, U. Böhme, K. Busch, F. Babonneau, H.-P. Martin, E. Müller, *Appl. Organomet. Chem.* **1997**, *11*, 71.

[3]   F. Babonneau, J. Maquet, C. Bonhomme, R. Richter, G. Roewer, D. Bahloul, *Chem. Mater.* **1996**, 1415.

[4]   F. Babonneau, R. Richter, C. Bonhomme, J. Maquet, G. Roewer, *J. Chim. Phys.* **1995**, *92*, 1745.

Supersilylated Bromodisilanes, Cyclotri-, and Cyclotetrasilanes

N. Wiberg*, H. Auer, Ch. M. M. Finger, K. Polborn
Institut für Anorganische Chemie
Ludwig-Maximilians-Universität München
Meiserstrasse 1, D-80333 München, Germany
Tel. Int. code + (89)5902230 – Fax: Int. code + (89)5902578
E-mail: niw@anorg.chemie.uni-muenchen.de

Keywords: Supersilyldisilanes / Bromosilanes / Cyclotrisilanes / Cyclotetrasilanes / Tetrasilatetrahedrane

Summary: Brominated 1,2-disupersilyldisilanes $(t\mathrm{Bu_3Si})_2\mathrm{Si_2H}_{4-n}\mathrm{Br}_n$ [$n = 1$, 2 (2 isomers), 3, 4] are formed by bromination of $t\mathrm{Bu_3Si}$–$\mathrm{SiH_2}$–$\mathrm{SiH_2}$–$\mathrm{Si}t\mathrm{Bu_3}$ with $\mathrm{Br_2}$ or $\mathrm{CBr_4}$. Reaction of these compounds with $t\mathrm{Bu_3SiNa}$ in THF at low temperature lead via Br/Na exchange to sodium compounds $(t\mathrm{Bu_3Si})_2\mathrm{Si_2H}_{4-n}\mathrm{Br}_{n-1}\mathrm{Na}$ which are transformed into cyclotrisilanes and cyclotetrasilanes ($n = 2$) or into a tetrasilatetrahedrane ($n = 4$) at higher temperature. Reaction mechanisms are discussed.

Introduction

Recently we gained tetrasupersilyl-*tetrahedro*-tetrasilane, the first molecular compound with a cluster of four silicon atoms located at the corners of a tetrahedron [1]. It can be made by reaction of tetrabromo-1,2-disupersilyldisilane with supersilylsodium in THF in more than 50% yield.

$$\mathrm{Si}t\mathrm{Bu_3} = \text{Tri-}t\text{-butylsilyl} = \text{"Supersilyl"}$$

Eq. 1.

In order to get more insight into mechanism of the formation of the tetrahedrane we studied the reaction of tetrabromo-1,2-disupersilyldisilane with supersilylsodium in presence of trapping reagents. We also reacted partially halogenated 1,2-disupersilyldisilanes with supersilylsodium, because in the latter case the dehalogenation reactions should stop at earlier stages.

Synthesis of Disilanes (tBu$_3$Si)$_2$Si$_2$H$_{4-n}$Br$_n$

Bromination of 1,2-disupersilyldisilane (**1**) with stoichiometric amounts of Br$_2$ in CH$_2$Cl$_2$ leads to the partially brominated disilanes **2**, **3a** and **4** (cf. Eq. 2). **2** and **3a** are accessible in high yield, while **4** is obtained only in mixture with **3a** and **5**. Compound **4** can be separated by preparative HPLC in moderate yield due to partial methanolysis by the eluent methanol (for a better synthesis of **4** cf. dehalogenation of **5**). With an excess of Br$_2$ the perbrominated disilane **5** is formed in quantitative yield. 1,1-Dibromo-1,2-disupersilyldisilane (**3b**) is formed together with all other brominated compounds by reaction of **1** and CBr$_4$, but has not yet been isolated. Compounds **1-5** are air-stable, but the partially brominated compounds slowly undergo solvolysis in methanolic solution.

$$(t\text{Bu}_3\text{Si})_2\text{Si}_2\text{H}_4 + n\,\text{Br}_2 \xrightarrow[n\,=\,1\,-\,4]{(\text{CH}_2\text{Cl}_2)} (t\text{Bu}_3\text{Si})_2\text{Si}_2\text{H}_{4\text{-}n}\text{Br}_n + n\,\text{HBr}$$

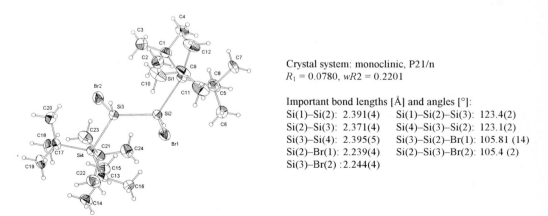

Eq. 2.

According to crystal structure analyses of 1,2-dibromo-1,2-disupersilyldisilane (**3a**) (Fig. 1) and tetrabromo-1,2-disupersilyldisilane (**5**) (Fig. 2) both compounds exhibit a *trans*-relationship of the two bulky supersilyl groups. The bromine atoms of **3a** are in *cis*-relationship.

Crystal system: monoclinic, P2$_1$/n
$R_1 = 0.0780$, $wR2 = 0.2201$

Important bond lengths [Å] and angles [°]:
Si(1)–Si(2): 2.391(4)   Si(1)–Si(2)–Si(3): 123.4(2)
Si(2)–Si(3): 2.371(4)   Si(4)–Si(3)–Si(2): 123.1(2)
Si(3)–Si(4): 2.395(5)   Si(3)–Si(2)–Br(1): 105.81 (14)
Si(2)–Br(1): 2.239(4)   Si(2)–Si(3)–Br(2): 105.4 (2)
Si(3)–Br(2) :2.244(4)

Fig. 1. Crystal structure of 1,2-dibromo-1,2-disupersilyldisilane (**3a**).

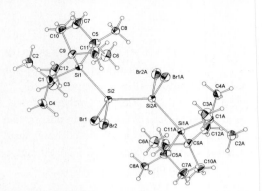

Crystal system: monoclinic, C2/c
$R_1 = 0.0663$, $wR2 = 0.1972$

Important bond lengths [Å] and angles [°]:
Si(1)–Si(2): 2.458(4)   Si(1)–Si(2)–Si(2A): 130.2(2)
Si(2)–Si(2A): 2.423(6)   Br(2)–Si(2)–Br(1):   103.70(13)
Si(2)–Br(1): 2.257(3)   Br(2)–Si(2)–Si(2A): 102.4(2)
Si(2)–Br(2): 2.244(4)   Br(1)–Si(2)–Si(2A): 103.5(2)

Fig. 2.   Crystal structure of tetrabromo-1,2-disupersilyldisilane (**5**).

Dehalogenation of tBu$_3$Si–SiHBr–SiH$_2$–SitBu$_3$ (2)

Dehalogenation of **2** with an equimolar amount of tBu$_3$SiNa in THF at -78°C leads to the isolable monosodium compound **6** in quantitative yield. A subsequent silylation of **6** by Me$_3$SiCl under formation of **7** is possible (Eq. 3). This reactions show that tBu$_3$SiNa is a suitable dehalogenation reagent for $(t$Bu$_3$Si$)_2$Si$_2$H$_{4-n}$Br$_n$.

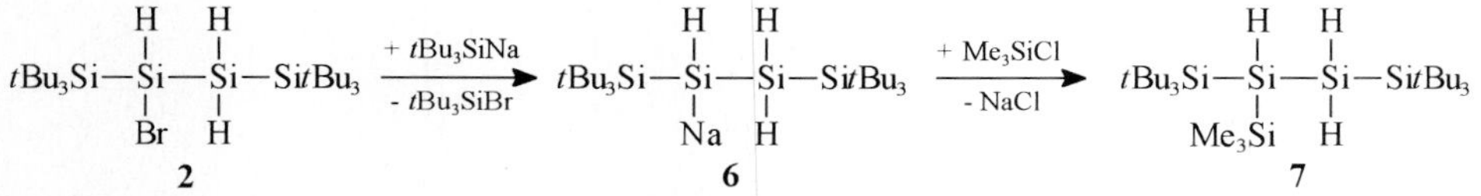

$$t\text{Bu}_3\text{Si}-\underset{\underset{\text{Br}}{|}}{\overset{\overset{\text{H}}{|}}{\text{Si}}}-\underset{\underset{\text{H}}{|}}{\overset{\overset{\text{H}}{|}}{\text{Si}}}-\text{Si}t\text{Bu}_3 \quad \xrightarrow[-\ t\text{Bu}_3\text{SiBr}]{+\ t\text{Bu}_3\text{SiNa}} \quad t\text{Bu}_3\text{Si}-\underset{\underset{\text{Na}}{|}}{\overset{\overset{\text{H}}{|}}{\text{Si}}}-\underset{\underset{\text{H}}{|}}{\overset{\overset{\text{H}}{|}}{\text{Si}}}-\text{Si}t\text{Bu}_3 \quad \xrightarrow[-\ \text{NaCl}]{+\ \text{Me}_3\text{SiCl}} \quad t\text{Bu}_3\text{Si}-\underset{\underset{\text{Me}_3\text{Si}}{|}}{\overset{\overset{\text{H}}{|}}{\text{Si}}}-\underset{\underset{\text{H}}{|}}{\overset{\overset{\text{H}}{|}}{\text{Si}}}-\text{Si}t\text{Bu}_3$$

                    2   **6**   **7**

Eq. 3.

Dehalogenation of tBu$_3$Si–SiHBr–SiHBr–SitBu$_3$ (3a)

Reaction of **3a** with an equimolar amount of tBu$_3$SiNa at -78°C with subsequent warming to room temperature leads to a mixture of various cyclosilanes. The cyclotrisilane **8** is the major component (59%), other products are the cyclotrisilane **9** (5%) and the cyclotetrasilanes **10** (8%), **11** (25%) and **12** (3%) [2]. The all-*cis*-cyclotetrasilane is not formed. At higher temperature (65°C) **3a** reacts with tBu$_3$SiNa mainly to the cyclotrisilane **9** (74 %) and not at all to the cyclotrisilane **8**. The yields of the cyclotetrasilanes are not significantly temperature-dependent.

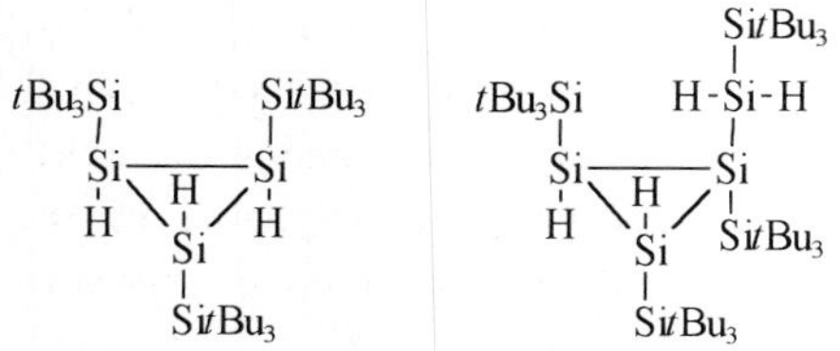

8 *cis, trans*   **9** *cis, trans*

Fig. 3.

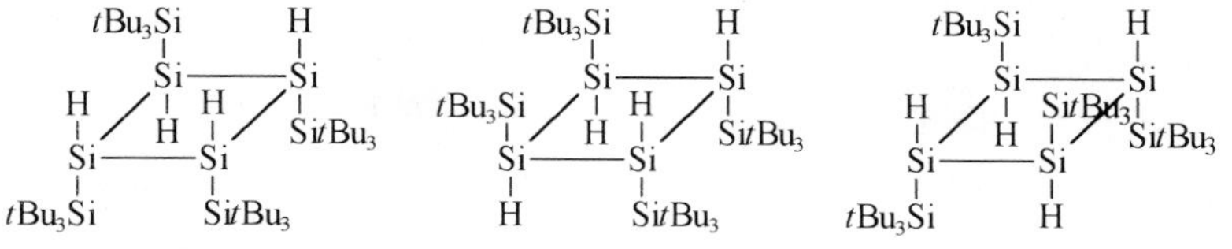

10 *cis-cis-trans* **11** *cis-trans-cis* **12** all-*trans*

Fig. 4.

The mechanism of this reaction has not yet been established in detail, but we were able to trap the disilene **13** with diphenylacetylene at 65°C (Eq. 4).

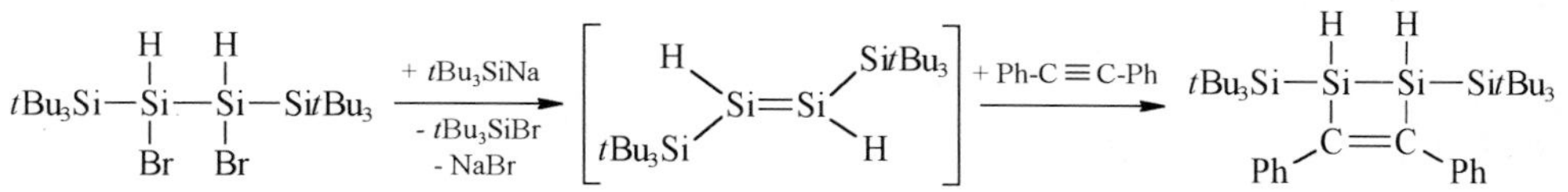

3a **13** **14**

Eq. 4.

Therefore, it might be possible that the disilene intermediate **13** dimerizes under formation of a diradical **15,** which undergoes cyclisation to the cyclotetrasilanes **10-12**, in the cyclotrisilane **8** and the silylene:SiH–Si*t*Bu_3 (which again can give **13** by dimerisation) or in the cyclotrisilane **9** (Eq. 5)

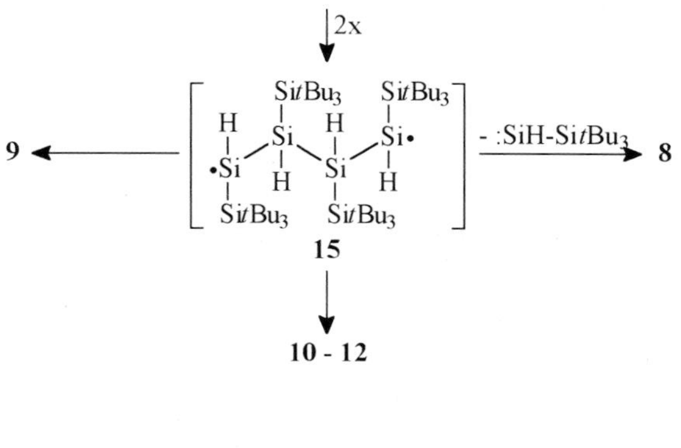

Eq. 5.

Dehalogenation of *t*Bu_3Si–SiBr_2–SiBr_2–Si*t*Bu_3 (5)

As previously mentioned we found that the dehalogenation of **5** with two equivalents of *t*Bu_3SiNa leads to the tetrahedrane **22** as a single product in remarkable high yield. [1, 2].

The reaction starts, in analogy to all dehalogenation reactions mentioned above, with a rapid Br/Na exchange at -100°C. This was shown by protonation of the sodium compound **16** (Eq. 6). The reaction of the **16** with a Me_3NHCl as protonating reagent is the best way to get **4** in high yield (MeOH as protonating agent leads to **17**).

Trapping experiments with triethylsilane and diphenylacetylene, which lead to **20** and **21**, proved the existence of the silylene **18** and the disilene **19** as reaction intermediates of the reaction in discussion. The reaction steps from **18** and **19** towards the formation of the tetrahedrane **22** are not yet established.

Eq. 6.

References:

[1]   N. Wiberg, C. M. M. Finger, K. Polborn, *Angew. Chem.* **1993**, *105*, 1140.

[2]   N. Wiberg, C. M. M. Finger, H. Auer, K. Polborn, *J. Organomet. Chem.* **1996**, *521*, 377.

Syntheses and ^{29}Si NMR Spectra of Halogenated Trisilanes and Cyclopentasilanes

Karl Hassler*, Wolfgang Köll, Ulrich Pöschl
Institut für Anorganische Chemie
Technische Universität Graz
Stremayrgasse 16, A-8010 Graz, Austria
E-mail: hassler@fscm1.tu-graz.ac.at

Keywords: Trisilanes / Cyclopentasilanes

Summary: The reaction of *p*-tolylphenyltrisilanes Si$_3$Ph$_n$*p*-Tol$_m$ (*n* + *m* = 8) with tri-fluoromethanesulfonic acid followed by a reduction with LiAlH$_4$ makes it possible to selectively substitute *p*-tolyl groups with hydrogen atoms. By using this strategy, we were able to synthesize 23 of the 29 possible aryltrisilanes Ar$_n$Si$_3$H$_{8-n}$ (Ar = Ph, *p*-Tol; *n* = 1-8). They are easily converted into the corresponding halogenated trisilanes X$_n$Si$_3$H$_{8-n}$ (X = Cl, Br, I) with either liquid or gaseous hydrogen halides. Usually, no further purification is necessary as the reaction proceeds without the formation of detectable amounts of byproducts.

In an analogous way, nonaphenyl- and octaphenylcyclopentasilane can be converted into nonahalo- and octahalocyclopentasilanes, the latter being an equimolar mixture of *cis*-and *trans*-isomers. All compounds were characterized with ^{29}Si NMR spectroscopy.

The preparation of halogenated oligosilanes by dearylation of arylsilanes offers many advantages over methods such as the halogenation of Si$_2$H$_6$ or Si$_3$H$_8$ by use of HX [1], Cl$_2$ [2], BCl$_3$ or BBr$_3$ [3] or SnX$_4$ [4]. The dearylation usually proceeds without byproducts and in yields up to 95%, and no further purification of the reaction products is necessary. A direct halogenation of silanes most always results in mixtures of isomers that are nearly impossible to separate on a preparative scale. Because of a decreasing thermal stability of halosilanes with increasing mass of the halogen atom and increasing number of silicon atoms in the oligosilane, each purification step dramatically reduces the achievable yields.

Halogenated trisilanes X$_n$Si$_3$H$_{8-n}$ (X = Cl, Br, I) are therefore most easily prepared from HX and the corresponding aryltrisilanes. For this purpose, we have synthesized a number of arylated trisilanes by

a) dearylation of phenylated trisilanes such as Si$_3$Ph$_8$ or (SiPh$_3$)$_2$SiH$_2$ with triflic acid (CF$_3$SO$_3$H), followed by reduction with LiAlH$_4$

b) reaction of Si$_2$Cl$_6$ or Ph$_2$ClSiSiClPh$_2$ with KSiPh$_3$, followed by dearylation and reduction as above

c) synthesis of *p*-tolylphenyltrisilanes Si$_3$Ph$_n$*p*-Tol$_m$ (*n*+*m*=8), followed by dearylation and reduction as above. Under suitable reaction conditions, *p*-tolyl groups can be substituted by hydrogen atoms without affecting the phenyl groups.

Table 1 summarizes the ^{29}Si NMR data of all trisilanes that have been prepared in our group and Scheme 1 presents a few examples of reactions that have been used.

Table 1. ^{29}Si NMR data (ppm, Hz) of aryltrisilanes $Si_3H_nAr_{8-n}$ (n = 0-8; Ar = Ph, p-Tol).

$Si_3H_nAr_{8-n}$	Trisilane	$\delta(Si)$	$\delta(Si^*)$	$\delta(Si^{**})$	$^1J_{SiH}$	$^1J_{Si^*H}$	$^1J_{Si^{**}H}$	$^2J, {}^3J$
Si_3Ar_8	$Ph_3Si–Si^*Ph_2–SiPh_3$	-18.5	-41.4					
	$Ph_3Si–Si^*Tol_2–SiPh_3$ [5]	-19.0	-42.7					
	$Tol_3Si–Si^*Ph_2–Si^{**}Ph_3$	-19.4	-42.4	-18.5				
	$Tol_3Si–Si^*Ph_2–SiTol_3$	-19.1	-42.5					
Si_3HAr_7	$Ph_2HSi–Si^*Ph_2–Si^{**}Ph_3$	-30.2	-42.2	-19.2	192.7			
	$Tol_2HSi–Si^*Ph_2–Si^{**}Ph_3$	-30.2	-41.4	-18.4	188.0			
	$Ph_3Si–Si^*HPh–SiPh_3$ [5]	-17.2	-64.2			170.6		
$Si_3H_2Ar_6$	$PhH_2Si–Si^*Ph_2–Si^{**}Ph_3$	-56.5	-42.3	-19.7	190.4			$^3J_{SiH}$ = 5.8
	$TolH_2Si–Si^*Ph_2–Si^{**}Ph_3$	-57.0	-42.3	-19.6	186.3			$^3J_{SiH}$ = 5.7
	$Ph_3Si–Si^*H_2–SiPh_3$ [5]	-15.3	-105.3			173.7		
	$Ph_2HSi–Si^*Ph_2–SiHPh_2$	-32.2	-42.6		189.5			
	$Tol_2HSi–Si^*Ph_2–SiHTol_2$	-32.5	-42.6		193.9			
	$Ph_2HSi–Si^*HPh–Si^{**}Ph_3$	-30.5	-65.1	-17.9	190.4	174.1		$^2J_{Si^*H}$ = 9.1
$Si_3H_3Ar_5$	$H_3Si–Si^*Ph_2–Si^{**}Ph_3$	-94.0	-41.4	-19.2	191.7			
	$PhH_2Si–Si^*HPh–Si^{**}Ph_3$	-57.6	-66.4	-18.2	193.8	178.6		
	$PhH_2Si–Si^*Ph_2–Si^{**}HPh_2$	-58.1	-42.3	-32.3	190.7		189.2	$^3J_{SiH}$ = 6.2
	$Ph_2HSi–Si^*H_2–Si^{**}Ph_3$	-30.8	-106.7	-16.0	193.7	177.6		$^2J_{Si^*H}$ = 10.7
	$Ph_2HSi–Si^*HTol–SiHPh_2$	-31.1	-67.5		191.9	177.4		
$Si_3H_4Ar_4$	$PhH_2Si–Si^*Ph_2–SiH_2Ph$	-58.2	-41.7		191.1			
	$TolH_2Si–Si^*Ph_2–SiH_2Tol$	-58.6	-41.9		190.0			$^2J_{Si^*H}$ = 6.0
	$Ph_2HSi–Si^*H_2–SiHPh_2$	-31.3	-108.3		194.4	181.3		$^2J_{SiH}$ = 9.8
	$H_3Si–Si^*Ph_2–Si^{**}HPh_2$	-96.4	-42.0	-32.1	192.7		190.1	
	$PhH_2Si–Si^*HPh–Si^{**}HPh_2$	-58.7	-67.7	-31.5	193.2	182.3	190.7	
$Si_3H_5Ar_3$	$PhH_2Si–Si^*HPh–SiH_2Ph$	-58.7	-68.1		193.3	185.5		
	$PhH_2Si–Si^*H_2–Si^{**}HPh_2$	-60.2	-109.3	-31.3	195.1	184.4	195.8	
	$H_3Si–Si^*HPh–Si^{**}HPh_2$	-97.8	-69.4	-31.1	195.2	185.0	190.9	
	$H_3Si–Si^*Ph_2–Si^{**}H_2Tol$	-95.6	-41.0	-56.9	192.9		191.0	
	$H_3Si–Si^*H_2–Si^{**}Ph_3$	-100.2	-111.2	-16.3	197.4	184.6		$^2J_{Si^*H}$ = 4.4
$Si_3H_6Ar_2$	$H_3Si–Si^*Ph_2–SiH_3$	-95.5	-42.1		194.1			$^3J_{SiH}$ = 3.5
	$H_3Si–Si^*HPh–Si^{**}H_2Ph$	-98.3	-70.1	-57.8	195.2	188.5	195.5	
	$PhH_2Si–Si^*H_2–SiH_2Ph$ [6]	-60.1	-110.6		196.9	188.0		$^2J_{Si^*H}$ = 7.3
Si_3H_8	$H_3Si–Si^*H_2–SiH_3$ [7]	-98.7	-116.1		198.4	193.6		$^2J_{SiH}$ = 3.0
								$^2J_{Si^*H}$ = 4.7

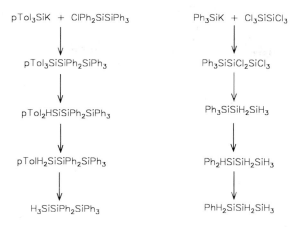

Scheme 1.

The aryltrisilanes are easily converted into the corresponding halotrisilanes with either liquid hydrogen halides at low temperatures or with gaseous hydrogen halides at ambient temperatures. The latter reaction has to be catalyzed by aluminum halides AlX_3 and is carried out in benzene. Table 2 gives the ^{29}Si chemical shifts (ppm vs. TMS) and coupling constants of chloro-, bromo-, and iodotrisilanes that have been prepared so far. Figs. 1 and 2 present the proton coupled ^{29}Si spectra of an arylated and a halogenated trisilane as typical examples.

Table 2. ^{29}Si NMR data (ppm, Hz) of halotrisilanes $Si_3H_nX_{8-n}$ (X = Cl, Br, I).

$Si_3X_{8-n}H_n$	Trisilane	δ(Si)	δ(Si*)	δ(Si**)	$^1J_{SiH}$	$^1J_{Si\cdot H}$	$^1J_{Si\cdot\cdot H}$	$^2J, ^3J$
Si_3X_7H	$HCl_2Si–Si^*Cl_2–Si^{**}Cl_3$	-7.0	-5.6	-3.8	285.6			$^2J_{Si\cdot H} = 235.3$
								$^3J_{Si\cdot\cdot H} = 6.1$
	$HBr_2Si–Si^*Br_2–Si^{**}Br_3$	-26.9	-25.0	-37.5	289.9			$^2J_{Si\cdot H} = 33.4$
								$^3J_{Si\cdot\cdot H} = 5.4$
	$HI_2Si–Si^*I_2–Si^{**}I_3$	-96.2	-90.4	-162.4	273.1			$^2J_{Si\cdot H} = 27.7$
	$I_3Si–Si^*IH–SiI_3$	-165.9	-81.0			235.0		$^2J_{SiH} = 20.9$
$Si_3X_6H_2$	$H_2ClSi–Si^*Cl_2–Si^{**}Cl_3$	-32.6	+1,0	-2.5	242.4			$^2J_{Si\cdot H} = 19.9$
	$H_2BrSi–Si^*Br_2–Si^{**}Br_3$	-45.0	-19.2	-35.8	246.9			$^2J_{Si\cdot H} = 20.3$
								$^3J_{Si\cdot\cdot H} = 4.5$
	$HBr_2Si–Si^*Br_2–SiBr_2H$	-25.5	-24.3		286.6			$^2J_{Si\cdot H} = 31.5$
								$^3J_{SiH} = 4.9$
	$H_2ISi–Si^*I_2–Si^{**}I_3$	-79.2	-84.0	-157.6	244.2			$^2J_{Si\cdot H} = 18.5$
								$^3J_{Si\cdot\cdot H} = 3.5$
	$I_3Si–Si^*H_2–SiI_3$ [8]	-165.8	-96.5			220.1		$^2J_{SiH} = 10.7$
	$HI_2Si–Si^*I_2–SiI_2H$	-92.6	-90.6		273.5			$^2J_{Si\cdot H} = 25.7$
								$^3J_{SiH} = 2.8$

Continuation Table 2.

Group	Compound							Coupling constants
	$HI_2Si–Si^*IH–Si^{**}I_3$	-96.9	-78.7	-163.5	271.1	236.2		$^2J_{SiH} = 15.3$ $^2J_{Si^*H} = 21.6$ $^2J_{Si^{**}H} = 19.6$
$Si_3X_5H_3$	$H_3Si–Si^*Br_2–Si^{**}Br_3$	-86.2	-7.7	-32.5	217.2			$^2J_{Si^*H} = 9.9$ $^3J_{Si^{**}H} = 3.2$
	$HBr_2Si–Si^*Br_2–Si^{**}BrH_2$	-25.6	-19.1	-43.8	285.6		246.2	$^2J_{Si^*H} = 19.5$ (t) $^2J_{Si^*H} = 30.8$ (d) $^3J_{SiH} = 4.1$
	$HCl_2Si–Si^*ClH–SiCl_2H$	-0.3	-31.1		279.5	232.8		$^2J_{SiH} = 14.3$ $^2J_{Si^*H} = 28.1$
	$HBr_2Si–Si^*BrH–SiBr_2H$	-22.3	-45.3		281.1	237.2		$^2J_{SiH} = 14.7$ $^2J_{Si^*H} = 25.7$ $^3J_{SiH} = 3.4$
	$H_3Si–Si^*I_2–Si^{**}I_3$ [8]	-84.1	-73.3	-150.7	217.0			$^2J_{Si^*H} = 10.0$
	$HI_2Si–Si^*I_2–Si^{**}IH_2$	-89.9	-88.6	-76.2	269.0		244.1	$^2J_{Si^*H} = 15.3$ (t) $^2J_{Si^*H} = 24.1$ (d) $^3J_{SiH} = 3.4$ $^3J_{Si^{**}H} = 3.3$
	$HI_2Si–Si^*IH–SiI_2H$	-94.9	-82.8		270.0	236.7		$^2J_{SiH} = 14.5$ $^2J_{Si^*H} = 20.4$ $^3J_{SiH} = 2.3$
$Si_3I_4H_4$	$HI_2Si–Si^*H_2–SiI_2H$	-88.3	-82.8		264.5	214.6		$^2J_{SiH} = 7.4$ $^2J_{Si^*H} = 16.9$
	$HI_2Si–Si^*IH–SiIH_2$	-92.5	-83.9	-81.8	267.8	232.3	241.0	$^2J_{SiH} = 14.1$ $^2J_{Si^*H} = 19.7$ (d) $^2J_{Si^*H} = 13.8$ (t) $^2J_{Si^{**}H} = 12.4$
	$H_3Si–Si^*I_2–Si^{**}I_2H$	-82.7	-80.9	-90.6	216.2		267.0	
$Si_3I_3H_5$	$H_2ISi–Si^*IH–SiIH_2$ [8]	-80.7	-88.3		236.4	227.5		$^2J_{SiH} = 10.3$ $^2J_{Si^*H} = 13.7$
	$HI_2Si–Si^*H_2–SiIH_2$	-91.9	-96.2	-78.2	268.0	202.8	230.5	$^2J_{SiH} = 13.8$ $^2J_{Si^*H} = 11.8$ $^2J_{Si^{**}H} = 11.2$
$Si_3I_2H_6$	$H_2ISi–Si^*H_2–SiIH_2$ [8]	-77.6	-99.0		230.6	205.3		$^2J_{Si^*H} = 11.1$

Dihydrooctaphenylcyclopentasilane undergoes isomerization to *cis-* and *trans-*$1,3$-$H_2Si_5X_8$ when treated with HX. Attempts to separate the isomers by fractional crystallization failed. ^{29}Si NMR data of halogenated cyclopentasilanes are presented in Tables 3 and 4.

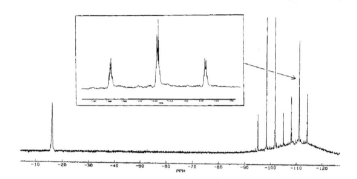

Fig. 1.   Proton coupled ^{29}Si spectrum of $SiH_3SiH_2SiPh_3$

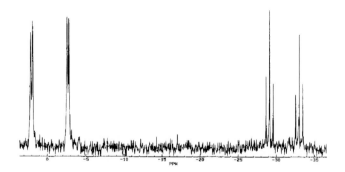

Fig. 2.   Proton coupled ^{29}Si spectrum of $HCl_2SiSiHClSiCl_2H$.

Table 3. ^{29}Si NMR data (ppm, Hz) of nonahalocyclopentasilanes HSi_5X_9 (X = Cl[9], Br, I).

Compound	Si(1)		Si(2,5)		Si(3,4)	
	δ(ppm)	J(Hz)	δ(ppm)	J(Hz)	δ(ppm)	J(Hz)
Si_5Cl_9H	-27.6	$^1J_{Si-H}$ 255.2	1.8	$^2J_{Si-H}$ 8.2	-2.1	$^3J_{Si-H}$ 7.7
Si_5Br_9H	-43.7	$^1J_{Si-H}$ 255.0	-24.6	$^2J_{Si-H}$ 9.1	-27.4	$^3J_{Si-H}$ 7.4
		$^1J_{Si-Si}$ 63.6		$^1J_{Si-Si}$ 63.6		$^1J_{Si-Si}$ 67.9
		$^2J_{Si-Si}$ 22.5		$^1J_{Si-Si}$ 67.9		$^2J_{Si-Si}$ 22.5
				$^2J_{Si-Si}$ 33.3		$^2J_{Si-Si}$ 33.4
Si_5I_9H	-75.8	$^1J_{Si-H}$ 246.0	-99.1	$^2J_{Si-H}$ 9.2	-97.8	$^2J_{Si-H}$ 5.5

Table 4. ^{29}Si NMR data (ppm, Hz) of 1,3-dihydrooctahalocyclopentasilanes 1,3-$H_2Si_5X_8$ (X = Cl, Br, I).

Compound	Si(1,3)		Si(2)		Si(4,5)	
	δ(ppm)	J(Hz)	δ(ppm)	J(Hz)	δ(ppm)	J(Hz)
$Si_5Cl_8H_2$	-28.7	$^1J_{Si-H}$ 253.7	6.8	$^2J_{Si-H}$ 8.8	4.0	$^2J_{Si-H}$ 9.0
		$^3J_{Si-H}$ 5.0		$^1J_{Si-Si}$ 75.0		$^3J_{Si-H}$ 9.0
		$^1J_{Si-Si}$ 75.3		$^2J_{Si-Si}$ 27.0		$^1J_{Si-Si}$ 75.4
		$^1J_{Si-Si}$ 27.0				$^2J_{Si-Si}$ 26.9
	-29.6	$^1J_{Si-H}$ 254.3	5.9	$^2J_{Si-H}$ 7.5	3.8	$^2J_{Si-H}$ 9.2
		$^3J_{Si-H}$ 5.5		$^1J_{Si-Si}$ 77.0		$^3J_{Si-H}$ 9.2
		$^1J_{Si-Si}$ 76.9		$^2J_{Si-Si}$ 28.3		$^1J_{Si-Si}$ 75.6
		$^2J_{Si-Si}$ 26.8				$^2J_{Si-Si}$ 27.7
$Si_5Br_8H_2$	-44.6	$^1J_{Si-H}$ 254.8	-20.7	$^2J_{Si-H}$ 9.3	-22.2	$^2J_{Si-H}$ 9.2
		$^3J_{Si-H}$ 4.6				$^3J_{Si-H}$ 9.2
	-45.5	$^1J_{Si-H}$ 255.0	-21.0	$^2J_{Si-H}$ 8.2	-22.8	$^2J_{Si-H}$ 9.7
		$^3J_{Si-H}$ 5.7				$^3J_{Si-H}$ 9.7
$Si_5I_8H_2$	-77.8	$^1J_{Si-H}$ 244.5	-99.8		-97.3	
	-79.0	$^1J_{Si-H}$ 247.4	-100.3		-97.6	

Acknowledgement: Financial support of the *Austrian Science Foundation* (FWF, project P 9378-CHE) is gratefully acknowledged.

References:

[1]   A. Stock, K .Somieski, *Ber. Dtsch. Chem. Ges.* **1920**, *53*, 759.
[2]   F. Fehér, P. Plichta, R. Guillery, *Inorg. Chem.* **1971**, *10*, 606.
[3]   C. H. Van Dyke, A. G. MacDiarmid, *J. Inorg. Nucl. Chem.* **1963**, *25*, 1503.
[4]   J. E. Bentham, S. Cradock, E. A. V. Ebsworth, *Inorg. Nucl. Chem. Lett.* **1971**, *7*, 1077.
[5]   K. Hassler, *Monatsh. Chem.* **1988**, *119*, 1051.
[6]   K. Hassler, U. Katzenbeisser, *J. Organomet. Chem.* **1991**, *421*, 151.
[7]   J. Hahn, *Z. Naturforsch.* **1980**, *35b*, 282.
[8]   K. Hassler, U. Katzenbeisser, *J. Organomet. Chem.* **1994**, *480*, 173.
[9]   K. Hassler, U. Pöschl, *J. Organomet. Chem.* **1996**, *506*, 93.
[10] H. Stüger, P. Lassacher, E. Hengge, *Z. Anorg. Allg. Chem.* **1995**, *621*, 1517.

Chlorination of Methylphenyloligosilanes: Products and Reactions

Christina Notheis*, Erica Brendler, Berthold Thomas
Institut für Analytische Chemie
Technische Universität Bergakademie Freiberg
Leipziger Straße 29, D-09599 Freiberg, Germany
Tel.: Int. code + (3731)392266 – Fax: Int. code + (3731)393666
E-mail: notheis@arsen.anch.tu-freiberg.de

Keywords: ^{29}Si NMR / Chloromethylphenyloligosilanes / Heating of Chloromethyloligosilanes

Summary: Several methylphenyltri- and tetrasilanes have been synthesized, characterized by ^{29}Si NMR and chlorinated. Tris(dimethylphenylsilyl)phenylsilane $(PhMe_2Si)_3SiPh$ and bis(dimethylphenylsilyl)methylphenylsilane $(PhMe_2Si)_2SiMePh$ were chlorinated stepwise to give partially chlorinated intermediates, which were also characterized by ^{29}Si NMR spectroscopy. In a last step, thermal treatment of tris(chlorodimethylsilyl)chlorosilane $(Cl_2MeSi)_3SiCl$, tris(chlorodimethylsilyl)chlorosilane $(ClMe_2Si)_3SiCl$ and bis(chlorodimethylsilyl)dichlorosilane $(ClMe_2Si)_2SiCl_2$ was carried out. Products and mechanistic suggestions are given.

Introduction

In the last few years, efforts to describe oligosilanes have been increasing and many systematic investigations on tri- and tetrasilanes have now been published. There are two principle synthetic methods that lead oligosilanes:

(1)   the controlled disproportionation of methylchlorodisilanes, followed by destillation [1],
(2)   the metallo condensation of monosilanes to oligosilanes followed by chlorination of the condensed products.

Since chlorination of the permethyloligosilanes leads to products with only one chlorine per silicon unit, we used another synthesis strategy: the condensation of chloromethylphenylsilanes to methylphenyloligosilanes followed by a phenyl/chloro exchange reaction.

The advantage of this method is that the number of chloro substituents per molecule depends directly on the number of phenyl groups and is, therefore, determined by the chloromethyl-phenylmonosilanes used for the condensation.

Results

We were able to characterize the following methylphenyloligosilanes, synthesized by the use of a slightly modified literature procedure [2].

Table 1. ^{29}Si chemical shift and $^{1}J_{SiSi}$ coupling constants of the synthesized methylphenyloligosilanes.

Silane	$\delta_{Si}{}^{A}$ [ppm]	$\delta_{Si}{}^{B}$ [ppm]	$^{1}J_{SiSi}$ [Hz][a]
(Ph$_2$MeSi^A)$_2$Si^{B}MePh	-19.7	-46.9	72.1
(PhMe$_2$Si^A)$_2$Si^{B}MePh[b]	-18.5	-46.5	72.9
(PhMe$_2$Si^A)$_2$Si^{B}Me$_2$[b]	-18.7	-48.2	72.8
(PhMe$_2$Si^A)$_2$Si^{B}Ph$_2$[c]	-19.2	-40.2	71.4
(Ph$_2$MeSi^A)$_3$Si^{B}Me	-16.8	-84.2	61.8
(PhMe$_2$Si^A)$_3$Si^{B}Me[d]	-15.4	-86.1	61.8
(PhMe$_2$Si^A)$_3$Si^{B}Ph	-15.9	-76.1	61.0
(PhMe$_2$Si^{A}Si^{B}MePh)$_2$	-17.7	-43.9	

[a] signs not determined; synthesis and characterization have been described – [b] in ref. [3] – [c] in ref. [4] – [d] synthesis has been described in [5].

Because we were especially interested in the chemical shifts of mixed substituted oligosilanes, tris(dimethylphenylsilyl)phenylsilane (PhMe$_2$Si)$_3$SiPh and bis(dimethylphenyl)methylphenylsilane (PhMe$_2$Si)$_2$SiMePh were – monitored by ^{29}Si-NMR – also partially chlorinated, the other methylphenyloligosilanes were completely chlorinated using acetylchloride and aluminiumtrichloride as described by LEHNERT et al. in [6]. Because of the higher viscosity and boiling points of the tri- and tetrasilanes, we used a solvent (hexane) in order to make isolation of the chlorinated products out of the reaction mixture possible. The identified oligosilanes are listed in Table 2.

Very interesting reactions were observed during heating of tris(chlorodimethylsilyl)chlorsilane (ClMe$_2$Si)$_3$SiCl in the presence of dichlorotetramethyldisilane Cl$_2$Me$_4$Si$_2$ as a silylene trap. This leads to products, which can formally be explained by a silylene generation out of a chlorine containing sidechain, and then insertion of the silylene into a silicon-chlorine bond of the disilane.

$$(ClMe_2Si)_3SiCl + ClMe_2SiSiMe_2Cl \quad \rightarrow \quad (ClMe_2Si)_2SiCl_2 \; + \; (ClMe_2Si)_2SiMe_2$$

$$(Cl_2MeSi)_3SiCl + Cl_2MeSi\text{-}SiMeCl_2 \quad \rightarrow \quad (Cl_2MeSi)_2SiCl_2 \; + \; (Cl_2MeSi)_2SiMeCl$$

$$(ClMe_2Si)_2SiCl_2 + ClMe_2Si\text{-}SiMe_2Cl \quad \rightarrow \quad Cl_3SiSiMe_2Cl \; + \; (ClMe_2Si)_2SiMe_2$$

Scheme 1.   Formal reactions of samples under heat.

During purification of tris(dichloromethylsilyl)chlorsilane ($Cl_2MeSi)_3SiCl$ and bis(chlorodimethylsilyl)chlorsilane ($ClMe_2Si)_2SiCl_2$ by destillation obviously analogous reactions took place. We were able to identify substances, which can be explained by a decomposition reaction as described above for ($ClMe_2Si)_3SiCl$. The reactions are listed in scheme 1. The substances printed in italics were confirmed by ^{29}Si NMR, the others are assumed to undergo very fast further reactions.

Table 2. ^{29}Si chemical shifts and $^1J_{SiSi}$ coupling constants of chlorinated samples.

Silane	$\delta_{Si}{}^A$ [ppm]	$\delta_{Si}{}^B$ [ppm]	$\delta_{Si}{}^C$ [ppm]	$^1J_{SiSi}$ [Hz][a]
$(PhMe_2Si^A)Si^BMePh(Si^CMe_2Cl)$	-19.0	-45.4	24.1	
$(ClMe_2Si^A)_2Si^BMePh$	22.8	-44.6		
$(ClMe_2Si^A)_2Si^BMeCl^{[b]}$	19.5	-0.7		89.5
$(PhMe_2Si^A)_2Si^BPh(Si^CMe_2Cl)$	-16.0	-74.1	27.7	
$(PhMe_2Si^A)Si^BPh(Si^CMe_2Cl)_2$	-16.3	-72.0	26.3	
$(ClMe_2Si^A)_3Si^BCl$	22.0	-24.6		71.3
$(ClMe_2Si^A)_2Si^BMeCl^{[b]}$	19.5	-0.7		89.5
$(Cl_2MeSi^A)_2Si^BMeCl^{[c]}$	23.7	-5.1		110.9
$(ClMe_2Si^A)_2Si^BMe_2{}^{[b]}$	24.4	-44.0		81.3
$(ClMe_2Si^A)_3Si^BMe^{[d]}$	26.4	-76.7		70.2
$(Cl_2MeSi^A)_3Si^BMe^{[c]}$	30.7	-63.6		86.4

[a] signs not determined – [b] see also ref. [3] – [c] see also ref. [7] – [d] see also [1].

Discussion

In spectra 1 to 3 (Fig. 1) all the synthesis steps carried out on model compounds ($PhMe_2Si)_3SiPh$ and ($ClMe_2Si)_3SiCl$ are shown. By comparing spectra 3 to 5, one can see that partially chlorinated intermediates appear along with the methylphenyl compound. Spectrum 4 shows the signals arising from the chlorinated tetrasilane. The exchange phenyl/chlorine at the *t*-silicon atom causes a large shift of the signal to lower field. The comparison of the value of $^1J_{SiSi}$ of ($ClMe_2Si)_3SiCl$ with the value of $^1J_{SiSi}$ of the related methylsubstituted tetrasilane tris-(chlorodimethylsilyl)methylsilane ($ClMe_2Si)_3SiMe$ shows a surprisingly small difference of about 1 Hz, but it is increasing with increasing number of chlorine substituents, as we expected from comparison with literature [1,4].

Spectrum 5 is typical for a heated sample in that the signals due to the „decomposition product" ($ClMe_2Si)_2SiMe_2$ appear along with the tetrasilane signals.

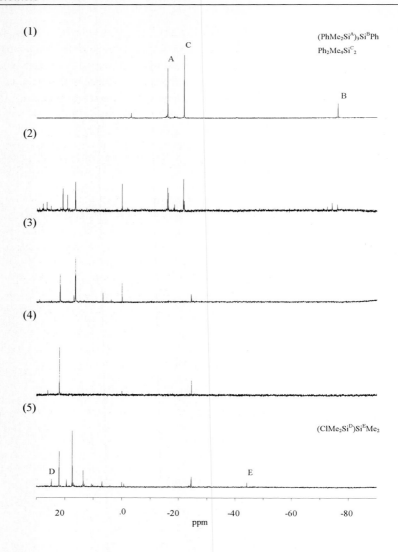

Fig. 1. Spectra of $(PhMe_2Si^A)_3Si^BPh$ along with $Ph_2Me_4Si^C_2$ (1), partially chlorinated intermediates and $(PhMe_2Si)_3SiPh$ (2), the chlorination reaction mixture (3), $(ClMe_2Si)_3SiCl$ (4), and the heated sample (5) containing $(ClMeSi^D)_2Si^EMe_2$.

The reaction mechanism for formation of this trisilane is not presently known. In the literature, different mechanisms are suggested. They have been established for the catalytic disproportionation of methylchlorodisilanes and have a common first step: a reversible nucleophilic attack of an electron-pair donor, enlarging the coordination sphere of one silicon from tetra- to pentacoordination. In the next step, the models differ.

One model states that a silylene is generated by an electron transfer such as an alpha-elimination, forming a monosilane and a donor-stabilized silylene. The other model favours the cleavage of the silicon-silicon bond along with the formation of an silanion and a donor-stabilized silicenium ion. To form a silylene, transfer of chloride from the silanion to the silicenium ion is necessary. Chloride might act as an electron donor itself. In the last step, the silylene inserts into the silicon-chlorine bond of another disilane molecule or oligosilane. However, the reaction conditions of the disproportionation reaction are quite different from our conditions since the presence of a lewis base like in [1] is excluded by the reaction conditions. We did not observe any reaction of the disilane that is comparable to the disproportionation reaction, but a decomposition of the tetrasilane (or trisilane). Therefore another explanation must be found.

We would like to suggest, that an attack of a chlorine takes place first at one of the silicon atoms of the tri- or tetrasilane under discussion to form a pentacoordinated silicon. In a second step, a silylene is generated which inserts into a silicon-chlorine bond of the disilane while the silicon skeleton of the oligosilane is shortened by one silicon atom. The source of the attacking chlorine is not yet known. An intramolecular reaction with a chlorine of a chloromethylsilyl-sidechain and the central silicon might occur, or remaining traces of the aluminium trichloride, used for the chlorination, might induce the reaction. This is possible since only catalytic amounts would be necessary to cause the decomposition.

Another possible explanation is that the larger number of chlorine substituents in the chloromethyltri- and tetrasilanes especially at the central silicon atom decreases the stability of these silanes (compared with that of the chloromethyldisilanes and the corresponding methylphenyldisilanes), such that the reaction temperature of 150°C was sufficient to cause thermal cracking of the silicon skeleton.

Further investigations are required to establish the reaction mechanism.

Acknowledgement: The authors express their gratitude to the *Deutsche Forschungsgemeinschaft* for financial support.

References:

[1]   U. Herzog, R. Richter, E. Brendler, G. Roewer, *J. Organomet. Chem.* **1996**, *507*, 221.
[2]   C. Notheis, E. Brendler, B. Thomas, *GIT Fachz. Lab.*, in press.
[3]   K. Schenzel, K. Hassler, *Spectrochim. Acta* **1994**, *50A*, 127.
[4]   K. Schenzel, K. Hassler, in: *Organosilicon Chemistry II: From Molecules to Materials* (Eds.: N.Auner, J. Weis), VCH, Weinheim, **1996**.
[5]   A. Sekiguchi, M. Nanjo, C. Kabuto, H. Sakurai, *J. Am. Chem. Soc.* **1995**, *117*, 4195.
[6]   R. Lehnert, M. Höppner H. Kelling, *Z. Anorg. Allg. Chem.* **1991**, *591*, 209.
[7]   G. Kolleger, K. Hassler, *J. Organomet. Chem.* **1995**, *485*, 233.

Selective Hydrogenation of Methylchlorooligosilanes

Uwe Herzog*, Gerhard Roewer
Institut für Anorganische Chemie
Technische Universität Bergakademie Freiberg
Leipziger Straße 29, D-09596 Freiberg, Germany
Tel: Int. code + (3731)393174 – Fax: Int. code (3731)394058
E-mail: herzog@merkur.hrz.tu-freiberg.de.

Keywords: Silanes / Oligosilanes / Stannanes / Hydrogenation / ^{29}Si NMR

Summary: Using a base-catalyzed hydrogenation of methylchlorooligosilanes (Si_3-Si_6) with trialkylstannanes, partially hydrogenated methylchlorooligosilanes can be prepared and identified by means of ^{29}Si and ^{1}H NMR spectroscopy.
Starting from methylchlorosilanes like $SiClMe_2$–$SiClMe$–$SiClMe_2$ certain partially hydrogenated compounds can be prepared selectively because of the different reactivity of the Si–Cl bonds towards hydrogenation by R_3SnH.

Introduction

The well known hydrogenation of methylchlorooligosilanes with $LiAlH_4$ in etheral solvents leads to the formation of the completely hydrogenated methylhydrogenoligosilanes only [1,2]. Until now nothing is known about tri- or higher oligosilanes simultaneously containing methyl, chlorine and hydrogen substituents.

Recently we investigated the partial hydrogenation of $SiCl_4$, methylchloromono- and -disilanes (especially $SiCl_2Me$–$SiCl_2Me$) [3,4] by trialkylstannanes in the presence of a Lewis base catalyst.

In general, the hydrogenation by trialkylstannanes takes place at the silicon atom with the highest acceptor strength. The acceptor strength increases with the number of silyl and chlorine substituents, e.g., in the order $SiCl_2Me$–$SiCl_2Me$ > $SiClMe_2$–$SiClMe_2$ > $SiCl_2Me_2$ > $SiClMe_3$. The same series can be derived from the stability of 2,2-bipy adducts of mono- and disilanes [5].

The first step of the overall reaction is the attack of the Lewis base catalyst at the silicon atom forming a hypervalent silicon atom and stretching the Si–Cl bond. In a second step this species reacts with the stannane to give the hydrogenated silane and trialkylchlorostannane:

Eq. 1.

Especially in the case of disilanes with $SiCl_2Me$ groups the product spectrum (i. e., amount of compounds with a SiHClMe group) can be influenced by a variation of the catalyst and the alkyl groups of the used stannane (0-32 % SiHClMe groups beside SiH_2Me and $SiCl_2Me$). A large amount of SiHClMe units has been obtained with Me_3SnH and Ph_3MePI as catalyst, whereas the combination of Bu_3SnH and Ph_3P produces SiH_2Me beside $SiCl_2Me$ groups only [6].

It should be possible to apply our method of selective hydrogenation also to higher oligosilanes as:

$SiClMe_2$–$SiMe_2$–$SiClMe_2$ (which is comparable with the hydrogenation of $SiClMe_2$–$SiClMe_2$),

$SiClMe_2$–$SiClMe$–$SiClMe_2$, $SiCl_2Me$–$SiClMe$–$SiCl_2Me$, $(SiCl_2Me)_2SiMe$–$SiClMe$–$SiCl_2Me$,

$SiMe(SiClMe_2)_3$ and $SiMe(SiMe_3)_i(SiMe_2Cl)_{3-i}$ $i = 1, 2$,

$Si(SiClMe_2)_4$ and $Si(SiMe_3)_i(SiClMe_2)_{4-i}$ $i = 1-3$,

$(SiClMe_2)_3Si$–$SiClMe$–$SiClMe_2$ and $(SiClMe_2)_2SiMe$–$SiMe(SiClMe_2)_2$.

Results and Discussion

Hydrogenation of oligosilanes with a $(Si)_2SiClMe$ unit

In such silanes the Si–Cl bond of this unit is hydrogenated at first (used catalyst: Ph_3MePI) and the so formed methylchlorohydrogenoligosilanes can be prepared in pure state:

$$SiClMe_2{-}SiClMe{-}SiClMe_2 \quad \xrightarrow[{+\ R_3SnCl}]{+\ R_3SnH} \quad SiClMe_2{-}SiHMe{-}SiClMe_2$$

Eq.2.

$$SiCl_2Me{-}SiClMe{-}SiCl_2Me \quad [7] \quad \xrightarrow[{+\ R_3SnCl}]{+\ R_3SnH} \quad SiCl_2Me{-}SiHMe{-}SiCl_2Me$$

Eq. 3.

$$(SiCl_2Me)_2SiMe{-}SiClMe{-}SiCl_2Me \quad \xrightarrow[{+\ R_3SnCl}]{+\ R_3SnH} \quad (SiCl_2Me)_2SiMe{-}SiHMe{-}SiCl_2Me$$

Eq. 4.

$$SiClMe_2{-}(SiClMe)_2{-}SiClMe_2 \quad \xrightarrow[{+\ 2\ R_3SnCl}]{+\ 2\ R_3SnH} \quad SiClMe_2{-}(SiHMe)_2{-}SiClMe_2$$

Eq. 5.

A hydrogenation of the other Si–Cl bonds occurs only if more than one (or two in Eq. 5) equivalents stannane was added. An excess of stannane leads to the completely hydrogenated species.

Table 1. ^{29}Si NMR chemical shifts of the investigated oligosilanes and their hydrogenation products (see also [6, 7]).

Verbindung	δ(Si$_A$)	δ(Si$_B$)	δ(Si$_C$)	δ(Si$_D$)	δ(Si$_E$)
Si$_A$ClMe$_2$–Si$_B$ClMe–Si$_A$ClMe$_2$	19.48	-0.68			
Si$_A$ClMe$_2$–Si$_B$HMe–Si$_A$ClMe$_2$	24.65	-68.51			
Si$_A$ClMe$_2$–(Si$_B$ClMe)$_2$–Si$_A$ClMe$_2$	20.19	2.19/3.22[b]			
Si$_A$ClMe$_2$–(Si$_B$HMe)$_2$–Si$_A$ClMe$_2$	26.10	-69.42/-70.83[b]			
Si$_A$Cl$_2$Me–Si$_B$ClMe–Si$_A$Cl$_2$Me	24.08	-4.65			
Si$_A$Cl$_2$Me–Si$_B$HMe–Si$_A$Cl$_2$Me	31.54	-60.38			
Si$_A$HClMe–Si$_B$HMe–Si$_C$Cl$_2$Me	0.70	-63.89	33.76		
Si$_A$H$_2$Me–Si$_B$HMe–Si$_C$Cl$_2$Me	-66.34	-66.34	35.94		
Si$_A$H$_2$Me–Si$_B$HMe–Si$_C$HClMe	-66.3	-71.78	3.98		
Si$_A$H$_2$Me–Si$_B$HMe–Si$_A$H$_2$Me	-64.83	-76.06			
(Si$_A$Cl$_2$Me)$_2$Si$_B$Me–Si$_C$ClMe–Si$_D$Cl$_2$Me	32.08	-63.66	4.45	25.22	
(Si$_A$Cl$_2$Me)$_2$Si$_B$Me–Si$_C$HMe–Si$_D$Cl$_2$Me	33.31	-66.59	-63.36	33.89	
(Si$_A$ClMe$_2$)$_3$Si$_B$–Si$_C$ClMe–Si$_D$ClMe$_2$	27.60	-109.92	10.60	20.82	
(Si$_A$ClMe$_2$)$_3$Si$_B$–Si$_C$ClMe–Si$_D$HMe$_2$	27.97	-110.17	14.78	-32.20	
(Si$_A$HMe$_2$)$_3$Si$_B$–Si$_C$HMe–Si$_D$HMe$_2$	-32.52	-138.86	-69.32	-34.54	
Si$_A$(Si$_B$ClMe$_2$)(Si$_C$Me$_3$)$_3$	-129.38	32.01	-9.83		
Si$_A$(Si$_B$ClMe$_2$)$_2$(Si$_C$Me$_3$)$_2$	-123.74	30.45	-9.64		
Si$_A$(Si$_B$ClMe$_2$)$_3$(Si$_C$Me$_3$)	-118.36	28.98	-9.34		
Si$_A$(Si$_B$ClMe$_2$)$_4$	-113.91	27.50			
Si$_A$(Si$_B$HMe$_2$)(Si$_C$Me$_3$)$_3$	-136.78	-33.49	-9.56		
Si$_A$(Si$_B$HMe$_2$)$_2$(Si$_C$Me$_3$)$_2$	-137.88	-33.44	-9.20		
Si$_A$(Si$_B$HMe$_2$)$_3$(Si$_C$Me$_3$)	-139.00	-33.40	-8.73		
Si$_A$(Si$_B$HMe$_2$)$_4$	-140.07	-33.35			
(Si$_A$ClMe$_2$)$_2$Si$_B$Me–Si$_B$Me(Si$_A$ClMe$_2$)$_2$	28.53	-73.78			
(Si$_A$HMe$_2$)(Si$_B$ClMe$_2$)Si$_C$Me–Si$_D$Me(Si$_E$ClMe$_2$)$_2$	-33.17	29.92	-79.54	-73.96	28.47/28.66[a]
(Si$_A$HMe$_2$)(Si$_B$ClMe$_2$)Si$_C$Me–Si$_C$Me(Si$_A$HMe$_2$)(Si$_B$ClMe$_2$)	-33.37/-33.52[b]	29.88/29.93 [b]	-79.12		
(Si$_A$HMe$_2$)$_2$Si$_B$Me–Si$_C$Me(Si$_D$HMe$_2$)(Si$_E$ClMe$_2$)	-32.95/-33.24[a]	-84.21	-78.48	-33.64	29.90
(Si$_A$HMe$_2$)$_2$Si$_B$Me–Si$_B$Me(Si$_A$HMe$_2$)$_2$	-33.2	-83.2			

[a] Diastereotopic silicon atoms – [b] Pair of enantiomers and meso species.

Hydrogenation of oligosilanes with a $(Si)_4Si$ unit

In contrast to linear silanes having a $(Si)_2SiMe_2$ unit (e. g., $SiClMe_2$–$SiMe_2$–$SiClMe_2$) or branched silanes with a $(Si)_3SiMe$ unit (e. g., $SiMe(SiClMe_2)_3$, $SiMe(SiCl_2Me)_3$), which can be hydrogenated by trialkylstannanes [6], the silanes $Si(SiMe_3)_x(SiClMe_2)_{4-x}$ ($x = 0$-3) with a quarternary $(Si)_4Si$ unit do not react with stannanes. Hydrogenation with $LiAlH_4$ / Et_2O leads to the completely hydrogenated species $Si(SiMe_3)_x(SiHMe_2)_{4-x}$ ($x = 0$-3). Probably steric hindrance prevents the attack of the Lewis base at the silicon atom of the $SiClMe_2$ group. This is confirmed by the selective single hydrogenation of the hexasilane **1** (which was prepared by chlorination of $[(SiMe_3)_2SiMe]_2$ with acetyl chloride and $AlCl_3$ at higher temperatures). Only the Si–Cl bond of the $SiClMe_2$ group which is not directly bonded to a quarternary Si atom is hydrogenated by stannanes:

Eq. 6.

But also by the hydrogenation of other silanes the formation of certain partially hydrogenated species is preferred (see Eq. 7, catalyst used: Ph_3MePI, Si* = chiral silicon atom).

```
        Cl              Cl
         \               \
         SiMe₂           SiMe₂
           \               /
           SiMe ——————— SiMe
           /               \
         SiMe₂           SiMe₂
         /               /
        Cl              Cl

                  │  Me₃SnH
                  ▼

        H               Cl
         \               \
         SiMe₂           SiMe₂
           \               /
           Si*Me ——————— SiMe
           /               \
         SiMe₂           SiMe₂
         /               /
        Cl              Cl
```

$$\swarrow \qquad \not\searrow$$

```
    H           Cl              H           Cl
     \           \               \           \
     SiMe₂       SiMe₂           SiMe₂       SiMe₂
       \           /               \           /
       Si*Me ——— Si*Me             SiMe ——— SiMe
       /           \               /           \
     SiMe₂       SiMe₂           SiMe₂       SiMe₂
     /           /               /           /
    Cl           H               H           Cl
```

```
        H               Cl
         \               \
         SiMe₂           SiMe₂
           \               /
           SiMe ——————— Si*Me
           /               \
         SiMe₂           SiMe₂
         /               /
        H               H

                  │
                  ▼

        H               H
         \               \
         SiMe₂           SiMe₂
           \               /
           SiMe ——————— SiMe
           /               \
         SiMe₂           SiMe₂
         /               /
        H               H
```

Eq. 7.

References:

[1] K. Schenzel, K. Hassler, *Spectrochim. Acta* **1994**, *50A*, 127.

[2] G. Kollegger, K. Hassler, *J. Organomet. Chem.* **1995**, *485*, 233.

[3] U. Pätzold, G. Roewer, U. Herzog, *J. Organomet. Chem.* **1996**, *508*, 147.

[4] U. Herzog, G. Roewer, U. Pätzold, *J. Organomet. Chem.* **1995**, *494*, 143.

[5] D. Kummer, A. Balkir, H. Köster, *J. Organomet. Chem.* **1979**, *178*, 29.

[6] U. Herzog, G. Roewer, E. Brendler, *J. Organomet. Chem.* **1996**, *511*, 85.

[7] U. Herzog, R. Richter, E. Brendler, G. Roewer, *J. Organomet. Chem.* **1996**, *507*, 221.

Electrochemical Formation of Cyclosilanes

Susanne Graschy, Christa Grogger, Edwin Hengge
Institut für Anorganische Chemie
Technische Universität Graz
Stremayrgasse 16, A-8010 Graz, Austria
Tel.: Int. code + (316)8738217 – Fax: Int. code + (316)8738701
E-mail: grogger@anorg.tu-graz.ac.at

Keywords: Cyclosilanes / Electrochemistry

Summary: Several dichlorosilanes R_2SiCl_2 (R = Me, Ph, *p*-Tol) were subjected to a systematic study on the correlation of the bulk of substituents and ring size of the products formed in the respective electrolyses. Furthermore, the influence of a change of anode material and electrolyte system on the electrolysis was investigated. Electrolyses of monohydrodichlorosilanes $RHSiCl_2$ (R = Ph, Me) were not successful in forming cyclic products, but gave linear polymers due to the high flexibility of the chain. Replacement of the organic group by the bulky cyclohexasilanyl substituent did not lead to the expected steric hindrance. Instead, electrolysis of $(Si_6Me_{11})HSiCl_2$ yielded bi(undecamethylcyclohexasilanyl) as a result of Si-Si bond cleavage and subsequent dimerization of two cyclohexasilanyl groups.

Introduction

Organosubstituted cyclosilanes are interesting starting materials for various purposes, e.g., metathesis of cyclotetrasilanes and formation of siloxene-like products. Employing the usual Wurtz-coupling procedure, the desired cyclosilanes very often are obtained, if at all, only in moderate yields [1].

Up to now, the main interest in electroreductive coupling of dichlorosilanes was the formation of linear oligo- and polysilanes [2]. But, as shown in our previous work [3], cyclic electrolysis products are not uncommon. Thus, we tested the applicability of electroreductive coupling for the *specific* formation of cyclotetra-, -penta-, and -hexasilanes.

To obtain cyclic products with functional side groups, our main interest was focused on phenylated and hydrogen containing silanes. As we were able to prove, diorganosilanes behave differently to hydrosilanes. For this reason the two groups are treated separately in the following discussion.

Replacement of HMPA

As reported earlier [4], best electrolysis results are obtained when using a mixture of THF and HMPA (hexamethylphosphoric triamide) with 0.02 M Et_4NBF_4 as solvent/electrolyte system.

However, with several hydrosilanes, e.g,. $PhHSiCl_2$, HMPA reacts to give an insoluble, white complex that cannot be electrolyzed. Therefore, it was necessary to find an alternative solvent/electrolyte system with sufficient specific conductivity. As THF is one of the rare solvents that shows a good stability in the cathodic as well as in the anodic region, we were intent on finding an appropriate supporting electrolyte for it. After disappointing experiments with Bu_4NCl and Bu_4NI, we finally found THF with 0.02 M Bu_4NBF_4 to give satisfying results not only in the electrolyses of hydrodichlorosilanes but also in that of diorganodichlorosilanes.

Diorganodichlorosilanes

Regarding the electrolyses of diorganodichlorosilanes, the expected correlation of the bulk of substituents and the ring size was confirmed. Obviously, the determinant factor is the steric condition on the electrode surface. Thus, the thermodynamically most stable six-membered ring was formed as main product when electrolyzing Me_2SiCl_2, whereas in the electrolysis of Ph_2SiCl_2 the kinetically preferred $(Ph_2Si)_4$ was formed (Table 1). Several approaches to shift the favored ring size – e.g., by addition of "Wurtz-proved" catalyst $Ph_3SiSiMe_3$ [5] – failed.

Perarylated silanes like Ph_2SiCl_2 always gave cyclotetrasilanes as the only reaction products, irrespective of the concentration of starting material [3]. Replacing the bulky aryl substituents by smaller methyl groups, ring formation is observed only at low silane concentrations. With rising silane concentration, a rise in the yield of polymeric products can be observed. To favor intramolecular reaction (which means ring formation) over intermolecular reaction, electrolyses must be carried out with silane concentrations below 0.5 mol/L. Thus, preparatively reasonable yields can be afforded only with large electrolysis cells. A kind of "technological" cell with hydrogen anodes and stainless steel cathodes with an overall surface area of about 1900 cm^2 is currently being used. With this cell we are confident of being able to carry out electrolyses on a larger scale and to yield isolable amounts of products.

Table 1.   Correlation of bulk of substituents and ring size.

Starting material	Main product	Byproduct
Me_2SiCl_2	$(Me_2Si)_6$	$(Me_2Si)_5$
$MePhSiCl_2$	$(MePhSi)_5$	$(MePhSi)_6$
Ph_2SiCl_2	$(Ph_2Si)_4$	-
p-TolPh $SiCl_2$	$(p$-TolPhSi$)_4$	-

To scan the influence of the electrolyte/anode system, electrolyses were carried out in three different systems:

1) Al-anode in THF / HMPA / Et_4NBF_4,
2) Mg-anode in THF / Bu_4NBF_4
3) H_2-anode in THF / Bu_4NBF_4

It appeared that a change in the electrolyte/anode system did not effect any change in the kind of reaction products. The only detectable difference was a change of yield, which was observed by GC/MS analysis. While the Al- and the Mg-system gave comparable results in each experiment, the hydrogen anode behaved differently. As reported earlier [4], phenylated chlorosilanes showed lower reduction potentials and additional overpotentials when electrolyzed with the hydrogen electrode, thus leading to a significant decrease in the current yield. In the case of p-TolPhSiCl$_2$, which was electrolyzed for the first time, no reaction products at all could be detected when using the hydrogen anode.

An additional experiment was carried out with Me$_2$SiCl$_2$, using Zn as sacrificial anode, as this metal is easily available and thus appropriate for large scale electrolyses. Unfortunately, it is too noble for this kind of electrolysis. The only detected cathodic reaction is the reduction of the anodically formed Zn^{2+} ions, thus covering the cathode surface with a Zn-coating.

Hydrochlorosilanes

The successful cathodic dimerization of the two hydrochlorosilanes MePhSiHCl and Ph$_2$SiHCl to the corresponding disilanes [6] under chloride elimination induced us to carry out electrolyses using hydrodichlorosilanes as substrates. In these cases Si-Si bond linkage should occur leading to interesting products with Si-H functional groups. Employing conventional chemical methods, such cyclic silanes are obtained only with difficulty and in very moderate yields.

By application of the reaction conditions described above it was not possible at all to synthesize cyclic organohydrosilanes even when using a concentration of the monosilane as low as 0.2 M. As a result of the low size of the hydrogen substituent, the formed oligosilane chain is too flexible for intramolecular coupling. Only intermolecular reaction takes place. The obtained electrolysis products turned out to be polymeric hydrosilane chains:

$$\text{n RSiHCl}_2 \xrightarrow{\text{I, } -2n \text{ Cl}^-} (\text{RHSi})_n$$

$$\text{R = Me, Ph}$$

Eq. 1.

The slight solubility of the poly(methylsilane) was sufficient enough to effect GPC investigations. The molecular weight distribution vs. polystyrene standard showed a relatively uniform chain length of fifteen Si units. Poly(phenylsilane) was completely insoluble, which made GPC analysis impossible, but calculations on base of elemental analysis yielded a chain length of more than 20.

Table 2. Chain lengths and current efficiencies.

Substrate	Product	Chain Length	% Current Yield
MeHSiCl$_2$	(MeHSi)$_n$	$n \approx 15$	38
PhHSiCl$_2$	(PhHSi)$_n$	$n > 20$	33

From our experience that cyclic silanes are formed more likely when using monosilanes with bulky substituents, we tried to electrolyze (undecamethylcyclohexasilanyl)dichlorosilane. However, in this case Si-Si bond cleavage took place and the only isolated product was bi(undecamethylcyclohexanyl), formation of which is obviously favored because of its remarkable stability.

Cyclic Voltammetry

Comparing the cyclic voltammograms of hydroorgano- and diorganodichlorosilanes in 0.1 M solution of tetrabutylammonium perchlorate (TBAP) in dimethoxyethane (DME), we found the former to show reduction at much lower potentials. Therefore, cathodic precipitation of magnesium occurs to a larger extent, thus leading to lower yields of poly(hydrosilanes).

Table 3. Reduction potentials of the cited silanes ($v = 0.1$ V/s).

Silane	E_{red} [V] vs. SCE
Me_2SiCl_2	0.5/-0.7
$MePhSiCl_2$	0.4/-0.5
Ph_2SiCl_2	-0.2/-0.9
$MeHSiCl_2$	-0.8
$PhHSiCl_2$	-0.9

The cyclic voltammetric investigations of $MeHSiCl_2$ and $PhHSiCl_2$ showed, in accordance to the results of non-hydrogenated silanes, a higher reduction potential for the methylated derivative than for the phenylated one [3]. In both cases the reduction is quasi-reversible, hence, with increasing potential sweep, the reduction waves shift to more negative values.

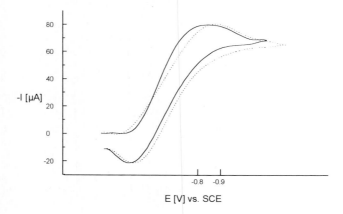

Fig. 1. Cyclic voltammograms of $MeHSiCl_2$ (—) and $PhHSiCl_2$ (···); $v = 0.1$ V/s.

Acknowledgment: We wish to thank *Wacker-Chemie GmbH* for support of this study.

References:

[1]   E. Hengge, R. Janoschek, *Chem Rev.* **1995**, *95*, 1495.

[2]   a) M. Bordeau, C. Biran, M.-P. Léger-Lambert, J. Dunoguès, *J. Chem. Soc., Chem. Commun.* **1991**, 1477.
      b) M. Umezawa, M. Takeda, H. Ichikawa, T. Ishikawa, T. Koizumi, T. Fuchigami, T. Nonaka, *Electrochim. Acta* **1990**, *35*, 1867.
      c) T. Shono, S. Kashimura, M. Ishifune, R. Nishida, *J. Chem. Soc., Chem. Commun.* **1990**, 1160.

[3]   Ch. Jammegg, S. Graschy, E. Hengge, *Organomet.* **1994**, *13*, 2397.

[4]   S. Graschy, Ch. Jammegg, E. Hengge, *Investigations Concerning the Electrochemical Formation of Di- and Polysilanes*, in: *Organosilicon Chemistry II: From Molecules to Materials* (Eds.: N. Auner, J. Weis), VCH, Weinheim, **1996**, p. 89.

[5]   R. West, *Organometallics* **1983**, *2*, 409.

[6]   E. Hengge, S. Graschy, Ch. Grogger-Jammegg, *"Electrochemical Behavior of Organosilanes with Si–H Functional Groups"*, XXVIIIth Organosilicon Symposium, Gainesville, **1995**.

Undecamethylcyclohexasilanyl Derivatives of Tin(IV) and Lead(IV)

Frank Uhlig*, Uwe Hermann
Lehrstuhl für Anorganische Chemie II
Universität Dortmund
Otto-Hahn-Str. 6, D-44227 Dortmund, Germany
Tel.: Int. code +(231)7553812
E-mail: fuhl@platon.chemie.uni-dortmund.de

Karl Klinkhammer
Anorganische Chemie
Universität Stuttgart
Pfaffenwaldring 55, D-70569 Stuttgart, Germany

Edwin Hengge
Institut für Anorganische Chemie
Technische Universität Graz
Stremayrgasse 16, A-8010 Graz, Austria

Keywords: Cyclosilanes / Stannylsilanes / NMR Spectroscopy

Summary: The synthesis and spectroscopic characterization of the first undeca-methylcyclohexasilanyl tin(IV) derivatives are reported. The crystal structure of bis(un-decamethylcyclohexasilanyl)dimethyltin has been determined. Reactions and stability of similar lead(IV) compounds are also discussed.

Introduction

Undecamethylcyclohexasilanyl potassium **1**, is a well known starting material for the synthesis of a variety of cyclic silane and silane transition metal derivatives [1-3]. A novel and convenient way to obtain **1** has been described recently [4, 5].

$$
\begin{array}{c}
\text{Me}_2\text{Si} \longrightarrow \text{SiMe}_2 \\
K \diagup \qquad \diagdown \\
\text{Si} \qquad \text{SiMe}_2 \\
\text{Me} \diagdown \qquad \diagup \\
\text{Me}_2\text{Si} \longrightarrow \text{SiMe}_2
\end{array}
$$

1

Using the above synthetic route we developed the synthesis of the undecamethylcyclohexasilanyl derivatives of tin(IV) and lead(IV).

Synthesis

Reaction of **1** with organotinmono- and dichlorides at -30 to -40°C led to derivatives **2a-f** as colorless solids in good yields (30-70%) (eq. 1-2).

Eqs. 1 and 2. Synthesis of **2**.

The products are stable against air and hydrolysis for up to a week. The compounds were characterized by NMR spectroscopy (especially by silicon-silicon and silicon-tin coupling constants, Table 1 and 3) and mass spectrometry. The crystal structure of **2c** is shown in Fig. 1.

Table 1. Definition of organo substituents R (CHS = undecamethylcyclohexasilanyl)

	R	R	R
2a	Me	Me	Me
2b	Ph	Ph	Ph
2c	Me	Me	CHS
2d	Me	Ph	CHS
2e	Ph	Ph	CHS
2f	*p*-Tol	*p*-Tol	CHS

Table 2. Selected bond lengths and bond angles of **2c**.

Bond Angles		Bond Lengths	
Si(11)–Sn–Si(21)	114.22°(11)	Si(21)–Sn	2.586(3) Å
C(3)–Sn–C(4)	101.5°(8)	Si(22)–Sn	2.587(3) Å
C(3)–Sn–Si(21)	110.5°(5)		

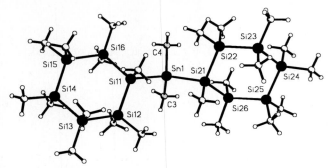

Fig. 1.   Structure of **2c**.

The reaction of **2c** with one or two equivalents of trifluoromethanesulfonic acid and lithium chloride yielded the chlorostannanes **4a** and **4b** (Eqs. 3 and 4).

Eqs. 3 and 4.   Reactions of **2c**.

However, the reaction of methyltrichlorotin with three equivalents of **1** failed to give the desired tris(undecamethylcyclohexasilanyl)methyltin **4a**.

Eq. 5.   Attempted synthesis of **4a**.

Compound **4a** was obtained by an alternative route, by reaction of **3a** with **1**. It was not isolated but unambigously identified by NMR spectroscopy.

Eq. 6. Synthesis of **4a**.

In contrast to the undecamethylcyclohexasilanyl tin derivatives **2**, most of the analogous lead compounds are unstable in solution at room temprature. Only compound **5a** could be isolated as a yellow-white solid (mp(dec.) 75-83°C). However, decomposition in solution at room temperature is observed within hours.

Eq. 7. Synthesis of **5a**.

Table 3. Selected ^{29}Si and ^{119}Sn NMR data of **2-5**.

	$\delta(^{119}Sn)$ [ppm]	$\delta(^{29}Si_{Me})$ [ppm]	$^{1}J_{Si-Sn119/117}$ [Hz]	Solvent
2a	-82	-76.9	348/330	C_6D_6
2c	-206	-71.5	204/195	THF/D_2O
3a	-171	-71.5	215/206	$CDCl_3$
4a	-285	-70.2	170/161	$CDCl_3$
5a	–	-72.0	648[a]	C_6D_6

[a] $^{1}J_{Si-Pb}$ [Hz]

Acknowledgement: The authors thank the *Deutschen Forschungsgemeinschaft* for financial support (DFG-project Uh74-3), *Wacker-Chemie GmbH* (Burghausen) and the *ASV-innovative Chemie GmbH* (Bitterfeld) for the generous gift of silanes. We are also grateful to Prof. Dr. K. Jurkschat (Dortmund) for supporting this investigation.

References:

[1]   E. Hengge, R. Janoschek, *Chem. Rev.* **1995**, *95*, 1495.

[2]   M. Ishikawa, M. Watanabe, J. Iyoda, H. Ikeda, *Organometallics* **1982**, *1*, 317.

[3]   E. Hengge, E. Pinter, M. Eibl, F. Uhlig, *Bull. Soc. Chim. Fr.* **1995**, *132*, 509.

[4]   F. Uhlig, P. Gspaltl, M. Trabi, E. Hengge, *J. Organomet. Chem.* **1995**, *493*, 33.

[5]   F. Uhlig, P. Gspaltl, E. Pinter, E. Hengge, in: *Organosilicon Chemistry II: From Molecules to Materials* (Eds.: N. Auner, J. Weis), VCH, Weinheim, **1996**, p. 109.

New Results in Cyclosilane Chemistry: Siloxene-like Polymers

Alois Kleewein, Uwe Pätzold, Edwin Hengge
Institut für Anorganische Chemie
Technische Universität Graz
A-8010 Graz, Stremayrgasse 16/IV, Austria
Tel.: Int. code + (316)8738200 – Fax: Int. code + (316)827685
E-mail: sekretariat@anorg.tu-graz.ac.at

Stefan Tasch, Günther Leising
Institut für Festkörperphysik
Technische Universität Graz
A-8010 Graz, Petersgasse 16/II, Austria
Tel.: Int. code + (316)8738470 – Fax: Int. code + (316)8738478
E-mail: if_graz@fscm1.dnet.tu-graz.ac.at

Keywords: Cyclosilanes / Siloxene / Polymers / Fluorescence Spectra

Summary: Several linear and cyclic silanes (four-, five- and six-membered rings) with silicon-halogen or silicon-triflate functions were prepared and hydrolyzed to polymeric structures similar to Wöhler siloxene and Kautsky siloxene. Optical investigations on the fluorescence of these polymers were carried out. The color and the fluorescence of the polymers are influenced by the ring size and the kind of substituents. Depending on the starting material the fluorescence maxima range from 400 to 550 nm.

Introduction

Calcium disilicide reacts with acids to form the Wöhler siloxene [1] or Kautsky siloxene [2], depending on the reaction conditions. These compounds show interesting optical behavior [3]. In the Kautsky formula, silicon six-membered rings are connected by oxygen atoms forming a polymeric layer (Fig. 1).

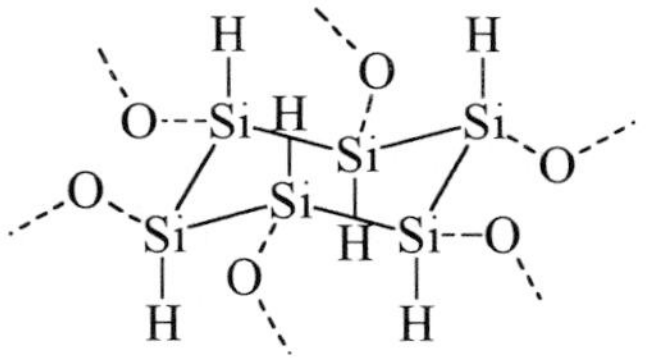

Fig. 1. Structure of siloxene.

To elucidate the question whether or not cyclosilane structures are essential for the color and fluorescence of the polymers, linear and cyclic silanes with silicon halogen or -triflate functions were prepared and hydrolyzed to siloxene-like structures.

Syntheses of Halogen Containing Linear and Cyclic Silanes

Several methods, all starting from appropriate phenylated or methylated silanes were used to prepare the following chlorinated linear silanes (Scheme 1).

$$Me_3Si\text{-}SiMe_3 \ + \ 2\ MeCOCl \ \xrightarrow{\ AlCl_3\ } \ ClMe_2Si\text{-}SiMe_2Cl$$
$$\textbf{1}$$

$$PhMeSiCl_2 \ + \ 2\ PhMe_2SiLi \ \longrightarrow \ PhMe_2Si\text{-}SiPhMe\text{-}SiMe_2Ph$$

$$PhMe_2Si\text{-}SiPhMe\text{-}SiMe_2Ph \ \xrightarrow{\ CF_3SO_3H\ } \ \xrightarrow{\ LiCl\ } \ ClMe_2Si\text{-}SiMeCl\text{-}SiMe_2Cl$$
$$\textbf{2}$$

$$(Ph_2MeSi)_2 \ \xrightarrow{\ 2CF_3SO_3H\ } \ \xrightarrow{\ 2PhMe_2SiLi\ } \ PhMe_2Si\text{-}(SiMePh)_2\text{-}SiMe_2Ph$$

$$PhMe_2Si\text{-}(SiMePh)_2\text{-}SiMe_2Ph \ + \ HCl \ \xrightarrow{\ AlCl_3\ } \ ClMe_2Si\text{-}(SiMeCl)_2\text{-}SiMe_2Cl$$
$$\textbf{3}$$

$$(Ph_2MeSi)_2 \ \xrightarrow{\ 2CF_3SO_3H\ } \ \xrightarrow{\ 2Ph_2MeSiLi\ } \ Ph_2MeSi\text{-}(SiMePh)_2\text{-}SiMePh_2$$

$$Ph_2MeSi\text{-}(SiMePh)_2\text{-}SiMePh_2 \ + \ HCl \ \xrightarrow{\ AlCl_3\ } \ Cl_2MeSi\text{-}(SiMeCl)_2\text{-}SiMeCl_2$$
$$\textbf{4}$$

Scheme 1.   Syntheses of chlorinated linear silanes.

The starting materials for the four-, five and six-membered ring systems, Si_4Ph_8, Si_5Ph_{10}, Si_6Ph_{12}, and $Si_6Me_6Ph_6$, were synthesised by well-known Wurtz-type reactions (Eqs. 1-2).

$$Ph_2SiCl_2 \ + \ Na/K \ \longrightarrow \ (SiPh_2)_n \quad (n = 4;\ 5;\ 6)$$

Eq. 1.

$$MePhSiCl_2 \ + \ Li \ \longrightarrow \ (SiMePh)_6$$

Eq. 2.

The triflic acid derivatives were prepared from the perphenylated cyclosilanes by dearylation with triflic acid (Eq. 3). The subsequent reaction of $Si_5Ph_8(CF_3SO_3)_2$ with lithium aluminium hydride and with hydrogen chloride in a second step led to 1,3-dihydrooctachlorocyclo-pentasilane (Eq. 4).

$$(SiPh_2)_n + n\,CF_3SO_3H \longrightarrow (SiPh(CF_3SO_3))_n \quad (n=4;\,5) \quad \textbf{5; 6}$$

Eq. 3.

$$(SiPh_2)_5 + 2\,CF_3SO_3H \longrightarrow 1,3\text{-}(CF_3SO_3)_2Si_5Ph_8 \xrightarrow{\ LiAlH_4\ } 1,3\text{-}H_2Si_5Ph_8$$
$$1,3\text{-}H_2Si_5Ph_8 + HCl \longrightarrow 1,3\text{-}H_2Si_5Cl_8 \quad \textbf{7}$$

Eq. 4.

The reaction of perphenylated cyclosilanes with hydrogen halides under pressure or with aluminum halides as catalysts allows the preparation of $(SiCl_2)_5$ and $(SiCl_2)_6$ (Eq. 5) as well as the preparation of $Si_5Ph_5I_5$ (Eq. 6).

$$(SiPh_2)_n + HCl \xrightarrow{\ AlCl_3\ } (SiCl_2)_n \quad (n=5;\,6) \quad \textbf{8; 9}$$

Eq. 5.

$$(SiPh_2)_5 + HI \longrightarrow (SiPhI)_5 \quad \textbf{10}$$

Eq. 6.

The methyl(chloro)cyclosilanes were achieved by the reaction of the corresponding phenyl-(methyl)cyclosilanes with hydrogen chloride (Eqs 7-8).

$$(SiPh(CF_3SO_3))_5 + 5\,MeLi \longrightarrow (SiPhMe)_5 \xrightarrow{\ HCl/AlCl_3\ } (SiMeCl)_5 \quad \textbf{11}$$

Eq. 7.

$$(SiMePh)_6 \xrightarrow{\ HCl/AlCl_3\ } (SiMeCl)_6 \quad \textbf{12}$$

Eq. 8.

Hydrolysis and Condensation

The hydrolysis of the halides was carried out by dropwise addition of an excess of aqueous tetra-hydrofuran (1:5) to a solution of the halides in THF. The solid (if any) was filtered off and washed with dry THF to remove the water and the remaining lower molecular fractions. The triflic acid derivatives were hydrolyzed in another way, because the triflic acid that was released in the hydrolysis step caused the tetrahydrofuran to form polymers. Thus toluene or dichloromethane solutions of the triflic acid derivatives were first stirred overnight with water. The aqueous layer was removed, tetrahydrofuran added and the remaining solids filtered off. Finally, the products were dried in vacuum at 50 °C for 5 hours. In order to investigate the condensation step, a sample of one of these polymers, namely the hydrolysis product of $(SiMeCl)_5$, was not subjected to the drying procedure. It remained partially soluble in THF and so we were able to record a GPC (size exclusion chromatogram) and a ^{29}Si NMR (INEPT pulse sequence). The polymer showed two distinct peaks at 672 g/mol (17%) and 1191 g/mol (13%). The main part (70%) was a very broad peak with a maximum at 1751 g/mol. Evaluation of the chromatogram resulted in 1912 g/mol as number average molecular weight (M_n) and 5554 g/mol as weight average molecular weight (M_w). The calculations refer to a poly(styrene) standard. A very broad signal, reaching from +7 to -20 ppm, with its maximum at -4.1 ppm was found in the ^{29}Si NMR spectrum, referred to TMS as standard. Interpreting these facts it seems to be, that the condensation already starts in the hydrolysis step and that oligomers with a chain length of n = 7 - 18 are formed. To obtain maximum condensation and insolubility in consequence a rigoros drying step is necessary. The polymers obtained by this way exhibit the following specifications (Table 1).

Table 1.   Summary of the hydrolysis and condensation products

No	Starting Material	Product Description	Fluorescence Maximum [nm]
1	$Si_2Me_4Cl_2$	slightly yellowish, thin oil	none
2	$Si_3Me_5Cl_3$	opaque, thick oil	none
3	$Si_4Me_6Cl_4$	white solid	very weak (399)
4	$Si_4Me_4Cl_6$	white solid	weak (413)
5	$Si_4Ph_4(OTf)_4$	dark yellow solid	553
6	$Si_5Ph_5(OTf)_5$	dark yellow solid	522
7	$1,3-H_2Si_5Cl_8$	pale orange solid	486
8	Si_5Cl_{10}	white solid	400
9	Si_6Cl_{12}	white solid	432
10	$Si_5Ph_5I_5$	dark yellow-green solid	540
11	$Si_5Me_5Cl_5$	slightly yellow-green solid	505
12	$Si_6Me_6Cl_6$	slightly yellow-green solid	436

Fluorescence Spectra

The fluorescence emission spectra were recorded with photoluminescence excitation-emission equipment, consisting of a Xe-arc source monochromatized by a grating double monochromator for the excitation. The fluorescence intensity from the samples was collected by lens optics into the entrance slit of a high resolution grating monochromator connected to a cooled photon-counting system. The fluorescence spectra presented here are uncorrected, since the spectral response of the emission setup is nearly constant in the wavelength region 450-800 nm. The liquid samples were measured in a 0.5 cm Infrasil cuvette, whereas the solid samples were prepared as a powder coat on paper. The preparation of the samples and the measurements were done using argon as protective gas.

The polymers obtained from linear starting materials (di-, tri and tetrasilane halides) exhibit neither a significant fluorescence emission nor noticeable optical absorption in the visible. However, a weak blue-ultraviolet fluorescence is observed, which we attribute to impurity emission centers. As expected, the products from cyclic silanes show quite strong fluorescence emission accompanied by an optical absorption in the visible range. The fluorescence spectra of the cyclic materials are depicted in Fig. 2 and Fig. 3.

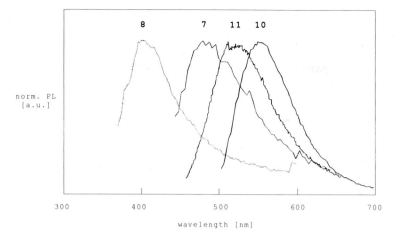

Fig. 2.   Photoluminescence spectra of the polymers of various cyclopentasilane derivatives.

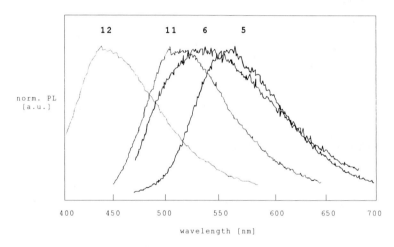

Fig. 3.   Photoluminescence spectra of the polymers of cyclosilanes with various ring size.

A bathochromic shift of the fluorescence maxima depending on the substituents takes place in the order of the starting materials (Fig. 2): $Si_5Cl_{10} < Si_5H_2Cl_8 < Si_5Me_5Cl_5 < Si_5Ph_5I_5$

A comparison of similar substitution products such as the methyl and phenyl compounds shows that the four-membered ring seems to be a stronger chromophore than the five- and six-membered rings. In our present knowledge only cyclic systems (and probably also cage structures) seem to be able to form fluorescent polymers in a matrix of oxygen.

Acknowledgements: The authors are very grateful to the *Fonds zur Förderung der wissenschaftlichen Forschung* (Wien) for financial support and to *Wacker-Chemie GmbH* (Burghausen) for gifts of silanes.

References:

[1]   F. Wöhler, *Lieb. Ann.* **1863**, *127*, 257.

[2]   H. Kautsky, *Z. Anorg. Allg. Chem.* **1921**, *117*, 209.

[3]   E. Hengge, A. Kleewein, in: *Tailormade Silicon-Oxygen Compounds – from Molecules to Materials* (Eds.: P. Jutzi, R. J. P. Corriu), Vieweg Verlag, Braunschweig, **1996**, pp. 89-98.

Stepwise Synthesis of Functional Polysilane Dendrimers

Christoph Marschner*, Edwin Hengge
Institut für Anorganische Chemie
Technische Universität Graz
Stremayrgasse 16, A-8010 Graz, Austria
Tel: Int. code + (316)8738202 – Fax: Int. code + (316)8738701
E-mail: marschner@anorg.tu-graz.ac.at

Keywords: Polysilanes / Dendrimers

Summary: Functional first generation dendrimers were synthesized starting from methyltrichlorosilane. Successive treatment with the lithium derivatives of dimethylphenylsilane and methyldiphenylsilane and conversion of the phenyl groups into bromides gave a stepwise construction of a branched polysilane.

Introduction

The study of dendritic polysilanes is a relatively new trend in silicon chemistry [1]. However, first results suggest interesting properties for these compounds. Lambert and coworkers pointed out that their polysilane dendrimer, which contains 27 heptasilane chains possesses a very similar λ_{max} value (272 nm) compared to the parent heptasilane (266 nm) but shows an extinction coefficient about ten times larger. Comparison of the number of silicon atoms in the molecules (13 *vs* 7) indicates an amplification of extinction per silicon atom of about 5 [1a].

This feature is not the only advantage of dendritic polysilanes. Due to their structural redundancy they are much better suited to retain their properties also in the case of Si-Si cleavage. In addition to this, the spherical shape of the dendrimer leads to enhanced resistance towards chain cleavage conditions. The results suggest that the interesting electronic properties attributed to polysilanes could be exploited much more effectively by the use of dendritic structures.

Synthesis

Contrary to the groups of Lambert, Sakurai, and Suzuki [1] we chose a synthetic route to dendritic polysilanes which makes use only of monosilanes. The reason for this strategy lies on the one hand in an improved accessibility of starting materials and on the other it allows a more flexible design of the desired structure.

Only dimethylphenylchlorosilane and methyldiphenlychlorosilane will be used as the source of methyl silyl groups once we started from methyltrichlorosilane. Lithium or potassium derivatives of the phenylsilanes are very easily accessible in one step from the reaction with lithium or C_8K [2].

Starting from methyltrichlorosilane the reaction with dimethylphenylsilyllithium leads to isotetrasilane **1**. This crystalline compound easily reacts with HBr at -80°C in essentially quantitative yield to give tribromide **2**. Reaction of the latter with methyldiphenylsilyllithium introduces a first branching point into the molecule. Reaction of neat **3** with hydrogen bromide gives hexabromide **4** under mild conditions in quantitative yield as a nicely crystalline compound. Treatment with lithium aluminum hydride leads to the terminal hydrosilane **5**.

Reaction of compound **4** with dimethylphenylsilyllithium leads to the first generation polysilane dendrimer **6**, which has been synthesized already by Sakurai et al [1b]. Again, treatment with HBr under the described conditions gives hexabromide **7**, also in a clean reaction as a crystalline compound. Reaction of **7** with lithium aluminum hydride finally gives hydrosilane **8**.

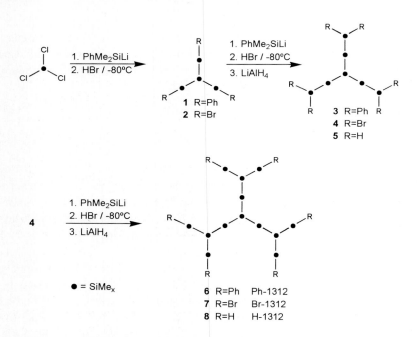

Scheme 1.   Synthesis of first generation dendritic polysilanes.

Tables 1 and 2 show some selected ^{29}Si and ^{13}C NMR data. Both series show a deshielding effect for the central unit. Both the core silicon atom as well as the attached methyl group experience a low field shift. This effect seems to be caused by steric reasons because compounds **6**, **7**, and **8**, which possess different substituents as end groups, show very similar values for $\delta(Si_1)$.

Another point which has also been reported by Sakurai et al. [1a] is the diastereotopicity of the outmost methyl groups which leads to two signals in the NMR spectrum for the diastereotopic groups. We observed this effect which appears in the carbon and proton spectra for compound **8**.

Table 1. ^{29}Si NMR data.

Compound	$\delta(Si_1)$	$\delta(Si_2)$	$\delta(Si_3)$	$\delta(Si_4)$	J
1 (Ph)	-86.1	-15.4			
2a (Br)	-72.9	19.7			
2b (TfO)	-82.4	46.5			
3 (Ph)	-69.1	-39.1	-18.7		
4 (Br)	-75.8	-29.7	25.5		
5 (H)	-77.0	-39.9	-61.8		t, 184 Hz
6 (Ph)	-64.7	-29.2	-79.1	-15.0	
7 (Br)	-64.4	-71.6	-29.7	21.8	
8 (H)	-65.3	-30.8	-65.4	-33.3	d, 178 Hz

Table 2. ^{13}C NMR data.

Compound	$\delta(Si_1)$	$\delta(Si_2)$	$\delta(Si_3)$	$\delta(Si_4)$	Phenyl groups
1 (Ph)	-11.4	-1.5			128.0, 128.6, 134.1, 140.5
3 (Ph)	-7.0	-1.2	-2.7		128.6, 129.5, 135.8, 138.2
4 (Br)	-8.4	-2.1	10.1		
5 (H)	-10.5	-2.5	-11.1		
7 (Br)	-8.2	2.0	0.0	5.9	
8 (H)	-6.2	1.0	-9.7	-3.4/-3.7	

Nomenclature

A proposal made recently by Lambert and coworkers [1d] for the nomenclature and structural description of polysilane dendrimers treats the problem of the high redundancy of very complex names of these compounds which possess a high degree of symmetry and might be described better by different ways.

We appreciate the introduced *gcsb*-system but would like to extend it to the point that substituents at the terminal silicon atoms different from methyl should be part of the description. For example compounds **6-8** would be described as Ph-1312, Br-1312 and H-1312, respectively.

Perspective

As already mentioned above we are currently investigating the synthesis of dendrimers with a different number of spacer groups. As a useful starting material for these studies we found compound **9** which is easily available from the reaction of methyltrichlorosilane with methyldiphenylsilyllithium.

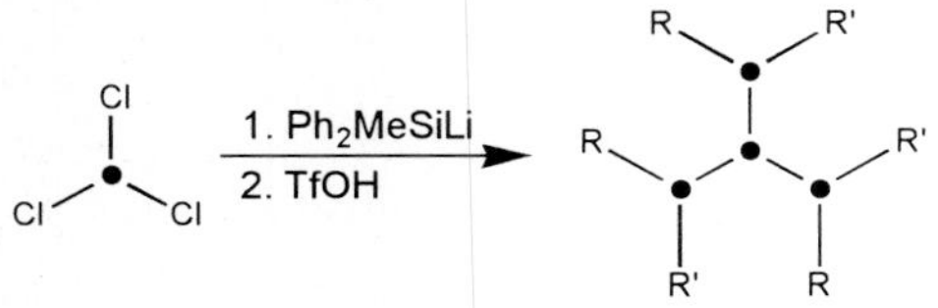

9 R=R'=Ph
10 R=Ph, R=OTf

Scheme 2.

The reaction of **9** with 3 equivalents of triflic acid in toluene gives the threefold monotriflated product, which can be used to obtain dendrimers not only with dimethylsilylene units as spacer but also with methylphenylsilylene groups.

References:

[1]   a) J. B. Lambert, J. L. Pflug, C. L. Stern, *Angew. Chem.* **1995**, *107*, 106.
       b) A. Sekiguchi, M. Nanjio, C. Kabuto, H. Sakurai, *J. Am. Chem. Soc.* **1995**, *117*, 4195.
       c) H. Suzuki, Y. Kimata, S. Satoh, A. Kuriyama, *Chem. Lett.* **1995**, 293.
       d) J. B. Lambert, J. L. Pflug, J. M. Denari, *Organometallics* **1996**, *15*, 615.
[2]   a) M. V. George, D. J. Peterson, H. Gilman, *J. Am. Chem. Soc.* **1960**, *82*, 403.
       b) A. Fürstner, H. Weidmann *J. Organomet. Chem.* **1988**, *354*, 15.

Synthesis and Reactivity of Novel Polysilynes and Branched Copolysilanes

Wolfram Uhlig

Laboratorium für Anorganische Chemie
Eidgenössische Technische Hochschule Zürich
ETH-Zentrum, CH-8092 Zürich, Switzerland
Tel.: Int. code (0)16325984 – Fax: Int. code (0)16321149

Keywords: Polysilyne / Copolysilane / Silyl triflate / Potassium-Graphite

Summary: Polyphenylsilyne and branched copolysilanes have been prepared by reductive coupling of the corresponding organochlorosilanes or halogenated oligosilanes with potassium-graphite. The reaction of the polymers with CF_3SO_3H leads to triflate-substituted derivatives. These compounds have been modified by consecutive reactions with nucleophiles, or by reductive coupling with C_8K. The simple syntheses allow variation in the properties of the preceramic polymers (molecular weights, content of carbon, solubility, functional groups). The polymers were characterized by ^{29}Si NMR spectroscopy.

Introduction

In recent years, much attention has been directed to silicon containing polymers as sources of novel materials [1]. Silyl or polysilyl groups are known to be involved in conjugation with unsaturated groups either through a $(d-p)_\pi$ overlap or a $(\sigma^*-p)_\pi$-type interaction. Therefore, the polymer systems can be expected to be polymeric conducters [2]. On the other hand, intense research has been focused on the elaboration of SiC, Si_3N_4, or Si/C/N-based materials from organosilicon precursors [3]. An idealized pre-ceramic polymer should possess a compromise of the following properties, sometimes incompatible:

- A molecular weight sufficiently high to prevent any volatilization of oligomers;
- presence of latent reactivity (functional substituents) to obtain thermosetting properties;
- a polymeric structure containing cages and rings to decrease the elimination of volatile fragments resulting from backbone cleavage;
- viscoelastic properties (fusibility, maleability, or solubility) to apply the polymer in the desired shape before the pyrolytic process;

- low organic group content to increase ceramic yield and avoid the production of undesired free carbon excess.

Polysilynes and branched copolysilanes are potential preceramic precursors. In this paper, we want to show that the properties of these polymers can be varied by relatively simple methods.

Results and Discussion

Phenyltrichlorosilane was reduced by potassium-graphite C_8K [4] in THF to polyphenylsilyne **1** ($\delta(^{29}Si)$ (CDCl$_3$): -69 ppm) in 83% yield (Eq.1) [5]. A molecular weight $M_w = 6400$ g/mol and a polydispersity $M_w/M_n = 2.0$ are observed using the "inverse reducing method" (addition of potassium-graphite in portions to the chlorosilane). The deficiency of the reducing agent leads to a decrease in the chain-initiation reactions. Therefore, molecular weights are higher than in the case of the "normal reducing method" (addition of the chlorosilane to the reducing agent) [6-8].

$$
Ph\!-\!\underset{\underset{Cl}{|}}{\overset{\overset{Cl}{|}}{Si}}\!-\!Cl \quad \xrightarrow[\substack{-24n\ C \\ -3n\ KCl}]{3n\ C_8K} \quad \left[Ph\!-\!\underset{|}{\overset{|}{Si}}\!-\! \right]_n \quad \mathbf{1}
$$

Eq.1.

Polysilynes [RSi]$_n$ with R = Me or other short-chain alkyl substituents cannot be obtained by this method. These polymers are insoluble in organic solvents and separation from graphite is impossible. Therefore, soluble polysilynes with a relatively small content of carbon must be copolymers. Copolysilynes and branched copolysilanes with alternating arrangements of the substituents have been obtained by reductive coupling of halogenated oligosilanes. The synthesis of the copolysilyne [(MeSi)$_{0.33}$(PhSi)$_{0.67}$]$_n$ **2** from the corresponding trisilane is shown in Scheme 1. The branched copolysilane [(Me$_2$Si)$_{0.75}$(PhSi)$_{0.25}$]$_n$ **3** has been prepared from a halogenated tetrasilane in an analogous way [9] (Scheme 2).

The synthesized polysilynes and branched copolysilanes could be modified by substitution of phenyl substituents by trifluoromethanesulfonic acid [10-12]. The reductive coupling of the triflate derivatives with potassium-graphite leads to polymeric networks with quaternary silicon atoms (Scheme 3). The functional substituted polysilynes **6-9** have been prepared by the reaction of the triflate derivatives with nucleophiles (Scheme 4). The polysilyne **10** was obtained by reduction of the triflate derivative with LiAlH$_4$.

The polymers **1-10** have been characterized by ^{29}Si NMR spectroscopy. The chemical shifts of the tertiary silicon atoms in polysilynes are observed in the same range as the shifts of the oligosilanes MeSi(SiMe$_3$)$_3$ ($\delta(^{29}Si) = -67.4$ ppm) and PhSi(SiMe$_3$)$_3$ ($\delta(^{29}Si) = -74.2$ ppm) and indicate a strainless tetra-hedral geometry of these silicon atoms. In a recent paper [5], we reported that polysilynes consist of cross-linked five- and six-membered rings. The width of the NMR signals (≈ 10 ppm) can be explained by the irregular structure. Table 1 summarizes the spectroscopic data of the polymers **1**-**10**.

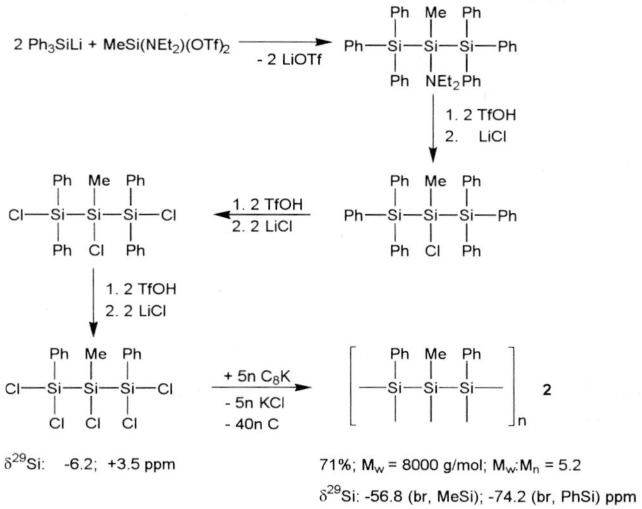

$$2\ Ph_3SiLi + MeSi(NEt_2)(OTf)_2 \xrightarrow[-\ 2\ LiOTf]{}$$

Ph—Si—Si—Si—Ph (with Ph, Me, Ph on top; Ph, NEt₂, Ph on bottom)

1. 2 TfOH
2. LiCl

Ph—Si—Si—Si—Ph (Ph Me Ph / Ph Cl Ph)

Cl—Si—Si—Si—Cl (Ph Me Ph / Ph Cl Ph)   ← 1. 2 TfOH 2. 2 LiCl

1. 2 TfOH
2. 2 LiCl

Cl—Si—Si—Si—Cl (Ph Me Ph / Cl Cl Cl)   + 5n C₈K / − 5n KCl / − 40n C → [—Si—Si—Si—]ₙ (Ph Me Ph) **2**

δ^{29}Si: −6.2; +3.5 ppm

71%; M_w = 8000 g/mol; M_w:M_n = 5.2
δ^{29}Si: −56.8 (br, MeSi); −74.2 (br, PhSi) ppm

Scheme 1.

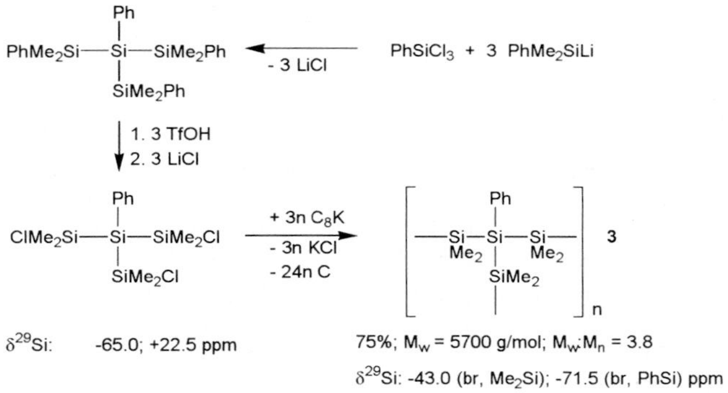

PhMe₂Si—Si—SiMe₂Ph (Ph on top, SiMe₂Ph on bottom) ← − 3 LiCl PhSiCl₃ + 3 PhMe₂SiLi

1. 3 TfOH
2. 3 LiCl

ClMe₂Si—Si—SiMe₂Cl (Ph on top, SiMe₂Cl on bottom) + 3n C₈K / − 3n KCl / − 24n C → [—Si—Si—Si— (Me₂ / Me₂), Ph on top, SiMe₂ on bottom]ₙ **3**

δ^{29}Si: −65.0; +22.5 ppm

75%; M_w = 5700 g/mol; M_w:M_n = 3.8
δ^{29}Si: −43.0 (br, Me₂Si); −71.5 (br, PhSi) ppm

Scheme 2.

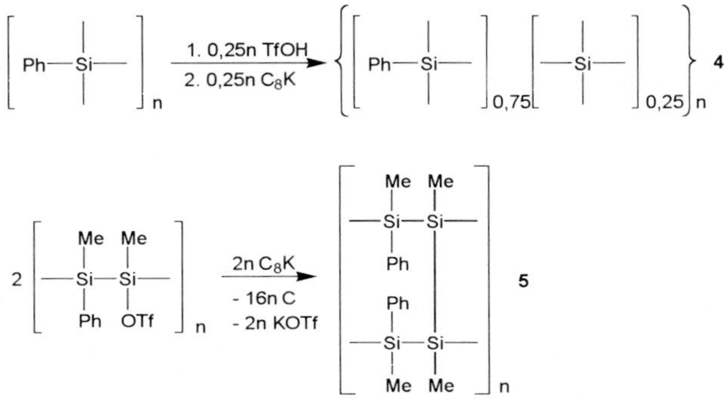

[Ph—Si]ₙ → 1. 0,25n TfOH 2. 0,25n C₈K → { [Ph—Si]₀,₇₅ [—Si—]₀,₂₅ }ₙ **4**

2 [—Si—Si— (Me Me / Ph OTf)]ₙ + 2n C₈K / − 16n C / − 2n KOTf → [—Si—Si— (Me Me) / Ph / Ph / —Si—Si— (Me Me)]ₙ **5**

Scheme 3.

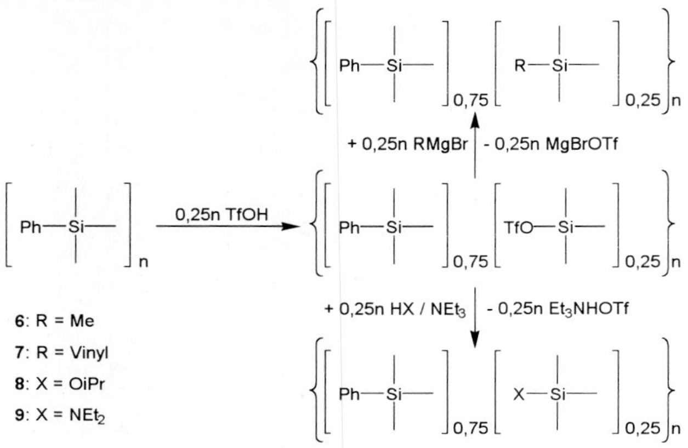

Scheme 4.

6: R = Me
7: R = Vinyl
8: X = OiPr
9: X = NEt$_2$

Table 1. Yields, molecular weights, polydispersities, and NMR data of the polymers **1-10**.

No	Formula	Yield [%]	M_w [g/mol]	M_w/M_n	$\delta(^{29}Si)$ [ppm]	$\delta(^{13}C)$ [ppm]	$\delta(^{1}H)$ [ppm]
1	[PhSi]$_n$	83	6400	2.0	-69.0 (PhSi))	134.0 (Ph) 138.5 (Ph)	7.0 (Ph)
2	[(MeSi)$_{0.33}$(PhSi)$_{0.67}$]$_n$	71	8000	5.2	-56.8 - (-74.2) (MeSi, PhSi)	-4.1 (Me) 136.1 (Ph)	-0.2 (Me) 6.9 (Ph)
3	[(Me$_2$Si)$_{0.75}$(PhSi)$_{0.25}$]$_n$	75	5700	3.8	-43.0 (Me$_2$Si) -71.5 (PhSi)	-3.9 (Me) 136.9 (Ph)	-0.3 (Me) 6.9 (Ph)
4	[(PhSi)$_{0.75}$(Si)$_{0.25}$]$_n$	85	12800	5.5	-70.0 (PhSi) -110.1 (Si)	133.1 (Ph)	6.9 (Ph)
5	[(MePhSi)$_{0.5}$(MeSi)$_{0.5}$]$_n$	86	14500	5.3	-46.6 (MePhSi) -59.1 (MeSi)	-1.9 (Me) 136.4 (Ph)	-0.1 (Me) 7.0 (Ph)
6	[(PhSi)$_{0.75}$(MeSi)$_{0.25}$]$_n$	74	6500	2.1	-54.1 - (-71.2) (MeSi, PhSi)	-3.5 (Me) 134.9 (Ph)	-0.2 (Me) 7.0 (Ph)
7	[(PhSi)$_{0.75}$(ViSi)$_{0.25}$]$_n$	77	5600	2.3	-61.9 - (-77.2) (ViSi, PhSi)	126.9 - 142.5 (Vi, Ph)	5.2 - 6.5 (Vi) 7.0 (Ph)
8	[(PhSi)$_{0.75}$(i-PrOSi)$_{0.25}$]$_n$	86	6900	2.2	-53.9 (i-PrOSi) -68.0 (PhSi)	27.0 (Me) 64.3 (CHO) 132.5 (Ph)	1.2 (Me) 4,4 (CHO) 7.0 (Ph)
9	[(PhSi)$_{0.75}$(Et$_2$NSi)$_{0.25}$]$_n$	82	6100	2.4	-57.5 (Et$_2$NSi) -69.6 (PhSi)	16.6 (Me) 44.0 (CH$_2$N) 136.1 (Ph)	1.1 (Me) 2.7 (CH$_2$N) 7.1 (Ph)
10	[(PhSi)$_{0.75}$(HSi)$_{0.25}$]$_n$	71	5400	2.5	-70.0 (PhSi) -86.1 (HSi)	137.0 (Ph)	3.5 (HSi) 6.9 (Ph)

Conclusions

The reductive coupling of halogenated oligosilanes by potassium-graphite leads to polysilynes and branched polysilanes with variable properties. The structure of the oligosilanes influences the degree of cross-linking, the content of carbon in the polymers, and therefore the solubility of the compounds. The protodesilylation of the phenyl derivatives by CF_3SO_3H and conversion with nucleophiles lead to numerous new functionally-substituted polymers. Therefore, the described syntheses offer convenient methods to vary the properties of preceramic polymers.

Acknowledgement: This work was supported by *Wacker-Chemie GmbH* (Burghausen). Furthermore, the author acknowledges Prof. R. Nesper for support of this investigation.

References:

[1]   H. R. Allcock, *Adv. Mater.* **1994**, *6*, 106.

[2]   R. J. P. Corriu, *Organometallics* **1992**, *11*, 2500.

[3]   M. Birot, J.-P. Pillot, J. Dunogues, *Chem. Rev.* **1995**, *95*, 1443.

[4]   A. Fürstner, H. Weidmann, *J. Organomet. Chem.* **1988**, *354*, 15.

[5]   W. Uhlig, *Z. Naturforsch.* **1995**, *50B*, 1674.

[6]   P. A. Bianconi, F. C. Schilling, T. W. Weidman, *Macromolecules* **1989**, *22*, 1697.

[7]   K. Furukawa, M. Fujino, N. Matsumoto, *Macromolecules* **1990**, *23*, 3423.

[8]   M. Sasaki, K. Matyjaszewski, *J. Polym. Sci., Part A: Polym. Chem.* **1995**, *33*, 771.

[9]   W. Uhlig, *Z. Naturforsch.* **1996**, *51b*, 703.

[10] W. Uhlig, *Chem. Ber.* **1996**, *129*, 733.

[11] D. A. Smith, C. A. Freed, P. A. Bianconi, *Chem. Mater.* **1993**, *5*, 245.

[12] W. Uhlig, *J. Organomet. Chem.* **1996**, *516*, 147.

Unusual Polyhedra by Lithiation of Silazanes

G. Becker*, S. Abele, U. Eberle, G. Motz, W. Schwarz

Institut für Anorganische Chemie
Universität Stuttgart
Pfaffenwaldring 55, D-70550 Stuttgart, Germany

Keywords: Lithiation / Silazanes / Crystal Structure Analyses

Summary: In reactions of bis(*tert*-butylamino)methyl-, bis(*tert*-butylamino)-, and tris-(*tert*-butylamino)silane with *n*-butyl lithium in a molar ratio of 1:2, 1:1, or 1:3, respectively, compounds are formed the structures of which may be deduced from a bicapped triangular dodecahedron on the one hand and a rhombic dodecahedron on the other. A single crystal X-ray structure determination of lithiated bis(*tert*-butyl-amino)methylsilane (**1**) shows the presence of strong agostic Li··H interactions in the solid, whereas the complex anion (**2**) obtained from bis(*tert*-butylamino)silane consists of Li–N bridged 1,3-di-*tert*-butyl-2-*t*-butylamido-4-(*N*-lithium-*t*-butylamido)cyclodisilazane and bis(*N*-lithium-*tert*-butylamido)silane fragments.

About ten years ago Bürger and co-workers prepared the lithiated silazanes $\{[Me_3C–N(Li)]_2SiMe_2\}_2$ (**I**) (Me = CH_3) [1] and $\{[Me_3Si–N(Li)]_3SiR\}_2$ (**II**) {R = CH_3, C_6H_5, $C(CH_3)_3$} [2] from *n*-butyl lithium and bis(*tert*-butylamino)dimethyl- or tris(trimethylsilyl)organylsilane, respectively, and determined their structures by X-ray analyses. The skeletons of these two dimeric molecules represent a bicapped triangular and a rhombic dodecahedron (Fig. 1).

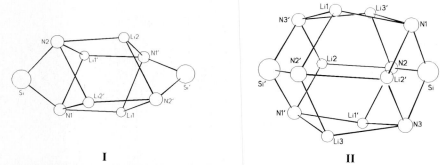

I   **II**

Fig. 1.   Structures of the dimeric molecules bis(*N*-lithium-*tert*-butylamido)dimethylsilane (**I**) and tris(*N*-lithium-trimethylsilylamido)organylsilane (**II**). For a better understanding the figures have been reduced to the molecular frameworks; *tert*-butyl and methyl groups of compound **I** and all trimethylsilyl substituents and the R groups of **II** are not shown.

At present these neutral complexes turn out to be the parent compounds of several related species being studied in our group. Lithiated silazanes of similar or different constitution have also been described by other authors [3-11].

Very recently we have demonstrated that the structure of solvent-free bis(*N*-lithium-*tert*-butyl-amido)methylsilane (**1**) [12] – a Si–H compound easily accessible by lithiation of the corresponding N–H derivative with *n*-butyl lithium in *n*-pentane – can be derived from polyhedron **I**, if several Li–N bonds of its skeleton are broken and two of those bicapped triangular dodecahedra are combined by additional Li···H and Li–N interactions (Fig. 2). In solution strong agostic Li···H interactions can be realized by considerably reduced $^1J_{Si-H}$ coupling constants (Table 1).

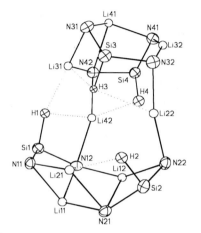

Characteristic structural parameters of **1**:

Si–N:	168 - 172 pm
Si–H:	140 - 149 pm
N–Li:	190 - 235 pm
Li···H:	197 - 232 pm
N–Si–N:	106 - 119°
N–Li–N:	81 - 168°
Li–N–Li:	66 - 150°

Fig. 2. Molecular framework of tetrameric bis(*N*-lithium-*tert*-butylamido)methylsilane (**1**) (thermal ellipsoids at 30 % probability level) together with characteristic ranges for bond lengths and angles. Dotted lines stand for strong agostic Li···H interactions. For δ^{29}Si NMR parameters and for further structural details see Tables 1 or 2, respectively.

Table 1. Survey of ^{29}Si NMR parameters (δ [ppm]; J [Hz]; d (doublet); t (triplet)) obtained from [D_6]-benzene solutions.

1[a]		**2**[a]		*trans*-3		*cis*-3		**4**[a]	
δ	$^1J_{Si-H}$	δ	$^1J_{Si-H}$	δ	$^1J_{Si-H}$	δ	$^1J_{Si-H}$	δ	$^1J_{Si-H}$
-27.11	149.5 (d)	-49.55	213.4 (d)	+4.76	–	+4.74	–	-41.06	189.9 (d)
-29.01	127.5 (d)	-54.33	235.9 (d)	-46.83	230.7 (d)	-55.79	223.3 (d)	-38.31	179.7 (d)
-29.57	[b]	-43.70	172.7 (t)						
-30.43	133.6 (d)								

[a] The NMR data are thought to characterise the compounds; the species present in solution are not known; [b] Correct determination not possible due to superposition.

In the meantime it has been shown that the reaction of bis(*tert*-butylamino)silane (Me$_3$C–NH)$_2$SiH$_2$ – now a silazane with two Si–H bonds – with an equimolar amount of an *n*-hexane solution of *n*-butyl lithium at -40°C in *n*-pentane leads to a product the structure of which could not be determined so far, which, however, contains cyclodisilazane **2b** most probably. Therefore, the compound obtained in well-shaped colorless crystals between 0 and -60°C from an *n*-pentane solution is treated with excess chlorotrimethylsilane (Eq. 1). Depending upon reaction temperature and solvent used (either *n*-pentane at room temperature or refluxing THF) *trans-* and *cis*-1,3-di-*tert*-butyl-2,4-bis[*tert*-butyl(trimethylsilyl)amino]cyclodisilazane (**3**) are formed selectively in a nearly quantitative yield. Both compounds, isolated as usual from the corresponding reaction mixtures, differ significantly in their reproducible melting points (149 vs. 204°C) and their δ^{29}Si NMR data (see Table 1) [12, 14].

Eq. 1.

When, however, the above mentioned colorless crystals isolated from *n*-pentane are redissolved in THF, on cooling the solution to -60°C a co-crystallizate precipitates (Eq. 2). It is built up of $[\{thf\}_4Li]^+$ cations, anions with a very complex framework (**2**), and two molecules of THF in the cavities of the ionic structure. In the anion a bis(*N*-lithium-*t*-butylamido)silane (**2a**) and a 1,3-di-*tert*-butyl-2-*tert*-butylamido-4-(*N*-lithium-*tert*-butylamido)cyclodisilazane fragment (**2b**) are linked together by mutual Li-N interactions (Fig. 3). Again, the structure can be derived from Bürger's triangular dodecahedron (**I**) by removing one lithium as a $[\{thf\}_4Li]^+$ cation and substituting a dimethylsilylene unit for a four-membered Si–N–Si–N ring.

With respect to the underlying mechanism it might be supposed that (*N*-lithium-*tert*-butylamido)-*tert*-butylaminosilane, the first step in the reaction course, is either subject to a second lithiation or eliminates hydrogen to form an electronical favored heteroallyl anion with an $-N=Si(H)-N^-$ fragment. This intermediate, however, is probably not stable enough to be isolated and dimerizes to give 1,3-di-*tert*-butyl-2,4-bis(*N*-lithium-*tert*-butylamido)cyclodisilazane (**2b**). In an alternative route lithium hydride elimination from bis(*N*-lithium-*tert*-butylamido)silane also seems to be possible, especially since the inorganic salt is highly reactive in statu nascendi and might act as an additional lithiation reagent of N–H groups evolving hydrogen (Eq. 2).

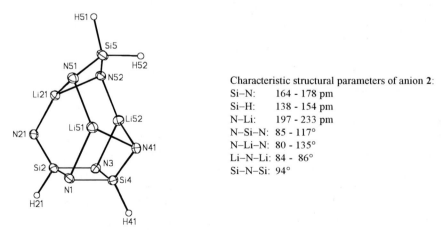

Eq. 2.

Characteristic structural parameters of anion **2**:

Si–N: 164 - 178 pm
Si–H: 138 - 154 pm
N–Li: 197 - 233 pm
N–Si–N: 85 - 117°
N–Li–N: 80 - 135°
Li–N–Li: 84 - 86°
Si–N–Si: 94°

Fig. 3.   Structure of anion **2** (thermal ellipsoids at 30 % probability level) together with characteristic ranges for bond lengths and angles. For a better insight into the structure, the *t*-butyl substituents at the nitrogen atoms have been omitted. δ^{29}Si NMR parameters and further structural details are given in Tables 1 or 2, respectively.

Finally, reaction of *n*-butyl lithium with tris(*tert*-butylamino)silane prepared in high yield from trichlorosilane and excess *tert*-butylamine, gives the completely lithiated silazane **4**. An X-ray structure determination shows the compound to be dimeric in the solid (Fig. 4). While six lithium and two silicon atoms build up the inner cube of a rhombic dodecahedron, six nitrogen atoms occupy the remaining vertices. Compounds with an analogous structure have been reported by several research groups [2, 7, 15].

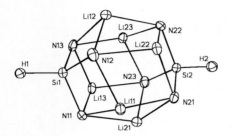

Characteristic structural parameters of **4**:
Si–N: 174 - 175 pm
Si–H: 143 pm
N–Li: 201 - 208 pm
N–Si–N: 106 - 107°
N–Li–N: 86 - 108°
Li–N–Li: 70 - 120°

Fig. 4.   Molecular structure of tris(*N*-lithium-*tert*-butylamido)silane (**4**) (thermal ellipsoids at 30 % probability level) together with characteristic ranges for bond lengths and angles. The *t*-butyl substituents at the nitrogen atoms are not shown; for δ^{29}Si NMR parameters and for further structural details see Tables 1 or 2, respectively.

Table 2.   Crystal data determined at -100 ± 3 °C and details of data collection for **1**, **2**, **3**, and **4**.

	1	**2**	***trans*-3**	***cis*-3**	**4**
Formula	$C_9H_{22}N_2Li_2Si^{[a]}$	$C_{48}H_{106}Li_4N_6O_6Si_3^{[b]}$	$C_{22}H_{56}N_4Si_4$	$C_{22}H_{56}N_4Si_4$	$C_{12}H_{28}N_3Li_3Si^{[a]}$
M	200.3	975.4	489.1	489.1	263.3
Space group	$P2_1/n$	$P2_1/n$	$P2_1/c$	$P2_1/c$	$P\bar{1}$
a [pm]	1176.9(3)	1686.5(2)	1096.1(2)	1017.6(3)	1006.1(2)
b [pm]	2392.3(7)	2108.4(3)	1351.5(3)	1115.0(3)	1667.4(3)
c [pm]	1923.9(5)	1745.8(2)	1020.0(2)	2771.9(5)	1716.2(3)
α [°]	90	90	90	90	62.68(3)
β [°]	100.91(1)	93.52(3)	100.97(3)	100.46(3)	86.85(3)
γ [°]	90	90	90	90	87.57(3)
Z	4 tetramers	4 formula units	2	4	6 formula units[c]
Unique reflections	6523	8428	2481	3815	7562
Unique F ≥ 4σ	6108	5776	1963	2968	6409
Parameters	835	814	248	492	566
Final Δρ [10^{30} e m^{-3}]	-0.35 / 0.47	-0.43 / 0.85	-0.38/0.48	-0.25/0.22	-0.40/0.57
wR$_2$ [13]	0.142	0.220	0.131	0.092	0.169

[a] Formula unit of the monomer; [b] Formula unit of the co-crystallizate consisting of cation, anion and two molecules of THF; [c] Half a centrosymmetric dimer and an acentric dimer in the asymmetric part of the unit cell.

Acknowledgement: We thank the Land *Baden-Württemberg* (Keramikverbund Karlsruhe-Stuttgart, KKS), the *Fonds der Chemischen Industrie*, and *Hoechst AG* (Frankfurt/Main) for generous financial support.

References:

[1]   D. J. Brauer, H. Bürger, G. R. Liewald, *J. Organomet. Chem.* **1986**, *308*, 119.

[2]   D. J. Brauer, H. Bürger, G. R. Liewald, J. Wilke, *J. Organomet. Chem.* **1985**, *287*, 305.

[3]   M. Veith, *Angew. Chem.* **1987**, *99*, 1; *Angew. Chem., Int. Ed. Engl.* **1987**, *26*, 1.

[4]   M. Veith, F. Goffing, V. Huch, *Chem. Ber.* **1988**, *121*, 943.

[5]   M. Veith, *Chem. Rev.* **1990**, *90*, 3.

[6]   M. Veith, M. Zimmer, P. Kosse, *Chem. Ber.* **1994**, *127*, 2099.

[7]   P. Kosse, E. Popowski, M. Veith, V. Huch, *Chem. Ber.* **1994**, *127*, 2103.

[8]   H. Chen, R. A. Bartlett, H. V. Rasika Dias, M. M. Olmstead, P. P. Power, *Inorg. Chem.* **1991**, *30*, 2487.

[9]   B. Tecklenburg, U. Klingebiel, D. Schmidt-Bäse, J. Organomet. Chem. **1992**, *426*, 287.

[10] L. H. Gade, N. Mahr, *J. Chem. Soc., Dalton Trans.* **1993**, 489.

[11] L. H. Gade, Ch. Becker, J. W. Lauher, *Inorg. Chem.* **1993**, *32*, 2308.

[12] G. Becker, S. Abele, J. Dautel, G. Motz, W. Schwarz, in: *Organosilicon Chemistry II: From Molecules to Materials* (Eds.: N. Auner, J. Weis), VCH, Weinheim, **1996**, p. 511.

[13] G. M. Sheldrick, Programm SHELXL93, Göttingen **1993**.

[14] In Fig. 5 of ref. [12] (p. 515) the structural data of the isomers **4b** and **4c** must be interchanged.

[15] L. Zsolnai, G. Huttner, M. Drieß, *Angew. Chem.* **1993**, *105*, 1549; *Angew. Chem. Int. Ed. Engl.* **1993**, *32*, 1439.

Isomeric Halosilylhydroxylamines:
Preparation and Thermal Rearrangements

R. Wolfgramm, U. Klingebiel*

Institut für Anorganische Chemie
Georg-August-Universität Göttingen
Tammannstr. 4, D-37077 Göttingen, Germany

Keywords: Fluorosilylhydroxylamines / Tris(silyl)hydroxylamines / Crystal Structures / Thermal Rearrangements

Summary: O-Fluorosilylhydroxylamines contain two functional groups. By treating with lithiumorganyls LiF elimination and ring closure are observed. The reaction of lithiated N,O-bis(trimethyl)silylhydroxylamine with fluorosilanes leads to the formation of N- or O-halosilylhydroxylamines depending on the substituents of the silanes. Bis(hydroxylamino)fluorosilanes are isolated in the reaction of the lithiated bis(trimethyl)silylhydroxylamine with fluorosilanes in the molar ratio of 2:1. Isomeric silylaminodisiloxanes are formed by the thermal rearrangement of tris(silyl)hydroxylamines. The rearrangement involves the insertion of a silicon moiety into the N–O bond with organyl group migration from the silicon to the nitrogen atom. The organyl-substituents show different tendencies toward migration.

Preparation of *O*-Fluorosilylhydroxylamines

The substitution of the hydroxyl H-atom of H_2NOH by fluorosilanes results from the reaction of $NH_2OH \cdot HCl$ with aminofluorosilanes **1**. If the substituents of the fluorosilanes are small the product cannot be isolated, because it polymerizes on removal of the solvent. However, when bulky substituents are used the product can be distilled at reduced pressure. These fluorosilylhydroxylamines **2** show no tendency to condense.

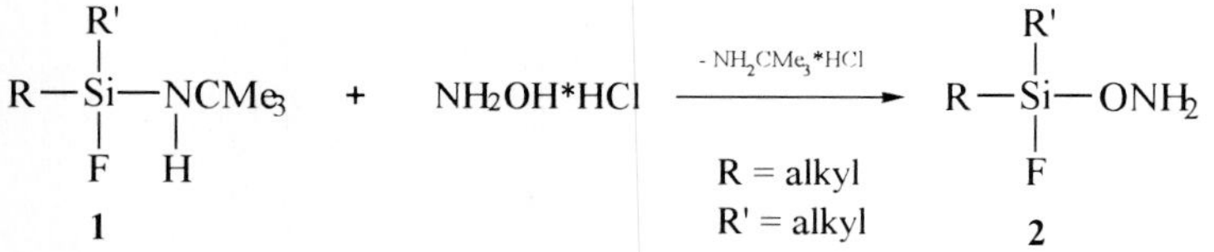

$$
\begin{array}{c}
\text{R'} \\
| \\
\text{R}-\text{Si}-\text{NCMe}_3 \\
| \quad | \\
\text{F} \quad \text{H} \\
\mathbf{1}
\end{array}
\quad + \quad NH_2OH*HCl
\quad \xrightarrow[\text{- NH}_2\text{CMe}_3\text{*HCl}]{}
\quad
\begin{array}{c}
\text{R'} \\
| \\
\text{R}-\text{Si}-\text{ONH}_2 \\
| \\
\text{F} \\
\mathbf{2}
\end{array}
$$

R = alkyl
R' = alkyl

Scheme 1.   Synthesis of **2**.

Fluorosilylhydroxylamines **2** contain two functional groups: N–H and Si–F. Treatment of fluorosilylhydroxylamines **2** with butyllithium results in ring closure to yield the cyclic silyl-hydroxylamines **3** containing a six-membered ring [1].

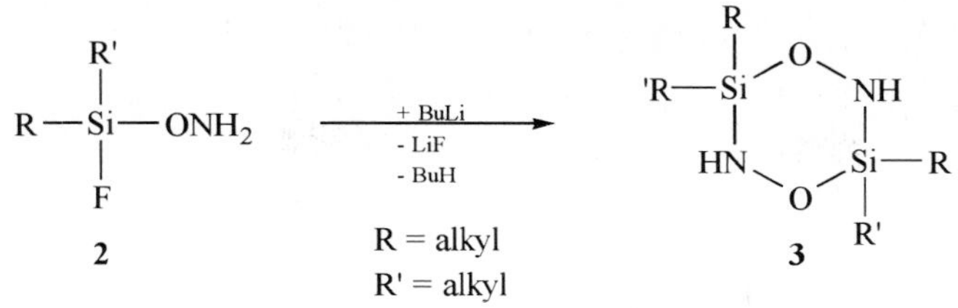

Scheme 2. Synthesis of **3**

The crystal structure of di-*t*-butylsilylhydroxylamine **4** indicates that the ring is not planar but rather has a twisted conformation [2].

Si(1)-N(1):	173.2 pm
Si(1)-O(1):	167.7 pm
N(1)-O(1):	147.6 pm
Si(1)-O(1)-N(2):	111.3°

Fig. 1. Crystal structure of **4**.

Preparation of Tris(silyl)hydroxylamines

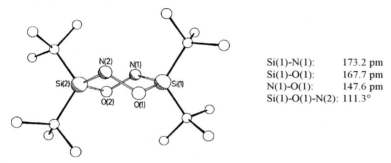

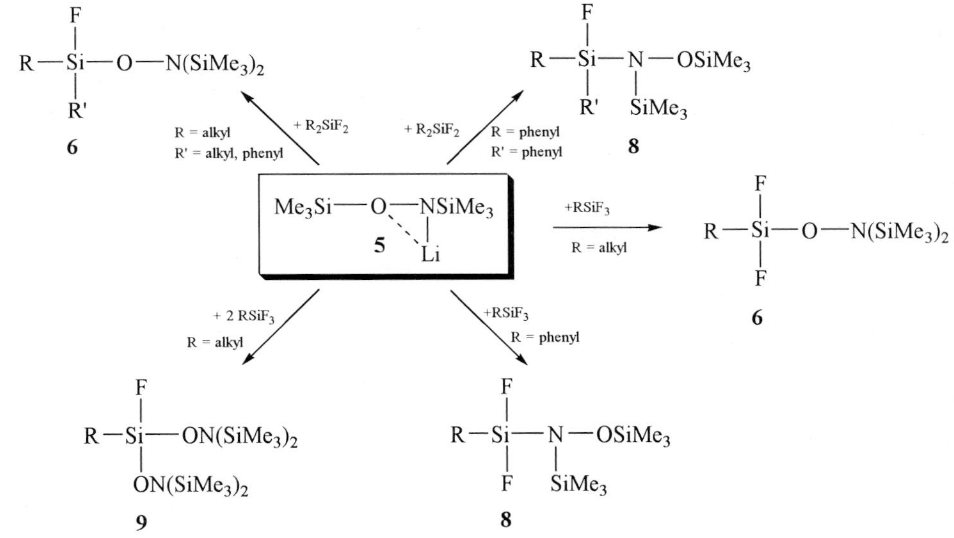

Scheme 3. Substitutionreactions of **5**.

By treating lithiated *N*,*O*-bis(trimethylsilyl)hydroxylamine **5** with alkylfluoro- or alkylarylfluoro-silanes *O*-fluorosilyl-*N*,*N*-bis(trimethylsilyl)hydroxylamines **6** are prepared. *O*-difluoro-2,4,6-tri-*t*-butylphenylsilyl-*N*,*N*-bis(trimethylsilyl)hydroxylamine **7** is available as crystalline solid, which is characterized by X-ray structure determination. The bond lengths and the angles are not significantly different from those in other silylhydroxylamines and the calculated geometries [3]. The angles at nitrogen prove a pyramidal geometry.

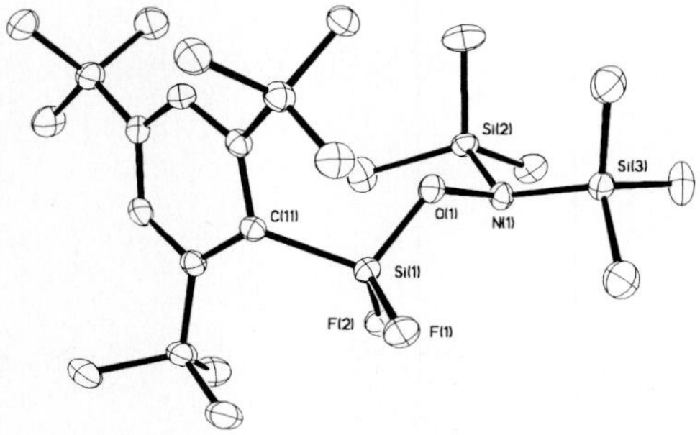

Si(1)-O(1):	163.0 pm
O(1)-N(1):	150.5 pm
N(1)-Si(2):	176.4 pm
Si(1)-O(1)-N(1):	117.6°
O(1)-N(1)-Si(3):	105.0°
O(1)-N(1)-Si(2):	106.2°
Si(2)-N(1)-Si(3):	129.6°
Σ< N(1) = 340°	

Fig. 2.   Crystal structure of **7**.

Arylfluorosilanes react with the lithiated *N*,*O*-bis(trimethylsilyl)hydroxylamine **5** under the same conditions to form *N*-fluorosilyl-*N*,*O*-bis(trimethylsilyl)-hydroxylamines **8**.

Both isomers, **6** and **8**, are obtained in the reaction with benzylfluorosilanes. By heating the mixture of the isomers in THF only the *N*-substituted compound **8** is isolated. Therefore, migration of one trimethylsilyl- and the fluorosilylgroup must occur.

Preparation of Bis(hydroxylamino)fluorosilanes

Bis(hydroxylamino)fluorosilanes are isolated in the reaction of the lithiated bis(trimethylsilyl)-hydroxylamine **5** with fluorosilanes in a 2:1 molar ratio. Similiar to the first substitution, the result depends on the substituents. Using alkylfluorosilanes the *O*-substitution products **9** are formed.

Thermal Rearrangement of Tris(organosilyl)hydroxylamines

Silylaminodisiloxanes are formed by heating tris(organosilyl)hydroxylamines at 200°C. They undergo a rearrangement involving the insertion of a silicon moiety into the nitrogen-oxygen bond and the transfer of one organyl group from silicon to nitrogen.

These observations suggest an intramolecular mechanism. The reaction takes place via the formation of pentacoordinated silicon atoms. Two different pathways, A and B, are proposed as possible mechanisms [4].

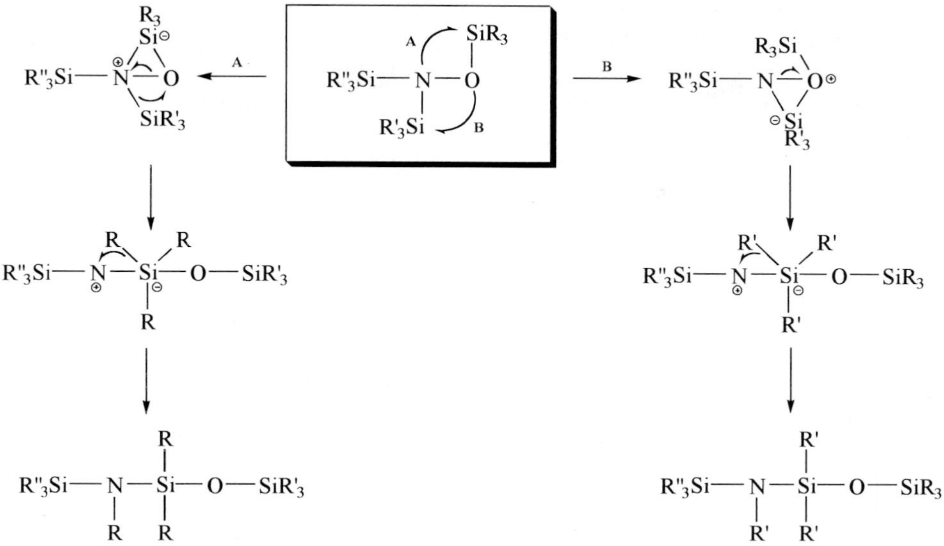

Scheme 4. Mechanism of thermal rearrangement.

We studied rearrangements of *O*-fluorosilylhydroxylamines **6** and found that the migration of the organyl groups and the insertion of the silyl groups depend on the substituents.

The formation of three isomers might be expected in the rearrangement of *O*-alkyl-fluorophenylsilyl-*N*,*N*-bis(trimethylsilyl)hydroxylamines **11**, if both mechanistic pathways have the same probability of occurence. However, only one isomer is isolated in this reaction. The alkylfluorosilyl group inserts into the N-O bond and the phenyl group migrates to the nitrogen atom. These products **12** can only be obtained by pathway A.

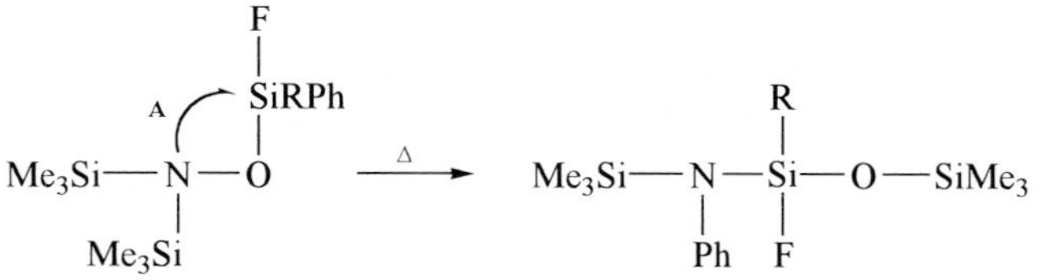

Scheme 5. Rearrangement of **11**.

This seems to indicate that aryl substituents have a stronger tendency to migrate than alkyl substituents, if they are bound to the same silicon atom.

The rearrangement of *O*-di-*tert*-butylfluorosilyl-*N*,*N*-bis(trimethylsilyl)hydroxylamine **13** also leads to one isomer, which can only be reached by pathway B.

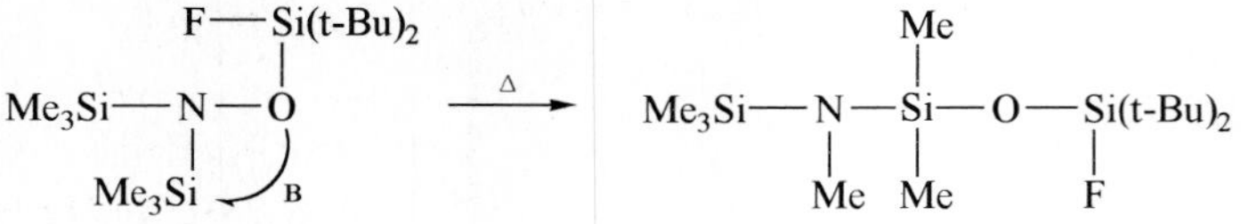

Scheme 6.   Rearrangement of **13**.

Other less bulky *O*-dialkylfluorosilyl-*N,N*-bis(trimethylsilyl)hydroxylamines form the expected isomer mixtures.

Acknowledgement: We thank the *Deutsche Forschungsgemeinschaft* und the *Fonds der Chemischen Industrie* for financial support.

References:

[1]   D. Bentmann, U. Klingebiel, *Z. Anorg. Allg. Chem.* **1981**, *477*, 90.

[2]   D. Bentmann, W. Clegg, U. Klingebiel, G. Sheldrick, *Inorg. Chim. Acta*, **1980**, *45*, L229.

[3]   N. Mitzel, M. Hofmann, E. Waterstradt, P. Schleyer, H. Schmidbauer, *J. Chem. Soc., Dalton Trans.* **1994**, 2503.

[4]   P. Boudjouk, R. West, *Intra Science Chem. Rept.* **1973**, *4*, 65.

Reactions of Hydridosilylamides

Kathrin Junge, Normen Peulecke, Katrin Sternberg, Helmut Reinke, Eckhard Popowski*

Fachbereich Chemie

Universität Rostock

Buchbinderstr. 9, D-18051 Rostock, Germany

Tel.: Int. code + (381)4981822 – Fax: Int. code + (381)4981763

Keywords: Hydridosilylamides / Reaction Behavior / Cyclodisilazanes

Summary: The hydridosilylamines $R_2(H)SiNHSiR_3$ (**1a**: R = $CHMe_2$, R' = Me; **1b**: R = Ph, R' = Me; **1c**: R = CMe_3, R' = Me; **1d**: R = Me, R' =Et; **1e**: R =Me, R' = Ph) and the corresponding hydridosilylamides **2a-2e** have been synthesized. Reaction of **2a-2e** with Me_3SiCl in *n*-hexane or *m*-xylene leads to mixtures of *N*-substitution products **3a-3e** and 1,3-cyclodisilazanes **4a-4e**. In THF **2a-2c** react only to give *N*-substitution products. A 1:1 mixture of $Me_2(H)SiNLiCMe_3$ and $Me_2(H)SiNLiSiMe_3$ gives the 1,3-cyclodisilazanes **5-7** and the noncyclic compounds **8** and **9** in the presence of Me_3SiCl in *n*-hexane. In the reaction of $Me_2(H)SiNLiCMe_3$ with $CH_2=C(Me)–CH=O$ in the presence of Me_3SiCl in *n*-hexane the imine $CH_2=C(Me)–CH=NCMe_3$ and the cyclosiloxanes $(Me_2SiO)_n$ (*n* = 3, 4) are formed.

Introduction

For the formation of cyclodisilazanes by salt elimination from silylamides =Si(X)–N(M)– (X = halogen, H; M = alkali metal) two routes are discussed: (i) intramolecular MX elimination and dimerization of the formed silanimine, (ii) intermolecular MX elimination to an intermediate =Si(X)–N(–)–Si(=)–N(M)– that immediately cyclizes with MX elimination [1-5]. The hydridosilylamides $Me_2(H)SiNLiR$ (R = $Me_2(H)Si$, Me_2CH, CMe_3) and $RR'(H)SiNLiSiMe_3$ (R, R' = Me; R = Me_3SiNH, Me_3SiNLi, R' = Me) react with Me_3SiCl in THF to give *N*-substitution products and in *n*-hexane to form 1,3-cyclodisilazanes, formation of which has been explained via route (i) [6, 7]. However, the silanimine has not been detected experimentally. Here we describe the synthesis of new hydridosilylamides and the reaction of these compounds and known hydridosilylamides with Me_3SiCl.

Results and Discussion

The hydridosilylamines $R_2(H)SiNHSiR'_3$ (**1a**: R = $CHMe_2$, R' = Me; **1b**: R = Ph, R' = Me; **1c**: R = CMe_3, R' = Me; **1d**: R = Me, R' =Et; **1e**: R =Me, R' = Ph) are obtained by coammonolysis of chlorosilanes $R_2(H)SiCl$ with Me_3SiCl (**1a**, **1b**) or by reaction of $(Me_3C)_2(H)SiNHLi$ with Me_3SiCl (**1c**) and $LiNHSiR'_3$ with $Me_2(H)SiCl$ (**1d**, **1e**), respectively, in *n*-hexane (Eqs. 1-3).

$$R_2(H)SiCl + 2\ Me_3SiCl \xrightarrow[\substack{-78 \rightarrow 20\ °C \\ -\ NH_4Cl}]{n\text{pentane, NH}_3\downarrow} R_2(H)SiNHSiMe_3 + 0.5\ (Me_3Si)_2N$$
$$\mathbf{1a}\ (67\ \%),\ \mathbf{1b}\ (66\ \%)$$

Eq. 1.

$$(Me_3C)_2(H)SiNHLi + Me_3SiCl \xrightarrow[\substack{1\ h\ 20\ °C \\ 2\ h\ reflux}]{THF} \mathbf{1c} + LiCl$$
$$(66\ \%)$$

Eq. 2.

$$LiNHSiR'_3 + Me_2(H)SiCl \xrightarrow[8\ h\ reflux]{n\text{hexane}} Me_2(H)NHSiR'_3 + LiCl$$
$$\mathbf{1d}\ (57\ \%),\ \mathbf{1e}\ (61\ \%)$$

Eq. 3.

Treatment of the silylamines **1a-1e** with n-butyllithium in equimolar ratio in n-hexane gives the hydridosilylamides $R_2(H)SiNLiSiR'_3$ **2a-2e** in nearly quantitative yield (Eq. 4). Compounds **2a-2e** are colorless, crystalline solids that can be recrystallized from n-alkanes.

$$\mathbf{1a\text{-}1e} + n\text{BuLi} \xrightarrow[16\ h\ 20\ °C,\ -\ n\text{BuH}]{n\text{hexane}} R_2(H)SiNLiSiR'_3$$
$$\mathbf{2a\text{-}2e}$$

Eq. 4.

Refluxing hydridosilylamides **2a-2e** dissolved in n-hexane or m-xylene does not lead to elimination of LiH.

Table 1. ^{29}Si NMR and IR data of hydridosilylamines and hydridosilylamides (δ in ppm, J in Hz, υ in cm^{-1})

Compound	$\delta^{29}Si(H)$	$\delta^{29}Si(C_3)$	$J(^{29}Si^1H)$	$\upsilon_{as}(SiNSi)$	$\upsilon(SiH)$
1a	1.7	3.8	188.2	928	2108
1b	-21.4	5.0	205.0	929	2119
1c	4.3	4.1	189.4	929	2112
2a	-1.2	-10.7	137.9	999	2033
2b	-26.6	-7.1	190.1	1014	2066
2c	2.9	-13.1	144.5	1078	1925

Compounds **2a-2e** react with Me$_3$SiCl in boiling n-hexane or m-xylene to form the N-substitution products **3a-3e** the 1,3-cyclodisilazanes **4a-4e**, and Me$_3$SiH (Scheme 1). For the same reaction time the conversion in m-xylene is clearly higher than that in n-hexane. Besides the N-substitution products and the 1,3-cyclodisilazanes the unreacted hydridosilylamides are found in the reaction mixture.

Compound	a	b	c	d	e
3	6 %	25 %	5 %	4 %	10 %
4	75 %	4 %	10 %	36 %	18 %

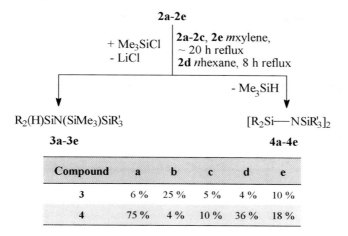

Scheme 1.

These results show that the substituents do not cause a principal change of the reaction behavior of the hydridosilylamides. Compounds **2a-2c** were also allowed to react in THF (7-15 h reflux). Only the *N*-substitution products are formed in good yield (**3a**: 97 %, **3b**: 86 %, **3c**: 95 %). The crystal structures of **4a** and **4b** have been determined (Fig. 1).

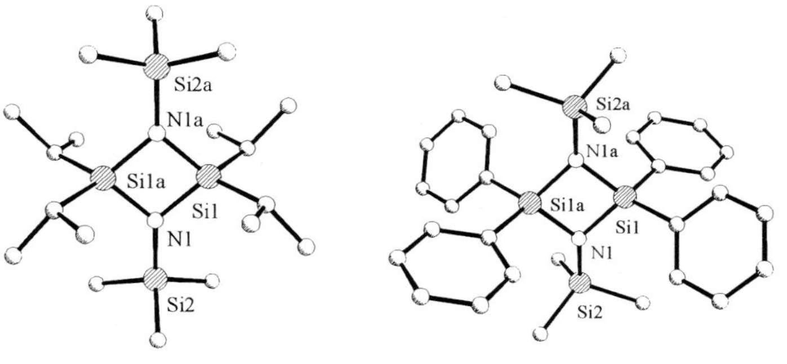

Fig. 1. Crystal structure of **4a** and **4b**; selected bond lengths and angles for **4a** [**4b**]: Si(1)–N(1) 175.6(3) [174.7(3)], Si(1)–N(1a) 175.0(3) [173.6(3)], Si(1a)–N(1) 175.0(3) [173.6(3)], Si(1a)–N(1a) 175.6(3) [173.6(3)] pm; N(1)–Si(1)–N(1a) 89.12(2) [89.24(2)], Si(1)–N(1)–Si(1a) 90.88(2) [90.76(2)]°

The four-membered ring SiNSiN is planar. The Si–N distances and the angles in the ring are comparable to values for other 1,2-cyclodisilazanes [8, 9]. In **4a** the sum of angles around the N(1) atom is 359.9 °, in **4b** 358.1 °.

A 1:1 mixture of the hydridosilylamides $Me_2(H)SiNLiCMe_3$ and $Me_2(H)SiNLiSiMe_3$ gives Me_3SiH, the expected 1,3-cyclodisilazanes **5-7** [6, 7] and surprisingly the noncyclic compounds **8** and **9** in the presence of Me_3SiCl in *n*-hexane under mild conditions (Scheme 2). The amides are stable in boiling *n*-hexane.

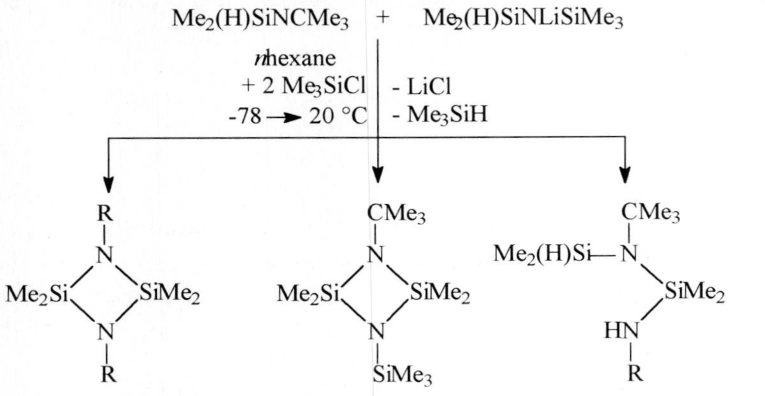

$$Me_2(H)SiNCMe_3 \quad + \quad Me_2(H)SiNLiSiMe_3$$

nhexane
+ 2 Me₃SiCl | - LiCl
-78 → 20 °C | - Me₃SiH

5: R=SiMe₃ (5 %)
6: R=CMe₃ (12 %)

7 (19 %)

8: R=SiMe₃ (5 %)
9: R=CMe₃ (9 %)

Scheme 2.

Lithiation of **8** and **9** in *n*-hexane and subsequent treatment with Me₃SiCl leads to **6** and **7**. The formation of **8** and **9** indicates that the 1,3-cyclodisilazanes are not produced via silanimine intermediates, but via the intermediates Me₂(H)SiN(CMe₃)SiMe₂NLiR (R = Me₃Si, Me₃C), probably formed by intermolecular elimination of LiH.

This is contrary to the result of the reaction of Me₂(H)SiNLiCMe₃ with methacrolein in presence of Me₃SiCl in *n*-hexane. In this case the amide behave like a silanimine. The imine CH₂=C(Me)–CH=NCMe₃, the cyclosiloxanes (Me₂SiO)ₙ (n = 3, 4) and Me₃SiH are obtained (Scheme 3). The imine and the cyclosiloxanes are produced via an oxaazasilacyclobutane intermediate, fomed by [2+2] cycloaddition of the silanimine Me₂Si=NCMe₃ with the aldehyde.

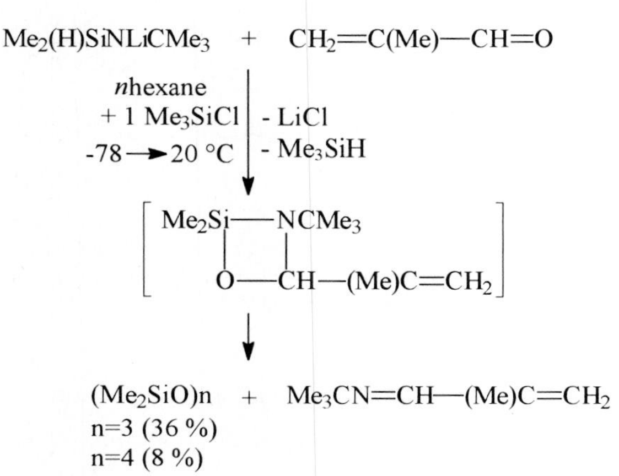

$$Me_2(H)SiNLiCMe_3 \quad + \quad CH_2{=}C(Me){-}CH{=}O$$

nhexane
+ 1 Me₃SiCl | - LiCl
-78 → 20 °C | - Me₃SiH

$$\left[\begin{array}{c} Me_2Si{-\!\!-}NCMe_3 \\ | \qquad\quad | \\ O{-\!\!-}CH{-\!\!-}(Me)C{=}CH_2 \end{array} \right]$$

(Me₂SiO)n   +   Me₃CN=CH—(Me)C=CH₂
n=3 (36 %)
n=4 (8 %)

Scheme 3.

Acknowledgement: We thank the *Fonds der Chemischen Industrie* for financial support.

References:

[1]   L. W. Breed, J. C. Wiley, *Inorg. Chem.* **1972**, *11*, 1634.

[2]   N. Wiberg, *J. Organomet. Chem.* **1984**, *273*, 141.

[3]   W. Clegg, U. Klingebiel, G. M. Sheldrick, D. Stalke, *J. Organomet. Chem.* **1984**, *265*, 17.

[4]   G. Raabe, J. Michl, in: *The Chemistry of Organosilicon Compounds* (Eds. S. Patai, Z. Rappoport), John Wiley & Sons, Chichester, **1989**, p. 1109.

[5]   K. A. Andrianov, V. M. Kopylov, C. M. Khananashvili, T. V. Nesterova, *Dokl. Akad. Nauk SSSR* **1967**, *176*, 85.

[6]   G. H. Wiseman, D. R. Wheeler, D. Seyferth, *Organometallics* **1986**, *5*, 146.

[7]   P. Kosse, E. Popowski, *Z. Anorg. Allg. Chem.* **1992**, *613*, 137.

[8]   L. Bihatsi, P. Hencsei, L. Parkanyi, *J. Organomet. Chem.* **1981**, *219*, 145.

[9]   S. Walter, U. Klingebiel, M. Noltemeyer, *Chem. Ber.* **1992**, *125*, 783.

Silylhydrazines:
Precursors for Rings, Hydrazones and Pyrazolones

Christian Drost, Uwe Klingebiel*, Henning Witte-Abel
Institut für Anorganische Chemie
Georg-August-Universität Göttingen
Tammannstraße 4, D-37077 Göttingen, Germany

Keywords: Silylhydrazines / Lithium derivatives / O-Silylpyrazolones / Crystal Structure

Summary: Mono(silyl)hydrazines condense to bis(silyl)hydrazines at higher temperature. The degree of oligomerization of mono- and dilithiated silylhydrazines in the crystal depends on the bulkiness of the substituents. Reactions of lithiated silylhydrazines with fluorosilanes lead, for example, to the formation of tetrakis(silyl)-hydrazines, six- , four- and five-membered rings. Formaldehyde derivatives of mono-(silyl)hydrazones are obtained in a reaction of mono(silyl)hydrazines with aqueous formaldehyde solution. O-Silylpyrazolones can be synthesized by treating mono(silyl)-hydrazines with acetoacetic ester. A dipyrazolonesilane is formed in the reaction of an O-silylpyrazolone with dichlorodimethylsilane.

Silylhydrazines

Stable mono(silyl)hydrazines are formed in reactions of lithiated hydrazine with bulky fluorosilanes [1,2]. They condense to form bis(silyl)hydrazines above 200 °C [2]:

Eq. 1.

Lithium Derivatives of Silylhydrazines

a) Monolithiated di-*t*-butylmethylsilylhydrazine **3** forms a hexameric unit in the crystal [3].

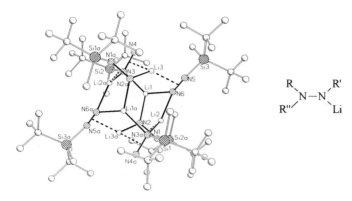

Fig. 1. Crystal structure of **3**.

b) The monolithium-derivative of bis(trimethylsilyl)hydrazine **4** reacts in ether with formation of a dimer composed of mono- and dilithiated units [4]:

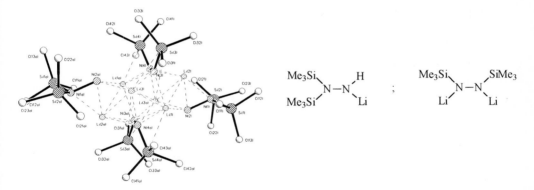

Fig. 2. Crystal structure of **4**.

Bis(*t*-butyldimethylsilyl)hydrazine **5** [2], tris(trimethylsilyl)hydrazine **6** [5, 6], and *N-t*-butyl-*N'*-(trimethylsilyl)hydrazine **7** [5] form dimers via Li–N bonds. In the case of **8** the silyl-*N*-atom is lithiated.

c) dilithiated silylhydrazines:

The oligomerization of dilithiumbis(silyl)hydrazides depends on the bulkiness of the substituents. For example, (Me₃SiNLi)₂ **(9)** forms a tetramer in the crystal [7, 8]. The dilithium derivative of bis(di-*t*-butylmethylsilyl)hydrazine **10** is the first isolated monomeric dilithium derivative of a silylhydrazine.

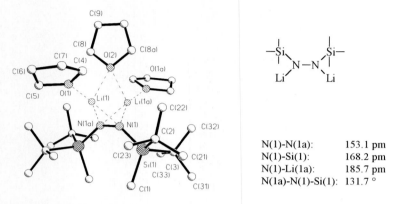

N(1)-N(1a):   153.1 pm
N(1)-Si(1):   168.2 pm
N(1)-Li(1a):   185.7 pm
N(1a)-N(1)-Si(1): 131.7 °

Fig. 3. Crystal structure of **10**.

Reactions

a) Dilithiated bis(silyl)hydrazines react with fluorosilanes, e. g., trifluorophenylsilane, to yield fluorofunctional tetrakis(silyl)hydrazines [1].

N–N: 149.2 pm
N–SiF$_2$: 169.3 pm
N–SiMe$_3$: 176.5 pm

Eq. 2.

The nitrogen atoms are nearly planar and the fluorine atoms are not equivalent because of a hindered rotation.

b) Elimination of lithium fluoride from lithiated fluoro(silyl)hydrazines leads to the formation of six-membered rings. Cyclization with formation of a four-membered ring has not been achieved by Elimination of LiF.

A quantitative fluorine-chlorine exchange occurs in the reaction of these lithium derivatives with trimethylchlorosilane; trimethylfluorosilane and lithium chloride are formed. The resulting silahydrazone is isolated as the [2+2]-cycloaddition product [8].

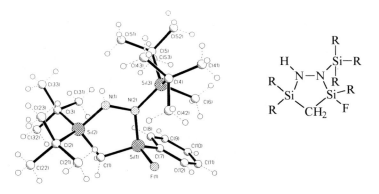

Eq. 3.

c) If the substituents are very bulky LiF-elimination leads to the formation of a carbon-silicon bond and a five-membered N_2Si_2C-ring **14** is formed [3].

Fig. 4. Crystal structure of **14**.

Silylhydrazones

The first stable monomeric formaldehyde derivatives are formed by reaction of mono(silyl)hydrazines with aqueous formaldehyde solution.

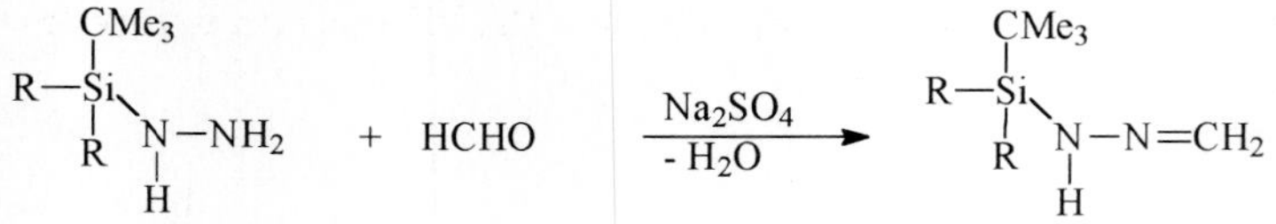

Eq. 4.

Silylpyrazolones

O-Silylpyrazolones can be obtained by treating mono(silyl)hydrazines with acetoacetic ester. A 1,3-silylgroup migration occurs at higher temperature.

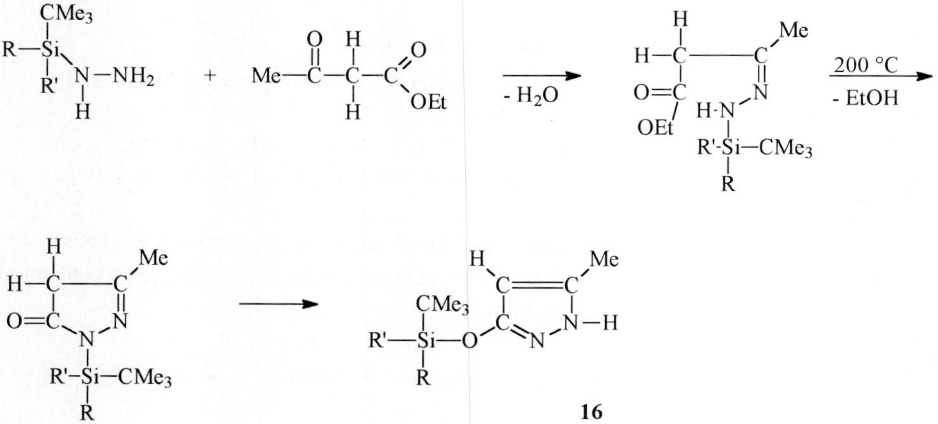

Eq. 5.

The *O*-silylpyrazolones obtained are water resistant.

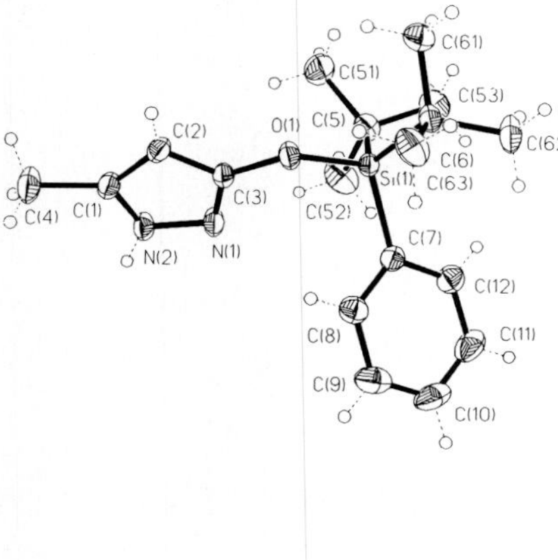

Fig. 5.   Crystal structure of **16**.

The reaction of two equivalents *O*-silylpyrazolone with dichlorodimethylsilane leads to the formation of the first dipyrazolonesilane:

R = OSi(CMe₃)₂Ph

Eq. 6.

References:

[1]   K. Bode, U. Klingebiel, *Adv. Organomet. Chem* **1996**, *40*, 1.

[2]   K. Bode, M. W. Gluth, R. Herbst-Irmer, U. Klingebiel, M. Noltemeyer, M. Schäfer, H. Witte-Abel, *Phosphorus, Sulfur, and Silicon,* **1996**, *108*, 121.

[3]   S. Dielkus, C. Drost, R. Herbst-Irmer, U. Klingebiel, *Angew. Chem.* **1993**, *105*, 1689.

[4]   K. Bode, C. Drost, C. Jäger, U. Klingebiel, M. Noltemeyer, Z. Zak, *J. Organomet. Chem.* **1994**, *482*, 285.

[5]   K. Bode, U. Klingebiel, M. Noltemeyer, H. Witte-Abel, *Z. Anorg. Allg. Chem.* **1995**, *621*, 500.

[6]   H. Hommer, H. Nöth, H. Sachdev, M. Schmidt, H. Schwenk, *Chem. Ber.* **1995**, *128*, 1187.

[7]   a) H. Nöth, H. Sachdev, M. Schmidt, H. Schenk, *Chem. Ber.* **1995**, *128*, 105.
      b) N. Metzler, H. Nöth, H. Sachdev, *Angew. Chem.* **1995**, *106*, 1838.

[8]   C. Drost, S. Freitag, C. Jäger, U. Klingebiel, M. Noltemeyer, G. M. Sheldrick, *Chem. Ber.* **1994**, *127*, 845.

Products from Multiple Insertion Reactions between Diisocyanates and Antiheteroaromatic 1,4-Bis(trimethylsilyl)-1,4-dihydropyrazine

Torsten Sixt, Fridmann M. Hornung, Anja Ehlend, Wolfgang Kaim*

Institut für Anorganische Chemie
Universität Stuttgart
Pfaffenwaldring 55, D-70550 Stuttgart, Germany
Tel.: Int. code + (711)6854170 – Fax: Int. code + (711)6854165
E-mail: Kaim@anorg55.chemie.uni-stuttgart.de

Keywords: Crystal Structure / Diisocyanates / Heterocycles / Insertion / Silicon-Nitrogen Bonds

Summary: Bifunctional polymethylenediisocyanates $O=C=N–(CH_2)_n–N=C=O$, $n = 4$ or 6, react with the extremely electron-rich 8π electron heterocycle 1,4-bis(trimethylsilyl)-1,4-dihydropyrazine **1** under insertion of the heterocumulene functions into the N–Si bonds. Of the expected electron-rich oligomeric structures the first isolated products are presented, including the crystal structure of the 1:2 addition product between $OCN–(CH_2)_4–NCO$ and **1**. This product contains two trimethylsilylurea functionalities.

Introduction

1,4-Bis(trimethylsilyl)-1,4-dihydropyrazine **1** is a thermally stable but very electron-rich molecule with 8 cyclically conjugated π-electrons in a virtually planar six-membered heterocyclic ring (antiheteroaromaticity) [1]. The biologically relevant [2] ring system can be stabilized by other 1,4-trialkylsilyl [3] or 1,4-acyl substituents [4]. In addition to outer-sphere electron transfer with acceptors such as TCNQ, TCNE, or C_{60} [3] and exchange reactions of the SiMe$_3$ substituents by SiMe$_2$SiMe$_3$, SiPh$_3$, or BMes$_2$ in ion-stabilizing solvents [3, 5] the insertion of heterocumulenes [6] such as CO_2, COS, CS_2, or R–NCO is a reaction that has been studied for **1** and related molecules [7-9]. The reaction with monoisocyanates R–NCO always yielded bis-insertion products [7] so that the use of bifunctional diisocyanates was expected to produce linear oligomers or polymers. After hydrolysis of the triorganosilyl groups the resulting polyurea species can be envisaged to form an aramide-like arrangement (Scheme 1).

Herein we report first results from the reaction between **1** and the simple diisocyanates $OCN–(CH_2)_n–NCO$ ($n = 4$ or 6).

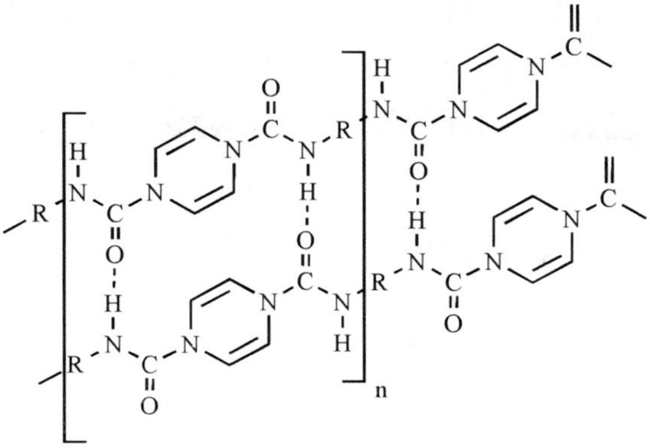

Scheme 1.

Results and Discussion

The reaction in dilute hexane solution between OCN–(CH$_2$)$_4$–NCO and **1** in either 1:1 or 1:2 molar ratio produced a series of oligomers [10] of which the pale-yellow 1:2 bis-insertion adduct **2** proved to be the major product.

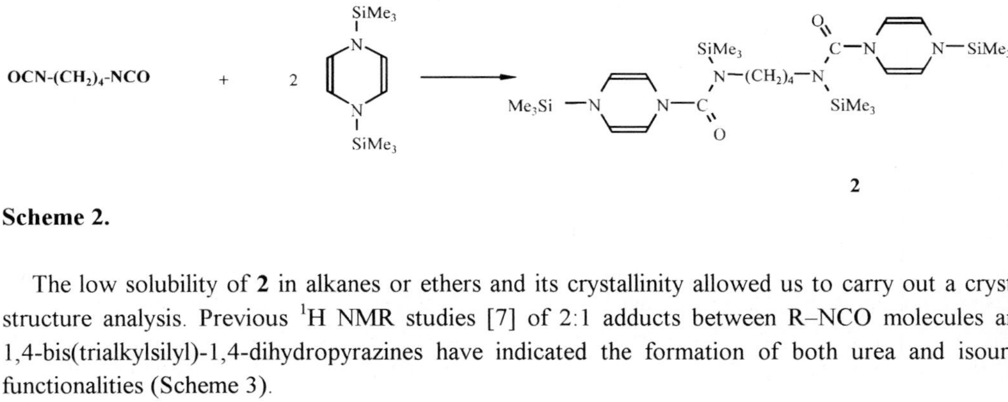

Scheme 2.

The low solubility of **2** in alkanes or ethers and its crystallinity allowed us to carry out a crystal structure analysis. Previous ^{1}H NMR studies [7] of 2:1 adducts between R–NCO molecules and 1,4-bis(trialkylsilyl)-1,4-dihydropyrazines have indicated the formation of both urea and isourea functionalities (Scheme 3).

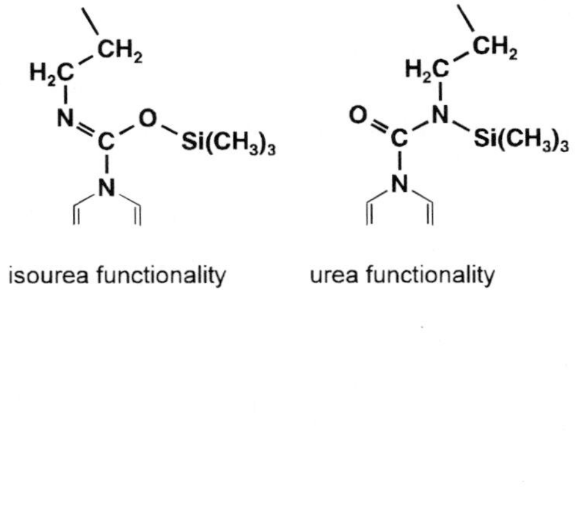

isourea functionality   urea functionality

Scheme 3.

The first crystal structure analysis of such a derivative should now settle the question [7-9] of which isomeric structure (Scheme 3) is formed between sterically unencumbered components.

A suitable single crystal of **2** for X-ray diffraction was obtained by the cooling to 3°C of a saturated solution in 1,2-dimethoxyethane [11]. Fig. 1 shows the structure of the centrosymmetric molecule in the crystal and the atomic numbering; the caption summarizes selected bond lengths and angles.

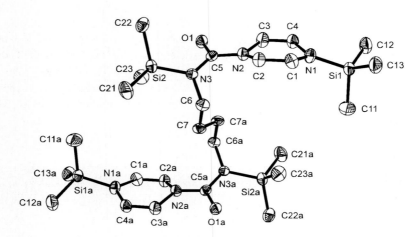

Fig. 1. Molecular structure of **2** in the crystal. Selected bond lengths (pm) and angles (°): C5–O1 123.5(4), C5–N2 137.1(5), C5–N3 137.5(5), C1–C2 132.5(5), C3–C4 130.7(5), N1–Si1 175.1(3), N3–Si2 177.5(3), C–Si (av.) 185.5; O1–C5–N2 119.9(3), O1–C5–N3 120.2(3), N2–C5–N3 119.9(3),C5–N2–C3 127.0(3), C5–N2–C2 118.6(3), C2–N2–C3 113.1(3), C5–N3–C6 120.7(3), C5–N3–Si2 113.0(2), Si2–N3–C6 121.5(2), C1–N1–C4 112.6(3), C1–N1–Si1 124.5(3), C4–N1–Si1 122.7(3).

The structure of **2** in the crystal confirms a "normal" *N*-silylurea situation following the primary insertion of the isocyanate function into one of the N–Si bonds of **1**. The bond lengths at the urea sites are also quite regular, including very similar (O)C–N(dihydropyrazine) and (O)C–N(SiMe$_3$/CH$_2$–) bond lengths.

The 1-trimethylsilyl-4-carbamoyl substituted 1,4-dihydropyrazine ring is no longer planar [1] but exhibits a distinct boat conformation [1, 3, 8]. The dihedral angle α [1] between the boat halves intersecting at N1···N2 lies at 158.6°, right between the virtually 180° of **1** and the approximately 140° observed for the 2,3,5,6-tetramethyl derivative [1]. The dihedral "bow" angles β [1] between the C1,C2,C3,C4 best plane and the C1,N1,C4 or C2,N2,C3 planes are higher at the carbamoylated side (19.2°) than at the silylated side (13.9°). Nevertheless, the mono-insertion and silylurea formation affects *both* sides of the conformationally very flexible [2] 1,4-dihydropyrazine ring.

In addition to the intramolecular effects there are weak but unambiguous π/π interactions between the 1,4-dihydropyrazine rings of different molecules of **2** in the crystal. The shortest such distance found between ring carbon atoms of different molecules was established at 352.2 pm (C1a-C4′).

For hexamethylenediisocyanate, OCN–(CH$_2$)$_6$–NCO, the reaction with **1** under various molar ratios and in several solvents always yielded the 2:2 adduct **3** as a yellow oil (Scheme 4).

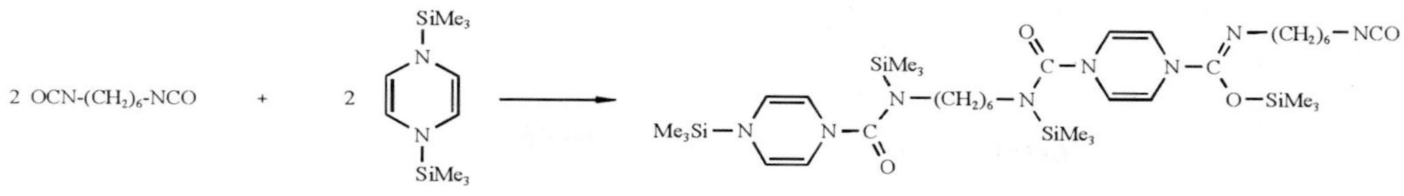

2 OCN-(CH₂)₆-NCO   +   2   [structure **1**]   →   [structure **3**]

Scheme 4.

In contrast to compound **2** with symmetrical insertion, the 2:2 linear adduct **3** exhibits two different unsymmetrical dihydropyrazine rings, including one with both urea and isourea functionalities (Scheme 4). This arrangement is supported by ¹H and ¹³C NMR data (Table 1) and is in agreement with the results for **2** and previous observations [7]. Apparently, the second insertion step at **1** requires greater activation and then proceeds to a different product, i.e., an isourea instead of a silyl-urea arrangement.

Table 1. ¹H NMR and ¹³C NMR data[a] for **2** and **3** in CDCl₃ solution at 300 K.

		SiCH	C–CH₂–C	N–CH₂–C	HC(Pz)	NCO	C=O	C=N
δ(¹H)	**2**	0.11[c] s 0.24[e] s	1.33 m,b	3.02 m,b	5.09[b] d 5.53[b] d			
	3	0.08[c] s 0.13[d] s 0.21[e] s 0.24[e] s	1.40 m,b	3.10 m,b	5.05 dd 5.14 m,b 5.21 m,b 5.38 m,b 5.49 d			
δ(¹³C)	**2**	-1.53[c] s 0.20[e] s	28.44 s	41.18 s	110.40 s 118.20 s		156.40[f] s	
	3	-1.80[c] s -0.04[d] s 0.22[e] s	26.27 d 29.05 d 30.96 m	45.02 d 47.11 d	108.00 b 110.10 s 118.00 s 119.90 b	121.9 b	149.80[g] d 156.10[f] s	161.1[g] b

[a] Broadened signal – [b] ³J = 6.3 Hz – [c] (Pz)N–SiCH – [d] (N=C)O–SiCH – [e] (O=C)N–SiCH – [f] Si(Pz)C=O – [g] N=C(Pz).

The limiting factor in the polymerization thus seems to be the diminished reactivity of the N–Si bond of the mono-insertion product; different conditions are currently being tested to obtain high molecular weight products.

Acknowledgement: This work was supported by the *Deutsche Forschungsgemeinschaft* and the *Fonds der Chemischen Industrie*.

References:

[1] H.-D. Hausen, O. Mundt, W. Kaim, *J. Organomet. Chem.* **1985**, *296*, 321.

[2] a) W. Kaim, *Angew. Chem.* **1983**, *95*, 201; *Angew. Chem., Int. Ed. Engl.* **1983**, *22*, 171.
b) W. Kaim, A. Schulz, F. Hilgers, H.-D. Hausen, M. Moscherosch, A. Lichtblau, J. Jordanov, E. Roth, S. Zalis, *Res. Chem. Intermed.* **1993**, *19*, 603.

[3] a) J. Baumgarten, C. Bessenbacher, W. Kaim, T. Stahl, *J. Am. Chem. Soc.* **1989**, *111*, 2126.
b) A. Lichtblau, A. Ehlend, H.-D. Hausen, W. Kaim, *Chem. Ber.* **1995**, *128*, 745.

[4] R. Gottlieb, W. Pfleiderer, *Liebigs Ann. Chem,* **1981**, 1451.

[5] A. Lichtblau, H.-D. Hausen, W. Schwarz, W. Kaim, *Inorg. Chem.* **1993**, *32*, 73.

[6] M. F. Lappert, B. Prokai, *Adv. Organomet. Chem.* **1967**, *8*, 243.

[7] W. Kaim, A. Lichtblau, T. Stahl, E. Wissing, in: *Organosilicon Chemistry: From Molecules to Materials* (Eds.: N. Auner, J. Weis), VCH, Weinheim, **1994**, p. 41.

[8] A. Ehlend, H.-D. Hausen, W. Kaim, *J. Organomet. Chem* **1995**, *501*, 283.

[9] A. Ehlend, H.-D. Hausen, W. Kaim, in: *Organosilicon Chemistry II: From Molecules to Materials* (Eds.: N. Auner, J. Weis), VCH, Weinheim, **1996**, p. 141.

[10] T. Sixt, *Diploma Thesis*, Universität Stuttgart, **1996**.

[11] Crystal data for **2**: $C_{26}H_{52}N_6O_2Si_4$, $M =593.1$ g/mol, monoclinic ($P2_1/c$; [No.14]) $a = 1129.2(2)$ pm, $b = 2357.2(3)$ pm, $c = 640.1(1)$ pm, $\beta = 94.18(11)°$, $V = 1699.3(5) \times 10^6$ pm³, $Z = 2$, $\rho_{calc} = 1.159$ g cm⁻³, $\mu(MoK_\alpha)= 2.06$ cm⁻¹, 4459 reflections (4106 independent, $3.46° < 2\Theta < 56.0°$; $h = $ -14 to 14, $k = $ -1 to 1, $l = $ -8 to 8) were collected at -90°C on a crystal of 0.6 x 0.2 x 0.15 mm size using monochromated Mo-K_α radiation; $F(000) = 644$, $R_1[I > 2\sigma(I)] = 0.0690$ [12a], $wR_2 = 0.2098$ [12b], GOF = 1.027 [12c]. The structure was solved by direct methods using the SHELXTL-PLUS package [13a], the refinement was carried out with SHELXL-93 [13b] employing full matrix least-squares methods. Anisotropic thermal parameters were refined for all non-hydrogen atoms. The hydrogen atoms were located and refined isotropically. Further information on the structure determination may be obtained from Fachinformationszentrum Karlsruhe GmbH, D-76344 Eggenstein-Leopoldshafen, Germany, on quoting the depository number CSD 405422, the names of the authors, and the book citation.

[12] a) $R = (\Sigma \, ||F_o| - |F_c|| \,)/\Sigma |F_o|$.
b) $wR_2 = \{\Sigma[w(\,|F_o|^2 - |F_c|^2)^2]/\Sigma[w(F_o^4)]\}^{1/2}$.
c) GOF = $\{\Sigma w(\,|F_o|^2 - |F_c|^2)^2/(n\text{-}m)\}^{1/2}$; $n = $ no. of reflections; $m = $ no. of parameters.

[13] a) G. M. Sheldrick, *SHELXTL-PLUS: An Integrated System for Solving, Refining and Displaying Crystal Structures from Diffraction Data*, Siemens Analytical X-Ray Instruments Inc., Madison, WI, U.S.A., **1989**.
b) G. M. Sheldrick, SHELXL-93, *Program for Crystal Structure Determination*, Universität Göttingen, Germany, **1993**.

Some Surprising Chemistry of Sterically Hindered Silanols

Paul D. Lickiss

Department of Chemistry
Imperial College of Science, Technology and Medicine
London SW7 2AY, UK
Fax: Int. code + (171)5945804
E-mail: p.lickiss@ic.ac.uk

Keywords: Silanols / Structure / Synthesis / Silanone / Steric Hindrance

Summary: The presence of a sterically bulky group such as $(Me_3Si)_3C$, $(PhMe_2Si)_3C$ or $(Me_3Si)_3Si$ attached to a silicon atom allows a range of silanols containing unusual functional groups to be prepared and their novel chemistry to be studied. Stable compounds such as the triol $(PhMe_2Si)_3CSi(OH)_3$ and the halosilanols $(Me_3Si)_3CSiHI(OH)$ and $(Me_3Si)_3CSiF(OH)_2$ may be prepared and their structures determined. The unusual hydrido silanols $(Me_3Si)_3CSiRH(OH)$ (R = H, F, Me, Et, nBu, Ph) may be prepared as stable solids and used as precursors to the silanone radical anions $(Me_3Si)_3CSiR(=O)^-$.

Introduction

Organosilanols are compounds of the general formula $R_{3-n}Si(OH)_n$ (R = alkyl, aryl, etc., n = 1, 2, or 3) and they are important intermediates in the formation of silicone polymers and silane coupling agents. Closely related alkoxysilanols $(RO)_{3-n}Si(OH)_n$ are important intermediates in sol-gel processes. The synthetic routes to silanols are commonly oxidation of Si–H containing species or hydrolysis of a halo-, amido-, or alkoxy-silanes. The most important reaction of silanols from an industrial point of view is their condensation resulting in the loss of water and the formation of siloxane linkages, Si–O–Si. Such condensation reactions are especially easy if the R substituents are small or if there are two or three OH groups attached to the same silicon atom. Silanols also tend to be thermally unstable with respect to condensation and both bases and acids catalyze condensation reactions. Thus, simple silanols such as $Me_2Si(OH)_2$ or $PhSi(OH)_3$ require considerable care in their preparation and isolation, and they readily condense to form polymeric materials. If, however, very bulky groups such as $(Me_3Si)_3C$, $(PhMe_2Si)_3C$, or $(Me_3Si)_3Si$ are attached to a silicon atom bearing an OH group then stable silanols may be readily prepared and isolated, and their chemistry investigated. The synthesis, structures and chemistry of some surprisingly stable silanols will be described below.

The Synthesis of Bulky Silanols

The bulky groups $(Me_3Si)_3C$, $(PhMe_2Si)_3C$, and $(Me_3Si)_3Si$ may readily be attached to a silicon atom by use of the lithium reagents $(Me_3Si)_3CLi$, $(PhMe_2Si)_3CLi$, and $(Me_3Si)_3SiLi$ which in turn may be prepared according to Eqs. 1-3 [1-3]. Even if prolonged heating and more than one equivalent of lithium reagent is used, only one $(Me_3Si)_3C$ or $(PhMe_2Si)_3C$ group may be attached to an individual silicon atom.

$$3\ Me_3SiCl\ +\ CHBr_3\ +\ 6\ Li\ \xrightarrow{\ THF\ }\ (Me_3Si)_3CH\ \xrightarrow{\ MeLi\ }\ (Me_3Si)_3CLi$$

Eq. 1.

$$3\ PhMe_2SiCl\ +\ CHCl_3\ +\ 6\ Li\ \xrightarrow{\ THF\ }\ (PhMe_2Si)_3CH\ \xrightarrow{\ MeLi\ }\ (PhMe_2Si)_3CLi$$

Eq. 2.

$$4\ Me_3SiCl\ +\ SiCl_4\ +\ 8\ Li\ \xrightarrow{\ THF\ }\ (Me_3Si)_4Si\ \xrightarrow[-\ SiMe_4]{\ MeLi\ }\ (Me_3Si)_3SiLi$$

Eq. 3.

The especially high degree of steric protection afforded by the $(Me_3Si)_3C$ group is readily appreciated when the stabilities of the trihalides $(Me_3Si)_3CSiX_3$ (X = Cl or Br) are considered. Neither compound, in sharp contrast to less hindered trihalosilanes, is hydrolyzed by atmospheric moisture. Indeed, the trichloride does not react with refluxing MeOH [4] and the tribromide may be recovered unchanged from refluxing aqueous ethanol [5]. The silicon-centerd analogue $(Me_3Si)_3SiSiCl_3$ however, is, readily hydrolyzed by moist diethyl ether [6], the longer central Si–Si bonds allowing nucleophilic attack to occur much more readily. Despite the low reactivity of $(Me_3Si)_3CSiCl_3$ towards water it does react with $LiAlH_4$ to give $(Me_3Si)_3CSiH_3$ [7], which may be used to prepare a range of interesting silanols.

A range of silanols that contain unusual combinations of functional groups at silicon can be prepared using the silane $(Me_3Si)_3CSiH_3$ as a starting material. Scheme 1 shows how the triol $(Me_3Si)_3CSi(OH)_3$ may be prepared *via* the novel $(Me_3Si)_3CSiHI(OH)$. An iodosilanol would normally undergo rapid self-reaction to give a siloxane but the presence of the bulky $(Me_3Si)_3C$ group prevents such bimolecular reactions, this is also the case for the other halosilanols shown in Scheme 1. The hydrolysis of $(Me_3Si)_3CSiHI(OH)$ directly to the triol $(Me_3Si)_3CSi(OH)_3$ without isolation of the diol $(Me_3Si)_3CSiH(OH)_2$ is thought to be due to the diol undergoing ionization to give a silanolate anion, which loses H^- to give the transient silanone $(Me_3Si)_3CSi(=O)OH$, which then rapidly undergoes addition of water to the double bond to give the observed triol [8]. The bromo- and the fluorosilanols shown in Scheme 1 are all air-stable compounds that may readily be isolated and characterized [9]. The structure of $(Me_3Si)_3CSiF(OH)_2$ is described below.

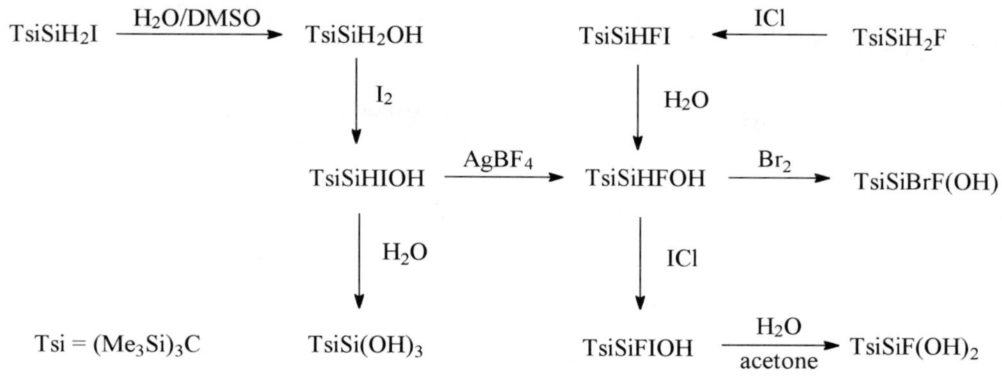

Scheme 1. Synthetic routes to silanols containing the bulky $(Me_3Si)_3C$ group.

The synthesis of a second series of compounds, the hydrido silanols $(Me_3Si)_3CSiRH(OH)$, is shown in Scheme 2 [8, 10-12]. These compounds are again unusual in that they have both H and OH as substituents at silicon, which, with smaller substituents than $(Me_3Si)_3C$ present, normally leads to species that undergo rapid self-condensation reactions. Such compounds are useful precursors to silanone radical anions as described below.

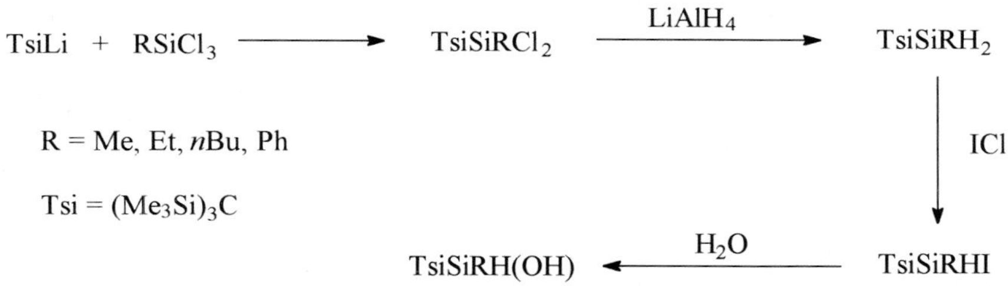

Scheme 2. Synthesis of bulky hydrido silanols.

The bulky lithium reagent $(PhMe_2Si)_3CLi$ is rather unreactive towards many chlorosilanes, for example it reacts neither with Me_3SiCl nor with Me_2SiCl_2 [13]. It does, however, react with the less bulky SiH_2Cl_2 to give $(PhMe_2Si)_3CSiH_2Cl$, which reacts readily with water to give the silanol $(PhMe_2Si)_3CSiH_2OH$ [9]. This is an unusual example of a stable silanol with two hydrogen substituents at the SiOH silicon, usually such species undergo condensation with the formation of siloxanes very readily. The silanol reacts with two equivalents of dimethyldioxirane to give the triol $(PhMe_2Si)_3CSi(OH)_3$, which is thermally stable up to 250 °C [9].

The Structures of Bulky Silanols

The structures of many silanols have been described in detail in a review [14]. The bulky silanol $(PhMe_2Si)_3CSiMeH(OH)$ had previously been shown by Eaborn and co-workers [15] to have a monomeric structure in which there was an intramolecular hydrogen bonding interaction between the

OH group and one of the aromatic rings. The less hindered $(PhMe_2Si)_3CSiH_2(OH)$ has a similar structure [9]. Intramolecular OH$\cdots\pi$ interactions are also seen in the triol $(PhMe_2Si)_3CSi(OH)_3$ [16], but in this case there are also OH$\cdots$OH hydrogen bonds and the structure comprises tetramers as shown in Fig. 1. It is not clear why the less hindered $(PhMe_2Si)_3CSiH_2(OH)$ and $(PhMe_2Si)_3CSiMeH(OH)$ should show no intermolecular OH$\cdots$OH hydrogen bonds but the more hindered triol does.

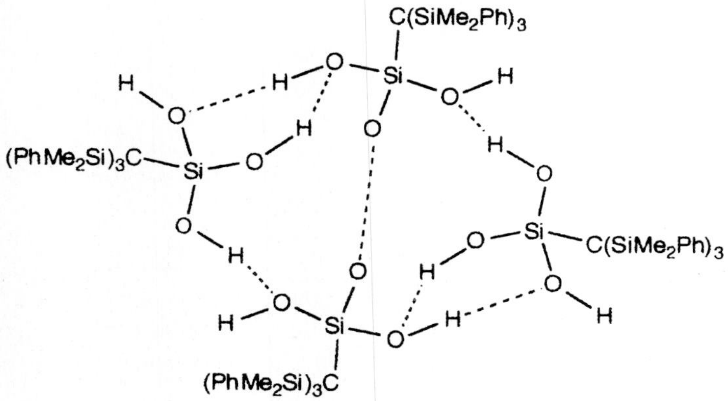

Fig. 1.   A representation of the hydrogen bonding in $(PhMe_2Si)_3CSi(OH)_3$. The hydrogen atoms on the two central oxygen atoms could not be located.

The fluorosilanol forms a hydrate if crystallised in solvent open to the atmosphere and the structure in the solid state is found to be the hexameric dihydrate $[(Me_3Si)_3CSiF(OH)_2]_6\cdot2H_2O$ as shown schematically in Fig. 2 [9].

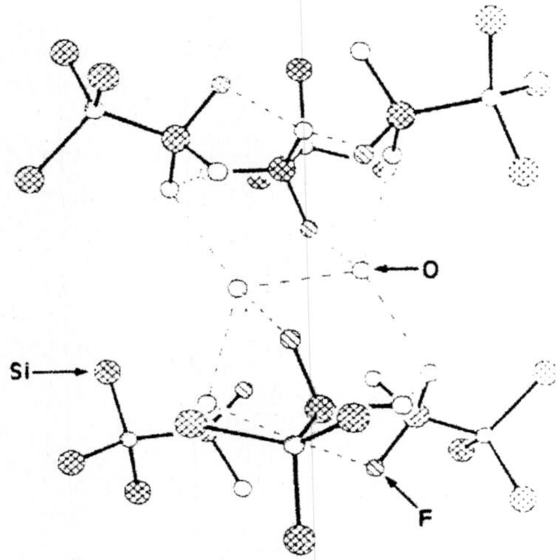

Fig. 2.   The hydrogen-bonded structure of $[(Me_3Si)_3CSiF(OH)_2]_6\cdot2H_2O$. The hydrogens atoms and the methyl groups have been omitted for clarity, hydrogen bonds are shown by dashed lines.

The structure comprises two hydrogen-bonded trimeric units that are connected together via the two water molecules both by OH···OH and OH···F hydrogen bonds. Unlike the hexameric $(Me_3Si)_3SiSi(OH)_3$ and most other silanols, $(Me_3Si)_3CSiF(OH)_2$ is not maximally hydrogen bonded, hydrogen bonding between the hexameric units to form stacks or layers being prevented by the hydrophobic Me_3Si exterior of the hexamer.

The triol $(Me_3Si)_3SiSi(OH)_3$ also forms a hexameric structure (very similar to that of $(Me_3Si)_3CSi(OH)_3$) but in this case it is maximally hydrogen bonded within the hexamer and there is no water incorporated into the structure [6].

Reactions of Sterically Hindered Silanols

Stable silanones, $R_2Si=O$ species, are not known but in 1992 Davies and Neville reported that the radical anion $tBu_2Si=O^-$ could be generated in solution and characterised by its EPR spectrum [17]. The silanone radical anion was generated by abstracting both hydrogens from the SiH(OH) group of $tBu_2SiH(OH)$ using a combination of $tBuOK$ and $tBuO^-$ in $tBuOH/(tBuO)_2$. As discussed above, compounds containing the SiH(OH) grouping are usually highly unstable towards self-condensation reactions, but the presence of an $(Me_3Si)_3C$ substituent does allow the synthesis of the surprisingly stable silanols $(Me_3Si)_3CSiRH(OH)$ as shown in Scheme 2. These silanols also react with $tBuOH/KOtBu/(tBuO)_2$ under UV irradiation to give the silanone radical anions $(Me_3Si)_3CSiR(=O)^-$ as shown in Scheme 3 [12]. EPR data for the radical anions are shown in Table 1. The spectra for the ethyl and the *n*-butyl species are of interest because they show coupling to only one of the two hydrogens in the α-CH$_2$ group; this is also found in the radical $(Me_3Si)_3CSi(Et)H^-$. This could be due either to the fact that the α-CH$_2$ hydrogens are diastereotopic or that rotation about the Si–Et bond is restricted by the presence of the bulky $(Me_3Si)_3C$ group.

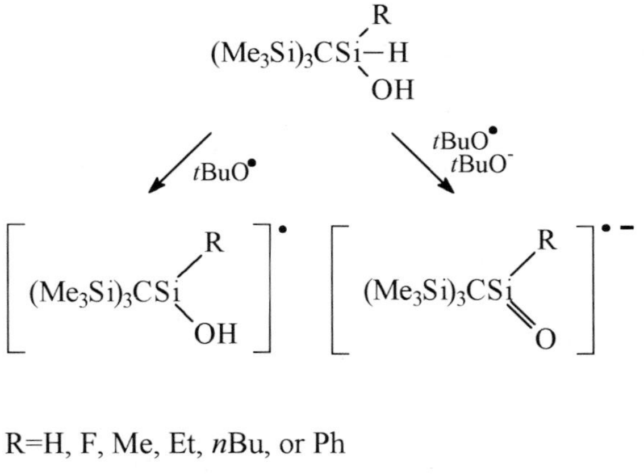

R=H, F, Me, Et, *n*Bu, or Ph

Scheme 3. The formation of silanone radical anions.

Calculations have shown [12] that, for the $(Me_3Si)_3CSi(Et)H^-$ radical, Fig. 3, the hydrogen forming a dihedral angle θ of 30° with the axis of the singly occupied sp^3 orbital should give a small, but non-zero, coupling (which is not observed in the spectra, although the linewidth is relatively large) while that forming an angle θ of 180° should give a large coupling, as observed experimentally. A similar magnetic inequivalence is observed in the ^{1}H NMR spectrum of $(Me_3Si)_3CSiEt(OH)H$, in which the SiH resonance is a doublet. A magnetic inequivalence is also seen in the NMR spectrum of $(Me_3Si)_3CSiEtH_2$ in which the CH_2 protons are not diastereotopic but which do give rise to a doublet of doublets for the Si–H proton [12]. This supports the suggestion that the inequivalence results from restricted rotation.

Table 1. EPR data for silanone radical anions.

Radical Anion	a(nH or F)/G	$a(^{29}Si\alpha)$/G	g
TsiSi(H)=O⁻K	32.90 (1 H)	186.0	2.002 85
TsiSi(Me)=O⁻K	4.25 (3 H)	172.1	2.002 72
TsiSi(Et)=O⁻K	10.06 (1 H)	164.7	2.002 65
TsiSi(Bu)=O⁻K	8.44 (1 H)	164.8	2.002 65
TsiSi(Ph)=O⁻K	0.11 (n H)[a]	< 75	2.002 65
TsiSi(F)=O⁻K	95.8 (1 F)	241	2.001 91

[a] The spectrum of the phenyl containing species was poorly resolved.

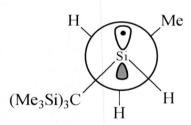

Fig. 3. The conformation of the $(Me_3Si)_3CSi(Et)H^-$ radical.

Conclusions

This work has shown that the presence of the very sterically demanding substituents $(Me_3Si)_3C$, $(PhMe_2Si)_3C$, and $(Me_3Si)_3Si$ at silicon may allow compounds of surprising stability to be formed. Thus, for example, the tribromide $(Me_3Si)_3CSiBr_3$ is stable to hydrolysis, the triol $(PhMe_2Si)_3CSi(OH)_3$ is stable to self-condensation and the halosilanol $(Me_3Si)_3CSiHI(OH)$ is stable towards self reaction. The remarkably stable hydridosilanols $(Me_3Si)_3CSiRH(OH)$ may also be used to generate a series of silanone radical anions including that of the first sila-aldehyde, $(Me_3Si)_3CSiH(=O)^-$, and the first sila-acylhalide, $(Me_3Si)_3CSiF(=O)^-$. The use of very bulky groups thus continues to provide compounds that exhibit a variety of surprising properties.

Acknowledgement: The work on silanone radical anions was carried out with Professors C. Eaborn (Sussex) and A. G. Davies (University College, London). Their collaboration is gratefully acknowledged. Crystallographic studies were carried out by Drs. P. B. Hitchcock, (Sussex), A. D. Redhouse (Salford) and F. Fronczek (Louisiana State, Baton Rouge) and the synthetic work by Y. Derouiche (Sussex), S. M. Whittaker (Salford) and A. G. Neville (University College, London). Financial support from the *EPSRC*, *The Royal Society*, and *The University of Salford* is also gratefully acknowledged.

References:

[1]   Z. H. Aiube, C. Eaborn, *J. Organomet. Chem.* **1984**, *269*, 217.

[2]   G. Fritz, B. Grunert, *Z. Anorg. Allg. Chem.* **1981**, *473*, 59.

[3]   G. Gutekunst, A. G. Brook, *J. Organomet. Chem.* **1982**, *225*, 1.

[4]   C. Eaborn, D. A. R. Happer, K. D. Safa, D. R. M. Walton, *J. Organomet. Chem.* **1978**, *157*, C50.

[5]   P. D. Lickiss, S. M. Whittaker, *unpublished results*.

[6]   S. S. Al-Juaid, N. H. Buttrus, R. I. Damja, Y. Derouiche, C. Eaborn, P. B. Hitchcock, P. D. Lickiss, *J. Organomet. Chem.* **1989**, *371*, 287.

[7]   S. S. Dua, C. Eaborn, D. A. R. Happer, S. P. Hopper, K. D. Safa, D. R. M. Walton, *J. Organomet. Chem.* **1979**, *178*, 75.

[8]   R. I. Damja, C. Eaborn, *J. Organomet. Chem.* **1985**, *290*, 267.

[9]   S. M. Whittaker, *Ph.D. thesis*, University of Salford, **1993**.

[10] C. Eaborn, D. E. Reed, *J. Chem. Soc., Perkin Trans. 2* **1985**, 1695.

[11] Z. H. Aiube, N. H. Buttrus, C. Eaborn, P. B. Hitchcock, J. A. Zora, *J. Organomet. Chem.* **1982**, *292*, 177.

[12] A. G. Davies, C. Eaborn, P. D. Lickiss, A. G. Neville, *J. Chem. Soc., Perkin Trans. 2* **1996**, 163.

[13] C. Eaborn, A. I. Mansour, *J. Chem. Soc., Perkin Trans. 2* **1985**, 729.

[14] P. D. Lickiss, *Adv. Inorg. Chem.* **1995**, *42*, 147.

[15] S. S. Al-Juaid, A. K. Al-Nasr, C. Eaborn, P. B. Hitchcock, *J. Chem. Soc., Chem. Commun.* **1991**, 1482.

[16] F. Fronczek, P. D. Lickiss, S. M. Whittaker, *unpublished results*.

[17] A. G. Davies, A. G. Neville, *J. Organomet. Chem.* **1992**, *436*, 255.

Silanetriols: Preparation and Their Reactions

Ramaswamy Murugavel, Andreas Voigt,
Mrinalini Ganapati Walawalkar, Herbert W. Roesky*
Institut für Anorganische Chemie
Universität Göttingen
Tammannstraße 4, D-37077 Göttingen, Germany
Tel.: Int. code + (551)393001 – Fax: Int. code + (551)393373
E-mail: hroesky@gwdg.de

Keywords: Silanetriols / Heterosiloxanes / Metallasiloxanes / Model Compounds / Catalysis

Summary: Synthetic routes have been developed in our laboratories for the preparation of a variety of silanetriols, $RSi(OH)_3$, where the central silicon atom is bonded to carbon, nitrogen or oxygen. In particular, the N-bonded (arylamino)silanetriols are air-stable in the solid state and are also soluble in a variety of organic solvents, thus making these compounds useful starting materials for the preparation of a variety of metallasiloxanes. The metallasiloxanes resulting from the silanetriols show very interesting structural features. The Si–O–M frameworks in these compounds are made up of either cyclic or three-dimensional polyhedral core structures. Both the silanetriols and the metallasiloxanes have been extensively characterized by means of IR and NMR spectroscopy as well as by single crystal X-ray diffraction studies in representative cases. Many of these metallasiloxanes contain hydrolyzable functionalities such as M–C, M–O, and Si–N bonds providing a possibility of using these compounds as starting materials in the preparation of supramolecular cage structures and synthetic zeolites under mild conditions.

Introduction

The major interests in scientific research have always been associated with finding the best possible solutions for the challenges offered by Nature. One such challenge has been an attempt to mimic naturally occurring processes by preparing soluble model compounds for naturally occurring insoluble substances such as silicate minerals. On the other hand, catalysis is the branch of chemistry dealing with the development of new catalysts which would help the large scale synthesis of important chemical substances. The development in the area of catalysis over the decades has transformed from multiphase heterogeneous catalytic systems to catalytically active compounds anchored on neutral substrates. However, the ultimate goal in this area would be preparing completely homogeneous catalytic systems having substantial turn-over numbers.

In an attempt to reach the aforementioned goals, we started our work in this area on the synthesis of soluble compounds which can offer a neutral base environment for active metal centers. Given the background that the silica-supported catalysts have been one amongst the most favorite systems for many decades [1], our first task was to synthesize the necessary silicon-based ligands. Silanols [2] have been tried out as ligands for a long time. Unfortunately, silanols containing a single –OH functionality give rise to monometallic products [3] and cannot be ideal catalysts, as most of the catalytic reactions are supported by more than one metal center. To overcome this problem, silanediols $R_2Si(OH)_2$ [4], disilanols $[R_2Si(OH)]_2O$ [5], other types of silanols [6, 7], and finally silanetriols $RSi(OH)_3$ [8] have been synthesized and tried out for their ligand capabilities. The latter class of compounds, namely silanetriols, have an advantage over the other types of ligands in the light of their use as ideal building-blocks for the formation of three-dimensional metal-containing super structures [8].

Towards this end, we have been recently successful in synthesizing a series of silanetriols with diverse substituents and studying their chemistry in detail. The various sections of this article summarize the following aspects of silanol chemistry: (a) various synthetic routes available for the preparation of soluble and stable silanetriols starting from commonly available starting materials, (b) common spectroscopic and structural properties of silanetriols, (c) reactions of silanetriols with various metal precursors leading to the preparation of novel polyhedral metallasiloxanes, (d) structures of metallasiloxanes, and (e) possible applications.

Preparation of Silanetriols

The preparation of stable silanetriols of the type $RSi(OH)_3$ has proven to be difficult, compared to the other types of silanols, in view of their well known tendency to self-condense and result in siloxane rings and polymers [9]. However, in recent years, the use of very bulky R groups on silicon and carefully chosen experimental conditions has led to the isolation of many stable silanetriols. While silanetriols such as $tBuSi(OH)_3$, $PhSi(OH)_3$, and $(c\text{-}C_6H_{11})Si(OH)_3$ have already been prepared [2], the smaller analogues such as $MeSi(OH)_3$ and $EtSi(OH)_3$ are yet to be synthetically realized.

In 1955, Tyler reported the preparation of the first silanetriol, $PhSi(OH)_3$ (**1**) starting from phenyltrimethoxysilane [10]. The acid assisted hydrolysis of $PhSi(OMe)_3$ at 10 °C resulted in the isolation of phenylsilanetriol in 75% yield (Scheme 1). This compound was found to be very unstable and decomposes on heating. Acid or base impurities also lead to the decomposition of this compound. Later, this synthetic methodology was used by other workers [11,12] to prepare a series of *ortho*-, *meta*-, and *para*-substituted phenylsilanetriols (**2**) starting from the respective aryltrimethoxysilanes (Scheme 1). Similarly, the hydrolysis of cyclohexyltrimethoxysilane in the presence of acetic acid leads to the isolation of $(c\text{-}C_6H_{11})Si(OH)_3$ (**3**) [13] (Scheme 1) whose crystal structure has been later determined by X-ray diffraction studies [14].

Silanetriols can also be prepared by amine-assisted hydrolysis of the respective trichlorosilanes. Phenylsilanetriol was thus prepared by Takiguchi by the hydrolysis of $PhSiCl_3$ with stoichiometric amounts of water in the presence of aniline as the HCl acceptor (Scheme 1) [15]. This mild synthetic procedure later proved to be very useful in the preparation of the tri-*t*-butoxysiloxysilanetriol $(tBuO)_3SiOSi(OH)_3$ (**4**) [16].

The groups of Eaborn [17-19] and Lickiss [2] reported on a series of sterically hindered silanetriols containing SiR_3 groups (Scheme 2). The hydrolysis of $(Me_3Si)_3SiSiCl_3$ employing the procedure of Takiguchi leads to the isolation of $(Me_3Si)_3SiSi(OH)_3$ (**5**) [19]. However, the similar triols $(Me_3Si)_3CSi(OH)_3$ (**6**) [17] and $(PhMe_2Si)_3CSi(OH)_3$ (**7**) [2] cannot be made directly from their trichlorides and require multi-step procedures. The synthetic routes used for the preparation of these silanetriols are shown in Scheme 2.

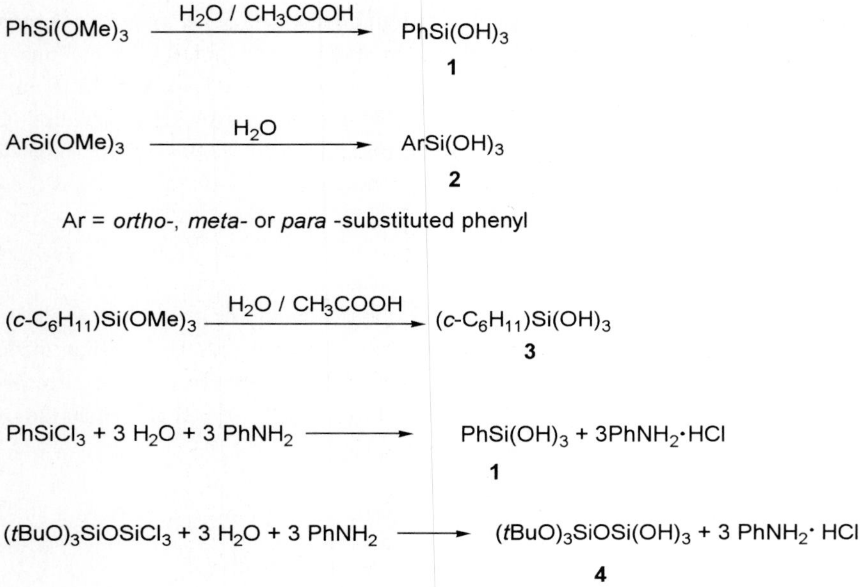

$$PhSi(OMe)_3 \xrightarrow{\text{H}_2\text{O / CH}_3\text{COOH}} PhSi(OH)_3$$

1

$$ArSi(OMe)_3 \xrightarrow{\text{H}_2\text{O}} ArSi(OH)_3$$

2

Ar = *ortho-*, *meta-* or *para* -substituted phenyl

$$(c\text{-}C_6H_{11})Si(OMe)_3 \xrightarrow{\text{H}_2\text{O / CH}_3\text{COOH}} (c\text{-}C_6H_{11})Si(OH)_3$$

3

$$PhSiCl_3 + 3\ H_2O + 3\ PhNH_2 \longrightarrow PhSi(OH)_3 + 3PhNH_2 \cdot HCl$$

1

$$(tBuO)_3SiOSiCl_3 + 3\ H_2O + 3\ PhNH_2 \longrightarrow (tBuO)_3SiOSi(OH)_3 + 3\ PhNH_2 \cdot HCl$$

4

Scheme 1.

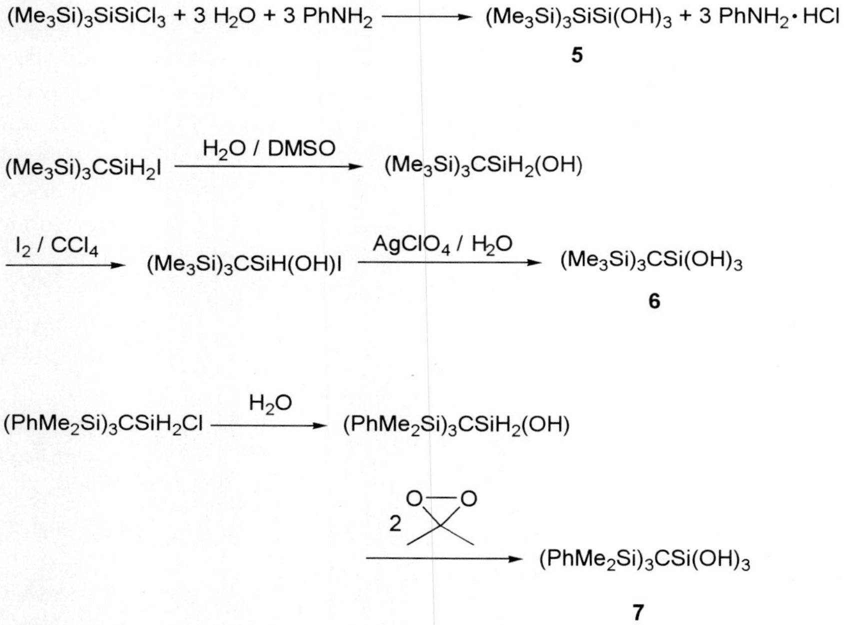

$$(Me_3Si)_3SiSiCl_3 + 3\ H_2O + 3\ PhNH_2 \longrightarrow (Me_3Si)_3SiSi(OH)_3 + 3\ PhNH_2 \cdot HCl$$

5

$$(Me_3Si)_3CSiH_2I \xrightarrow{\text{H}_2\text{O / DMSO}} (Me_3Si)_3CSiH_2(OH)$$

$$\xrightarrow{\text{I}_2 / \text{CCl}_4} (Me_3Si)_3CSiH(OH)I \xrightarrow{\text{AgClO}_4 / \text{H}_2\text{O}} (Me_3Si)_3CSi(OH)_3$$

6

$$(PhMe_2Si)_3CSiH_2Cl \xrightarrow{\text{H}_2\text{O}} (PhMe_2Si)_3CSiH_2(OH)$$

$$\xrightarrow{2 \ \text{O—O}} (PhMe_2Si)_3CSi(OH)_3$$

7

Scheme 2.

Recently, there have been reports on the synthesis of silanetriols with an M–Si bond. Rickard et al. reported the synthesis of the osmium containing silanetriol $OsCl(CO)(PPh_3)_2Si(OH)_3$ (**8**) starting from the corresponding trichlorosilane and aq. KOH (Scheme 3) [20]. Similarly, Malisch and co-workers prepared the group 6 metals [21] and iron [22] containing silanetriols (**9-11**) starting from the corresponding metallatrihydridosilanes (Scheme 3).

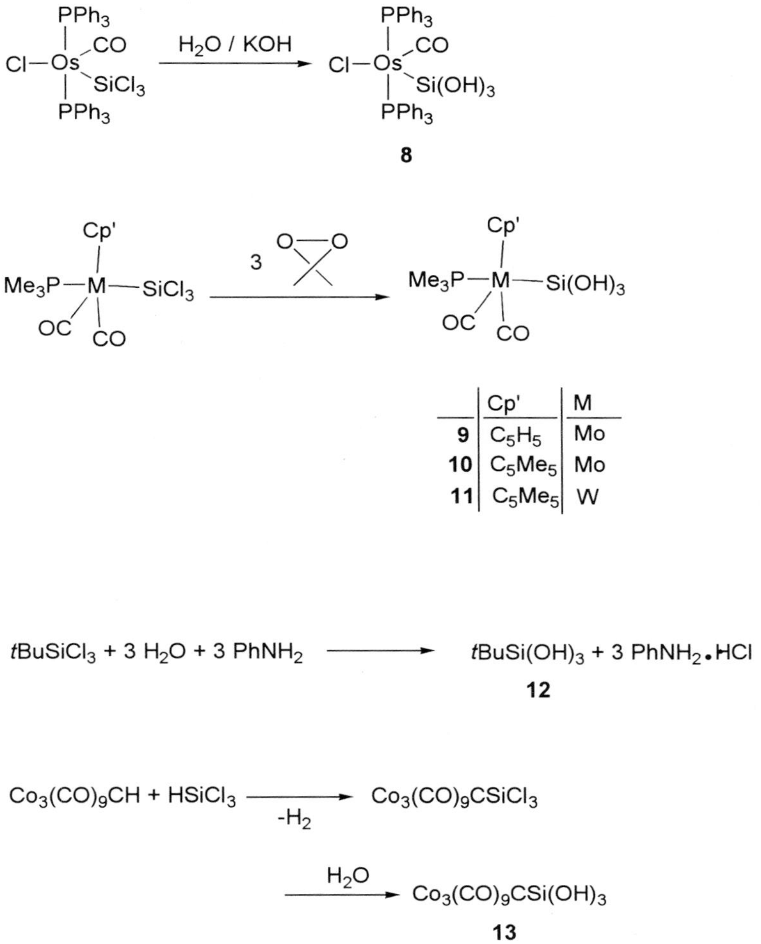

	Cp'	M
9	C_5H_5	Mo
10	C_5Me_5	Mo
11	C_5Me_5	W

$$tBuSiCl_3 + 3\ H_2O + 3\ PhNH_2 \longrightarrow tBuSi(OH)_3 + 3\ PhNH_2 \cdot HCl$$

12

$$Co_3(CO)_9CH + HSiCl_3 \xrightarrow{-H_2} Co_3(CO)_9CSiCl_3$$

$$\xrightarrow{H_2O} Co_3(CO)_9CSi(OH)_3$$

13

Scheme 3.

Our interest in this area has been in synthesizing a variety of silanetriols where silicon is bonded to carbon, nitrogen, or oxygen in order to assess their relative stabilities as well as the differences in their reactivities. The C-bonded silanetriol $tBuSi(OH)_3$ (**12**) was synthesized from the commercially available $tBuSiCl_3$ (Scheme 3) [23]. The bulky t-butyl group afforded the desired steric protection needed to avoid the self-condensation. However, it should be mentioned that the hydrolysis of $tBuSiCl_3$ in the presence of KOH leads to the isolation of the primary condensation product of the t-butylsilanetriol, $[tBuSi(OH)_2]_2O$ [2].

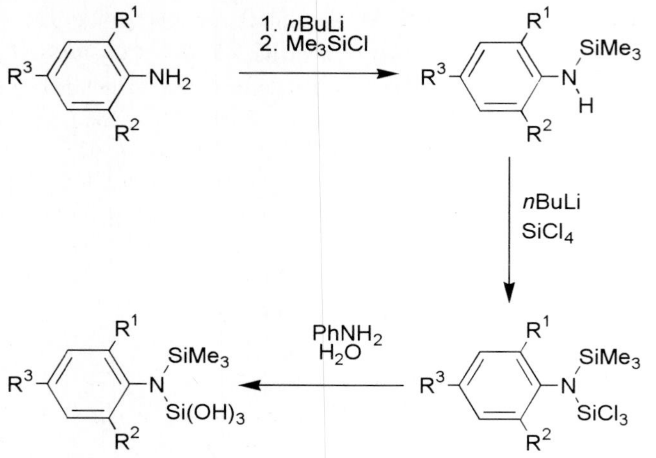

	R^1	R^2	R^3
14	Me	Me	H
15	*i*Pr	*i*Pr	H
16	Me	Me	Me
17	Me	*i*Pr	H

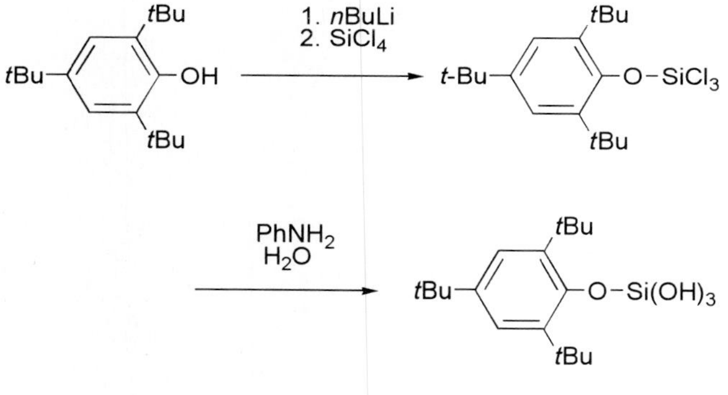

18

Scheme 4.

In order to obtain a catalytically useful silanetriol, we have adapted Seyferth's procedure [24] to prepare the cobalt carbonyl cluster anchored silanetriol $Co_3(CO)_9CSi(OH)_3$ (**13**) (Scheme 3) [25].

Subsequently, we reasoned that the silanetriols with a somewhat weak Si–N or Si–O linkage would have hydrolyzable functionalities and therefore would prove useful in building supramolecules based on metallasiloxanes. For the synthesis of stable nitrogen based silanetriols, sterically hindered primary aromatic amines proved to be attractive targets. We chose *ortho* disubstituted anilines as starting materials and substituted one of the amino hydrogen atoms by a $SiMe_3$ group to impart desired solubility and stability properties of the resulting silanetriols. The detailed synthetic methodology used in preparing the arylaminosilanetriols $RN(SiMe_3)Si(OH)_3$ (**14-17**) is shown in Scheme 4 [26, 27]. These silanetriols are perfectly air-stable in the solid state for extended periods of time and are soluble in a wide range of organic solvents including pentane.

We have also prepared the *O*-bonded silanetriol $(2,4,6\text{-}tBu_3C_6H_2)OSi(OH)_3$ (**18**) starting from 2,4,6-tri-*t*-butylphenol (Scheme 4) [27]. However, the synthesis of other *O*-bonded silanetriols proved to be difficult due to the ready cleavage of the Si–O(aryl) bond.

Spectroscopic Properties and Structures of Silanetriols

There are two characteristic –OH absorptions observed for most of our silanetriols in the region between 3400 and 3700 cm^{-1} (Fig. 1) [28]. The strong and broad absorption appearing around 3360 cm^{-1} is attributable to the OH groups which are involved in hydrogen bonding. Another somewhat less intense, but sharp absorption appearing at around 3600 cm^{-1} arises due to the free hydroxy groups present in these compounds.

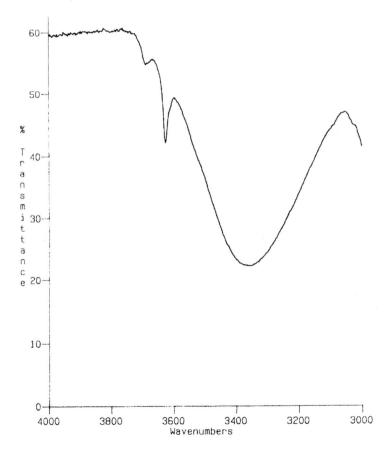

Fig. 1. Infrared spectrum of the silanetriol $(2,4,6\text{-}Me_3C_6H_2)N(SiMe_3)Si(OH)_3$ (**16**) as a CCl_4 solution in the –OH region (1.8×10^{-4} M).

The relative intensity of these two absorptions reveals the extent of the hydrogen bonding present in these molecules. This observation is also consistent with the O–H···O hydrogen bonding interactions found in the X-ray crystal structure of the silanetriol $(2,4,6-Me_3C_6H_2)N(SiMe_3)Si(OH)_3$ (16). Moreover, the observed spectral pattern of the (arylamino)silanetriols does not vary significantly on changing the medium of the spectral measurement. For example, the relative intensity of these two absorptions in the OH region of the silanetriol $(2,4,6-Me_3C_6H_2)N(SiMe_3)Si(OH)_3$ obtained as a KBr pellet, Nujol mull, or as a carbon tetrachloride solution (different concentrations) remains almost the same (Fig. 1). In the case of the metal-based silanetriol $OsCl(CO)(PPh_3)_2Si(OH)_3$ (8), only one vibration is observed at 3616 cm^{-1} indicating the possibility of no O–H···O interactions in this compound. The single crystal X-ray determination of this compound also reveals that this compound is devoid of any intra- or inter-molecular hydrogen bonds [20].

The ^{29}Si NMR chemical shifts for the silanetriols are summarized in Table 1 [28]. In general, the resonances for the Si atoms having three hydroxy groups are high-field shifted. Moreover, there are significant variations observed for this resonance depending upon the nature and the magnetic shielding effect of the fourth substituent on silicon (e.g., Si, C, N, or O). For example, the $\delta(SiO_3)$ for C-bonded silanetriols $tBuSi(OH)_3$ and $Co_3(CO)_9CSi(OH)_3$ (13) are -36.8 and -56.0 ppm, respectively. The corresponding resonance for the N-bonded silanetriols is shifted further upfield and appears around -65 ppm.

Table 1.   Infrared and ^{29}Si NMR Spectroscopic Data for Silanetriols.

No.	Compound	$\nu_{(OH)}$ [cm^{-1}]	δ(Si) [ppm]	Ref.
1	$PhSi(OH)_3$	3150	–	[10]
3	$(cC_6H_{11})Si(OH)_3$	3160	–	[13]
4	$(tBuO)_3SiOSi(OH)_3$	3430	–	[16]
5	$(Me_3Si)_3SiSi(OH)_3$	3402	-10.3	[19]
6	$(Me_3Si)_3CSi(OH)_3$	3425	–	[17]
8	$OsCl(CO)(PPh_3)_2Si(OH)_3$	3616	–	[20]
9	$Mo(C_5H_5)(CO)_2(PMe_3)Si(OH)_3$	3490	–	[21]
10	$Mo(C_5Me_5)(CO)_2(PMe_3)Si(OH)_3$	3645	–	[21]
11	$W(C_5Me_5)(CO)_2(PMe_3)Si(OH)_3$	3620	–	[21]
12	$tBuSi(OH)_3$	3100	-36.8[b]	[23]
13	$Co_3(CO)_9CSi(OH)_3$	3640, 3390	-56.0[a]	[24, 25]
14	$(2,6-Me_2C_6H_3)N(SiMe_3)Si(OH)_3$	3628, 3339	-66.2[a]	[26]
15	$(2,6-iPr_2C_6H_3)N(SiMe_3)Si(OH)_3$	3574, 3344	-67.3[b]	[26]
16	$(2,4,6-Me_3C_6H_2)N(SiMe_3)Si(OH)_3$	3592, 3400	-65.8[a]	[26]
17	$(2-iPr-6-MeC_6H_3)N(SiMe_3)Si(OH)_3$	3590, 3361	-65.3[a]	[26]
18	$(2,4,6-tBu_3C_6H_2)OSi(OH)_3$	3500	-83.7[b]	[27]

[a] in CDCl$_3$ – [b] d$_6$-DMSO.

In the case of the *O*-bonded silanetriol $(2,4,6-t\text{Bu}_3\text{C}_6\text{H}_2)\text{OSi(OH)}_3$ (**18**) containing a SiO_4 tetrahedron, this resonance is observed at -83.7 ppm. For the only Si-bonded silanetriol $(\text{Me}_3\text{Si})_3\text{SiSi(OH)}_3$ (**5**) this resonance appears at -25.8 ppm. Thus, the high-field shifts for the silanetriols increase in the following order : $\text{Si–SiO}_3 < \text{C–SiO}_3 < \text{N–SiO}_3 < \text{O–SiO}_3$.

Solid State Structures

By virtue of the presence of three OH groups, the silanetriols in the solid state tend to associate themselves through an extensive network of intra- and intermolecular hydrogen bonds resulting in interesting supramolecular motifs. The association through hydrogen bonds can be considered as one of the factors for their stability especially in the solid state. Although, so far only a very few X-ray crystal structures of silanetriols have been determined, the H-bonded network in these compounds shows considerable diversity. On the basis of their H-bond network, the known X-ray structures for silanetriols can be classified into several types [8]:

(1) *A double-sheet structure* in which the molecules arrange themselves in a head-to-head and tail-to-tail fashion. The alkyl groups and the OH groups form alternating hydrophobic and hydrophilic double sheets respectively. This type of arrangement is observed in $t\text{BuSi(OH)}_3$ [23] and $(c\text{-C}_6\text{H}_{11})\text{Si(OH)}_3$ [14].

(2) *Hexameric cage structures* are found for sterically more hindered silanetriols $(\text{Me}_3\text{Si})_3\text{CSi(OH)}_3$ [19] and $(\text{Me}_3\text{Si})_3\text{SiSi(OH)}_3$ [19]. These structures are formed from six extensively hydrogen bonded triol molecules. The polyhedral cage as defined by this hexamer comprises two equilateral triangles each of which is surrounded by three six-membered rings in a boat conformation and six five-membered rings in an envelope conformation. There are no hydrogen bonds between these discrete cages.

(3) In a related sterically hindered silanetriol, $(\text{Me}_2\text{PhSi})_3\text{CSi(OH)}_3$ [2], a *tetrameric structure* is observed.

(4) An *octameric cage structure* is found for the cobalt carbonyl cluster anchored silanetriol, $\text{Co}_3(\text{CO})_9\text{CSi(OH)}_3$ [25].

(5) The N-bonded silanetriol $(2,4,6-\text{Me}_3\text{C}_6\text{H}_2)\text{N(SiMe}_3)\text{Si(OH)}_3$ [26] organizes itself in a tubular form which is made up of four linear columns. Further, these columns are displaced with respect to each other by a $90°$ rotational relationship. As a result of this arrangement there is an interesting formation of *silanetriol tubes* in the crystal which contain a *hydrophilic interior* and a *hydrophobic exterior*.

(6) In a recently structurally characterized silanetriol, $\text{OsCl(CO)(PPh}_3)_2\text{Si(OH)}_3$, where the silicon atom is bonded to a metal atom (Os), no intermolecular O–H···O interaction is observed [20].

In this context, it should be mentioned that there are reports on the theoretical studies of the model silanetriol HSi(OH)_3. This molecule has a near *gauche* arrangement around all the three Si–O bonds, indicating a possible role of an anomeric effect in stabilizing this geometry [29].

Reactions of Silanetriols

While the chemistry of other types of silanols has been explored in somewhat detail [30], the reactivity studies of silanetriols, until recently, have mainly been limited to a few esterification reactions [31]. The recent objective in this area is to see whether the three OH groups on silicon can be reacted in concert with suitable metal and metalloid precursors to result in three-dimensional polyhedral cages containing a high metal to silicon ratio.

Among the silanetriols discussed in the preceeding section, only the (arylamino)silanetriols **14-16** have been extensively used for this purpose and the following sub-sections mainly describe the reactions of these three silanetriols with various metal precursor complexes.

Group 4 Derivatives

Reactions of simple silanediols and disilanols with titanium orthoesters, titanium halides and titanium amides proceed to give cyclic titanasiloxanes [30]. On the other hand, the silanetriols with three functional OH groups would prove appropriate synthons for constructing three-dimensional titanasiloxanes which would in turn serve as model compounds for catalytically useful Ti-doped zeolites [32]. The synthesis of cubic titanasiloxanes has been achieved in two ways.

The titanasiloxanes $[RSiO_3Ti(OR')]_4$ (**19-24**) shown in Scheme 5 are more readily accessible via a facile and efficient one-step synthesis involving titanium orthoesters and silanetriols [33]. In this reaction the driving force is the elimination of the corresponding alcohol, which results in the subsequent assembly of the three dimensional Si–O–Ti frameworks.

Since the $SnMe_3$ group is known to be an excellent leaving group, particularly in the reactions with metal halides, the tris-$OSnMe_3$ derivative $tBuSi(OSnMe_3)_3$ has been synthesized from the silanetriol $tBuSi(OH)_3$ (**12**). This trimethylstannoxy derivative reacts smoothly with $(\eta^5-C_5H_4Me)TiCl_3$ eliminating three equivalents of Me_3SnCl to yield the polyhedral titanasiloxane $[RSiO_3Ti(\eta^5-C_5H_4Me)]_4$ (**25**) (Scheme 5) [27].

Interestingly, the direct reaction of the (amino)silanetriol **15** with Cp^*TiCl_3 (Cp^* = C_5Me_5) in the presence of NEt_3 does not lead to the formation of the expected cubic titanasiloxane, and instead forms the novel Ti-siloxane cluster $[R_6Si_6O_{16}(OH)_6Ti_4 \cdot 2THF]$ (**26**) containing a Ti_2O_2 four-membered ring (Scheme 5) [34]. There are two types of Ti centers and two types of silanetriol moieties in this molecule. The Cp^* group of the starting material Cp^*TiCl_3 is lost during the course of this reaction.

In order to elucidate the reactivity of silanetriols with metal substrates containing only *two* reactive centers, the reactions of silanetriol **15** with Cp_2TiCl_2 and Cp_2ZrCl_2 (Cp = C_5H_5) have been carried out [35]. These reactions leave unreacted OH groups on silicon. While the reaction with Cp_2TiCl_2 affords the acyclic derivative $RSi(OH)_2OTiCp_2Cl$ (**27**) containing two free hydroxyl groups on the silanetriol, the eight-membered metallasiloxane $[RSi(OH)(O)(OZrCp_2)]_2$ (**28**) was obtained when the reaction was carried out between **15** and Cp_2ZrCl_2.

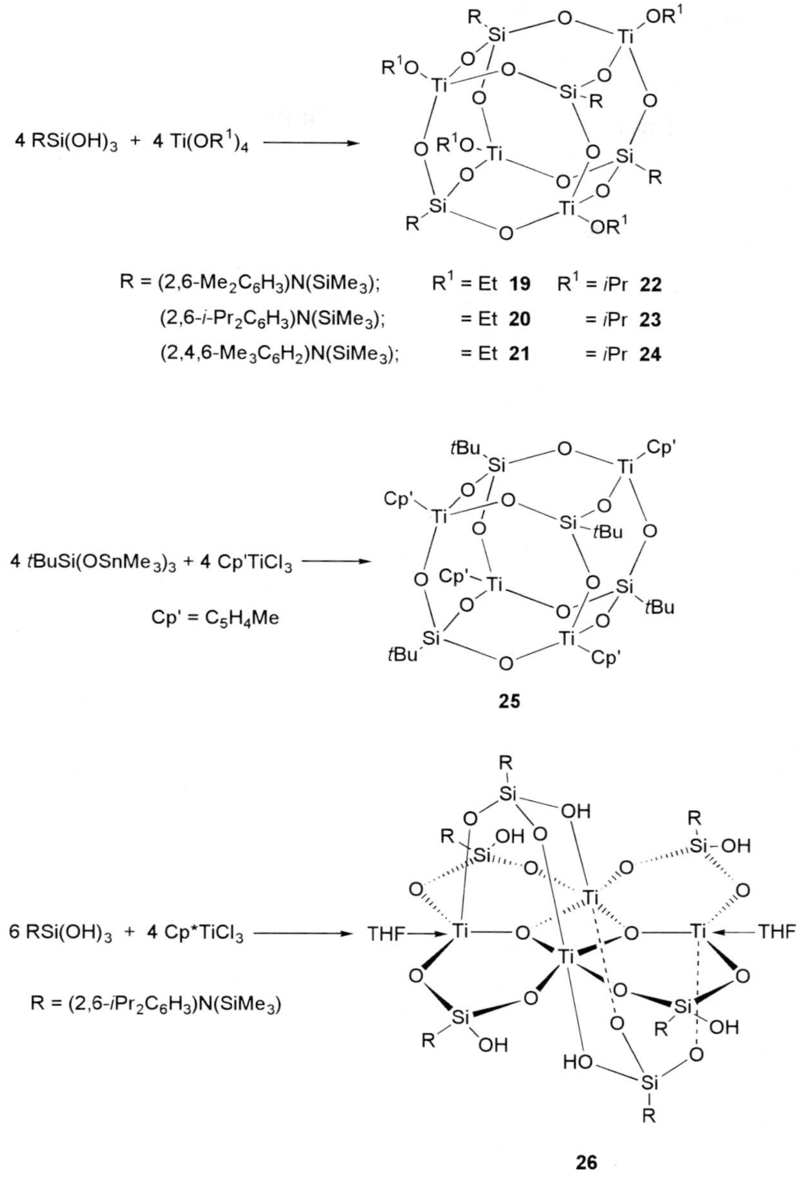

4 RSi(OH)$_3$ + 4 Ti(OR^1)$_4$ ⟶

R = (2,6-Me$_2$C$_6$H$_3$)N(SiMe$_3$);   R^1 = Et **19**   R^1 = *i*Pr **22**

(2,6-*i*-Pr$_2$C$_6$H$_3$)N(SiMe$_3$);   = Et **20**   = *i*Pr **23**

(2,4,6-Me$_3$C$_6$H$_2$)N(SiMe$_3$);   = Et **21**   = *i*Pr **24**

4 *t*BuSi(OSnMe$_3$)$_3$ + 4 Cp'TiCl$_3$ ⟶

Cp' = C$_5$H$_4$Me

25

6 RSi(OH)$_3$ + 4 Cp*TiCl$_3$ ⟶

R = (2,6-*i*Pr$_2$C$_6$H$_3$)N(SiMe$_3$)

26

Scheme 5.

Group 7 Derivatives

Unlike the reactions of silanediols and disilanols towards metal oxides, the silanetriol *t*BuSi(OH)$_3$ reacts with Re$_2$O$_7$ to yield the eight-membered siloxane [*t*BuSi(O)(OReO$_3$)]$_4$ (**29**) (Scheme 6) [23]. This compound contains a Si$_4$O$_4$ siloxane ring on which four ReO$_4$ fragments are anchored. All the four ReO$_4$ groups in **29** are attached to the same side of the siloxane ring making this compound a suitable model for silica supported metal oxide catalytic systems.

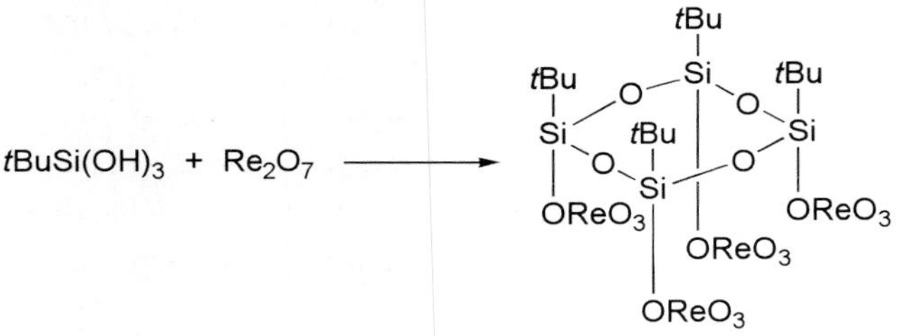

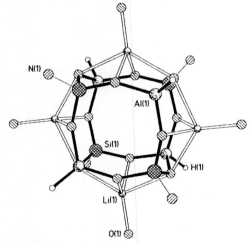

$$tBuSi(OH)_3 \; + \; Re_2O_7 \longrightarrow$$

29

Scheme 6.

Group 13 Derivatives

One of the most important use of the silanetriols is their ability to act as synthons to generate soluble analogues of naturally occurring aluminosilicates [36]. A number of reactions of silanetriols with aluminum precursors such as AlMe$_3$, Al-iBu$_2$H, LiAlH$_4$, and NaAlEt$_2$H$_2$ have been studied. The reactions of (amino)silanetriols **14-16** with Al-iBu$_2$H or AlMe$_3$ in a 1:1 stoichiometry at the reflux temperatures of hexane proceed via elimination of i-butane and hydrogen gas or methane to afford in quantitative yields the aluminosiloxanes [RSiO$_3$Al·dioxane]$_4$ (**30-32**) (Scheme 7) [37, 38]. The reaction of the cobalt carbonyl cluster anchored silanetriol **13** with AlMe$_3$ under similar reaction conditions yields the cubic aluminosiloxane [Co$_3$(CO)$_9$CSiO$_3$Al·THF]$_4$ (**33**) [39]. These compounds contain an Al/Si ratio of 1 and represent the first successful synthesis of soluble aluminosiloxanes having Al$_4$Si$_4$O$_{12}$ cage frameworks. The same structural unit is found in the smallest building blocks of the zeolite A. Moreover, the aluminosiloxane **33** has been found to be a useful hydroformylation catalyst (*vide infra*).

Fig. 2. Core structure of the cubic framework showing the side-on coordination of Li ions in [(2,6-iPr$_2$C$_6$H$_3$)N–(SiMe$_3$)SiO$_3$AlH]$_4$[Li•THF]$_4$ (**34**).

On the other hand, the naturally occurring aluminosilicates contain anionic aluminum centers [36]. In order to generate soluble derivatives containing such anionic Al centers, the reactions of LiAlH$_4$ or NaAlEt$_2$H$_2$ with silanetriol **15** have proved to be useful. These reactions lead to the isolation of anionic aluminosilicates [RSiO$_3$AlH]$_4$[Li·THF]$_4$ (**34**) or [RSiO$_3$AlEt]$_4$[Na·THF]$_4$ (**35**), respectively, in high yields (Scheme 7) [37]. The tetra-anionic Al/O/Si cubic core in these molecules is surrounded by four Li or Na cations. The cations are coordinated by the endocyclic oxygen atoms in a crown ether type coordination (Fig. 2). It should be noted that a large number of natural and synthetic zeolites contain alkali metal counterions in addition to the anionic aluminosilicate framework. The above compounds probably are the simplest model compounds for naturally occurring aluminosilicates.

$$4\ RSi(OH)_3\ +\ 4\ Al\text{-}iBu_2H \xrightarrow[65\ ^\circ C]{\substack{1,4\text{-dioxane}\\ \text{or thf,}\\ \text{hexane}}}$$

$$\text{or}$$

$$4\ AlMe_3$$

13 - 16

S = 1,4-dioxane or thf

R = (2,6-Me$_2$C$_6$H$_3$)N(SiMe$_3$)   **30**

(2,6-*i*Pr$_2$C$_6$H$_3$)N(SiMe$_3$)   **31**

(2,4,6-Me$_3$C$_6$H$_2$)N(SiMe$_3$)   **32**

Co$_3$(CO)$_9$C   **33**

$$4\ RSi(OH)_3\ +\ 4\ M[AlR_2{}'R_2{}''] \xrightarrow[65\ ^\circ C]{\substack{\text{thf}\\ \text{hexane}}} 4\ [M \cdot thf]$$

15

R = (2,6-*i*Pr$_2$C$_6$H$_3$)N(SiMe$_3$)

	R'	R"	M
34	H	H	Li
35	Et	H	Na

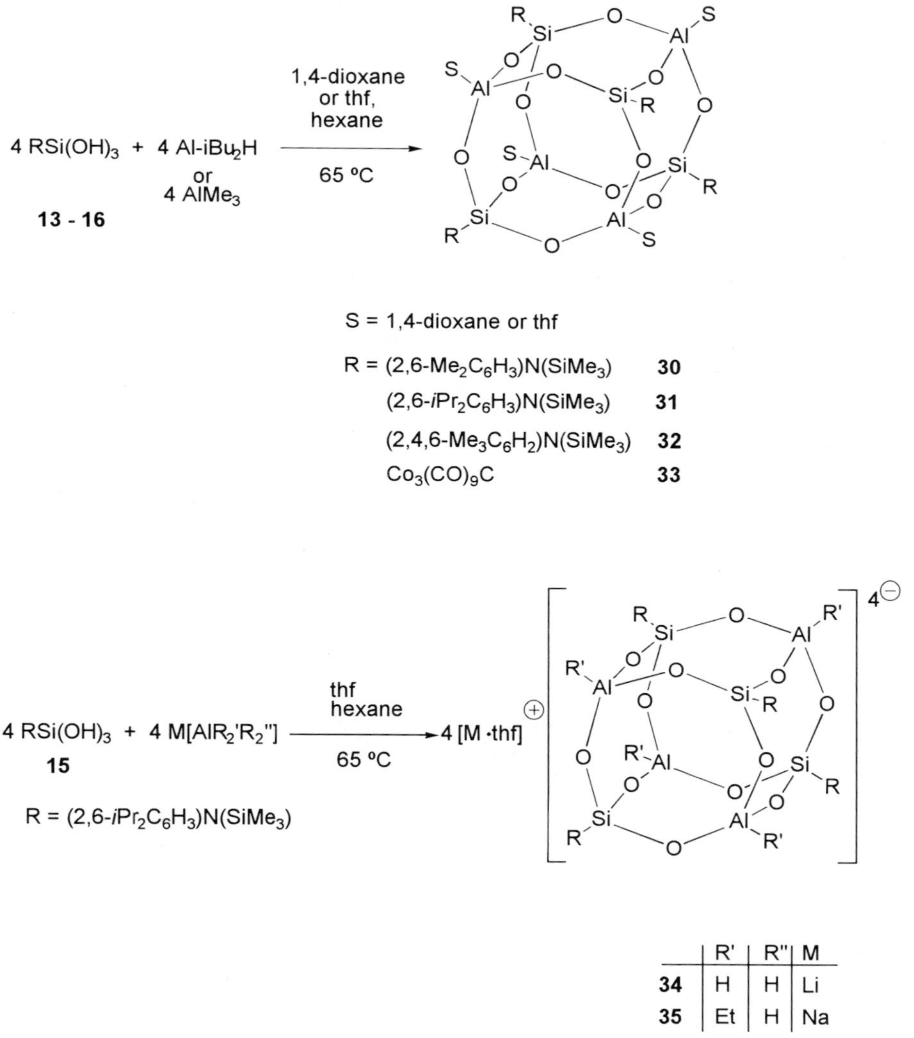

Scheme 7.

Modification of the reaction conditions in terms of temperature, stoichiometry and steric control on silanetriols leads to other interesting aluminosiloxanes [40]. Thus for example, the reaction between silanetriol 15 and Al-iBu$_2$H in a 1:1 molar ratio at -78 °C leads to the isolation of the eight-membered Si$_2$Al$_2$O$_4$ ring system 36 with one unreacted hydroxy group on each silicon (Scheme 8) [40]. In the mineral gismondine, (CaAl$_2$Si$_2$O$_8$(H$_2$O)$_4$)$_n$, similar eight-membered Si$_2$Al$_2$O$_4$ rings are known to be present [36]. The polyhedral aluminosiloxane 37 with a drum shaped core is isolated as the only product when the reaction is carried out in a 1:2 molar ratio of 15 and Al–iBu$_2$H (Scheme 8).

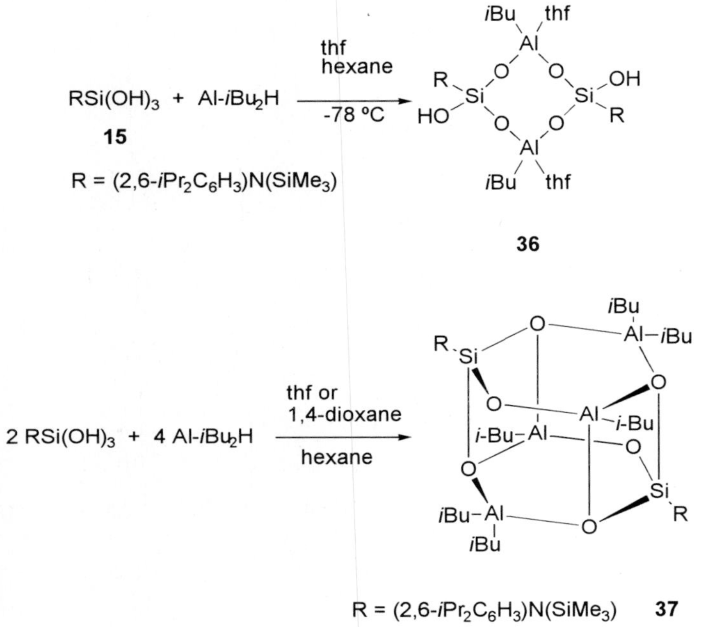

Scheme 8.

The products of the reactions of silanetriols with gallium [41] and indium alkyls [42] are very similar to those obtained from aluminum alkyls described above. The interest in Ga containing siloxanes stems from the known catalytic activity of Ga-doped zeolites in the dehydrogenation reactions of alkanes. The reactions of silanetriols 15 and 16 with GaMe$_3$ or InMe$_3$ in refluxing hexane/1,4-dioxane lead to the cubic Ga/In siloxanes [RSiO$_3$M·THF]$_4$ (M = Ga 38, 39; In 40, 41), respectively. In the resulting products, the Ga and In centers are coordinated to a dioxane solvent molecule. These compounds have a very similar structure to that of aluminosiloxanes 30-33 shown in Scheme 7.

The anionic cubic gallium and indium containing siloxanes have been prepared starting from LiMMe$_4$ (M = Ga or In) [42]. Thus, when the reactions of 15 and 16 are carried out with LiMMe$_4$ in refluxing hexane-THF in 1:1 reactant ratio, the anionic cages [RSiO$_3$MMe]$_4$[Li·THF]$_4$ (M = Ga 42, 43; In 44, 45) are obtained. The molecular structures of these Ga and In siloxanes have been determined by X-ray diffraction studies. Here the metal centers retain one of the methyl groups and the whole cage compound is a tetravalent anion. The charge is counterbalanced by four solvated lithium cations which are also coordinated by four oxygen atoms of the cubic framework.

The reactions of **15** and **16** with GaMe₃ or InMe₃ in 1:2 ratio at room temperature lead to the isolation of first examples of polyhedral gallium and indium containing siloxanes [RSiO(OMMe₂)(OMMe)]₂ (M = Ga **46**, **47**, In **48**, **49**) with M₄Si₂O₆ frameworks [41, 42]. The molecular structures of **46**, **47**, and **48** have been determined by single crystal X-ray diffraction studies (Fig. 3). The core structure of these compounds is similar to the drum shaped aluminosiloxanes described above. The crystal structures reported for these compounds represent the first structure determinations of soluble compounds containing a Si–O–M (M = Ga or In) linkage.

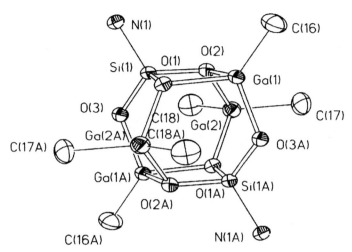

Fig. 3.   Core structure of [(2,6-*i*Pr₂C₆H₃)N(SiMe₃)SiO(OGaMe₂)(OGaMe)]₂ (**47**) showing the drum polyhedron.

Group 14 Derivatives

The silanetriols **14-16** react with trimethylsilyl chloride in a 1:1 ratio to yield the products RSi(OH)₂(OSiMe₃) (**50-52**) respectively [43]. These compounds show monosubstitution of the trisilanol by a trimethylsilyl group. The reaction of silanetriol **15** with GeMe₃Cl in a 1:2 ratio yields the acyclic siloxane RSi(OH)(OGeMe₃)₂ (**53**). The reactions of the silanetriols **14-16** with trimethyltin chloride in 1:3 ratio give the acyclic stannasiloxanes RSi(OSnMe₃)₃ (**54-56**) [44].

The reactions of silanetriol **15** and **16** with R₂SnCl₂ (R = Me, Ph) have been studied. The reaction of dimethyl and diphenyltindichloride proceeds with all the OH groups of silanetriol reacting and yielding the bicyclic compounds [RSi(OSnR₂O)₃SiR] (**57-60**) [44]. The molecular structure of these compounds have been deduced by X-ray diffraction (Fig. 4). These compounds represent a rare class of bicyclic compounds obtained from the reaction of silanetriols. Formation of these products does not appear to depend on the stoichiometry of the reactants. While these compounds are formed in extremely good yields in a 2:3 ratio by reaction of silanetriol and R₂SnCl₂, they are also formed in other stoichiometric ratios, albeit in varying yields.

The reactions between silanetriols and many trichlorometal derivatives normally lead to the isolation of unidentifiable products. On the contrary, the reaction between silanetriols **14** and **15** with phenyltintrichloride in the presence of NEt₃ is neat and yields the cubic stannasiloxanes [RSiO₃SnPh]₄ (**61**, **62**) [44].

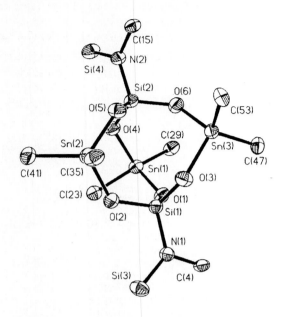

Fig. 4.   Bicyclic core structure of the stannasiloxane [RSi(OSnPh₂O)₃SiR] (R = (2,6-Me₂C₆H₃)(SiMe₃)N) **(59)**.

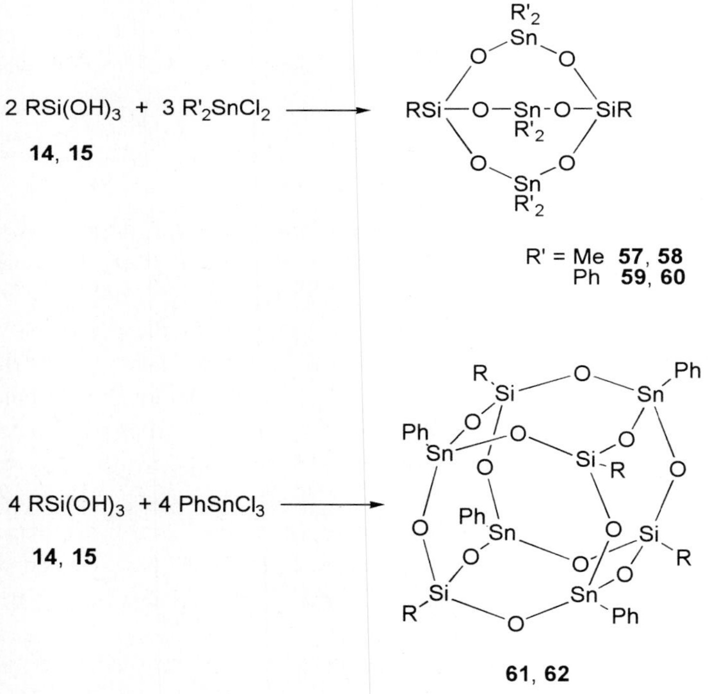

$$2\ RSi(OH)_3\ +\ 3\ R'_2SnCl_2\ \longrightarrow$$

14, 15

R' = Me **57, 58**
Ph **59, 60**

$$4\ RSi(OH)_3\ +\ 4\ PhSnCl_3\ \longrightarrow$$

14, 15

61, 62

Scheme 9.

Properties of Metallasiloxanes

Although the structures of most of the metallasiloxanes described above have been unambigously determined by single crystal X-ray structural studies, ^{29}Si NMR provides a convenient tool in deducing their structures (Table 2). All the resonances of the silicon atoms in the Si–O–M units are shifted upfield with respect to the chemical shifts of the parent silanetriols. As an example, in the titanasiloxanes, an upfield shift of around 32 ppm is observed. Similar shifts are found for other cubic metallasiloxanes. The most significant shift is observed for the anionic cubic aluminosiloxanes, where the SiO$_3$ signal appears at δ = -112.0 ppm. It should be noted that this value is very similar to that of the solid-state NMR chemical shifts for a variety of aluminosilicate gels.

Table 2. Infrared and ^{29}Si NMR Spectroscopic Data for Some Representative Metallasiloxanes.

No.	Compound	ν$_{(SiOM)}$ [cm^{-1}]	δ(Si) [ppm]	Δδ(Si) [ppm]	Ref.
23	[R^{1}SiO$_3$Ti(O-iPr)]$_4$	968	-96.5[c]	29.2	[27]
25	[tBuSiO$_3$Ti(C$_5$H$_4$Me)]$_4$	974	-68.6[b]	31.8	[17]
27	R^{1}Si(OH)$_2$(OTiCp$_2$Cl)	973	-71.8[b]	4.5	[34]
28	[R^{1}Si(OH)O(OZrCp$_2$)]$_2$	968	-71. 8[e]	4.5	[34]
31	[R^{1}SiO$_3$Al•dioxane]$_4$	–	-80.3[b]	13.0	[29]
33	[R^{2}SiO$_3$Al•THF]$_4$	1018	-73.6[b]	17.6	[15b]
35	[R^{1}SiO$_3$AlEt]$_4$[Na•THF]$_4$	–	-112.0[d]	44.7	[28]
36	[R^{1}Si(OH)O(OAl-iBu•THF)]$_2$	963	-77.0[e]	9.7	[30]
37	[R^{1}SiO(OAl-iBu$_2$)(OAl-iBu)]$_2$	1053	-65.3[b]	-2.0	[30]
38	[R^{1}SiO$_3$Ga•dioxane]$_4$	1052	-77.9[b]	10.6	[31]
40	[R^{1}SiO$_3$In•dioxane]$_4$	1046	-77.9[c]	10.6	[32]
46	[R^{1}SiO(OGaMe$_2$)(OGaMe)]$_2$	1025	-65.4[b]	-1.9	[31]
48	[R^{1}SiO(OInMe$_2$)(OInMe)]$_2$	1014	-65.9[c]	-1.4	[32]
51	R^{1}Si(OH)$_2$(OSiMe$_3$)	974	-72.9[e]	5.6	[33]
55	R^{1}Si(OSnMe$_3$)$_3$	965	-71.8[c]	4.5	[34]
59	[R^{1}Si(OSnPh$_2$O)$_3$SiR^1]	960	-67.9[e]	0.6	[34]

R^1 = (2,6-iPr$_2$C$_6$H$_3$)N(SiMe$_3$), R^2 = Co$_3$(CO)$_9$C; [a] $\Delta\delta_{Si}$ = δ_{Si}(metallasiloxane) – δ_{Si}(free silanetriol) – [b] in C$_6$D$_6$ – [c] in CDCl$_3$ – [d] in d$_8$-THF – [e] in d$_8$-DMSO.

In all the cubic metallasiloxanes, a M$_4$Si$_4$O$_{12}$ polyhedron is present. The alternate corners of the cube in these compounds are occupied by M (M = Ti, Al, Ga, or In) and Si. The edges of the cube contain the μ^2-bridging O atoms which link the metal and silicon atoms. The average Si–O–Ti angles are larger in comparison with the corresponding Si–O–Al angles. The sides of the cubic framework comprise six M$_2$Si$_2$O$_4$ eight-membered rings which adopt an approximate C_4 crown conformation. The O–Si–O angles in all the compounds remain largely tetrahedral. Both in titana- and aluminosiloxanes, the exocyclic M–O bond lengths are longer compared to the framework M–O bond lengths. This difference is considerable in the case of Al compounds (0.17 Å), owing to the difference in the nature of the interaction between Al and the exocyclic ligand (THF or dioxane). In the case of the cubic anionic aluminosiloxanes, the Li^+ or Na^+ counter ions lie adjacent to the four faces of the cube and are coordinated by the four oxygen atoms of the siloxane framework. The anionic gallium and indium containing cubic siloxanes have a very similar structural framework.

In the drum compounds, the $M_4Si_2O_6$ (M = Al, Ga, or In) polyhedron is made up of two six-membered rings in the top and bottom, and two six- and four-membered rings on the sides. All the six-membered rings are in a boat conformation while the four-membered rings are planar. The M–O distances associated with μ^2-oxygens are considerably shorter than those of μ^3-oxygens. The gallium and indium containing drum compounds also display very similar structural features. The ring aluminosiloxane contains an eight-membered $Al_2Si_2O_4$ unit with bridging oxygen atoms. The ring has a chair conformation with two oxygen atoms lying above the plane formed by the other six atoms (about 0.4 Å).

While several metallasiloxanes have already been synthesized using these silanetriols, the late transition metals and lanthanides containing metallasiloxanes remain virtually unexplored. In general, the reactions of silanetriols with metal alkyls seem to be straightforward leading to quantitative yields of the metallasiloxanes. On the other hand, the reactions of silanetriols or their lithium salts with the metal halides often do not produce desired products.

Applications

Hydroformylation reactions are important from the industrial point of view and the two commonly used hydroformylation catalysts are either Rh or Co based. We thought it would be interesting to anchor a SiO_3 unit on a cobalt cluster via hydrosilylation. This would be a close model to a silica-supported cobalt cluster. Secondly, since the reactions of silanetriols have been demonstrated to afford three-dimensional metallasiloxanes, we anticipated that this silanetriol would react with substrates such as trialkylaluminums, affording cobalt carbonyl cluster anchored aluminosiloxanes. Such compounds would resemble a modified zeolite having on its surface catalytically active cobalt carbonyl moieties and might inspire the preparation of actual zeolite systems with these modifications.

We reacted the silanetriol $Co_3(CO)_9CSi(OH)_3$ (**13**) with MMe_3 (M = Al, Ga, or In) in a 1:1 ratio and obtained the cubic cobalt carbonyl cluster containing group 13 heterosiloxane $[Co_3(CO)_9CSiO_3M]_4$ (M = Al, Ga, or In) [39]. The gross structure of the aluminosiloxane is similar to that of the other cubic aluminosiloxanes described *vide supra*. The interesting aspect of this soluble heterosiloxane is the presence of four $Co_3(CO)_9C$ cluster units on each silicon of the cubic framework. In order to test the catalytic viability of this model compound and as well as the starting silanetriol **13** we used them in typical hydroformylation reactions involving 1-hexene. In the case of **13** and the corresponding aluminosiloxane, the main product is heptanal, showing a regioselectivity of over 60%. In view of this encouraging result, from what are essentially first generation aluminosiloxane-based catalysts, one is inclined to believe that this synthetic path can prove appropriate for the design of a new generation of hydroformylation catalysts by suitable modifications.

Acknowledgement: This work was supported by the *Deutsche Forschungsgemeinschaft* and the *Hoechst AG.*

References:

[1]   a) Y. I. Yermakov, B. N. Kuznetsov, V. A. Zakharov, *Catalysis by Supported Complexes*, Elsevier, Amsterdam, **1981**.
      b) T. Seiyama, K. Tanabe, *New Horizons in Catalysis*, Elsevier, Amsterdam, **1980**.

[2]   P. D. Lickiss, *Adv. Inorg. Chem.*, **1995**, *42*, 147.

[3]   M. G. Voronkov, E. A. Maletina, V. K Roman, *Heterosiloxanes*, Soviet Scientific Review Supplement, Series Chemistry, Vol. 1 (Ed.: M. E. Vol'pin, transl.: K. Gingold), Harwood, Academic, London, **1988**.

[4]   M. Shakir, H. W. Roesky, *Phosphorus, Sulfur, and Silicon* **1994**, *93/94*, 13.

[5]   I. Abrahams, M. Motevalli, S. A. A. Shah, A. C. Sullivan, *J. Organomet. Chem.* **1995**, *492*, 99; and references cited therein.

[6]   Y. T. Struchkov, S. V. Lindeman, *J. Organomet. Chem.* **1995**, *488*, 9.

[7]   F. J. Feher, T. A. Budzichowski, *Polyhedron*, **1995**, *14*, 3239; and references cited therein.

[8]   R. Murugavel, V. Chandrasekhar, H. W. Roesky, *Acc. Chem. Res.* **1996**, *29*, 183.

[9]   a) E. G. Rochow, *Silicon and Silicones*, Springer, Berlin, **1987**.
      b) W. Noll, *Chemie und Technologie der Silikone*, VCH, Weinheim, **1968**.
      c) B. J. Aylett, *Organometallic Compounds, Vol. 1, Part 2*, 4th ed., Chapman and Hall, London, **1979**, p. 1.

[10] L. J. Tyler, *J. Am. Chem. Soc.* **1955**, *77*, 770.

[11] a) K. A. Andrianov, A. A. Zhdanov, *Zh. Obsh. Khim.* **1957**, *27*, 156.
      b) I. I. Lapkin, T. N. Povarnitsina, *Zh. Obsh. Khim.* **1964**, *34*, 1202.

[12] Z. Michalska, Z. Lasocki, *Bull. Acad. Polon. Sci., Ser. Sci. Chim.* **1971**, *19*, 757.

[13] H. Ishida, J. L. Koenig, *J. Poly. Sci. Polym. Phys. Ed.* **1979**, *17*, 1807.

[14] H. Ishida, J. L. Koenig, K. C. Gardner, *J. Chem. Phys.* **1982**, *77*, 5748.

[15] T. Takiguchi, *J. Am. Chem. Soc.* **1959**, *81*, 2359.

[16] Y. Abe, I. Kijima, *Bull. Chem. Soc. Jpn.* **1969**, *42*, 118.

[17] R. I. Damja, C. Eaborn, *J. Organomet. Chem.* **1985**, *290*, 275.

[18] N. H. Buttrus, R. I. Damja, C. Eaborn, P. B. Hitchcock, P. D. Lickiss, *J. Chem. Soc., Chem. Commun.* **1985**, 267.

[19] S. S. Al-Juaid, N. H. Buttrus, R. I. Damja, Y. Derouiche, C. Eaborn, P. B. Hitchcock, P. D. Lickiss, *J. Organomet. Chem.* **1989**, *371*, 287.

[20] C. E. F. Rickard, W. R. Roper, D. M. Salter, L. J. Wright, *J. Am. Chem. Soc.* **1992**, *114*, 9682.

[21] W. Malisch, R. Lankat, S. Schmitzer, J. Reising, *Inorg. Chem.* **1995**, *34*, 5701.

[22] W. Malisch, S. Schmitzer, G. Kaupp, K. Hindahl, H. Käb, U. Wachtler, in: *Organosilicon Chemistry: From Molecules to Materials* (Eds.: N. Auner, J. Weis), VCH, Weinheim, **1994**, p. 185.

[23] N. Winkhofer, H. W. Roesky, M. Noltemeyer, W. T. Robinson, *Angew. Chem.* **1992**, *104*, 670; *Angew. Chem., Int. Ed. Engl.* **1992**, *31*, 599.

[24] a) D. Seyferth, C. N. Rudie, M. O. Nestle, *J. Organomet. Chem.* **1979**, *178*, 227.
b) D. Seyferth, C. L. Nivert, *J. Am. Chem. Soc.* **1977**, *99*, 2359.

[25] U. Ritter, N. Winkhofer, H.-G. Schmidt, H. W. Roesky, *Angew. Chem.* **1996**, *108*, 591; *Angew. Chem., Int. Ed. Engl.* **1996**, *35*, 524.

[26] R. Murugavel, V. Chandrasekhar, A. Voigt, H. W. Roesky, H.-G. Schmidt, M. Noltemeyer, *Organometallics.* **1995**, *14*, 5298.

[27] N. Winkhofer, A. Voigt, H. Dorn, H. W. Roesky, A. Steiner, D. Stalke, A. Reller, *Angew. Chem.* **1994**, *106*, 1414; *Angew. Chem., Int. Ed. Engl.* **1994**, *33*, 1352.

[28] A. Voigt, R. Murugavel, U. Ritter, H. W. Roesky, *J. Organomet. Chem.* **1996**, *521*, 279.

[29] a) A. E. Reed, C. Schade, P. v. R. Schleyer, P. V. Kamath, J. Chandrasekhar, *J. Chem. Soc., Chem. Commun.* **1988**, 67.
b) Y. Apeloig, in. *The Chemistry of Organic Silicon Compounds, Part 1*; (Eds.: S. Patai, Z. Rappoport), Wiley, Chichester, **1989**, p. 79.

[30] R. Murugavel, A. Voigt, M. G. Walawalkar, H. W. Roesky, *Chem. Rev.* **1996**, *96*, 2205.

[31] S. S. Al-Juaid, C. Eaborn, P. B. Hitchcock, P. D. Lickiss, *J. Organomet. Chem.* **1992**, *423*, 5.

[32] (a) A. Bhaumik, R. Kumar, *J. Chem. Soc., Chem. Commun.* **1995**, 869.
(b) R. Joseph, T. Ravindranathan, A. Sudalai, *Tetrahedron Lett.* **1995**, *36*, 1903; and references cited therein.

[33] A. Voigt, R. Murugavel, V. Chandrasekhar, N. Winkhofer, H. W. Roesky, H.-G. Schmidt, I. Usón, *Organometallics* **1996**, *15*, 1610.

[34] A. Voigt, R. Murugavel, M. L. Montero, A. Wessel,F.-Q. Liu, H. W. Roesky, I. Usón, T. Albers, E. Parisini, *Angew. Chem.* **1997**, *109*, 1020; *Angew. Chem. Int. Ed. Engl.* **1997**, *36*, 1001.

[35] A. Voigt, R. Murugavel, H. W. Roesky, H.-G. Schmidt *J. Mol. Structure*, in press.

[36] F. Liebau, *Structural Chemistry of Silicates*, Springer, Berlin, **1985**, p 244.

[37] M. L. Montero, A. Voigt, I. Usón, H. W. Roesky, *Angew. Chem.* **1995**, *107*, 2761; *Angew. Chem., Int. Ed. Engl.* **1995**, *34*, 2504.

[38] V. Chandrasekhar, R. Murugavel, A. Voigt, H. W. Roesky, H.-G. Schmidt, M. Noltemeyer, *Organometallics* **1995**, *15*, 918.

[39] U. Ritter, N. Winkhofer, R. Murugavel, A. Voigt, D. Stalke, H. W. Roesky, *J. Am. Chem. Soc.* **1996**, *118*, 8580.

[40] M. L. Montero, I. Usón, H. W. Roesky, *Angew. Chem.* **1994**, *106*, 2198; *Angew. Chem., Int. Ed. Engl.* **1994**, *33*, 2103.

[41] A. Voigt, R. Murugavel, E. Parisini, H. W. Roesky, *Angew. Chem.* **1996**, *108*, 823; *Angew. Chem. Int. Ed. Engl.* **1996**, *35*, 748.

[42] A. Voigt, M. G. Walawalkar, R. Murugavel, H. W. Roesky, E. Parisini, T. Lubini, *Angew. Chem.*, in press.

[43] R. Murugavel, A. Voigt, V. Chandrasekhar, H. W. Roesky, H.-G. Schmidt, M. Noltemeyer, *Chem. Ber.* **1996**, *129*, 391.

[44] R. Murugavel, A. Voigt, H. W. Roesky, *Organometallics* **1996**, *15*, 5097.

Silsesquioxanes as Crown Ether Analogs

Uwe Dittmar, Heinrich C. Marsmann*, Eckhard Rikowski
Fachbereich Chemie und Chemietechnik
Universität-GH Paderborn
Warburger Straße 100, D-33095 Paderborn, Germany
Tel.: Int. code + (5251)602571

Keywords: Silsesquioxanes / Alkali metal ion complexes / Extraction constants

Summary: The siloxane skeletons of oligomeric silsesquioxanes consist of simple cage structures formed by concatenated tri-, tetra-, and pentameric siloxane rings. They are in some respects molecular models of the zeolites. The lone pairs of the oxygen atoms of the rings can act as donors to alkali ions, as they do in crown ethers. To test the capabilities of the silsesquioxanes in this direction, the distribution equilibria of alkali picrates between an aqueous solution and silsesquioxanes dissolved in CH_2Cl_2 were studied by UV/VIS spectroscopy. The extraction constant is a function of the size of the siloxane rings and the radii of the alkali ions. In the trimethyl siloxy derivates, the exocyclic oxygen atoms are also involved in the bonding of the alkali ions.

Introduction

Silsesquioxane have the general formula $(RSiO_{3/2})_n$ with R = H, halogen, organic residue, etc. The siloxane skeletons of the oligomeric species form cages and isomers are possible [1, 2]. A schematic representation of the silsesquioxane skeletons studied here is found in Fig.1. With the exception of $(RSiO_{3/2})_6$, the siloxane skeleton is formed by concatenated tetra- and pentameric siloxane rings.

Results

The lone pairs of the oxygen atoms of siloxanes are capable of donating to alkali cations. This is true for cyclic poly(dimethylsiloxanes) [3-5] and for silicic acid esters where a relation between the ring size and the radius of the alkali cation determines the stability of the complexes [6].

To test the ability of the silsesquioxanes to function as ligands to alkali cations, a 10^{-4} m aqueous solution of picric acid with 2×10^{-4} mol alkali hydroxide was mixed with a 10^{-4} m solution of the silsesquioxane in CH_2Cl_2 according to [7]. The concentration of the picric acid anions was determined by measuring the extinction of the absorption band at 370 nm and using the Beer-Lambert law.

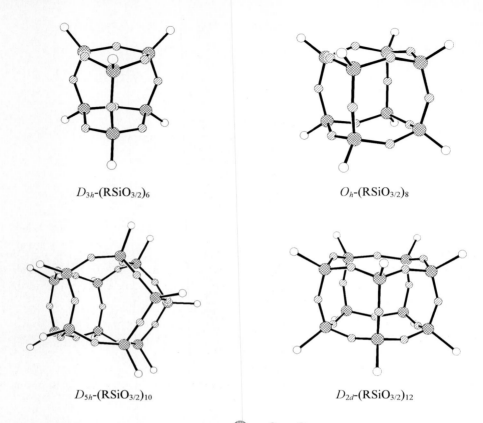

$D_{3h}\text{-}(\text{RSiO}_{3/2})_6$

$O_h\text{-}(\text{RSiO}_{3/2})_8$

$D_{5h}\text{-}(\text{RSiO}_{3/2})_{10}$

$D_{2d}\text{-}(\text{RSiO}_{3/2})_{12}$

Fig. 1.   Siloxane skeletons of simple silsesquioxanes; ⬤ Si, ◐ O, ○ R ;
Abbreviations: N = $(\text{CH}_3)_3\text{Si}(\text{CH}_2)_3$, M = $\text{R}_3\text{SiO}_{1/2}$, D = $\text{R}_2\text{SiO}_{2/2}$, T = $\text{RSiO}_{3/2}$, Q = $\text{SiO}_{4/2}$, Hex = nHexyl-.

The extraction constants can be calculated using the following assumptions:

a) Because of the hydrophobic nature of the silsesquioxanes their concentration in the aqueous phase is negligible.

b) The alkali picrates do not dissolve in the organic phase without complexion.

c) There is no multiple complexion.

Then the equilibrium

$$M^+_{\text{H}_2\text{O}} + A^-_{\text{H}_2\text{O}} + L_{\text{CH}_2\text{Cl}_2} \rightleftharpoons (M^+LA^-)_{\text{CH}_2\text{Cl}_2}$$

M^+ = Alkali metal ion, A^- = pricrate anion, L = silsesquioxane

exists and the extraction constant K_E is given by the following equation.

$$K_E = \frac{[M^+LA^-]_{CH_2Cl_2}}{[M^+]_{H_2O} \cdot [A^-]_{H_2O} \cdot [L]_{CH_2Cl_2}}$$

The values so obtained are collected in Table 1.

Table 1. Extraction constants K_E [L^2/mol^2].

Silsesquioxane	Li	Na	K	Rb	Cs
Hex_8T_8	451	2286	–	–	–
T_8N_8	328	–	–	–	–
Q_8M_8	1121	531	–	2655	4316

No measurable transfer occurred with the silsesquioxanes Q_6M_6, $Q_{10}M_{10}$ and $Q_{12}M_{12}$ (Q = $SiO_{3/2}$, M = $(CH_3)_3SiO–$). So it seems that the cubic silsesquioxanes only are capable of bonding alkali ions, although the other silsesquioxanes also contain tetrameric rings. Even then the degree of transfer of alkali picrates is considerably lower than in the cases of cyclic siloxanes [3-5] and of cyclic silicic acid esters [6]. To explain this effect, one has to consider the structure of silsesquioxanes in more detail. X-Ray structures exist for hexa- [8], octa- [9-17], deca- [18], and dodecasilsesquioxanes [19]. Assuming that the structure of the silsesquioxane skeleton does not depend on the side chain in a larger extent, the positions of the oxygen and the silicon atoms in the rings building the silsesquioxanes are as depicted in Fig. 2. Here a covalent radius of 1.18 Å is used for the oxygen and 0.73 Å for the silicon [20].

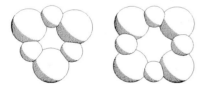

Typical pore diameter of the six membered ring:
1.14 Å for the Si atoms
1.67 Å for the O atoms

Fig. 2a. Sketches of the siloxane rings constituting the silsesquioxanes; Surface structure units of Q_6M_6.

Typical pore diameter of the eight membered ring:
2.00 Å for the Si atoms
2.38 Å for the O atoms
Distance of the exocyclic O atoms: 3.2-3.4 Å

Fig. 2b. Surface structure units of Q_8M_8 and R_8T_8 (the figure for the exocyclic Oatoms is only for Q_8M_8).

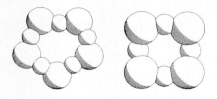

Typical pore diameter of the ten membered ring:
2.91 Å for the Si atoms
2.96 Å for the O atoms

Fig. 2c.   Surface structure units of $Q_{10}M_{10}$ and $D_{2d}-Q_{12}M_{12}$

Inspection of Fig. 2 reveals that – except for the octasilsesquioxanes – the oxygen atoms are also above and below the plane of the ring given by the four silicon atoms. Therefore, not all oxygen atoms can be in contact with an alkali ion. Thus, the octasilsesquioxanes are the only ones to have enough stability to afford a transport into the organic phase. The degree of transport of the different ions in the organic phase depends on their size. The sodium ion ($d = 2.32$ Å) fits exactly in the middle of the ring formed by the four oxygen atoms in the octasilsesquioxanes. The larger Rb^+ ($d = 3.50$ Å) and Cs^+ ($d = 3.76$ Å) ions can be fixed only with the help of the exocyclic oxygen atoms of the $-OSi(CH_3)_3$ side groups. No extraction happens with the silsesquioxanes Hex_8T_8 and T_8N_8 lacking such oxygen atoms. The potassium ion ($d = 3.04$ Å) is too large to be complexed by the ring oxygen atoms alone and too small to be kept in place with the additional help of exocyclic oxygen atoms. It is not transported in any case. The small lithium ion ($d = 1.8$ Å) is probably found inside the octasilsesquioxane cage.

References:

[1]   M. G. Voronkov, V. I. Lavrent'yev, *Top. Curr. Chem.* **1982**, *102*, 199.

[2]   H. Bürgy, G. Calzaferri, D. Herren, A. Zhadanov, *Chimia* **1991**, *45*, 3.

[3]   C. J. Oliff, P. Ladbrook, *Bioelectrochem. Bioenergetics* **1979**, *6*, 105.

[4]   C. J. Oliff, G. R. Pickering, K. J. Rutt, *J. Inorg. Nucl. Chem.* **1980**, *42*, 288.

[5]   C. J. Oliff, G. R. Pickering, K. J. Rutt, *J. Inorg. Nucl. Chem.* **1980**, *42*, 1201.

[6]   H. C. Marsmann, M. Seifert, *Z. Naturforsch.* **1991**, *46b*, 693.

[7]   H. K. Frensdorf, *J. Am. Chem. Soc.* **1971**, *93*, 4684.

[8]   Yu. I. Smolin, Yu. F. Shepelev, A. S. Ershov, D. Hoebbel, W. Wieker, *Kristallografiya* **1984**, *29*, 712.

[9]   K. Larsson, *Ark. Kemi* **1960**, *16*, 203.

[10]   V. E. Shklover, Yu. T. Struchov, N. N. Makarova, K. A. Andrianov, *Zh Strukt. Khim.* **1977**, *19*, 1107.

[11]   I. A. Baidina, N. V. Podberezskaya, V. I. Alekseev, T. N. Martynova, S. V. Borisov, A. N. Kanev, *Zh. Strukt. Khim.* **1979**, *20*, 648.

[12]   N. V. Podberezskaya, I. A. Baidina, V. I. Alekseev, S. V. Borisov, T. N. Martynova, *Zh. Strukt. Khim.* **1981**, *22*, 116.

[13]   N. V. Podberezskaya, S. A. Magarill, I. A. Baidina, S. V. Borisov, L. E. Gorsh, A. N. Kanev, T. N. Martynova, *Zh. Strukt. Khim.* **1982**, *23*, 120.

[14]   Yu. I. Smolin, Yu. F. Shepelev, R. Pomes, *Khimiya Silikatov i Oksidov* **1982**, 68.

[15]   V. W. Day, W. G. Klemperer, V. V. Mainz, D. M. Millar, *J. Am. Chem. Soc.* **1985**, *107*, 8262.

[16]   F. J. Feher, T. A. Budzichowski, *J. Organomet. Chem.* **1989**, *373*, 153.

[17] T. B. E. Auf der Heyde, H.-B. Bürgi, H. Bürgy, K. W. Törnroos, *Chimia* **1991**, *45*, 38.

[18] A. Baidina, N. V.Podberezskaya, S. V.Borisov, V. I.Alekseev, T. N.Martynova, A. N.Kanev, *Zh. Strukt. Khim.***1980**, *21*, 125.

[19] W. Clegg, G. M. Sheldrick, N. Vater, *Acta Cryst.* **1980**, *B36*, 3162.

[20] J. E. Huheey, *Anorganische Chemie*, Walter de Gruyter, Berlin, **1987**, p. 78, p. 278.

Azomethine-Substituted Organotrialkoxysilanes and Polysiloxanes

Frank Mucha, Gerhard Roewer*
Institut für Anorganische Chemie
Technische Universität Bergakademie Freiberg
Leipziger Straße 29, D-09596 Freiberg, Germany
Tel.: Int. code + (3731)393194 – Fax.: Int. code + (3731)394058

Keywords: Silicon / Azomethine Compounds

Summary: During the last few years intense research has been focussed on the elaboration of synthetic materials [1-4] based on modified silicon esters. Products with interesting properties were formed by synthesis of a novel group of derived silicon esters. The condensation of carbonyl- and amino groups leads to organotrialkoxysilanes containing azomethine bonds. From these compounds three-dimensional, cross-linked silicon polymers were synthesized by hydrolysis/condensation.

The monomeric precursors were formed according to Eq. 1. by the condensation of 3-aminopropyltrialkoxysilanes and carbonyl-substituted 2-hydroxybenzaldehydes.

Eq. 1.   Structure with numbered carbon atoms for correspondence to Table 1.

The water formed was captured in situ using zeolithes to prevent any hydrolytic cleavage of Si–OR"-bonds. The products are viscous, moisture-sensitive liquids, which cannot be distilled without decomposition even in vacuo. Table 1 contains a selection of analytical data.

Table 1. NMR data and GC/MS molpeaks of new azomethine compounds (Numeration of carbon atoms according to Eq. 1.)

No	R'	R''	^{29}Si NMR [ppm]	^{13}C NMR [ppm]					^{1}H NMR PhOH [ppm]	GC/MS Molpeak
				C1	C7	C8	C9	C10		
I	H	CH$_3$	-42.1	161.5	165.2	61.7	24.4	7.2	13.5	283
II	CH$_3$	CH$_3$	-42.1	164.0	171.5	51.5	23.9	7.5	16.1	297
III	C$_6$H$_5$	CH$_3$	-41.9	163.9	174.0	50.5	24.2	8.3	15.9	dec.
IV	H	C$_2$H$_5$	-45.9	161.5	164.9	61.9	24.6	8.0	13.5	325
V	CH$_3$	C$_2$H$_5$	-45.7	164.1	171.4	51.7	24.2	8.2	16.2	339
VI	C$_6$H$_5$	C$_2$H$_5$	-45.9	163.5	173.9	53.9	24.5	8.2	15.5	401

By a thermal treatment of the monomer for some hours in vacuo the compounds undergo an intermolecular elimination of the alcohol R''OH. Instead of the alkoxy groups the phenolic oxygen atoms take part in the formation of Ph–O–Si-bridged oligomers (n = 2-4) (Fig. 2.). They are dark red, highly viscous liquids or solids, readily soluble in common organic solvents. It is possible to complete the cleavage of alkoxy groups by addition of water to the solution. The resulting products coincide with the cross-linked polymers, which are described below.

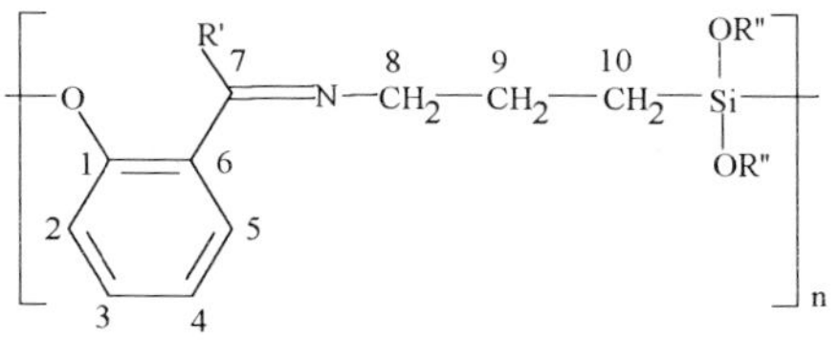

Fig. 2. General structure of the oligomers (R' = H, CH$_3$, C$_6$H$_5$; R'' = CH$_3$, C$_2$H$_5$).

In Table 2 some spectral data of the oligomer formed from monomer **IV** in Table 1 (R' = H, R'' = C$_2$H$_5$) are given for an example.

Table 2. Characteristic NMR and IR data of an oligomer

^{29}Si NMR [ppm]		^{1}H NMR [ppm] PhOH	characteristic IR bands [cm^{-1}]		
terminal unit	connecting link		azomethine terminal	azomethinec onnecting	Si–C
-45,2	-44,7	missing	1.629	1.602	763

By adding water to the solution in polar organic solvents the alkoxysilylazomethines were converted to three-dimensional cross-linked polymers. This overall reaction proceeds in two stages: The silicon-alkoxy bonds undergo hydrolysis, giving silanol groups fallowed by condensation to form Si–O–Si-bridged structures. To complete the second stage an elevated temperature is required.

Depending on the degree of hydrolysis various substituted silicon atoms are detectable in the polymers by ^{29}Si NMR. Three of these various chemical environments are exhibited in Fig. 3.

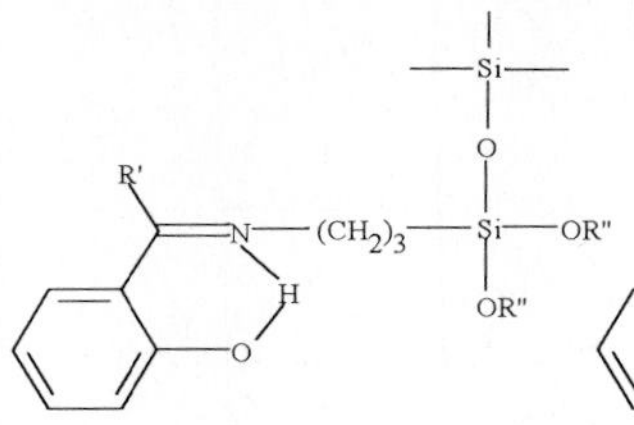

Fig. 3. Various environments of silicon in the hydrolysis/condensation polymers (R' = H, CH$_3$, C$_6$H$_5$; R'' = CH$_3$, C$_2$H$_5$).

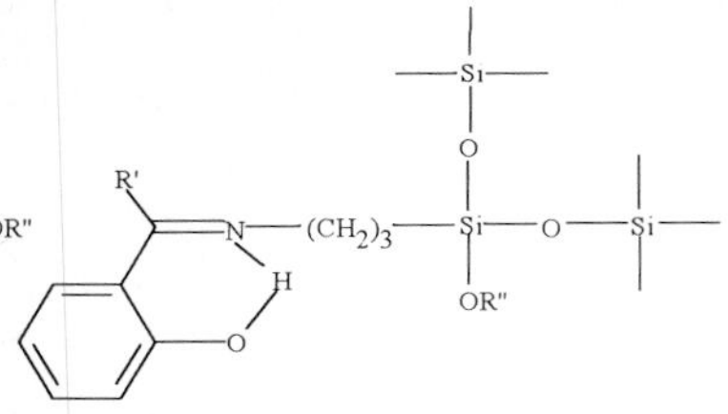

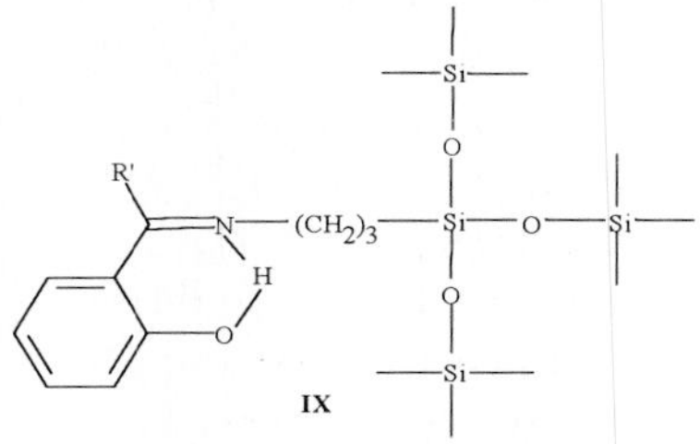

Table 3 contains a selection of significant analytical data for two polymers resulting from the monomers **IV** and **V** in Table 1.

Table 3. Spectral data of two polymers; numeration of carbon atoms analogous to Eq. 1.

R'	R''	^{29}Si NMR [ppm]			^{13}C NMR [ppm]			IR bands [cm^{-1}]		
		VII	VIII	IX	C1	C7	C10	C=N	Si–O–Si	Si–O–C
H	C$_2$H$_5$	-54.3	-9.1	-65.4	161.4	164.9	9.2	1.633	1.030	1.120
CH$_3$	C$_2$H$_5$	-61.7	-68.2	-71	164.8	171.5	11.2	1.615	1.028	1.127

Because of their properties, such as hardness, low brittleness, and high resistance to chemical attack some of the synthesized polymers are potentially useful for practical applications.

References:

[1] *Sol-Gel Processing and Applications* (Ed.: Y. A. Attia), Plenum Press, New York, London, **1994**.

[2] C. J. Brinker, G. W. Scherer, *Sol-Gel Science – The Physics and Chemistry of Sol-Gel Processing*, Academic Press, London, San Diego, **1990**.

[3] *Sol-Gel Optics: Processing and Applications* (Ed.: L. C. Klein), Kluwer Academic Publisher, Boston, **1994**.

[4] E. P. Plueddemann, USP 3819675 **1974**, Dow Corning.

On the Reaction of (tBu$_2$SnO)$_3$ with Organochlorosilanes. Simple Formation of [(tBu$_2$SnO)$_2$(tBu$_2$SiO)]

Jens Beckmann, Klaus Jurkschat*

Fachbereich Chemie, Lehrstuhl für Anorganische Chemie II
Universität Dortmund
D-44221 Dortmund, Germany
Tel.: Int. code + (231)7553800 – Fax: Int. code + (231)7553797
E-mail: kjur@platon.chemie.uni-dortmund.de

Dieter Schollmeyer

Institut für Organische Chemie
Universität Mainz
Saarstraße 21, D-55099 Mainz, Germany
Tel.: Int. code + (6131)395320 – Fax: Int. code + (6131)394778
E-mail: scholli@uacdr0.chemie.uni-mainz.de

Keywords: Mixed Diorganotin / Diorganosilicon Oxide / ^{29}Si NMR / ^{119}Sn NMR / X-Ray Analysis

Summary: The reaction of (tBu$_2$SnO)$_3$ with tBu$_2$SiCl$_2$ led to formation of [(tBu$_2$SnO)$_2$(tBu$_2$SiO)] (**3**) and tBu$_2$Sn[OSi(tBu$_2$)O]$_2$SntBu$_2$ (**4**) whereas treatment of (tBu$_2$SnO)$_3$ with tBu$_2$SiF$_2$ afforded tBu$_2$Sn[OSi(F)tBu$_2$]$_2$ (**8**). The compounds were characterized by means of NMR spectroscopy, mass spectrometry, and in case of **4** also by X-ray analysis.

Introduction

Recently we and others have shown that (tBu$_2$SnO)$_3$ [1] acts as a convenient anhydrous source of O^{2-} in reactions with organotin halides allowing the controlled synthesis of new organotin clusters, e. g.,. {[RClSn(CH$_2$)$_3$SnClR]O}$_4$ [2] and {[R(O)Sn]$_2$CMe$_2$}$_2$ [3].

As part of a current study on (tBu$_2$SnO)$_3$ as an anhydrous source of O^{2-} in reactions with organoelement halides, such as BX$_3$, RBX$_2$, RSnX$_3$, RPbX$_3$, R$_2$PbX$_2$, RTiX$_3$, R$_2$TiX$_2$, etc. (X = halogen, R'COO) we report here results on its reaction with organochlorosilanes.

Silicon is known to have a high oxophilicity and we thought that the use of (tBu$_2$SnO)$_3$ in reactions with R$_2$SiCl$_2$ and RSiCl$_3$, respectively, would allow the controlled synthesis of Si–O rings and cages of defined size. However, for small R groups, such as Me and Ph, mixtures of various products were obtained, which have not been separated yet.

In contrast, the reaction of tBu$_2$SiX$_2$ (X = F, Cl) with (tBu$_2$SnO)$_3$ provided some unexpected results, which are shown below.

Results and Discussion

The reaction of $(tBu_2SnO)_3$ (**1**) with tBu_2SiCl_2 (**2**) in boiling toluene afforded a clear solution from which a precipitate **A** was filtered off after cooling to room temperature (Scheme 1).

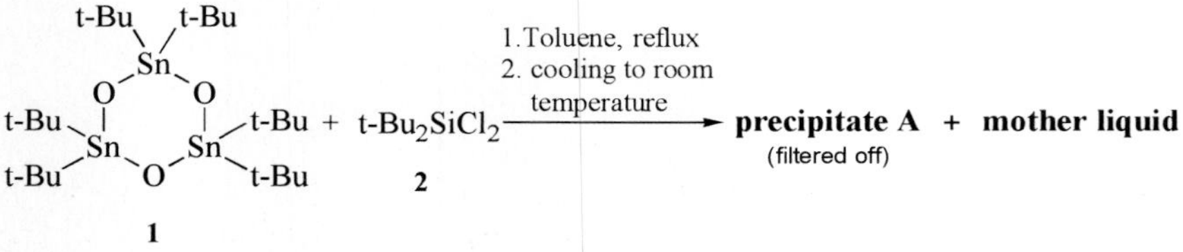

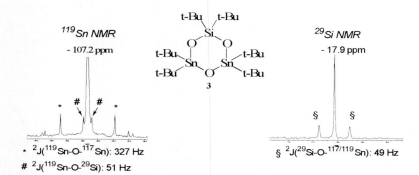

Scheme 1.

The ^{119}Sn NMR spectrum of **A** in $CDCl_3$ displayed three signals of different integral ratio at -107.2 ppm ($\approx$ 92 %), $^2J(^{119}$Sn–O–^{17}Sn) 327 Hz, -178.5 ppm ($\approx$ 6 %) and -82.9 ppm ($\approx$ 2 %), $^2J(^{119}$Sn–O–^{117}Sn) 360 Hz, respectively, with the latter signal unambiguously assigned to **1**.

Recrystallization of precipitate **A** from toluene gave crystalline **3** in about 80 % yield. The ^{119}Sn and ^{29}Si NMR spectra of **3** are shown in Fig. 1.

Fig. 1. ^{119}Sn and ^{29}Si NMR spectra of **3**.

The chemical shifts, the coupling patterns, and the signal to coupling integral ratios confirmed unambiguously the structure of **3**. The same holds for ^{1}H NMR ($CDCl_3$; δ tBuSi 1.16 ppm, δ tBuSn 1.49 ppm, $^3J(^{119}$Sn–^{1}H) 94.7 Hz, integral ratio 1:2), ^{13}C NMR ($CDCl_3$; δ SiC 22.2 ppm, δ Si–C–CH_3 29.2 ppm, δ SnC 39.7 ppm, $^1J(^{119}$Sn–^{13}C) 497 Hz, δ Sn–C–CH_3 30.6 ppm) and mass spectrometry. The ^{119}Sn CP MAS NMR spectrum displayed a singlet at -106 ppm indicating the solid state and solution structures of **3** to be identical.

The signal at -178.5 ppm in the ^{119}Sn NMR spectrum of precipitate **A** was tentatively assigned to the eight-membered ring **4**. In order to verify this assumption **4** was synthesized in high yield from $tBu_2Si(OLi)_2$ [4] and tBu_2SnCl_2 (Eq. 1).

2 t-Bu₂Si(OLi)₂ + 2 t-Bu₂SnCl₂ →[hexane, − LiCl]

$$\text{(structure 4)}$$

4

Eq.1.

The NMR data taken from a CDCl₃ solution of **4** are fully consistent with the eight-membered ring structure (¹H NMR: δ *t*BuSi 1.02 ppm, δ *t*BuSn 1.37 ppm, $^{3}J(^{119}\text{Sn}-^{1}\text{H})$ 98.9 Hz, integral ratio 1:1; ¹³C NMR: δ SiC 22.3 ppm, δ Si–C–CH3 29.6 ppm, δ SnC 39.3 ppm, $^{1}J(^{119}\text{Sn}-^{13}\text{C})$ 540 Hz, δ Sn–C–CH₃ 30.4 ppm; ²⁹Si NMR: -25.7 ppm, $^{2}J(^{29}\text{Si}-\text{O}-^{119}\text{Sn})$ 98 Hz; ¹¹⁹Sn NMR: -178.5 ppm, $^{2}J(^{119}\text{Sn}-\text{O}-^{29}\text{Si})$ 98 Hz). The molecular structure of **4** is shown in Fig. 2. It proves the eight-membered ring found in solution also for the solid state.

The ¹¹⁹Sn CP MAS NMR chemical shift of **4** amounts to -232 ppm, which is 53.5 ppm to lower frequency in comparison with the solution chemical shift. We tentatively attribute this to different conformations in the solid state and in solution. To the best of our knowledge **3** and **4** represent the first completely characterized mixed diorganosilicon/diorganotin oxides.

Both mass spectrometry and differential thermoanalysis show for **3** and **4** complete loss of the *t*-butyl groups, making these compounds potentially interesting as precursors for well-defined mixed silicon-tin oxides.

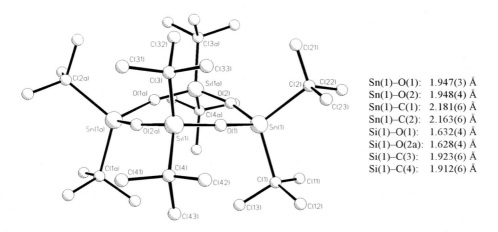

Sn(1)–O(1): 1.947(3) Å
Sn(1)–O(2): 1.948(4) Å
Sn(1)–C(1): 2.181(6) Å
Sn(1)–C(2): 2.163(6) Å
Si(1)–O(1): 1.632(4) Å
Si(1)–O(2a): 1.628(4) Å
Si(1)–C(3): 1.923(6) Å
Si(1)–C(4): 1.912(6) Å

Fig. 2. Molecular structure of **4**.

The signals in the ²⁹Si and ¹¹⁹Sn NMR spectra of the mother liquid **B** are tentatively assigned to species **5** and **7** (Eq. 2). This is especially supported by the intensity of the ¹¹⁷/¹¹⁹Sn satellites in the ²⁹Si spectrum proving unambiguously a 2:1 tin-silicon ratio in **5** and a 1:1 tin-silicon ratio in **7**. The formation of **6** and **7** is the result of hydrolysis of **5** by atmospheric moisture.

Compound **6** it is known from the literature[5] and was isolated and completely characterized.

$$\text{5, } \delta\ ^{119}\text{Sn : -67.9 ppm}$$
$$\delta\ ^{29}\text{Si : -23.5 ppm}$$
$$^2\text{J}(^{119}\text{Sn-O-}^{29}\text{Si) 105 Hz}$$

$$\text{7, } \delta\ ^{119}\text{Sn : -56.8 ppm}$$
$$\delta\ ^{29}\text{Si : -14.7 ppm}$$
$$^2\text{J}(^{119}\text{Sn-O-}^{29}\text{Si) 85 Hz}$$

Eq. 2.

The reaction of **1** with *t*Bu$_2$SiF$_2$ yielded **8** almost quantitatively (Eq. 3). Compound **8** is stable against atmospheric moisture and does not react with excess **1**.

$$\delta\ ^{119}\text{Sn: - 164.7 ppm, } ^2\text{J}(^{119}\text{Sn-O-}^{29}\text{Si) 83 Hz}$$
$$\delta\ ^{29}\text{Si: - 15.0 ppm, } ^1\text{J}(^{29}\text{Si-}^{19}\text{F) 312 Hz}$$
$$^2\text{J}(^{29}\text{Si-O-}^{119}\text{Sn) 84 Hz}$$

Eq. 3.

Further studies on this project are currently underway.

Acknowledgement: We thank the *Deutsche Forschungsgemeinschaft* and the *Fonds der Chemischen Industrie* for financial support.

References:

[1]   H. Puff, W. Schuh, R. Sievers, W. Wald, R. Zimmer, *J. Organomet. Chem.* **1984**, *260*, 271.

[2]   D. Dakternieks, K. Jurkschat, D. Schollmeyer, H. Wu, *Organometallics* **1994**, *13*, 4121.

[3]   B. Zobel, M. Schürmann, K. Jurkschat, in preparation.

[4]   S. Schütte, U. Klingebiel, D. Schmidt-Bäse, *Z. Naturforsch.* **1993**, *48b*, 263.

[5]   H. Puff, H. Hevendehl, K. Höfer, H. Reuter, W. Schuh, *Organomet. Chem.* **1985**, *287*, 163.

Silanols and Siloxanes Substituted with the Chiral Iron Fragments Cp(OC)(RPh$_2$P)Fe (R = Ph, (H)(Me)(Ph)C(Me)N) [1]

Wolfgang Malisch*, Michael Neumayer, Klaus Perneker,
Norbert Gunzelmann, Konrad Roschmann
Institut für Anorganische Chemie
Universität Würzburg
Am Hubland, D-97074 Würzburg, Germany
Tel.: Int. code + (931)8885277 – Fax: Int. code + (931)888-4618
E-mail: anor142@rzbox.uni-wuerzburg.de

Keywords: Ferrio-Silanes / Stereogenic Metal Centers / Diastereoselective Synthesis

Summary: Condensation of the diastereomerically pure ferrio-silanol *(RS,SR)*-Cp(OC)(Ph$_3$P)Fe–Si(Me)(Ph)(OH) (**1**) with the organochlorsilanes ClSiMe$_2$R (R = H (**2a**), Me (**2b**), H$_2$C=CH (**2c**)) leads to the ferrio-disiloxanes *(RS,SR)*-Cp(OC)(Ph$_3$P)Fe–Si(Me)(OSiMe$_2$R)(Ph) (R = H (**3a**), Me (**3b**), CH=CH$_2$ (**3c**)). Irradiation of Cp(OC)$_2$Fe–SiCl$_2$R (R = Me, Mes (**4a,b**)) in the presence of Ph$_3$P yields the phosphane-substituted derivatives Cp(OC)(Ph$_3$P)Fe–SiCl$_2$R (R = Me (**5a**), Mes (**5b**)), which are converted to the ferrio-silanediols Cp(OC)(Ph$_3$P)Fe–Si(OH)$_2$R (R = Me (**6a**), Mes (**6b**)) by Cl/OH-exchange. Analogously the ferrio-silanediols *(R$_{Fe}$S$_C$)*- and *(S$_{Fe}$S$_C$)*-Cp(OC)[(H)(Me)(Ph)C(Me)NPh$_2$P]Fe–Si(OH)$_2$Ph (**8**) are obtained. Condensation of **6a** with R$_2$SiCl$_2$ (R = Me, Ph) generates Cp(OC)(Ph$_3$P)Fe–Si(OSiR$_2$Cl)$_2$Me (R = Me (**9a**), Ph (**9b**)). **8** and the chlorosilanes RMe$_2$SiCl (R = H, H$_2$C=CH) yield the ferrio-trisiloxanes *(R$_{Fe}$S$_C$)*- and *(S$_{Fe}$S$_C$)*-Cp(OC)[(H)(Me)(Ph)C(Me)NPh$_2$P]Fe–Si(OSiMe$_2$H)$_2$Ph (**10**) or the ferrio-disiloxanols *(R$_{Fe}$S$_C$S$_{Si}$)*- and *(R$_{Fe}$S$_C$R$_{Si}$)*-Cp(OC)[(H)(Me)(Ph)C(Me)NPh$_2$P]Fe–Si(OH)(OSiMe$_2$CH=CH$_2$)Ph (**11**), respectively, isolated enantiomerically pure.

Optically active organometallic compounds, especially pseudotetrahedral half-sandwich complexes of the type Cp(OC)(Ph$_3$P)Fe–R with iron as the center of chirality have been extensively investigated in the past. However, synthetic utilization of stereocontrol by the chiral iron has been limited to systems with σ-bonded carbon ligands [2]. In contrast, silicon-iron complexes have not yet found analogous application. In context with our studies concerning metallo-silanols we have established simple routes to isolate diastereomerically pure derivatives with a chiral iron fragment.

Transformation of the ferrio-silanol Cp(OC)(Ph₃P)Fe–Si(Me)(OH)(Ph) (*RS,SR*-1) to the ferrio-disiloxanes Cp(OC)(Ph₃P)Fe–Si(Me)(OSiMe₂R)(Ph) (R = H (*RS,SR*-3a), Me (*RS,SR*-3b), CH=CH₂ (*RS,SR*-3c))

Recently we have demonstrated that the ferrio-silanol (*RS,SR*)-1 is available, starting either from the diastereomerically pure ferrio-chlorosilane (*RR,SS*)-Cp(OC)(Ph₃P)Fe–Si(Me)(Cl)(Ph) or the ferrio-silane (*RS,SR*)-Cp(OC)(Ph₃P)Fe–Si(Me)(H)(Ph). In both cases, the transformation to the ferrio-silanol **1** – Cl/OH-exchange at silicon or insertion of oxygen into the Si–H bond using dimethyldioxirane – occurs stereospecifically [3].

The condensation of metallo-silanols with organochlorosilanes [4] is a common method of preparing metallo-siloxanes. It has now been applied to gain diastereomerically pure ferrio-siloxanes (*RS,SR*)-3a-c by interaction of the ferrio-silanol (*RS,SR*)-1 with the organochlorosilanes 2a-c in ether or benzene at room temperature (Eq. 1).

Eq. 1.   Synthesis of **3a-c**.

Neither iron nor silicon is involved in the condensation. Therefore, the chiral ferrio-siloxanes **3a-c** show the same configuration with respect to the two stereogenic centers as the starting material **1**.

Ferrio-silanediols   Cp(OC)(Ph₃P)Fe–Si(OH)₂R   [R   =   Me,   Mes   (6a,b)]   and Cp(OC)[H(Me)(Ph)C*(Me)N–Ph₂P]Fe–Si(OH)₂Ph (6a,b,$S_{Fe}S_C$-8): Synthesis, Structure and Reaction with Chlorosilanes

For Ph₃P-substituted ferrio-silanedioles of the type Cp(OC)(R'₃P)Fe–Si(OH)₂R the iron represents the center of chirality. Selective reaction of one of the two diastereotopical (OH)-ligands at the silicon should create a new chiral center, indicated by the formation of diastereomers.

Synthesis of the ferrio-silanediols **6a,b** can be performed according to the substitution route described for **1**. It starts with the ferrio-dihalogenosilanes **4a,b**, which, on irradiation in the presence of an excess of Ph₃P, yield the phosphane-substituted ferrio-dichlorosilanes **5a,b** (Eq. 2). Introduction of the OH groups at the silicon of **5** can be attained with KOH in the two-phase system THF/H₂O. (Eq. 2).

A sequence analogous to Eq. 2 can be accomplished by using the chiral phosphane Ph₂P–N(Me)C(H)(Me)(Ph) as a ligand. In this case the ferrio-dichlorosilane **7** is obtained as a (1 : 1)-mixture of the epimers ($S_{Fe}S_C$)-7 and ($R_{Fe}S_C$)-7 shown in scheme 1, which are separated by fractional crystallization. Both epimers have been converted in the described manner to the corresponding ferrio-silanediols ($S_{Fe}S_C$)-8 and ($R_{Fe}S_C$)-8 (Scheme 1).

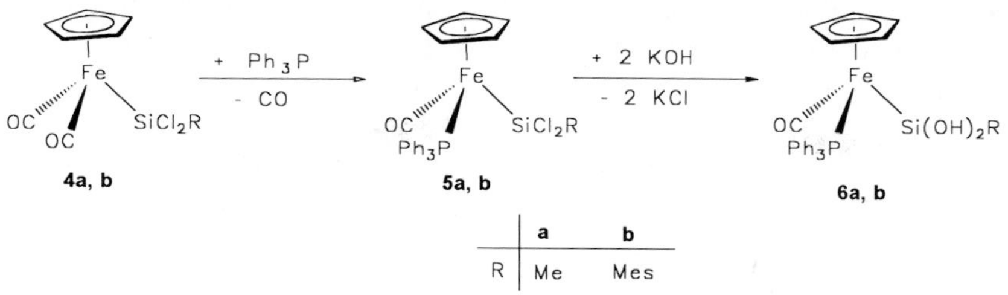

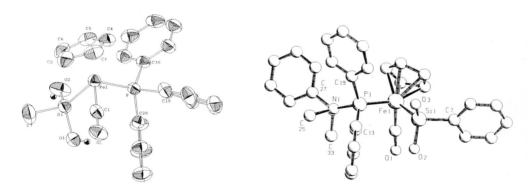

Eq. 2. Synthesis of **5a,b** and **6a,b**.

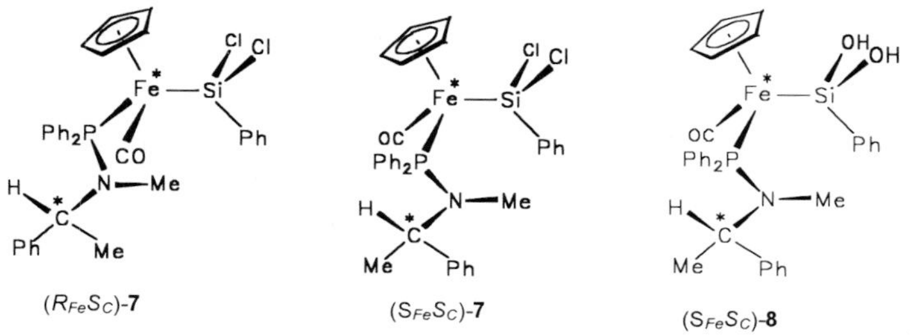

	a	b
R	Me	Mes

Scheme 1. Ferrio-silanes $(R_{Fe}S_C)$-**7**, $(S_{Fe}S_C)$-**7**, and $(S_{Fe}S_C)$-**8**.

Structure analysis of **6a** and $(S_{Fe}S_C)$-**8** (Fig. 1) reveals hydrogen bonding in the first case indicated by the oxygen-oxygen distances of 2.863 Å [5]. It is, however, limited to the formation of chair-like Si_2O_4-dimers in the crystal.

Fig. 1. Molecular structure of $Cp(OC)(Ph_3P)Fe–Si(OH)_2Me$ (**6a**) and $(S_{Fe}S_C)$-$Cp(OC)[(H)(Me)(Ph)C(Me)NPh_2P]Fe–Si(OH)_2Ph$ (**8**).

$(S_{Fe}S_C)$-**8** shows no hydrogen bonding due to the sterically more demanding transition metal group. In context with our studies concerning stereocontrolled reactions of silicon-iron complexes, the selective condensation of one of the Si(OH)-units in **6a** or $(S_{Fe}S_C)$-**8**/$(R_{Fe}S_C)$-**8** leading to a chiral silicon atom is attractive, especially with respect to asymmetric induction by the chiral metal fragment.

The condensation of **6a** with the diorganodichlorosilanes R_2SiCl_2 (R = Me, Ph) directly leads to the ferrio-trisiloxanes **9a,b**, irrespective of the stereochemistry (Scheme 2). Analogously, $(S_{Fe}S_C)$-**8** or $(R_{Fe}S_C)$-**8** reacts with $Me_2(H)SiCl$ yielding the ferrio-trisiloxanes **10**. $(CH_2=CH)Me_2SiCl$, however, affords the two expected diastereomers of the ferrio-disiloxanol **11** (ratio 1.5 : 1) in a yield of 78 % (Scheme 2). The separation is conducted by column chromatography.

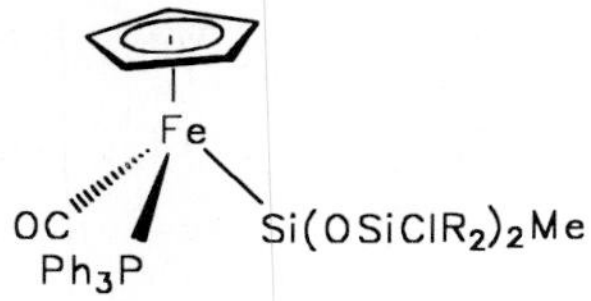

	9	**a**	**b**
	R	Me	Ph

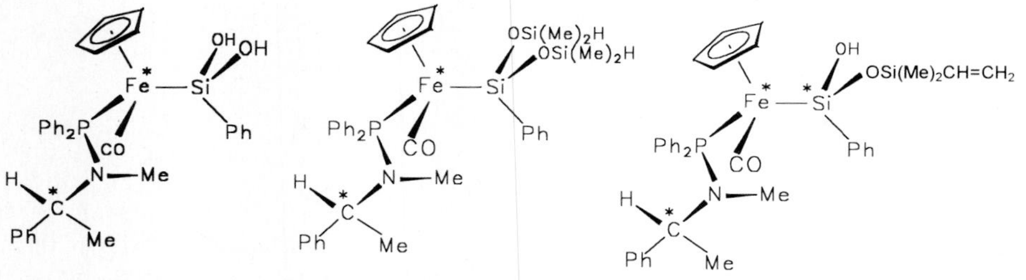

$(R_{Fe}S_C)$-**8** $(R_{Fe}S_C)$-**10** $(R_{Fe}S_C S_{Si})$-**11**

Scheme 2.   Ferrio-silanes **9a,b**, $(R_{Fe}S_C)$-**8**, $(R_{Fe}S_C)$-**10**, and $(R_{Fe}S_C)$-**11**.

$(R_{Fe}S_C S_{Si})$-**11** and $(S_{Fe}S_C S_{Si})$-**11** represent the first ferrio-silanes with three defined stereocenters, which offer interesting perspectives with respect to reactions involving the (OH)Si- or the $(CH_2=CH)Si$-group.

Acknowledgement: This work has been generously supported by the *Deutsche Forschungsgemeinschaft* (SFB 347: "*Selektive Reaktionen Metall-aktivierter Moleküle*"; Schwerpunktprogramm "*Spezifische Phänomene in der Siliciumchemie*").

References:

[1]   Part 15 of the series "*Metallo-silanols and Metallo-siloxanes*". In addition, Part 41 of the series "*Synthesis and Reactivity of Silicon Transition Metal Complexes*". Part 14/40, see W. Malisch, J. Reising, M. Schneider, in: *Organosilicon Chemistry III: From Molecules to Materials* (Eds.: N. Auner, J. Weis), VCH, Weinheim, **1997**, p. 415.

[2]   a) S. G. Davies, J. M. Dordor-Hedgecock, K. H. Sutton, J. C. Walker, C. Bourne, R. H. Jones, K. Prout, *J. Chem. Soc., Chem. Commun.* **1986**, 607.

b) S. G. Davies, *Aldrichimica* **1990**, *23*, 31.

c) S. G. Davies, K. S. Holland, K. H. Sutton, J. P. McNally, *Isr. J. Chem.* **1991**, *31*, 25.

[3]   W. Malisch, M. Neumayer, O. Fey, W. Adam, R. Schumann, *Chem. Ber.* **1995**, *128*, 1257.

[4]   a) S. Möller, O. Fey, W. Malisch, W. Seelbach, *J. Organomet. Chem.* **1996**, *507*, 239.

b) W. Malisch, S. Schmitzer, R. Lankat, M. Neumayer, F. Prechtl, W. Adam, *Chem. Ber.* **1995**, *128*, 1251.

c) R. Lankat, W. Malisch, S. Schmitzer, *Inorg. Chem.* **1995**, *34*, 5701.

[5]   a) A. G. Brook, M. Hesse, K. M. Baines, R. Kumarathasan, A. J. Lough, *Organometallics* **1993**, *11*, 4259.

b) R. West, E. K. Pham, *J. Organomet. Chem.* **1991**, *403*, 43.

c) P. D. Lickiss, *Adv. Inorg. Chem.* **1995**, *42*, 147.

Si–H-Functionalized Ferrio-Trisiloxanes
$C_5R_5(OC)_2Fe–Si(Me)(OSiMe_2H)_2$ (R = H, Me) [1]

Joachim Reising, Wolfgang Malisch*, Reiner Lankat
Institut für Anorganische Chemie
Universität Würzburg
Am Hubland, D-97074 Würzburg, Germany
Tel.: Int. code + (931)8885277 – Fax: Int. code + (931)8884618
E-mail: anor142@rzbox.uni-wuerzburg.de

Keywords: Metallo-Siloxanes / Photoinduced Cyclization / Photoinduced Metallation

Summary: Condensation of the metallo-silanediols $C_5R_5(OC)_2Fe–Si(Me)(OH)_2$ [R = H (**1a**), Me (**1b**)] with $Me_2(H)SiCl$ leads to the ferrio-trisiloxanes $C_5R_5(OC)_2Fe–Si(Me)(OSiMe_2H)_2$ (**2a,b**), which undergo photo-induced rearrangement to the cyclo(ferra)siloxanes $C_5R_5(OC)(H)Fe–SiMe_2O–Si(H)(Me)–OSiMe_2$ (**3a,b**). Conversion to the dinuclear derivative $C_5Me_5(OC)(H)Fe–SiMe_2O–Si(Me)[Fe(CO)_2Cp]–OSiMe_2$ (**4a**) is achieved in the case of **3b** by photolysis in the presence of $Cp(OC)_2Fe–CH_3$. The isomeric cyclo(ferra)siloxane $C_5Me_5(OC)_2Fe–Si(Me)–OSiMe_2–[Fe(H)(CO)Cp]–SiMe_2O$ (**4b**) is obtained by photoreaction of the ferrio-trisiloxane $C_5Me_5(OC)_2Fe–Si(Me)(OSiMe_2H)_2$ (**2b**) with $Cp(OC)_2Fe–CH_3$.

Metal-fragment substituted silanols are known to be stable towards self-condensation due to the strongly reduced acidity of the Si–OH proton [2]. However, these species show ready reaction with organochlorosilanes R_3SiCl, which gives access to metallo-siloxanes [2], constituting attractive models for transition metal complexes anchored on silica surfaces.

Introduction of Si–H functionality at the γ-silicon of these metallo-siloxanes promises a rich area concerning consecutive reactions, especially oxidative addition to coordinatively unsaturated metal centers [3].

This possibility has now been used to synthesize novel mono- and dinuclear cyclo(metallo)siloxanes starting with the ferrio-trisiloxanes **2a,b** generated from the ferrio-silanediols **1a,b** by condensation with dimethylchlorosilane in the presence of triethylamine (Scheme 1, step a). Si–H-functionality of **2a,b** is crucial for the photochemically induced ring closure reaction yielding the six-membered cyclo(ferra)siloxanes **3a,b** after loss of CO (Scheme 1, step b). The rearrangement of **2** to **3**, resulting in the interchange of the position of the α- and γ-siliconatoms, involves consecutive oxidative addition of both Si–H functions at iron after CO elimination. The SiH unit of **3b** created at γ-silicon is susceptible towards photoinduced introduction of a $Cp(OC)_2Fe$-fragment using $Cp(OC)_2Fe–CH_3$ to produce the dinuclear derivative **4** (Scheme 1, step c).

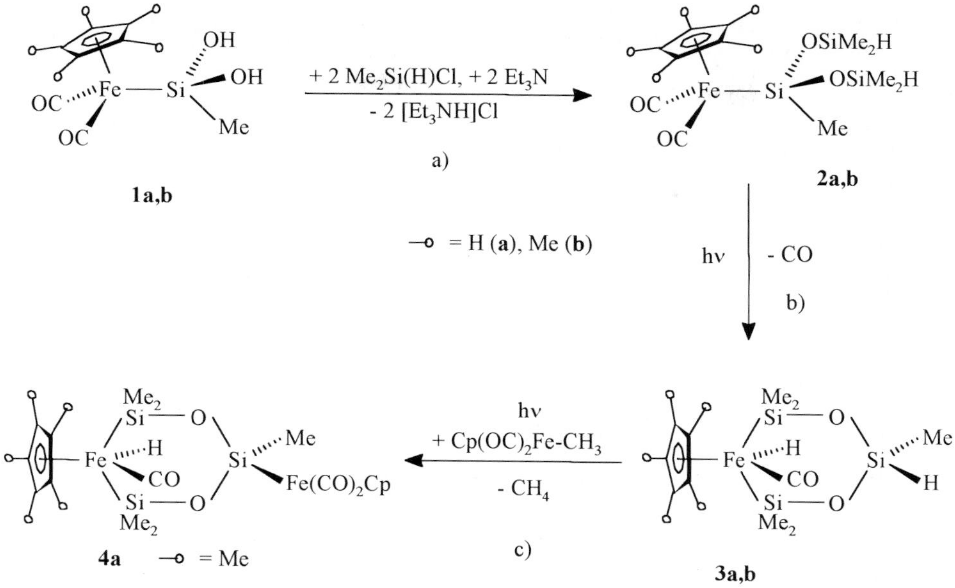

Scheme 1. Synthesis of **2**, **3**, and **4a**.

The cyclo(ferra)trisiloxanes **3a,b** and **4a** are obtained as a mixture of two isomers, resulting from different geometrical arrangements of the ligands at the γ-silicon and the iron, incorporated in the ring.

Dinuclear cyclo(metalla)siloxanes containing both an endocyclic and an exocyclic metal fragment, in addition, can be realized by the photochemical reaction of metallo-trisiloxanes with the methyl-iron complex Cp(OC)$_2$Fe–CH$_3$, as demonstrated recently for the molybdenum complexes C$_5$Me$_5$(OC)$_2$(Me$_3$P)Mo–Si(R)(OSiMe$_2$H)$_2$ (R = Me, Ph) [3]. According to this route, the dinuclear iron complex **4b** is obtained by irradiation of the ferrio-trisiloxane C$_5$Me$_5$(OC)$_2$Fe–Si(Me)(OSiMe$_2$H)$_2$ **2b** in the presence of Cp(OC)$_2$Fe–CH$_3$ (Scheme 2).

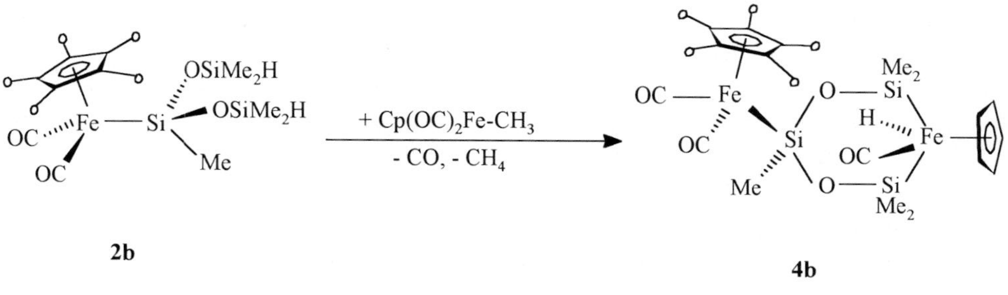

Scheme 2. Synthesis of **4b**.

Compound **4b** is isomeric with **4a**, differing in the position of the iron fragments. While **4a** has the C_5Me_5-substituted iron incorporated in the ring, in the case of **4b** an exocyclic position is adopted.

Acknowledgement: This work has been generously supported by the *Deutsche Forschungsgemeinschaft* (SFB 347: *"Selektive Reaktionen Metall-aktivierter Moleküle"*; Schwerpunktprogramm: *"Spezifische Phänomene in der Siliciumchemie"*).

References:

[1]   Part 13 of the series *"Metallo-Silanols and Metallo-Siloxanes"*. In addition, Part 39 of the series *"Synthesis and Reactivity of Silicon Transition Metal Complexes"*. Part 12/38: see Ref. [2a].

[2]   a) W. Malisch, R. Lankat, J. Reising, *Inorg. Chem.* **1995**, *34*, 5701.
b) W. Malisch, S. Schmitzer, R. Lankat, M. Neumayer, F. Prechtl, W. Adam, *Chem. Ber.* **1995**, *128*, 1251.
c) W. Malisch, S. Möller, R. Lankat, J. Reising, S. Schmitzer, O. Fey, in: *Organosilicon Chemistry II: From Molecules to Materials* (Eds.: N. Auner, J. Weis), VCH, Weinheim, **1996**, p. 575.
d) S. Möller, O. Fey, W. Malisch, W. Seelbach, *J. Organomet. Chem.* **1996**, *507*, 239.

[3]   W. Malisch, R. Lankat, O. Fey, J. Reising, S. Schmitzer, *J. Chem. Soc., Chem. Comm.* **1995**, 1917.

Novel Siloxy-Bridged Di-, Tri-, and Tetranuclear Metal Complexes from Ferrio- and Tungsten-Silanols [1]

Wolfgang Malisch*, Joachim Reising, Martin Schneider

Institut für Anorganische Chemie
Universität Würzburg
Am Hubland, D-97074 Würzburg, Germany
Tel.: Int. code + (931)8885277 – Fax: Int. code + (931)8884618
E-mail: anor142@rzbox.uni-wuerzburg.de

Keywords: Metallo-Siloxanes / Condensation

Summary: Metallo-silanols of the type $L_nM–SiMe_2OH$ (L_nM = $Cp(OC)_2Fe$ (**1**), $C_5R_5(OC)_2(Me_3P)W$, R = H (**5a**), Me (**5b**)) undergo condensation with the electron deficient metal halides $Cp_nM'Cl_{4-n}$ (n = 2, M' = Ti, Zr; n = 1, M' = Ti) yielding hetero-di-, tri-, and tetranuclear metallo-siloxanes. While the ferrio-dimethylsilanol **1** shows stepwise condensation with titanocene- and zirconocenedichloride to give the dinuclear $Cp(OC)_2Fe–SiMe_2O–M(Cl)Cp_2$ (M = Ti (**3a**), Zr (**3b**)) and trinuclear complexes $[Cp(OC)_2Fe–SiMe_2O]_2–MCp_2$ (M = Ti (**4a**), Zr (**4b**)), the corresponding tungsten-silanols **5a,b** yield exclusively the dinuclear species $C_5R_5(OC)_2(Me_3P)W–SiMe_2O–Ti(Cl)Cp_2$ (R = H (**6a**), Me (**6b**)) in the case of titanocenedichloride and the trinuclear complexes $[C_5R_5(OC)_2(Me_3P)W–SiMe_2O]_2–ZrCp_2$ (R = H (**7a**), Me (**7b**)) with zirconocenedichloride, respectively. Moreover, the reaction of **1** with $CpTiCl_3$ succeeds in the substitution of all chlorine atoms to yield $[Cp(OC)_2Fe–SiMe_2O]_3–TiCp$ (**9**), while **5a,b** exclusively produce the trinuclear compounds $[C_5R_5(OC)_2(Me_3P)W–SiMe_2O]_2–Ti(Cl)Cp$ (R = H (**10a**), Me (**10b**)).

Recently, it has been demonstrated that transition metal fragment substituted silanols are both stable with respect to self-condensation and reactive towards various chlorosilanes. Due to these properties metallo-silanols provide interesting starting materials for the synthesis of special siloxane arrangements at metal centers [2].

Condensation of organosilanols with transition metal halides is well established [3], but comparable reactions of metallo-silanols, which should provide a convenient entrance into the field of siloxy-bridged multinuclear complexes, has not been realized. In this context we have carried out the reaction of ferrio- and tungsten-silanols $L_nM–SiMe_2OH$ (L_nM = $Cp(OC)_2Fe$ (**1**); $C_5R_5(OC)_2(Me_3P)W$, R = H (**5a**), Me(**5b**)) with the electron deficient metal halides Cp_nMCl_{4-n} (n = 2, M = Ti, Zr; n = 1, M = Ti).

Titanocene- or zirconocenedichloride perform stepwise condensation with the ferrio-dimethylsilanol **1** in the presence of triethylamine to give the mono- or bis(ferrio-siloxy)-substituted derivatives **3a,b** or **4a,b**, respectively (Eq. 1).

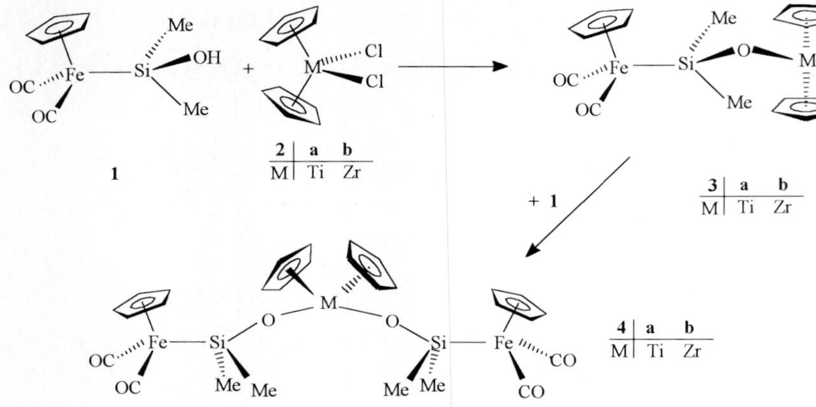

Eq. 1.   Synthesis of **3a,b** and **4a,b**.

Compared with the ferrio-silanol **1**, the complex **4b** shows a high field shift of the ^{29}Si NMR resonance ($\Delta\delta = 11.02$ ppm).

The tungsten-dimethylsilanols **5a,b** display different behavior in so far as for the reaction with titanocenedichloride **2** formation of only one Ti–O–Si-unit is observed (**6a,b**), while zirconocene dichloride directly yields the trinuclear species **7a,b** as a result of an exchange of both chlorine atoms (Fig. 1).

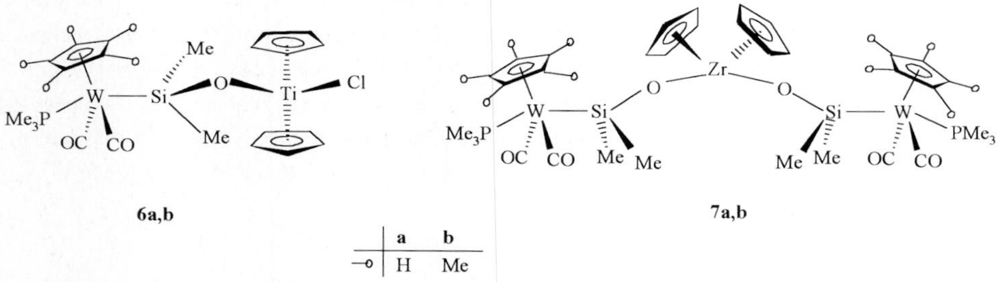

Fig. 1.

6a,b and **7a,b** are obtained in good yields as red-orange microcrystalline powders. **6a** shows a ^{29}Si NMR shift of 55.41 ppm, with the appropriate coupling constants $^{1}J_{WSi} = 41.58$ Hz and $^{2}J_{PWSi} = 16.46$ Hz. The trinuclear species **7a** is characterized by a high field shift of the ^{29}Si NMR resonance ($\delta = 40.25$ ppm) with coupling constants $^{1}J_{WSi} = 44.28$ Hz and $^{2}J_{PWSi} = 15.34$ Hz.

Treatment of the ferrio-silanol **1** with cyclopentadienyltrichlorotitanium **8** yields, under the conditions of Eq. (1), the tris(ferrio-siloxy)-substituted complex **9**. In contrast, the analogous reaction of the tungsten-silanols **5a,b** does not succeed in the substitution of all chloro atoms, but leads to the incorporation of only two metallo-siloxy groups to give the trinuclear metal complexes **10a,b**, irrespective of the molar ratio (Fig. 2).

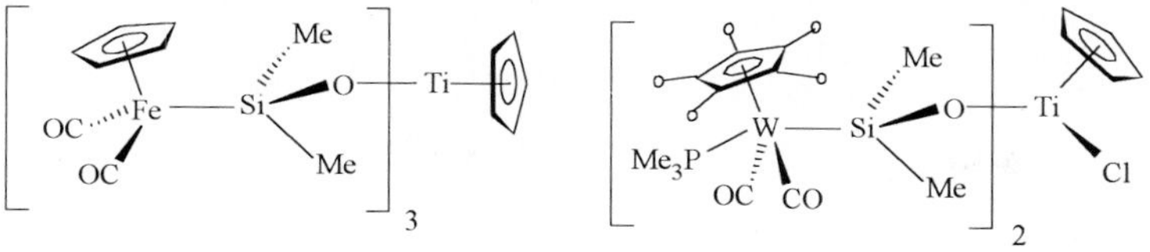

Fig. 2

The different behaviour of iron- and tungsten-silanols presumably is best interpreted in terms of the higher steric demand of the tungsten fragment. Support for this idea is expected from the structural investigations of the siloxy-bridged polynuclear compounds, presented above.

Acknowledgement: This work has been generously supported by the *Deutsche Forschungsgemeinschaft* (Schwerpunktprogramm: "*Spezifische Phänomene in der Siliciumchemie*").

References:

[1]   Part 14 in the series "*Metallo-silanols and Metallo-siloxanes*". In addition, Part 40 of the series "*Synthesis and Reactivity of Silicon Transition Metal Complexes*". Part 13/39 see J. Reising, W. Malisch, R. Lankat, in: *Organosilicon Chemistry III: From Molecules to Materials* (Eds.: N. Auner, J. Weis), VCH, Weinheim, **1997**, p. 412.

[2]   W. Malisch, S. Möller, R. Lankat, J. Reising, S. Schmitzer, O. Fey, in: *Organosilicon Chemistry II: From Molecules to Materials* (Eds.: N. Auner, J. Weis), VCH, Weinheim, **1996**, p. 575; W. Malisch, R. Lankat, J. Reising, *Inorg. Chem.* **1995**, *34*, 5701; W. Malisch, S. Schmitzer, R. Lankat, M. Neumayer, F. Prechtl, W. Adam, *Chem. Ber.* **1995**, *128*, 1251; S. Möller, O. Fey, W. Malisch, W. Seelbach, *J. Organomet. Chem.* **1996**, *507*, 239; W. Malisch, R. Lankat, O. Fey, J. Reising, S. Schmitzer, *J. Chem. Soc., Chem. Comm.* **1995**, 1917.

[3]   D. W. von Gudenberg, H.-C. Kang, W. Massa, K. Dehnicke, C. Maichle-Mössmer, J. Strähle, *Z. Anorg. Allg. Chem.* **1994**, *620*, 1719.

The 2-Dimethylaminomethyl-4,6-dimethylphenyl Substituent: A New Intramolecular Coordinating System with High Steric Demand

Uwe Dehnert, Johannes Belzner*
Institut für Organische Chemie
Georg-August-Universität Göttingen
Tammannstr. 2, D-37077 Göttingen, Germany
Tel./Fax: Int. code + (551)399475
E-mail: jbelzne@gwdg.de

Keywords: Aryllithium Compounds / Chelating Substituents / Highly Coordinated Silicon Compounds / Sterically Hindered Silicon Centers

Summary: The 2-dimethylaminomethyl-4,6-dimethylphenyl substituent **2** is introduced at silicon via the structurally characterized lithio compound **2-Li**. In this manner 2-dimethylaminomethyl-4,6-dimethylphenyl substituted silanes, such as trichlorosilane **3**, difluorosilane **4**, and dihydridosilane **5** are available. The structural analyses of these silanes in the solid as well as in solution show **3** and **4** to be highly coordinated silicon compounds.

The mesityl substituent **1** and the 2-(dimethylaminomethyl)phenyl substituent **1a** have shown their ability to stabilize unsaturated silicon centers in a kinetic and thermodynamic manner, respectively [1, 2]. The 2-dimethylaminomethyl-4,6-dimethylphenyl substituent **2** combines the steric demand of **1** and the chelating amino side-chain of **1a** (Fig. 1). Along the synthetic pathway to potential precursors of unsaturated silicon compounds, 2-dimethylaminomethyl-4,6-dimethylphenyl substituted silanes have been synthesized and structurally characterized.

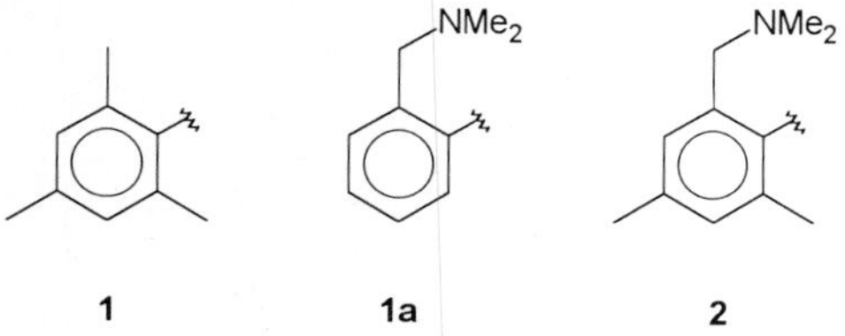

Fig. 1.

Stirring a solution of an equimolar mixture of **2a** and *n*BuLi in Et₂O with subsequent crystallization at -32°C gave rise to the formation of (**2-Li**)₂·Et₂O as pale yellow crystals (50-56 %, Eq. 1).

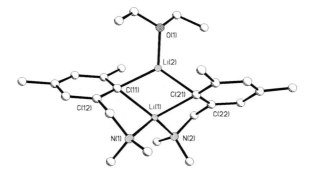

Eq. 1. Synthesis of 2-dimethylaminomethyl-4,6-dimethylphenyllithium (**2-Li**).

X-Ray analysis shows that **2-Li** forms an approximately C_2-symmetric dimer. Interestingly, Li-1 is chelated by both amino donors, whereas Li-2 is coordinated by just one external molecule of the solvent resulting in a threefold coordination (Fig. 2.).

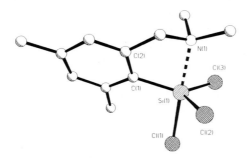

Fig 2. Structure of (**2-Li**)$_2$·Et$_2$O in the solid; selected interatomic distances [pm] and angles [°]: Li(1)···Li(2) 237.0(6), Li(1)–C(11) 225.6(5), Li(2)–C(11) 217.0(5), Li(1)–C(21) 221.6(5), Li(2)–C(21) 214.4(5), Li(1)–N(1) 207.1(5), Li(1)–N(2) 205.7(5), O(1)–Li(2) 196.1(5); Li(1)–C(11)–Li(2) 64.7(2), Li(1)–C(21)–Li(2) 65.8(2), C(11)–Li(1)–C(21) 111.5(2), C(11)–Li(2)–C(21) 117.9(2), N(1)–Li(1)–C(11) 88.0(2), N(2)–Li(1)–N(1) 125.2(2), C(11)–Li(2)–O(1) 118.8(2), C(21)–Li(2)–O(1) 123.2(2).

Fig 3. Structure of **3** in the solid; selected interatomic distances [pm] and angles [°]: Si(1)–Cl(1) 218.26(9), Si(1)–Cl(2) 209.36(8), Si(1)–Cl(3) 207.99(9), Si(1)–C(1) 188.3(2), Si(1)···N(1) 211.7(2), Si(1)···N(1)–Cl(1) 177.02(6); C(1)–Si(1)–Cl(3) 126.83(7), Cl(3)–Si(1)–Cl(2) 116.28(4), Cl(2)–Si(1)–C(1) 115.40(7).

In solution, no different lithium nuclei could be observed by ^{7}Li NMR spectroscopy, even at -80°C in nonetheral solvents such as [D$_8$]-toluene, which should slow down exchange processes. In addition, no splitting due to a ^{13}C^{7}Li coupling was detectable at this temperature in the ^{7}Li and ^{13}C NMR spectra.

Reaction of **2-Li** with SiCl$_4$ and SiF$_4$ results in formation of trichlorosilane **3** and difluorosilane **4**, respectively. The dihydridosilane **5** is synthesized by reaction of **2-Li** with 0.5 equiv. HSiCl$_3$ and subsequent reduction with LiAlH$_4$ in an overall yield of 52 % (Scheme 1).

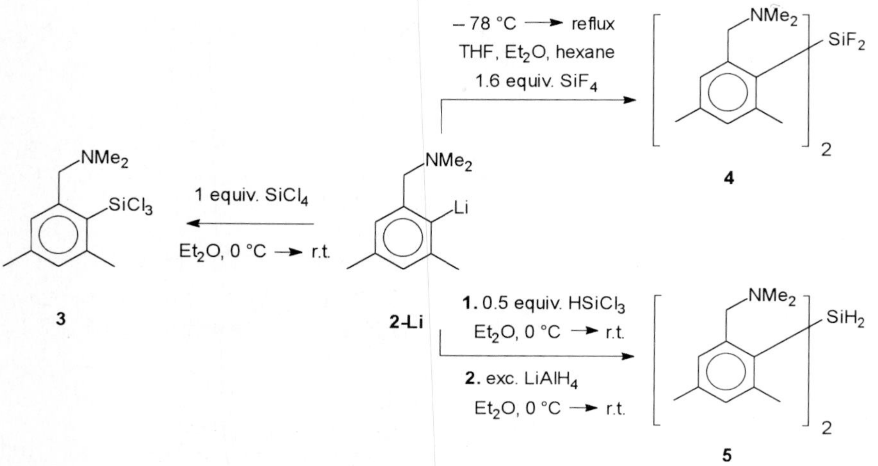

Scheme 1.

Trichlorosilane **3** is a pentacoordinated silicon species. The solid state structure of **3** shows a short Si–N distance of 212 pm (sum of the van der Waals radii of Si and N: 360 pm), a nearly ideal trigonal bipyramidal arrangement around the silicon center and an enlongated Si–Cl bond length to the apical chlorine substituent Cl(1), which are features of the Si–N interaction (Fig. 3) [3]. The position of the silicon atom is 29.8 pm outside the plane of the arene, probably due to the steric influence of the 6-methyl group of the aromatic ring.

The structure of **3** in solution at room temperature is dynamically pentacoordinated, showing a singlet for both the benzylic protons and the NMe$_2$ group in the ^{1}H NMR spectrum. A significant upfield shift in the ^{29}Si NMR spectrum in comparison to mesityltrichlorosilane indicates a highly coordinated silicon compound (Table 1).

Disubstitution of a silicon center by the 2-dimethylaminomethyl-4,6-dimethylphenyl substituent may result in hexacoordinated silicon compounds. This is manifested in the X-ray structure of difluorosilane **4**, which shows a weak Si–N interaction of both amino side chains resulting in Si–N distances of 282 pm and 287 pm (Fig. 4). Both fluorine substituents occupy positions *trans* to the attacking nitrogen donor atoms, thereby forming N–Si–F angles of 173° and 172°. Thus, the silicon center is weakly hexacoordinated showing a coordination sphere, that may be described as a bicapped distorted tetrahedron [4].

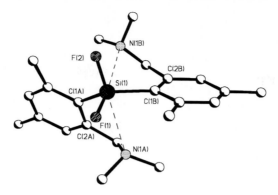

Fig. 4. Structure of **4** in the solid; selected interatomic distances [pm] and angles [°]: Si(1)–F(1) 160.2(2), Si(1)–F(2) 160.3(2), Si(1)–C(1a) 187.8(2), Si(1)–C(1b) 188.3(2), Si(1)···N(1a) 286.9, Si(1)···N(1b) 281.8; F(2)–Si(1)···N(1a) 172.9, C(1a)–Si(1)–C(1b) 124.60(9), C(1b)–Si(1)–F(2) 105.85(8), C(1a)–Si(1)–F(1) 104.25(9), F(1)–Si(1)–F(2) 97.62(9).

The ^{29}Si NMR spectrum of **4** in CDCl$_3$ shows a signal at -46.7, which is shifted upfield in comparison with dimesityldifluorosilane (Table 1). The ^{1}H NMR spectrum of **4** at room temperature shows singlets for both the benzylic protons and the NMe$_2$ groups, thus indicating a dynamic penta- or hexacoordinated structure in solution.

Table 1. Selected ^{29}Si NMR data for **3-6** and reference compounds.

	δ	Δδ	$^1J_{Si-F}$ [Hz]	$^1J_{Si-H}$ [Hz]
3 MesSiCl$_3$[a]	-67.1 -3.4	-63.7		
ArSiCl$_3$[b] PhSiCl$_3$	-61.7 -0.8	.60.5		
4 Mes$_2$SiF$_2$ [a]	-46.7 -23.3	-23.4	279 299	
Ar$_2$SiF$_2$[b] Ph$_2$SiF$_2$	-53.9 -29.1	-24.8	270 303	
6 **5** Mes$_2$SiH$_2$[a]	-59.5 -61.3 -61.6	-1.8 +0.3		198 204 195

[a] Mes = 2,4,6-trimethylphenyl – [b] Ar = 2-(dimethylaminomethyl)phenyl.

The influence of the polarizability [3] of the substituents attached to the silicon center on Si–N coordination is clearly evidenced by dihydrosilane **5**. The X-ray structure analysis shows at best *one* weak Si–N interaction which is manifested in an Si–N distance of 298 pm. Thus, the geometry around the silicon center is that of a capped distorted tetrahedron (Fig. 5) [4].

Apparently, no Si–N interaction exists in solution. The chemical shift of **5** in the ^{29}Si NMR spectrum is comparable with that of dimesitylsilane or compound **6**, in which the donor properties of the amino groups towards silicon are blocked by coordination to AlMe$_3$. In addition, NMR measurements at various temperatures in the range -40 °C to +90 °C do not lead to a significant change in that value. In the case of an intramolecularly coordinated silicon compound one would expect a significant upfield shift with a decrease in temperature [5].

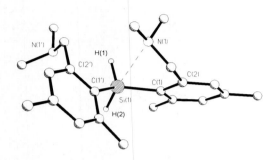

Fig. 5.   Structure of **5** in the solid; selected interatomic distances [pm] and angles [°]: Si(1)–H(1) 138.6, Si(1)–H(2) 141.3, Si(1)–C(1) 190.0(2), Si(1)–C(1′) 188.2(2), Si(1)···N(1) 298.4, Si(1),N(1′) 322.0, H(2)–Si(1)···N(1) 171.7; C(1)–Si(1)–H(2) 104.9, C(1′)–Si(1)–H(2) 103.5, C(1′)–Si(1)–C(1) 120.6(1), C(1′)–Si(1)–H(1) 115.4, C(1)–Si(1)–H(1) 105.9, H(1)–Si(1)–H(2) 105.0.

In conclusion, in spite of its steric demand, the 2-dimethylaminomethyl-4,6-dimethylphenyl substituent is able to form highly coordinated silicon compounds. Investigations currently in progress are directed towards the synthesis of unsaturated silicon compounds stabilized by this substituent.

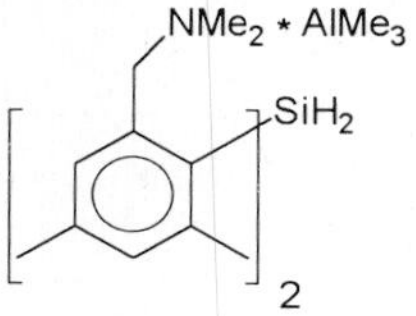

Fig. 6.

Acknowledgement: This work was supported by the *Deutsche Forschungsgemeinschaft* and the *Fonds der Chemischen Industrie*. We thank Dr. M. Noltemeyer for determination of the X-ray structures and Dr. A. Frenzel for support by synthesis of **4**.

References:

[1]   R. West, M. J. Fink, *Science* **1981**, *214*, 1343.

[2]   P. Arya, J. Boyer, F. Carré, R. J. P. Corriu, G. Lanneau, J. Lapasset, M. Perrot, C. Priou, *Angew. Chem.* **1989**, *101*, 1069.

[3]   C. Chuit, R. J. P. Corriu, C. Reyé, J. C. Young, *Chem. Rev.* **1993**, 93, 1371; and references cited therein.

[4]   R. R. Holmes, *Chem. Rev.* **1996**, *96*, 927.

[5]   B. J. Helmer, R. West, R. J. P. Corriu, M. Poirier, G. Royo, A. de Saxce, *J. Organomet. Chem.* **1983**, *251*, 295.

Reaction Behavior of Hypervalent Silanes

Heinrich Lang*, Eduard Meichel, Markus Weinmann, Michael Melter
Institut für Chemie, Lehrstuhl Anorganische Chemie
Technische Universität Chemnitz
Straße der Nationen 62, D-09107 Chemnitz, Germany
Fax: Int. code + (371)5311833
E-mail: heinrich.lang@chemie.tu-chemnitz.de

Keywords: Hypervalent Silanes / Heterobutadienes / Oligomers / Polymers

Summary: The synthesis of the pentacoordinated silane $(2\text{-}Me_2NCH_2C_6H_4)$-$(CH=CH_2)Si(H)_2$ (**1**) is described. A comparison of the chemical behavior of hypervalent **1** with tetravalent silicon compounds is carried out.

Introduction

Silicon compounds with coordination number larger than four are the object of many studies first with respect to their application as catalysts in organic and inorganic syntheses and second as starting materials for the preparation of a broad variety of organosilicon compounds [1]. Additionally, hypervalent silicon hydride compounds can successfully be used as model compounds to study, for instance, the mechanism of nucleophilic substitution reactions, which is of great interest since the silicon atom is able to easily extend its coordination number [1]. Moreover, hypervalent silanes are suitable as starting materials for the synthesis and stabilization of low-valent silanediyl transition metal complexes [2-5].

In this context, we here describe the synthesis and reaction behavior of the pentacoordinated vinyl-functionalized silane $(2\text{-}Me_2NCH_2C_6H_4)(CH=CH_2)Si(H)_2$.

Results and Discussion

Synthesis

The pentacoordinated silane $(2\text{-}Me_2NCH_2C_6H_4)(CH=CH_2)Si(H)_2$ (**1**) can be prepared by reacting equimolar amounts of $(2\text{-}Me_2NCH_2C_6H_4)Li$ with $(CH_2=CH)SiCl_3$ in diethyl ether solution at ambient temperature, followed by treatment of $(2\text{-}Me_2NCH_2C_6H_4)(CH=CH_2)Si(Cl)_2$ with $LiAlH_4$ [6].

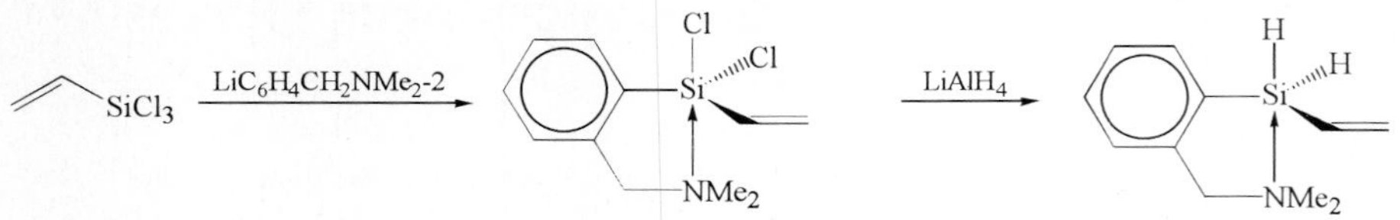

Eq. 1.

Reaction Chemistry

The highly reactive pentacoordinated silane $(2\text{-}Me_2NCH_2C_6H_4)(CH=CH_2)Si(H)_2$ (**1**) reacts with alcohols ROH (R = CH_3, iC_3H_7, $CH_2C\equiv CH$) to produce $(2\text{-}Me_2NCH_2C_6H_4)(CH=CH_2)Si(OR)_2$ (**2a**: R = CH_3, **2b**: R = iC_3H_7, **2c**: R = $CH_2C\equiv CH$).

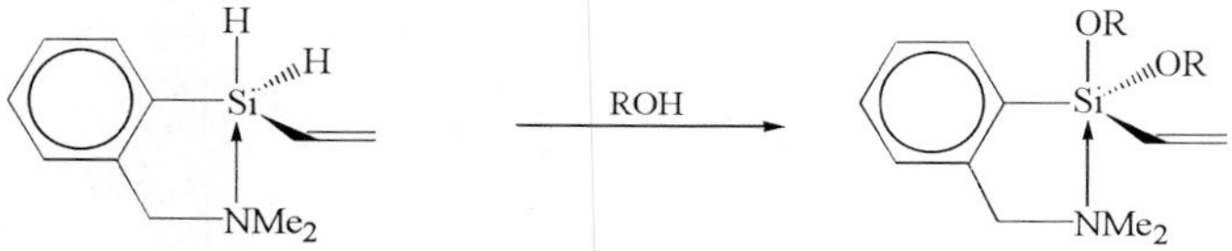

Eq. 2.

Hydrolysis of **1** yields the oligomeric siloxane $[(2\text{-}Me_2NCH_2C_6H_4)(H_2C=CH)SiO]_n$ (**3**).

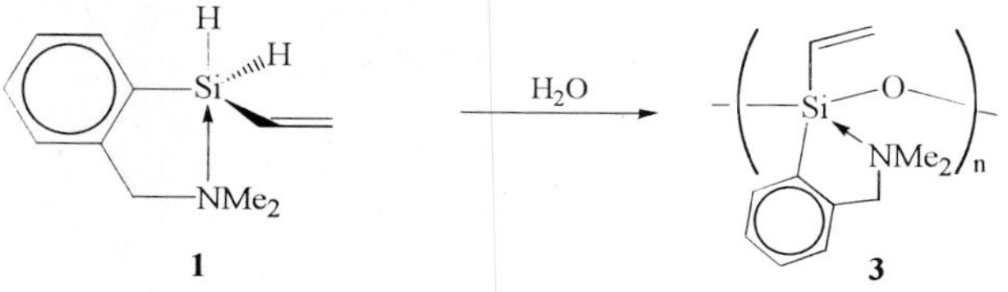

Eq. 3.

Replacing the weaker Broensted acids by stronger carboxylic acids makes the substitution of the hydrogen atoms in compound **1** more rapid. In the absence of any catalyst hypervalent **1** reacts spontaneously with benzoylic acid under evolution of dihydrogen to produce the benzoyloxysilane **4** [6]. Similar observations were made by Corriu et al. [6, 7]. On heating **4** to 120°C elimination of benzaldehyde occurs and oligomeric **3** is formed.

Moreover, on further addition of benzoylic acid to compound **4** benzanhydride and **3** are formed [6].

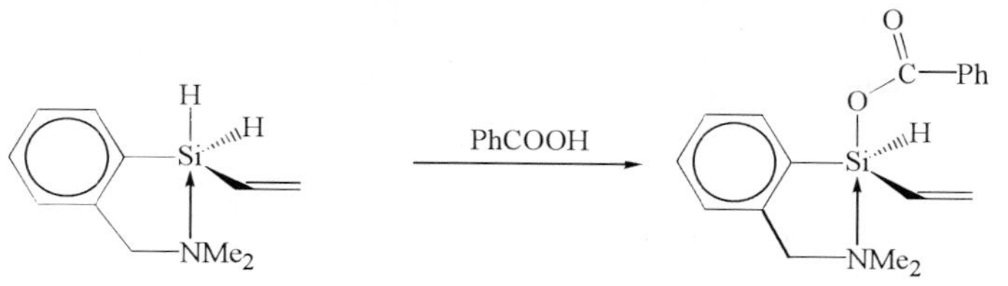

Eq. 4.

One possibility to synthesize oligomeric hypervalent carbosilanes represents the head-to-tail oligomerization of compound **1** in the presence of catalytic amounts of hexachloroplatinic acid and 2-ethylhexanol. This hydrosilylation is performed on heating neat **1** slowly to 200°C.

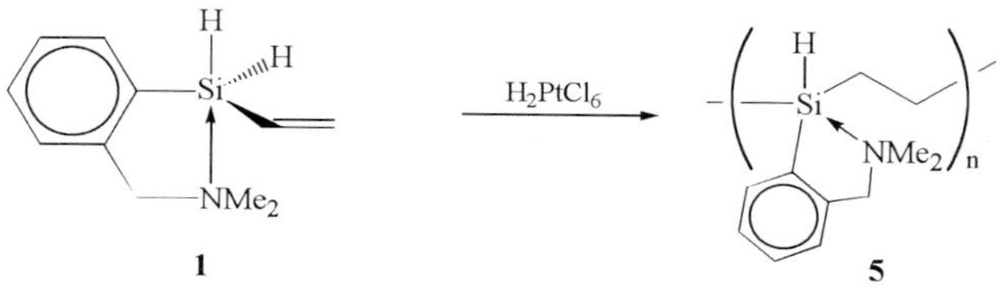

Eq. 5.

A further hydrosilylation takes place if silane **1** is treated with stoichiometric amounts of the heterocumulenes $PhN=C=X$ ($X = S, O$) in tetrachloromethane at 25°C [6b, 8]. The corresponding mono-substituted thioformamide and formamide silanes $(2\text{-}Me_2NCH_2C_6H_4)(H_2C=CH)Si(H)\text{-}(PhNCHX)$ (**6a**: $X = S$, **6b**: $X = O$) are formed in high yields.

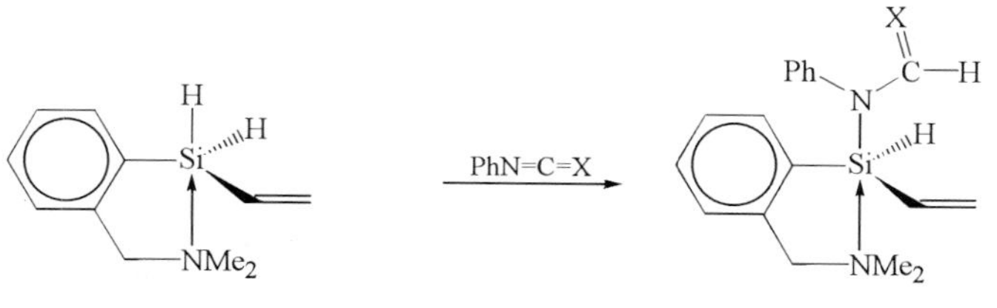

Eq. 6.

While molecule **6a** is stable in solution for weeks, it is found that the appropriate isocyanate-substituted derivative **6b** rearranges to form the 1,3,5-triphenyl-1,3,5-hexahydrotriazine **7**, along with oligomeric $[(2\text{-}Me_2NCH_2C_6H_4)(H_2C=CH)SiO]_n$ (**3**). A possible reaction mechanism for the formation of **3** and **7** from **6b** is presented in Scheme 1.

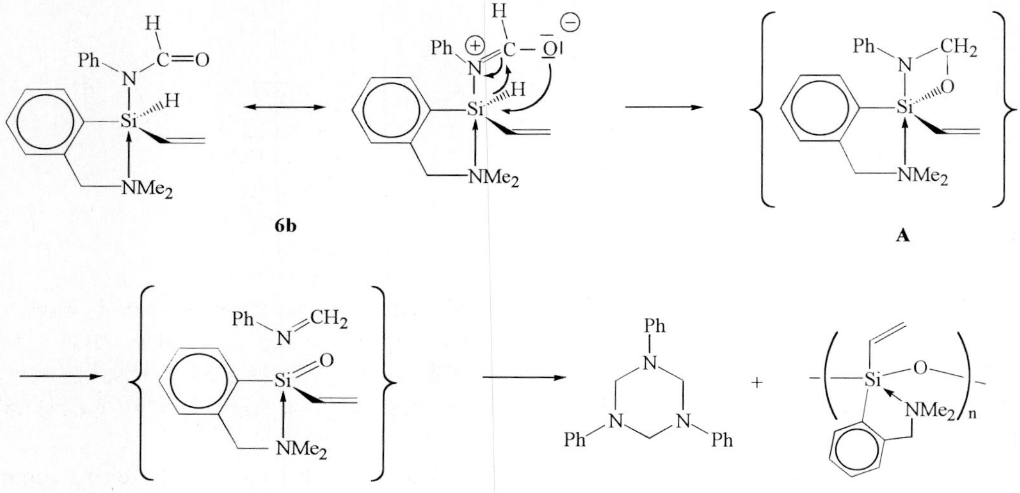

Scheme 1.   Formation of **3** and **7** from **6b** [6b].

Presumably in the first step the 1-sila-2-oxa-4-azetane (type **A** molecule) is formed. By means of a [2+2] retro-cycloaddition reaction intermediate **A** yields PhN=CH$_2$ and (2-Me$_2$NCH$_2$C$_6$H$_4$)-(CH$_2$=CH)Si=O (Scheme 1). The latter is not stable under the reaction conditions and **7** is formed by trimerization of PhN=CH$_2$, while oligomerization of (2-Me$_2$NCH$_2$C$_6$H$_4$)(CH$_2$=CH)Si=O yields the siloxane **3**.

Attempts to initiate a desirable second insertion of PhN=C=X into the remaining silicon-hydrogen bond in compounds **6a** or **6b** failed. However, if a 10-fold excess of PhN=C=X (X = S, O) is used, trimers of phenylisothiocyanate (**8a**) or phenylisocyanate (**8b**) are formed. The latter can also be obtained in much better yields by direct reaction of **1** with a 10-fold excess of PhN=C=X.

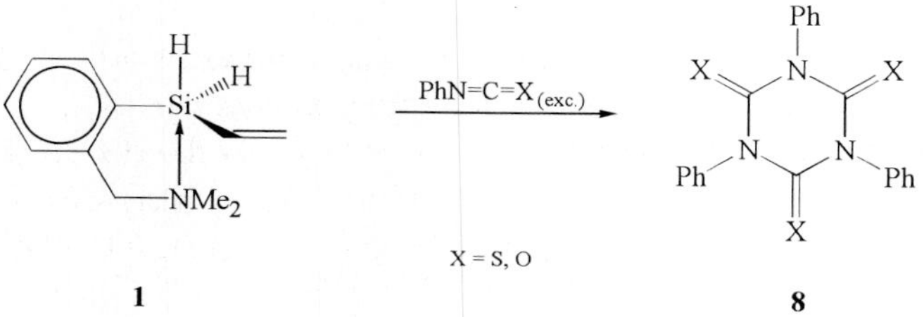

Eq. 7.

The hypervalent silane **1** can also be used to prepare donor-stabilized low-coordinated silicon compounds, such as 1-thia-2-sila-1,3-dienes or 1-metalla-2-sila-1,3-dienes [3, 6b, 9].

On treatment of the vinyl-functionalized hypervalent silane (2-Me$_2$NCH$_2$C$_6$H$_4$)(CH=CH$_2$)Si(H)$_2$ (**1**) with sulfur in chloroform at 25 °C the formation of the 1-thia-2-sila-1,3-diene **9** is observed. It appears that **9** is extremely sensitive to moisture: By elimination of H$_2$S oligomeric **3** is formed [6].

Eq. 8.

The formation of heterobutadiene **9** can be explained by an insertion-elimination reaction: Insertion of sulfur into the silicon-hydrogen bond yields in the first step (2-Me$_2$NCH$_2$C$_6$H$_4$)(CH=CH$_2$)Si(SH)$_2$ as intermediate. This species eliminates H$_2$S, producing the 1-thia-2-sila-1,3-diene **9**, which is one of the rare examples possessing a silicon-sulfur double-bond unit [9].

Applying the method described by Corriu and coworkers [3], 1-metalla-2-sila-1,3-dienes can be synthesized by photochemical reaction of hypervalent **1** with, e.g., Cr(CO)$_6$. The organometallic heterobutadiene **10** is formed in 43 % yield.

Eq. 9.

The identity of all compounds **1 - 10** presented is confirmed by analytical and spectroscopic (IR, ^{1}H, ^{13}C, ^{29}Si NMR, MS) data. Additionally, the solid state structures of **1-10** were established by X-ray diffraction [6b].

In conclusion, it could be shown that the pentacoordinated silane (2-Me$_2$NCH$_2$C$_6$H$_4$)-(CH=CH$_2$)Si(H)$_2$ (**1**), in comparison with tetravalent analogues, shows an increased reactivity towards several substrates. This expresses itself, for instance, in the alcoholysis or hydrolysis of hypervalent **1** without the presence of added catalysts. In a similar manner, the synthesis of pentacoordinated silyl carboxylates, such as (2-Me$_2$NCH$_2$C$_6$H$_4$)(H$_2$C=CH)Si(H)[OC(O)Ph] and (2-Me$_2$NCH$_2$C$_6$H$_4$)(H$_2$C=CH)Si[OC(O)Ph}$_2$, can be achieved. The high reactivity of the pentacoordinated silane **1** can be used for the preparation of oligomeric carbosilanes, as well as *N*-functionalized formamide and thioformamide compounds. In addition, molecule **1** can serve as a catalyst for the cyclotrimerization of heterocumulenes, such as PhN=C=X (X = S, O). Finally, the pentacoordinated silane **1** gives access to novel intramolecular donor-stabilized heterobutadienes like (2-Me$_2$NCH$_2$C$_6$H$_4$)(H$_2$C=CH)Si=S and (2-Me$_2$NCH$_2$C$_6$H$_4$)(H$_2$C=CH)Si=Cr(CO)$_5$.

Acknowledgements: We are greatful to the *Deutsche Forschungsgemeinschaft*, the *Fonds der Chemischen Industrie*, and *Wacker-Chemie GmbH* (Burghausen) for financial support of this work. We thank Dr. Ch. Limberg and Ms S. Ahrens for many discussions.

References:

[1]   See, for example: C. Chuit, R. J. P. Corriu, C. Reye, J. C. Young, *Chem. Rev.* **1993**, *93*, 1371; and refs. cited therein.

[2]   a) C. Zybill, *Nachr. Chem. Tech. Lab.* **1989**, *37*, 248.
       b) C. Zybill, *Top. Curr. Chem.* **1991**, *160*.
       c) C. Zybill, H. Handwerker, H. Friedrich, *Adv. Org. Chem.* **1994**, *36*, 229.

[3]   a) R. J. P. Corriu, B. P. S. Chauhan, G. F. Lanneau, *Organometallics* **1995**, *14*, 1646.
       b) B. P. S. Chauan, R. J. P. Corriu, G. F. Lanneau, Ch. Priou, N. Auner, H. Handwerker, E. Herdtweck, *Organometallics* **1995**, *14*, 1657.
       c) R. J. P. Corriu, B. P. S. Chauan, G. F. Lanneau, *Organometallics* **1995**, *14*, 4014.

[4]   H. Handwerker, C. Leis, R. Probst, P. Bissinger, A. Grohmann, P. Kiprof, E. Herdtweck, J. Blümel, N. Auner, C. Zybill, *Organometallics* **1993**, *12*, 2162.

[5]   a) T. D. Tilley, *Acc. Chem. Res.* **1993**, *26*, 22.
       b) S. D. Grumbine, R. K. Chadha, T. D. Tilley, *J. Am. Chem. Soc.* **1992**, *114*, 1518.
       c) D. A. Strauss, S. D. Grumbine, T. D. Tilley, *J. Am. Chem. Soc.* **1990**, *112*, 7801.

[6]   a) H. Lang, M. Weinmann, A. Gehrig, B. Schiemenz, B. Nuber, *Organometallics*, submitted.
       b) M. Weinmann, *Ph.D. thesis*, Universität Heidelberg, **1994**.

[7]   a) P. Arya, J. Boyer, R. J. P. Corriu, G. F. Lanneau, M. Perrot, *J. Organomet. Chem.* **1988**, *346*, C11.
       b) R. J. P. Corriu, G. F. Lanneau, M. Perrot, *Tetrahedron Lett.* **1987**, *34*, 3941.

[8]   R. J. P. Corriu, G. F. Lanneau, M. Perrot-Petta, V. D. Mehta, *Tetrahedron Lett.* **1990**, *31*, 2585.

[9]   H. Suzuki, N. Tokitoh, R. Okazaki, *J. Am. Chem. Soc.* **1994**, *116*, 11578; and refs. cited therein.

Investigations of Nucleophilic Substitution at Silicon: An Unprecedented Equilibrium between an Ionic and a Covalent Chlorosilane

D. Schär, J. Belzner*

Institut für Organische Chemie
Georg-August-Universität Göttingen
Tammannstr. 2, D-37077 Göttingen, Germany
Tel.: Int. code + (0551)393285
E-mail: jbelzne@gwdg.de

Keywords: Nucleophilic Substitution / Siliconium Compounds / Silicon Halides

Summary: A series of halosilanes $R_2SiH(Hal)$ (R = 2-$(Me_2NCH_2)C_6H_4$) has been synthesized. The influence of the leaving group ability of the halogen substituent, the polarity of the solvent, temperature, and concentration on the formation of intramolecularly coordinated silyl cations has been investigated.

Intramolecularly coordinated silyl cations are known [1, 2], but the factors favoring the formation of these complexes are not fully understood. We have prepared a series of monohalogenated silanes bearing the 2-(dimethylaminomethyl)phenyl (R = 2-$(Me_2NCH_2)C_6H_4$) substituent in order to obtain more insight into this question.

Iodosilane **2-I** and bromosilane **2-Br** were obtained quantitatively by treating silane **1** with bromine or iodine. Chlorosilane **4** [3] was synthesized by reaction of lithio compound **3** (2 equiv.) with trichlorosilane. The chlorosilane **4** was transhalogenated to yield fluorosilane **5** by using $AgBF_4$.

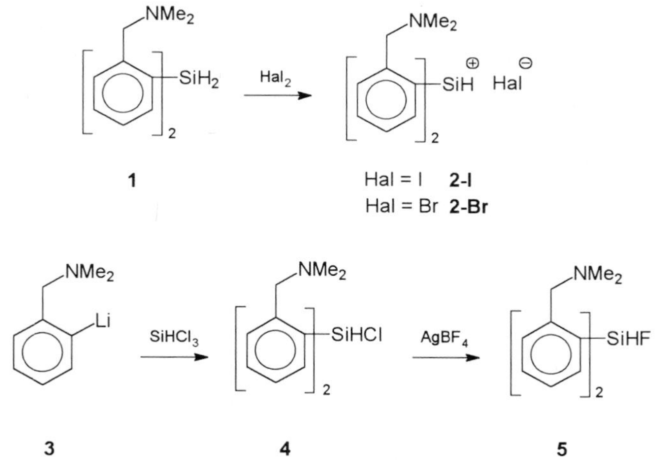

Scheme 1. Synthesis of bis[2-(dimethylaminomethyl)phenyl]halosilanes.

The covalent character of the silicon fluorine bond in **5** is evidenced by the observation of an Si–F coupling constant (see Table 1).

Table 1. ^{29}Si NMR Data of [2-(dimethylaminomethyl)-phenyl] Substituted Fluorosilanes in C_6D_6.

	δ	$^1J_{Si-H}$ [Hz]
R_2SiMeF	-4.4	271
R_2SiHF (**5**)	-38.5	276
R_2SiF_2	-50.8	271

R = 2-(Me$_2$NCH$_2$)C$_6$H$_4$.

In contrast, the similarity of the ^{1}H, ^{13}C, and ^{29}Si NMR spectra of the bromine and iodine compounds (**2-Br**, **2-I**) to the spectra of R$_2$SiH$^+$OTf (**2-OTf**), which is known to form a separated ion pair in the solid [4] (see Fig. 1c) and in solution [5], proves that these compounds form cationic complexes.

Table 2. ^{1}H NMR and ^{29}Si NMR Data of Bis[2-(dimethylaminomethyl)phenyl]-silanes in CDCl$_3$.

	^{1}H NMR			^{29}Si NMR	
	NMe$_2$	CH$_2$N	SiH	δ	$^1J_{Si-H}$ [Hz]
4 (mod. II)	2.16	3.60, 3.74	5.80	-58.1	289
4 (mod. I)	2.69, 2.92	4.35, 4.44	4.63	-51.8	272
2-Br	2.35, 2.71	4.42	4.64	-51.5	269
2-I	2.71, 2.93	4.41, 4.47	4.74	-51.5	273
2-OTf	2.56, 2.76	4.27, 4.33	4.60	-51.6	272
2-Al	2.62, 2.83	4.25, 4.32	4.61	-51.5	272

The ^{1}H, ^{13}C, and ^{29}Si NMR spectra of chloro compound **4** are highly dependent on the solvent: In C$_6$D$_6$ one set of signals is observed for the aryl substituent and for the silicon bound hydrogen (Figure 1a), whereas in polar solvents such as CDCl$_3$ two sets of signals are formed, indicating the existence of two modifications (mod. **I**, mod. **II**) of **4** (Fig. 1b). The inverse ^{1}H,^{1}H NOESY spectrum of **4** shows that both species are in rapid equilibrium with each other.

The signal set of modification **I** is very similar to the spectrum of cationic complexes such as **2-OTf** (compare Figs. 1b and 1c), thus identifying this modification as a pentacoordinated siliconium compound. Moreover, addition of 1 equiv. ClAlMe$_2$ to a solution of **4** in CDCl$_3$, thereby transforming the chloride of **4** into the less nucleophilic Me$_2$AlCl$_3^{2-}$ anion, reduces the complex ^{1}H NMR spectrum (Fig. 1b) to that of cationic complexes (compare Fig. 1c).

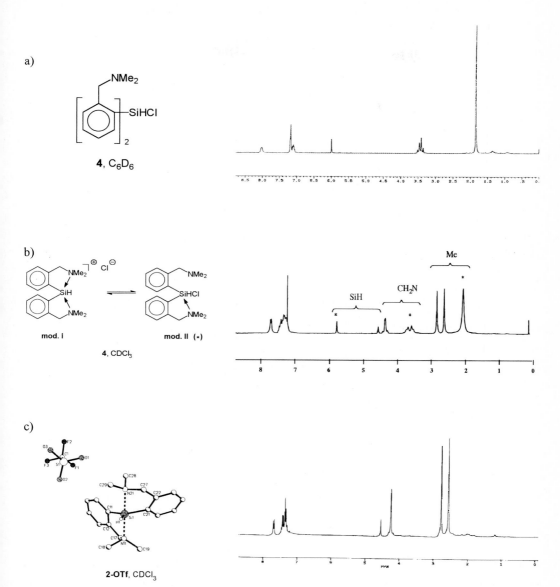

Fig. 1. ¹H NMR spectra of **4** in C₆D₆ (a) and CDCl₃ (b) (signals of mod. **II** of **4** are marked with an asterisk) and the ¹H NMR spectrum of **2-OTf** (c) in CDCl₃.

The second set of signals of the NMR spectrum of **4** in CDCl₃ resembles that of **4** in less polar C₆D₆ (compare Figs. 1a and 1b), which indicates that the second modification (mod. **II**) is the covalent chlorosilane. There is further evidence for assuming that mod. **II** is a non-dissociated species:

- The X-ray structure of **4** clearly shows that **4** is a covalent pentacoordinated chlorosilane in the solid state (Fig. 2).

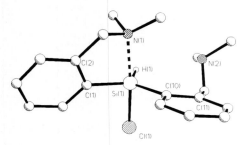

Fig. 2.   Structure of **4** in the solid state; distances [pm] and angles [°]: Si(1)-Cl(1) 225.7(2), Si(1)···N(1) 218.8(5), C(10)-Si(1)-H(1) 122(2), C(1)-Si(1)-H(1) 121(2), C(10)-Si(1)-C(1) 117.2(2), N(1)···Si(1)-Cl(1) 173.6(1).

- A dilute solution of **4** in CH$_2$Cl$_2$, in which the equilibrium between both species is shifted towards the (assumed) covalent modification **II**, reveals a significantly lower equivalent conductivity than a solution of ionic **2-OTf**.
- We have found a linear correlation of the Si–H coupling constant with the Si–H stretching frequency for covalent silicon compounds bearing the 2-(dimethylamino-methyl)phenyl substituent (A-L) [6] (Fig. 3). The values of v_{SiH} (obtained in the solid state) and $^1J_{SiH}$ (obtained in C$_6$D$_6$) of **4** (point M) fit this correlation.

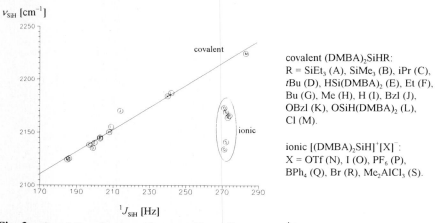

covalent (DMBA)$_2$SiHR:
R = SiEt$_3$ (A), SiMe$_3$ (B), iPr (C),
*t*Bu (D), HSi(DMBA)$_2$ (E), Et (F),
Bu (G), Me (H), H (I), Bzl (J),
OBzl (K), OSiH(DMBA)$_2$ (L),
Cl (M).

ionic [(DMBA)$_2$SiH]$^+$[X]$^-$:
X = OTf (N), I (O), PF$_6$ (P),
BPh$_4$ (Q), Br (R), Me$_2$AlCl$_3$ (S).

Fig. 3.   Correlation of wave number v_{SiH} with coupling constant $^1J_{SiH}$.

The equilibrium between ionic modification **I** and covalent modification **II** is influenced by polarity of the solvent, concentration and temperature. Polar solvents (such as CD$_3$CN, CDCl$_3$ and CD$_2$Cl$_2$), high concentration and low temperature favor the formation of the cationic complex.

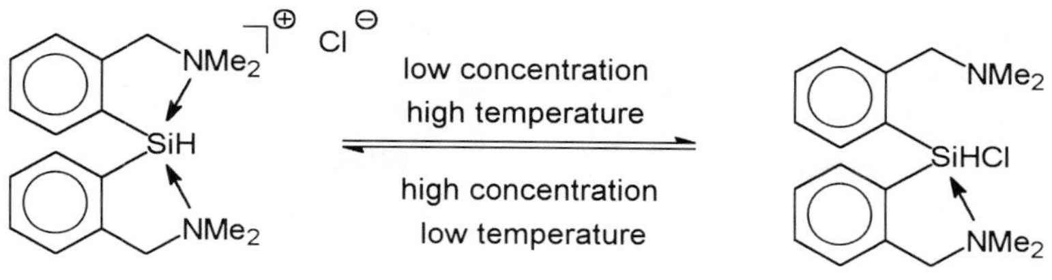

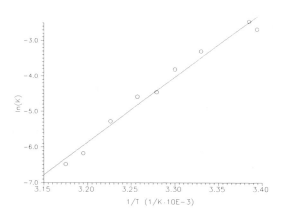

Scheme 2. Influence of concentration and temperature on the equilibrium between the modifictions **I** and **II** of **4**.

Assuming that the potential intermediates of dissociation reaction from mod. **II** to mod. **I**, such as contact ion pairs, are of very low concentration in the observed temperature range, the thermodynamic data for the dissociation reaction can be obtained using a plot of $\ln(K)$ against $1/T$ (see Fig. 4 and Table 3).

Table 3. Thermodynamic data for the dissociation reaction (**II→I**).

ΔH	ΔS	ΔG_{300}
-150 kJmol^{-1}	-530 Jmol^{-1}K^{-1}	8.2 kJmol^{-1}

Fig. 4. Correlation between $\ln(K)$ and $1/T$.

The large entropy change is noteworthy, reflecting the loss of degrees of freedom on going from the covalent compound, which, according to NMR studies, is dynamically pentacoordinated, to the rigid framework of the siliconium compound (compare Fig. 1c). However, we cannot exclude at this point that an appreciable change of solvent polarity due to the formation of ionic species is the cause of the unusually large entropy effects.

Acknowledgement: This work was supported by the *Deutsche Forschungsgemeinschaft* (financial support, fellowship to J.B.), *Fonds der Chemischen Industrie* (financial support) and *Graduiertenförderung des Landes Niedersachsen* (fellowship to D.S.). We thank Dr. R. Herbst-Irmer, Dr. M. Noltemeyer and Dipl.-Chem. B. O. Kneisel for the determination of X-ray structures and Prof. Dr. H. Schneider for the conductivity measurements.

References:

[1]   C. Chut, R. J. P. Corriu, A. Mehdi, C. Reyé, *Angew. Chem.* **1993**, *105*, 1372; *Angew. Chem., Int. Ed. Eng.* **1993**, *33*, 1311.

[2]   V. A. Benin, J. C. Martin, M. R. Willcott, *Tetrahedron Lett.* **1994**, *35*, 2133.

[3]   **4** was obtained independently by Auner et al.: N. Auner, R. Probst, F. Hahn, E. Herdtweck, *J. Organomet. Chem.* **1993**, *459*, 25.

[4]   J. Belzner, D. Schär, *Organometallics* **1995**, *14*, 1840.

[5]   J. Belzner, D. Schär, *"Highly Coordinated Silicon Compounds - Hydrazino Groups as Intramolecular Donors"*, in: *Organosilicon Chemistry II: From Molecules to Materials* (Eds.: N. Auner, J. Weis), VCH, Weinheim, **1996**, p. 459.

[6]   Similar correlations are known: P. Grove, *C. R. Acad. Sci., Paris* **1996**, *262*, 815; H. Bürger, W. Kilian, *J. Organomet. Chem.* **1996**, *18*, 299.

Ligand Exchange Mechanism in Novel Hexacoordinate Silicon Complexes

Daniel Kost*, Sonia Krivonos, Inna Kalikhman
Department of Chemistry
Ben Gurion University of the Negev
Beer Sheva 84105, Israel
Tel.: Int. code + (9727)6461192 – Fax: Int. code + (9727)6472943
E-mail: Kostd@bgumail.bgu.ac.il

Keywords: Hexacoordinate Silicon / Multinuclear NMR / Stereodynamics

Summary: Novel neutral hexacoordinate silicon complexes **1-6** were prepared and the mechanism of ligand exchange studied. Two simultaneous nondissociative N-methyl exchange processes were found. On the basis of persistence of $^{1}J_{Si-F}$ and $^{2}J_{N-Si-H}$ in **5** and **1a**, respectively, dissociative exchange was ruled out. The two consecutive processes were assigned (in order of increasing activation energy) to (X, Cl)- and (O, O)-1,2-shifts, based on the low-temperature NOESY and Saturation Transfer spectrum. The first observation of $^{15}N^{1}H$ coupling across the Si–N coordinative bond is reported.

The stereodynamic behavior of hypervalent silicon complexes continues to present a challenge to silicon chemists and to attract interest among mechanistic enthusiasts due to its complexity [1-3]. Both penta- and hexacoordinate ligand exchange phenomena have been observed and discussed in terms of various mechanistic pathways, including inter- and intramolecular, dissociative and nondissociative exchange reactions [4-7]. In this paper we present our continued effort to elucidate the mechanism of ligand-site exchange in a series of novel hexacoordinate silicon complexes, possessing two identical bidentate chelate ligands.

$$2\ Me_2NN{=}C(R)OSiMe_3\ +\ XSiCl_3\ \xrightarrow[-\ 2\ Me_3SiCl]{}$$

1a X = H; R = Me	**3a** X = Ph; R = Me		
1b X = H; R = Ph	**3b** X = R = Ph		
1c X = H; R = CF$_3$	**3c** X = Ph; R = CF$_3$		
2a X = R = Me	**3d** X = Ph; R = CH$_2$Ph		
2b X = Me; R = Ph	**4a** X = Cl; R = Me		
2c X = Me; R = CF$_3$	**4b** X = Cl; R = Ph		
	4c X = Cl; R = CF$_3$		

Eq. 1.   Synthesis of hexacoordinate silicon complexes.

Complexes **1-4** were prepared as outlined in Eq. 1. An X-ray crystallographic study of **2b** confirmed the general octahedral structure depicted in Eq. 1 in the solid state. In solution, the ^{1}H, ^{13}C, and ^{29}Si NMR spectra indicated the presence of only one diastereomer, out of the possible six different arrangements of these ligands in an octahedral geometry [9]. The hexacoordination at silicon and the associated octahedral geometry were confirmed for the solution structure by the ^{29}Si chemical shifts, ranging between δ = -121 to -147 ppm, which are substantially shifted upfield relative to tetra- and pentacoordinate analogs.

The NMR spectra of all compounds were temperature-dependent in various solvents, indicating exchange phenomena at the NMR time scale. At the slow exchange limit temperatures, in several solvents, the four *N*-methyl groups of compounds **1-3** gave rise to four singlets. In the equally substituted dichloro compounds (**4**), only two *N*-methyl signals are observed, proving the C_2 structure with the chlorine ligands in *cis* position relative to each other. In addition, the C–Me signals in **1a-3a** and other ring substituents were also diastereotopic at low temperature, and indicated the nonequivalence of the two chelate rings.

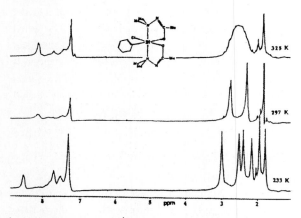

Fig. 1.   Variable temperature ^{1}H NMR spectra of **3a** in [D$_6$] acetone.

Increase of the temperature results in signal broadening and eventual coalescence of signals due to diastereotopic groups. The four *N*-methyl singlets coalesce, in *two different and well resolved* processes, initially to two lines, and finally to one singlet (Fig. 1). In some of the solvents, such as CD$_2$Cl$_2$ and [D$_6$] acetone, the two barriers were resolved, while in others ([D$_8$] toluene, CCl$_4$) the coalescence of signals due to both processes overlapped partly or completely. In the latter the determination of rate processes and activation barriers was done with the aid of a line-shape analysis program, using two non-zero rate constants. The activation free energies are collected in Table 1 for [D$_8$] toluene solutions.

The evidence shows that the lower of the two barriers is associated with the exchange of the chelate rings, such that pairs of diastereotopic groups on both rings become equivalent, and not the geminal *N*-methyl groups on each of the rings. This is demonstrated for **3b** by the following NOE and Saturation Transfer experiment (Fig. 2): at 240K in [D$_6$] acetone solution the *N*-methyl region of the ^{1}H NMR spectrum of **3b** shows four singlets. At this temperature the second, high-barrier rate process is too slow to be effective; the low-barrier process is sufficiently rapid to permit saturation transfer in a typical experiment.

Table 1. Free energies of activation for exchange in **1-3** in [D$_8$] toluene solution and for **4** in [D$_5$] nitrobenzene.

Compound	X	R	$T^{[a]}$ [K]	$\Delta G_1^{*[b]}$ [kcal/mol]	$\Delta G_2^{*[b]}$ [kcal/mol]	$\Delta G_1^{*[c]}$ [kcal/mol]
1a	H	Me	340	17.0	17.8	16.5 (C*Me*)
1b	H	Ph	345	17.5	18.5	
1c	H	CF$_3$	345	17.5	18.0	
2a	Me	Me	280	13.8	14.8	
2b	Me	Ph	285	13.7	14.8	
2c	Me	CF$_3$	283	13.7	14.6	13.8 (CF$_3$)[d]
3a	Ph	Me	337	15.9	16.9	15.8 (C*Me*)
3b	Ph	Ph	337	15.8	16.4	15.8 (*o*-H)
3c	Ph	CF$_3$	335	15.9	16.3	15.4 (*o*-H)
3d	Ph	CH$_2$Ph	335	15.9	16.9	15.8 (*o*-H)
4a[e]	Cl	Me	346	17.9		
4b[e]	Cl	Ph	363	19.5		
4c[e]	Cl	CF$_3$	393	20.7		

[a] Temperature near the coalescence at which the corresponding rate constants were obtained by spectra simulation. –[b] Barriers obtained from the coalescence of *N*-methyl signals by line-shape analysis and the Eyring Equation. – [c] Barriers at the coalescence temperature of indicated groups. – [d] Free energies of activation obtained from ^{19}F NMR. – [e] Measured in [D$_5$] nitrobenzene solution. Due to the higher symmetry only two *N*-methyl singlets and a single coalescence process are observed.

The low-field signal in Fig. 2 was selectively irradiated at low power for 3s, after which a 90° acquisition pulse was applied. Saturation transfer from the irradiated site is evident to the other negative signal, and identifies the latter as the exchange partner of the irradiated methyl group.

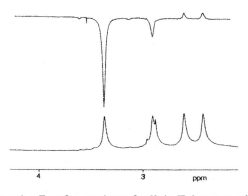

Fig. 2. Difference NOE and Saturation Transfer experiment for **3b** in [D$_6$] acetone solution at 240 K (*N*-methyl region shown). Upper trace: difference spectrum; lower trace: reference spectrum. The low field signal was selectively irradiated for 3s prior to the observation pulse.

However, in addition to the kinetic saturation transfer, two small but distinct positive NOE signals are seen. Since significant NOE interaction can only take place in this molecule between geminal *N*-methyl groups, it is evident that the exchange-partner and the geminal partner of the irradiated signal are different groups. It follows that at this temperature exchange takes place between *N*-methyl groups on different chelate rings, and *not* between geminal methyl groups.

This conclusion is also supported by evidence for another complex in the series: **3a**. The ^{1}H spectra (Fig. 1) as well as the ^{13}C NMR spectra show that simultaneously with the first (low barrier) process exchanging *N*-methyl groups also the *C*-methyl groups, which reside each on a different chelate ring, exchange and coalesce. It is thus evident that this first process effects an exchange of the chelate rings.

Information concerning the higher-barrier process is obtained from the variable temperature ^{1}H NMR spectra of **3d** (Fig. 3). In **3d** the prochiral benzyl groups attached to the chelate rings can "sense" chirality in their vicinity: the benzyl methylene protons are diastereotopic as long as the silicon is chiral; rapid inversion of configuration at the silicon center would render these protons enantiotopic. Examination of the spectra in Fig. 3 reveals that the initial low barrier process, which exchanges the chelate rings and converts the four *N*-methyl singlets into two, also effects coalescence of the initial two AB quartets due to the benzyl methylene protons into one. The fact that after the first rate process the benzyl methylene protons are still diastereotopic and give rise to an AB quartet (at 286K, Fig. 3), means that no configuration-inversion at silicon could have occurred. However, further increase of temperature brings about coalescence of the quartet into a singlet, in accord with rapid inversion at the silicon chiral center. We may conclude that while the first process exchanges chelate rings, the second inverts the configuration at silicon.

The question whether the exchange mechanisms are inter- or intramolecular can be answered without difficulty: we found the coalescence temperatures and barriers to be independent of the complex concentration, within a range of eight-fold dilution. We are thus left with a choice of *intramolecular* exchange mechanisms. These can be divided into two main categories: dissociative and nondissociative ligand exchange reactions. We next focus our attention to answer this question.

The most likely dissociation processes for the hexacoordinate complexes at hand are Si–N dative bond cleavage or Si–Cl ionic dissociation. The observation in **3d** that in the high-barrier process both *N*-methyl and the benzyl methylene protons exchanged simultaneously, by the same process and with the same activation barrier, suggests that N–Si does not open. Exchange of the geminal methyl groups by Si–N cleavage, followed by rapid rotation about the N–N bond and reclosure of the chelate ring, cannot effect simultaneous exchange of the methylene protons. This process must hence be ruled out. However, the possibility that Si–N cleavage does occur, and is followed by rapid inversion at the silicon center by pseudorotation or a similar process, may be considered.

More conclusive evidence against Si–N cleavage during the exchange processes was obtained from ^{15}N^{1}H coupling data. The ^{15}N NMR spectra of **1a** were run in CDCl$_3$ solution at varions temperatures, and $^2J_{\text{N}\rightarrow\text{Si-}\underline{\text{H}}}$ coupling between Si–$\underline{\text{H}}$ and nitrogen was observed *across the Si–N coordination* (Fig. 4). To the best of our knowledge this is the first report of proton to nitrogen coupling through N→Si coordination. The coupling was measured $^2J_{\text{N}\rightarrow\text{Si-}\underline{\text{H}}} = 12$ Hz. The striking observation, however, was that the doublet obtained for the *N*-methyl nitrogen *persisted* up to 335K, whereas in the ^{1}H spectrum in this solvent at ambient temperature the *N*-methyl signals had already coalesced. It thus is evident that Si–N cleavage does not take place on this time scale and that both rate processes observed for **1-4** do not involve Si–N cleavage.

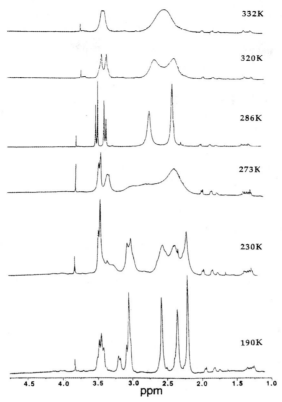

Fig. 3. ^{1}H DNMR spectra of **3d** in CD$_2$Cl$_2$ solution. The benzyl methylene proton region at low temperature is limited in resolution, but coalescence to an AB quartet is evident at 286K.

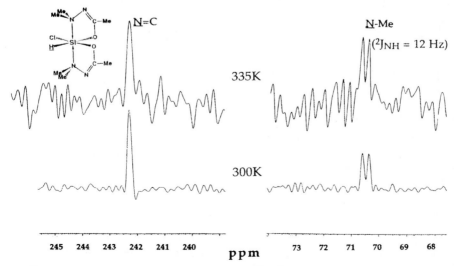

Fig. 4. ^{15}N NMR spectra of **1a** in CDCl$_3$ solution, showing ^{15}NSi^{1}H coupling.

For the second possible dissociation process, Si–Cl cleavage, the fluoro analog **5** was prepared. The very strong one-bond coupling between silicon and fluorine [$^1J_{Si-F}$ = 273 Hz] was observed in the ^{29}Si NMR spectrum (Fig. 5). In addition, coupling between fluorine and the *ipso*-carbon of the Si–P̲h̲ group was observed in the ^{13}C spectrum [$^2J_{F-Si-C}$ = 43.6 Hz]. Both of these coupling constants were monitored over a range of temperature, and were found to persist unchanged upon heating up to 350K, at which other diastereotopic groups had already coalesced. This proves the stability of the Si–F bond under the conditions of fast *N*-methyl exchange, and hence rules out Si–F dissociation as the process causing this exchange. It might be argued that Si–F behavior does not necessarily reflect on the Si–Cl bond stability in analogous complexes. However, the observation of similar NMR spectra (^{1}H, ^{29}Si), and in particular similar stereodynamic behavior of **5** and the chloro complexes **1**-**4**, assures that they undergo the same types of exchange reactions, and hence it may be concluded that ionic dissociation does not occur in **1**-**4** either.

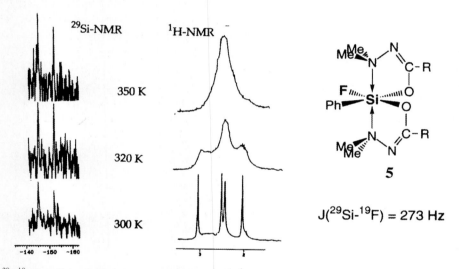

Fig. 5. ^{29}Si^{19}F Coupling at various temperatures, compared to ^{1}H coalescence of *N*–Me signals.

In the absence of a dissociative process, the remaining mechanistic options are only intramolecular nondissociative of the pseudorotation family. The nondissociative site-exchange reactions commonly discussed for octahedral complexes are the Ray-Dutt [10] and Bailar twist mechanisms [11], by which three ligands are rotated relative to the other three via a trigonal prism transition state. None of these processes effects a topomerization in one step. At least two consecutive such twists are necessary in order to bring about the observed topomerizations. However, with these mechanisms it would be difficult to rationalize the observation of two distinct barriers in **1**-**4**.

The mechanism which seems to fit best all of the current observations is an intramolecular interchange of adjacent ligands via the so-called bicapped tetrahedron transition state or intermediate [12]. There are two such exchanges, the (X,Cl)-1,2-shift and the (O,O)-1,2-shift (Fig. 6), which account for the observation of two distinct rate processes in the NMR spectra. Examination of these exchange processes shows that they precisely generate the observed NMR spectral changes: the (X,Cl)-shift interchanges the two chelate rings; the one *trans* to Cl becomes *trans* to X, and vice versa, as observed by the NOE and other experiments. The (O,O)-1,2-shift also effects an interchange of the chelate rings, but in addition inverts the chiral configuration at silicon.

We prefer the 1,2-shift mechanism over consecutive Ray-Dutt and Bailar twists for several reasons: (a) these two 1,2-shift processes effect the observed changes directly, each in one step [13]. They nicely account for the two observed barriers. (b) The transition states for these reactions are bicapped tetrahedron structures: such geometries have recently been observed as stable *ground state* structures in similar hexacoordinate silicon complexes, by Corriu and his group [14]. It is thus very likely to expect low energy intermediates or transition states with this geometry, by contrast to the trigonal prism transition state required for each of the twist mechanisms. (c) Additional support for this mechanism comes from the observation of a significant solvent effect on the barriers for topomerization: the twisting of two adjacent ligands out of the central plane of the octahedral skeleton to form the bicapped tetrahedron geometry involves steric strain, and requires that the coordinated nitrogen atoms be pushed away from silicon [13]. This constitutes partial (but not complete!) cleavage of the Si–N bonds, and requires activation energy. However, the bond-lengthening may be assisted by the solvent: in hydrogen-bond donor solvents (CHCl$_3$) the NMe$_2$ groups may be hydrogen bonded to solvent molecules, and hence the activation barrier associated with Si–N lengthening is reduced. Likewise, π-acceptor solvents tend to complex with the donor dimethylamino group and thereby reduce the activation barrier. This expectation is, in fact, borne out by experiment: the barriers for **3b** have been measured in a series of solvents, and are listed in Table 2. A rather dramatic solvent effect on barriers is found, in the same trend suggested above.

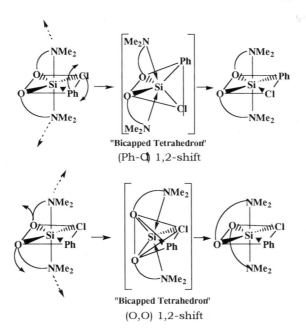

"Bicapped Tetrahedron"
(Ph-Cl) 1,2-shift

"Bicapped Tetrahedron"
(O,O) 1,2-shift

Fig. 6. Ligand exchange mechanism by 1,2-shift of adjacent Ph, Cl (upper part) and O, O (lower part) via bicapped tetrahedron transition state or intermediate.

Table 2. Solvent effect on activation free energies for *N*-methyl exchange in **3b**.

Solvent	Number of signals at 300K	T_c [K]	ΔG_1^* [kcal/mol]	ΔG_2^* [kcal/mol]
CCl$_4$	4	340	16.4	16.9
CD$_3$C$_6$D$_5$	4	337	15.8	16.4
C$_6$D$_6$	4	340	16.3	16.7
C$_4$Cl$_6$	4	345	16.6	16.9
C$_5$D$_5$N	2	292	13.9	
	2	280	14.0	
	2	340		16.1
C$_6$D$_5$NO$_2$	2	270	13.2	
	2	340		16.2
(CD$_3$)$_2$CO	2	278	13.8	
	2	286	13.7	
	2	315		15.0
CDCl$_3$	2	325		15.5
CD$_2$Cl$_2$	2	216	10.6	
	2	230	10.9	
	2	315		15.0
CD$_3$NO$_2$	2	<240	≈10.6	
	2	335		16.0

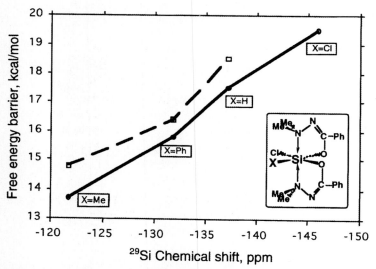

Fig. 7. Activation barriers for *N*-methyl exchange plotted against ^{29}Si chemical shifts.

Further evidence, compatible with lengthening of the Si–N dative bond during the exchange, comes from a remarkable correlation found between the barrier heights and the ^{29}Si chemical shifts (Fig. 7). The chemical shifts change as a result of the X group electronegativities. The effect of X on the barrier may be rationalized as follows: the more electron withdrawing the X group, the stronger is the coordination of nitrogen to silicon. As a result, a greater activation energy is required to lengthen the Si–N bond and facilitate the 1,2-shift.

On the basis of the evidence presented so far it is still impossible to distinguish between the two exchange reactions: (X,Cl)- and (O,O)-shift. In order to resolve this question we prepared the analogous complex **6**, containing an enantiomerically pure chiral ligand, R = CH(Me)Ph, using the optically pure (*R*)-(-)-2-phenylpropionic acid. Compound **6** can have two diastereomers, *RRR* and *RSR*, differing in the configuration at silicon. Indeed, the slow exchange limit NMR spectrum showed eight *N*-methyl singlets, in two unequally populated groups of four, with an isomer ratio of 4:3.

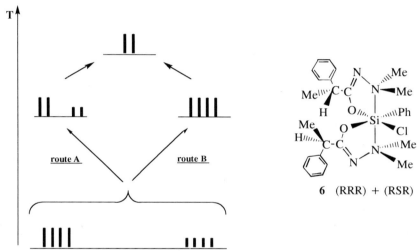

Fig. 8.   Schematic presentation of two possible routes for the *N*-methyl coalescence in **6**.

An analysis of the two exchange reactions for **6** shows two possible NMR routes, depicted in Fig. 8. If (X,Cl) exchange is the low energy process, since it involves *only* interchange of chelate rings, no interchange between the diastereomers is expected and the reaction would follow route A. However, if (O,O) exchange is the low energy process, route B would be followed, since this process involves epimerization at silicon and hence the interchange: *RRR* ⇌ *RSR*

The 2D NOESY and Saturation Transfer experiment (Fig. 9) readily distinguishes between the routes A and B: the negative NOE signals identify the geminal *N*-methyl pairs, while the positive cross signals connect exchange-related signal pairs. It is immediately apparent that exchange takes place only *within* each diastereomer, according to route A, i.e., signals of the major isomer (labeled 1) exchange only with those of the same isomer, and signals labeled 2 exchange only with others labeled 2. It is thus unambiguously concluded that (X,Cl) exchange precedes (O,O) exchange upon increase of the temperature.

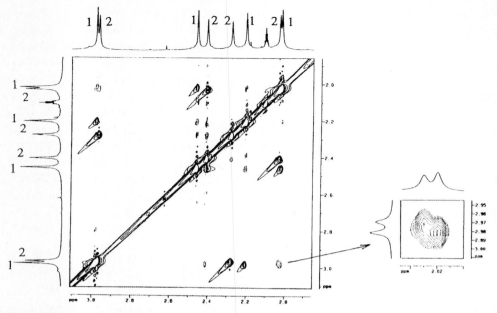

Fig. 9. 2D NOESY and Saturation Transfer spectrum for **6** at 250K in [D$_8$] toluene solution. Exchange (positive) signals: solid lines; NOE (saturation transfer, negative) signals: dotted lines.

The high temperature ^{1}H NMR spectra of **6** apparently also show a third process, the silicon-nitrogen bond cleavage: as is shown in Fig. 8, at the fast exchange limit temperature the geminal *N*-methyl groups should still be diastereotopic and nonequivalent, due to the proximity of the carbon chiral center. In practice, however, the coalescence and exchange phenomena of the two rate processes extend over a large temperature range before the signals narrow down again. Up to 380K in [D$_5$] nitrobenzene solution the *N*-methyl signals are still too broad, and above this temperature the signal narrows into *one* broad singlet. This could indicate that coalescence of the geminal methyl groups takes place via Si–N cleavage followed by rotation about the N–N bond, or it might be the result of accidental equivalence of the geminal methyl groups.

In conclusion, two consecutive ligand-site exchange processes were measured in hexacoordinate complexes **1-4**, and were assigned to (X,Cl) and (O,O)-1,2-shifts (in order of increasing activation barrier). Possibly also Si–N cleavage follows in these compounds at higher temperature. ^{15}N^{1}H coupling was observed for the first time across the N→Si coordinative bond.

Acknowledgement: Financial support from the *Israel Science Foundation*, administered by the *Israel Academy of Sciences and Humanities*, and from the *Israel Ministry of Sciences and Arts* is gratefully acknowledged.

References:

[1] S. N. Tandura, N. V. Alekseev, M. G. Voronkov, *Top. Curr. Chem.* **1986**, *131*, 99.

[2] R. R. Holmes, *Chem. Rev.* **1990**, *90*, 17.

[3] C. Chuit, R. J. P. Corriu, C. Reye, J. C. Young, *Chem. Rev.* **1993**, *93*, 1371.

[4] F. Carre, R. J. P. Corriu, A. Kpoton, M. Poirier, G. Royo, J. C. Young, *J. Organomet. Chem.* **1994**, *470*, 43.

[5] W. B. Farnham, J. F. Whitney, *J. Am. Chem. Soc.* **1984**, *106*, 3992.

[6] C. Breliere, F. Carre, R. J. P. Corriu, W. E. Douglas, M. Poirier, G. Royo, M. Wong Chi Man, *Organometallics* **1992**, *11*, 1586.

[7] F. Carre, C. Chuit, R. J. P. Corriu, A. Fanta, A. Mehdi, C. Reye, *Organometallics* **1995**, *14*, 194.

[8] A. O. Mozzhukhin, M. Yu. Antipin, Yu. T. Struchkov, B. A. Gostevskii, I. D. Kalikhman, V. A. Pestunovich, M. G. Voronkov, *Metaloorg. Khim.* **1992**, *5*, 658; *Chem. Abst.* **1992**, *117*, 234095w.

[9] D. Kost, I. Kalikhman, M. Raban, *J. Am. Chem. Soc.* **1995**, *117*, 11512.

[10] J. C. Bailar, *J. Inorg. Nucl. Chem.* **1958**, *8*, 165.

[11] P. Ray, N. K. Dutt, *J. Indian Chem. Soc.* **1943**, *20*, 81.

[12] R. Hoffmann, J. M. Howell, A. R. Rossi, *J. Am. Chem. Soc.* **1976**, *98*, 2484.

[13] A. Rodger, B. F. G. Johnson, *Inorg. Chem.* **1988**, *27*, 3061.

[14] C. Breliere, F. Carre, R. J. P. Corriu, M. Poirier, G. Royo, J. Zwecker, *Organometallics* **1989**, *8*, 1831.

Ligand Exchange via Coordinative Si–N Bond Cleavage and Pseudorotation in Neutral Pentacoordinate Silicon Complexes

Inna Kalikhman, Daniel Kost*

Department of Chemistry

Ben Gurion University of the Negev

Beer-Sheva 84105, Israel

Tel.: Int. code + (9727)6461192 – Fax: Int. code + (9727)6472943

E-mail: Kostd@bgumail.bgu.ac.il

Keywords: Pentacoordinate / Multinuclear NMR / Stereodynamics

Summary: Introduction of a chiral ligand to a series of pentacoordinate silicon complexes led to the assignment of two intramolecular rate processes: Si–N cleavage and pseudorotation. Linear correlations between ^{15}N and ^{29}Si chemical shifts of the complexes, as well as between the latter and the N-methyl exchange barriers were attributed to variation in strength of N→Si coordination.

A convenient method for the preparation of penta- and hexacoordinate silicon complexes was recently developed and reported [1]. The synthesis was based on an exchange reaction between poly-halogenated silanes with the *O*- or *N*-trimethylsilylated acylhydrazines and led to the formation of (O–Si)- and (N–Si)- mono and bis- chelate complexes (Scheme 1).

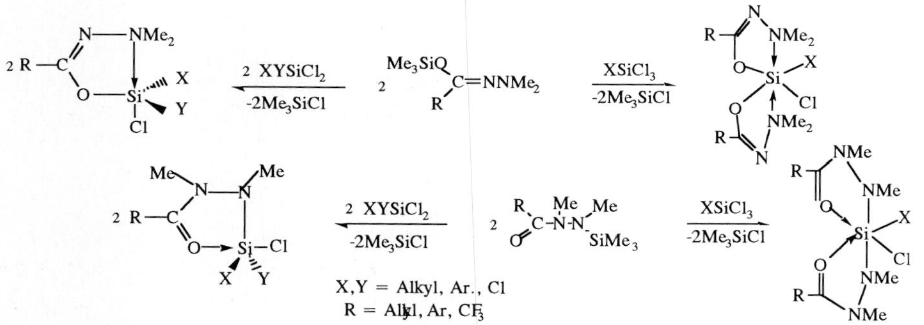

Scheme 1. Synthesis of penta- and hexacoordinated silicon complexes.

Penta- and hexacoordinate silicon complexes often show very complex ligand exchange reactions, which have been the subject of intense studies [2-5]. In the present paper we analyze the ligand exchange processes and ^{1}H, ^{15}N, and ^{29}Si NMR chemical shifts of a series of pentacoordinate complexes in a search for a better understanding of the exchange mechanisms. Compounds **2a-2h** were readily prepared by the reaction shown in Eq. 1.

Eq. 2. Synthesis of chelates **2**.

The trigonal bipyramid geometry of **2a-2h** was demonstrated by an X-ray crystallographic structure of one of the complexes, **2d** [6]. The Si–N and Si–Cl bond-lengths (2.264 and 2.192Å, respectively) are characteristic of pentacoordination at silicon.

Table 1. ^{29}Si and ^{15}N NMR chemical shifts and coordinative shifts for complexes **2**.

No	R	δ^{29}Si [ppm]	Δ^{29}Si[a] [ppm]	δ^{15}NMe[b] [ppm]	Δ^{15}NMe[a] [ppm]	ΔG^* [kcal mol^{-1}]
2a	Me	-32.8	-51.9	66.5	7.6	10.9
2b	PhCH$_2$	-31.8	-52.2	66.4	7.5	11.0
2c	PhMeCH	-33.8	-54.2	66.4, 65.6[c]	7.8	11.4
2d	Ph	-29.0	-47.8	66.8	6.6	10.6
2e	4-Me–Ph	-30.0	-48.2			10.7
2f	4-NO$_2$–Ph	-23.0	-44.5			9.4
2g	3,5-(NO$_2$)$_2$–Ph	-14.7	-38.5	68.2	3.7	8.9
2h	CF$_3$	-3.2	-30.0	64.7	2.0	< 7

[a] Coordinative shift: difference between pentacoordinate (**2a-2h**) and tetracoordinate precursor (**1a-1h**) shifts – [b] Shifts are in ppm relative to external NH$_4$Cl – [c] Shifts are given for two diastereomers.

The solution structure of **2a-2h** can be determined by the ^{29}Si and ^{15}N NMR chemical shifts, which are sensitive measures of coordination. In Table 1 are listed the ^{29}Si and ^{15}N chemical shifts for **2a-2h**, along with the "coordinative" shifts: the increment added to the chemical shift upon complexation, relative to the tetracoordinate precursor **1**. The large upfield shifts of the ^{29}Si resonances in **2** relative to **1**, as well as in comparison to Me$_2$(Cl)SiO(R)C=NNMe$_2$, (δ^{29}Si = 15.9 ppm), are evidence for the pentacoordination in complexes **2**.

Likewise, the downfield shifts of the ^{15}N signals due to *N*-methyl groups reflect electron donation from nitrogen to silicon and indicate the presence of a dative bond. Similar ^{15}N coordinative shifts have only been reported previously for silatranes and analogous triazasilatranes [7,8].

Since both the ^{15}N and ^{29}Si coordinative shifts appear to reflect the same phenomenon, i.e., the strength of nitrogen to silicon coordination, it is not surprising to find that they are linearly correlated, as depicted in Fig. 1. It may be of interest to note the rather significant dependence of these chemical shifts upon the nature of the remote R-ligand group.

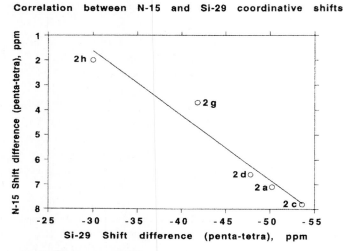

Fig. 1. Δ^{15}N as a function of Δ^{29}Si coordinative shifts.

2a-2h have a chiral silicon center which causes the two *N*-methyl groups to be diastereotopic. The exchange of the *N*-methyl sites in the NMR spectra of **2a-2h** provides a tool for monitoring structural changes in solution. The activation free energies for interchange of the *N*-methyl groups in **2a-2h** were measured by variable temperature ^{1}H NMR spectroscopy in CD$_2$Cl$_2$ solution, and are shown in Table 1. The site exchange observed by NMR may result from several possible dynamic processes: it could be an intermolecular reaction, or it could be an intramolecular, dissociative or nondissociative reaction [3, 9, 10]. The barriers were found to be independent of complex concentration, within a ten-fold dilution range, and thus an intermolecular process may be excluded.

Within the intramolecular mechanistic range, the two most likely *N*-methyl exchange mechanisms are either a Berry type pseudorotation, or Si–N dative bond cleavage followed by rotation about the NN bond and reclosure of the chelate ring [3, 9, 10]. These two mechanisms can be distinguished for **2c**, taking advantage of the chiral ligand (R = CHMePh) attached to the chelate ring. In this compound the presence of the permanent carbon chiral center and a silicon chiral center should generate two diastereomers, observable by NMR at the slow exchange limit temperature, as long as epimerization at silicon is slow. Indeed the low temperature ^{1}H NMR spectrum of **2c** (Fig. 2, 200 K) shows two sets of signals for each group, in accord with the presence of two diastereomers in an equilibrium concentration ratio 4:5. The diastereotopic *N*-methyl groups give rise to four singlets, two for each diastereomer.

Gradual increase of the temperature brings about broadening and eventual coalescence of each pair of *N*-methyl signals, from four to two, *without concomitant coalescence of other signals*. This can only be the result of Si-N bond cleavage, while no epimerzation at silicon and no exchange of diastereomers take place. The activation free energy for Si-N cleavage in **2c** is 11.4 kcal mol^{-1}. This barrier falls well within the range of barriers measured for the other complexes **2** (Table 1). It may hence be concluded that for all of these complexes the measured *N*-methyl exchange barrier represents the same type of process, i.e., Si–N cleavage, followed by rotation about the NN bond and reclosure of the chelate ring.

When a similar compound (**3**) was studied, in which silicon is not chiral, but the carbon ligand is, we see similar behavior: nonequivalence of geminal *N*-methyls in this compound can result only from the chirality at carbon, and coalescence can only result from Si–N cleavage. The spectra at various temperatures indicate such Si–N bond cleavage, and the barrier measured is again in the same range as for the other compounds: 10.6 kcal mol^{-1}.

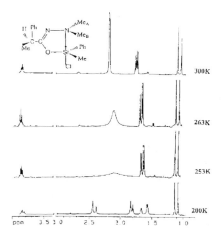

A remarkable linear correlation was found between the barriers for Si–N bond cleavage and the corresponding ^{29}Si chemical shifts (Fig. 3). The correlation reflects the variance in coordination strengths in the various complexes: the higher barriers measured for electron-releasing R groups indicate stronger complexation, which, in turn, is associated with a more pentacoordinate nature of the silicon and hence a greater upfield shift.

Fig. 2.   Low range variable temperature ^{1}H NMR spectra of **2c** in CD$_2$Cl$_2$ solution.

Since, in addition to Si–N cleavage, pseudorotation is also expected in pentacoordinate complexes, the high temperature NMR spectra of **2c** were further studied in nitrobenzene-D$_5$ solution (Fig. 4). As the figure shows, the doubling of all signals due to the diastereomers disappears at higher temperature (activation free energy: 18.7 kcal mol^{-1}). The coalescence of signals due to diastereomers must be due to epimerization at the silicon center, very likely through Berry pseudorotation. The rather high barrier for pseudorotation, relative to literature values [3], may result from one of the following reasons (or both): in order to effect an epimerization at silicon, several consecutive pseudorotation processes must occur. During these exchanges, the electronegative Cl ligand is forced out of the apical position and is replaced by the less electronegative carbon ligand. This change involves substantial increase in energy. The second phenomenon is the change of chelate binding from apical-equatorial to equatorial-equatorial, which is associated with considerable ring strain.

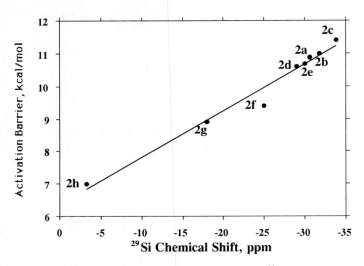

Fig. 3. Activation free energies for N-methyl exchange in **2** plotted against ^{29}Si chemical shifts.

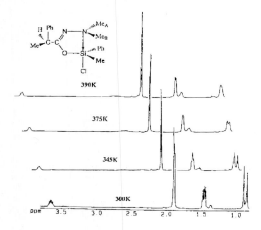

Fig. 4. High range variable temperature ^{1}H NMR spectra of **2c** in $C_6D_5NO_2$ solution.

Acknowledgments: We thank the *Fund for Basic Research*, administered by the *Israel Academy of Sciences and Humanities*, and the *Israel Ministry of Sciences* for support.

References:

[1] I. D. Kalikhman, B. A. Gostevskii, O. B. Bannikova, V. A. Pestunovich, M. G. Voronkov, *Metalloorg. Khim.* 1**989**, *2*, 937; *Metalloorg. Khim.* 1**989**, *2*, 701; *Metalloorg. Khim.* 1**989**, *2*, 205; *CA* **1990**, *112*: 1188926r; *CA* **1990**, *112*: 139113p; *CA* **1990**, *112*: 77291j.

[2] C. Chuit, R. J. P. Corriu, C. Reye, J. C. Young, *Chem. Rev.* **1993**, *93*, 1371.

[3]   F. Carré, R. J. P. Corriu, A. Kpoton, M. Poirier, G. Royo, J. C. Young, *J. Organomet. Chem.* **1994**, *470*, 43.

[4]   D. Kost, I. Kalikhman, M. Raban, *J. Am. Chem. Soc.* **1995**, *117*, 11512.

[5]   K. Tamao, T. Hayashi, Y. Ito, *Organometallics* **1992**, *11*, 2099.

[6]   I. D. Kalikhman, *Ph. D. thesis*, Irkutsk, **1990**.

[7]   V. A. Pestunovich, S. N. Tandura, B. Z. Shterenberg, V. I. Baryshok, M. G. Voronkov, *Izv. Akad. Nauk SSSR, Ser. Khim.* **1979**, 2459.

[8]   E. Kupce, E. Liepins, A. Lapsina, G. Zelcans, E. Lukevics, *J. Organomet. Chem.* **1987**, *333*, 1.

[9]   R. J. P. Corriu, A. Kpoton, M. Poirier, G. Royo, J. Y. Corey, *J. Organomet. Chem.* **1984**, *277*, C25.

[10] G. Klebe, *J. Organomet. Chem.* **1987**, *332*, 35.

Phosphine Coordination to Silicon Revisited

Gerhard Müller*, Martin Waldkircher, Andreas Pape
Fakultät für Chemie
Universität Konstanz
Universitätsstr. 10, D-78464 Konstanz, Germany
Fax: Int. code + (7531)883140
E-mail: xanorg@vg10.chemie.uni-konstanz.de

Keywords: Phosphorus Coordination / Donor-Acceptor Complexes / Structure

Summary: Silicon is coordinated only reluctantly by phosphines. Thus, silicon may conveniently serve to test the ability of phosphorus ligands to coordinate to main group elements. In this article the main classes of phosphine complexes of silicon are briefly reviewed. For each class of compounds the latest results are also given.

Introduction

Phosphorus coordination to main group elements by simple phosphines is rare. The most notable exceptions are the notoriously electron-deficient group 13 elements as acceptor atoms, in particular boron, aluminum, gallium, and indium. For the group 14 elements the ability to form donor-acceptor complexes with phosphines is well known to decrease from tin to silicon [1]. Tetravalent silicon seems to be a true borderline case, which has only a greatly reduced affinity for additional phosphorus donors. The reluctance of silicon to form hypervalent compounds with phosphorus donor atoms is nicely reflected by the virtual absence of such compounds in pertinent comprehensive review articles [2]. Thus, silicon may serve as a convenient acceptor center to test the ability of phosphine ligands to coordinate to main group elements. This article is intended to give a brief overview of the main classes of phosphine complexes of silicon with particular emphasis on structural aspects. Due to the limited amount of space this review cannot be comprehensive and the citations are only intended to be leading references. In addition to the reviewed literature, the latest developments in the field are also included.

Silicon Coordination by Neutral Phosphines

Complexes of tertiary organophosphines with silicon tetrahalides, in particular SiF_4, $SiCl_4$, and $SiBr_4$, have been prepared already more than 30 years ago [3]. Most of these complexes are of the general composition $(R_3P)_2SiX_4$ and contain hexacoordinated silicon. This work was later extended [4] and the 1:2 complexes could be shown by vibrational spectroscopy to contain *trans*-octahedral silicon [5, 6]. This was proven by X-ray crystallography for $(Me_3P)_2SiCl_4$ [7] but its molecular structure was rather ill-defined due to poor crystal quality and apparent crystal twinning.

We prepared the 3,5-dimethylbenzyl-dimethylphosphine complex *trans*-(3,5-Me$_2$C$_6$H$_3$CH$_2$-PMe$_2$)$_2$SiCl$_4$ (**1**) by simple mixing of the phosphine [8] with SiCl$_4$ in the appropriate molar ratio and determined its crystal and molecular structure [9, 10]. The resulting structural parameters (Fig. 1), especially the Si–P and Si–Cl bond lengths, should be reliable reference values for this class of compounds. In CDCl$_3$ solution, complex **1** maintains its *trans* hexacoordination of the central silicon atom as follows from its characteristic ^{29}Si and ^{31}P NMR resonances ($\delta(^{29}$Si) = -209.4 ppm (t), $^1J_{\text{Si–P}}$ = 240 Hz; $\delta(^{31}$P) = 5.9 ppm (s) with ^{29}Si satellites).

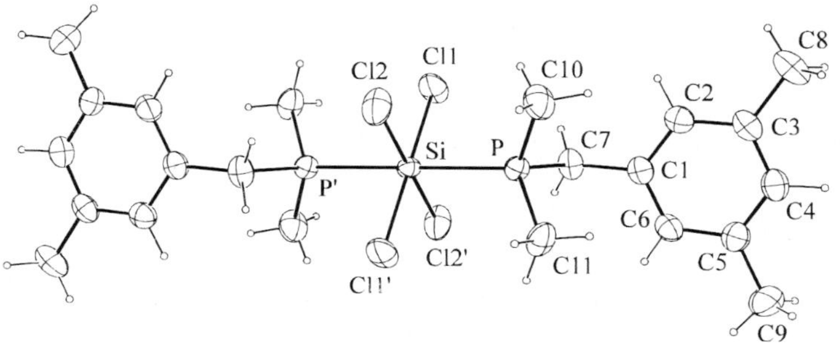

Fig. 1. Molecular structure of **1** in the crystal and atomic numbering scheme adopted [9] (ORTEP, displacement ellipsoids at the 50% probability level; H atoms with arbitrary radii). Important bond distances (Å) and angles (deg.): Si–P 2.359(1), Si–Cl1 2.205(1), Si–Cl2 2.212(1); P–Si–Cl1 87.02(3), P–Si–Cl2 90.73(2), Cl1–Si–Cl2 89.89(3). Due to the crystallographic $\bar{1}$ (C_i) symmetry of **1**, the *trans* angles at silicon are strictly 180°.

Silicon Coordination by Anionic Phosphines

Phosphine coordination to main group elements is greatly improved when anionic phosphines are used. This has been amply demonstrated by Karsch [11] with phosphinomethanides of the general type **A** as ligands. They have the carbanionoid center directly attached to the phosphine P atom, which renders the latter particularly electron-rich. Anionic phosphines of type **B**, where the negatively charged center is separated from the phosphorus donor atom by more than one bond, have been used much less for the synthesis of stable main group element phosphine complexes. They should have less electron-rich phosphorus atoms and should resemble much more "normal" (neutral) phosphines. In the following, examples for silicon complexes of both classes of anionic phosphines are given.

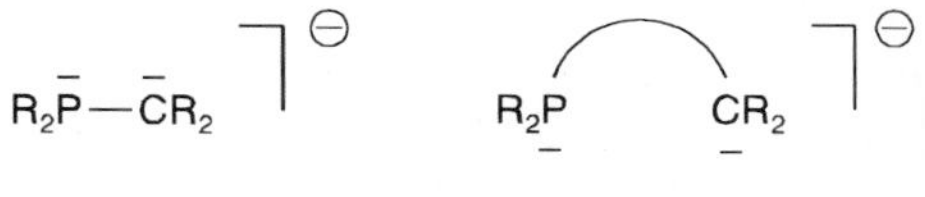

$$\left[R_2\bar{P}-\bar{C}R_2 \right]^{\ominus} \qquad \left[R_2P \overset{\frown}{\underset{-}{}} CR_2 \right]^{\ominus}$$

$$\textbf{A} \qquad\qquad \textbf{B}$$

Phosphinomethanide Complexes of Silicon

The Si^{IV} phosphinomethanide complex Cl$_2$Si[(PMe$_2$)$_2$CSiMe$_3$]$_2$ (**2**) with a Cl$_2$P$_4$ coordination sphere at silicon is shown below (Fig. 2) [12]. In this complex the chlorine atoms are arranged *trans* to each other as are pairwise the phosphorus atoms.

The respective dimethyl silicon compound Me$_2$Si[(PMe$_2$)$_2$CSiMe$_3$]$_2$ (3) is *cis* configured in the solid state [13]. In both complexes 2 and 3 the central silicon atom is hexacoordinated again. When comparing the structural details of 2 with those of 3, one notes that in the *cis* complex 3 the methyl groups exert a pronounced *trans* influence on the respective Si–P bonds opposite to them. These bonds are 2.868(2)/ 2.962(3) Å long, whereas the other two Si–P bonds, which are *trans* to each other, have bond lengths of 2.300(1) and 2.306(1) Å, respectively [13]. Alternatively stated, one may say that the methyl substituents in 3 render the central silicon atom less capable of accepting additional phosphorus donor bonds. This situation is releaved to some extent, when the silicon atom is part of a cyclobutane ring. Now the (intra-ring) C–Si–C angle is forced to be closer to the octahedral standard of 90°, which results in smaller differences between the "axial" and "equatorial" Si–P bond lengths [14]. In the meantime also phosphinomethanide complexes of Si^{IV} containing pentacoordinate silicon have been prepared, which have geometries intermediate between trigonal bipyramidal and square pyramidal [14].

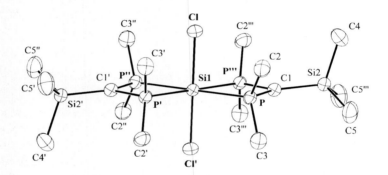

Fig. 2.   Structure of 2 in crystals of 2·THF [12]. The molecule has crystallographically imposed 2/*m* (*C*$_{2h}$) symmetry which renders all Si–P and Si–Cl bonds, respectively, to be equivalent (Si–P 2.348(1), Si–Cl 2.272(1) Å).

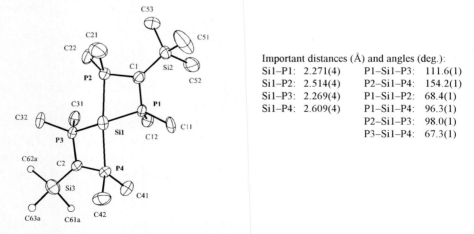

Important distances (Å) and angles (deg.):

Si1–P1:	2.271(4)	P1–Si1–P3:	111.6(1)
Si1–P2:	2.514(4)	P2–Si1–P4:	154.2(1)
Si1–P3:	2.269(4)	P1–Si1–P2:	68.4(1)
Si1–P4:	2.609(4)	P1–Si1–P4:	96.3(1)
		P2–Si1–P3:	98.0(1)
		P3–Si1–P4:	67.3(1)

Fig. 3.   Structure of 4 in the crystal [15]. The molecule has a distorted pseudo trigonal bipyramidal geometry with a (stereochemically active) lone pair in the equatorial plane. Of the disordered methyl groups at Si3 only one orientation is shown.

The extraordinary potential of phosphinomethanide ligands to form Si–P bonds (and to stabilize main group elements in low oxidation states) has been demonstrated by the preparation of the first σ-bonded silylene (silandiyl) complex $Si[(PMe_2)_2CSiMe_3]_2$, **4**, containing divalent silicon (Fig. 3) [15].

Other Anionic Phosphines as Ligands to Silicon

A fairly large number of silyl-substituted phosphines of type **B** has been prepared [16], but only in a few cases a (not necessarily intramolecular) phosphorus silicon donor-acceptor interaction has been diagnosed [17]. Using specifically the *ortho*-metallated benzylphosphines **C** and **D** as ligands, we investigated whether these anionic phosphines are able to enforce high coordination numbers at Si^{IV} by (preferably intramolecular) P→Si complexation. *ortho*-Metallated benzylphosphines are especially suitable for chelating complexation because of their P,C difunctionality and the formation of fivemembered rings upon complexation. Their potential for the complexation of main group elements has been shown already in some detail [18]. The compounds prepared by us are shown below [9]. They were synthesized from the lithiated ligands [18e, 19] and the respective silanes, $SiCl_4$ and $Si(OMe)_4$, in the appropriate molar ratios. Closely related compounds have been reported recently [16k].

With ligands **C** (R = Ph, Me) we observed no phosphorus coordination to silicon at all. This applies both to the solution and to the solid state. Fig. 4 shows the solid state structure of **10** [9, 20], while ref. [21] contains that of **6**. Characteristic for the situation in solution of compounds **5-12** are ^{29}Si NMR signals which appear at much lower field, as compared, e.g., to the hexacoordinated silicon in **1**, while the ^{31}P resonances of the uncoordinated phosphorus atoms in **5-12** are slightly shifted to higher field with respect to those in **1** [22]. Coupling between ^{29}Si and ^{31}P in compounds **5-12** is generally much smaller than in **1**, if it is observed at all. The smaller coupling clearly takes place through the ligand. In **6** also a small $^5J_{P-C}$ coupling between phosphorus and the methyl carbon atoms at silicon is observed [21].

Note that in contrast to the complexes of SiCl$_4$ with neutral phosphines, as shown above, the complexes with the carbanionic ligands **C** have at least one silicon-carbon bond. Apparently, this is enough to effectively make an additional P→Si donor-acceptor bond impossible. It is somewhat surprising that this is the case also in solution where an intramolecular P–Si bond should be particularly favored by the chelate effect.

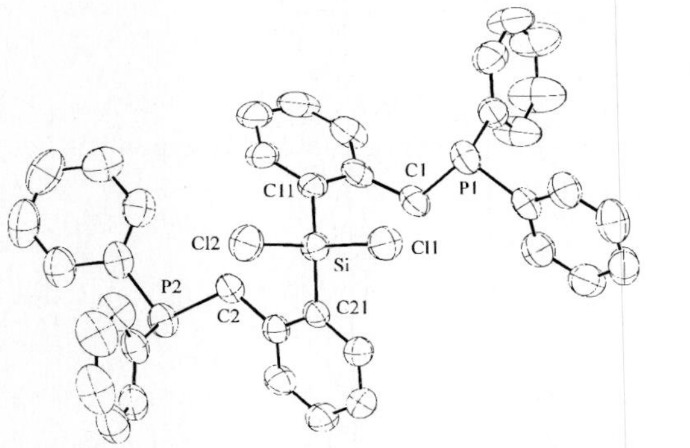

Selected distances (Å):
Si–Cl1:   2.052(1)
Si–Cl2:   2.047(1)
Si–C11:   1.868(4)
Si–C21:   1.857(3)

Fig. 4.   Solid state structure of **10** [9].

Compound **10** has been used for the preparation of chromium silylene (silandiyl) complexes containing coordinatively (and presumably also electronically) unsaturated Si^{II} [23]. Not surprisingly, in this complex an additional phosphorus coordination to silicon is observed (d(Si–P) = 2.380(1) Å), whereby silicon expands its coordination number from three to four.

Finally, with the potentially tridentate (CP$_2$) ligand **D** we were able to prepare a 1:1 complex with SiCl$_4$, **13**, in which the central silicon atom is octahedrally hexacoordinated again. Its coordination sphere consists of one carbon and three chlorine atoms in one plane and two phosphorus atoms *trans* to each other above and below this plane (Fig. 5) [24, 25]. The two five-membered chelate rings formed upon complexation of **D** are close to planar in **13**. This is clearly a consequence of the Si–P bond lengths, which are slightly longer than 2.3 Å. Planar chelate rings in complexes of **D** had been predicted previously if the intra-ring P→Element donor bonds are in this bond length range [26]. Not surprisingly, the octahedral coordination of **13** persists in solution, as indicated again by the NMR data (δ(^{29}Si) = -186 ppm (t), $^1J_{Si-P}$ = 138 Hz; δ(^{31}P) = -17 ppm (s)). The fact that hexacoordination at silicon is possible with ligand **D** clearly must be a consequence of the ligand shape, which strongly imposes a meridional octahedral complex geometry, thereby overcoming the reluctance of silicon to accept additional phosphorus complexation, as was evident in compounds **5-12**.

In summary, we conclude that in contrast to phosphinomethanides **A**, anionic phosphines of type **B** can only form intra- or intermolecular phosphorus silicon donor-acceptor complexes under particularly favorable circumstances. These are seen primarily in suitable ligand geometries. Multidentate phosphine ligands which strongly impose coordination numbers higher than four to silicon seem to be especially suitable. In accord with previous findings [4], chlorine atoms as further silicon substituents favor the phosphine coordination additionally. With some caution we also state that hypervalent silicon with phosphorus donor atoms seems to prefer hexacoordination over pentacoordination.

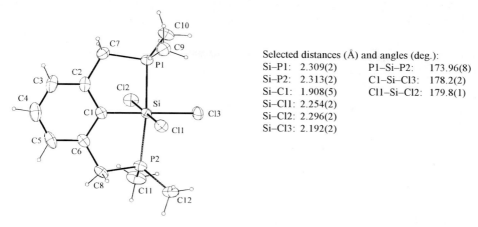

Selected distances (Å) and angles (deg.):

Si–P1:	2.309(2)	P1–Si–P2:	173.96(8)
Si–P2:	2.313(2)	Cl1–Si–Cl3:	178.2(2)
Si–C1:	1.908(5)	Cl1–Si–Cl2:	179.8(1)
Si–Cl1:	2.254(2)		
Si–Cl2:	2.296(2)		
Si–Cl3:	2.192(2)		

Fig. 5.   Molecular structure of **13** in the solid state [24].

Acknowledgements: We are grateful to the *Deutsche Forschungsgemeinschaft*, and the *Fonds der Chemischen Industrie* for continuous generous support of our work.

References:

[1]   N. C. Norman, N. L. Pickert, *Coord. Chem. Rev.* **1995**, *145*, 27; W. Levason, C. A. McAuliffe, *Coord. Chem. Rev.* **1976**, *19*, 173.

[2]   C. Chuit, R. J. P. Corriu, C. Reye, J. C. Young, *Chem. Rev.* **1993**, *93*, 1371; S. N. Tandura, M. G. Voronkov, N. V. Alekseev, *Top. Curr. Chem.* **1986**, *131*, 99.

[3]   K. Issleib, H. Reinhold, *Z. Anorg. Allg. Chem.* **1962**, *314*, 113.

[4]   I. R. Beattie, G. A. Ozin, *J. Chem. Soc. (A)*, **1969**, 2267.

[5]   I. R. Beattie, G. A. Ozin, *J. Chem. Soc. (A)*, **1970**, 370.

[6]   For phosphine adducts of hexachlorodisilane containing pentacoordinated silicon see: L.-P. Müller, W. W. du Mont, J. Jeske, P. G. Jones, *Chem. Ber.* **1995**, *128*, 615.

[7]   H. E. Blayden, M. Webster, *Inorg. Nucl. Chem. Lett.* **1970**, *6*, 703.

[8]   G. Müller, M. Waldkircher, M. Winkler, *Z. Naturforsch.* **1994**, *49b*, 1606.

[9]   M. Waldkircher, unpublished results.

[10]   Crystal structure data of **1**: Enraf Nonius CAD4 diffractometer, $Mo_{K\alpha}$ radiation, $\lambda = 0.71069$ Å, graphite monochromator. $C_{22}H_{34}Cl_4P_2Si$, $M_r = 530.362$, triclinic space group $P\bar{1}$ (No. 2), $a = 8.164(2)$, $b = 8.494(3)$, $c = 10.470(3)$ Å, $\alpha = 98.37(1)$, $\beta = 109.34(1)$, $\gamma = 103.45(1)°$, $V = 646.38$ Å^3, $Z = 1$, $d_{calcd} = 1.362$ g cm^{-3}, $\mu(Mo_{K\alpha}) = 6.4$ cm^{-1}, $F(000) = 278$ e, $R(F) = 0.048$, $wR(F^2) = 0.120$ for 2922 unique F^2 and 133 refined parameters (SHELXL93).

[11] H. H. Karsch, *Russ. Chem. Bull.* **1993**, *42*, 1937; H. H. Karsch, A. Appelt, G. Müller, *Organometallics* **1986**, *5*, 1664; H. H. Karsch, A. Appelt, J. Riede, G. Müller, *Organometallics* **1987**, *6*, 316; H. H. Karsch, B. Deubelly, G. Hanika, J. Riede, G. Müller, *J. Organomet. Chem.* **1988**, *344*, 153; H. H. Karsch, K. Zellner, P. Mikulcik, J. Lachmann, G. Müller, *Organometallics*, **1990**, *9*, 190, and refs. cited therein; H. H. Karsch, K. Zellner, G. Müller, *J. Chem. Soc., Chem. Commun.* **1991**, 466; H. H. Karsch, K. Zellner, S. Gamper, G. Müller, *J. Organomet. Chem.* **1991**, *414*, C39; H. H. Karsch, G. Baumgartner, S. Gamper, J. Lachmann, G. Müller, *Chem. Ber.* **1992**, *125*, 1333; H. H. Karsch, G. Grauvogl, P. Mikulcik, P. Bissinger, G. Müller, *J. Organomet. Chem.* **1994**, *465*, 65.

[12] H. H. Karsch, B. Deubelly, U. Keller, O. Steigelmann, J. Lachmann, G. Müller, *Chem. Ber.* **1996**, *129*, 671.

[13] H. H. Karsch, B. Deubelly, U. Keller, F. Bienlein, R. Richter, P. Bissinger, M. Heckel, G. Müller, *Chem. Ber.* **1996**, *129*, 759.

[14] H. H. Karsch, R. Richter, E. Witt, *"Novel Sila-Phospha-Heterocycles and Hypervalent Silicon Compounds with Phosphorus Donors"*, in: *Organosilicon Chemistry III: From Molecules to Materials* (Eds.: N. Auner, J. Weis), VCH, Weinheim, **1997**, p. 460.

[15] H. H. Karsch, U. Keller, S. Gamper, G. Müller, *Angew. Chem.* **1990**, *102*, 297; *Angew. Chem., Int. Ed. Engl.* **1990**, *29*, 295.

[16] a) H. G. Ang, P. T. Lau, *J. Organomet. Chem.* **1972**, *37*, C4.

b) A. A. Oswald, L. L. Murrel, L. J. Boucher, *Div. Pet. Chem., Am. Chem. Soc.* **1974**, *19*, 155.

c) L. J. Boucher, A. A. Oswald, L. L. Murrel, *Div. Pet. Chem., Am. Chem. Soc.* **1974**, *19*, 161.

d) M. Capka, J. Schraml, H. Jancke, *Collect. Czech. Chem. Commun.* **1978**, *43*, 3347.

e) J. Grobe, G. F. Scheuer, *Z. Anorg. Allg. Chem.* **1978**, *443*, 83.

f) J. Grobe, J. Hendriock, G. F. Scheuer, *Z. Anorg. Allg. Chem.* **1978**, *443*, 97.

g) R. D. Holmes-Smith, R. D. Osei, S. R. Stobart, *J. Chem. Soc., Perkin Trans.* **1983**, 861.

h) J. Grobe, N. Krummen, D. Le Van, *Z. Naturforsch.* **1984**, *39b*, 1711.

i) H. G. Ang, B. Chang, W. L. Kwik, *J. Chem. Soc., Dalton Trans.* **1992**, 2161.

j) F. L. Joslin, S. R. Stobart, *Inorg. Chem.* **1993**, *32*, 2221.

k) R. A. Gossage, G. D. McLennan, S. R. Stobart, *Inorg. Chem.* **1996**, *35*, 1729.

[17] See especially: J. Grobe, W. Hildebrandt, R. Martin, A. Walter, *Z. Anorg. Allg. Chem.* **1991**, *592*, 121.

[18] a) G. Müller, J. Lachmann, A. Rufinska, *Organometallics* **1992**, *11*, 2970.

b) G. Müller, J. Lachmann, *Z. Naturforsch.* **1993**, *48b*, 1248.

c) G. Müller, J. Lachmann, *Z. Naturforsch.* **1993**, *48b*, 1544.

d) M. Winkler, M. Lutz, G. Müller, *Angew. Chem.* **1994**, *106*, 2372; *Angew. Chem., Int. Ed. Engl.* **1994**, *33*, 2279.

e) A. Pape, M. Lutz, G. Müller, *Angew. Chem.* **1994**, *106*, 2375; *Angew. Chem., Int. Ed. Engl.* **1994**, *33*, 2281.

[19] H.-P. Abicht, K. Issleib, *Z. Anorg. Allg. Chem.* **1976**, *422*, 237; *Z. Anorg. Allg. Chem.* **1978**, *447*, 53.

[20] Crystal structure data of **10**: $C_{38}H_{32}Cl_2P_2Si$, M_r = 649.618, monoclinic space group $P2_1/c$ (No. 14), a = 15.521(3), b = 8.834(1), c = 25.547(4) Å, β = 106.18(1)°, V = 3364.07 Å^3, Z = 4, d_{calcd} = 1.282 g cm^{-3}, $\mu(Mo_{K\alpha})$ = 3.5 cm^{-1}, $F(000)$ = 1352 e, $R(F)$ = 0.059, $wR(F)$ = 0.045 for 4253 unique F_o with $F_o > 4.0 \, \sigma(F_o)$ and 388 refined parameters (SHELX76).

[21]   G. Reber, J. Riede, G. Müller, *Z. Naturforsch.* **1988**, *43b*, 915.

[22]   Selected NMR data (C_6D_6, RT; chemical shifts in ppm): 7: $\delta(^{29}Si)$ = -1.7 (s), $\delta(^{31}P)$ = -39.3 (s); **8**: $\delta(^{29}Si)$ = -5.9 (ddd), $^4J_{Si-P}$ = 10.0, $^1J_{Si-H}$ = 293 Hz, $^3J_{Si-HA}$ = 8 Hz; $^4J_{Si-HB}$ = 2 Hz, $\delta(^{31}P)$ = -38.5 (s); **9**: $\delta(^{29}Si)$ = -54.2 (d), $^4J_{Si-P}$ = 1.1 Hz, $\delta(^{31}P)$ = -41.3 (s).

[23]   H. Handwerker, M. Paul, J. Blümel, C. Zybill, *Angew. Chem.* **1993**, *105*, 1375; *Angew. Chem., Int. Ed. Engl.* **1993**, *32*, 1313.

[24]   A. Pape, unpublished results.

[25]   Crystal structure data of **13**: $C_{12}H_{19}Cl_3P_2Si$, M_r = 359.678, orthorhombic space group *Pbca* (No. 61), a = 11.022(1), b = 13.085(1), c = 22.889(4) Å, V = 3301.12 Å^3, Z = 8, d_{calcd} = 1.447 g cm^{-3}, $\mu(Mo_{K\alpha})$ = 8.0 cm^{-1}, $F(000)$ = 1488 e, $R(F)$ = 0.140, $wR(F^2)$ = 0.125 for 3054 unique F^2 and 207 refined parameters (SHELXL93).

[26]   G. Müller, J. Lachmann, J. Riede, *Z. Naturforsch.* **1992**, *47b*, 823.

Novel Sila-Phospha-Heterocycles and Hypervalent Silicon Compounds with Phosphorus Donors

Hans H. Karsch*, Roland Richter, Eva Witt

Anorganisch-Chemisches Institut

Technische Universität München

Lichtenbergstr. 4, D-85747 Garching, Germany

Tel./Fax: Int. code + (89)32093132

E-mail: Hans.H.Karsch@lrz.tu-muenchen.de

Keywords: Phosphinomethanides / Silaheterocycles / Pentacoordination / Hexacoordination

Summary: The reaction of two equivalents of lithium phosphinomethanides with di- or trifunctional chlorosilanes yields novel five- and six-membered heterocycles by multistep rearrangements or transmetallation reactions. Silaethene intermediates and hypervalent intermediates are likely to be involved. The reaction of one or two equivalents of lithium phosphinomethanide {Li[C(PMe$_2$)$_2$(SiMe$_2$Ph)]}·TMEDA with pTolSiCl$_3$ and CH$_2$CH$_2$CH$_2$SiCl$_2$ yields novel penta- and hexacoordinated silicon complexes. Both are the first examples of truly hypervalent organosilicon species with phosphorus donors characterized by X-ray structure determination.

1. Introduction

The ambidentate nature of phosphinomethanides **I** is well documented. They may react with electrophiles either via the carbon or the phosphorus atom, forming heteroelement substituted methanes or phosphorus ylides. The reactivity of the phosphinomethanides can be tuned by the choice of P and C substituents R, X, and Y and by the specific reaction conditions.

R = Me, Ph

X,Y = H, PR$_2$, SiR$_3$

Using *di*phosphinomethanides (Y = PR$_2$), e.g., Li[C(PMe$_2$)$_2$(SiMe$_3$)], the ylides obtained show structural in solution [1]. PhSiCl$_3$ reacts with Li[C(PMe$_2$)$_2$(SiMe$_3$)] to give a non-rigid, pentacoordinated compound with Si–P bond formation. The isolated colorless oil is characterized by NMR spectroscopy. Storage of **1** for several weeks at ambient temperature results in the formation of PhSiCl$_2$–C(SiMe$_3$)=PMe$_2$–PMe$_2$ (**3**). The formation of **3** may be understood by an initial rearrangement to the tetraheteroatom substituted methane derivative and subsequently to the ylide **3** involving a P–P bond formation. By similar rearrangement reactions, even heterocycles are formed [2].

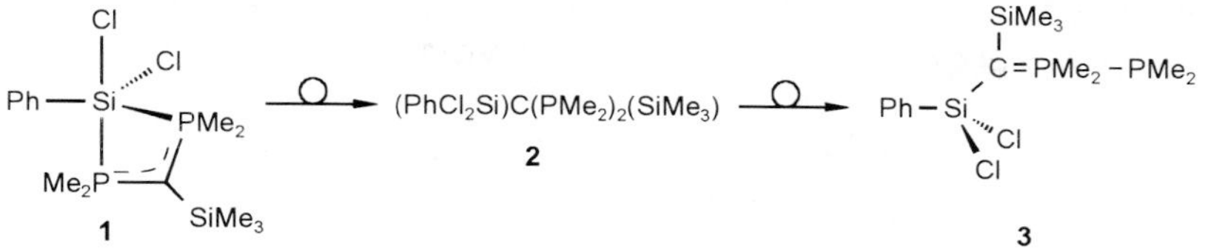

Eq. 1.

The reaction of $SiCl_4$ with two equivalents of the not completely heteroatom substitituted lithium *di*phosphinomethanide $Li[CH(PMe_2)_2]$ leads to Si–C bond formation [3].

$$SiCl_4 \quad + \quad 2\ LiCH(PMe_2)_2 \longrightarrow Cl_2Si[CH(PMe_2)_2]_2$$
$$\mathbf{4}$$

Eq. 2.

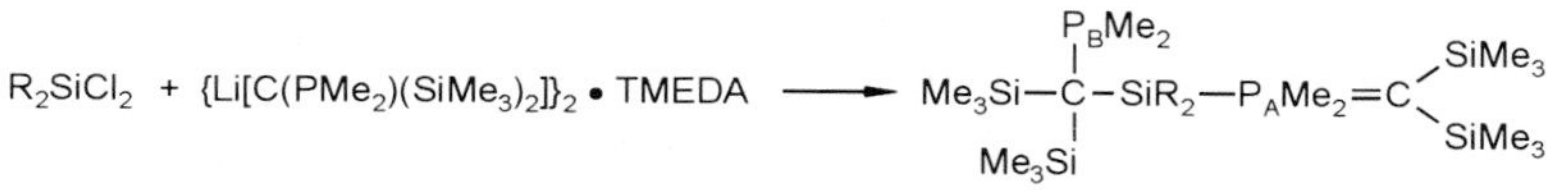

$$R = Cl\ (\mathbf{5a})$$
$$R = Me\ (\mathbf{5b})$$

Eq. 3.

In contrast, with $\{Li[C(PMe_2)(SiMe_3)_2]\}_2 \cdot TMEDA$, both Si–C and Si–P bond formation is observed.

With $SiCl_4$ and Me_2SiCl_2 two equivalents of $Li[C(PMe_2)_2(SiMe_3)]$ react under Si–P bond formation exclusively, yielding hexacoordinated silicon compounds [4]. $Me_2Si[(PMe_2)_2C(SiMe_3)]_2$ shows "true" hexacoordination only in solution, whereas in the solid state, only weak $Si-P_{eq}$ interactions add to a molecular framework on the borderline between tetrahedral and (distorted) octahedral. $Me_2Si[(PMe_2)_2C(SiMe_3)]_2$ decomposes in solution after a few days at room temperature, forming a mixture of mainly three different rearrangement products, which can be characterized by ^{31}P NMR spectroscopy [5].

Results

Reaction of Ph₂SiCl₂ with two equivalents of Li[CH(PMe₂)₂]

In contrast to the reaction of $SiCl_4$ with two equivalents of $Li[CH(PMe_2)_2]$, Ph_2SiCl_2 reacts with $LiCH(PMe_2)_2$ under formation of a six-membered heterocycle.

Steric hinderance obviously renders a second substitution step more difficult than a transmetallation reaction. *Di*phosphinomethane $CH_2(PMe_2)_2$ is formed and identified by NMR spectroscopy. A subsequent LiCl elimination occurs either intra- (via a silaethene derivative) or intermolecularly. In both cases dimerization leads to the formation of a novel six-membered heterocycle. Formally, in **6** two phosphinocarbene ligands bridge two diphenyl silylenes via carbon and phosphorus atoms. The $^{31}P\{^{1}H\}$ NMR spectrum shows an AA'BB'-type pattern. The structure of **6** unambiguously has been confirmed by an X-ray structure determination.

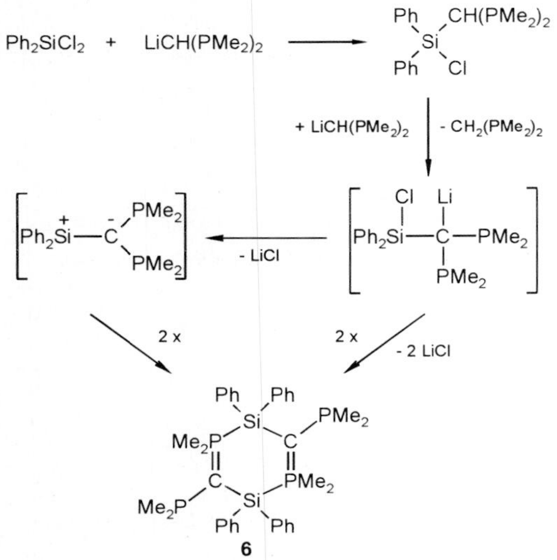

Scheme 1. Proposed pathway for the formation of **6**.

Reaction of PhSiCl₃ with {Li[C(PMe₂)(SiMe₃)₂]}₂·TMEDA

As described in Eq. 3, SiCl₄ reacts with {Li[C(PMe₂)(SiMe₃)₂]}₂·TMEDA under formation of a disubstitution product. An analogous reaction sequence (monosubstitution: Si–C bond formation, **7**; disubstitution: Si–P bond formation, **8**) can be observed in the reaction of PhSiCl₃ with {Li[C(PMe₂)(SiMe₃)₂]}₂·TMEDA. On storage of pure **8** at ambient temperature conversion to a novel heterocycle **13**, which is obtained as very air-sensitive colorless crystals, is observed.

The reaction pathway depicted in Scheme 2 seems reasonable. An isomerization, comparable to that in Eq. 1, initiates a multistep rearrangement. The sterically overcrowded **9** is not stable and releases steric strain by Me₂PCl elimination, thus generating a silaethene intermediate **10**. Me₂P–PMe₂=C(SiMe₃)₂ can be identified as a byproduct and thus indicates a reaction of Me₂PCl with {Li[C(PMe₂)(SiMe₃)₂]}₂·TMEDA. The silaethene **10** is not detected: it isomerizes by SiMe₃ group migration to give **11**, which in turn suffers a methyl group migration, assisted by Si–P bond formation to give **13**. This last step is reminiscent of a known type of reaction at an aluminum center [6].

In line with the proposed structure of **13** is the single line $^{31}P\{^{1}H\}$ NMR resonance at $\delta = -24.45$ ppm, which is accompanied by two sets of satellites that are mirrored in the $^{29}Si\{^{1}H\}$ NMR spectrum. An X-ray analysis of **13** confirms the structure in solution being present also in the solid state.

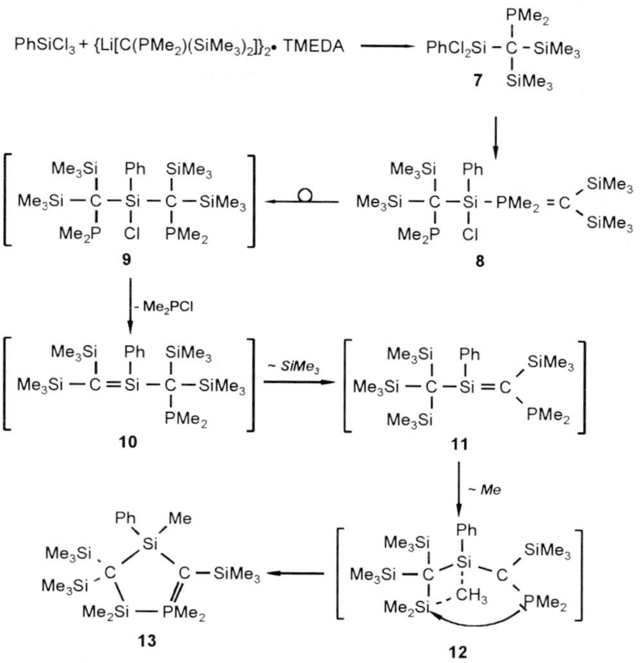

PhSiCl$_3$ + {Li[C(PMe$_2$)(SiMe$_3$)$_2$]}$_2$• TMEDA ⟶ PhCl$_2$Si – C – SiMe$_3$

7

[Scheme 2 reaction diagram showing structures **9**, **8**, **10**, **11**, **13**, **12** with intermediate transformations]

9 ⟵ O ⟵ **8**

– Me$_2$PCl

10 ⟶ ~ SiMe$_3$ ⟶ **11**

~ Me

13 ⟵ **12**

Scheme 2.   Proposed reaction scheme for the formation of **13**.

Hypervalent Silicon Phosphinomethanide Complexes

The aim of obtaining a crystalline complex with pentacoordinated silicon was achieved as follows: In order to get a better crystallinity compared whit that of PhSiCl$_2$[(PMe$_2$)$_2$C(SiMe$_3$)], the phenyl group was replaced by a *p*-tolylgroup and the SiMe$_3$ by a SiMe$_2$Ph group.

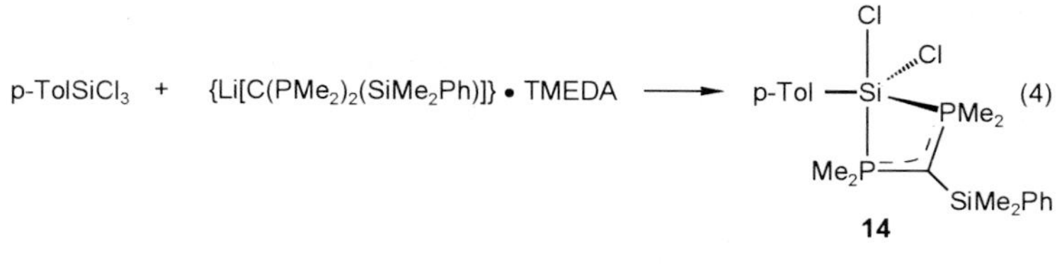

p-TolSiCl$_3$ + {Li[C(PMe$_2$)$_2$(SiMe$_2$Ph)]} • TMEDA ⟶ [structure **14**: p-Tol—Si with Cl, Cl, PMe$_2$, Me$_2$P, SiMe$_2$Ph]   (4)

14

Eq. 4.

Treatment of *p*TolSiCl$_3$ with {Li[C(PMe$_2$)$_2$(SiMe$_2$Ph)]}·TMEDA in an ethereal solvent leads to the isolation of the monosubstitution product. *p*TolSiCl$_2$[(PMe$_2$)$_2$C(SiMe$_2$Ph)] is obtained as a colorless crystalline solid. Only one single resonance in the ^{31}P{^{1}H} NMR spectrum of **14** at ambient temperature (δ^{31}P 39.08 ppm) and -100°C (δ^{31}P 39.33) indicates conformational unrigidity with rapid equilibration of the phosphino groups (pseudorotation). The singlet resonance is accompanied by silicon satellites (^1J(PSi) = 77.3 Hz), which is reflected by a triplet resonance for the central silicon nucleus in the ^{29}Si{^{1}H} NMR spectrum. Pentacoordination is also indicated by the high field shift of this nucleus (δ^{29}Si -75.98). The ^{1}H and ^{13}C NMR spectra of **14** are in full accord with the given structure, which has been confirmed by X-ray analysis.

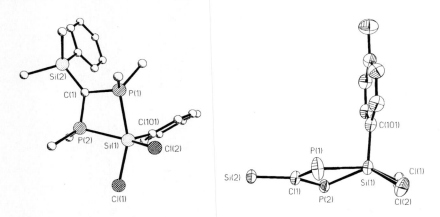

Fig. 1.   Molecular structure of **14** (H atoms omitted).

In the solid state (Fig. 1), the molecule adopts a geometry at the central Si(1) atom, which can be described as intermediate between a trigonal bipyramid and a tetragonal pyramid. The deviation from the tbp-geometry is documented by the (comparatively) large Cl(2)–Si(1)–P(2) angle (134.58°) and the small Cl(1)–Si(1)–P(1) angle (158.64(4)°) and is caused by the very small "bite-angle" of the diphosphinomethanide ligand with 69.82(3)° (P(1)···P(2) nonbonding distance = 2.697 Å). In comparison with "equatorial" bond distances in tbp-structures, the "axial" distances are usually longer by ca 8 %, which roughly fits the actual difference in Si–P$_{ax}$ (2.407(1) Å) and Si–P$_{eq}$ (2.304(1) Å) bond lenghts.

The aim of an optimized stabilization of an octahedral environment at silicon compared with Me$_2$Si[(PMe$_2$)$_2$C(SiMe$_3$)]$_2$ was achieved by replacing the Me$_2$Si moiety by a silacyclobutane moiety. As above, the SiMe$_3$ group of the ligand was replaced by a SiMe$_2$Ph group for better crystallinity.

$$\text{silacyclobutane-SiCl}_2 \; + \; 2\,\{\text{Li[C(PMe}_2)_2(\text{SiMe}_2\text{Ph})]\} \cdot \text{TMEDA} \; \longrightarrow \; \mathbf{15} \qquad (5)$$

Eq. 5.

Two equivalents of {Li[C(PMe$_2$)$_2$(SiMe$_2$Ph)]}·TMEDA react with $\overline{\text{CH}_2\text{CH}_2\text{CH}_2\text{SiCl}_2}$ in diethyl ether to give **15** as a colorless, crystalline solid. The ^{31}P{^{1}H} NMR spectrum of **15** shows a broad singlet resonance at +20°C. Cooling to -60°C establishes an AA'BB'-spin system, as expected for a *cis*-conformation. In contrast to Me$_2$Si[(PMe$_2$)$_2$C(SiMe$_3$)]$_2$, which decomposes in solution in a few days at room temperature, the thermal stability of **15** is noteworthy: heating **15** for four weeks (80°C) leaves **15** unchanged besides a minor amount of HC(PMe$_2$)$_2$(SiMe$_2$Ph) (hydrolysis) [5].

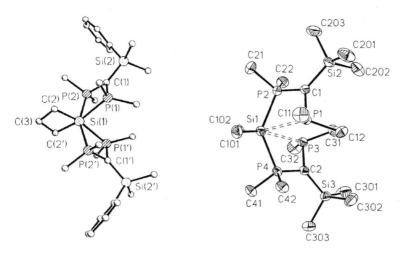

Fig. 2. Molecular structure of **15** and of Me$_2$Si[(PMe$_2$)$_2$C(SiMe$_3$)]$_2$ for comparison (H atoms omitted).

The molecular structure (Fig. 2) of **15** shows three planar four-membered ring systems (sum of the angles in each case 360.0°), one of the silacyclobutane moiety and two of the *di*phosphinomethanide chelate rings. The difference between the Si–P$_{ax}$ and Si–P$_{eq}$ bond lenghts (2.374(1)/2.481(1) Å) is within the expected range considering the respective *trans* influences and clearly distinguishes this "true" hexacoordination from the case of Me$_2$Si[(PMe$_2$)$_2$C(SiMe$_3$)]$_2$ (Si–P$_{ax}$: 2.300/2.306 Å; Si–P$_{eq}$: 2.868/2.962 Å). The dihedral angle between the silacyclobutane plane and the plane through P(1)–Si(1)–P(1)' of 17° indicates the deviation from an ideal octahedral geometry. The small P(2)–Si(1)–P(2)' angle of 155° is a consequence of the ring strain within the four membered rings with a ligand bite angle of 68.4°, which also causes a small P(1)···P(2) nonbonding distance of 2.729 Å.

References:

[1] H. H. Karsch, R. Richter, A. Schier, *Z. Naturforsch.* **1993**, *48b*, 1533.

[2] H. H. Karsch, R. Richter, A. Schier, M. Heckel, R. Ficker, W. Hiller, *J. Organomet. Chem.* **1995**, *501*, 167.

[3] H. H. Karsch, R. Richter, B. Deubelly, A. Schier, M. Paul, M. Heckel, K. Angermaier, W. Hiller, *Z. Naturforsch.* **1994**, *49b*, 1798.

[4] a) H. H. Karsch, B. Deubelly, U. Keller, O. Steigelmann, J. Lachmann, G. Müller, *Chem. Ber.* **1996**, *129*, 671.

b) H. H. Karsch, B. Deubelly, U. Keller, F. Bienlein, R. Richter, P. Bissinger, M. Heckel, G. Müller, *Chem. Ber.* **1996**, *129*, 759.

[5] R. Richter, *Dissertation*, Technische Universität München, **1996**.

[6] H. H. Karsch, K. Zellner, G. Müller, *Organometallics* **1991**, *10*, 2884.

Germanium Analogues of Zwitterionic Spirocyclic $\lambda^5 Si$-Silicates

Joachim Heermann, Reinhold Tacke *

Institut für Anorganische Chemie
Bayerische Julius-Maximilians-Universität zu Würzburg
Am Hubland, D-97074 Würzburg, Germany

Peter G. Jones

Institut für Anorganische und Analytische Chemie
Technische Universität Carolo-Wilhelmina zu Braunschweig
Hagenring 30, D-38023 Braunschweig, Germany

Keywords: Germanium / Zwitterions / $\lambda^5 Si$-Silicates / Crystal Structure

Summary: The novel zwitterionic (molecular) spirocyclic $\lambda^5 Ge$-germanates **1b-7b** have been synthesized and characterized. Compounds **1b-7b** are germanium analogues of the already known zwitterionic $\lambda^5 Si$-silicates **1a-7a** (Ge/Si exchange). Syntheses and properties of the Si/Ge analogues are reported. In addition, the crystal structures of **4b** and **5b·H₂O** are described.

Introduction

Over the past few years, a series of zwitterionic spirocyclic $\lambda^5 Si$-silicates with an SiO_4C framework have been synthesized and structurally characterized [1]. The $\lambda^5 Si$-silicates **1a-7a** are typical examples of this particular type of compound (Fig. 1). In this paper, we report on the analogous zwitterionic spirocyclic $\lambda^5 Ge$-germanates **1b-7b** (Fig. 1).

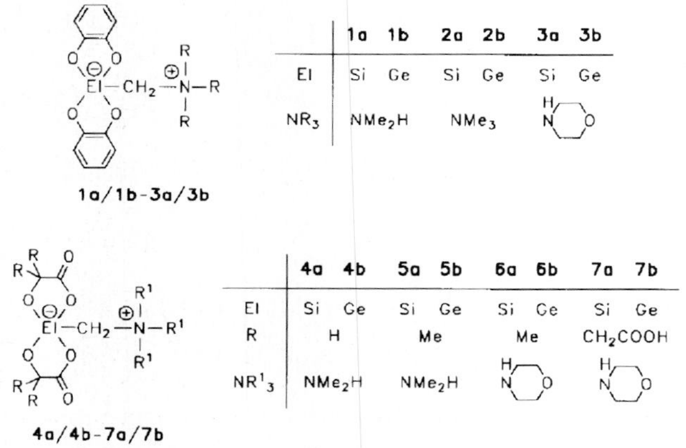

Fig. 1.

Compounds **1a-7a** and **1b-7b** are characterized by the presence of a pentacoordinate (formally negatively charged) silicon or germanium atom and a tetracoordinate (formally positively charged) nitrogen atom. Syntheses and properties of the Si/Ge analogues **1a/1b-7a/7b** are reported. In addition, the crystal structures of the $\lambda^5 Ge$-germanates **4b** and **5b·H$_2$O** are described.

Results and Discussion

Following the strategy used for the syntheses of the $\lambda^5 Si$-silicates **1a** [1a], **2a** [2], and **3a** [1c], the zwitterionic $\lambda^5 Ge$-germanates **1b-3b** were synthesized according to Scheme 1. Compound **1b** was obtained by reaction of the germane $(MeO)_3GeCH_2NMe_2$ with two mole equivalents of 1,2-dihydroxybenzene. The derivative **3b** was synthesized analogously, starting from trimethoxy-(morpholinomethyl)germane. *N*-Quaternization of $(MeO)_3GeCH_2NMe_2$ with methyl iodide and subsequent reaction of the resulting ammonium salt $[(MeO)_3GeCH_2NMe_3]I$ with two mole equivalents of 1,2-dihydroxybenzene yielded compound **2b**.

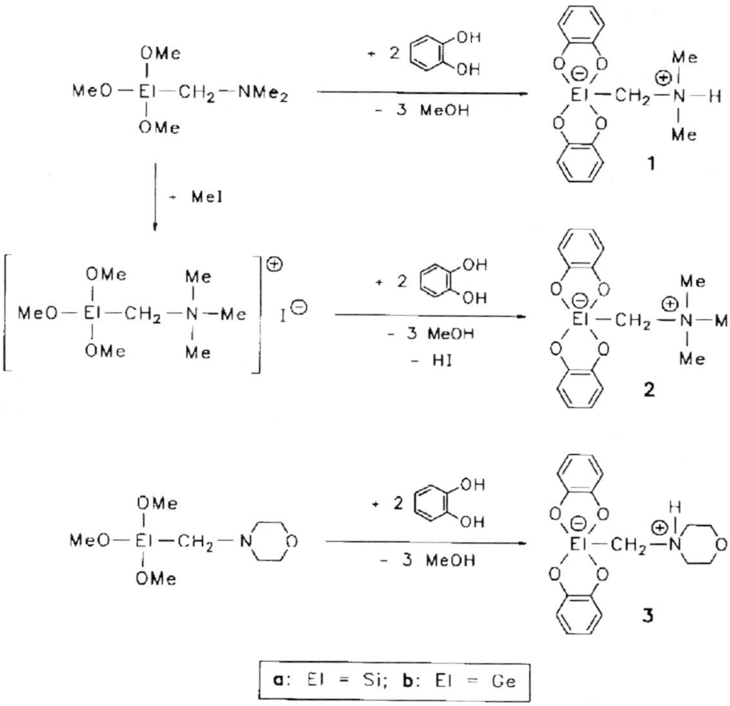

Scheme 1.

The $\lambda^5 Ge$-germanates **4b-7b** could also be prepared by analogy to the syntheses of the related $\lambda^5 Si$-silicates **4a** [3], **5a** [3], **6a** [4], and **7a** [5] (Scheme 2). Compounds **4b** and **5b** were obtained by reaction of the germane $(MeO)_3GeCH_2NMe_2$ with two mole equivalents of glycolic acid and 2-methyllactic acid, respectively. The derivatives **6b** and **7b** were obtained analogously by treatment of trimethoxy(morpholinomethyl)germane with 2-methyllactic acid and citric acid, respectively. The

λ^5Ge-germanates **4b** and **5b** were additionally synthesized by an alternative method involving a Ge–C bond cleavage reaction (Ge–Ph cleavage; formation of benzene), starting from the germane Ph(MeO)$_2$GeCH$_2$NMe$_2$.

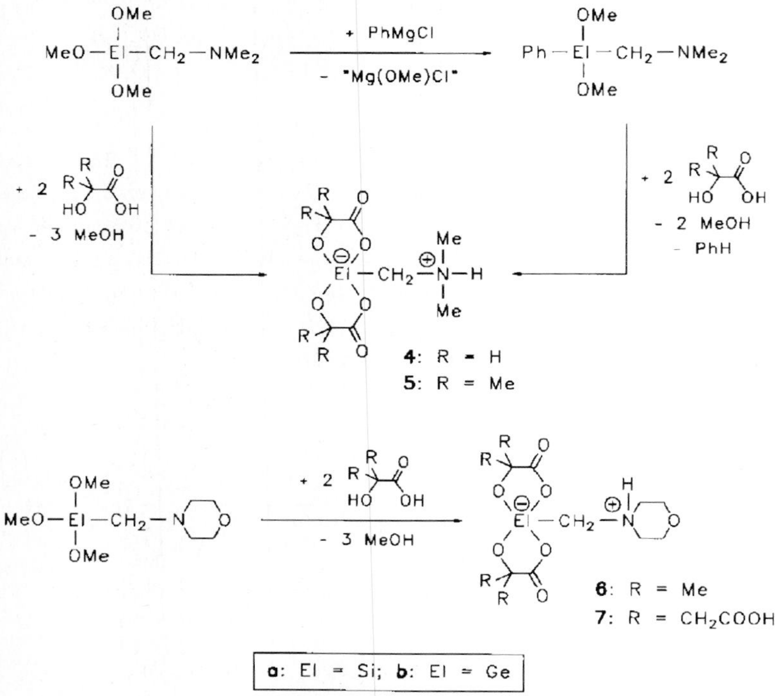

Scheme 2.

The λ^5Ge-germanates **1b-7b** were prepared analogously to the related silicon compounds **1a-7a** in acetonitrile at room temperature and isolated in good yield as crystalline solids. Their identities were established by elemental analyses, NMR experiments (^{1}H, ^{13}C; [D$_6$]DMSO), and mass spectrometry. In addition, the crystal structures of **4b** and **5b**·H$_2$O (obtained by recrystallization of **5b** from water) were determined by single-crystal X-ray diffraction.

As may be expected from their zwitterionic nature, the λ^5Si-silicates **1a-7a** and the λ^5Ge-germanates **1b-7b** are high-melting crystalline solids. All compounds are almost insoluble in unpolar solvents and exhibit very poor solubility in polar organic solvents.

The structures of the zwitterions in the crystal of **4b** and **5b**·H$_2$O are depicted in Fig. 2. The coordination polyhedra surrounding the germanium atoms in these compounds are distorted trigonal bipyramids with the carboxylate oxygen atoms in the axial positions. Compound **5b**·H$_2$O is isostructural with its silicon analogue **5a**·H$_2$O [1k]. Selected geometric parameters for the coordination polyhedra of **4b** and **5b**·H$_2$O are given in Table 1. In general, the molecular structures of **4b** and **5b** (see Fig. 2) are very similar to those observed for related zwitterionic λ^5Si-silicates.

Table 1. Selected geometric parameters for **4b** and **5b·H$_2$O** (distances in pm, angles in deg).

	Ge–O(1)	Ge–O(2)	Ge–O(3)	Ge–O(4)	Ge–C(1)	O(1)–Ge–O(3)
4b	195.29(13)	178.60(14)	191.74(14)	177.26(14)	195.4(2)	171.51(6)
5b·H$_2$O	192.32(11)	178.01(11)	192.06(11)	179.20(11)	194.7(2)	170.76(5)

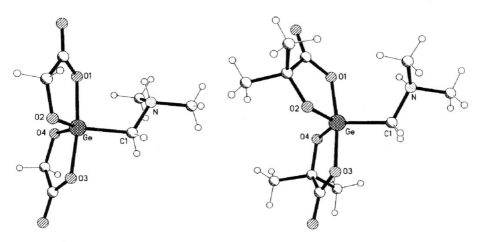

Fig. 2. Molecular structure of the zwitterions in the crystal state of **4b** (left) and **5b·H$_2$O** (right). In this context, see Table 1.

In conclusion, there are many similarities in the synthetic and structural chemistry of the title compounds and their corresponding silicon analogues (in this context, see also refs. [1j, 6]).

Acknowledgement: Financial support of this work by the *Deutsche Forschungsgemeinschaft* and the *Fonds der Chemischen Industrie* is gratefully acknowledged.

References:

[1]   a) R. Tacke, A. Lopez-Mras, J. Sperlich, C. Strohmann, W. F. Kuhs, G. Mattern, A. Sebald, *Chem. Ber.* **1993**, *126*, 851.

b) R. Tacke, A. Lopez-Mras, W. S. Sheldrick, A. Sebald, *Z. Anorg. Allg. Chem.* **1993**, *619*, 347.

c) J. Sperlich, J. Becht, M. Mühleisen, S. A. Wagner, G. Mattern, R. Tacke, *Z. Naturforsch.* **1993**, *48b*, 1693.

d) R. Tacke, A. Lopez-Mras, P. G. Jones, *Organometallics* **1994**, *13*, 1617.

e) R. Tacke, M. Mühleisen, P. G. Jones, *Angew. Chem.* **1994**, *106*, 1250; *Angew. Chem. Int., Ed. Engl.* **1994**, *33*, 1186.

f) M. Mühleisen, R. Tacke, *Chem. Ber.* **1994**, *127*, 1615.

g) M. Mühleisen, R. Tacke, *Organometallics* **1994**, *13*, 3740.

h) R. Tacke, M. Mühleisen, A. Lopez-Mras, W. S. Sheldrick, *Z. Anorg. Allg. Chem.* **1995**, *621*, 779.

i) R. Tacke, J. Becht, O. Dannappel, M. Kropfgans, A. Lopez-Mras, M. Mühleisen, J. Sperlich, in: *Progress in Organosilicon Chemistry* (Eds.: B. Marciniec, J. Chojnowski), Gordon and Breach Publishers, Amsterdam, **1995**, p. 55.

j) R. Tacke, O. Dannappel, M. Mühleisen, in: *Organosilicon Chemistry II: From Molecules to Materials* (Eds.: N. Auner, J. Weis), VCH, Weinheim, **1996**, p. 427.

k) R. Tacke, O. Dannappel, in: *Tailor-made Silicon-Oxygen Compounds – From Molecules to Materials* (Eds.: R. Corriu, P. Jutzi), Vieweg-Verlag, Braunschweig/Wiesbaden, **1996**, p. 75.

[2]   J. Heermann, R. Tacke, unpublished results.

[3]   O. Dannappel, *Ph. D. thesis*, Universität Karlsruhe, **1995**; O. Dannappel, R. Tacke, unpublished results.

[4]   J. Becht, *Ph. D. thesis*, Universität Karlsruhe, **1994**; J. Becht, R. Tacke, unpublished results.

[5]   M. Mühleisen, *Ph. D. thesis*, Universität Karlsruhe, **1994**; M. Mühleisen, R. Tacke, unpublished results.

[6]   R. Tacke, J. Sperlich, B. Becker, *Chem. Ber.* **1994**, *127*, 643.

PART II

SILICON BASED MATERIALS

INTRODUCTION

Norbert Auner
Humboldt-Universität
Berlin, Germany

Gordon Fearon
Dow Corning Corporation
Midland, MI, USA

Johann Weis
Wacker-Chemie GmbH
München, Germany

Among Silicon Based Polymers silicones are a unique class of materials featuring even contradictory properties such as

heat resistant	low temperature resistant
hydrophobic	hydrophilic
adhesion	release
profoaming	defoaming
electroconductive	(electro)insulating
thermoconductive	(thermo)insulating
transparent	pigmented

due to an impressing variability in chemistry.

With respect to their Si–O-backbone and the organic substituents attached to the silicon atoms they are hybrids between pure inorganic and organic polymers. Covering a tremendous range in molecular weight they are used in nearly every industry. This versatility is also reflected in scientific publications and patents.

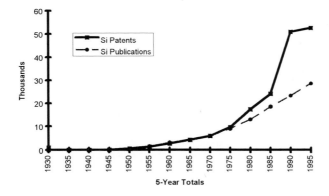

Fig. 1.

The figure above shows the number of papers covering synthetic silicon-based materials which have been published in recognized scientific journals for five-year periods starting in 1930. This figure also shows the number of worldwide patents which have been granted covering these same materials during the same time period. These data points have been collected for many years by the same information specialist using the same search procedures, so whereas the absolute numbers can be challenged, the trends are correct. Publications are tracked through Chemical Abstracts and Patents through the Derwent Patent Database. Between 1950 and 1955, approximately 100 publications were listed, between 1985 and 1995 this number exceeds 30,000. During this same time period the number of patents issued increased from less than 200 to greater than 50,000. The publications provide a measure of scientific knowledge and this continues to be expanded rapidly. Patents give some measure of conversion of this science to useful technology. During the last five-year period growth of patents appear to have slowed significantly, in fact, the number peaks at ~8000 per year in 1993 and since then has decreased. If this number continues to decrease, then there is real cause for concern because it will indicate that we are no longer converting this new scientific knowledge to goods and services at historic rates.

Dimethylpolysiloxanes are by far the most important silicon-based materials from a commercial standpoint. Business based on silicon science and technology is also growing impressively based on over 8000 products sold into most market segments in all geographic areas. Worldwide sales of silicon-based materials exceeded $ 6 billion dollars in 1995. Many of these materials are sold as key intermediates which are incorporated into final products, so the ulitmate business supported or impacted is very much larger.

The direct, copper catalyzed reaction of methyl chlorid with silicon, the "Direct Process", is the most important reaction for the production of methylsilicon intermediates. A similar reaction with hydrogen chloride yields trichlorosilane, which in turn is the basis for trifunctional intermediates and also for polycrystalline silicon, the raw material for the semiconductor industry. This Direct Process has been practiced for fifty years and much progress has been made by many towards understanding and optimizing the process, but much still remains to be learned about this complex series of heterogeneous reactions.

Past kinetic studies show that coadsorbed reactants on the silicon surface, methylchloride, methyl radicals and chlorine radicals compete with products for surface sites, but much is still unknown about these surface chemical processes. This reaction is carried out at elevated pressures so conventional high vacuum surface science and techniques cannot be used to uncover details of the reaction path. To circumvent this problem, Bent [1] and co-workers have developed low temperature techniques which permit direct observation of the interactions of reactants and products with the silicon surface. This type of understanding is key to continuous improvement of the Direct Process.

But the Direct Process has now been practiced with little fundamental change for over fifty years. Conversion of silica (sand) to diorganopolysiloxane such as dimethylpolysiloxane is accomplished through a series of reactions very simplisticly represented as follows:

$$SiO_2 + C \rightarrow Si + CO_2$$

Eq. 1.

Silica is reduced to silicon in a highly endothermic process normally carried out in an arc furnace.

$$Si + CH_3Cl \rightarrow (CH_3)_nSiCl_{4-n}$$

Eq. 2.

Silicon is reacted with methyl chloride by a process discovered by Mueller & Rochow over fifty years ago. This "Direct Process" gives a mixture of mono-, di-, trimethyl chlorosilanes together with small amounts of polysilanes, but normally the reaction is operated to maximize the yield of dimethyldichlorsilane.

$$n\,(CH_3)_2SiCl_2 + n\,H_2O \rightarrow \{(CH_3)_2Si(OH)_2\} \rightarrow -[(CH_3)_2SiO]_n- + 2n\,HCl$$

Eq. 3a.

$$m\,(CH_3)_2SiCl_2 + (m+1)\,H_2O \rightarrow HO[(CH_3)_2SiO]_mH + 2m\,HCl$$

Eq. 3b.

Dimethyldichlorosilane ist then hydrolyzed to yield intermediate silanol functional species which are normally unstable under the conditions of hydrolysis so condense to form an oligomer mixture consisting of cyclic dimethylsiloxanes (3a) and hydroxylterminated linear dimethylsiloxanes (3b). Hydrogen chloride is liberated in this reaction.

$$HCl + CH_3OH \rightarrow CH_3Cl + H_2O$$

Eq. 4.

Hydrogen chloride is carefully recovered and reacted with methanol to yield methylchloride which is then recycled to reaction #2. The key point is that the chlorine atom is recycled. Chlorine is money! So, all manufacturers work hard to contain and recycle every atom.

On paper the direct reaction of an organic species with silica can be accomplished without reduction to silicon. At high temperatures, silica is known to form a highly reactive species referred to as silicon monoxide. Recent investigation by Cannady [2] and others have shown that silicon monoxide will react with organic species such as ethane to give detectable amounts of methylsiloxane species. Of course, many will point out that silicon monoxide is only stable at very high temperatures, and that the conditions employed, high vacuum, plasma and rapid cooling of intermediates are far from practical. But a small window has perhaps been opened!

Polymer structures are often described by using the capital letters M (mono-), D (di-), T (tri-), and Q (quadruple-) indicating the functionality of the respective monomer units. Thus linear trimethylsilyl terminated silicone fluids are characterized by $M(D)_xM$. Linear polyorganosiloxanes can be produced either from cyclic organosiloxanes by ring-opening polymerization, which is promoted by both anionic or cationic catalysts [3, 4], or from oligomeric hydroxyterminated dimethylsiloxanes by polycondensation in the presence of acid catalysts [5, 6].

The polymerization reaction leads to an equilibrium mixture of linear polysiloxanes and cyclic siloxanes, wheras the equilibrium of the polycondensation reaction can be shifted to the polymer side by removing the condensation water.

Silicone polymers exhibit significantly lower surface tension than most solids. Exemptions are only polyolefins and polytetrafluoroethylene. Therefor it is not surprising that much of today's business for silicon based materials depends on their surface activity.

Adhesion, release, profoaming, defoaming, surface protection, all depend in one way or another on the properties of silicon(e) interfaces. Many research groups, both academic and industrial in all parts of the world are constantly providing new insights into the structure and properties of these molecular interfaces. This work is often lead by, or drives, the discovery and development of new analytical techniques, both static and dynamic, to characterize the interfaces. Based on this improved understanding, new families of silicon-based materials are being developed with specifically tailored structure and performance characteristics. This broad field is the topic of many excellent and comprehensive reviews. Following are but a few highlights.

Major new insights into the surface science of silicone materials were catalyzed by the work of deGennes, whose theories of polymer wetting and adhesion have inspired many to explore his provocative themes [7]. These explorations focused attention on the need for a polymeric material that is well characterized, liquid over a wide molecular weight range, with controlled molecular weight distribution and crosslinkable in a controlled fashion. Dimethylpolysiloxane is the best available candidate and has become central to a revolution in polymer surface physics.

Much polymer surface chemistry assumes equilibrium behaviors but many important applications for surface active materials, e.g. foaming, defoaming, emulsification, demulsification are dynamic and require new theories and techniques to study phenomena such as surface restructuring and viscoelastic properties of monolayers. Reactions at silicon surfaces such as the direct process studies mentioned before have sparked new understanding of adsorption and desorption behaviors of reactive intermediates. Careful examination of the release of adhesives from silicone and other release coatings has revealed that this process is strongly influenced by rheological as well as surface forces which explains why silicone release coatings are superior to fluorocarbon material, even though they have higher surface tension. Intertwined with these new insights are new developments in microscopy, ultra high vacuum surface analysis, techniques to measure surface forces and to characterize surfactant/polymer microstructures. All this work [8] has lead to the discovery of families of new surface active materials, nanocomposites, block copolymers, novel surfactants, fluorosilicones with very low surface tensions as well as providing the understanding for new applications of well established material. A fertile field indeed!

However, especially with those applications where silicones are used as auxiliaries or process aids, they are often released into the environment, which causes public concern sometimes. In this context, a brief exploration of silicon impact on the environment may be useful.

Despite their lipophilic nature and stability in their intended application, they do not bioaccumulate and are degraded in the environment [9]. High molecular weight polydimethylsiloxanes are quickly absorbed onto soil or particulate matter where they undergo a biotic degradation to short chain water soluble silanols. These low molecular weight materials undergo further photodegradation to carbon dioxide, water and polysilicic acid. Polysilicic acid is a naturally occurring material present in all groundwater at ~10 ppm, and all evidence to-date suggests that this is the primary building block for incorporation of silicon into diatoms, plants and other life forms. Low molecular weight polysiloxanes, often referred to as volatile methyl siloxanes, quickly volatilize into the atmosphere where they undergo rapid oxidative chemical degradation to silica, carbon dioxide and water, all naturally occurring in the environment. Research groups are working hard in many parts of the world today to continually improve and expand our knowledge of silicon impact on the environment. As this scientific understanding [10] is expanded it continues to support the belief that dimethylsiloxane-based materials are among the most environmentally acceptable synthetic materials available today.

Despite of the silicon breast implant campaign which is a serious issue in the United States but evidently not in the rest of the world, the above statement on the environmental impact of silicones is also applicable to the impact on health of silicon medical devices and silicon implants, which are daily used in hospitals worldwide due to their excellent biocompatibility to human tissue and blood [11-13] There is also a long lasting experience clinically documented over decades on their reliability in specific applications.

In the case of the silicone breast implant a claim was brought up that it effects the human immune system to cause a whole variety of connective tissues disorders. In fact, all documented scientific investigations, including several recent results from well recognized investigators, continue to show that there is no association between silicone and either typical or atypical tissue disease [14, 15].

Today all available evidence suggests that dimethylsiloxane is one of the most acceptable polymeric materials known from the standpoint of health and environmental impact. But not all silicon based compounds are inert. For instance, families of physiologically active silicon-based materials are well known today and several very active research programs focus on the discovery of effective silicon based pharmaceuticals [16, 17]. All evidence today suggests that there is a natural "silicon cycle" which incorporates varying amounts of silicon into all life forms, plant or animal [18]. So a major challenge is to educate the public that all silicon-based materials are not the same and certainly all of them are not inert!

Silicon is the most abundant solid element in the earth's crust and probably in the universe as a whole [19]. Silicon inorganic materials are all around us, sand, rocks, glasses, cements and mortars all have high silicon content and either occur naturally or are produced by well established processes which have often evolved through thousands of years. During the last few decades new classes of synthetic silicon inorganic materials have been discovered, developed and commercialized. All of the families are based on new chemical processes. The production of polycrystalline silicon with controlled purity by the chemical vapor deposition of trichlorosilane is probably one of the first examples of carefully synthesized silicon inorganic material. This silicon is the basis of the semiconductor industry.

The chemical vapor deposition of trichlorosilane is significantly endothermic ($\Delta H = 964$ kJ) and requires high reaction temperatures ($\sim 1150°C$). This reaction also yields considerable quantities of tetrachlorosilane as a byproduct:

$$4 \ SiHCl_3 + 2 \ H_2 \quad \xrightarrow{\Delta T} \quad 3 \ Si + SiCl_4 + 8 \ HCl$$

Eq. 5.

Tetrachlorosilane is partly converted back to trichlorosilane:

$$SiCl_4 + H_2 \quad \rightleftharpoons \quad SiHCl_3 + HCl$$

Eq. 6.

$$Si + 3 \ SiCl_4 + 2 \ H_2 \quad \rightleftharpoons \quad 4 \ SiHCl_3$$

Eq. 7.

The remaining tetrachlorosilane is predominantly used for the production of fumed silica [20], which finds wide use in a variety of industrial applications. Most important are reinforcement of silicone elastomers as active filler and thickening of liquids as a rheological additive. A highly dispersed particle structure, high surface area and surface energy are the main characteristics of fumed silica.

Silica optical fibers produced by the chemical vapor deposition of silicon intermediates are now revolutionizing communication networks. Photovoltaic cells based on silicon, again produced primarily by chemical vapor deposition processes are now important power generators for electrical and electronic devices.

The use of well characterized polymers with controlled silicon to carbon, nitrogen, oxygen or other heteroatom ratio first reported by Yajima [21] is now emerging as a flexible and important route to silicon carbide, silicon nitride, silicon oxide and other silicon-based inorganic materials and structures. Typically a soluble thermoplastic, polymer with the correct silicon to heteroatom ratio is extruded or coated in the desired shape or form, then converted to a ceramic by heating.

Based on this approach, families of silicon carbide fiber and silicon ceramic composites are now being routinely produced based on polycarbosilane precursors [22]. These new materials are finding a wide range of new applications, for example, as a hot zone component in the next generation turbojets where silicon carbide composite components now routinely service at operating temperatures well in excess of 1000°C under high static thrusts (up to 50,000 psi) and high sonic pressure (up to 800 DB). No metallic components survive under these conditions.

New silicon oxide ceramic coatings based on hydrogensilsesquioxane are now finding applications as interlayer dielectrics in new generation of multilayer semiconducting devices. These materials have excellent gap filling characteristics, low dielectric constants and are relatively simple to apply [23].

Silicon solgel science and technology is probably one of the fastest growing fields of silicon material research today [24]. Most approaches depend on the hydrolysis and condensation of tri- or tetrafunctional silanes, generally alkoxy silanes in solution to provide families of highly branched silanol functional systems. On removal of the solvent further condensation of the silanol functionality occurs to give crosslinked networks with low or no carbon content, depending upon the composition of the starting silanes. Other inorganic or organic monomers or polymers are often incorporated, sometimes in colloidal form to give a vast range of compositions of matter which are variously referred to as Ceramers, Ormosils, Ormocers or Polycerams to list but some of the names applied.

To date the major commercial impact of these materials is as thin films; hard glass-like abrasion resistant coatings for plastics, corrosion protective coatings for metals, surface and interlayer coatings for semiconductors are all important applications for this type of material. But this research field of new silicon polymer science, based on quite readily available monomers, is just beginning to grow. Progress today is limited by our understanding of these systems. Much is not known about the kinetics and mechanism of hydrolysis and condensation of tri- and tetra-functional silane intermediates. The final materials are highly branched amorphous systems which are not easily characterized. New analytical techniques must be discovered and applied to probe these complex systems. In summary, the material science and technology of silicon inorganic materials needs much more attention!

References:

[1] D. H. Sun, A. Gurevick, L. Kaufman, B. Bent, A. P. Wright, B. Naasz, *Proc. 42nd National Symposium*, American Vaccuum Society, **1995**, Mineapolis.

[2] J. P. Cannady, *Organosilicon Symposium*, Evanston, Illinois, **1996**.

[3] T. C. Kendrick, B. Parbhoo, J. W. White, in: *The Chemistry of Organic Silicon Compounds*, Wiley, New York, **1989**, p. 1289.

[4] P. V. Wright, in: *Ring-Opening Polymerisation, Vol 2* (Eds.: K. I. Ivin, T. Saegusa), Elsevier, New York, **1984**, chap. 14

[5] Wacker, EP 0 258 640, **1986**.

[6] Wacker, EP 0 208 285, **1985**.

[7] P. G. de Gennes, *Reviews of Modern Physics* **1995**, *57*, 827.

[8] M. J. Owen, in: *Frontiers of Polymers and Advanced Material* (Ed.: P.N. Prasad), Plenum Press, New York, **1994**.

[9] G. Chandra, *Proc. Amer. Association of Textile Chemists and Colorists*, **1995**, *27*, 21.

[10] *The Handbook of Environmental Chemistry: Organosilicon Materials* (Eds.: O. Hutzinger, G. Chandra), Springer Verlag, Berlin, **1997**.

[11] R. Alastair Winn, P. Redinger, K. Williams, *J. Biomater. Appl.* **1989**, *3*, 645.

[12] J. K. Quinn, M. J. Courtney, *Br. Polym. J.* **1988**, *20*, 25.

[13] N. Kossovsky, J. P. Heggers, C. M. Robson, *C R C Crit. Rev. Biocompat.* **1987**, *3*, 53.

[14] B. G. Silverman, S. L. Brown, R. E. Bright, R. G. Kaczmarck, J. B. Anowsmith-Lowe, D. A. Kessler, *Ann Intern Med* **1996**, *124*, 744.

[15] L. L. Perkins, B. D. Clark, P. J. Klein, R. R. Cook, *Annals of Plastic Surgery* **1995**, *35*, 561.

[16] R. Tacke, D. Reichel, M. Kropfgans, P. G. Jones, E. Mutschler, J. Gross, X. Hou, M. Waelbroeck, G. Lambrecht, *Organometallics* **1995**, *14*, 251.

[17] R. Tacke, D. Terunuma, A. Tafel, M. Muhleisen, B. Forth, M. Waelbroeck, J. Gross, E. Mutschler, T. Friebe, G. Lambrecht, *J. Organomet. Chem.* **1995**, *501*, 145.

[18] J. D. Birchall, *Chem Soc Review* **1995**, 351.

[19] E. G. Rochow, *Silicon & Silicones*, Springer Verlag, Berlin, **1987**.

[20] H. Barthel, L. Rösch, J. Weis, in: *Organosilicon Chemistry II: From Molecules to Materials* (Eds.: N. Auner, J. Weis), VCH, Weinheim, **1996**, p. 761.

[21] S. Yajima, *Amer Ceram Soc Bull* **1983**, *62*, 893.

[22] H. O. Davis, R. D. Petrak, *Journal of Nuclear Materials* **1995**, 219, 26.

[23] D. Pramanik, J. V. Tietz, K. Schiebert, *VMIC* **1993**, *93*, 329.

[24] U. Schubert, N. Husing, A. Lorenz, *Chem. Mater* **1995**, 7, 2010.

The Direct Process to Methylchlorosilanes: Reflections on Chemistry and Process Technology

B. Pachaly*, J. Weis
Wacker-Chemie GmbH
Geschäftsbereich S – Werk Burghausen
Johannes-Heß-Straße 24, D-84489 Burghausen, Germany

Keywords: Direct Process / Methylchlorosilanes / Copper Catalyst / Chloromethane / Silicon

Summary: The Direct Process discovered by Rochow and Müller around 1940 is the basic reaction used to produce methylchlorosilanes, which are the monomeric intermediates used for production of silicones. An understanding of the elementary reactions, the nature of active sites and the action of promotors does not nearly come close to the performance level of the industrial process and the economic importance. The silylene-mechanism is a useful model to understand the complex product mixture from the reaction of silicon with chloromethane.

Introduction

The Direct Process is the reaction of silicon with chloromethane to form methylchlorosilanes (Eq. 1). This reaction is unique, in that it is the only solid-catalyzed gas-solid reaction applied in the chemical industry. The Direct Process was first discovered by Rochow [1] and independently Müller [2] around 1940.

$$Si + CH_3Cl \rightarrow (CH_3)_x SiCl_{4-x} \,, \, x = 0\text{-}4$$

Eq. 1. The reaction of silicon with chloromethane to methylchlorosilanes.

The reaction is carried out in fluidized-bed reactors utilizing copper, copper alloys or copper compounds as catalyst and zinc, zinc alloys or zinc compounds as promotors. The most widely used catalyst is partially or completely oxidized copper. As copromotors minor amounts of tin, antimony or phosphorous are used. The temperature applied is typically 260-310°C. The methylchlorosilanes (MCS) are the monomeric intermediates for a remarkable number of silicone products based on polymethylsiloxanes. The most important methylchlorosilanes are dimethyldichlorosilane, methyltrichlorosilane and trimethylchlorosilane, from which more than 90 % of the silicone products are derived. Major silicone products produced from these methylchlorosilanes are RTV- and HTV-rubbers, methylfluids and fluid emulsions, agents for paper coating, textile finishing, antifoaming or release agents, resins and masonry water repellents.

The Silicones Industry

The first industrial production of methylchlorosilanes using the Direct Process started in 1947 in the USA (General Electric, Waterford), in Germany between 1951 and 1955 three companies entered into it. During the intervening five decades the Direct Process became a worldwide utilized process in the fast growing world of silicones and in 1993 the production of methylchlorosilanes passed 1 000 000 t per year in the Western World. In 1995 the production of methylchlorosilanes was about 1 250 000 t, for which 1 000 000 t chloromethane and 300 000 t silicon were consumed. The vale of the silicones market in 1995 was about 6 billion US$. These figures emphazise the economic importance of the Direct Process in the silicones industry.

Table 1. Major producers of methylchlorosilanes and production volume by regions in 1995.

Region	MCS-Production (tpy)	Producer
USA	490 000	Dow Corning, General Electric
Europe	470 000	Wacker, Dow Corning, Rhone-Poulenc, Bayer, Hüls
Japan	190 000	Shin-Etsu, Toshiba, Toray
ROW	100 000	CIS, PR China
Total	1 250 000	

The Formation of Methylchlorosilanes

The copper-catalyzed reaction of silicon with chloromethane leads to a complex mixture of varions monomeric and dimeric methylchlorosilanes termed crude silane.

Table 2. Methylchlorosilanes in the crude silane.

Compound	Typical Yield	bp [°C]	Heat of Formation [kJ/mole][3, 4]
$(CH_3)_2SiCl_2$	80-90	70	-466
CH_3SiCl_3	3-15	66	-569
$(CH_3)_3SiCl$	2-5	57	-354
CH_3HSiCl_2	0.5-4	40	-393
$(CH_3)_2HSiCl$	0.1-0.5	35	-282
$SiCl_4$	<0.1	58	-663
$(CH_3)_4Si$	0.1-0.3	26	-233
$HSiCl_3$	<0.1	32	-500
$(CH_3)_xSi_2Cl_{6-x}$, x=2-5	2-8	>140	–

The disilanes with $x = 2$ and 3, which is the main portion of the high-boiling fraction, are converted to monosilanes by subsequent cleavage with hydrogen chloride.

The performance targets, such as reactivity of the reaction mixture, selectivity, extension of a production campaign or reproducibility, are controlled by quite a number of parameters. The most important parameters are temperature, pressure, catalyst and promoters, poisons and inhibitors, silicon composition and structure, particle size distribution of solids, dust removal from fluidized-bed reactor, homogeneity of fluidized-bed and the purity of chloromethane.

The reaction of silicon and chloromethane proceeds phenomenologically in a way that the copper catalyst forms precipitates on the most activated areas of the silicon surface, which are the areas around intermetallic phases, grain boundaries or defects. In these areas one observes the fastet consumption of the silicon particle, also the general surface reacts, but much slower [5].

The reaction mechanism is not clarified yet. Older proposals by Hurd and Rochow, Bazant, Klebansky, Fikhtengolts and Voorhoeve, or Golubtsov are today replaced by the silylene mechanism [6], which provides a good explanation for the products obtained in the Direct Process.

$$Cu^{\delta+} - Si^{\delta-} + CH_3Cl \; \rightarrow \; Cl^{\delta-} - CH_3^{\delta+} \; \rightarrow \; Cu + [CH_3SiCl]$$

$$/ \qquad /$$

$$Cu^{\delta+} - Si^{\delta-}$$

Diffusion replaces consumed silicon

$$[CH_3SiCl] + CH_3Cl \; \rightarrow \; (CH_3)_2SiCl_2$$

Insertion of silylenes into C-Cl bond forms methylchlorosilanes

Scheme 1.   The silylene-mechanism for the formation of methylchlorosilanes.

The silylene-mechanism alone does not explain the formation of Si–H bonds, higher aliphatic groups attached to silicon (e.g., ethyl and propyl) and the formation of hydrocarbons, mainly methane or elemental carbon.

Chemisorbed methyl-groups undergo α-elimination to form hydrogen and methylene groups. The hydrogen then leads to methane by reductive elimination or remains bound to silicon. Methylene groups insert to form C_2- or higher aliphatic groups or further a-elimination finally leads to formation of carbon. More than 50 different C_1- to C_8- hydrocarbons formed by complex coupling and insertion reaction have been identified in the crude silane. In the state-of-the-art industrial production about 0.5-1% of the crude silane are hydrocarbons, the dominant one is methane with a formation rate of about 4 mg per g of MCS. Hydrocarbon formation can be catalyzed by trace elements of silicon, like Ni, Mn, Ti, or Cr, therefore the purity of silicon is an important parameter for high selectivity. Copper is believed to form a Cu_3Si-phase in the active sites [7]. The Cu_3Si-phase facilitates the reaction of silicon with chloromethane, the silicon is much more reactive than in silicon metal itself.

The action of promoters is not understood yet. Copper is the catalyst but is not sufficient for economic production of MCS. Zinc is used as the promoter and in addition minor amounts of either tin, antimony or phosphorous are also used. Obviously, Zn can lower the appearent activation energy for desired silanes [8].

Table 3. Apparent activation energies [kJ mol^{-1}] for varying Zn-content in reaction mixture.

Compound	0 % Zn	0.4 % Zn	1.2 % Zn
$(CH_3)_2SiCl_2$	130	84	50
CH_3SiCl_3	88	134	84

Raw Materials

Silicon is produced in rotating submerged arc furnaces by reduction of quartz with carbon. Carbon sources are coal, coke or wood. Typically, for 1000 kg of silicon 2500 kg SiO_2, 1600 kg C and 11500 kWh electrical energy are consumed [9].

The liquid silicon is tapped from the furnace into ladles. For the application in the Direct Process refining by oxygen blowing or slag addition is needed to obtain the desired content of foreign elements, like Al or Ca, for controlled activation. After refining the silicon is solidified by casting into iron moulds. To use it in the Direct Process the lumpy silicon has to be crushed, ground and optionally sieved to a particle size of 0-500 μm.

Chloromethane can be produced by either chlorination of methane or reaction of methanol with hydrogen chloride. In an integrated production of silicones hydrogen chloride obtained by hydrolysis of methylchlorosilanes is recycled to the Direct Process via a chloromethane synthesis. The losses are compensated by make-up chloromethane or hydrogenchloride. Impurities, like water, methanol, dimethyl ether or oxygen, must be kept at as low a level as possible.

Direct Process Reaction System

The production of MCS is performed in fluidized-bed reactors with chloromethane as the fluidizing agent and reactant. The contact mass, which is the mixture of silicon powder, copper catalyst and promoters, is charged to the reactor according to the silicon consumption. The crude silane is leaving the reactor via cyclones to remove solids and after condensation and filtration (or vice versa) it goes to the distillation. Unconsumed chloromethane is recycled.

Even if one can control steady-state concentration of catalyst and foreign elements in the fluidized-bed, the selectivity decreases over time to finally reach a point at which shut-down is more cost-efficient than proceeding with the run. Therefore, the Direct Process is a semicontinuous process with extension of production campaigns of several days to several weeks depending on process control and silicon utilization.

Dust removal is crucial for long production runs with a high selectivity level. The fines removed via the cyclones contain a lot of the copper catalyst. For economic reasons and low environmental impact it is neccessary to recover the copper catalyst from solid byproducts and utilize unconverted silicon in external processes, like the production of ferroalloys [10].

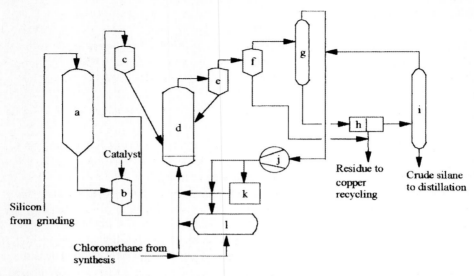

a) Silicon silo; b) Contact mass preparation; c) Contact mass silo; d) Fluidized bed reactor; e) Main cyclone to recycle silicon; f) Cyclone to recover
catalyst; g) Scrubber to condense crude silane; h) Filter; i) Chloromethane distillation column; j) Compressor; k) Chloromethane purification; l) Chloromethane storage tank

Fig. 1.   Direct process reaction system.

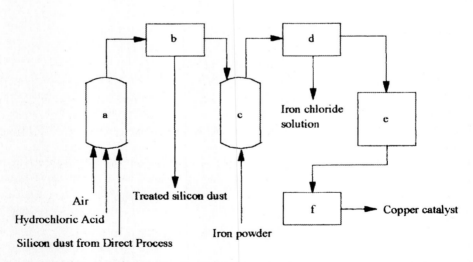

a) Suspension vessel to dissolve copper; b) Filter; c) Cementation vessel; d) Filter; e) Drying unit for drying and oxidation of copper powder; f) Jet mill

Fig. 2.   Wacker process for copper reycling and treatment of silicon byproducts.

New Developments and Outlook

The observation that the formation of active sites are preferred at intermetallic phases and grain boundaries triggered efforts to increase these in silicon by more rapid solidification procedures. Atomization is one approach to produce rapidly solidified silicon [11]. Since 1993, water granulation of silicon is utilized on a commercial scale [12, 13].

The growth of the silicones industry is expected to continue with about 5 % p.a. increase of MCS-production.

Even after five decades of industrial utilization no alternative method is coming in sight to substitute the Direct Process. Neither the reaction in liquid phase nor the insertion reaction of silicon monoxide SiO can compete with the way we produce MCS today.

Also, chlorine-free production, which was attempted from the very beginning, is the less interesting, the more the silicone producers succeed in recycling chlorine in integrated silicones production [14].

The focus for future developments has to be on reduction of production costs for silicon, more than 75 % of the MCS-production costs are due to raw material input.

The most important goal, however, must be to close the significant gap between the performance level of the industrial MCS-Production by the Direct Process on the one hand and the poor understanding of the elementary reactions, the nature of the active sites and the action of promoters on the other.

References:

[1]   General Electric, US 2 380 995, **1941** (E. Rochow).

[2]   VEB Siliconchemie, DD 5 348, **1942** (R. Müller).

[3]   R. Walsh, in: *The Chemistry of Organosilicon Compounds*, Wiley, New York, **1989**.

[4]   R. Walsh, *J. Chem. Soc. Fraday Trans.* **1983**, *79*, 2233.

[5]   H. M. Rong, *Silicon for the Direct Process to Methylchlorosilanes*, Diploma thesis, NTH Trondheim, **1992**.

[6]   K. M. Lewis, D .G .Rethwisch, *Catalyzed Direct Reactions of Silicon*, Elsevier, Amsterdam, **1993**.

[7]   R. J. Voorhoeve, *Organosilanes Precursors to Silicones*, Elsevier, Amsterdam, **1967**.

[8]   J. P. Agarwala, J. L. Falconer, *Int.J.Chem.Kin.* **1987**, *19*, 519.

[9]   T. Margaria, J. C. Anglezio, C. Servant, *Proc. 6th Int. Ferroalloys Congress*, Cape Town, **1992**, 209.

[10]   B. Pachaly, H. Straußberger, W. Streckel, *Proc.Conf. Silicon for Chemical Industry II*, Loen, **1994**, 235.

[11]   M. Schulze, E. Licht, *Proc. Conf. Silicon for Chemical Industry I*, Geiranger, **1992**, 131.

[12]   B. Pachaly, *Proc. X Int. Symposium on Organosilicon Chemistry*, Poznan, **1993**, 236.

[13]   L. Nygaard, H. Brekken, H. U. Lie, T. E. Magnussen, A. Sveine, *Proc. 7th Int. Ferroalloys Congress*, Trondheim, **1995**, 665.

[14]   B. Pachaly, U. Goetze, K. Mautner, *Proc. XXVIIIth Organosilicon Symposium*, Gainesville, **1995**, B-9.

On the Nature of the Active Copper State and on Promoter Action in Rochow Contact Masses

Heike Ehrich, Dietrich Born, Jürgen Richter-Mendau, Heiner Lieske*

Institut fürAngewandte Chemie Berlin-Adlershof e.V.

Abteilung Katalyse

Rudower Chaussee 5, D-12484 Berlin, Germany

Tel.: Int. code + (30)63924372 – Fax: Int. code + (30)6392 4392

E-mail: lieske@aca.fta-berlin.de

Keywords: Rochow Reaction / Active Copper Species / Promoter / Zinc / SEM / EDX

Summary: The question of the nature of the catalytically active copper species in Rochow contact masses has been investigated using the SEM/EDX technique. The results do not support the hypothesis of active η-Cu_3Si, but they provide more direct evidence for the existence and the catalytic action of X-ray amorphous Cu–Si surface species, i.e., extremely dispersed particles or even two-dimensional species like Cu–Si surface compounds, which we proposed recently. The investigation of zinc-promoted and non-promoted contact masses on basis of the pure and technical-grade silicon showed that the mode of operation of the famous Rochow promoter zinc can be understood rather as a moderation than as a real acceleration. By moderating the initial reaction rate, the promoter enables a sufficiently high stationarity of the reaction.

Introduction

In spite of decades of industrial practice, of its enormous economic importance and of numerous relevant scientific papers, see [1-3], the Rochow synthesis has been poorly understood so far, in a scientific sense. This situation has often been discussed and is surely due to special features, which make this reaction unique in the field of heterogeneous catalysis. One of the two reacting substances, silicon, plays the role of a catalyst support at the same time. Within the reacting system, several technical-grade solids (silicon, various copper species, promoters, carbonaceous deposits) interact, each of them with a broad variety of structure, morphology, and chemical composition. Thus, the number of possible influences on the reaction is hardly to be surveyed. This extreme complexity implies at the same time a poor methodological accessibility of the reaction system. The surface of the "contact mass", i.e., the mixture of silicon, copper species, promoters, impurities, and carbonaceous deposits, is of corresponding complexity. As a further aggravation, the reaction is restricted to μm-sized "pits" on the contact mass surface, which have a fine structure (see below). Hence, highly spatially resolving methods are necessary to characterize the reacting surface selectively. Up to now, it has been difficult to overcome these problems.

With this paper, we want to return to two basic questions of Rochow synthesis, with which we have dealt in the last years and about which we have reported preliminary results:

- the nature of the catalytically active copper species in Rochow contact masses, and
- the mode of action of promoters of the Rochow reaction, taking the industrially important promoter zinc as an example.

In the following, the present knowledge about these points, inclusive our own recent experiences, will shortly be outlined once more, and it will be shown what we could learn about active copper species and zinc promoter action by combining catalytic tests with the spatially resolving SEM/EDX technique.

Materials and Methods

Silicon: Three types of Si powders with grain sizes between 71 and 250 μm were used: monocrystalline and polycrystalline pure silicon, "$Si_{pure, m}$" and "$Si_{pure, p}$" as well as the technical-grade silicon "Silgrain" from ELKEM, Norway, "Si_{tech}", with 0.20% Al, 0.13% Fe, 0.03 % Ca, 0.015% Ti, < 20 ppm V, Cr, Mn, Ni, P.

Copper Components ("catalyst"): Anhydrous copper(II) chloride and copper(II) oxide p.a., from MERCK, Germany, were used.

Methyl Chloride: CH_3Cl with a content of >99.6% was dried over molecular sieve 4A.

Zinc Promoter: Pure metal powder <45 μm; mostly 0.5 wt.% Zn related to contact mass weight.

Catalytic Experiments: Catalytic measurements were carried out by means of a vibrating glass microreactor [4]. The rotating vibration of the reactor simulates fluidized bed conditions. The composition of the reaction products and the methyl chloride conversion was measured by on-line gas chromatography. Under the assumption of a differential behavior of the reactor, initial reaction rates were calculated from the methyl chloride conversion as a measure of the catalytic activity of the contact masses. The selectivity of the reaction is not a matter of discussion in this paper and will be mentioned only exceptionally.

SEM/EDX Analysis: The morphology and the element composition and distribution on the contact mass surface were investigated with a Scanning Electron Microscope (Cambridge Instruments S 360) combined with an Energy Dispersive X-ray spectrometer using a Delta Class Analyzer 8000 (KEVEX). For the scanning electron micrographs, a beam energy of 15 kV and magnifications from 1000 to 5000 were applied. The lateral resolving power of EDX point analysis has been about 2 μm. The scanning electron micrographs to be seen in this paper represent typical observations on the respective contact mass samples.

On the Nature of the Catalytically Active Copper Species

The Present Knowledge on Active Copper Species

The Cu–Si phase η-Cu$_3$Si has frequently been found in contact masses and has been discussed as a catalytically active component for decades. The idea was originated by Trambouze et al. [5], who observed correlations between η-phase content and catalytic performance of contact masses and proposed a dissociative adsorption of methyl chloride (CH$_3$→Cu;Cl→Si) on the Cu$_3$Si surface as an important reaction step. Klebansky and Fikhtengolts [6] and Voorhoeve et al. [1] established a mechanism with the reverse polarization (CH$_3$→Si; Cl→Cu). The clearness of their mechanistic and kinetic model might have contributed to its widespread acceptance. But, from the very beginning, there were also doubts. Müller and Gümbel [7] could not confirm the importance of η-Cu$_3$Si. A russian school, e.g., Turetskaya et al. [8], also rejected this hypothesis, and ascribed to the η-phase a role in copper redistribution processes within the contact mass. These authors assumed surface chlorinated Si species as active sites. Recent SEM [9] and SEM/XPS/Auger [10] investigations by Banholzer et al. neither decided the question whether η-Cu$_3$Si is the active species in real contact masses nor left doubts about the importance of this species. The group of Falconer indeed could demonstrate that methylchlorosilanes can be produced on massive η-Cu$_3$Si from methyl chloride, e.g., Frank et al. [11]. But, in accordance with our preliminary notice [12], this finding does not necessarily mean that η-Cu$_3$Si is the active component also in real, powder-like contact masses, where various copper species of possible catalytic relevance can be formed from the original copper component during the induction period. Furthermore, Frank et al. [13] proposed that the η-Cu$_3$Si surface is not active itself, but only delivers an underlayer for the formation of active sites, which were assumed to contain Si–Cl, Si–Cu and Si–C bonds. Finally, Floquet et al. [14] studied the Rochow reaction on Si(100) model surfaces in the presence of several copper catalysts. The authors discussed a catalytic role of η-Cu$_3$Si, but did not rule out the reaction without this phase as an intermediate.

Despite the immense dissemination of the η-phase hypothesis among the experts, we felt it necessary to check this idea and came to the following statements [12, 15]. There are types of contact masses, which contain η-Cu$_3$Si, but are not active (e.g., some non-promoted types or contact masses within the induction period). On the other hand, there are contact mass types, which are active, but do not contain any X-ray detectable amount of η-Cu$_3$Si (e.g., some types with copper powder as copper component). Hence, η-Cu$_3$Si seems to be neither a sufficient nor a necessary precondition of the catalytic activity of contact masses.

A copper balance within the copper powder containing contact masses mentioned above [15], based on gravimetric and X-ray analysis, showed that these samples practically contained only metallic copper with a particle size > 500 nm, which can by no means be the active catalyst, already because of its much too low surface area. On the other hand, other copper species than metallic copper, which could be assumed as catalytically active, could have been present only in rather small amounts in these samples, corresponding to the statistical errors of the analytical methods used. If so, the small amount of catalytically active copper must be in a highly dispersed state, in order to exhibit a surface area, which is large enough to bring about the rather high catalytic performances observed.

Our preliminary conclusions have been summarized in Fig. 1, taken from [12]. We ascribed the catalytic activity to so-called "Cu–Si surface species". These have been assumed to be extremely small copper silicide particles or even "two-dimensional" Cu–Si surface compounds. Such species would not be detectable by X-rays, because of their low content and their low dimensions. There had already been speculations about the existence and catalytic activity of Cu–Si surface compounds [16, 17]. From a logical point of view, we could not exclude an additional action of *highly dispersed* η-Cu₃Si *in some cases*, see the upper part of Fig. 1, but the renunciation of this assumption leads to a more simple working hypothesis, i.e., to only *one* active species and is therefore to be preferred for the time being. The plausibility of our hypothesis is additionally supported by the consideration of the circumstances of solid diffusion within the contact mass, outlined in Fig. 1.

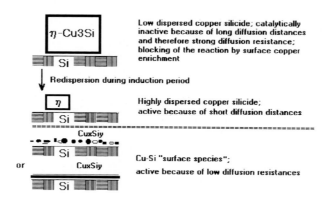

Fig. 1. Scheme of possible catalytically active copper species and diffusion circumstances in Rochow contact masses, from [12].

Results and Discussion on Active Copper Species

The catalytic activity of the Cu–Si surface species explained above was a logical, but only an indirect derivation from our former experiments. For some time, we were not able to present more direct evidence. The reason was the complicated surface topography of a working contact mass. According to all experiences, the Rochow reaction exclusively takes place in sharply demarcated pits on the silicon surface, the shape of which usually corresponds to the respective crystallographic planes. The inner surface of these pits mostly has an own structure, i.e., it is partly covered by particles of copper species, often doubtless by η-Cu₃Si particles [15]. But, the characteristic dimensions of these structures are generally below the resolving power of appropriate surface methods like the combination SEM/EDX.

It was by chance that we found a contact mass, by means of which we could overcome this problem for the first time. Fig. 2 depicts a typical surface structure of the sample "CuCl₂/Si_tech/Zn". This sample consisted of technical-grade silicon, 5 wt.% Cu as CuCl₂ and 0.5 wt.% zinc promoter. Its especially simple surface structure, see below, was surely due to its short reaction time: the sample reacted for only 20 min with methyl chloride. But, already after this short time, it had reached its full activity, see second section, and also its full selectivity for the main product dimethyldichlorosilane (about 80%).

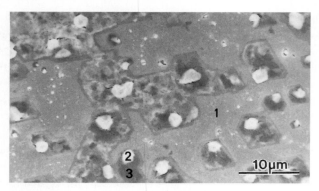

Fig. 2.   Typical surface image of a contact mass sample $CuCl_2/Si_{tech}/Zn$ after 20 min reaction time, and EDX analysis: 1) non-attacked Si surface, < 0.5 wt.% Cu, 96 wt.% Si; 2) copper-containing particle, 52 wt.% Cu, 41 wt.% Si; 3) surface within a pit beside a copper-containing particle, 2 wt.% Cu, 90 wt.% Si.

In Fig. 2, the pit-like reactive areas of this short-time sample of $CuCl_2/Si_{tech}/Zn$ are well visible. For the problem under discussion, the structure of these pits is of special interest. This structure is especially simple. Each pit contains only one relatively large copper-containing particle. The X-ray analysis showed that practically all the $CuCl_2$ added to the contact mass was present as η-Cu_3Si. Consequently, the large particles must consist of η-Cu_3Si. The particles are surrounded by large "free" areas within the pits. These areas must have been the scene of the synthesis reaction, because here silicon was removed and converted into methylchlorosilanes. As further information, Fig. 2 gives the EDX results of three points on the contact mass surface, in detail (1) on non-attacked silicon surface, (2) on a copper-containing particle, and (3) on a "free" area beside a copper containing particle, within the reactive area. The EDX results demonstrate that remarkable copper concentrations were only measured on the copper (η-Cu_3Si)-containing particles. On the free area beside the particle, the copper concentration was hardly higher than on the non-attacked silicon surface. *Just this is the point, which was to be demonstrated:* The Cu–Si surface species explained above are highly dispersed, practically two-dimensional, and can consequently contain only low amounts of copper. If they really are identical with the catalytically active surface, it is obvious that the EDX tool, which penetrates deeply (few μm) into the material, can detect only very low copper concentrations in the places where the reaction proceeds. This has been confirmed.

With our view of the active copper state, we do not exclude that η-Cu_3Si can play a role in the reacting system. η-Cu_3Si is probably not the main carrier of the catalytic activity, but possibly it plays a role as a copper reservoir. In the period after 20 min reaction time, i.e., after the state in Fig. 2, redispersion of η-Cu_3Si can be observed. Already after 60 min, the simple structure of the pits is destroyed and a lot of small particles are to be observed within the pit. (The surface then looks like that in Fig. 6, which is discussed only in the second section.) This redispersion process, probably comparable to our observations [12, 15], demonstrates a certain involvement of η-Cu_3Si in the reaction, which is perhaps linked with a minor catalytic activity. It can be imagined that the η-Cu_3Si particles act as a copper pool for the formation of new Cu-Si surface species. From such point of view, η-Cu_3Si could be regarded as a precursor of the active species. If so, however, there should exist also other ways to form these species: there are also active contact masses without any detectable η-phase.

On trial, one could discuss an alternative of the above model. One could assume that the synthesis reaction proceeds on the surface of the η-Cu₃Si particles and that the stationary state of the reaction is maintained by silicon diffusion from the free surface within the pits to the η-phase particles. This diffusion would be necessary in order to compensate the silicon deficiency on the particle surface, which would be generated by the silicon consuming reaction. However, such an interpretation must, in our opinion, be excluded. No driving force is imaginable that would be able to manage the necessary detachment of silicon atoms out of the very stable Si crystal lattice at the usual reaction temperatures. Moreover, nowadays it is well known that the diffusion coefficient of Si in the Cu–Si system is orders of magnitude lower than that of Cu. On the other hand, it is well imaginable that Cu, which is essentially more mobile in the Cu–Si system and can easily diffuse on and into silicon surfaces [18-20], diffuses from the η-Cu₃Si particles onto the silicon surface, forming the Cu–Si surface species as a precondition of the reaction. We will come back to this question at the end of the following discussion on the action of the zinc promoter.

On the Mode of Action of the Promoter Zinc

The Present Knowledge on Promoter Action

The mode of action of promoters has been discussed in very different ways, with a number of them being summarized [1-3]. In the sixties, the action of trace amounts of the promoters Al, In, Ga, P, Sb, Bi and Pb was interpreted in terms of the electronic theory of heterogeneous catalysis [1]. The promoters Zn, Al, and Sb were assumed to take part in the surface chlorination of silicon [21]. According to [8], Zn and Al could catalyze the formation of the intermediate CuCl from copper and the alkyl chloride. Zn may act as a methylating agent [22] or it may direct the polarity of dissociative adsorption of methyl chloride [23]. Sn was suggested to promote the reaction synergistically with Zn or Al, by reducing the surface tension and by increasing the dispersion of metal alloys [22]. In presence of Zn, the silicon diffusion is no longer rate-limiting [11, 24] and carbidic carbon is converted into graphitic carbon, freeing active sites in this way [13]. According to [25], promoters decrease the segregation energy of silicon in copper silicides and cause silicon enrichment on the surface. Finally, promoters could influence the crystal growth of copper silicides [26].

Some of the papers quoted propose a more or less direct interference of the promoters on the reaction mechanism, e.g., a methylating action or an influence on methyl chloride adsorption. Of course, such actions cannot be excluded from a logical point of view and, moreover, it cannot be excluded that one promotor can act in more than in one way.

However, we have already reported results [27] that gave rise to reflect on the necessity of such a hypothesis. In Fig. 3, reaction rates on contact masses composed of three different silicon qualities, non-promoted and zinc-promoted with CuCl₂ and CuO as the copper source, are compared. It is to be seen that the catalytic performances are essentially influenced not only by the zinc promoter, but also by the silicon quality and by the nature of the copper source. But, above all, it is obvious that also *the efficiency of the promoter significantly depends on silicon quality and copper source*. In case of the CuO-containing systems, zinc exhibits a strong promoter action with each of the three silicon qualities. On the other hand, considering the CuCl₂ containing systems, with CuCl₂/Si_{tech} the promoter action of zinc is a remarkable one, too, whereas there is hardly an influence of the promoter on the catalytic activity of CuCl₂/Si_{pure, m} under the conditions chosen. Following the mentioned hypothesis of an interference of the promoter into the mechanism of the chemical reaction, an accelerating action should be observable in each case. Hence, doubts are justified, whether this hypothesis is necessary at all.

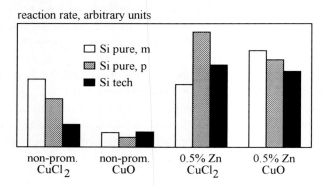

reaction rate, arbitrary units

Fig. 3   Steady state reaction rates after 3 h at 340°C on non-promoted and zinc-promoted contact masses on basis of $CuCl_2$ and CuO as copper sources and of the silicon types $Si_{pure, m}$, $Si_{pure, p,}$ and Si_{tech}.

Results and Discussion on Promoter Action.

What else could be the role of the zinc promoter? In the following, we will try to give a new explanation, based on a comparison of the catalytic and the surface structural properties of the contact masses $CuCl_2/Si_{tech}$, $CuCl_2/Si_{tech}/Zn$ and $CuCl_2/Si_{pure, p}$.

In literature on Rochow synthesis, frequently it is usual to compare stationary state values of reaction rate and selectivities, or mean values over a certain reaction time. Fig. 4 shows that such practice can lead to completely misleading conclusions. In this figure, the catalytic activities of the contact masses $CuCl_2/Si_{tech}$, $CuCl_2/Si_{tech}/Zn$, and $CuCl_2/Si_{pure, p}$ after 20, 60, 120, and 180 min are compared.

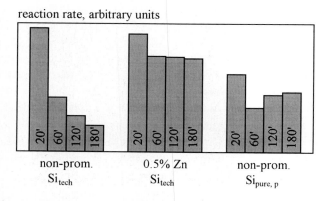

reaction rate, arbitrary units

Fig. 4.   Reaction rates on $CuCl_2/Si_{tech}$, $CuCl_2/Si_{tech}/Zn$ and $CuCl_2/Si_{pure, p}$ at 340°C after 20, 60, 120, and 180 min.

In the case of $CuCl_2/Si_{tech}$, the reaction starts very quickly, reaches a high maximum value already after 20 min, but comes to only a low stationary level after 180 min. Unlike this behavior, with

$CuCl_2/Si_{tech}/Zn$ the reaction rate reaches stationarity at a high level immediately after the short induction period. Hence, it would be wrong to state that the non-promoted system was of low reactivity. Rather, it seems that its reactivity was temporarily even higher than of the zinc-promoted system in the very beginning. (The frequency of GC analysis might have been to low in order to catch the real maximum of reactivity.) But, there must have been a mechanism responsible for the detrimental loss of activity observed and, possibly, there is even a connection between the high initial activity and the following activity decline.

An answer to these questions is given by the results of the SEM/EDX investigations of the two samples. In Fig. 5, a typical surface image of the non-promoted $CuCl_2/Si_{tech}$ system is depicted. It is to be seen that, except e.g., point 1, practically the whole of the silicon surface had been attacked by the reaction and is covered by an overlayer. The EDX analysis of this overlayer gave rather high copper concentrations in the range 50-80 %, see, e.g, point 2. The surface of $CuCl_2/Si_{tech}/Zn$, Fig. 6, significantly differs from $CuCl_2/Si_{tech}$: the reaction exclusively took place in the typical pit-like reactive areas surrounded by unattacked silicon surface.

Fig. 5.   Surface image of non-promoted $CuCl_2/Si_{tech}$ after 3 h reaction, and EDX analysis: 1) non-attacked surface, 0.5 wt.% Cu, 98 wt.% Si; 2) reacted surface covered by an overlayer, 77 wt.% Cu, 20 wt.% Si.

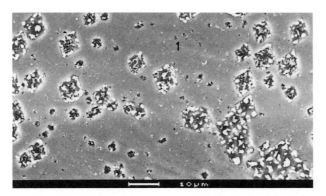

Fig. 6.   Surface image of zinc-promoted $CuCl_2/Si_{tech}$ after 3 h reaction, and EDX analysis: 1) non-reacted surface, 0.5 wt.% Cu, 98 wt.% Si; 2) reacted area with bright particles, 15 wt.% Cu, 68 wt.% Si.

In the following, these observations shall be interpreted in terms of a model, which in truth ascribes to the zinc promoter the role of a moderator of the Rochow reaction.

As can be concluded from Fig. 4, the initial reactivity of the surface of Si_{tech} is very high. The relevant copper species present, possibly copper chlorides, which can easily form copper-silicidic phases by reaction with silicon, can easily attack the whole of the surface and react with surface silicon, resulting in Cu–Si species and finally metallic copper, e.g., [26]. Due to this very fast formation of catalytically active Cu–Si species and of precursors thereof, the reaction becomes very fast already after a short time. But, on the other hand, this overall attack on the silicon surface gives copper species the possibility to be deposited practically over the whole of the silicon grain. This means in terms of our model of catalytically active Cu–Si surface species, explained in the first section of this paper, that there is a lack of still free silicon surface area, which is needed in order to form the active "two-dimensional" Cu–Si species. The surface is simply blocked by thick copper-containing layers. As consequence, the reaction goes down after a short time and the contact mass reaches only low stationary activity.

The initial activity of the technical-grade silicon is higher than of pure silicon, see Fig. 4. This difference should be due to the presence of impurities in the former. The impurities seem to act as a catalyst for the attack of copper species on silicon, i.e., for the destruction of the native SiO_2 overlayer on the silicon surface. If this assumption was true, the image of the reacting surface of a highly pure silicon should differ from that of technical grade silicon. This is confirmed by Fig. 7, which depicts the surface of the non-promoted $CuCl_2/Si_{pure, p}$ contact mass. Here we see the usual picture of distinct reactive areas on a silicon surface, which is only partly attacked by the reaction. Because of the presence of free surface, active Cu–Si surface species could be formed and the reaction rate reached a stationary level above $CuCl_2/Si_{tech}$.

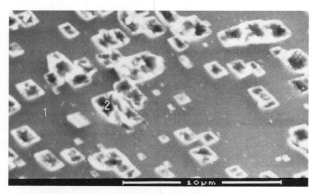

Fig. 7.   Surface image of non-promoted $CuCl_2/Si_{pure, p}$ after 3h reaction, and EDX analysis: 1) non-attacked surface, 0.5 wt.% Cu, 98 wt.% Si; 2) reacted surface (pit), 19 wt.% Cu, 71 wt.% Si.

The surface image of $CuCl_2/Si_{tech}/Zn$ qualitatively corresponds to $CuCl_2/Si_{pure, p}$: also here distinct reactive pits are typical, i.e. a continuous formation of Cu–Si surface species was possible. This results in a relatively high stationary level of activity. If so, the role of the zinc promoter in truth was not a really accelerating one, but was a moderation, leading to a sufficiently high stationary reaction rate. The zinc promoter succeeded in keeping free potentially active surface area, which otherwise would have been blocked by copper species. We can only speculate about the mechanism of this action. Following our above argumentation, it seems plausible that the promoter interacted with the silicon impurities. Zinc, probably as zinc chloride, should have eliminated the catalytic, but finally detrimental, action of the impurities, which consist of oxides of several metals. It is interesting to remember that zinc chloride is used as a flux in soldering procedures, because of its ability to remove oxide layers from metal surfaces [28].

On the whole, it seems that the famous promoter zinc acts in truth not as a real accelerator, as has been believed up to now, but as a kind of moderator of the initial reaction rate of the Rochow reaction and helps to maintain a sufficiently high stationary activity of a contact mass in this way. It is, of course, an open question, whether this mode of action also applies to other promoters and whether the promoter zinc can act also in other ways.

How to explain contact mass exhaustion? The following discussion is based on the results of both of the previous two sections. It is a general experience that the activity of a contact mass charge, e.g, in a discontinuous lab-scale experiment, decreases after a certain stationary period, before all the added silicon is consumed. There have been attempts to generally explain this behavior [1], besides the generally accepted damaging influences of coke deposition and enrichment of silicon impurities. Our concept of catalytically active Cu–Si surface species and of blocking copper containing overlayers seems to be relevant in this respect and allows a simple interpretation of contact mass exhaustion: As soon as the available free silicon surface area of a contact mass decreases too much, the reaction rate will decrease. The decrease in free surface area may be caused by extensive silicon consumption or by extensive deposition of copper-containing layers, the consequence is practically the same. It is interesting to compare this concept with the model of Voorhoeve [1] from the sixties. Voorhoeve also proposed the existence of free silicon surface as a precondition of the synthesis reaction. His model was based on the assumption of a catalytically active η-Cu_3Si surface and of a lateral diffusion of silicon towards η-Cu_3Si. The latter process was thought to maintain a sufficiently high silicon concentration at the η-Cu_3Si surface, which reacts steadily with methyl chloride. In this model, the reaction breaks down when the silicon grains are abraded by the synthesis reaction to such a degree that most of the grain surface is covered by η-Cu_3Si.

Unlike Voorhoeves model, in our concept XRD undetectable Cu–Si surface species with a very low total copper content steadily react with methyl chloride and can be steadily restored by copper diffusion from a copper source like η-Cu_3Si onto the silicon surface. The plausibility of this idea was demonstrated in Fig. 1. Both Voorhoeve's and our model predict a break-down of the reaction, as soon as the silicon grain surface is essentially covered by high amounts of copper-rich species. As a matter of fact, the models only differ in the assumption about the relative mobility of the atoms involved. The high mobility of copper atoms in the Cu/Si system [19, 20], which has been shown only after Voorhoeves work, seems to favor our proposal.

Conclusions

- Non-promoted and zinc-promoted Rochow contact masses on the basis of pure and of technical-grade silicon were investigated by catalytic tests and by the spatially resolving SEM/EDX technique.

- The investigations confirmed the existence and the catalytic action of X-ray amorphous Cu–Si surface species, i.e., extremely dispersed Cu–Si particles or even two-dimensional species like Cu–Si surface compounds, which we proposed recently. Ascribing the catalytic activity of Rochow contact masses to such species allows to explain contradictions in the literature concerning the role of XRD detectable Cu–Si phases like η-Cu₃Si. The role of η-Cu₃Si as a catalytically active species seems to be only a minor one, but it could play a role as a copper reservoir for the formation of active Cu–Si surface species.

- The promoter zinc seems to act rather as a moderator than as a real accelerater of the reaction. It enables a sufficient stationarity of the catalytic activity. It seems that the promoter prevents an overall attack of the reaction on the whole of the silicon surface, catalyzed by silicon impurities, which otherwise would lead to a general blockage of the surface by copper species. Possibly, the promoter is able to bind catalyzing impurities. In this way, the active Cu–Si surface species mentioned above can continuously be formed by lateral diffusion of copper species onto still free silicon surface.

Acknowledgement: This work was supported by the *Bundesminister für Wirtschaft* (Germany), within the project No. 253D of the *Arbeitsgemeinschaft industrieller Forschungsvereinigungen e.V.* (Köln), Forschungsvereinigung *Dechema* (Frankfurt/Main). Moreover, the authors are grateful to *Hüls-Silicone* (Nünchritz, Germany) and to *Wacker-Chemie* (Burghausen, Germany) for generous supply of materials.

References:

[1]   R. J. H. Voorhoeve, *Organohalosilanes, Precursors to Silicones*, Elsevier, New York, **1967**.
[2]   M. P. Clarke, *J. Organomet. Chem.*. **1989**, *376*, 165.
[3]   K. M. Lewis, D. G. Rethwisch, *Catalyzed Direct Reactions of Silicon*, Elsevier, New York, **1993**.
[4]   B. I. Baglaj, K. M. Weisberg, M. F. Mazitov, R. M. Masagutov, *Kinet. Katal.* **1975**, *16*, 804.
[5]   P. Trambouze, *Bull. Soc. Chim. France* **1956**, 1756.
[6]   A. L. Klebansky, V. S. Fikhtengolts, *J. Gen. Chem. USSR* **1957**, *27*, 2693.
[7]   R. Müller, H. Gümbel, *Z. Anorg. Allg. Chem.* **1964**, *327*, 302.
[8]   R. A. Turetskaya, K. A. Andrianov, I. V. Trofimova, E. A. Chernyshev, *Usp. Khim.* **1975**, *44*, 444.
[9]   W. F. Banholzer, N. Lewis, W. Ward, *J. Catal.* **1986**, *101*, 405.
[10] W. F. Banholzer, M. C. Burrel, *J. Catal.* **1988**, *114*, 259.
[11] T. C. Frank, K. B. Kester, J. L. Falconer, *J. Catal.* **1985**, *91*, 44.
[12] H. Lieske, H. Fichtner, I. Grohmann, M. Selenina, W. Walkow, R. Zimmermann, *"Characterization of Rochow Contact Masses by Catalytic Results, XRD, Chemisorption and XPS"*, in: *Proceedings of the Conference on Silicon for Chemical Industry*, (Eds.: H. A. Øye, H. Rong), June 16-18, **1992**, Geiranger, Norway, Institute of Inorganic Chemistry, Trondheim, **1992**, p. 111.

[13] T. C. Frank, K. B. Kester, J. L. Falconer, *J. Catal.* **1985**, *95*, 396.

[14] N. Floquet, S. Yilmaz, J. L. Falconer, *J. Catal.* **1994**, *148*, 348.

[15] H. Lieske, H. Fichtner, U. Kretzschmar, R. Zimmermann, *Appl. Organomet. Chem.* **1995**, *9*, 657.

[16] A. I. Gorbunov, A. P. Belyi, G. G. Filippov, *Usp. Khim.* **1974**, *43*, 683.

[17] A. Sh. Varadashvili, L. M. Khananashvili, N. I. Tsomaya, N. P. Lobusevich, E. S. Starodubtsey, V. M. Kopylov, B. M. Kipnis, Yu. N. Novikov, M. E. Vol'pin, *Soobshch. Akad. Nauk Gruz. SSR* **1989**, *134/3*, 565.

[18] F. A. Veer, B. H. Kolster, W. G. Burgers, *Trans. Metall. Soc. AIME* **1968**, *242*, 669.

[19] Shih-Hung Chou, A. J. Freeman, S. Grigoras, T. M. Gentle, B. Delley, E. Wimmer, *J. Chem. Phys.* **1988**, *89*, 5177.

[20] Shih-Hung Chou, A. J. Freeman, B. Delley, E. Wimmer, S. Grigoras, T. M. Gentle *"Chemisorption and catalytic activity of Cu and Ag atoms on Si(111) surfaces"*, in: *Catalyzed Direct Reactions of Silicon*, (Eds.: K. M. Lewis and D. G. Rethwisch), Elsevier, New York, **1993**, p. 299.

[21] M. G. R. T. de Cooker, J. W. de Jong, P. J. van den Berg, *J. Organomet. Chem.* **1975**, *86*, 175.

[22] L. D. Gasper-Galvin, D. M. Sevenich, H. B. Friedrich, D. G. Rethwisch, *J. Catal.* **1991**, *128*, 468.

[23] Jong Pal Kim, D. G. Rethwisch, *J. Catal.* **1992**, *134*, 168.

[24] J. P. Agarwala, J. L. Falconer, *Int. J. Chem. Kin.* **1987**, *19*, 519.

[25] K. M. Lewis, D. McLeod, B. Kanner, *"The significance of Si and Zn surface enrichment in the Rochow reaction"*, in: *Catalysis 1987* (Ed.: J. W. Ward), Elsevier, Amsterdam, **1988**, p. 415.

[26] G. Weber, D. Viale, H. Souha, B. Gillot, P. Barret, *Solid State Ionics* **1989**, *32/33*, 250.

[27] H. Lieske, U. Kretzschmar, R. Zimmermann, *"Action of Promoters in Rochow Contact Masses"*, in: *Proceedings of Silicon for the Chemical Industry II* (Eds.: H. A. Øye et al.), June 8-10, 1994, Loen, Norway, Tapir Forlag, Trondheim, **1994**, p. 147.

[28] A. F. Hollemann, E. Wiberg, *Lehrbuch der Anorganischen Chemie*, Walter de Gruyter, Berlin-New York, **1985**, p. 1038.

On the Acid- and Base-Catalyzed Reactions of Silanediols and Siloxanediols in Water

H. Kelling*, W. Rutz, K. Busse, Ch. Wendler, D. Lange

Fachbereich Chemie

Universität Rostock

Buchbinderstraße 9, D-18051 Rostock, Germany

Tel.: Int. code + (381)4981755 – Fax: Int. code + (381)4981763

Keywords: Silanediols / Siloxane-1,1-diols / Condensation / Distribution / Kinetics

Summary: The acid- and base-catalyzed reaction of some partially water-soluble silanediols $XMeSi(OH)_2$ and siloxane-1,1-diols have been investigated preferably in the two-phase system water/toluene. The diols react in water only whereas the less water-soluble products of the primary reactions are dissolved in the inert toluene phase. The silanediols give condensation only, the siloxanediols also show Si–O–Si-cleavage. The distribution rates of the model compounds and the rate constants of the primary reactions in water have been investigated.

Introduction

Silane- and siloxanediols are important intermediates in silicone synthesis. During the technical chloro- or alkoxysilane hydrolysis the primary-formed silanols and siloxanols react preferably in the aqueous phase due to their considerable water solubility. To get more information on these reactions in water we investigated some silane- and siloxanediols using a method already described [1].

For some silanediols $XMeSi(OH)_2$ kinetic data and mechanistic considerations already exist for the condensation in organic solvents (dioxane, methanol) [2-4], which are interesting for comparision with results for the reaction in water.

The two-phase experiments should have the advantage that only the starting products are sufficiently water soluble, whereas most reaction products are transfered in the toluene phase. There are consecutive reactions not to be expected due to the insolubility and the inactivity of the catalysts in toluene.

Results and Discussion

Model Compounds and their Distribution in Water / Toluene

The silanediols $XMeSi(OH)_2$ and the siloxane-1,1-diols (Table 1) have been prepared by hydrolysis of the corresponding dichlorosilanes. We succeeded in the preparation of the previously unknown hexamethylcyclotetrasiloxan-1,1-diol , ($D_3D^{(OH)}_2$), (Fig. 1.).

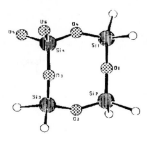

Fig. 1.   Structure of $D_3D^{(OH)_2}$

This relatively stable compound can be recrystallized from water and has a water solubility of $1,3 \times 10^{-2}$ mol l^{-1} at $25°C$. The X-ray-data show two molecules together with two molecules of water in an asymmetric unit. Further investigations on H-bonding in the crystal lattice are in progress.

For all model compounds the distribution ratios α have been determined according to (Eq. 1) measuring the absorption of the $\delta(CH_3)_s–Si$ IR band in toluene before and after distribution (see Table 1).

$$\alpha = c(\text{in toluene}) / c(\text{in water})$$

Eq. 1.

The also determined α-values for some primary condensation products $[XMeSi(OH)]_2O$ are greater than 50 except for X = Me ($\alpha = 0.3$).

The α-values increase slighly with increasing temperature due to the decreasing water solubility with decreasing H-bond interaction.

The Main Reactions of the Model Compounds in Water

For all $XMe_2Si(OH)_2$ and for $MM^{(OH)_2}$ the condensation corresponding to (Eq. 2.) is the primary reaction.

$$2 \ XMeSi(OH)_2 \ \text{-------}\!\!\!\succ \ [XMeSi(OH)]_2O \ + \ H_2O$$

Eq. 2.

Except for $Me_2Si(OH)_2$ no consecutive reaction have been found. For $Me_2Si(OH)_2$, also hexamethylcyclotrisiloxane (D_3), octamethylcyclotetrasiloxane (D_4), and octamethyltetrasiloxane-α,ω-diol have been detected in the toluene phase due to the water solubility of the primary condensation product $[Me_2Si(OH)]_2O$ and its condensation with $Me_2Si(OH)_2$ or with itself.

In constrast to the very fast condensation of $MM^{(OH)_2}$, no condensation product was found for $M_2D^{(OH)_2}$. In agreement with a first-order concentration decrease and the determination of hexamethyldisiloxane only in toluene is siloxane bond cleavage the prefered reaction under both acidic and basic conditions.

Also for the cyclotetrasiloxanediol $D_3D^{(OH)_2}$ the siloxane bond cleavage is dominating, at least at the low starting concentration due to the limited water solubiliy.

Table 1.   Distribution ratios α and rate constants k_a and k_b of the acid-and base-catalyzed condensation.

Silanediol	$\alpha = c(\text{in toluene}) / c(\text{ in water})$		$10^2\,k_a$ $[\text{L}^2\text{mol}^{-2}\text{s}^{-1}]$	k_b $[\text{Lmol}^{-1}\text{s}^{-1}]$
	25°C	40°C		
Me$_2$Si(OH)$_2$	≤0.001	≤0.001	32.00	0.20
EtMeSi(OH)$_2$	0.013	0.025	2.70	0.012
VinMeSi(OH)$_2$	0.005	0.006	3.60	0.18
BrCH$_2$MeSi(OH)0$_2$	0.20	0.24	0.57	
ClCH$_2$MeSi(OH)$_2$	0.29	0.36	1.21	0.22
Me$_3$SiOSiMe(OH)$_2$ MM$^{(OH)}$$_2$	0.14	0.27	58.00	0.0014
(Me$_3$SiO)$_2$Si(OH)$_2$ M$_2$D$^{(OH)}$$_2$	3.57	4.90	0.01[a]	0.10[a]
D$_3$D$^{(OH)}$$_2$(see Fig. 1)	14.9		0.40[a]	1.76[a]

[a] Rate constants for Si–O–Si bond cleavage, k (Lmol^{-1}s^{-1}).

Kinetic Data

To follow the reaction in the aqueous phase a toluene solution with a definite diol concentration was agitated with an equal volume of water. After measuring the equilibrium concentration of the distribution the reaction was started by addition of the catalyst (10^{-3}-10^{-2} m HCl or NaOH, respectively). At definite reaction times probes of the toluene phase were analyzed by GC or (and) IR. The reaction of the cyclosiloxanediol D$_3$D$^{(OH)}$$_2$ was investiated in water directly by HPLC.

The acid-catalyzed condensation was found to be of second order with respect to the diol and of first order to the catalyst. The rate constants k_a corresponding to (Eq. 3) are given in Table 1.

$$-dc(\text{diol}) / dt = k_a \cdot c^2(\text{diol}) \cdot c(\text{Kat.})$$

Eq. 3.

As already known for the silanol condensation in organic solvents [3, 5] and for other nucleophilic reactions at Si–O compounds, the reaction rates decrease with increasing electronegativity of the substituents X at Si and for more bulky substituents X. This is in agreement with a mechanism characterized by a fast protonation preequilibrium (Eq. 4)

$$\text{XMeSi(OH)}_2 + \text{H}_3\text{O}^+ \text{-------}\!\!\!\!\rightarrow \text{XMeSi(OH)(OH}_2)^+ + \text{H}_2\text{O}$$

Eq. 4.

The equilibrium concentration of the protonated silanol is rate-determining in the condensation reaction (Eq. 5).

$$\text{XMeSi(OH)}_2 + \text{XMeSi(OH)(OH}_2)^+ \text{-------}\!\!\!\!\rightarrow [\text{XMe(OH)Si]}_2 + \text{H}_3\text{O}^+$$

Eq. 5.

The acid and the base catalyzed siloxane cleavage of $M_2D^{(OH)_2}$ and $D_3D^{(OH)_2}$ is first-order with respect to both the diol and the catalyst concentration.

The base-catalyzed condensation of the silanediols and $MM^{(OH)_2}$ was found to be of first order with respect to the silanediol and the catalyst concentration. The rate constants k_b (Table 1) corresponding to (Eq. 6) increase slightly with increasing electronegativity of X in competition with a decrease for more bulky substituents X.

$$-dc(\text{diol}) / dt = k_b \cdot c(\text{diol}) \cdot c(\text{Kat.})$$

Eq. 6.

This is in agreement with analog results for the base-catalyzed condensation in organic solvents [4]. The catalyst base OH⁻ is converted nearly quantitative in silanolate. The silanolate concentration is approximatly constant during the main part of the reaction. Differences in the reaction rates are mainly caused by substituent effects on the electrophilicity of the silanol.

Acknowledgement: This work was supported by the *Fonds der Chemischen Industrie*.

References:

[1]   W. Rutz, D. Lange, H. Kelling, *Z. Anorg. Allg. Chem.* **1985**, *528*, 98.
[2]   S. Chrzczconowicz, Z. Lasocki, *Bull. Acad. Pol. Sci., Ser. Sci. Chim.* **1962**, *10*, 4.
[3]   Z. Lasocki, *Bull. Acad. Pol. Sci., Ser. Sci. Chim.* **1964**, *12*, 4.
[4]   B. Dejak, Z. Lasocki, W. Mogilnicki, *Bull. Acad. Pol. Sci., Ser. Sci. Chim.* **1969**, *17*, 1.
[5]   S. Bilda, D. Lange, E. Popowski, H. Kelling, *Z. Anorg. Allg. Chem.* **1987**, *550*, 186.

Trace Analysis of Mono- and Trifunctional Groups in Polydimethylsiloxanes Using Reaction Headspace GC

Jürgen Graßhoff

Hüls Silicone GmbH

Analytical Department

D-01612 Nünchritz, Germany

Tel.: Int. code + (35265)73510 – Fax: Int. code + (35265)73503

Summary: Polydimethylsiloxanes (PDMS) were cleaved using boron trifluoride etherate under defined conditions in a headspace sampler, and the resulting fluorosilanes were quantified by gas chromatography. The method makes it possible to determine traces of mono- and trifunctional organosiloxy groups in PDMS in concentrations down to < 10 mg/kg.

Introduction

Various attempts have been made in the past to determine the contents of various functionalities in silicone polymers by cleavage of the siloxane linkage and analysis of the monomers formed thereby [1]. Considerable experimental problems, especially the difficulty of reliably preventing the cleavage of Si–C bonds and of achieving the sensitivity necessary for trace analysis, have resulted in these methods not being used in routine analysis [2].

These limitations can be overcome by reacting the polymer with boron trifluoride etherate under defined conditions in a headspace sampler. The reaction results in a mixture of gaseous fluorosilanes that is transferred directly, without further workup, into a gas chromatograph, where they components are separated and quantified by means of a flame ionization detector (FID) or mass selective detector (MSD).

Fluorosilane Derivatization

Smith [3] has summarized various methods for cleaving polysiloxanes using boron trifluoride etherate.

$$3 \equiv Si - O - Si \equiv + 2BF_3 * \left(C_2H_5\right)_2 O \rightarrow 6 \equiv Si - F + 2\left(C_2H_5\right)_2 O + B_2O_3$$

Eq. 1.

Reaction of polydimethylsiloxanes in accordance with the above equation takes place sufficiently rapidly at moderately elevated temperature to give the corresponding fluorosilanes. On the other hand, the cleavage of polysiloxanes with a high trifunctional content is reported to be frequently incomplete [3].

Table 1. Selection of organofluorosilanes as products of the BF$_3$ etherate cleavage.

No.	Name	Formula	*bp* [°C]	MW [g/mol]
1	Methyldifluorosilane	CH$_3$(H)SiF$_2$	-35.6	82
2	Methyltrifluorosilane	CH$_3$SiF$_3$	-30.2	100
3	Vinyltrifluorosilane	C$_2$H$_3$SiF$_3$	-25	112
4	Dimethylfluorosilane	(CH$_3$)$_2$(H)SiF	-9	78
5	Ethyldifluorosilane	C$_2$H$_5$(H)SiF$_2$	-7.5	96
6	Ethyltrifluorosilane	C$_2$H$_5$SiF$_3$	-4.2	114
7	Dimethyldifluorosilane	(CH$_3$)$_2$SiF$_2$	2.7	96
8	Trimethylfluorosilane	(CH$_3$)$_3$SiF	6.4	92
9	Methylvinyldifluorosilane	CH$_3$(C$_2$H$_3$)SiF$_2$	25.5	108
10	Ethylmethyldifluorosilane	C$_2$H$_5$(CH$_3$)SiF$_2$	38	110
11	Dimethylvinylfluorosilane	(CH$_3$)$_2$C$_2$H$_3$SiF	43	104
12	Ethyldimethylfluorosilane	C$_2$H$_5$(CH$_3$)$_2$SiF	51	106

However, the reactivity and the wide boiling range of these compounds have caused considerable problems in analytical workup of the reaction mixtures, so that frequently only qualitative and semiquantitative results are recorded in the literature [2, 4].

Optimization of the Separation and Identification of the Cleavage Products by GC-MS

Optimal results for the separation of fluorosilanes (Table 1) were achieved with thick-film capillary columns of immobilized 3,3,3-trifluoropropylmethylpolysiloxane, e.g., Rtx-200 (RESTEK) or DB-210 (J&W). Besides the chemical stability of the separating phase towards the reagent, particularly in favor of their use were the good selectivity (Fig. 1) for many organofluorosilanes of industrial interest and a drastic reduction in adsorption effects.

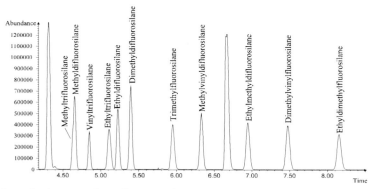

Fig. 1. TIC of a qualitative organofluorosilane test mixture.

The latter was the precondition, in conjunction with suitably sensitive detectors, for lowering the detection limit. The cleavage products were identified by comparison of retention times and examination of the mass spectra. For this purpose, appropriate model siloxane systems were prepared and reacted with boron trifluoride etherate.

Technical Implementation

Besides the requirement for adequate analytical reliability, crucial for routine use are robustness and the possibility of substantial automation. For these reasons, we used a HS 40 headspace sampler (Perkin-Elmer) directly coupled to an HP 5890 series II gas chromatograph (Hewlett-Packard). This combination avoids adsorptive and reactive surfaces in the sample path and allows the temperature to be controlled from cleavage to detection. This makes even trace analysis very reliable and reproducible.

Several series of tests were carried out to optimize the amounts of sample and reagent, and the reaction time and temperature. Completeness of reaction was confirmed by GPC analyses. To shorten the analysis cycle time and increase the useful life of the FID, back-flushing of high-boilers was implemented. The fluorosilanes can be substantially separated by gas chromatography using two wide-bore Rtx-200 capillary columns ($df = 1$ mm, $l = 60$ m, $di = 0.53$ mm) connected in series at 45°C isothermal.

Quantification can take place by the normalized percent method or by standard addition using the integrated peak areas of the fluorosilanes, taking account of the substance-specific correction factors. Calibration and checks of linearity were carried out using mixtures of siloxanes of known purity, e.g., methyltris(trimethylsiloxy)silane (M_3T) in octamethylcyclotetrasiloxane (D_4).

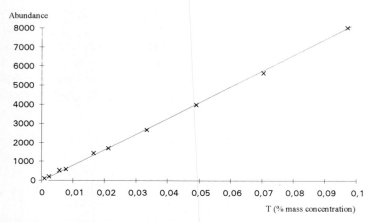

Fig. 2.   Calibration plot for the concentration of methyltrisiloxy groups (T) in polydimethylsiloxanes.

Conclusions

The described method makes it possible to determine trace amounts of T- and M-groups in polydimethylsiloxanes down to 1 mg/kg and is thus suitable for quality control of such products of the technical qualities currently available. However, the precondition for achieving these detection limits is adequate conditioning of the system and purity of the reagents and materials.

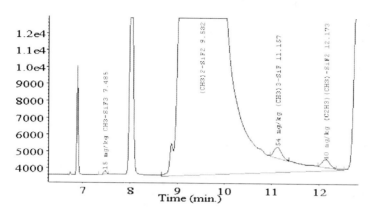

Fig. 3. Gas chromatogram of a fluorosilane mixture from the cleavage of a polydimethylsiloxane (DDS hydrolysate).

No adsorption effects or Si–C cleavage are observed even in the trace range. The idea that an unacceptable blank is produced by even a small amount of cleavage of dimethyldifluorosilane to methyltrifluorosilane during the depolymerization can thus be ruled out.

Similarly good results were obtained in reactions of polydimethylsiloxanes with methylvinyl and dimethylvinylsiloxy groups. No interfering side reactions were observed in this case either. For Si–H-containing siloxanes, we were able to confirm the partial replacement of hydride hydrogen by fluorine on the silicon atom which was observed by Smith [3]. Under standard conditions, this reaction occurred to the extent of 35-65 %, depending on the other substituents. However, these organosiloxy groups can also be determined by previous derivatization or independent determination of the hydrogen bonded to silicon.

References:

[1]   A. L. Smith, *The Analytical Chemistry of Silicones*, Wiley, New York, **1990**, p. 209.
[2]   J. Franc, K. Placek, *J. Chrom.* **1972**, *67*, 37.
[3]   A. L. Smith, *Analysis of Silicones*, Wiley, New York, **1974**, p. 159.
[4]   H. Hahnewald, H. Rotzsche, VEB Chemiewerk Nünchritz, **1989**, unpublished method.

Synthesis and Investigation of the Surface Active Properties of New Silane Surfactants

S. Stadtmüller*, K.-D. Klein, K. Köppen, J. Venzmer

Th. Goldschmidt AG

Goldschmidtstr. 100, D-45127 Essen, Germany

Keywords: Hydrolysis / Silanes / Surface Activity / Surfactants / Trisiloxanes

Summary: Organomodified trisiloxanes show remarkable surface activity. Thus they can be used as additives in detergents, foaming agents or agrochemicals. However, this class of compounds has a limited range of applications, since it is susceptible to hydrolytic decomposition in aqueous solution. The hydrolysis occurs at the silicon-oxygen-bonds present within the molecules. Recently, a new class of silane surfactants free of Si–O– bonds has been developed. These silane surfactants proved to be hydrolytically stable even under extreme pH. Furthermore they exhibit surfactant properties comparable to those typical of trisiloxane surfactants.

Introduction: Trisiloxane Surfactants

Today, a broad range of organosilicone compounds is commercially available. Because of their inherent structural variability they find industrial applications in such diverse fields as cosmetics or hair care, textiles-, and the automobile industry, floor polishes, and paints.

Among the organomodified silicones in particular hydrophilically-substituted trisiloxane derivatives serve as adjuvants in agricultural formulations. Aqueous solutions of such trisiloxanes show outstanding surface activity, such as low dynamic surface tension and excellent wetting and spreading capabilities. Those so-called "superspreaders" allow water to spread on a hydrophobic surface within tens of seconds when added in a small amount (about 0.1 wt.%) [1]. They enhance the performance of herbicidal formulations since they also promote good adhesion of spray droplets, stomatal infiltration, and fastness to rain.

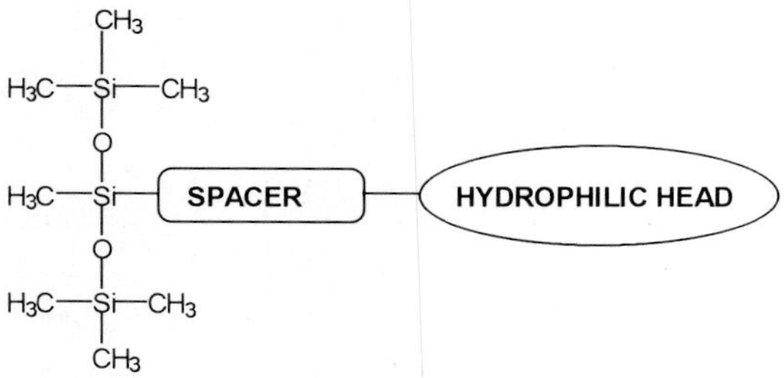

Fig. 1. Schematic structure of organomodified trisiloxanes.

In general, these organomodified trisiloxanes are based on a common structural principle. They consist of a lyophobic silicone backbone containing an alkyl spacer group connected by a silicon-carbon bond. The hydrophilic moiety, which can be either ionic or nonionic, is attached to the alkyl spacer.

In the presence of hydroxy- or hydronium ions trisiloxanes are prone to hydrolytical instability caused by a degradation process starting at the Si–O bonds of the silicone backbone [2, 3]. The resulting oligomers and hexamethyldisiloxane exhibit only poor surface activity (Fig. 2). Depending on the kind of application this decomposition in aqueous media limits the use of this class of compounds.

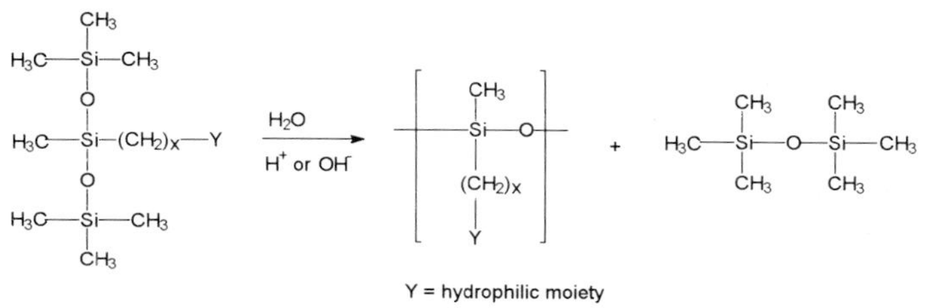

Y = hydrophilic moiety

Fig. 2.   Hydrolytic decomposition of organomodified trisiloxanes.

The ideal solution to overcome this problem is to find a compound that is hydrolytically stable and shows the same excellent surfactant properties as the trisiloxanes. Obviously, such a material must not contain Si–O bonds in its structure.

Synthesis of Trimethylsilane Surfactants

A first approach to obtain silicon-containing surfactants without Si–O bonds was made by Dow Corning [4, 5]. They reacted carbosilanes with several α-olefins containing reactive (e.g., epoxy-functional) moieties, catalyzed by platinum complexes. The surfactant properties and high stabilities of these compounds indicated a promising solution. However, the corresponding carbosilanes as precursors were synthesized via a Grignard reaction, so these silane surfactants were difficult and costly to produce.

The availability of trimethylsilane in larger amounts and reproducible quality was our key to silane surfactant chemistry [6]. The successful reduction of chlorotrimethylsilane with hydrides, such as magnesium hydride in etheral solution, gave the corresponding hydrogen silane in excellent yields. Further transformations with α,β-unsaturated olefins carrying a hydrophilic moiety opened up exciting new opportunities for the synthesis of Si-surfactants (Fig. 3) [7].

The Pt-catalyzed hydrosilylation of trimethyl silane and alkenols or alkenyl-polyethers lead to nonionic silane surfactants, whereas the addition of allylglycidyl ether to trimethylsilane results in a precursor for ionic derivatives. The epoxy group is highly reactive towards nucleophilic agents and can be easily transformed into quaternary ammonium, betaine, or sulfonate complexes. Additionally, cation-anion complexes can be formed by the transformation of two equivalents of epoxy silane with one equivalent of trialkyl ammonium hydrogen sulfite. The reaction of hydroxyalkyltrimethylsilane with sulfamic acid followed by transamination with a primary amine is a route to anionic silane sulfonate surfactants.

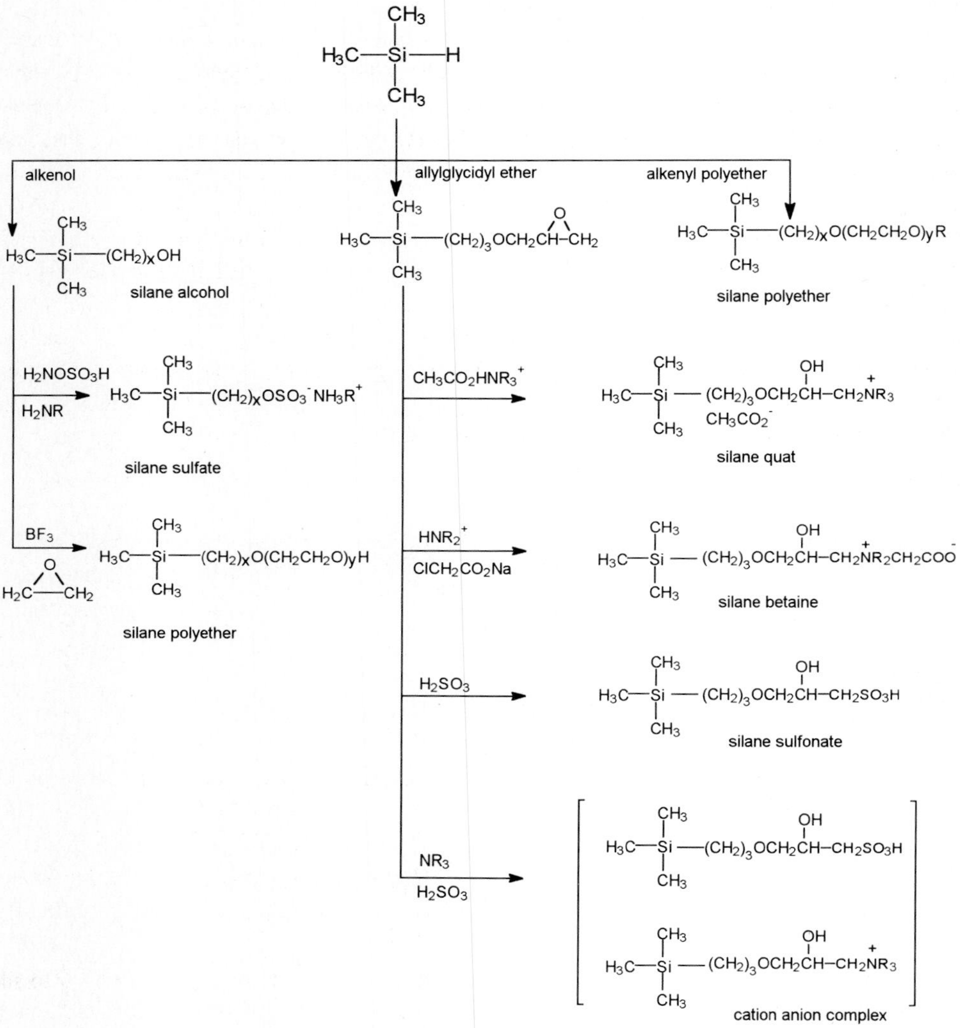

Fig: 3.   Synthetic route to amphiphilic trimethyl silane derivatives.

Trimethylsilanes connected to a polyether as the hydrophilic moiety can be readily synthesized through hydrosilylation of an alkenol followed by ethoxylation. By this method a variety of silane polyethers has been synthesized differing in the length of the alkyl spacer (C_3 to C_{11}) and the average number of ethylene oxide units (y = 2 to 10). Among these surfactants the one containing a hexyl spacer group and carrying a polyether with on average four ethylene oxide units (K 4237) exhibits the best surface activity according to preliminary results. Tests according to the OECD method 301

D revealed that this surfactant is 88 % biodegradable. Nevertheless, it is also hydrolytically stable, as shown by some investigations of surface activity as discussed below.

Surface Activity of Trimethylsilane Surfactants

Several methods have been employed to study the surface active behavior of the nonionic trimethylsilane surfactant K4237 with an hexyl spacer and four ethylene oxide groups. The static surface tension has been determined as a function of surfactant concentration using the du Noüy ring method. It decreases continously to a value of 23 mN/m at 0.03 wt.% and remains constant upon further increase in concentration. In the case of conventional surfactants, such a breakpoint in the surface tension versus concentration plot indicates the onset of micelle formation (*cmc*). Here, this point coincides with the occurence of a slight turbidity as in the case of trisiloxane surfactants [8]. In aqueous solution above this critical aggregation concentration a birefringent lyotropic liquid crystalline phase separates. An aqueous dispersion of this compound consists of vesicles, as determined by Video Enhanced Contrast Microscopy [9]. Such a dispersed surfactant-rich phase has been reported to be a requirement for superspreading of trisiloxane surfactants [10]. Indeed, aqueous solutions of nonionic as well as ionic silane surfactants exhibit excellent spreading characteristics on hydrophobic surfaces such as polypropylene film.

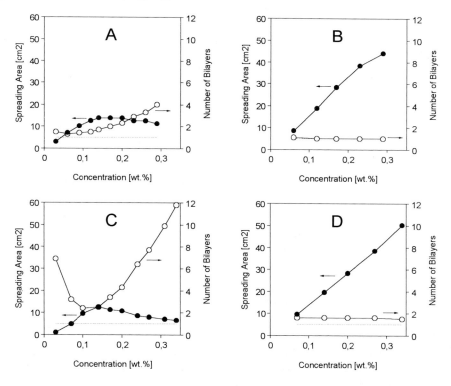

Fig. 4. Spreading area (absolute values [cm²]: ●; relative to number of surfactant molecules expressed in number of bilayers: ○) as a function of concentration and relative humidity. Trisiloxane surfactant (Tegopren 5840) at **(A)** 50 % and **(B)** 100 % relative humidity; Silane surfactant (K 4237) at **(C)** 50 % and **(D)** 100 % relative humidity.

In the case of nonionic polyether trisiloxane surfactants such as Tegopren 5840 spreading usually decreases at concentrations above 0.1 % (Fig. 4A). By eliminating evaporation, i.e., spreading in an atmosphere of 100 % relative humidity, the spreading area is exactly proportional to the amount of surfactant (Fig. 4B). Considering the size and number of surfactant molecules, the final structure after the spreading process corresponds to a single bilayer. From measurements of the spreading kinetics it can be concluded that the non-linearity observed during spreading in air is the result of a local formation of a non-spreading viscous liquid crystalline phase at the air/water interface caused by evaporation. Spreading experiments using the optimized trimethylsilane surfactant K 4237 revealed quite a similar behavior (Figs. 4C and D). Again, the spreading area decreases with increasing the concentration above a certain limit when the experiment is performed at laboratory atmosphere (50 % relative humidity), whereas it is exactly proportional to the amount of surfactant when spreading is carried out at 100 % relative humidity. However, the spreading area relative to the number of surfactant molecules is smaller, indicating a different type of aggregation behavior at the interface.

To test its hydrolytic stability this silane surfactant was stored as aqueous solutions (0.1wt.%) over a range of pH values from 2 to 12. The diameters obtained from spreading of a 50 μL droplet was unchanged over a period of more than six months. This is convincing proof that no significant decomposition takes place. This extraordinary stability under both acidic and alkaline conditions allows the use of such silane surfactants in applications where long term storage is required.

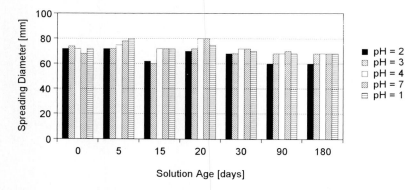

Fig. 5. Spreading of a nonionic silane surfactant (K 4237) at varions values of pH as a function of solution age.

Conclusion

By using a recently developed magnesium hydride technology, the trisiloxane lyophobic part in superspreading surfactants can be substituted by a trimethylsilane moiety. This synthetic route leads to both nonionic and ionic silane surfactants, which are hydrolytically stable even under extreme pH. Aqueous solutions of these new surfactants exhibit surface tension and wetting properties comparable to the traditional organomodified trisiloxane surfactants. The combination of hydrolytic stability and biodegradability offers chance for the widespread application of these silane based surfactants.

References:

[1] S. Zhu, W. G. Miller, L. E. Scriven, H. T. Davis, *Colloids Surf.* **1994**, *90*, 63.

[2] K.-D. Klein, D. Schaefer, P. Lersch, *Tenside Surf. Det.* **1994**, *31*, 115.

[3] G. Feldmann-Krane, W. Höhner, D. Schaefer, S. Silber, *DE Pat. Anm.* **1993**, 4317605.4.

[4] A. R. L. Colas, A. A. D. Renauld, G. C. Sawicki, *GB Pat.* **1988**, 8819567.

[5] M. J. Owen, *GB Pat.* **1974**, 1520421.
     K.-D. Klein,W. Knott, G. Koerner, *DE Pat. Anm.* **1993**, 43 13130.1.

[6] K.-D. Klein, W. Knott, G. Koerner, *DE Pat.* **1994**, 4320920.3; *DE Pat.* **1994**, 4330059.6.

[7] K. P. Ananthapadmanabhan, E. D. Goddard, P. Chandar, *Colloids Surf.* **1990**, *44*, 281.

[8] H. Leonhard, H. Rehage, J. Venzmer, unpublished results.

[9] M. He, R. M. Hill, Z. Lin, L. E. Scriven, H. T. Davis, *J. Phys. Chem.* **1993**, *97*, 8820.

Carbohydrate-Modified Siloxane Surfactants: The Effect of Substructures on the Wetting Behavior on Non Polar Solid Surfaces

R. Wagner*, L. Richter, Y. Wu
Max-Planck-Institut für Kolloid- und Grenzflächenforschung
Rudower Chaussee 5, D-12489 Berlin, Germany
Tel.: Int. code + (30)63923160 – Fax: Int. code + (30)6923102
E-mail: wagner@mpikg.fta-berlin.de

J. Weißmüller, J. Reiners
Geschäftsbereich Pflanzenschutz
Bayer AG
Alfred-Nobel-Str. 50, D-40789 Monheim, Germany

K.-D. Klein, D. Schaefer, S. Stadtmüller
Theodor Goldschmidt AG
Goldschmidtstr. 100, D-45127 Essen, Germany

Keywords: Si-Surfactants / Wetting Tension / Surface Tension / Contact Angle / Spreading

Summary: The wetting behavior of liquid siloxanes and aqueous solutions of carbohydrate-modified siloxane surfactants on perfluorinated surfaces has been investigated. Siloxanyl moieties in surfactants level off to a large extent the influences of other structural elements. The donor-acceptor portions of the surface tension and the interfacial tension solid/liquid converge and amount to about 1-2 mN/m. The contact angle is not a linear function of the surface tension. It results from the superposition of surface tension and interfacial tension solid/liquid, both independent of each other.

Introduction

Low molecular weight siloxane surfactants represent a unique class of substances combining superior surface activity and compatibility of the permethylated backbone with hydrocarbon phases. For these reasons aqueous solutions of certain siloxane surfactant types rapidly wet low energy surfaces (polypropylene, polyethylene, natural waxes) [1-3]. Nevertheless, the influences of the surfactant molecule substructures on the wetting process are not well understood. Therefore, a combined investigation of the energy situation at the interfaces vapor-aqueous surfactant solution and low energy solid-aqueous surfactant solution has been carried out. Additionally, we investigated the wetting behavior of various liquid siloxane precursors.

Materials and Methods

For that purpose carbohydrate-modified siloxane surfactants bearing four independent structural elements (i) siloxanyl moiety (si), (ii) spacer (sp), (iii) carbohydrate unit (ch) and (iv) modifying element (mo) have been synthesized [4, 5, 6].

```
        ┌──────┐
        │  si  │
        └──────┘
           │
        ┌──────┐    ┌──────┐
        │  sp  │────│  mo  │
        └──────┘    └──────┘
           │
        ┌──────┐
        │  ch  │
        └──────┘
```

Fig.1.   Structure of the synthesized surfactants.

In order to investigate the influence of the siloxanyl structure a series of disaccharide derivatives has been synthesized according to the following sequence (Eq. 1).

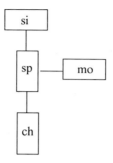

Eq. 1.   Synthesis of carbohydrate-modified siloxane surfactants.

In order to determine the contact angles (Θ) (Eq. 2) of aqueous surfactant solutions ($c > cmc$) on an apolar perfluorinated surface (FEP®, Du Pont, solid surface tension γ_{sv} 18.9 mN/m) the surface tension (γ_{lv}) and the wetting tension (α) have been measured [7].

$$\frac{\alpha}{\gamma_{lv}} = \frac{\gamma_{sv} - \gamma_{sl}}{\gamma_{lv}} = \cos\Theta$$

Eq. 2.   Young's equation.

From wetting experiments on apolar surfaces ($\gamma_{sv}^{LW} = \gamma_{sv}$) the Lifshitz-van der Waals portions of the surface tension (γ_{lv}^{LW}) (Eq. 3) and interfacial tension solid/liquid (γ_{sl}^{LW}) (Eq. 4) can be determined [8,9].

$$\gamma_{lv}^{LW} = \frac{\gamma_{lv} * (1 + \cos\Theta)^2}{2\sqrt{\gamma_{sv}^{LW}}}$$

Eq. 3.

$$\gamma_{sl}^{LW} = \gamma_{sv}^{LW} + \gamma_{lv}^{LW} - 2\sqrt{\gamma_{sv}^{LW} * \gamma_{lv}^{LW}}$$

Eq. 4.

The interfacial energy data of aqueous solutions of surfactants synthesized according to Eq. 1 are summarized in Table 1 [10].

Table 1.   Interfacial energy properties of selected surfactants.

No.	Si-structure	cmc (mol/L)	γ_{lv} (mN/m)	γ_{lv}^{LW} (mN/m)	$\gamma_{lv}^{+/-}$ (mN/m)	γ_{sl} (mN/m)	γ_{sl}^{LW} (mN/m)	$\gamma_{sl}^{+/-}$ (mN/m)	Θ
1	$(CH_3)_3SiOSi(CH_3)_2-$	6.4×10^{-4}	21.9	20.7	1.2	1.2	0.0	1.2	35
2	$(CH_3)_3Si[OSi(CH_3)_2]-$	4.4×10^{-4}	22.2	21.0	1.2	1.2	0.1	1.1	37
3	$[(CH_3)_3SiO]_2(CH_3)Si-$	7.5×10^{-4}	21.5	19.0	2.5	2.4	0.0	2.4	40
4	$[(CH_3)_3SiO]_3Si-$	2.8×10^{-4}	22.6	20.4	2.2	2.2	0.0	2.2	42
5	$[(C_2H_5)_3SiO]_2(CH_3)Si-$	4.0×10^{-4}	26.9	25.5	1.4	1.9	0.5	1.4	47
6	$(CH_3)_3SiCH_2Si(CH_3)_2-$	4.0×10^{-4}	23.4	22.7	0.7	0.8	0.2	0.6	39

The properties of the liquid precursors (Table 2) were investigated by wetting experiments on a perfluoroalkyl methacrylate (FC 722®, 3M) covered glass plate (solid surface tension γ_{sv} 12.5 mN/m) [11].

Table 2. Interfacial energy properties of liquid precursors.

No.	liquid	γ_{lv} (mN/m)	γ_{lv}^{LW} (mN/m)	$\gamma_{lv}^{+/-}$ (mN/m)	γ_{sl} (mN/m)	γ_{sl}^{LW} (mN/m)	$\gamma_{sl}^{+/-}$ (mN/m)	α (mN/m)	Θ
7	$(CH_3)_3SiOSi(CH_3)_2H$	15.7	14.6	1.1	1.2	0.1	1.1	11.3	44
8	$(CH_3)_3Si[OSi(CH_3)_2]H$	17.0	15.6	1.4	1.6	0.2	1.4	10.9	50
9	$[(CH_3)_3SiO]_2(CH_3)SiH$	16.9	15.3	1.6	1.8	0.1	1.7	10.7	51
10	$[(CH_3)_3SiO]_3SiH$	17.5	16.2	1.3	1.5	0.2	1.3	11.0	51
11	$[(C_2H_5)_3SiO]_2(CH_3)SiH$	23.8	21.9	1.9	3.2	1.3	1.9	9.3	67
12	$(CH_3)_3SiCH_2Si(CH_3)_2H$	18.8	17.7	1.1	1.5	0.4	1.1	11.0	54

Results

The data in Table 2 show that for methylated liquids the surface tension (γ_{lv}) (interface liquid/vapour) increases from di- to tetrasiloxanes (**7→9→10**). This is mainly due to an increase of the Lifshitz-van der Waals portion (γ_{lv}^{LW}). Surprisingly, we did not find a polarity increase ($\gamma_{lv}^{+/-}$) for the change from the trisiloxane (two oxygen atoms) to the tetrasiloxane (three oxygen atoms). At the interface solid/liquid these molecules behave slightly differently. Intensive interactions of the Lifshitz-van der Waals type with the condensed perfluorinated matter yield a dramatic reduction of this interfacial tension portion (γ_{sl}^{LW}). On the other hand, this strictly apolar surface cannot compensate donor-acceptor portions ($\gamma_{sl}^{+/-}$). These values are practically identical to those of $\gamma_{lv}^{+/-}$.

The change from methyl-to ethyl-substituted siloxanes (**9→11**) yields a considerable surface tension increase. Again, this behavior is mainly due to an increase of the Lifshitz-van der Waals portion. Considering the data for the interfacial tension solid/liquid and its portions the energy difference between methyl and ethyl groups at interfaces becomes even more transparent. For the methylated liquid **9** the donor-acceptor portion $\gamma_{sl}^{+/-}$ is not accompanied by a significant γ_{sl}^{LW} portion. Ethyl groups make siloxanes less flexible [12] and cause a considerable γ_{sl}^{LW} portion.

The same trends are observed for the change from siloxanes to carbosilanes (**7→12**). The higher surface tension of the surprisingly polar carbosilane is caused by an increased γ_{lv}^{LW} portion. Obviously methyl groups attached to flexible siloxane backbones are energetically distinctly different from those bonded to rigid structures. Expectedly a significant γ_{sl}^{LW} portion occurs at the interface solid/liquid for compound **12**.

From an interfacial energy point of view the main difference between liquid siloxanes and siloxane surfactants in aqueous solution consists in the enhanced orientation of the latter at interfaces. The surfactants hydrophilic moiety is oriented to the water phase, the hydrophobic part to the air (at the liquid/vapour interface) or to the condensed perfluorinated matter (at the solid/liquid interface).

From this background the diverging results for the isomeric siloxanes **7→8** on the one hand and surfactants **2→3** on the other hand become reasonable. The data sets for the liquid siloxanes **7** and **8** show minor differences. Obviously, the position of the Si–H bond is almost irrelevant for the energy balance at both interfaces. In contrast for the surfactants **2** and **3** the position of substitution has a major influence on the energy situation.

For the straight-chain structure **2** a higher surface tension (γ_{lv}) but lower interfacial tension solid/liquid (γ_{sl}) were determined. Interestingly, the net surface tension increase results from the superposition of a moderately decreased donor-acceptor portion ($\gamma_{lv}^{+/-}$) and a considerably increased Lifshitz-van der Waals portion (γ_{lv}^{LW}). Straight-chain structures improve the chances for intermolecular interactions of the Lifshitz-van der Waals type [13]. On the other hand we assume that only one of the two oxygen atoms of the straight-chain trisiloxanyl moiety is sufficiently close to the water surface to influence the donor-acceptor portion. As to expect from the results for pure liquids only donor-acceptor portions $\gamma_{sl}^{+/-}$ were detected at the interface solid/liquid. The values are identical to those for $\gamma_{lv}^{+/-}$ indicating a common source for the donor-acceptor portions. The increased $\gamma_{sl}^{+/-}$ value for the branched structure **3** indicates that in this particular case both oxygen atoms are exposed to the perfluorinated surface.

Acknowledgement: This work was supported by the *German Ministry for Research and Technology* (reg. No. 0310317 A/B) and the *Deutsche Forschungsgemeinschaft* (reg. No. WA 1043/1-1).

References:

[1] S. Zhu, W. G. Miller, L. E. Scriven, H. T. Davis, *Colloids Surfaces* **1994**, *90*, 63.

[2] F. Tiberg, A. M. Cazabat, *Europhys. Lett.* **1994**, *25*, 205.

[3] F. Tiberg, A. M. Cazabat, *Langmuir* **1994**, *10*, 2301.

[4] R. Wagner, R. Wersig, G. Schmaucks, L. Richter, B. Weiland, A. Hennig, A. Jänicke, J. Reiners, W. Krämer, J. Weißmüller, W. Wirth, DE 4318536, **1994**.

[5] R. Wagner, L. Richter, R. Wersig, G. Schmaucks, B. Weiland, J. Weißmüller, J. Reiners, *Appl. Organomet. Chem.* **1996**, *10*, 421.

[6] R. Wagner, L. Richter, B. Weiland, J. Reiners, J. Weißmüller, *Appl. Organomet. Chem.* **1996**, *10*, 437.

[7] T. Young, *Phil. Trans. Roy. Soc.* **1805**, *95*, 65.

[8] G. Czichocki, B. Gilsenbach, M. Olschewski, L. Richter, *Plaste und Kautschuk* **1987**, *34*, 445.

[9] D. K. Owens, R. C. Wendt, *Appl. Polym. Sci.* **1969**, *13*, 1741.

[10] R. Wagner, L. Richter, J. Weißmüller, J. Reiners, K.-D. Klein, D. Schaefer, S. Stadtmüller, *Appl. Organomet. Chem.* **1997**, *11*, 617.

[11] R. Wagner, L. Richter, Y. Wu, J. Weißmüller, J. Reiners, E. Hengge, A. Kleewein, K. Hassler, *Appl. Organomet. Chem.* **1997**, in press.

[12] S. Grigoras, T. H. Lane, *Silicon-Based Polymer Science: A comprehensive resource*, ACS, **1990**, p. 125.

[13] M. J. Rosen, *J. Am. Oil Chem. Soc.* **1972**, *49*, 293.

Synthesis and Application of
ω-Epoxy-Functionalized Alkoxysilanes

G. Sperveslage, K. Stoppek-Langner, J. Grobe*
Anorganisch-Chemisches-Institut
Westfälische Wilhelms-Universität
Wilhelm-Klemm-Str. 8, D-48149 Münster, Germany

Keywords: Epoxysilanes / Alkenylsilanes / Surface Modification / Diffuse Reflectance Infrared Fourier Transformation

Summary: The development of materials with well-designed surface properties becomes increasingly important in many technological applications. Our studies particularly focus on the preparation and application of organosilicon compounds $(EtO)_n Me_{3-n}Si–(CH_2)_m–CH-CH_2–O$ as versatile tools for obtaining functionalized surfaces via SA-technique. Vibrational spectroscopic studies of silica surfaces silanized with the ω-epoxysilanes suggest that the labile oxirane ring can only be preserved if the reaction is carried out at ambient temperature. On the contrary, silylation at $T > 50°C$ quantitatively leads to ring opening most likely due the attack of ethanol formed during the anchoring process.

Surface modification with organosilanes

Organosilanes are widely used to modify the surface properties of various inorganic substrates. The most important criterions for effective silylation are (i) a covalent fixation of the molecule and (ii) the introduction of functional groups with high synthetic potential for further derivatization. Considering these aspects, ω-epoxysilanes of the general type $(EtO)_n Me_{3-n}Si-(CH_2)_m–CH–CH_2–O$ represent a promising class of compounds. First, the EtO-Si unit is capable of interacting with suitable reactive groups on the material surface (e.g., surface OH groups) and, second, the labile oxirane ring allows a variety of opening processes to yield tailor-made functional groups.

However, the high reactivity of the epoxy unit requires special preparative routes to maintain the function during the stepwise synthesis of the silane molecule. In addition, a systematic study of the silane/surface interaction is necessary to guarantee the availability of the oxirane unit for further reactions.

Preparative Routes to ω-Epoxy-Functionalized Alkoxysilanes

In Fig.1, two possible pathways to ω-epoxy-functionalized alkoxysilanes are displayed: In case of route I, the first step consists of the partial hydrosilylation of α,ω-dienes with $HSiMe_nCl_{3-n}$ to give the corresponding ω-alkenylchlorosilanes in 53-79 % yield. Although this reaction proceeds smoothly in the presence of H_2PtCl_6 catalyst, the formation of isomers due to double bond migration is observed. This is a particular problem for the hydrosilylation of long chain α,ω-dienes with trichlorosilane $HSiCl_3$ yielding 20 % (hexadiene) and 48 % (octadiene) of the product with internal double bonds, respectively. In the case of 1,9-decadiene, this isomer is obtained exclusively. Note that the yields given in this paper refer to the mixture of both isomers; further separation was not carried out.

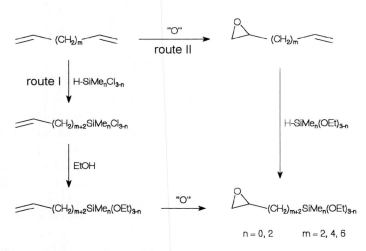

Fig. 1.   Synthetic routes to ω–epoxyalkoxysilanes.

The alcoholysis of the ω-alkenylchlorosilanes (second step of route I) is conveniently carried out either in the presence of NEt_3 (dimethylchlorosilanes) or via removing HCl with an inert gas flow (trichlorosilanes). The product yields of the ω-alkenylethoxysilanes are in the range between 62 and 92 %. In the last step epoxidation of the terminal double bond by m-chloroperbenzoic acid in CH_2Cl_2 was performed affording the ω-epoxyfunctionalized ethoxysilanes in reasonable yields of 50-75 %.

Using route II [1], the desired silanes are accessible in a two-step synthesis: The ω-epoxyalkenes are obtained by partial epoxidation of the corresponding α,ω-dienes with m-chloroperbenzoic acid. The lower product yields (47-50 %) compared to the direct epoxidation of ω-alkenylethoxysilanes are caused by polymerization of the dienes in a side-reaction. The subsequent hydrosilylation requires the ethoxysilane $HSi(OEt)_3$ as educt in order to exclude ring opening during the otherwise nescessary alcoholysis step. The lower reactivity of $HSi(OEt)_3$ compared to chlorosilanes significantly reduces the formation of isomers but, on the other hand, considerably decreases the product yields (31-70 %).

Table 1 summarizes the obtained ω-epoxyalkylsilanes prepared via routes I and II. In general, the three-step route I allows preparation in higher yields, however, at the expense of different isomers formed during the hydrosilylation step. Furthermore, tri- as well as monoethoxysilanes can be obtained by this pathway, thus allowing the preparation of silane multi- and monolayers.

Table 1. ω-Epoxyalkoxysilanes via route I and route II.

Monoethoxysilanes	Triethoxysilanes
EtOMe$_2$Si–(CH$_2$)$_4$–CH–CH$_2$–O	(EtO)$_3$Si–(CH$_2$)$_4$–CH–CH$_2$–O
EtOMe$_2$Si–(CH$_2$)$_6$–CH–CH$_2$–O	(EtO)$_3$Si–(CH$_2$)$_6$–CH–CH$_2$–O
EtOMe$_2$Si–(CH$_2$)$_8$–CH–CH$_2$–O	(EtO)$_3$Si–(CH$_2$)$_8$–CH–CH$_2$–O

Surface Silylation with *ω*-Epoxyalkylethoxysilanes

The silylation of hydroxylated substrates has been discussed in numerous studies [2] and proceeds via condensation with reactive surface OH groups. Two examples are given in Fig.2: In case of the silica substrate, the absorption band due to isolated Si–OH groups (3745 cm^{-1}) [3] disappears after application of the epoxysilylester, whereas on the alumina surface isolated Al–OH centers (3695 cm^{-1}) [4] are involved in the reaction.

In order to evaluate whether or not the epoxy group is preserved during anchoring of the silane molecule, a systematic study of silylated silica and alumina substrates was performed. In a first series, the reactions were carried out at elevated temperatures using different procedures: Pure silane without solvent (50°C), silylation in water at pH 3-4 and 75°C [5], and silylation in toluene under reflux [5,6], (further experimental details in [7]). DRIFT spectroscopy was applied to follow the reaction sequence.

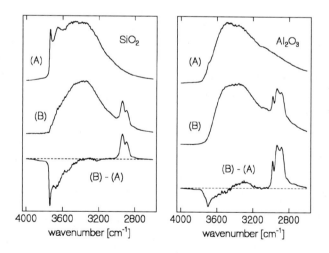

Fig. 2. DRIFT spectra (in Kubelka-Munk units) of silylated SiO$_2$ and Al$_2$O$_3$ substrates.

Fig.3 shows the obtained spectra of the silica sample after silylation with (EtO)$_3$Si–(CH$_2$)$_4$–CH–CH$_2$–O, (EtO)$_3$Si–(CH$_2$)$_6$–CH–CH$_2$–O and (EtO)$_3$Si–(CH$_2$)$_8$–CH–CH$_2$–O, respectively:

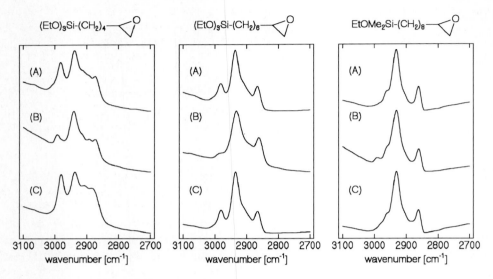

Fig. 3. ν(CH) range of different epoxysilanes SA-layers on silica (absorbance given in Kubelka-Munk units); (A) pure silane; (B) silane in H₂O; (C) silane in toluene.

Characteristic vibrations of the oxirane unit (3047 cm⁻¹) are not detectable suggesting ring opening due to the presence of water or ethanol. Closer inspection of the ν(CH)-range underlines that the hydrolysis of the EtO–Si groups proceeds much slower if the reaction is carried out in toluene, showing up in the considerably higher intensity of the νCH vibration due to the ethoxy-methyl group.

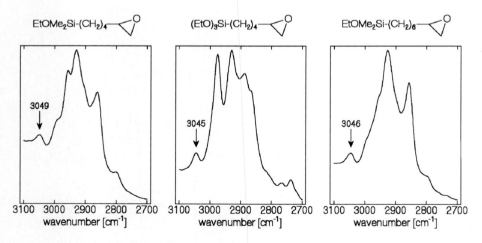

Fig. 4. ν(CH) range of different epoxysilanes SA-layers on silica prepared under ambient conditions (absorbance given in Kubelka-Munk units).

Using the same procedure similar results are obtained with monoethoxysilanes: Again, the typical vibration of the oxirane at 3047 cm⁻¹ is missing. Furthermore, the DRIFT spectra show bands of EtO groups (2985 cm^{-1}) which suggest that the three-membered ring is cleaved by ethanol formed during the anchoring step. In case of the Al$_2$O$_3$ substrate, similar DRIFT spectra are obtained. According to Posner [8], these effects can be explained in terms of oxirane ring opening catalyzed by the alumina surface.

Since the interaction of ω-epoxysilanes with oxide surfaces at elevated temperatures exclusively leads to the loss of the epoxy function, additional silylation experiments were carried out under ambient conditions. Fig.4 shows some diffuse reflectance infrared spectra of the silica material. In all cases, an additional absorption band around 3046 cm^{-1} characteristic of the intact oxirane fragment is detected.

However, the spectra of the monoethoxysilanes again indicate the presence of EtO groups (2980 cm^{-1}) suggesting that the labile epoxy ring is partially cleaved by ethanol. On the other hand, it is obvious that the extent of ring opening under these conditions is much smaller compared with silylation at higher temperatures.

Conclusions

ω-Epoxy-functionalized alkoxysilanes can be obtained in several synthetic steps from α,ω–dienes utilizing conventional organic reactions like epoxidation and hydrosilylation. Reactions of the silylating agents with oxides result in condensation with surface hydroxyl groups (Si–OH, Al–OH) to form covalent bonds. At $T > 50°C$, the alcohol formed during the anchoring step or water present on the oxide surface lead to a quantitative cleavage of the oxirane unit and loss of functionality. At ambient temperature, however, the epoxy groups are partially preserved. Future work, therefore, will focus on derivatization reactions to generate a variety of suitable functional centers attached to the oxide substrate.

References:

[1] M. Yamaya, M. Yanagisawa, T. Yamazaki, EP Pat. No. 0485985 A1, **1992**.
[2] W. Hertl, *J. Phys. Chem.* **1968**, *72*, 1248; W. Hertl, *J. Phys. Chem.* **1968**, *72*, 3993; G. A. Pasteur, H. Schonhorn, *Appl. Spectrosc.* **1975**, *29*, 512; H. O. Finklea, R. Vithange, *J. Phys. Chem.* **1982**, *86*, 3621; J. D. Miller, H. Ishida, *J. Phys. Chem.* **1987**, *86*, 1593.
[3] R. K. Iler, *The Chemistry of Silica*, Wiley-Interscience, New York, **1975**.
[4] H. Knözinger, P. Ratnasamy, *Catal. Rev. – Sci. Eng.* **1978**, *17*, 31; G. Busca, V. Lorenzelli, G. Ramis, R. J. Willey, *Langmuir* **1993**, *9*, 1492.
[5] H. Weetall, in: *Methods in Enzymology, 44*, Academic Press, New York, **1976**.
[6] I. Haller, *J. Am. Chem. Soc.* **1978**, *100*, 8050.
[7] G. Sperveslage, *Diploma thesis*, Universität Münster, **1995**.
[8] G. H. Posner, D. Z. Rogers, *J. Am. Chem. Soc.* **1977**, *99*, 8208.

Diffuse Reflectance IR and Time-of-Flight SIMS Investigation of Methoxysilane SA-Layers on Silica and Alumina

K. Stoppek-Langner*
Anorganisch-Chemisches Institut
Wilhelm-Klemm-Straße 8
Westfälische Wilhelms-Universität, D-48149 Münster,Germany

K. Meyer, A. Benninghoven
Physikalisches Institut
Wilhelm-Klemm-Straße 10
Westfälische Wilhelms-Universität, D-48149 Münster,Germany

Keywords: Methoxysilanes / SA-layer / Diffuse Reflectance Infrared / Time-of-flight SIMS

Summary: In the present study, the formation of SA-layers from alkylmethoxysilanes $RSi(OCH_3)_3$ (R = CH_3, C_2H_5, nC_3H_7, iC_4H_9) on silica and alumina substrates was investigated. DRIFT measurements indicate the presence of residual methoxy groups in the hydrolysis product. Due to a more acidic character of the silica surface, the contribution of free OCH_3 groups is significantly larger compared to the alumina substrate. TOF-SIMS studies on the outmost layers of the polymeric network suggest the formation of completely hydrolyzed condensation products for both substrates. High resolution data can be obtained by extraction of the silanized oxides with anhydrous acetone prior to monolayer preparation on etched silver targets.

Introduction

The application of organosilanes R_nSiX_{4-n} offers a large variety of possibilities to modify the surface characteristics of solid state substrates. In the field of chromatography [1], alkylchlorosilanes (X = Cl) are used to enhance the separation properties of suitable adsorbents. Recent studies also focus on the generation of a tailor-made surface architecture via self-assembled monolayers utilizing both functionalized and long chain alkylsilanes [2]. Our studies primarily focus on the application of alkoxysilanes to generating SA-layers on oxide materials. The water-repellant properties of the polysiloxanes formed by hydrolytic polycondensation decrease the wettability without significantly altering the given physico-mechanical parameters of, e.g., building materials [3]. In order to understand the surface modification process, a closer inspection utilizing analytical techniques is required with particular interest on the surface characteristics of the molecular layer. In this contribution, two spectroscopic techniques with different information depths (diffuse reflectance infrared fourier transform spectroscopy, DRIFT, time-of-flight secondary ion mass spectrometry, TOF-SIMS) were combined to analyze the silanized substrate.

DRIFT Investigation

In Fig. 1, the $\nu(CH)$ range of the methoxysilanes on silica and alumina are displayed. The characteristic vibration located at 2850 cm^{-1} indicates residual CH_3OSi groups in the polysiloxane.

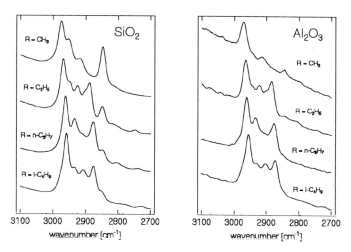

Fig. 1. $\nu(CH)$ ranges in the DRIFT spectra of silica and alumina after silylation with alkyltrimethoxysilanes $RSi(OCH_3)_3$

Provided that particle size, packing density effects and scattering coefficients for the samples given in Fig. 1 are constant, the relative peak intensities in the spectra serve as a measure to obtain concentration information [4]. In the present study, the intensities of the $\nu_{sym}(OCH_3)$ and $\nu_{asym}(CH_3)$ absorptions were calculated via NLLS curve fitting of the baseline corrected spectra in the 3100-2700 cm^{-1} range. Demonstrated by the normalized band intensitiy ratio R (Fig. 2), it is obvious that the contribution of unreacted Si–OCH$_3$ is significantly larger for the silica, suggesting that the hydrolysis process is less complete on this substrate. This is especially true for the shorter chain alkylsilanes $CH_3Si(OCH_3)_3$ and $C_2H_5Si(OCH_3)_3$, showing the strongest influence of the substrate.

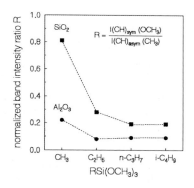

Fig. 2. Normalized band intensity ratios R for alkyltrimethoxysilane SA-layers on silica and alumina.

This behavior can be explained in terms of different oxide surface acidities. Whereas silica (pH ≈ 4) exhibits acidic properties, alumina surfaces shows a much more basic character (pH ≈ 8) [5] and, therefore, allows further hydrolysis of the alkoxysilanes. The influence of the surface pH on the characteristics of anchored molecular structures has also been reported by Deo and Wachs [6] for the preparation of supported vanadia catalysts via incipient wetness.

TOF-SIMS Investigation

In a previous study [7], we reported on the TOF-SIMS characteristics of hydroxylated substrates chemically modified with reactive methylsilanes CH_3SiX_3 (X = Cl, OCH_3, OC_2H_5). The mass spectra exhibit the same peak patterns regardless of the substrate (alumina, silica) or the silane used, indicating a quantitative hydrolysis in the uppermost monolayers. The higher C_2-C_4 alkylmethoxysilanes behave in the same way: as depicted in Fig. 3 for the $iC_4H_9Si(OCH_3)_3$, very similar spectra are obtained.

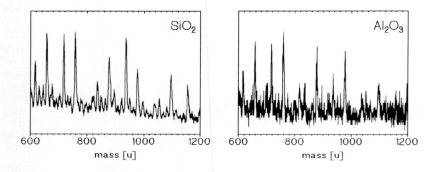

Fig. 3.   Positive TOF-SIMS spectra of $iC_4H_9Si(OCH_3)_3$ on SiO_2 and Al_2O_3.

Using an acetone extraction, small amounts of white, pale yellow crystals can be harvested during the solvent treatment; for the isobutylsilane, the obtained material contains some oily components as well as white crystalline compounds. Fig. 4 displays the positive TOF-SIMS spectra of the methyl and ethyl extraction products, respectively. The dominating peak clusters for the methylpolysiloxane correspond to T_n silsesquioxanes (n even ≥ 6) and homo-silsesquioxanes T_nOH (n odd ≥ 5) with spherical Si–O frameworks [7]. In contrast to this, the higher alkylpolysiloxanes (C_2-C_4) exclusively show mass lines due to T_nOH silsesquioxanes. The absence of T_n clusters is not related to the extraction process; T^R_n cage molecules with higher alkyl groups R exhibit much higher solubilities compared to the analogous methyl compounds [8].

However, it is unlikely that the T^R_nOH siloxanes with R = C_2H_5, nC_3H_7, and iC_4H_9 have the same spherical Si–O framework found for the corresponding methylsiloxanes. First, T_n cage molecules for the higher alkyl precursors are not detected in the spectra. Although polyhedral siloxanes with larger alkyl groups are known to be formed during hydrolytic polycondensation of suitable monomers $RSiX_3$, the T_n types are always generated in much higher quantities than the T_nOH silsesquioxanes. Selective formation of spherical T_nOH siloxanes under the conditions used in the present study can be excluded.

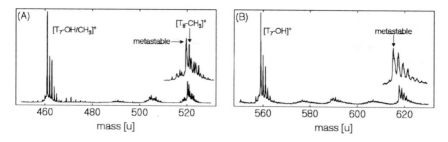

Fig. 4. Positive TOF-SIMS spectra of extracted polysiloxanes on silver; (A): methylsiloxane, (B): ethylsiloxane.

Secondly, the TOF-SIMS spectra show distinct differences in the fragmentation behavior of the higher T_nOH molecules ($n > 7$) (Fig. 5). In case of the methyl derivatives, the contribution of the fragment ion $[T_nOH–OH]^+$ increases with the number of repeating units in the molecule, whereas for ethyl, *n*-propyl, and *i*-butyl species the loss of the alkyl group is favored. This furthermore suggests the presence of isomeric T_nOH siloxanes.

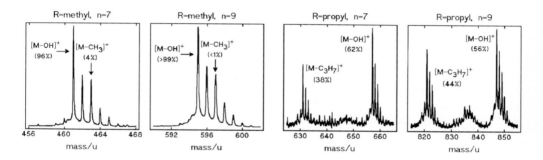

Fig. 5. Fragmentation patterns of the silsesquioxanes T^R_nOH; monolayer preparation on silver.

The formation of these isomers can be explained in terms of different pathways during the hydrolytic polycondensation process. At present, two different hydrolysis processes are proposed in the literature. According to Sprung and Guenther [9], the polysiloxane build-up from methylalkoxysilanes $CH_3Si(OR)_3$ involves the stepwise formation of linear → cyclic → polycyclic → polyhedral compounds with randomly increasing molecular size. For higher alkyltrichlorosilanes (C_6-C_9) a similar mechanism is assumed [10].

On the other hand, GC/MS studies on the polycondensation of $C_2H_5SiCl_3$ [11] suggest a two step mechanism, including rapid hydrolysis of the monomer and formation of linear, cyclic, and polycyclic products. The polyhedral silsesquioxanes are generated during a second and considerably slower step via cleavage and rebuilding from (strained) polysiloxane molecules.

Accordingly, polycyclic siloxanes can be considered as the direct precursors to polyhedral silsesquioxanes and, therefore, represent possible sources for T_n and T_nOH isomers, respectively. The following figure schematically shows the formation of polyhedral silsesquioxanes via intramolecular condensation. $T_n(OH)_4$ siloxanes lead to different $T_n(OH)_2$ isomers with symmetrical (**I**) and anti-symmetrical (**II**) position of the remaining reactive groups (–●) [12]. The formation of spherical T_n silsesquioxanes, however, can only occur from the type (**I**) siloxanes.

In case of $T_n(OH)_3$ ladder molecules, analogous mechanisms either lead to T_nOH siloxanes with regular polyhedral structure (IV), or allow the formation of structural isomers (III). Considering the data obtained from the mass spectra, the polysiloxanes generated from the C_2-C_4 alkyltrimethoxysilanes are likely to be of this structural type.

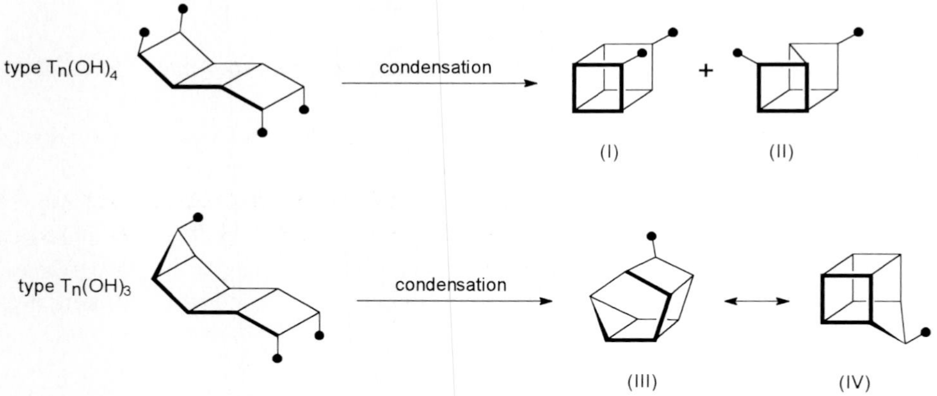

Fig. 6.   Schematic model of intramolecular condensation processes in ladder-like polysiloxanes.

Conclusion

Molecular and structural characteristics of methoxysilane SA-layers deposited on inorganic subtrates can be obtained using DRIFT spectroscopy in combination with TOF-SIMS. The hydrolysis rates are strongly influenced by the acid/base properties of the oxide surface. In accord to this, the contribution of residual $Si–OCH_3$ units is much lower for SA-layers on the basic alumina. The outmost layers of the polysiloxanes consist of completely hydrolyzed species. In case of methylsilanes, the SIMS spectra indicate the presence of T_n and T_nOH silsesquioxanes whereas for higher alkylsilanes the formation of T_nOH isomers is likely. This hypothesis has to be addressed in future studies with particular focus on structural aspects.

Acknowlegdements: Part of this work was funded by the *Deutsche Forschungsgemeinschaft*.

References:

[1]   J. Nawrocki, J. Buszewski, *Chromatogr. Rev.* **1988**, *449*, 1; R. K. Gilpin, M. F. Burke, *Anal. Chem.* **1973**, *1383*, 45.
[2]   D. G. Kurth, T. Bein, *Langmuir* **1993**, *2965*, 9; P. Silberzahn, L. Leger, D. Ausserre, J. J. Benattar, *Langmuir* **1991**, *1647*, 7.

[3] S. Brandriss, S. Mergel, *Langmuir* **1993**, *1232*, 9: J. Grobe, K. Stoppek-Langner, W. Müller-Warmuth, S. Thomas, A. Benninghoven, B. Hagenhoff, *Nachr. Chem. Tech. Lab.* **1993**, *1233*, 41.

[4] M. P. Fuller, P. R. Griffiths, *Anal. Chem.* **1978**, *1906*, 50; J. P. Blitz, R. S. Shreedhara Murthy, D. E. Leyden, *Appl. Spectrosc.* **1986**, *829*, 40.

[5] J. R. Anderson, *Structure of Metallic Catalysts*, Academic Press, London, **1975**.

[6] G. Deo, I. Wachs, *J. Phys. Chem.* **1991**, *5889*, 95.

[7] B. Hagenhoff, A. Benninghoven, K. Stoppek-Langner, J. Grobe, *Adv. Mater.* **1994**, *142*, 6.

[8] K. Olsson, *Arkiv Kemi* **1958**, *367*, 13.

[9] M. M. Sprung, F. O. Guenther, *J. Am. Chem. Soc.* **1955**, *3996*, 77.

[10] K. A. Andrianov, B. A. Izmailov, *J. Organomet. Chem.* **1967**, *435*, 3.

[11] V. I. Lavrent'yev, V. M. Kovrigin, G. G. Treer, *Zh. Obshch. Khim.* **1981**, *124*, 51.

[12] M. M. Sprung, F. O. Guenther, *J. Polym. Sci.* **1958**, *17*, 28; J. F. Feher, D. A. Newman, J. F. Walzer, *J. Am. Chem. Soc.* **1989**, *1741*, 111.

Ether-Substituted Triethoxy- and Diethoxymethylsilanes: Precursors for Hydrophilic, Elastic Consolidants for Natural Stones

Roland Fabis, Christian Zeine, Joseph Grobe*
Anorganisch-Chemisches Institut
Westfälische Wilhelms-Universität Münster
Wilhelm-Klemm-Str. 8, 48149 Münster, Germany
Tel.: Int. code + (2518)33111 − Fax: Int. code + (2518)33108

Keywords: Preservation of Historical Monuments / Stone Consolidants / Hydrophobation / (3-Alkoxypropyl)triethoxysilanes / (3-Alkoxypropyl)-diethoxymethylsilanes

Summary: In this paper, (3-alkoxypropyl)triethoxy- and (3-alkoxypropyl)diethoxymethylsilanes are introduced as a possible basis for hydrophilic, elastic stone consolidants. The preparation of these compounds is described. Addition of (3-alkoxypropyl)diethoxymethylsilanes to commercial stone consolidants reduces the number of shrinking fissures of the resulting SiO_2-gel. Application of (3-alkoxypropyl)triethoxysilanes without any additives even offers the opportunity to produce hydrophilic, elastic consolidants for natural stones.

Introduction

In the preservation of historical monuments and other works of art, consolidation of weathered and damaged natural stone is of particular importance. The use of silicic acid esters, especially tetraethoxy-silane (TEOS) and its oligo-condensates, often generates brittle fillings with strong shrinking fissures at the end of the condensation process (Fig. 1a), thus reducing the consolidating effect considerably.

Wendler has tried to solve this problem by adding diethoxydimethylsilane (DEDMS) to TEOS in order to establish elastic properties in the SiO_2-gel [1]. In fact, the addition of DEDMS reduces the shrinking, but also introduces hydrophobic properties into the polycondensate. Consequently, water-based treatments of the strengthened stone area, e. g., plasterwork, is more difficult.

We have investigated the application of ether-substituted triethoxy- and diethoxysilanes ((3-alkoxypropyl)triethoxy-; (3-alkoxypropyl)diethoxymethylsilanes) for the preparation of hydrophilic, elastic stone consolidants.

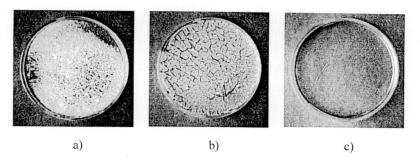

a)   b)   c)

Fig. 1. Photographs of stone consolidants after polycondensation: (a) unmodified stone consolidant; (b) modified stone consolidant, elastificated with (3-methoxypropyl)diethoxymethylsilane; (c) (3-methoxypropyl)triethoxysilane.

Results and Discussion

Preparation of (3-Alkoxypropyl)triethoxy- and (3-Alkoxypropyl)diethoxymethylsilanes [2, 3]

Because of numerous applications, increasing demand and simple access, the industrial production of γ-functionalized alkyltriethoxysilanes has gained growing importance. The products are used as adhesive media, filling compounds and surface-modifying agents in various branches of industry.

Regarding this development, it is surprising that so far only one γ-alkoxy-functionalized propyltriethoxysilane ((3-ethoxypropyl)triethoxysilane) has been described in the chemical literature [4, 5]. It was prepared from sodium ethoxide and (3-chloropropyl)triethoxysilane in an ether synthesis according to Williamson. This procedure is not suited for the synthesis of other alkoxypropyltriethoxysilanes because of the possible exchange of alkoxy groups [6].

Our route to a variety of (3-alkoxypropyl)triethoxysilanes is based on a hydrosilylation, as shown in Scheme 1.

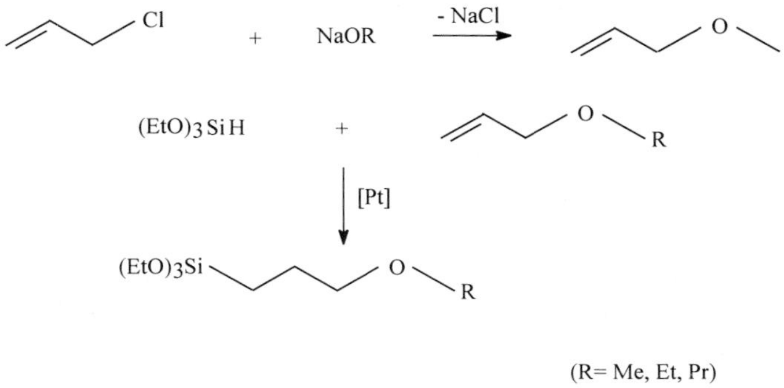

$$\text{Cl} \quad + \quad \text{NaOR} \quad \xrightarrow{- \text{NaCl}} \quad \text{O}$$

$$(EtO)_3SiH \quad + \quad \text{O}{\diagdown}R$$

$$\Big\downarrow [Pt]$$

$$(EtO)_3Si \diagup\diagdown\diagup\diagdown \text{O}{\diagdown}R$$

(R= Me, Et, Pr)

Scheme 1. Preparation of (3-alkoxypropyl)triethoxysilanes.

The alkyl allyl ethers used for the catalytic process were obtained from the corresponding sodium alkoxides and allylchloride by the Williamson ether synthesis. Hydrosilylation using triethoxysilane and Speier´s catalyst turned out to be a straightforward reaction with yields of 65-85 %, depending on the allyl ethers applied.

Scheme 2 shows the analogous preparation of the (3-alkoxypropyl)diethoxymethylsilanes, starting with the catalytic addition of dichloromethylsilane to the corresponding alkoxy allyl ethers followed by nucleophilic substitution of the chlorine atoms by ethoxy groups.

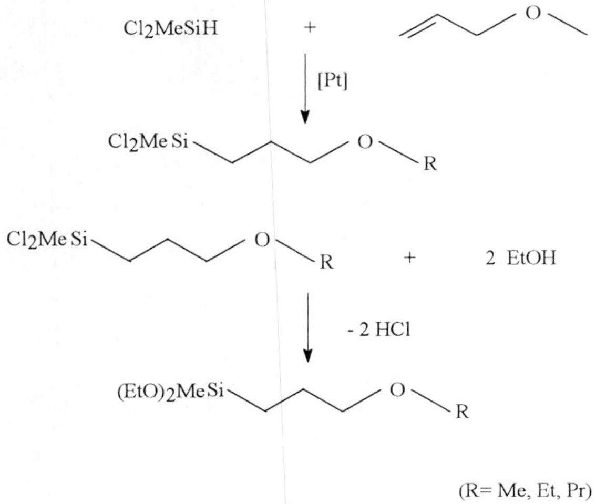

(R= Me, Et, Pr)

Scheme 2.   Preparation of (3-alkoxypropyl)diethoxymethylsilanes.

Depending on the length of the alkyl groups the hydrosilylation can be performed with a yield of 55-80 %, while the substitution of Cl by OEt gives the final products in 85-90 % yield.

Investigation of the Hydrophobic and Elastic Properties of Ether-Functionalized Silylester Condensates

Compared with alkylethoxysilanes of different chain length the ether-substituted derivatives have proved to be less hydrophobic in a series of applications [3, 7, 8]. Table 1 shows the water uptake of Burgpreppacher sandstone treated with DEDMS and the prepared (3-alkoxypropyl)diethoxy-methylsilanes after under water storage for 24 h.

Table 1.   Water uptake of treated Burgpreppacher sandstone after storing under water for 24 h.

Compound	water uptake, %
Untreated	100
Dimethyldiethoxysilane	24
Methoxypropyldiethoxymethylsilane	79
Ethoxypropyldiethoxymethylsilane	77
Propoxypropyldiethoxymethylsilane	76

However, for sandstone cubes impregnated with alkylsilanes a still lower water absorption is observed than for samples treated with the alkoxypropylsilanes. An interesting result is obtained (Table 2) by regarding the water uptake of Sander sandstone treated with alkoxypropyl-triethoxysilanes and alkyltriethoxysilanes of comparable chain length. The less hydrophobic character of the ether-substituted silanes is obvious.

Table 2. Water uptake of treated Sander sandstone after storing under water for 24 h.

Compound	water uptake, %
Untreated	100
Methoxypropyltriethoxysilane	86
Ethoxypropyltriethoxysilane	89
Propoxypropyltriethoxysilane	87
Pentyltriethoxysilane	26
Hexyltriethoxysilane	18
Heptyltriethoxysilane	17

To investigate the elastic properties resulting from ether-functionalized silylester condensates, the diethoxymethyl derivatives were used in 1:1-mixtures with the commercial stone consolidant F510 of Remmers Bauchemie to prepare thick siloxane films in Petri dishes. Fig. 1b shows a photograph of the polycondensate obtained from F510 modified with (3-methoxypropyl)diethoxymethylsilane. In comparison with the film produced by hydrolysis and condensation of the unmodified consolidant (Fig. 1a), considerably less shrinking fissures are observed.

The ether-substituted triethoxysilanes were brought to polycondensation without addition of stone consolidants yielding films like the one shown in Fig. 1c for the polycondensate of (3-methoxypropyl)triethoxysilane which contains almost no shrinking fissures and therefore is very probably well suited as an elastic, nonhydrophobic consolidant.

References:

[1] E. Wendler, D. D. Klemm, R. Snethlage, *Consolidation and Hydrophobic Treatment of Natural Stone*, Fifth International Conference on Durability of Building Materials and Compounds, Brighton **1990**.

[2] R. Fabis, *Ph. D. thesis*, Westfälische Wilhelms-Universität Münster, **1994**.

[3] C. Zeine, *Diploma thesis*, Westfälische Wilhelms-Universität Münster, **1994**.

[4] E. Z. Koval, A. A. Pashchenko, V. A. Sviderski, *Mikrobiol. Zh.* **1982**, *44*, 88.

[5] Z. W. Kornetka, J. Bartz, *Chem. Abstr.* **1985**, *103*, 142175.

[6] W. Simmler, *"Kieselsäureester"*, in: *Methoden der Organischen Chemie (Houben-Weyl)*, Band VI/2, Thieme, Stuttgart, **1963**, p. 77.

[7] I. Krull, *Diploma thesis*, Westfälische Wilhelms-Universität Münster, **1990**.

[8]   M. Wessels, J.Grobe, *"Untersuchungen zur Hydrophobierung von Sander Schilfsandstein mit Mischungen aus Alkyl-triethoxysilanen"*, in: *Jahresberichte Steinzerfall-Steinkonservierung 1993* (Ed.: R. Snethlage), Verlag Ernst, Berlin, **1995**, 77.

Organosilicon Compounds for Stone Impregnation – Long-Term Effectivity and Weathering Stability

Christian Bruchertseifer, Stefan Brüggerhoff
Zollern-Institut beim Deutschen Bergbau-Museum/DMT
Herner Str. 45, D-44787 Bochum, Germany
Tel.: Int. code + (0234)9684032 – Fax: Int. code + (0234)9684040
E-mail: bruchcbd@rz.ruhr-uni-bochum.de

Karl Stoppek-Langner, Joseph Grobe
Anorganisch-Chemisches Institut
Westfälische Wilhelms-Universität Münster
Wilhelm Klemm Str. 8, D-48149 Münster, Germany

Martin Jursch, Hans-Jürgen Götze
Lehrstuhl für Analytische Chemie
Ruhr-Universität Bochum
Universitätsstr. 150, D-44780 Bochum, Germany

Keywords: Hydrophobization / Alkylalkoxysilanes / Propylsilane / Octylsilane / Organosilicon Coating

Summary: First results of a large field exposure study with organosilicon hydrophobic agents applied to natural stone are presented. Remaining qualities of polysiloxane coating after eight years of exposure turned out to be highly different regarding the stone materials as well as the protective agents.

Introduction

Central effort of building conservation in our times is the improvement of effectivity and long-term stability of protective agents. Application of impregnating systems is aimed at a drastic reduction of moisture uptake of usually porous building materials (e.g., natural stone, concrete, brick, etc.) to prevent weathering processes caused by the presence of water [1]. An important assumption to fulfill the tasks of long-term impregnation and consolidation of building materials is the research on correlations between the structure of protective polycondensates and its interaction with the substrate including the degradation behavior of the coating. With respect to the hydrophobization of mineral surfaces, organosilicon compounds derived from alkylalkoxysilanes $X–(CH_2)_nSi(OR)_3$ have taken a leading position in the spectrum of protective agents.

Field Exposure with Organosilicon Hydrophobic Agents

Alkylalkoxysilane and -siloxane based agents are successfully used in the protection of building materials [2]. In order to elucidate the effects of weathering on organosilicon coatings, a large field exposure study including ten different types of freshly quarried natural stone, prepared as prisms (300x300x150 mm), was initiated by the Zollern-Institut in 1986 [3]. Besides general aspects concerning the effectivity and durability of water repellent agents, the campaign should yield a database for a tailor-made treatment of weathered stone materials used in famous monuments. Eleven different commercially available types of organosilicon water repellents (in organic solvents) were applied:

I. Silanes

1. C_4 silane + hydrolized consolidant,
2. C_4 silane + catalyst,
3. C_4 silane + consolidant + catalyst,
4. C_3/C_8 silane mixture (7:1) + catalyst.

(consolidant: Tetraethoxysilane)

II. Siloxanes

1. Consolidant + C_1 siloxane,
2. C_1 siloxane + catalyst,
3. C_1 siloxane,
4. C_1/C_8 siloxane mixture + catalyst,
5. C_1/C_8 siloxane mixture + catalyst + consolidant.

III. Silicones

1. C_1 silicone,
2. C_1/C_8 silicone mixture.

Application was carried out by soaking the stone material for one minute. Two samples of each variety of stone remained untreated. The prisms were exposed at six different sites in Germany (among them Dortmund and Duisburg) covering a wide range of climatic and immission situations. One set of samples was stored in the laboratory for comparative investigation.

Organosilicon Compounds: Long-Term Efficiency and Durability

After eight years of exposure, an extensive examination was started in 1994 to evaluate the effectivity and durability of the water repellent treatments [4]. Macroscopic tests as water uptake measurements were carried out to determine the remaining effect of the protective organosilicon layer. Due to the fact that surface information from hydrophobic treatment of mineral surfaces is supplied by surface sensitive measuring techniques, TOF-SIMS and additional DRIFT-studies on treated and exposed material were performed.

In order to get information about the effectivity and durability of the complete agent spectrum, the two exemplary different types of natural stone, Schleeriether (SRS) and Obernkirchener (OKS) sandstone, were chosen for investigation. The SRS represents a relatively weak material with claylike mineral bonding. On the contrary, OKS, due to its quartzitic mineral bonding, represents a strong material. After careful preparation of the prisms water absorption measurements were carried out by complete water storage of treated and corresponding untreated material (SRS: 192 h, OKS: 768 h). In coincidence with usual practical advance and for simplification here only water *uptake* as efficiency criterion was taken as a basis. Fig.1 presents an overview on the remaining hydrophobic effectivity of the organosilicon protective systems applied to SRS- and OKS-prisms eight years ago, stored in the laboratory (REF), and exposed in Dortmund (DO) and Duisburg (DU), respectively. The hydrophobic effectivity is expressed as a relative value ([rel. %]) of the difference in the water uptake of treated and untreated material (normalized to 100 %).

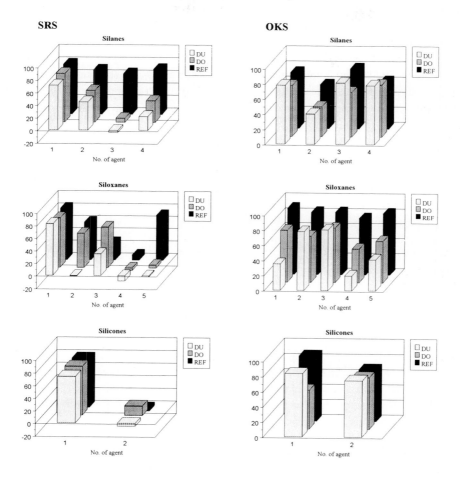

Fig. 1. Hydrophobic effectivity of organosilicon protective agents applied to SRS- (left column) and OKS-samples (right column) stored in the lab (REF) and exposed (DO, DU).

Effectivity and durability of the water repellent treatments turn out to be partly quite different considering the stone materials as well as the organosilicon compounds, but in general it is found that long-term hydrophobization of OKS is distinctly more stable as compared to SRS. Considering SRS (Fig. 1, left column), the differences in long-term effects within the agent spectrum must be categorized as extreme. This generalizing aspect relates to the reference samples as well as to the weathered material. A drastic example is given by the hydrophobic effectivity of the silicones (bottom). Comparison of the siloxane based agents No. 1 and 4 reveals similar results (middle). In case of the combination product (No. 1) addition of the (highly concentrated) consolidant obviously leads to an improved bonding of the water-repellent siloxane. Water absorption of the substrates treated with the silanes No. 1 and 3 (top) also reveals an interesting aspect: Though composition of the agents is very similar, the polysiloxane formed from product No. 1 shows a nearly unchanged hydrophobic effectivity, whereas protective properties of the polysiloxane formed from silane No. 3 have decreased distinctly by weathering.

In case of the silane treated OKS (Fig. 1, right column) hydrophobic effectivity of the exposed material after 768 h is in some cases even slightly *higher* than that of the laboratory-stored prisms (silanes No. 1, 3, and 4, exposed in DU, top). This effect is most likely due to progressive polycondensation of the applied agents during exposure. In spite of the mainly satisfying results of OKS-treatment, long-term effectivity of the siloxane based agent No. 1 (comparing REF to DO, DU, middle) decreases significantly (in contrast to SRS) probably due to the formation of micro-cracks in the consolidant network caused by weathering. Hydrophobic effectivity of the prisms treated with siloxane No. 4 and 5 is reduced in the same sequence (REF > DO > DU). The silicones both exhibit comparable durability (bottom).

Surface Investigation of Impregnating Mixtures

Secondary ion mass spectrometry (SIMS) in the time-of-flight mode (TOF) is a very powerful measuring technique for analyzing the surface properties of polymeric structures [5]. In combination with DRIFT (Diffuse Reflectance Infrared Fourier Transform) spectrometry, which enables elucidation of the bulk properties of the polycondensate formed, valuable details about the protective layer can be obtained. The material chosen for surface investigation of the protective agents relevant in the exposure study was OKS. TOF-SIMS spectra of the systems containing C_1 and C_1/C_8 siloxane/silicone applied to stone slabs (19x10x2 mm) prior to a reaction time of four weeks show the same peak pattern as pure C_1 silane, therefore suggesting the formation of spherical silsesquioxanes and their homoderivatives (Fig. 2) [6].

Mass peaks due to condensation products of the C_8 component are absent. Additional DRIFT investigation of the fine pulverized stone slabs show the presence of the polysiloxane only in case of the pure C_1 systems (siloxanes No. 1, 2 and 3, silicone No. 1) whereas the spectra of the impregnating mixtures (siloxanes No. 4 and 5, silicone No. 2) give evidence *neither* of the C_1 nor of the C_8 component. In order to obtain detailed information about the condensation process of mixtures, SIMS- and DRIFT- analysis on silane No. 4 was carried out. Fig. 3 shows the mass spectra (positive mode) of the relevant single systems (a) C_3 silane (methoxy), (b) C_8 silane (ethoxy) and (c) C_3/C_8 silane (7:1). Similar to the result obtained from SIMS analysis of the siloxane mixtures only mass peaks due to polysiloxanes formed from the C_3 component (shorter alkyl group) are detected. The observation of high mass peaks in the SIMS spectrum of pure C_8 silane, however, is unexpected [7]. Catalytic effects of the stone surface seemingly lead to the formation of spherical C_8 silsesquioxanes. Corresponding DRIFT-spectra are shown in Fig. 4.

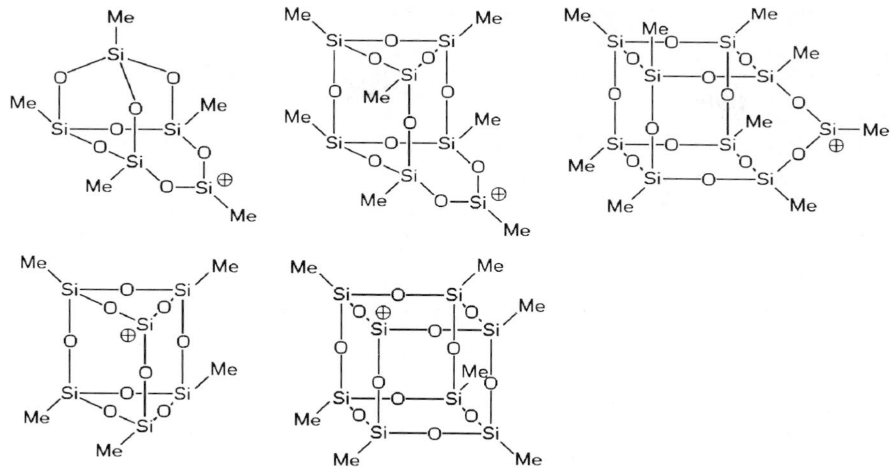

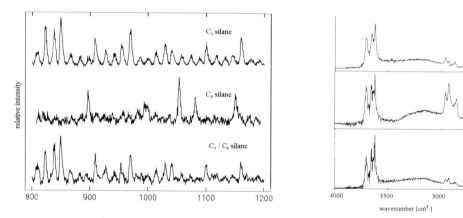

Fig. 2.   Structures of C_1 silsesquioxanes (below) and homosilsesquioxanes (above).

Fig. 3.   TOF-SIMS spectra of C_3 (methoxy) and C_8 (ethoxy) silane and the mixture.

Fig. 4.   DRIFT spectra of C_3 (methoxy) and C_8 (ethoxy) silane and the mixture.

Diffuse reflectance signals at 3600-3750 cm^{-1} are observed in each spectrum and are caused by the stretching vibration of surface silanol groups (OH-bond) from clay minerals. The spectrum of C_3 silane exhibits a typical peak pattern characterized by signals due to the CH stretching vibration (C_3 group at 2900-3000 cm^{-1}, methoxy group at 2850 cm^{-1}). In case of C_8 silane the signal caused by the ethoxy group (2975 cm^{-1}) is covered by the band due to the C_8 group. In contrast to the siloxane and silicone mixtures, the DRIFT spectrum of the silane mixture gives evidence of the protective polysiloxane.

Surface Studies on Treated and Exposed Material

Investigation of changes in surface properties of the organosilicon coating caused by the influences of weathering were carried out on OKS samples showing most significant differences in the remaining protective effectivity of the applied agent (silane No.4: low water uptake, siloxane No.4: high water uptake, both exposed in DU, Fig.2.). Stone slabs were cut from the most affected side of the prisms. According to the results of water absorption measurement, SIMS depth profiling of the material treated with the silane mixture gives evidence of C_3 silsesquioxanes. Considering the siloxane mixture, however, analogous peak groups due to the C_1 component are detected in the reference but *not* in the exposed stone. Coincidently, additional DRIFT studies on the exposed material treated with the two different agents show the presence of the polycondensate only in case of the silane system (Fig.5).

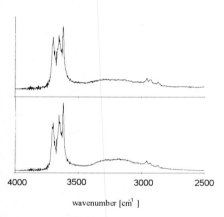

wavenumber [cm^1]

Fig. 5.   DRIFT spectra of silane No. 4 applied to OKS (reference: above, exposed sample: below).

Conclusions

TOF-SIMS studies on impregnating mixtures suggest enrichment of the more reactive component in the uppermost polysiloxane layer. Furthermore, the results obtained from combination of TOF-SIMS and DRIFT indicate stronger interaction between stone surface and the polycondensate formed from the silane mixture as compared to the siloxane system. In order to confirm the trends, surface analysis will be extended to larger numbers of samples. In general, combination of different analytical techniques for systematic studies on the structure of polysiloxane coating and its interaction with mineral surfaces will be a main subject of future investigation.

References:

[1] J. Grobe, K. Stoppek-Langner, *Forschungsjournal der Westfälischen Wilhelms-Universität Münster* **1992**, *1*, 36.

[2] M. Roth, *Baugewerbe* **1982**, *2*, 38.

[3] P. W. Mirwald, *Bautenschutz + Bausanierung* **1986**, *6*, 24.

[4] Ch. Bruchertseifer, S. Brüggerhoff, J. Grobe, K. Stoppek-Langner, *Proceedings of the 1st International Symposium of "Surface treatment of building materials with water repellent agents"*, Delft, the Netherlands, **1995**, p. 27.1.

[5] A. Benninghoven, F. G. Rüdenauer, H. W. Werner, *Secondary Ion Mass Spectrometry*, Wiley, New York, **1987**.

[6] B. Hagenhoff, *Ph. D. thesis*, Westfälische Wilhelms-Universität, Münster, **1993**.

[7] K. Stoppek-Langner, *Ph. D. thesis*, Westfälische Wilhelms-Universität, Münster, **1991**.

Studies on the Regioselectivity of the Hydroformylation with Alkenylalkoxysilanes

Manfred Wessels, Joseph Grobe*

Anorganisch-Chemisches Institut
Westfälische Wilhelms-Universität Münster
Wilhelm-Klemm-Str. 8, D-48149 Münster, Germany

Keywords: Hydroformylation / Aldehyde-Functionalized Silanes / Regioselectivity

Summary: The hydroformylation of alkenylalkoxysilanes catalyzed by rhodium complexes allows the preparation of aldehyde-functionalized silanes. It is possible to improve the regioselectivity of this reaction by adding an excess of triphenylphoshine and by temperature variation.

Introduction

The hydroformylation of alkenylalkoxysilanes, accessible by using known methods of organosilicon chemistry, allows the synthesis of a variety of aldehyde-functionalized organosilanes differing in chain length as well as in kind and number of alkoxygroups. These compounds have proved to be superior to combinations of aminoalkylsilanes and glutaraldehyde for the functionalization of silica surfaces with aldehyde groups. They can be used as a new class of spacer molecules for the immobilization of biomolecules in many biotechnological processes [1, 2]. For example, the aldehyde group is suited to attach enzymes directly to solid surfaces, without the aid of a bifunctional crosslinking compound, because the free amino groups of the enzyme can react with the carbonyl function of the silane to form an azomethine linkage (Fig. 1).

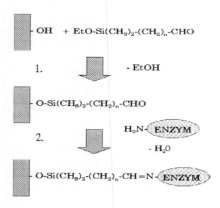

Fig. 1. Immobilization using aldehyde-functionalized silanes.

While the conditions of regioselectivity in the hydroformylation of alkenes have been thoroughly studied [3], only little is known about the analogous reaction of alkene groups in organosilicon compounds. Takeuchi and Sato have investigated the mechanism and regioselectivity of the hydroformylation of vinyltrimethylsilane in 1990 [4], with the result that in this special case the regioselectivity depends mainly on type and concentration of the catalyst, on the CO pressure and the reaction temperature.

Since the position of the aldehyde function on the alkyl chain will probably affect the quality of a fixation, we have studied the regioselectivity of the synthesis of aldehyde-functionalized mono- and trialkoxysilanes, using rhodium-hydrido-carbonyl-tris(triphenylphosphine) as catalyst. In general, the hydroformylation of the terminal double bond in organoalkenylsilanes results in the formation of isomeric aldehydes (Eq. 1).

$$(RO)_{3-m}(CH_3)_m Si(CH_2)_n-CH=CH_2$$
$$+$$
$$\rightarrow$$
$$(RO)_{3-m}(CH_3)_m Si(CH_2)_n-CH_2-CH_2-CHO \; (n)$$
$$+$$
$$CO + H_2$$
$$(RO)_{3-m}(CH_3)_m Si(CH_2)_n-CH(CH_3)-CHO \; (iso)$$

Eq. 1. $n = 0, 1, 2, 4, 6, 8; m = 0, 2.$

Experimental Results

Our investigation was aimed at studying the effects of different substituent patterns in the silyl fragment, variation of the chain length of the alkenyl group, the reaction temperature and the concentration of additional PPh$_3$ on the yield and regioselectivity of the hydroformylation. The results obtained for the hydroformylation of different types of allylsilanes are listed in Table 1.

Table 1.   Effect of the substituent patterns in the silyl fragment[a].

Silane	Product	Yield (%)	*n/iso*
allyltrimethylsiane	trimethylsilylbutanal	95	78/22
allyldimethylmethoxysilane	dimethylmethoxysilylbutanal	92	75/25
allyldimethylethoxysilane	dimethylethoxysilylbutanal	96	60/40
allytrimethoxysilane	trimethoxysilylbutanal	88	54/46
allytriethoxysilane	triethoxysilylbutanal	91	51/49

[a] RhH(CO)(PPh$_3$)$_3$-catalyzed hydroformylation: Silane (10 mmol), Toluene (10mL), CO (40 bar), H$_2$ (40 bar), 10 h 80°C.

The data show that in general excellent product yields are obtained, but that the regioselectivity decreases from *n/iso* = 78/22 for trimethylsilylbutanal to 51/49 for triethoxysilylbutanal.

The length of the alkenyl chain is only of minor influence on the regioselectivity (Table 2), at least there is no systematic tendency.

Table 2. Effect of the alkenyl chain length[a].

Silane	Product	Yield (%)	n/iso
allyldimethylethoxysilane	dimethylethoxysilylbutanal	96	60/40
1-hexenyldimethylethoxysilane	dimethylethoxysilylhepanal	86	62/38
1-octenyldimethylethoxysilane	dimethylethoxysilylnonanal	82	58/42
allyltriethoxysilane	triethoxysilylbutanal	91	52/48
1-butenyltriethoxysilane	triethoxysilylpentanal	85	54/46
1-hexenyltriethoxysilane	triethoxysilylheptanal	92	52/48
1-octenyltriethoxysilane	triethoxysilylnonanal	84	55/45
1-decenyltriethoxysilane	triethoxysilylundecanal	56	48/52

[a] $RhH(CO)(PPh_3)_3$-catalyzed hydroformylation: Silane (10 mmol), Toluene (10mL), CO (40 bar), H_2 (40 bar), 10 h, 80-120 °C.

of vinyltrimethylsilane hydroformylation [4]. At 50°C the *n*-aldehydes were obtained in an *n/iso* ratio of 75/25, which changed to 45/55 on raising the reaction temperature to 120°C.

We observed the same effect for the hydroformylation of vinyltrimethoxysilane (Table 3), for which the optimal ratio of 65/35 was found at 60°C. Reaction at 120°C leads to the inverse result of 30% *n*- and 70% *iso*-aldehyde.

Table 3. Effect of the reaction temperature[a].

Silane	Temp. (°C)	Product	Yield (%)	*n/iso*
vinyltrimethoxysilane	60	trimethoxysilylpropanal	60	65/35
vinyltrimethoxysilane	80	trimethoxysilylpropanal	83	52/48
vinyltrimethoxysilane	120	trimethoxysilylpropanal	80	30/70

[a] $RhH(CO)(PPh_3)_3$-catalyzed hydroformylation: Silane (10 mmol), Toluene (10mL), CO (40 bar), H_2 (40 bar), 10 h.

In a preliminary study we have also investigated the influence of additional triphenylphosphine on the hydroformylation process and its regioselectivity. The results for $ViSiMe_3$, $ViSi(OMe)_3$, and $allylSi(OMe)_3$ (Table 4) are in accord with those of the Japanese group [4]:

Table 4. Effect of triphenylphoshine addition[a].

Silane	Temp. (°C)	Product	Yield (%)	*n/iso*
vinyltrimethylsilane	60	trimethylsilylpropanal	88	95/5
vinyltrimethoxysilane	60	trimethoxysilylpropanal	95	80/20
allyltrimethoxysilane	60	trimethoxysilylbutanal	93	71/29

[a] $RhH(CO)(PPh_3)_3$-catalyzed hydroformylation: Silane (10 mmol), Toluene (10mL), CO (40 bar), H_2 (40 bar), 10 h, excess of PPh_3.

Both the overall yield and the *n*-selectivity are considerably improved. By combining the effect of lower temperatures and higher PPh₃ concentrations enhanced yields and *n*-selectivities are attainable.

Discussion

To interpret the observed effects on the regioselectivity a modified model of the hydroformylation cycle is presented in Fig. 2 for the special case of vinylsilane precursors.

Regioselectivity in hydroformylation is influenced by electronic and steric effects [4, 5]. Thus the formation of the C α–Rh bond is favored over that of the C β–Rh bond by the well known β-silicon effect (Fig. 3), which stabilizes a positive charge on the β–C atom. From the resulting intermediate **Ia** the *iso*-product should form predominantly. On the other hand, steric effects induced by bulky substituents on silicon or rhodium would favor the sterically less hindered normal alkyl rhodium complex with the C β–Rh intermediate **IIa** as the precusor to the *n*-aldehyde. The observed *n/iso*-ratios very close to 1:1 for the Rh-catalyzed hydroformylation of vinyltrimethylsilane indicate that the electronic β-effect obviously is canceled out by the steric demand of the Me₃Si-groups. Since addition of PPh₃ will favor an active complex with a larger number of bulky phosphine ligands (L = PPh₃ in Fig. 2), the formation of the linear alkylrhodium complex intermediate **IIa** to **IId** is prefered [6].

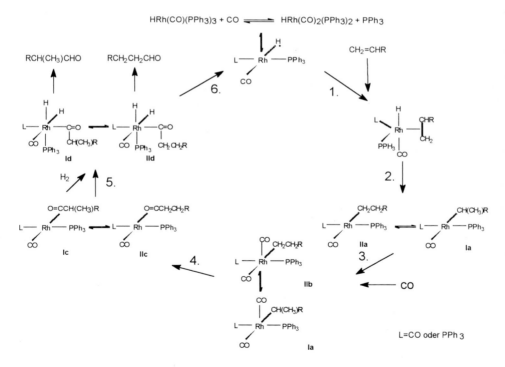

Fig. 2. Modified catalytic cycle for the hydroformylation of vinylsilanes (R=SiR'₃): Intermediates **Ia** to **Id** are favored by the β-silicon effect (Fig. 3), small ligands L like CO, and higher temperature, **IIa** to **IId** by a bulky ligand, e. g., L = PPh₃.

A plausible picture of the β-silicon effect [5] on the hydroformylation of vinylsilanes R'₃SiCH=CH₂ with HRh(CO)(PPh₃)₃ as catalyst is shown in Fig. 3.

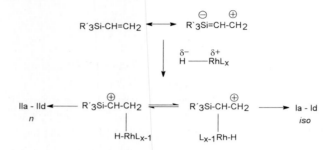

Fig. 3.   Schematic description of the β-silicon effect on the regioselectivity of vinylsilane hydroformylation.

Conclusion

This preliminary study allows to improve yield and regioselectivity of alkenylsilane hydroformylation by carrying out the process at low temperature and with high concentration of triphenylphosphine. Further investigations are necessary to support the mechanistic model.

References:

[1]   C. Brüning, J. Grobe, *J. Chem. Soc., Chem. Commun.* **1995**, 2323.
[2]   J. Grobe, C. Brüning, M. Wessels, *"Aldehyde-functionalized Silanes: New Compounds to Improve the Immobilization of Biomolecules"*, in: *Organosilicon Chemistry II: From Molecules to Materials* (Eds.: N. Auner, J. Weis), VCH, Weinheim, **1996**, p. 243.
[3]   G. Dümbgen, D. Neubauer, *Chem. Ing. Techn.* **1969**, *41*, 974.
[4]   R. Takenuchi, N. Sato, *J. Organomet. Chem.* **1990**, *293*, 1.
[5]   M. R. Ibrahim, W. Jorgenson, *J. Am. Chem. Soc.* **1989**, *111*, 819.
[6]   C. K. Brown, G. Wilkinson, *J. Chem. Soc.* **1970**, 2753.

Novel Precursors for Inorganic-Organic Hybrid Materials

Stefan Kairies, Klaus Rose*

Fraunhofer-Institut für Silicatforschung
Neunerplatz 2, D-97082 Würzburg, Germany

Keywords: Hybrid Materials / Precursors / Triazines

Summary: Novel alkoxysilyl-functionalized triamino triazines are obtained by nucleophilic addition of alkoxysilane compounds bearing isocyanato groups. Conversion of NH_2 groups to the corresponding sodium amide enables both nucleophilic addition to epoxysilanes or α,β-unsaturated carbonyl silane compounds in addition to nucleophilic substitution of chloropropylsilanes. An alternative pathway comprises the formation of the triamino-substituted triazine ring with simultaneous introduction of a hydrolysable organoalkoxysilyl group, via reaction of 3-aminopropyltriethoxysilane with 2,4,6-trichloro-1,3,5-triazine. Reaction of the cyclic phosphornitrilic chloride trimer $[-Cl_2P=N-]_3$ with nitrogen nucleophiles such as 3-aminopropyltriethoxysilane facilitates the introduction of the silane component in a single step. Alternatively, allylic alcohol is used as a nucleophilic *O*-alkyl source, followed by hydrosilylation of the residual olefinic C=C-bond.

Introduction

Progress in almost all fields of advanced technology is largely dependent upon the development of new and advanced materials. Extensive efforts have therefore been directed towards the production of materials possessing the desirable properties of both organic polymers and glasses or ceramics [1]. One approach employed towards syntheses of inorganic-organic hybrid materials involves the sol-gel processing of organo-alkoxysilane precursor compounds of the general type $R'Si(OR)_3$, often in combination with alkoxides of Al, Ti, Zr [2]. Hydrolysis and condensation reactions facilitate the formation of an inorganic Si–O–Si network, within which organic functionalities are linked by a stable chemical bond. An organic polymeric network may be formed additionally using reactive organic moieties [2].

The ultimate objective of this investigation is the synthesis of advanced and thermally stable organically modified siloxanes by incorporation of 6-membered ring structures of triamino triazine **A** and cyclophosphazene **B** (Fig. 1.). It is therefore necessary to synthesize novel monomeric precursor compounds for sol-gel processing, which possess 6-membered ring structures chemically linked to alkoxysilyl moieties.

Fig. 1.   Structure of triamino triazine **A** and cyclophosphazene **B**.

The new compounds bearing these ring structures were characterized by elemental analysis and spectroscopic methods.

Nucleophilic Reactions of the Triamino Triazine Ring

High yields of alkoxysilyl functionalized triamino triazines are obtained by nucleophilic addition of the NH_2-group of 2,4,6-triamino-1,3,5-triazine to 3-isocyanatopropyltriethoxysilane to yield the modified urea-bridged product. The general reaction scheme, following a literature procedure [3], is depicted in Eq. 1 and the products formed given in Table 1.

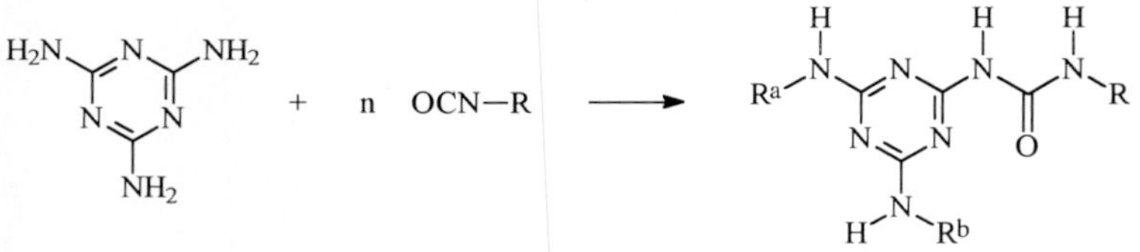

Eq. 1.   Synthesis of urea bridged modified triazines.

Table 1.   Products of the isocyanate addition to triamino triazine.

n	R	R^a	R^b	Product	Yield [%]
1	$(CH_2)_3Si(OEt)_3$	H	H	**1**	90
2	$(CH_2)_3Si(OEt)_3$	CONHR	H	**2**	80
3	$(CH_2)_3Si(OEt)_3$	CONHR	CONHR	**3**	94

Control of reaction stoichiometry of the precursor compounds allows the number of alkoxysilane substituents in the final product to be varied over the range from one to three. The solubility in organic solvents increases with number of substituents (one to three). High yields up to 94 % were obtained for the trisubstituted product **3**. Reactions with $n = 2$ and 1 gave yields of 80 % and 90 % for products **2** and **1** respectively, which were purified by recrystallisation.

Glycidoxypropyltrimethoxysilane and methacryloxypropyltrimethoxysilane are also potential compounds for the introduction of hydrolysable alkoxysilylgroups to the triazine ring system. As melamine is a weak nucleophile and does not react with epoxides or α,β-unsaturated carbonyls, it is necessary to increase its nucleophilicity by converting NH_2 groups to the corresponding sodium amide [4]. This sodium amide derivative reacts by nucleophilic addition with glycidoxysilane or methacryloxysilane, as outlined in Eq. 2.

4

5

R = (CH$_2$)$_3$Si(OMe)$_3$

Eq. 2.   Addition of triazine amide to glycidoxypropyltrimethoxysilane and methacryloxypropyltrimethoxysilane.

After workup with ammonium chloride (liberation of NH_3 and NaCl) and extraction with hot dimethylsulfoxide products **4** and **5** can be isolated as white solids in yields of ca. 30 %. The low yield of the desired product observed for the epoxide addition reaction is attributed to an anionic epoxide polymerisation process [5].

Following an alternative reaction pathway, the triazine sodium amide derivative undergoes nucleophilic substitution with chloropropyltrimethoxysilane at temperatures of 120 °C (Eq. 3). Under these reaction conditions an alkoxy/amide exchange was observed, resulting in low yields of ca. 30 % for product **6**.

6

Eq. 3.   Substitution reaction of triazin amide with chloropropyltrimethoxysilane.

Attempts to introduce more than one substituent bearing an alkoxysilyl group via this reaction pathway were unsuccessful. Alkylation of the remaining NH_2 groups, in order to increase solubility, was not accomplished due to the reactivity of the alkoxysilyl groups.

Reaction of Cyanuric Chloride with Amine Compounds [6]

The synthesis of alkoxysilyl-modified melamine derivatives comprised the formation of the triamino substituted triazine ring from trichloro triazine and simultaneous introduction of a hydrolysable organoalkoxysilyl group. Thus, aminopropyltriethoxysilane or *N*-methylaminopropyltrimethoxysilane reacts with 2,4,6-trichloro-1,3,5-triazine or with partially *N,N*-diethylamine-substituted triazine to give the triazine derivatives **7** to **12** (Eq. 4.; Table 2).

$$\left[-N=C \underset{|}{\overset{Cl}{\vphantom{|}}} - \right]_3 \xrightarrow[\text{2. HRN}-(CH_2)_3-Si(OR')_3]{\text{1.}\qquad HNEt_2} \left[-N=C(NEt_2)_m(NR-(CH_2)_3-Si(OR')_3)_n- \right]_3$$

Eq. 4.   Reaction of trichloro triazine with amines.

Table. 2. Products from the reaction of trichloro triazine with amines.

R	R'	m	n	Product	Yield [%]
H	Et	$^2/_3$	$^1/_3$	7	93
Me	Me	$^2/_3$	$^1/_3$	8	91
H	Et	$^1/_3$	$^2/_3$	9	92
Me	Me	$^1/_3$	$^2/_3$	10	90
H	Et	0	1	11	92
Me	Me	0	1	12	95

The degree of substitution is determined by reaction temperature [7]. The first substituent is introduced in an exothermic reaction at 0°C, the second at 40-45°C and complete substitution required temperatures of 80-100°C. Tertiary amines were employed as base in order to remove hydrochloric acid liberated during the reaction. More satisfactory results (i.e., reaction time and yield) were obtained by employing a two-fold mole increase of amine-nucleophile, thereby obtaining products 7 to 12 as clear, viscous and distillable liquids in yields greater than 90 %.

Phosphornitrilic Chloride and *N*-Nucleophiles

The inorganic –P=N–backbone in phosphazenes possess good thermal and remarkable radiation stability [8]. Many attempts were made to prepare phosphazene-hybrid materials from polyphosphazenes [9]. Our approach was to synthesize small phosphazene moieties for the use in the sol-gel process and to combine both inorganic materials, siloxane and phosphazene.

The precursor compound, phosphornitrilic chloride trimer, reacts with nitrogen and oxygen nucleophiles to form stable products [10]. Reaction of the trimeric phosphazene chloride with nitrogen nucleophiles, such as 3-aminopropyltriethoxysilane, facilitates the introduction of the silane component in a single step resulting in the modified compounds **13** and **14** in yields of about 90 % (Eq. 5.; Table 3).

$$\left[-N=P \underset{|}{\overset{Cl}{\underset{Cl}{\vphantom{|}}}} - \right]_3 \xrightarrow[\text{2. } H_2N \diagup\diagdown Si(OEt)_3]{\text{1.}\qquad H_2NBu} \left[-N=P(NHBu)_m(NH-(CH_2)_3-Si(OEt)_3)_n- \right]_3$$

Eq. 5.   Reaction of phosphornitrilic chloride with amines.

Table 3. Products of the reaction of phosphornitrilic chloride with amines.

m	n	Product	Yield [%]
0	2	13	90
1	1	14	92

For the synthesis of compounds bearing different amine substituents, it is convenient to introduce first the unsilylated amine to avoid side reaction at the Si-alkoxy group during extended reaction times. The number of substituents is controlled by stoichiometry where geminal, in addition to non-geminal, arrangement of substituents at one phosphorous of the phosphazene is possible [11].

During the reaction of phosphornitrilic chloride with secondary amines such as *N*-methylaminopropyl-trimethoxysilane, a 6-fold substitution does not occur and the compounds are hydrolytically unstable due to remaining chloro groups.

Phosphornitrilic Chloride Trimer and *O*-Nucleophiles

In a two-step process allylic alcohol is used as a nucleophilic *O*-alkyl source, followed by hydrosilylation of the residual olefinic C=C-bond to yield the alkoxysilyl-substituted phosphazene precursor for sol-gel chemistry.

As the nucleophilicity of alcohols is insufficient for substitution reactions on chlorophosphazenes, the corresponding sodium salts were used. Employing this type of reaction, a variety of compounds can be synthesized as shown in Eq. 6 and Table 4.

Eq. 6. Reaction of phosphornitrilic chloride trimer with sodium alkoxides followed by hydrosilylation.

Table 4. Products of the reaction of phosphornitrilic chloride with sodium alkoxides.

R	X	m	n	p	Product	Yield/%
Et	OEt	1	1	3	15	40
CH_2CF_3	Cl	1	1	2	16	90
	Cl	2	0	1	17	83

The products obtained with two different substituents where $m = 1$ $n = 1$ are mixtures of isomers with a ratio of allyloxy to ethoxy (e.g., trifluorethoxy) one to one. Separation of those isomers for further investigations was not necessary.

The hydrosilylation reaction was conducted with catalytic amounts of hexachloroplatinic acid in THF. It was observed that triethoxysilane gave only low yields of product **15** (ca. 40 %). More satisfactory results were obtained with chlorosilanes instead of alkoxysilanes. Thus, mixed isomers **16** were obtained as clear liquids in a yield of 90 % and compound **17** was isolated after vacuum distillation in a yield of ca. 83 %.

Thermal Stability of Triamino Triazine and Phosphazene Modified Polysiloxanes

In preliminary investigations 6-membered ring modified polysiloxanes were synthesized via sol-gel processing and analysed by DTA. Materials obtained from the threefold substituted urea bridged melamine derivative **3** decompose at 210°C. Higher thermal stability is achieved for polysiloxanes prepared from compounds **7-12**. It is apparent that high degrees of alkoxysilyl substitution increase T_{dec} of the corresponding modified polysiloxane at 225-270°C as observed in compounds **7, 9, 11** possessing one to three alkoxysilyl substituents.

N- and *O*-bridged phosphazene siloxanes synthesized to date exhibit lower thermal stability in the range 190-230°C compared with melamine siloxanes. The influence of trifluoroethoxy groups on the higher thermal stability of the ternary silyl substituted compound **16** is demonstrated clearly by the $T_{dec} = 230°C$. Data obtained in the literature [12], would suggest the *N*-substituted phosphazene siloxanes decompose at lower temperatures (190-200°C) as *O*-bridged materials (210-230°C).

Conclusion

The results presented here demonstrate that a great variety of monomeric alkoxysilanes bearing 6-membered rings of melamine and cyclophosphazene is available. Thermal analysis of polysiloxanes derived from these precursor compounds exhibit remarkable stability.

Acknowledgment: We thank the *Deutsche Forschungsgemeinschaft* for financial support.

References:

[1]   B. M. Novak, *Adv. Mat.* **1993**, *5*, 422.

[2]   U. Schubert, N. Hüsing, A. Lorenz, *Chem. Mater.* **1995**, *7*, 2010.

[3]   J. E. Herweh, W. Y. Whitmore, *J. Chem. Eng. Data* **1970**, *15*, 593.

[4]   R. Livine, W. C. Fernelius, *Chem. Rev.* **1954**, *54*, 508.

[5]   I. Yoshio, I. Shimichi, *Bull. Chem. Soc. Japan* **1966**, *39*, 2490.

[6]   H. Moscher, F. Whitmore, *J. Am. Chem. Soc.* **1945**, *67*, 662.

[7]   D. W. Kaiser, J. T. Thurston, J. R. Dudley, F. C. Schaefer, I. Hechenbleickner, D. Holm-Hansen, *J. Am. Chem. Soc.* **1951**, *73*, 2984.

[8]   H. R. Allcock, *Adv. Mat.* **1994**, *6*, 106.

[9]   H. R. Allcock, *Chem. Mat.* **1994**, *6*, 1476.

[10] S. S. Krishnamurthy, A. C. Sau, M. Woods, *Adv. Inorg. Chem. Radiochem.* **1978**, *21*, 41.

[11] C. W. Allen, *Chem. Rev.* **1991**, *91*, 119.

[12] H. R. Allcock, *Acc. Chem. Res.* **1979**, *12*, 351.

Mesomorphic Properties of Poly(diphenylsiloxane)

B. R. Harkness*, M. Tachikawa, I. Mita

Dow Corning Asia Ltd. Research Center
603 Kishi, Yamakita, Kanagawa 258-01 Japan
Fax: Int. code + (46576)4422
E-mail: usdccnrl@ibmmail.com

Keywords: Poly(diphenylsiloxane) / Mesophase / Oligomers

Summary: Poly(diphenylsiloxane) (PDPhS) has been reported to form a liquid crystalline phase above an apparent melting temperature of 260°C. To gain further insight into the structure of the high temperature mesophase, the morphologies of both the high molecular weight polymer and low molecular weight oligomers of PDPhS have been examined by DSC, high temperature optical microscopy, X-ray diffraction and dynamic mechanical analysis. The results suggest that the high temperature phase of PDPhS is more representative of a conformationally disordered crystalline phase than a liquid crystalline phase.

Introduction

A survey of the literature reveals a wealth of information concerning the physical properties of poly(dimethylsiloxane) with a few reports focusing on the properties of poly(dialkylsiloxane)s with slightly longer alkyl side-chains such as ethyl and propyl, however a member of the siloxane family that has received relatively little attention is poly(diphenylsiloxane) (PDPhS). For pure PDPhS, most research to date has focussed on evaluating its bulk properties, yet even these remain to be thoroughly examined. An interesting observation with regards to the structure of PDPhS has been the unusual temperature dependent change that occurs in the X-ray diffraction pattern [1] above 230°C, where all of the crystalline reflections disappear with the exception of an intense small angle reflection at $2\theta = 8.8°$ and a broad halo at $2\theta = 20°$. That the $2\theta = 8.8°$ reflection remained and actually increased in intensity on heating, combined with the simplicity of the diffraction pattern, suggested a possible transition to a mesomorphic state on melting of the crystalline phase.

That certain flexible polydialkylsiloxanes form a mesophase above the melting point of the crystalline phase is well known [2-4], however, the state of order of the flexible chains in the mesomorphic structure has been the focus of some debate. Early reports have referred to the mesophase formed by poly(dialkylsiloxane)s as a conformationally disordered crystal or 'condis crystal', which is defined as a crystalline structure in which the flexible chains are able to undergo dynamic conformational change within the confines of a three-dimensional lattice [5]. Alternatively, it has recently been argued that the mesophase formed by polydialkylsiloxanes is a hexagonal columnar liquid crystalline phase with positional order in only two dimensions [6]. The state of order for mesomorphic PDPhS has not been examined in detail although it has been suggested in a recent report that the high temperature phase of PDPhS is a nematic liquid crystal as based upon microscopic textures and X-ray diffraction patterns [7].

To extend the basic property profile of the bulk polymer and help better understand the high temperature mesomorphic state, high molecular weight PDPhS and its low molecular weight oligomers have been examined by X-ray diffraction analysis, DSC, dynamic mechanical analysis and high temperature optical microscopy. The results of this examination and interpretation of the results are presented.

Results and Discussion

Polydiphenylsiloxane was prepared by a lithium silanolate initiated ring-opening polymerization of hexaphenylcyclotrisiloxane in diphenyl ether at 180°C [7]. Since the polymerization is kinetically controlled, careful monitoring of the reaction is required to minimize depolymerization processes that lead to the formation of the thermodynamically more stable octaphenylcyclotetrasiloxane. The reaction is terminated and the high molecular weight polymer stabilized by end-blocking with a small amount of a chlorosilane. Cooling of the reaction mixture below 150°C results in the precipitation of the polymer. A characteristic of PDPhS is its inability to dissolve in organic solvents below 150°C, thus limiting analysis of the dissolved polymer to high temperatures in high boiling solvents. Hence, important characteristics of PDPhS such as molecular weight and hydrodynamics have not been examined to date. In this light the focus continues to be on the characterization of the bulk polymer.

The DSC heating and cooling scans for a thermally equilibrated sample of PDPhS are shown in Fig. 1. On heating, the transition from the crystalline phase to the mesophase is represented by a broad peak centered at 260°C. Heating to 400°C revealed no additional transitions. There was no evidence for a glass transition in the DSC between 20 and 400°C.

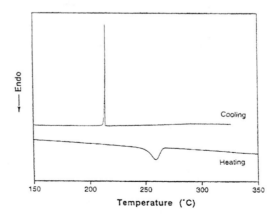

Fig. 1.   DSC heating and cooling scans for a thermally equilibrated sample of PDPhS (heating rate = 10°C min^{-1}).

On cooling, the transition from the mesophase to the crystalline phase is represented by a sharp peak at 218°C ($\Delta H = 7$ kJ mol^{-1}; $\Delta S = 14$ JK^{-1}·mol^{-1}). In performing the above DSC studies it was important to consider the preparative and thermal history of the polymer sample. PDPhS prepared by rapidly quenching a hot solution of the polymer into a cold non solvent showed poorly resolved peaks in the DSC scan. Heat cycling of this sample in the DSC between 100 and 300°C (rate = 10°C/min) resulted in a sharpening of the DSC peaks coupled with a continuous increase in the enthalpy of fusion over 10 cycles. This suggests that an equilibrium condition with respect to the level of crystallinity may not be assumed upon heating above the 260°C transition followed by cooling.

To gain greater insight into the 260°C transition, the storage modulus (G') and tan δ of the equilibrated polymer were recorded to 400°C (Fig. 2). To prepare samples for mechanical testing, the polymer was compression molded into flat test plates at 300°C. Even under these conditions it was apparent that the material did not flow smoothly, giving brittle plates that resembled a fused solid. On heating, the glass transition (T_g) of the polymer appears as a small reduction in the modulus between 40 and 80°C, with a maximum in tan δ at 65°C. As the temperature of the sample is further increased, a second transition appears above 230°C accompanied by a drop in the modulus by one order of magnitude and a second maximum in tan δ at 265°C. The fall in the storage modulus at 265°C to a plateau value of 10^8 dyne/cm² is significantly less than what one would expect for a transition from a crystalline phase to a flowable liquid crystalline phase. In fact, at these high temperatures the polymer has a leathery texture and can be cut with a knife without brittle fracture. Continued heating of the sample above 265°C produces a plateau region in G' similar to what one might expect for a crosslinked polymer. This is followed by a possible third transition above 400°C that may represent the onset of a high temperature melting process.

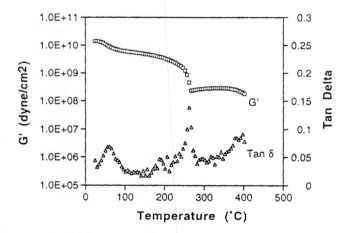

Fig. 2. Storage modulus (G') and tan δ as a function of temperature for PDPhS.

To further test the flowability of the polymer at high temperatures and test for molecular orientation, a small amount of the sample was placed between a glass plate and a cover glass and heated to 360°C in a Mettler hot-stage while viewing the sample through the crossed polars of an optical microscope. On heating no significant change in the birefringence was observed upon passing through the 265°C transition up to a maximum temperature of 360°C. At 360°C the sample was subjected to shear by sliding the top and bottom glass plates in opposite directions by approximately 1 mm, yet this failed to induce any change in the birefringence. In addition, significant resistance of the sample to the mechanical shear was indicative of a very high viscosity as might be expected given the high storage modulus recorded at this temperature.

Given that high molecular weight PDPhS has a high isotropization temperature ($T > 400°C$) and an onset of thermal decomposition in the same temperature range implies that it is not possible to microscopically observe the transition from the isotropic state to the mesomorphic state that exists above 400°C. A potential solution to this problem was to reduce the molecular weight of the polymer, which would function to suppress the temperature for transition to the mesophase and the transition to the isotropic melt. By carefully controlling the ratio of initiator to monomer ratio in the polymerization reaction and controlling the reaction temperature and polymerization time it was possible to prepare oligomers of diphenylsiloxane with number average molecular weights ranging from approximately 2200 to 3450 g/mol. With these samples a phase diagram on heating was constructed (Fig. 3), from which a critical molecular weight for mesophase formation of approximately 2400 was determined. Oligomers with M_n 2400 and higher showed both a mesomorphic phase and an isotropic phase below 360°C. This created an opportunity to observe for the first time the formation of the mesophase from the isotropic state for a diphenylsiloxane polymer. Interestingly, on cooling the isotropic melt the mesophase was found to grow slowly as three-dimensional crystalline bodies. Shearing of the partially formed mesophase failed to induce shear orientation in the isotropic phase nor deform or change the birefringence of the crystallites. This implied that the mesophase has crystalline characteristics and a three-dimensional lattice structure.

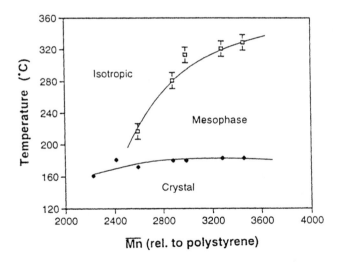

Fig. 3.   Phase diagram on heating for diphenylsiloxane oligomers.

In conclusion, the experimental evidence acquired to date strongly suggests that the high temperature mesophase of poly(diphenylsiloxane) is more crystalline in character than liquid crystalline and is perhaps representative of a conformationally disordered crystalline phase.

References:

[1]   D. Y. Tsvankin, V. Y. Levin, V. S. Papkov, V. P. Zhukov, A. A. Zhdanov, K. A. Andrianov, *Polym. Sci. USSR* **1980**, *21*, 2348.

[2]   Yu. K. Godovski, V. S. Papkov, *Adv. Polym. Sci.* **1989**, *88*, 129.

[3]   Yu. K. Godovski, V. S. Papkov, *Makromol. Chem. Macromol. Symp.* **1986**, *4*, 71.

[4]   G. J. J. Out, A. A. Turetski, M. Möller, D. Oelfin, *Macromolecules* **1994**, *27*, 3310.

[5]   B. Wunderlich, M. Möller, J. Grebowicz, H. Baur, *Adv. Polym. Sci.* **1988**, 87, 1.

[6]   G. Ungar, *Polymer* **1993**, 34, 2050.

[7]   M. K. Lee, D. J. Meier, *Polymer* **1993**, *34*, 4882.

Applications of Silicone Elastomers for Electrical and Electronic Fields

Masaharu Takahashi
Silicone-Electronics Materials Research Center
Shin-Etsu Chemical Co., Ltd.
Tel.: Int. code + (27384)5310 – Fax: Int. code + (27384)5305
1-10, Hitomi, Matsuida, Gunma 379-02, Japan

Keywords: Silicone elastomers / Coat / Transparent / Conductivity / Electrical

Summary: Applications of silicone elastomers for electrical and electronic fields are described. Applications are separated into five categories which involve applications concerning the volume resistivity, coating materials, technology and materials of transparency, other applications for these fields and issues of contact failure with low molecular weight siloxane.

Introduction

More than 50 years have passed since the Silicone Industry was born. Silicone elastomers have been researched and developed constantly for various industrial fields, especially electrical and electronic fields. The annual production of chlorosilanes in 1995 reached about 200 000 tons (110 000 tons in polysiloxane base) in Japan.

In the beginning electrical and electronic uses of silicone elastomer covered the needs of insulation materials, wire and cable, but nowadays they are spreading to wider needs in these fields and other new fields. For instance, silicone elastomers are used for coating, potting, insulating, sealing for semiconductor device encapsulation, opto-electronic materials and so on. The development of silicone elastomers in these fields depends on not only the improvement of qualities such as electric properties, heat resistance, flame resistance, purity, but also the innovation of fabricating technology such as curing methods and various fabricating equipments.

In this paper, applications of silicone elastomers for electrical and electronic fields are described.

Applications for Electrical and Electronic Fields

Applications are separated into the following five categories. The first group is applications concerning the volume resistivity of silicone elastomers. These are insulation stocks, electrically conductive stocks and semiconductive stocks. The second group is the coating materials used for electrical and electronic fields, for example, the conformal coating and the junction coating resin. The third group is the technology and materials of transparency which are the radiation cure with ultra-violet ray, the optical fiber coat and silicone gel. And fourth is about other applications relating to electrical and electronic fields such as rubber contact, thermally conductive rubber and rubber isolator. Finally the contact failure issue with low molecular weight siloxane is discussed.

Applications Concerning Volume Resistivity

The market for silicone elastomer products of this decade has expanded dramatically with increase of business machines and home electronics such as personal computers, copy machines, printers, TV sets, video cameras and so on. Various silicone products are used for electrical and electronic fields.

Based on the electrical conductivity, there are three grades of silicone rubbers. Insulating stocks are used for anode caps used for the cathode ray tube of TV sets, or plug boots and oil bleed connectors for automobiles, for instance. Electrically conductive silicones are used in products like rubber contact switches for TV remote controllers, handy phones, zebra connectors, EMI (electron magnetic interference) shield gasket and so on. And semiconductive silicone elastomers used for semiconductive rollers for plain paper copy machines and page printers, and antistatic materials. These relatively new materials have the volume resistivity of 10^5-10^{10} ohm-cm, and may be used in many other industrial fields.

Insulative Silicone Rubber

At first, the anode cap material is singled out as a typical application. It is well known that the screen size of a TV sets tends to expand year after year. With that tendency, the anode cap size became bigger and bigger. These days its size is 60 to 70 mm in diameter. As a result, the breakdown voltage (BDV) of material was improved. And it is necessary to produce the high grade material which has a break down voltage higher than 100 kV/mm in direct current (Fig. 1).

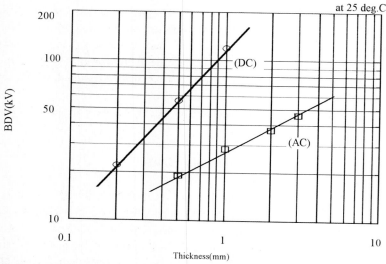

Fig. 1. Breakdown voltage of insulative silicone.

And it is also very important to give the material flame retardancy as well as high breakdown voltage. The mechanism of burning and combustion of silicone elastomers is very different from that of other synthetic polymers [1-4]. Normally the synthetic polymers produce inflammable gases and water vapor in burning. Silicone elastomers also produce inflammable gases such as cyclo-siloxanes. However, silicone makes a three-dimensional structure by the oxidation reaction of siloxane side chain. Furthermore, after the combustion it makes the ash ; silica. This makes silicones different from other polymers.

In the case of flame retardant silicone elastomer, many ingredients such as silica, platinum, and other flame retardant agents are incorporated into the base siloxane polymer. But there is no need to use the halogenated flame retardant agent, for example, bromine or chlorine compounds. This difference is an advantage of silicones compared with other synthetic polymers in terms of health and safety.

Electrically Conductive Silicone Rubber

Electrically conductive rubber is usually made by mixing electro-conductive materials into siloxane polymer. For example carbon black, graphite, electro-conductive metal powder such as silver, nickel, metal oxide such as tin oxide or electro-conductive fiber. The electro-conductivity depends on the content, configuration and structuring of these electro-conductive materials.

As actual applications of electrically conductive silicone, there are several examples: the gasket for EMI shields and electro-conductive rolls for business machines. The special fabricating case of electro-conductive silicone is the zebra connector.

As shown in Fig. 2, the electro-conductive silicone rubber is also now used for the connectors of liquid crystal displays in computers and digital watches. The zebra connector is a composite made by alternating silicone insulator layers and conductive silicone layers, and both edges are supported with silicone rubber. The zebra connector was developed for connection of electrodes formed on flat surfaces. The connector is especially suitable if soldering is difficult.

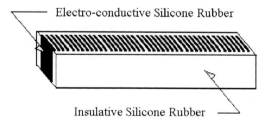

Electro-conductive Silicone Rubber

Insulative Silicone Rubber

Fig. 2.   Zebra connector.

Semiconductive Silicone Rubber

The semiconductive silicone is currently a topic of great interest. Fig. 3 displays silicone rubber rolls used in a laser beam printer. There are several issues of electric devices surrounding OPC drum. Until now the wire coron system has been popular and has been used for electro-charging system onto OPC drum, but this system has a problem producing ozone gas. Recently, the semiconductive rubber roll has become popular due to requirements for lower ozone gas generation and for more compact design. Also, a semiconductive rubber roll is being used as a developing roll in place of a magnetic roller. Attention is being paid to semiconductive rubber rolls because they are superior in getting a clear image of halftone. The required volume resistivity is in the range 10^5-10^7 ohm-cm.

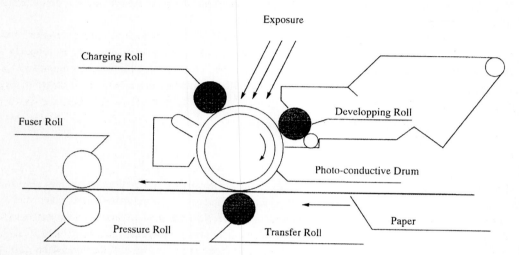

Fig. 3.   Silicone rubber rollers used in laser beam printer.

The technology of compounding design has also been developed to make the semiconductive rubber. In the past, electro-conductive carbon black was merely mixed into silicone rubber. However this method did not provide stable semiconductivity. To solve this difficult problem several ideas were tried [5-10]. One was incorporating the semiconductive metal oxide filler instead of carbon black. Another idea was to develop a localized configuration of carbon black in compound matrix; for instance, the method of polymer blending technology of silicone and carbon black-loaded EPDM. In this case, more than any thing else the compounding process is very important. The carbon black is first dispersed into the EPDM, and then the silicone is blended into this compound. The carbon black has limited migration into the silicone phase after blending of the silicone into the EPDM / carbon because of the better compatibility of the carbon with the EPDM. It is thought that this formation of macrodispersion stabilizes the resistivity.

Coating Materials

Two types of coating materials are used in electrical and electronic equipments. One is conformal coating, the other is junction coating resin. These coating materials also occupy an important place in the electrical and electronic fields.

Conformal Coating

As the printed circuit boards used in electrical equipment for automobiles and airplanes are exposed to severe environments involving coarse particulates, rapid temperature variations, high temperature and high humidity, for example, there is the possibility of mechanical damage or the occurrence of electrical problems in board components. An environmentally resistant conformal coating is used to protect board surfaces.

Certainly conformal coating has less impact resistance than conventional potting or molding, but has the following advantages: small size and light weight, easy repair, good heat radiation properties, minimal thermal expansion damage and good operatability. Silicone conformal coating, which has excellent resistance to heat, cold and weather, is used in electrical equipment for automobiles and airplanes, instead of other conformal coating materials like acrylic or urethane. Recently, non-solvent type products for silicone conformal coating were developed because of dangers of solvents in terms of health and air pollution.

Junction Coating Resin

Junction coating resin is used like a conformal coating to protect semiconductors from mechanical shock and electrical interface. These resins are available in forms which are soft and flexible, or rigid and hard when cured. A variety of curing systems to meet a variety of applications is available.

Fig. 4. shows a typical application of junction coating resin. The junction coating resin is used as a protective coating of diodes and transistors. In this figure these dotted areas are the junction coating resin. Recently various types of coating materials are developed: gel type, rubber type, rigid type and elast-plastic type. For instance, one is an application for hybrid IC chip coating and another is for thermal head coating.

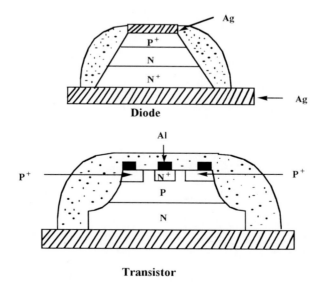

Fig. 4.   Junction cating resin.

Transparent Materials

Third category is related to the subject of transparent materials and technology. First is the cure technology by UV radiation. Second is optical fiber coat technology and third is potting gel technology.

Curing Systems

Usually four types of curing system are used for silicone elastomers. One is condensation reaction type using the hydrolyzable radical, second is addition reaction type, which uses hydrosilation reaction. The third type is organic peroxide curing and the last is radiation reaction using EB cure and UV cure.

Fast cure technology, improved handling and non-corrosion characteristics are needed for these curing systems. Nowadays the high processability of production is in demand, so there is a need for fast curing systems in place of the former systems. Applications of the UV curing system are like these; adhesion and temporary fixing devices, gap filling or potting of optical machine and equipment, and hard coat or sealing available for low moisture permeability need. However, its usage is restricted because of a lack of cure in shadow areas. So, a double cure system (Scheme 1), which includes radiation cure and moisture cure, has been developed [11]. For example, in the upper curing system, radiation cure is a photo-addition reaction between isopropenoxy group and mercapto group. The moisture cure is a condensation reaction of isopropenoxy group. Therefore, even in shadow areas the coating will cure by the condensation reaction. In the future, the importance of a double cure system will be increasing.

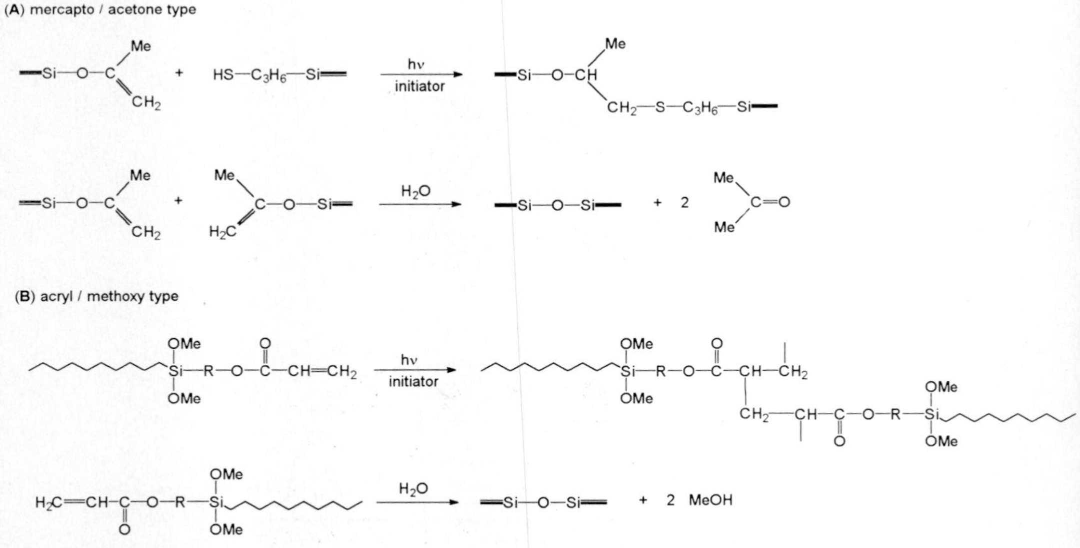

(A) mercapto / acetone type

(B) acryl / methoxy type

Scheme 1.   Double cure system.

Coating for Optical Fibers

The next topic is related to optical fiber coating applications. Optical fiber communication has been developing rapidly and will be a key technology in the future. Silicones are used to coat and protect optical fibers while further enhancing their overall transmission properties. Actually, silicone elastomers are being used for the primary and buffer coat. These have the following quality requirements: external stress relief, easy control of refractive index, high transparency and high coating speed.

Further more silicone is noted as a good core material. Plastic fiber is superior to glass fiber in short distance communications. It is easier to handle and lower in cost. At present polymethacrylate is used as a core material. But its usable conditions are limited because of low heat resistance. Good heat resistance of silicone makes it a noted core material, but its transmission loss is high because of moisture absorption. The glass transition temperature of silicone is -120°C. Under typical working conditions of common fiber, the siloxane chain exhibits micro-Brownian motion. Therefore moisture permeability of the silicone is large. Newly developed silicone core material possesses high glass transition temperature and high light quantity retention by the introduction of T units and phenylene units (Table 1).

Table 1. Silicone-core optical fiber.

Core Materials	Tg [°C]	Light Quantity Retention [%][a]
$\left(\bigcirc - SiO_{1.5} \right)_x \left(MeViSiO \right)_y \left(Me_2SiO \right)_z$ $HMe_2Si - \bigcirc - O - \bigcirc - SiMe_2H$	90	91
$ViMe_2SiO - \left(MeViSiO \right)_x - SiMe_2Vi$ $HMe_2SiO - \left(Me_2SiO \right)_x \left(MeHSiO \right)_y - SiMe_2H$	-120	0

[a] After exposure 60°C: 90 % RH 1000 h

Silicone Gel

Let's look at silicone gel, a multi-functional material, for electronic applications. This material has excellent physical properties, typical for organopolysiloxanes and also other unique properties which are derived from the low crosslinking density of the gel. Addition reaction type curing mechanism is used for silicone gels, because it offers excellent thermal stability, low shrinkage and ease of molecular design.

Table 2 shows the physical properties of some potential encapsulants. Silicone gel has a high thermal expansion coefficient compared with that of other organic resins, but the generated thermal stress is much lower due to its low modulus. Therefore, if silicone gel is used as a potting material, it can prevent electrical failures caused by the thermal stresses.

Table 2. Physical properties of silicone gel.

Resin	Modulus [kgf/mm^2]	Coeff. of Linear expansion [x 10^{-5}/deg]
Epoxy	200-500	5-8
Polyester	200-500	5-10
Polyurethane	7-300	10-20
Polyimide	100-200	1-5
Nylon	5-300	1-8
Silicon Gel	0.001-0.01	20-35

Other Applications

Rubber Switches

Keyboard switches are integrated composition of insulate silicone rubber and conductive silicone rubber filled with carbon black. They are commonly called rubber contacts, which serve as an electrical contact and a mechanical spring. Contacts made of silicone rubber were developed in response to demands for more compact size, mass production and for lower cost.

Rubber contacts were made for calculators at early stages of development. These contacts are finding new applications not only in calculators but also in TV touch channel changers, personal computer keyboards and in personal handy phone and so on. These are required to have very high contact durability. The keyboard for computers is required to exceed a durability of over ten million cycles.

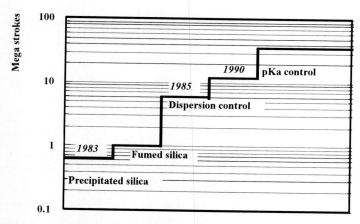

Fig. 5. Progress of fatigue life of silicone rubber.

Fig. 5. shows the technology progress of the extension fatigue durability of silicone rubber. At present, silicone rubber achieves a fatigue durability of about 50 million cycles by using improved dispersion uniformity of the reinforcing silica filler [12].

Thermally Conductive Silicones

It is becoming very important to reduce the heat from electronic devices which tend to be downsized year after year. The thermal conductive silicone sheets show the good performance because of good heat resistance and flame retardancy and so on. An example of application is the thermally conductive silicone used for power transistors.

Silicone Rubber Isolator

One application of silicone isolator is the pick up attenuator for the laser beam systems of compact disk player.

Butyl rubber is well known as a good vibration absorbing material. But its performance is not good enough because of the large dependency of its isolating properties on temperature. The reason is that tan δ changes around the glass transition temperature of the polymer. On the contrary, silicone rubber has a steady tan δ in wide range of temperature (Fig. 6).

It is thought that the good vibration absorption properties of silicone rubber are obtained from the energy dissipating macrostructure that is comprised of filler and wetting agent [13]. So, silicone rubber has stable damping properties in the range -50 to 100 °C.

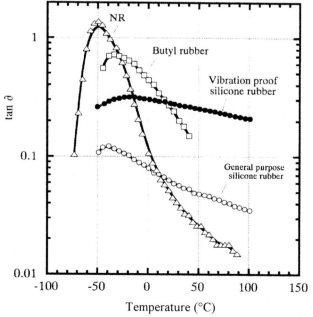

Fig. 6. Vibration absorbing properties of elastomers.

Low Molecular Siloxane Issue

Downsizing of electronic products is making rapid progress and better reliability is a serious concern. Some failures of these devices can be attributed to the volatility of constituents in silicones used as adhesives and potting materials, etc. Low molecular weight silicone is volatilized and burnt by arc energy, occurring from the electrical contacts. This is the reason why low molecular weight siloxanes are changed into silica, and deposited on the contact surfaces. Silica works as an insulator and abrasive and thus results in contact failure.

The contact failure appears within a limited current and voltage range [14]. It is well known that there is a relationship between silicone vapor concentration and the relay life. To get the long life of switching cycles, a reduction of the low molecular weight siloxane included in products is needed.

Fig. 7 shows the content of low molecular weight siloxanes in the silicone polymer. As shown in the figure, almost all of the low molecular weight siloxanes can be removed from the original polymer through refining process. Nowadays, refined polymers are used for RTV rubber for electrical and electronic applications. Refined silicone sealants are used as clean room sealants in the field related to semiconductor manufacturing.

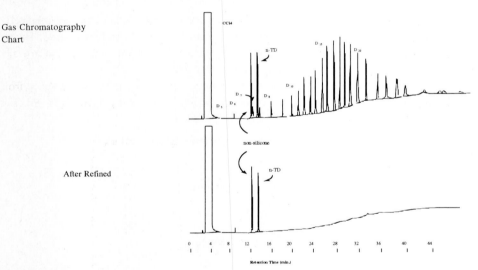

Fig. 7.   Gas chromatography chart.

Future Needs to Silicone Elastomers

As mentioned above, silicone elastomer is the active material and is yet developing in electrical and electronic fields. And furthermore, silicone based-polymers are noted as a new functional materials for these fields of the future. The research and development on silicon-based polymers has been continued by many researchers and many companies in the world.

So we hope that silicone, including silicone elastomers, will be prosperous in these fields in future.

Acknowledgement: I thank K. Itoh and M. Ikeno for advising, and T. Nakamura for supporting the manuscript.

References:

[1]   M. G. Noble, J. R. Brower, USP 3514424.

[2]   M. R. MacLaury, *J. Fire and Flammability* **1979**, 10, 175.

[3]   J. W. Harder, USP 3652488.

[4]   K. Itoh, T. Yoshida, Jpn. Tokkyo Kouhou, JP 51-24302.

[5]   M. Takahashi, T. Nakamura, K. Numata, USP 5209872.

[6]   T. Fukuda, K. Mita, Jpn. Tokkyo Kouhou, JP 5-64993.

[7]   M. Takahashi, T. Nakamura, K. Numata, Jpn. Tokkyo Koukai, JP 3-195749.

[8]   M. Takahashi, T. Nakamura, K. Numata, Jpn. Tokkyo Koukai, JP 3-195752.

[9]   M. Takahashi, T. Nakamura, K. Numata, Jpn. Tokkyo Koukai, JP 3-190964.

[10] Shimizu, Jpn. Tokkyo Koukai, JP 63-251464.

[11] M. Arai, S. Sato, K. Fujioka, Jpn. Tokkyo Kouhou JP 93 20455.

[12] N. Omura, M. Takahashi, T. Nakamura, *Presented at the Meeting, Rubber Div., ACS,* **1990**.

[13] T. Nakamura, T. Fukuda, *Automotive Polymer & Design,* **1989**, 25.

[14] M. Aramata, *Silicone News, Shin-Etsu Chemical Co.* **1989**, *38*, 9.

Cyclic Liquid Crystalline Siloxanes: Chemistry and Applications

F.-H. Kreuzer*, N. Häberle, H. Leigeber, R. Maurer, J. Stohrer, J. Weis
Consortium für elektrochemische Industrie GmbH
Zielstattstr. 20, D-81379 München, Germany
Wacker-Chemie GmbH
Hans-Seidel-Platz 4, D-81737 München, Germany

Keywords: Siloxanes / Polymers / Cholestric reflectors / Crosslinking

Summary: Cyclic siloxanes substituted with mesogenic groups, which are connected to the backbone by aliphatic spacers, exhibit liquid crystalline (LC) phases as the classic calamitic liquid crystals.

Two types of materials were synthesized during the last few years, crosslinkable and non crosslinkable compounds. Compared with their linear analogues they exhibit low viscosity as a result of the low degree of polymerization, combined with high glass transition temperatures. Applications of non crosslinkable species use both of these properties. The crosslinkable types do not require a high glass transition temperature as the vitrifying concept. Nevertheless, the low viscosity is essential for processing. The orientation of the mesogenic groups in particular requires a low bulk viscosity.

The most important applications are as follows: retardation plates, cholesteric reflectors and filters, storage materials for optical informations, agents for preventing unauthorized copying of documents, and pigments for iridescent and polarizing coatings.

Introduction

Cyclic liquid crystalline siloxanes can be considered as a class of the liquid crystalline (LC) side-chain polymers with a low degree of polymerization.

Although the three LC types have quite different structures, many properties are dominated by the mesogens employed. This is valid especially for the optical properties, for example double refraction, and static electrical properties, for example the dipole moment. Dynamic properties, the switchability by electric fields, the dielectric loss factor, and, in general, properties depending on temperature, are strongly influenced by the coupling forces between the mesogenic groups. Coupling forces result from dipole moments, dispersion forces, chemical links via spacers, their lengths, and the number of spacers leading off the mesogenic groups. Last but not least the embedding of the mesogens in a system of backbone coils, networks, or a higher or less ordered lattice may play an important role.

Thus, side-chain systems can exhibit many properties in between, well-oriented and solid materials. Many applications for cholesteric, nematic, and smectic cyclic siloxanes have been proposed. Most of them use cholesterics. Cholesteric liquid crystals (n*) or tilted smectic phases reflect the incident light in a specific wavelength range and with circular polarization. The wavelength of reflection is determined by the pitch and by the helicity of the cholesteric phase. The reflected light has the same sense of helicity as that of the cholesteric phase. The transmitted beam – with the same wavelength as the reflected one – has the opposite sense [1]. Monomeric cholesterics exhibit a strong temperature dependence of the reflection wavelength. This property has been used for optical temperature measurements [2], but is not acceptable when the reflective properties of cholesterics are to be used in optical filters. Sophisticated temperature compensation can also be achieved with low mass systems, but they suffer from the liquid properties, which require encapsulation in glass cells [3]. In contrast to non-glassy, low molecular mass systems nearly constant optical properties over a broad range of temperature can be achieved if materials with a glass transition temperature above the application temperature are used.

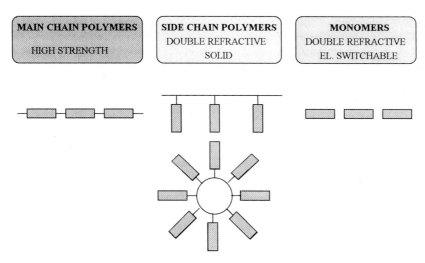

Fig. 1.   Architecture of various types of liquid crystals.

Another approach to fix the LC properties is crosslinking after orientation. The versatility of the silicone chemistry provides materials tailored for specific applications.

Synthesis

Linear LC siloxanes have been well known since the end of the seventies [4] and an enormous variety of compounds, including networks, have been synthesized in the meantime [5]. These side-chain polymers are usually synthesized by a well-established standard procedure, that is, the coupling of ω-alkenyl substituted mesogenic groups to H-siloxanes with the aid of a Pt catalyst. This concept has been extended to various geometric and chemical classes of backbone siloxanes, linear [6], cyclic [7, 8] and cage-like siloxanes [9], or siloxanes with a great variety of substituents [10].

Synthesis of Non-Crosslinkable Materials

As a first approach we tried to design a material with, processing characteristics like those of thermoplastics or hot melts. The advantages of such a system are clear: one component system, simple processing, no shelf-life problems and no material losses by irreversible crosslinking. To reach this goal a material with a T_g above RT and good orientation properties had to be synthesized.

One of the most important properties of polydimethylsiloxanes is their low T_g. Therefore, with respect to the glass transition, siloxanes are handicapped compared with other backbone systems, such as poly(meth)acrylates.

Starting with linear siloxane systems and using the C-3 spacer due to the low cost of allyl chloride and using commercially available H-siloxane with DP 35, the resulting materials had a very poor orientability. To improve this property, the bulk viscosity was consequently reduced by using H-siloxanes with short chain lengths. As a result, the T_g also dropped dramatically as the influence of the trimethylsilyl end groups became stronger and stronger.

Table 1. Comparison of linear and cyclic LC siloxanes.

Therefore, we switched to cyclic systems without end groups and with higher T_g [7, 11]. A systematic study on T_g of linear and cyclic LC siloxanes was done by Richards et al. [12] and confirmed this concept.

Basically, a glass transition temperature of up to 80°C can be achieved [10], but these materials are limited with respect to variations of the mesogenic groups. The general route to obtain these materials is shown in Scheme. 1.

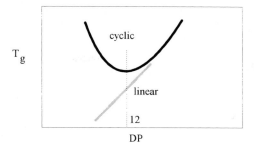

Fig. 2. Comparison of the T_g of linear cyclic LC siloxanes. If the degree of polymerization is low, the cyclic systems show a higher glass transition temperature than the linear analogues. Scheme after Richards et al. [12]

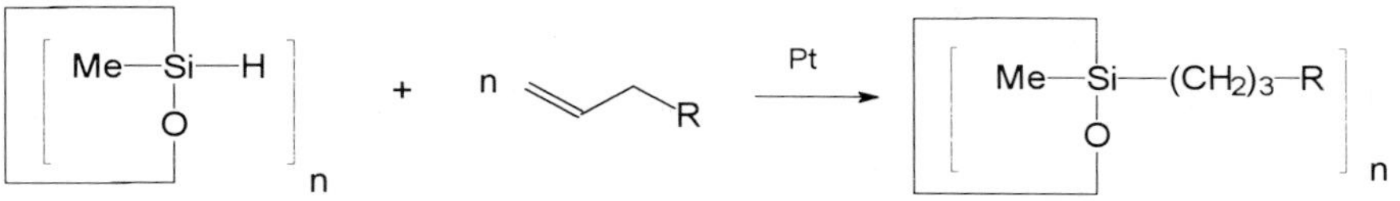

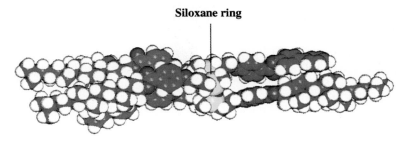

Scheme 1. Reaction of an ω-unsaturated mesogen with a H-cyclosiloxane.

A molecule with three cholesterol units and one biphenyl unit according to Scheme 1 connected to a tetrasiloxane shows an elongated structure (Fig. 3). Molecules of this shape can be arranged in nematic, smectic, or cholesteric phases.

Siloxane ring

Tetrasiloxane with 1 biphenyl unit and 3 cholesterol units

Fig. 3. Van der Waals plot of a cyclic siloxane with 4 Si units (yellow), 1 mesogenic group II, and 3 mesogenic groups I (Modeling program Insight / Discover).

Fig. 3 suggests that tetrasiloxanes with two mesogenic groups can only be realized with the ratios 1:0, 3:1, 1:1, 1:3, and 0:1. Pure materials consisting of molecules with only one defined structure can be obtained easily if only one mesogen is employed. From the technical point of view instead of the pure substances bulk materials with the desired liquid crystalline phases are of interest. Fortunately in most cases the various substituted oligomers are miscible in all ratios, last but not least a consequence of the oligomeric character. Thus each desired composition can be obtained by simultaneous or stepwise addition of two or more mesogenic groups to the siloxane backbone.

An example of such a system with the ring size 4 and the mesogenic groups from Table 1 is given in Fig. 4.

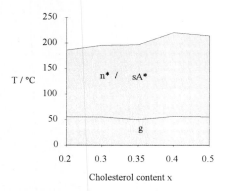

Fig. 4. Phase behavior of a non crosslinkable cyclic LC siloxane system (*n* = 4) with mesogens I and II.

To obtain the mixtures with the requested molar ratio of side chains the reaction is carried out using the cyclosiloxane and a mixture of mesogens – if the reactivity of the mesogens is comparable. Otherwise a two step reaction is preferred. The product distribution is defined by a binomial distribution [9].

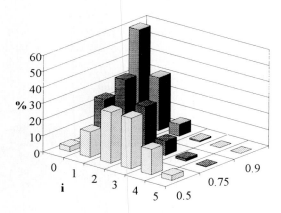

Fig. 5. Statistical distribution (molar fraction) of single copentamers $A_i B_{5-i}$ (cyclopentasiloxane with *i* = 0...5 mesogens A and 5-*i* mesogens B) for a randomly substituted pentamer $A_x B_{5-x}$ with three average fractions *x* of mesogen A.

If allyloxy spacers are used a splitting reaction at the oxygen can be observed. Thus the number of different substituents is multiplied. The product distribution, even for such complex systems, can be calculated on the basis of NMR and GPC data and can be confirmed by MALDI-TOF mass spectrometry.

Synthesis of Crosslinkable Materials

Due to the limitations of non crosslinkable cyclic liquid crystalline siloxanes resulting from the fact that the T_g is dominated by the mesogenic groups connected to the cyclic siloxane backbone and, therefore, other aspects for the selection of the mesogens are limited, a crosslinkable system had to be established. This strategy requires the presence of polymerizable moieties within the side chain groups, such as epoxide, cinnamate or (meth)acrylate groups. Methacrylates are preferred due to their reasonably fast photopolymerization and thermal stability.

To obtain liquid-crystalline films with excellent optical properties, the crosslinkable cyclic siloxane should also be of low viscosity. Therefore, small-ring cyclic siloxanes with 4-7 siloxane units are used for these materials. Linear siloxanes can also be used because the argument to meet the condition $T_g >> $ RT is no longer essential.

The crosslinking moiety can be introduced into the siloxane backbone by two pathways: by coupling of an OH-protected unit to the siloxane followed by methacrylation of the deprotected OH group [13] or by direct coupling of the methacrylate-bearing unit to the siloxane backbone chain [14].

The second method should be feasible, at least for small and medium-sized siloxanes, as the reactivity of ω-alkenes and methacrylates towards hydrosilylation differs by a factor of at least 10, thus minimizing premature crosslinks via hydrosilylation.

These two synthetic approaches have been applied to the cooligomers shown in Scheme 1, employing a chiral mesogen I and an achiral methacrylate-bearing mesogen III.

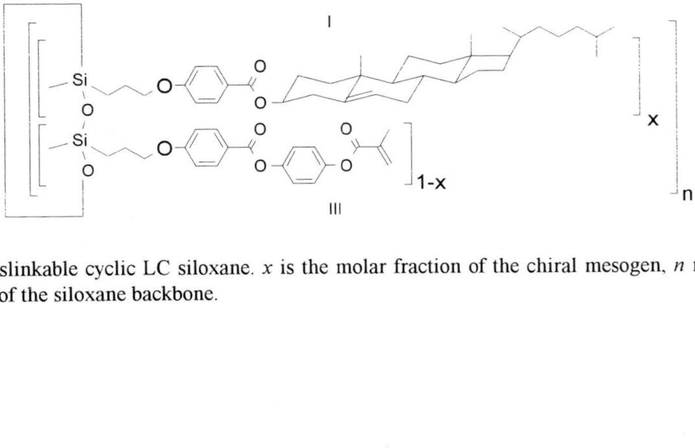

Scheme 2.   Crosslinkable cyclic LC siloxane. *x* is the molar fraction of the chiral mesogen, *n* represents the ring size of the siloxane backbone.

Properties

Phase Behavior

The phase behavior of cyclic LC siloxanes is dominated by the mesogens used. The ring size of the siloxane backbone also plays an important role. As an example, the phase diagram of a system according to Scheme 2 is given in Fig. 6.

In non crosslinkable materials an s_A* phase occurs for high pitch pentamers [15]. All cooligomers show homogeneous phases. This is easily rationalized via the statistical occurrence of at least 6 differently substituted pentamers (see Fig. 5 for n = 5), 2 being homooligomers and 4 being cooligomers with 1-4 chiral side chains.

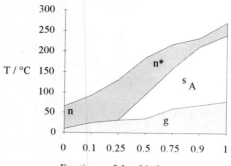

Fig. 6.   Mesomorphic properties of a series of crosslinkable LC siloxane cooligomers (*n* = 5) depending on the fraction of two mesogens (Scheme 1).

Viscosity

For viscosity adjustment of crosslinkable siloxane oligomers reactive monomers are added. Thus, the viscosity of the siloxane oligomer at 80°C can be reduced from 15 Pa s to 1 Pa s (Fig. 7).

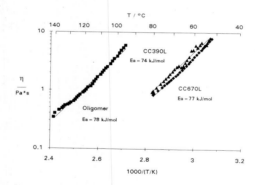

Fig. 7.   Viscosity of crosslinkable cyclosiloxane (n = 4.5, x = 0.45) and the crosslinkable cholesteric LC siloxane mixtures CC390L and CC680L.

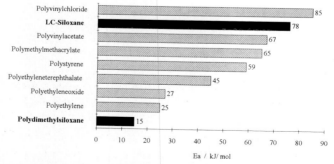

Fig. 8.   Activation energy of viscous flow. Data adapted from D. W. Van Krevelen, P. J. Hoftyzer, Properties of Polymers, Elsevier, **1976**.

Fig. 8 shows that the activation energy for viscous flow is of the order of 70 kJ mol^{-1}. This is totally different from dimethylsiloxanes and is a hint that the temperature dependence of the flow behavior is determined by the mesogenic groups attached to the LC siloxanes.

Optical Properties

Beside the double refraction, the property used most often at the moment is the characteristic reflection behavior of the cholesteric phase.

From theory the wavelength of reflection (λ_r) is inversely proportional to the concentration of chiral component, yielding the following relation:

$$\lambda_r = \text{const.} / C_{\text{Chiral}}$$

Eq. 1. C_{Chiral} is the concentration of chiral component in the mixture.

Thus, LC siloxanes with reflection wavelengths in the UV, in the visible, and IR regions can be obtained by variation of the concentration of the chiral compound. Additionally, structural properties like the ring size influence the wavelength of reflection.

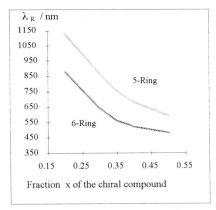

Fig. 9.   The wavelength of reflection depends on the cholesterol content and the size of the siloxane ring.

By substituting the cholesterol moiety by dihydrocholesterol, the range of reflection wavelengths can be extended to the IR. Right handed materials were obtained by using doristerol derivatives instead of cholesterol compounds [16].

Tuning the Reflection Wavelength by Mixing

As mentioned above LC siloxanes are mutually miscible. This is also valid for mixtures containing reactive monomers for adjusting the viscosity.

By mixing of two LC siloxane mixtures CC670L (red) and CC390L (blue), the amount of the chiral component in the mixture is changed. Accordingly the wavelength of reflection λ_r can be varied between 390 nm and 670 nm (see Fig. 10).

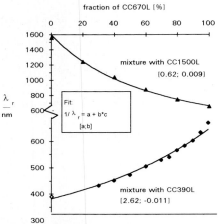

Fig. 10. Reflection wavelengths of blends: Dependence of reflection wavelength on amount of CC670L in the binary mixtures of CC670L and CC390L, resp. CC1500L. The experimental values were fitted by a hyperbola according to equation (3) with the coefficients given in the figure.

The concentration of the chiral component in the mixture is composed of the respective concentrations in CC670L (C_1) and CC390L (C_2).

$$C_{Chiral} = x \cdot C_1 + (1\text{-}x) \cdot C_2$$

Eq. 2. x is the fraction (w/w) of CC670L in the mixture.

Combining equations (1) and (2) allows description of the relation between wavelength of reflection and the fraction of CC670L in the mixture.

$$\lambda_r = \frac{1}{a + x * b}$$

Eq. 3.   The coefficients a and b were determined to match the experimental data (see Fig. 6). The same procedure was applied to mixtures of CC1500L and CC670L.

The versatility of this procedure may be demonstrated by using right-and left-handed LC siloxanes both reflecting in the blue at 390 nm. Starting from these two components all wavelengths of reflection (right-and left-handed) up to reflection in the IR region can be obtained including a nematic material ($\lambda_r = \infty$).

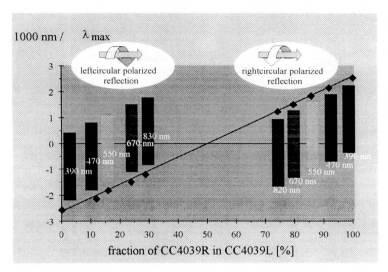

Fig. 11. Mixing of right and left handed materials.

Temperature Dependence of λ_r

Temperature Dependence of Non-Crosslinkable LC Siloxanes

It is obvious that we must distinguish the two temperature regions above and below T_g. Below T_g the order parameter of the mesogens in the glassy state is not influenced by variation in temperature. Only thermal expansion or contraction is responsible for small variations of the cholesteric pitch. On heating the pitch should increase and, therefore, λ_r should shift to longer wavelength. Above T_g a more significant temperature dependence is observed. All LC siloxanes synthesized until now containing cholesterol or doristerol derivatives exhibit a shift of λ_r towards shorter wavelength.

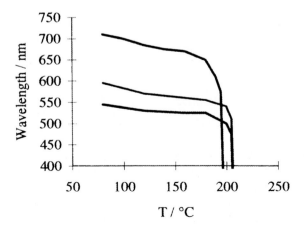

Fig. 12. Temperature dependence of non-crosslinkable cyclic LC siloxanes with different λ_r.

Temperature Dependence of λ_r of Crosslinkable LC Siloxanes

Crosslinkable LC siloxanes in the non-crosslinked state have a similar temperature dependence of λ_r as the non-crosslinkable materials. Mixtures with monomers are more sensitive. The temperature-dependent λ_r can be frozen in by polymerization. Thus different λ_r values can be obtained with the same material, crosslinked at different temperatures.

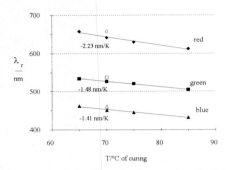

Fig. 13. Influence of curing temperature on λ_r. The reflection wavelength of the cured material was measured at 20°C. A linear relationship holds for the examined temperature range 65-85°C with a gradient of about -2 nm/K. The open symbols show λ_r before crosslinking.

Temperature Dependence of λ_r of the Cured / Crosslinked Material

After curing λ_r becomes nearly independent of temperature up to 140°C (see Fig. 14).

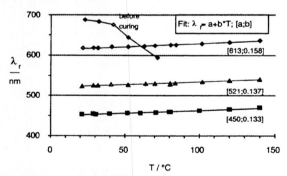

Fig. 14. Temperature dependence of the reflection wavelength for cured layers of LC siloxanes. The slope of the linear behavior correlates well with that determined from thermal expansion. For comparison the temperature dependence before curing is also shown for a red reflecting LC siloxane.

The very small residual increase (less than 0.2 nm/K) can be attributed to thermal expansion of the cured polymer. This was verified by measuring the coefficient of thermal expansion via the change of density by heating pieces of cured solid film from 20°C to 50°C. A value of $\Delta\lambda/\Delta T = (0.11^{+}; -0.04)$ nm/K was calculated from this measurement, which is in good agreement with the observed wavelength shift. Preservation of the reflection band was observed down to -196°C. This sounds trivial for a polymer but should be mentioned for comparison with monomeric liquid crystalline materials, which tend to crystallize at low temperature.

Refractive Index and Bandwidth

The refractive indices n_p and n_o of LC siloxanes before curing show the typical liquid crystalline behavior of vanishing double refraction at the clearing point (compare Fig. 15).

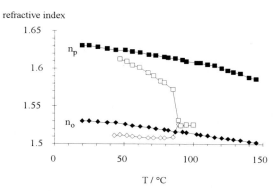

Fig. 15. Refractive indexes, n_p and n_o, before and after curing at 70°C for a green reflecting mixture.

After curing a nearly constant double refraction of $\Delta n = 0.1$ is observed in the interval of the measured temperatures. This indicates the preservation of the liquid crystalline order when the temperature is increased. From the relation between double refraction and bandwidth of the cholesteric reflection, the bandwidth of a green reflecting LC siloxane may be calculated [2]:

$$\Delta\lambda = \lambda_r \cdot \Delta n/n = 530 \text{ nm} \cdot \textbf{Fehler!} = 33.5 \text{ nm}$$

Eq. 4.

This is in good agreement with spectra recorded from a small area of homogeneous color with a spectrometer attached to a microscope. In larger samples the measured bandwidths often vary by 30-50 nm due to variations of sample thickness or imperfect orientation.

Applications

Applications were proposed and developed with respect to the specific properties of the LC silicone properties. For all applications the solid state at RT, achieved by vitrification or chemical curing is essential. The required specific optical properties are a consequence of the LC phase.

Processing

The application of non-crosslinkable LC siloxanes is very simple: The material can be applied by contacting a hot (90°C $< T <$ 150°C) substrate and subsequent shearing with a doctor blade or with a second substrate layer, e.g., glass. The application can also be done with the doctor blade technique starting from the melt.

Processing of the crosslinkable materials (mixtures) starts with the same techniques as before and additional curing with UV irradiation at 90°C employing 1-2.5% photoinitiator (Irgacure 907).

Applications Using the Nematic and Smectic Phase

Retardation plates or foils are widely used as optical elements. Mostly used applications are $\lambda/4$ or $\lambda/2$ plates. Their purpose is to convert linear polarized light to circular polarized light or circular to linear polarized light. As a special application retardation plates with high damage resistance are requested for high energy lasers. For this purpose a nematic LC silicone material was studied [17].

If crosslinkable materials are used, patterned retardation plates can be realized using the temperature dependence of Δn. Starting from high Δn values at low temperature the Δn for higher temperature can be fixed by photopolymerization up to $\Delta n = 0$ above T_i.

A more sophisticated retardation plate is a film consisting of the perpendicular arranged combination of two optical uniaxial materials, e.g., uniaxially stretched polycarbonate foil with a homeotropic oriented LC siloxane layer [18]. The purpose of such a foil is the improvement of the optical properties of LC displays, especially the viewing angle dependence of the contrast.

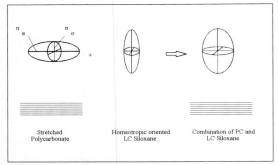

Fig. 16.   Combination of two optical uniaxial materials: The figure shows the optical indicatrix of an uniaxially stretched plastic foil with horizontal direction of n_e and the indicatrix of a homeotropic oriented LC siloxane with horizontal direction of n_o (schematically).

To demonstrate the effect of such a combination a three-layer system between crossed polarizers may be considered: The first picture was calculated as model for a LC display operating by change of Δn by application of an electric field, the second is the combination with a color-compensating foil and the third is the additional application of a homeotropic oriented layer of a LC siloxane.

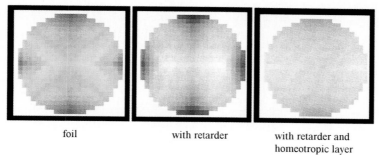

| foil | with retarder | with retarder and homeotropic layer |

Fig. 17.   Optical effect of a homeotropic oriented layer of LC siloxane.

Applications Using the Cholesteric Phase

Basic types of the cholesteric layers are summarized in Fig. 18. Well known is the unperturbed oriented arrangement of the helical axis perpendicular to a substrate (A). Beside this classical helix structure perturbed species exist (B, C) [19-22]. (A) exhibits only a single reflection band due to the constant pitch and undistorted helical director distribution. In (B) the pitch is still constant but the helical director distribution is distorted, resulting in reflections of higher order. The additional bands can be explained by higher order components of the periodically distorted helical structure. In (C) the director is again distributed helically, but the pitch changes from one side of the layer to the other, either by a continous pitch gradient, as shown in (C), or stepwise by a multilayer. As a result, such a system can reflect light over a broad range of the spectrum.

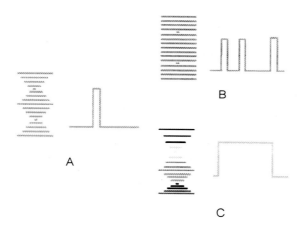

	Pitch	**Helical Director Distribution**
A	const	not distorted
B	const	periodically distorted
C	not const.	not distorted

Fig. 18.   Helix structures and their resulting spectra.

Cholesteric Filters and Reflectors

Filters and reflectors using the cholesteric phase may be classified by their application although all use the reflection and polarization properties. They can be made of crosslinkable or non crosslinkable LC siloxanes [23, 24].

The simplest devices are homogeneous planar oriented cholesteric or s_A^* layers. They can be used in the transmission or reflection mode as color selecting and polarizing plates. Depending on the service temperature non-crosslinked or crosslinked material may be employed. Beside these planar cholesteric layers spherical devices can also be made [25]. These devices operate as color selecting and polarizing reflectors as well as real optical elements like spherical mirrors. Reflectors for all parts of the electromagnetic spectrum can be made using cholesterics reflecting in the UV or IR.

Cholesteric reflectors for all wavelengths of the visible spectrum [26-29] will be used as reflectors and polarizers for back lights of LC displays to get a higher yield of the emitted light: Classical dyes containing polarizers absorb 50% of the polarized light in contrast to the cholesteric systems based on a non-absorptive interference mechanism.

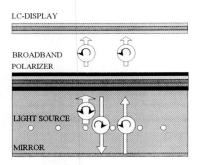

Fig. 19.   Application of a cholesteric broadband polarizer.

Color LCDs using back lights suffer from the high energy consumption for illumination of the display. The broadband polarizer technique would bring some improvements but the best way would be to have a system working only with ambient light. A possible approach to reach this goal are structured color reflecting and polarizing devices. LC siloxanes can be structured by a lithographic process if their temperature dependence of λ_r is high enough (Fig. 20).

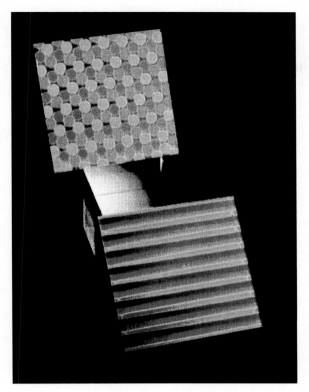

Fig. 20.   Lithographic process for patterned cholesteric RGB reflectors. T1<T2<T3<T_i<T4. At the different
temperatures the color is fixed by photocrosslinking.

Retardation Plates

IR or UV reflecting cholesterics are colorless in the visible region of the spectrum. Therefore,
retardation plates can be realized for STN displays using a long pitch material. Also UV reflecting
LC siloxanes are of interest for retardation plates because they exhibit behavior like an optical
negative uniaxial material.

Optical Storage of Information [30]

There is a lot of definitions of optical information. Apart from writing analog or digital signals into
an information storage layer an optical marker can also be assumed as information. Thus, copy
protection devices for documents or bank notes using cholesteric decorations with strong color
change depending on the angle of observation have been proposed [31, 38].

Thermorecording is possible with materials that have T_g > RT and a moderate clearing
temperature T_i. With the aid of a laser beam the sample is locally heated above T_i and the isotropic
non-reflecting state is preserved by rapidly cooling below T_g. The reflecting state can be recovered
again by slowly heating above T_g.

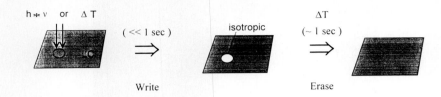

h ✳ v   or   Δ T   (≪ 1 sec)   isotropic   ΔT (~ 1 sec)

Write   Erase

Fig. 21.   Principle of a process for thermorecording.

An example of such an array of written and erased dots is shown in Fig. 22.

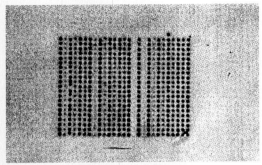

Fig. 22.   Array of written and erased dots in a yellow reflecting cholesteric matrix.

Although this principle looks very simple, the technical realization with high data rates is rather difficult. A totally different approach is based on the optical rotation diffusion [32,33]. For this process mostly azo dye containing liquid crystals are used: The azo dye molecule undergoes cis / trans isomerization when irradiated. The process is effective if the optical transition moment of the azo compound is parallel to the polarization plane of the incident light. Thermal relaxation happens simultaneously but this process occurs statistically, so no direction is preferred. The resulting process is characterized by movement of the light triggered azo molecules to a position with a minimum of irradiation density. During this process the other mesogens are also shuffled by the azo molecules, more or less in the same direction. If linear polarized light is used, the direction of the azo and the other moieties in the final state is perpendicular to the polarization plane. If non-polarized light is used the preferred direction is parallel to the incident beam. LC siloxanes applied for this purpose were cholesterics, reflecting in the IR, so the properties in the visible range of the spectrum were not influenced by λ_r.

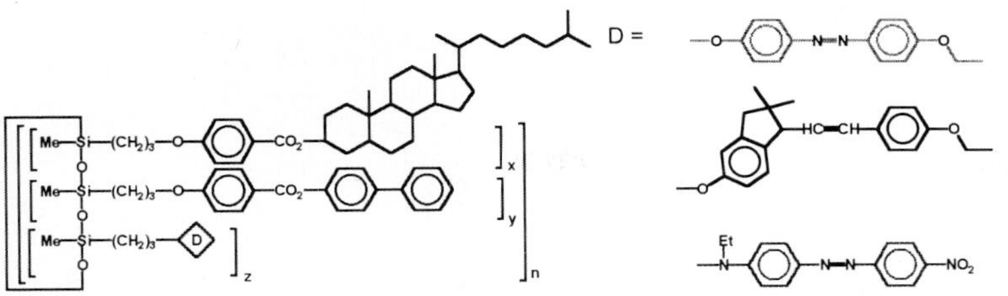

Scheme 3.   Structures used for optical rotation diffusion experiments.

Retardation plates structured by light using the rotation diffusion mechanism are based on non-crosslinkable cholesteric cyclic siloxanes [34,35]. The following samples demonstrate light-induced variation of Δn.

The observation of the specific effects of a distorted helix resulting in unusual overtones of the cholesteric reflection was made using a non-crosslinkable material with the cholesteric reflection in the near IR.

With these materials of Scheme 3 the deformed helix B from Fig. 18 was realized. A spectrum of an IR reflecting cholesteric siloxane in the non-irradiated and irradiated states is shown in Fig. 24. The portrait shown on the right is a result of the higher order reflection.

Fig. 23.   Retardation generated by irradiation of azo-(**a**, **b**) and stilbene-(**c**, **d**) containing LC siloxanes; **a** is the sample after irradiation without polarizers; **b** between crossed polarizers; **c** is between polarizers (angle 45°); **d** with - 45°.

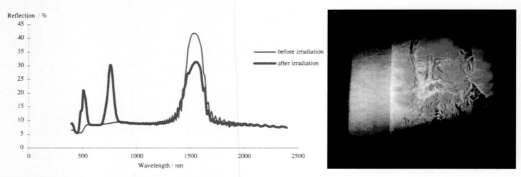

Fig. 24. Reflections of higher order by irradiation with linear polarized light.

Pigments and Coatings

One of the first attempts to apply LC siloxanes for decorative applications [36] was made with the non-crosslinkable materials. At room temperature this material is hard and brittle and melts above T_g as low viscous liquid crystals, which orient spontaneously on the substrate. To improve the orientation a spatula is recommended. The application temperature is 100-150°C.

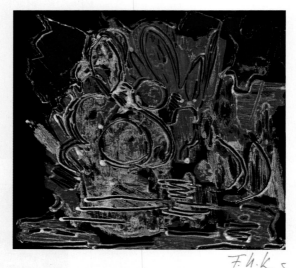

Fig. 25.   Small painting on black paper with aid of LC siloxanes.

Beside the application of the glassy cholesteric materials as meltable paints a very promising application is being developed from the disintegration of oriented, crosslinked cholesteric layers: Small, insoluble platelets can be used as iridescent pigments [37-39]. Such platelets can be suspended in inorganic media or organic lacquers and can be applied by a simple spray technique. The color effects are very strong if a black substrate is used for absorption of the non-reflected light. On the other hand very delicious colors result from combinations with colored substrates or mixtures with traditional pigments [40].

Fig. 26.   Racing car coated with green LC siloxane pigments.

References:

[1]   H. I. De Vries, *Acta Cryst.* **1951**, *4*, 219.

[2]   H. Kelker, R. Hatz, *Handbook of Liquid Crystals*, VCH, Weinheim, **1980**.

[3]   M. Schadt, J. Fünfschilling, *Jap. J. Appl. Phys.* **1990**, *29*, 1974.

[4]   H. Finkelmann, G. Rehage, *Makromol.Chem., Rapid Comm.*, **1980**, *1*, 31.

[5]   C. Burger, F.-H. Kreuzer, *"Polysiloxanes and Polymers Containing Siloxane Groups"*, in: *Silicon in Polymer Synthesis*, (Ed.: H. R. Kricheldorf), Springer, Berlin, **1996**, p.113.

[6]   B. Krücke, M. Schlossarek, H. Zaschke, *Acta Polym.* **1988**, *39*, 607.

[7]   F.-H. Kreuzer, M. Gawhary, R. Winkler, H. Finkelmann (Consortium für elektrochemische Industrie GmbH), EP 060 335, **1981**.

[8]   R. D. C. Richards, W. D. Hawthorne, J. S. Hill, M. S. White, D. Lacey, J. A. Semlyen, G. W. Gray, T. C. Kendrick, *J. Chem. Soc., Chem. Commun.* **1990**, 95.

[9]   F.-H. Kreuzer, R. Maurer, P. Spes, *Makromol. Chem. Makromol. Symp.* **1991**, *50*, 215.

[10] F.-H. Kreuzer, D. Andrejewski, W. Haas, N. Häberle, G. Riepl, P. Spes, *Mol. Cryst. Liq. Cryst.* **1991**, *199*, 345.

[11] F.-H. Kreuzer, *Proceedings 11. Freiburger Arbeitstagung Flüssigkristalle*, **1981**, 5.

[12] R. D. C. Richards, W. D. Hawthorne, J. S. Hill, M. S. White, D. Lacey, J. A. Semlyen, J. W. Grey, T. C. Kendrick, *J. Chem. Soc., Chem. Commun.* **1990**, 95.

[13] D. Andrejewski, M. Gawhary, H.-J. Luckas, R. Winkler., F.-H. Kreuzer (Consortium für elektrochemische Industrie GmbH), EP 358 208, **1988**.

[14] J. Stohrer, H.-J. Luckas, F.-H. Kreuzer, DE 4 440 209, **1994**.

[15] T. J. Bunning, H. E. Klei, E. T. Samulski, R. L. Crane, R. J. Linville, *Liq. Cryst.* **1991**, *10*, 445.

[16] H.-P. Weitzel, F.-H. Kreuzer, R. Maurer (Consortium für elektrochemische Industrie GmbH) DE 4 234 845.5, **1992**; *Chem. Abstr.* **1994**, *122*, 201399.

[17] E. M. Korenic, S. D. Jacobs, J. K. Houghton, A. Schmid, F.-H. Kreuzer, *Appl. Opt.* **1994**, *33*, 1889.

[18] T. Ohnishi, T. Nogushi, M. Kuwabara, K. Higashi, M. Namioka, A. Shimizu (Sumitomo Chemical Company), EPA 617 111 A1, **1994**.

[19] R. B. Meyer, *Appl. Phys. Lett.* **1969**, *14*, 208.

[20] R. Dreher, *Solid State Commun.* **1973**, *12*, 519.

[21] S. Mazkedian, S. Melone, F. Rustichelli, *Le J. de Physique*, **1976**, *37*, 731.

[22] L. H. Hajdo, A. C. Eringen, *J. Opt. Soc. Am.* **1979**, *69*, 1017.

[23] R. Maurer, D. Andrejewski, F.-H. Kreuzer, A. Miller, *SID Int. Symp. Dig. Tech. Papers*, **1990**, *21*, 110.

[24] N. Häberle, H. Leigeber, R. Maurer, A. Miller, J. Stohrer, R. Buchecker, J. Fünfschilling, M. Schadt, *SID Int. Display Res. Conf. San Diego*, **1991**, 57.

[25] R. Maurer, S. Beiergrößlein, F.-H. Kreuzer (Consortium für elektrochemische Industrie GmbH), EP 631 157 B1, **1993**.

[26] R. Maurer, D. Andrejewski, F.-H. Kreuzer, A. Miller, *SID 90 DIGEST*, **1990**, 110.

[27] L. Li, S. M. Faris, *SID 96 DIGEST*, **1996**, 111.

[28] D. J. Broer, J. Lub (Philips), EP 606 940 A3, **1993**.

[29] D. Coates, M. J. Goulding, Greenfield, J. M. W. Hanmer, S. A. Marden. O. L. Parri, *SID 96 APPLICATIONS DIGEST*, **1996**, 67.

[30] F.-H. Kreuzer, Ch. Bräuchle, A. Miller, A. Petri, *"Cyclic Liquid-Crystalline Siloxanes as Optical Recording Media"*, in: *Polymers as Electrooptical and Photooptical Active Media* (Ed.: V. Shibaev), Springer, Berlin, **1996**, p. 111.

[31] Ch. Heckenkamp, G. Schwenk, J. Moll (GAO Gesellschaft für Automation und Organisation mbH) DE 3 942 663, **1989**.

[32] J. Michl, E. W. Thulstrup, *Spectroscopy with Polarized Light*, VCH, Weinheim, **1986**.

[33] M. Eich, J.-H. Wendorff, B. Reck, H. Ringsdorff, *Makromol. Chem., Rap. Commun.* **1987**, *8*, 59.

[34] A. Petri, Ch. Bräuchle, H. Leigeber, A. Miller, H.-P. Weitzel, F.-H. Kreuzer. *Liq. Cryst.* **1993**, *15*, 113.

[35] T. J. Bunning, F.-H. Kreuzer, *TIP*, **1995**, *3*, 318.

[36] D. Makow, *Color Res. Appl.* **1986**, *11*, 205.

[37] S. Faris (Reveo Inc.), USP 5 364 557, **1991**.

[38] Ch. Müller-Rees, J. Stohrer, R. Maurer, F.-H. Kreuzer, S. Jung, F. Csellich (Consortium für elektrochemische Industrie GmbH) EP 601 483 A1, **1992**.

[39] E. M. Korenic, S. D. Jacobs, S. M. Faris, L. Li, *Proc. of the IS&T/SID 1995 Col. Im. Conf.: Col. Sci., Sys. Appl.* **1995**, 60.

[40] H.-J. Eberle, A. Miller, F.-H. Kreuzer, *Liq. Cryst.* **1989**, *5*, 907.

Modified Polydimethylsilsiloxanes with Fluorescent Properties

B. Strehmel[#]*, C. W. Frank[#]
Institut für Physikalische und Theoretische Chemie
Humboldt-Universität zu Berlin
Bunsenstraße 1, D-10117 Berlin, Germany
Fax.: Int. Code + (30)20932365
E-mail: bernds@chemie.hu-berlin.de
[#]Department of Chemical Engineering Stanford University
Stanford, CA 94305-5025, USA
Fax.: Int. Code + (415)7239780
e-mail: curt@chemeng.stanford.edu

W. Abraham, M. Garrison
Cygnus Therapeutic Systems
400 Penobscot Drive, Redwood City, CA 94063, USA
Fax: Int. Code + (415)3695318

Keywords: Fluorescence Probes / Networks / Mobility / Polarity

Summary: The fluorescence behavior of covalent bonded probes was investigated in linear and crosslinked polydimethylsiloxanes. The probe shows a dual emission that depends on the network density and the concentration on swelling agent. Investigations of linear polydimethylsiloxanes with a bonded probe as a function of THF concentration indicate the existence of ground state complexes between the solvent and the probe molecule.

Introduction

In the last years, the fluorescence probe technique has become an increasingly importance in many fields of material science because this technique opens up the possibility to examine the microscopic environment of different materials. First, this method has been successfully applied in the field of polymer science to study the polymer building process [1], the relaxation behavior of linear [2] and crosslinked polymers [3], polymer blends [4], and the interaction between polymer chains and various solvents[5].

The goal of this work was to show how the fluorescence probe with the general structure **I** (Scheme 1) behaves in swollen polydimethylsiloxane networks and how the solvation can be examined of linear polydimethylsiloxanes in the presence of covalent bonded probes. This molecule consists of a donor part, e.g., a dialkylamino group, and an acceptor, which is usually an ester group. The following points are important to discuss for a better understanding of the fluorescence behavior of **I** in polydimethylsiloxanes.

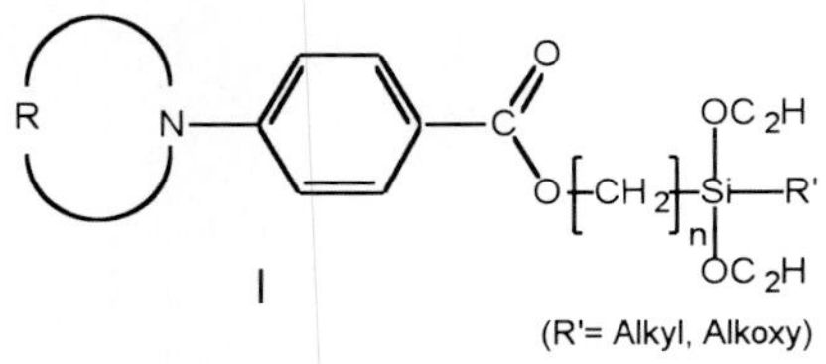

Scheme 1.   General Structure of the Fluorescence Probe Attached at Polydimethylsiloxanes.

(1) It was described earlier [6] that this class of probes is able to show dual fluorescence. This means that two bands exist in the fluorescence spectrum and the long wavelength band is caused by a charge transfer emission. The position of the band is strongly dependent on the microscopic polarity around the probe. Furthermore, the ratio of the two fluorescence bands depends on the mobility of the probe in the material investigated. These both values are therefore a quantity for the examination of the fluorescence behavior of **I**.

(2) Compounds with the general structure of **I** can possess a different spacer length n. This allows us to consider the fluorescence behavior at different lengths from the polymer chain. It was described previously that the length of the spacer group is important for the fluorescence behavior of the probe [6, 7].

(3) The functionality of alkoxy groups at the silicon allows us to localize the probe in the polymer. Trifunctional compounds are located at the network junction of a crosslinked polydimethylsiloxane while bifunctional probe molecules are placed on the main chain during polymer formation.

(4) Probes without a functionalized silane group cannot besuccessfully applied in this study because they cannot be localized at a special position in the network. This is important for the study of the swelling process of crosslinked polydimethylsiloxanes. If the probe is not attached it will be completely extracted during the swelling process.

Study of Crosslinked Polydimethylsiloxanes

The networks were synthesized by Sn(II)-catalyzed reaction between hydroxyl terminated polydimethylsiloxanes and methyltriethoxysilane. Attachment of the fluorescence probe occurs as described in Fig. 1.

The networks were prepared using a different molecular weight of hydroxyl terminated polydimethylsiloxane (PDMS–OH) and in different crosslinker concentrations. The swelling results are summarized in Table 1. From Table 1 we can conclude that both the molecular weight of the hydroxylterminated polydimethylsiloxane and the crosslinker concentration have an important influence on the emission of the probe. The self reaction of the crosslinker and the flexibility of the network are important quantities to describe the emission behavior of the probes applied in this work.

As can be seen from Table 1, the parameter v_2 increases with increasing molecular weight of PDMS–OH because the flexibility of the polymer chains increase in the same direction. Furthermore, the ratio CT/LE shows higher values with increasing molecular weight of PDMS–OH because of an increase in mobility. Eq. 1 can explain this behavior of the probe molecule.

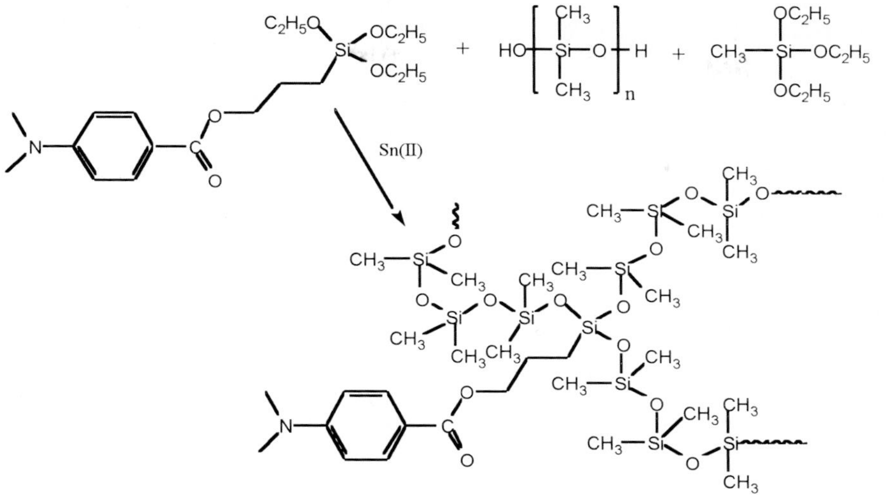

Fig. 1. Attaching of a fluorescence probe on the basis of a dialkylaminobenzoic ester at the network junction during the crosslinking process

$$\frac{CT}{LE} = k_{rot} \cdot C$$

Eq. 1. CT = area of the CT-band, LE = area of the locally excited state; C = constant that contains all photophysical parameters for both the charge transfer and the locally excited state; k_{rot} = rate constant for the formation of the CT -state from the LE-state, this value is strongly dependent on the probe mobility.

Typical spectra obtained for the probed attached at the network junction are shown in Figs. 2 and 3. The probe in THF possesses the highest ratio for CT/LE. An increased intensity for the LE was measured for the swollen sample. This effect can be explained by different polarity /mobility of the probe. One can assume that covalent bonded probes possess another probe mobility than free dissolved probe molecules. Furthermore, the covalent bonded probe molecule that shows a higher polarity in comparison to the siloxane chains is located at the network junction. The attached probe molecule is surrounded mainly by siloxane chains of the network. Addition of polar swelling solvents leads to an increase of the CT-emission and the ratio CT/LE is mainly influenced by the composition of polymer and swelling agent (compare spectra for dried and swollen N1 samples in Fig. 2). Therefore, the covalently bonded probe shows another fluorescence behavior in comparison to the free dissolved probes that can be surrounded also by solvent molecules.

Table 1. Summary of the swelling results in dependence on the crosslinker concentration and the molecular weight of the siloxane used, swelling solvent tetrahydrofurane THF.

network	M_w(PDMS-OH)/(g/mole)	SiOH/ SiOC$_2$H$_5$[a]	CT/LE[b]	E(CT)/cm^{-1}[c]	v_2[d]
N1	4200	1:3.0	1.55	22763	0.304
N2	18000	1:3.0	1.90	22579	0.265
N3	18000	1:1.7	2.55	22572	0.184
N4	18000	1:1.0	3.04	22520	0.108
N5	4200	1:2.3	2.70	22397	0.293
N6	4200	1:1.7	2.79	22383	0.272
N7	4200	1:1.0	3.31	22431	0.285

[a] SiOH/SiOC$_2$H$_5$ = ratio of the mass ratio between the hydroxyl terminated polydimethylsiloxane and the crosslinker– [b] CT/LE = ratio of the charge transfer band (CT) to the band of the locally excited state (LE) after band deconvolution, this ratio is strongly dependent on the mobility of the probe– [c] E(CT) = energy of the CT-state – [d] v_2 = volume fraction of polymer at the swelling equilibrium.

This effect becomes also clear in Fig. 3 where the fluorescence behavior was investigated as a function of crosslinker concentration. An increase of crosslinker concentration accelerates self crosslinking reactions of the crosslinking agent. The resulting networks contain flexible parts obtained by reaction of the PDMS–OH with the crosslinker as well as rigid components that are formed by the self reaction of the crosslinker. The probe is located in both parts of the resulting network. Changes of the fluorescence in a swollen sample can only be detected in flexible parts because the rigid one does not swell. The results obtained are summarized in Table 1.

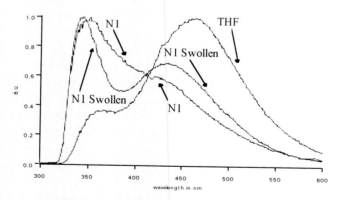

Fig. 2. Fluorescence spectra of dimethylaminobenzoic acid esters in swollen and dried (N1) polydimethyl-siloxane networks, as a reference the spectrum of the probe in THF is plotted, the intensity is plotted in a. u.

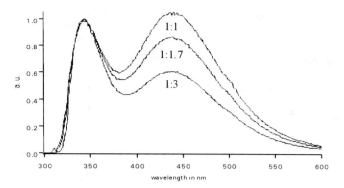

Fig. 3.   Fluorescence spectra of dimethylaminobenzoic acid esters in swollen polydimethylsiloxane networks as a function of the crosslinker concentration; corresponding [–Si–OH]:[–Si–O–C$_2$H$_5$] ratios are drawn in the figure, the intensity is plotted in a.u.

Fluorescence Studies of the Main Chain

For a better understanding of the probe behavior in PDMS networks it was necessary to investigate the fluorescence behavior of a probe covalently bonded to a soluble polydimethylsiloxane chain. Therefore, a dialkoxysilane containing 4-piperidinobenzoic acid ester was allowed to react with a hydroxyl-terminated polydimethylsiloxane giving a soluble polymer with a covalently bonded probe at the main chain. This reaction occurs in the presence of Sn(II) octoate. The reaction scheme is drawn in Fig. 4. It was shown by GPC (UV detection) and NMR spectroscopy that the probe reacted with the hydroxyl groups of the polydimethylsiloxane chains.

Moreover, the addition of THF to a solution consisting of polydimethylsiloxane and small amounts of the polymer described in Fig. 4 yields a decrease of the fluorescence intensity. The spectra obtained are drawn in Fig. 5.

As can be seen from Fig. 5, the addition of THF leads to a quenching of the fluorescence. In the next step the spectra were deconvoluted and the results obtained for the decrease of the LE band (high energy band) were used to quantify the quenching experiment. Furthermore, it was found that the quenching behavior can be only described by a mixture of both static and dynamic quenching. The following equation was applied for the calculation of the rate constant for quenching (k_{sv}) and the equilibrium constant between the probe and the quencher (K).

$$\frac{\frac{I_0}{I} - 1}{[quencher]} = \tau_0 \cdot k_{SV} + K + \tau_0 \cdot k_{SV} \cdot K \cdot [quencher]$$

Eq. 2. τ_o = fluorescence lifetime of the probe without quencher, k_{SV} = rate constant for quenching (dynamic process), K = equilibrium constant between probe and quencher, I = integral of the quenched fluorescence, I_0 = integral of the fluorescence without quencher.

Eq. 2 allows one to determine the parameters K and k_{SV}. The determination of k_{SV} yields a value of $3.2 \cdot 10^8$ s^{-1}($\tau_o = 3$ ns [8]). Furthermore, the equilibrium constant K results a value of about 0.5. This leads to the conclusion that about 2 solvent molecules form a complex with a probe molecule in the ground state.

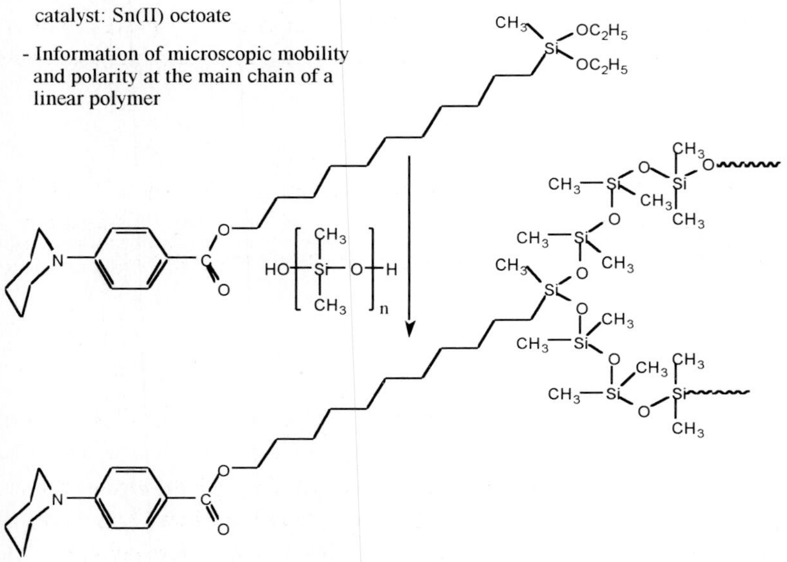

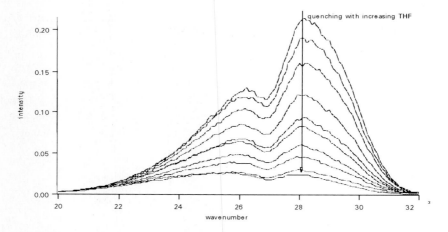

catalyst: Sn(II) octoate

- Information of microscopic mobility and polarity at the main chain of a linear polymer

Fig. 4.   Reaction Scheme of the Formation of a Linear Polydimethylsiloxane Containing Covalent Bonded 4-Piperidinobenzoic Acid Ester.

Fig. 5.   Fluorescence spectra of a linear polydimethyl-siloxane containing covalent bonded 4-piperidinobenzoic acid ester in polydimethyl-siloxane as a function of different contents on THF (unit for the wavenumber: x 1000 cm^{-1}; spectra for the corresponding quencher concentrations from the top to the bottom in M: 0, 0.079, 0.208, 0.473, 0.736, 1.004, 1.409, 1.833, 3.033, 4.1859)

Conclusions

The probes applied in this work are able to detect differences in both the mobility and polarity of the matrix. Selective attachment of the probe to the polymer opens the opportunity to study the behavior of the probe at different points fixed at the polymer. A variation of the spacer length would give selective information about the mobility and polarity, respectively. In general, covalently bonded probes are reasonable tools to examine swollen samples because they cannot be extracted during the swelling procedure.

Acknowledgement: B.S. would like to acknowledge the *German Academic Exchange Service* for a fellowship that opened the opportunity to work at Stanford University. Furthermore, he wishes to thank the *Deutsche Forschungsgemeinschaft* for a Habilitation-fellowship.

References:

[1]   B. Strehmel, M. Younes, V. Strehmel, S. Wartewig, *Progress in Colloid & Polymer Science* **1992**, *91*, 83.
[2]   B. Strehmel, V. Strehmel, H.-J. Timpe, K. Urban, *Eur. Polym. J.* **1992**, *28*, 525.
[3]   B. Strehmel, D. Anwand, H. Wetzel, *Advances in Resist Technology and Processing XI, Proceedings SPIE – The International Society for Optical Engineering* **1994**, *2195*, 801.
[4]   C. W. Frank, M. A. Gashgari, *Macromolecules* **1979**, *12*, 163.
[5]   K. Ficht, K. Fischer, H. Hoff, C. D. Eisenbach, *Makormol. Chem., Rapid Commun.* **1993**, *14*, 515.
[6]   W. Rettig, *Angew. Chem.* **1986**, *98*, 969.
[7]   B. Strehmel, C. W. Frank, *Polymer Preprints* **1995**, *36*(1), 385.
[8]   W. Rettig, unpublished results.

Sensitized Cationic Photocrosslinking of α,ω-Terminated Disiloxanes: Cation Formation in Nonpolar Media

U. Müller*, A. Kunze

Institut für Organische Chemie
Martin-Luther-Universität Halle-Wittenberg
Geusaer Straße, D-06217 Merseburg, Germany
Tel.: Int. code + (3461)462009 – Fax: Int. code + (3461)462080

Ch. Herzig, J. Weis

Wacker-Chemie GmbH
Geschäftsbereich S - Werk Burghausen
Johannes-Heß-Str. 24, D-84489 Burghausen, Germany
Tel.: Int. code + (8677)834833 – Fax: Int. code + (8677)3093

Keywords: Photocrosslinking / Kinetics / Cation Formation

Summary: The photoinduced cationic crosslinking of α,ω-disiloxanes was investigated by real time measurements. Moreover, the photoinduced cation formation was studied using photochemical methods in nonpolar solvents. A combination of a lipophilic substituted iodonium salt with several sensitizers was used as initiator system. The effiiciency of cation formation is a function of media and sensitizer used. The crosslinking of the disiloxanes can be described by means of a cationic chain process. The crosslinking reaction is a function of light intensity and sensitizer used. Moreover, the efficiency of cation formation determines the crosslinking kinetics.

Introduction

Vinyl ether and epoxy chemistry established more than 50 years ago finally found its way into radiation curing technology. These compounds have been demonstrated to be reactive monomers in curing processes photoinitiated by iodonium and sulfonium salts. In particular, vinyl ethers are known to be among the most reactive monomers in photocuring chemistry.

In silicone chemistry, there has been recently a resurgent interest in the chemistry and technology of photocurable compositions derived from epoxy and vinyl ether silicones. Several types of silicones with vinyl and epoxy groups were designed [1, 2] which differ in functionality, spacer group and monomer content. The cured products have excellent physical properties and are of commercial interest in a number of areas. The crosslinking chemistry of such modifided silicones involves the well known cationic crosslinking/polymerization of vinyl ethers. The silicone unit acts basically as an internal solvent. The initiating species, mainly protons, were formed in a photochemical process from photoinitiators, that are dissolved in the silicones.

Iodonium salt/sensitizer combinations are important initiators of cationic photopolymerization [3]. There have been numerous investigations of their sensitized photoreactions in polar solvents [3-5]. As has been shown, under these circumstances the quenching mechanism is electron transfer to give the radical cation of the sensitizer and the neutral radical of the onium salt and all in all a proton, which can initiate the cationic polymerization. However, our application of the sensitizer-iodonium salt combinations takes place in α, ω-terminated disiloxanes (experimental products of *Wacker-Chemie* Burghausen/Germany, see Scheme 2). Yet, so far, no mechanistic investigations of electron transfer sensitized iodonium salt photolysis in such media and with technically important onium compounds have been reported.

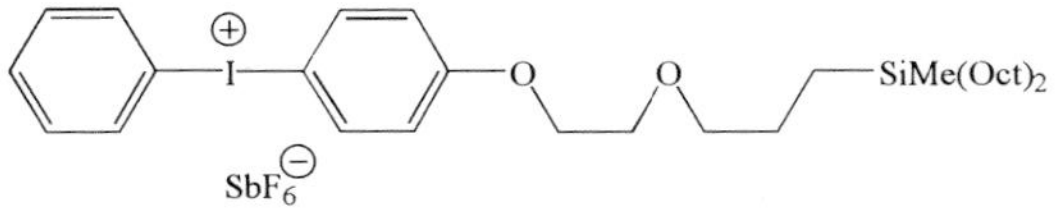

Scheme 1.

In this work, we therefore study the photoreactions of lipophilic substituted iodonium salts ($I_{(1)}$, experimental products of *Wacker-Chemie* Burghausen/Germany) with electronically excited sensitizers in various solvents. Several reactions were studied in a mixture of dimethoxyethane (DME) and hexamethyldisiloxane (HMDS) to simulate the situation in the real media, see Scheme 2.

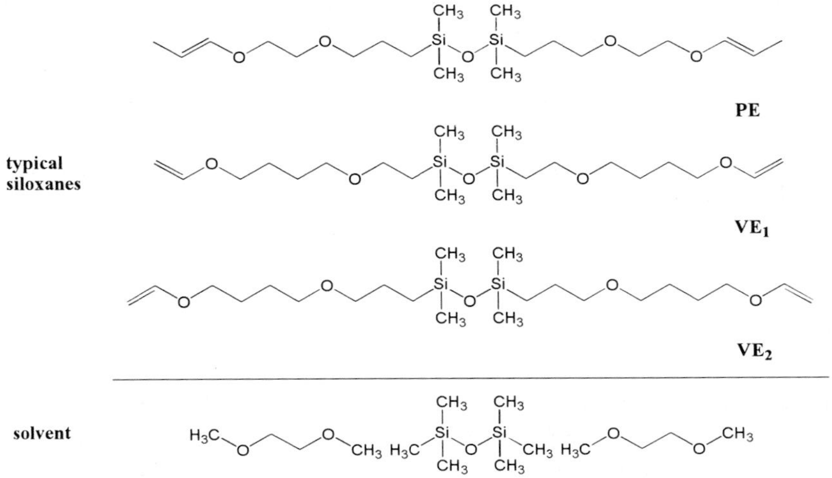

Scheme 2.

Additionally, the electron transfer reactions of Ph_2I^+ SbF_6^- ($I_{(2)}$) were studied to compare the results in these solvents with those from the lipophilic product $I_{(1)}$. Moreover, we compare these results with the results of crosslinking chemistry of model siloxanes terminated with reactive propenyl ether and vinyl ether.

General Crosslinking Kinetics

Kinetic studies have shown that the photoinitiated cationic crosslinking of such α,ω-terminated disiloxanes (see Scheme 2) can be considered as a polymerization process [6]. Due to the specific kinetic situation in bulk, Eq. 1 must be used to describe the rate of the photocrosslinking process (R_P) under stationary irradiation conditions [7]. Only this general expression or the reduced form (2),

$$R_P = -d[M]/dt = -(dx/dt)\cdot[M]_o = k(x)\cdot[M]^{\alpha}\cdot I_o^{\beta}$$

Eq. 1.

$$R'_P = -dx/dt = k(x)\cdot[M]^{\alpha}\cdot I_o^{\beta}$$

Eq. 2.

where [M] is the molar concentration of double bonds, t is the time, $k(x)$ is a conversion (x) dependent quantity, I_o is the intensity of the incident light, α and β are exponents, reflects the real situation in the polymeric systems investigated in all details, because the conditions for the reaction partners are changed at each time.

The kinetics of the radical crosslinking corresponds to a thermal reaction with the exception of the initiation step. According to Eq. 1, the dependence of R_P upon I_o can be deduced only from the rate of cation formation Eq. 3 (see also [8]),

$$R_i = \eta \Phi_{H^+}\cdot \eta_{abs}\cdot I_o$$

Eq. 3.

where R_i is the rate of the cation formation, η_{abs} fraction of the absorbed light, Φ_{H^+} is the quantum yield of the primary cation formation, η is the addition efficiency of the primary formed cation to the monomer. Under this assumption one can formulate Eq. 4:

$$R'_P = k'(x)\cdot[M]^{\alpha}(\eta \Phi_{H^+}\cdot \eta_{abs}\cdot I_o)^{\beta}$$

Eq. 4.

where $k'(x)$ is a conversion dependent quantity.

According to Eq. 4, it is also to be expected that the photoinitiator used influences the rate of crosslinking via Φ_{H^+}, η and η_{abs} [8]. On the other hand, from Eqs. 3 and 4 it follows that only R_i determines R'_P and this proportion this relation depends on the initiator used. Moreover, the initiator used also influences the inhibition time (see Eq. 5),

$$t_i = [Q]/R_i = [Q]/\eta \Phi_{H^+}\cdot \eta_{abs}\cdot I_o$$

Eq. 5.

where [Q] is the quencher concentration.

Sensitized Cation Formation in Nonpolar Solvent

Using fluorescence and laser flash techniques one can show that the lipophilic substituted iodonium hexafluoroantimonates react efficiently with singlet and triplet states of several sensitizers. Furthermore, one observes the decay of the electron donor, which was investigated with 9,10-diphenylanthracene (DPA) and 9,10-dimethylanthracene as sensitizers, which act as electron donor. Using the CIDNP-technique and laser flash photolysis one can prove that electron transfer takes place to the iodonium salts [9], which results in formation of a proton (see Scheme 3).

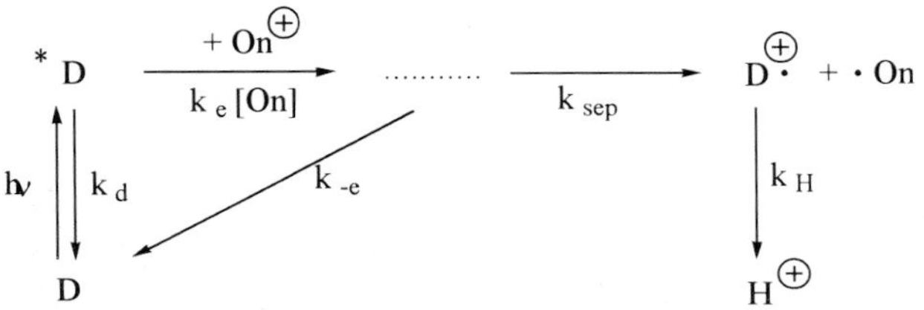

Scheme 3.   Photosensitized electron transfer.

The importance of acid formation requires a determination of the efficiency of this process. We have adapted an indicator dye method (see Scheme 4) to the measurements of quantum yields for acid generation. The results obtained using several sensitizers and $I_{(1)}$ are summarized in Fig. 1.

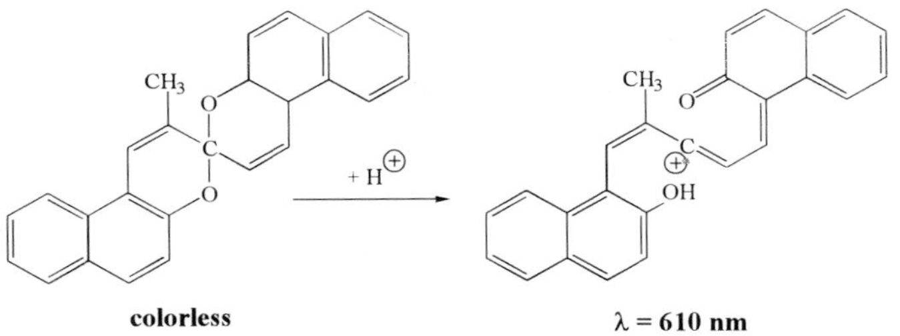

Scheme 4.   Indicator dye method for proton formation

From Fig. 1 one can deduce that the senzitizer possesses a great influence on the efficiency of acid formation. Highest values are found using thioxanthone, phenothiazine and 1-chlorohioxanthone as sensitizer. Assuming the standard scheme for photoinduced electron transfer, Scheme 3, one can write for the quantum yield of acid formation Eq. 6:

$$\Phi_{H^+} = \Phi_{-D} = \eta_{sep} \cdot k_e[On]/(1/\tau + k_e[On])$$

Eq. 6.

$$1/\Phi_{H^+} = 1/\Phi_{-D} = 1/(\tau \cdot \eta_{sep} \cdot k_e[On]) + 1/\eta_{sep}$$

Eq. 7.

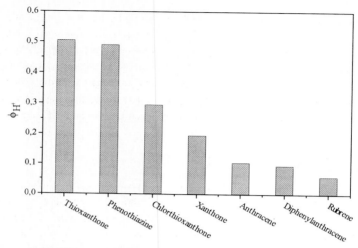

Fig. 1.   Quantum yield of proton formation using several sensitizer and $I_{(1)}$ as onium salt (λ = 365 nm, [Sens] = 10^{-4} mol/l, [On] = 10^{-2} mol/l, solvent: DME/HMDS-mixture (1:1))

where Φ_{-D} is the quantum yield of senstizer decay, η_{sep} is the separation efficiency of the primary formed intermediate product, k_e is the quenching constant of the excited sensitizer by the onium salt, [On] is the onium salt concentration, τ is the life time of the excited sensitizer. A relationship linear in [On] is obtained by inverting Eq. 6.

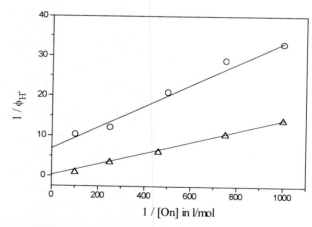

Fig. 2.   Plot $1/\Phi_H^+$ vs $1/$[On] for phenothiazine (triangle) and DPA (circles), experimental details see Fig. 1.

Fig. 2 shows for several sensitizers the plot $1/\Phi_H^+$ vs $1/$[On]. For all sensitizers used linear relationships were found. From Fig. 2 one can see that the slope and the intercept of the plot are a function of the sensitizer. By dividing and inverting of the slope of the plot by the intercept one can obtain the $k_e \cdot \tau$-value, which indicates to a certain extent the efficiency of the electron transfer reaction. The values obtained differ by a factor of 3. The highest value was obtained using DPA as sensitizer. The lowest value was obtained using phenothiazine. Nevertheless, the efficiency of proton formation is much higher using the phenothiazine as sensitizer than using the anthracene derivative. The quantum yield of proton formation was determined by η_{sep} and this value is a function of the intermediate formed. Finally, Eq. 7 suggests that η_{sep} can be determined by inverting the intercepts. Owing to the fact that these are much less accurate than the slopes, this procedure led to a comparatively large scatter of data. The values obtained for η_{sep} with this method changes from 2.9 (phenothiazine) over 0.65 (thioxanthone) to 0.13 (9,10-diphenylanthracene). Values > 1 are impossible with the above model, which we again ascribe to the propagation of errors from the determination of the slope and of the intercept. Nevertheless, it is possible that values > 1 indicate perhaps a chain reaction of acid formation.

Photolysis of the Sensitizer

According to Scheme 3 proton formation is combined with consumption of the sensitizer. For the mechanism of Scheme 3, the quantum yield Φ_{-D} of sensitizer photolysis should be equal to Φ_{H^+}. Using 9,10-diphenylanthracene as sensitizer Φ_{-D} was measured as function of the onium salt concentration (see Fig. 3). In Table 1 the values for $k_e \cdot \tau$ and η_{sep} are summarized with $I_{(1)}$ and $I_{(2)}$ as onium compounds. The slopes of plots according to Eq. 5 have also been compiled in Table 1. In a given solvent these slopes are fairly similar for both iodonium salts, and they show a systematic and consistent trend towards higher values with decreasing solvent polarity. These results are in agreement with the Stern-Volmer constant $k_q \cdot \tau$ determined using fluorescence quenching, which described the overall reaction of the excited sensitizer with the iodonium salt. Furthermore, the solvent polarity determines mainly the intercept, which is a size of the separation efficiency η_{sep} of the primary formed intermediate, the quantum yield Φ_{-D} of sensitizer photolysis and the quantum yield Φ_{H^+} of proton formation. The intercepts differ for the unsubstituted and the lipophilic products in the solvents used. Identical values were obtained only in methanol (MEOH) and the mixture MeOH/DME (4:1). Coherently the values for $k_e \cdot \tau$, determined by dividing and inverting the slope by the intercept, differ for $I_{(1)}$ and $I_{(2)}$. This result is surprising, as the values for $k_q \cdot \tau$ determined using the fluorescence quenching technique are nearly identical for both [9]. Moreover, the values for $k_e \cdot \tau$ are some times larger than the $k_q \cdot \tau$-values. These facts are indigestible with the above model, which permits only $k_e \cdot \tau < k_q \cdot \tau$. Presumable, the sensitizer (or a derivative with nearly identical absorption) were formed, which leads to lower Φ_{-D} and to an erroneous intercept and slope.

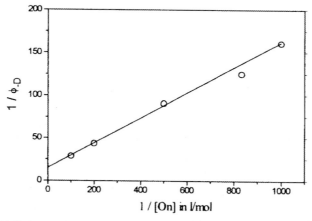

Fig. 3. Plot $1/\Phi_{-D}$ vs $1/$[On] (system : **DPA/I$_{(1)}$** experimental details see Fig. 1)

Table 1. Results of measurements of the sensitizer photolysis Φ_{-D} in several solvents (irradiation wavelength = 365 nm; [**DPA**] = 10^{-4} mol/L; [**On**] = $10^{-2}= 10^{-3}$ mol/L, see also [9])

solvent	$I_{(1)}$		$I_{(2)}$	
	$k_{e^\bullet\tau}$ [a]	η_{sep}	$k_{e^\bullet\tau}$ [a]	η_{sep}
MeCN	530 (95) 0.017	0.11	166 (89) 0.018	0.33
MeOH	71 (72) 0.043	0.33	81 (60) 0.037	0.33
MeOH/DME (4:1)	62 (58) 0.049	0.33	81 (40) 0.037	0.33
MeOH/DME (1:4)	133 (37) 0.060	0.125	48 (36) 0.063	0.33
DME	188 (39) 0.161	0.033	50 (30) 0.080	0.25
DME/HMDS (1:1)	100 (33) 0.143	0.07	[b]	[b]
heptane	245 (22) 1.074	0.0038	[b]	[b]

[a] Second value slope of the plot $1/\Phi_{-D}$ vs $1/$[On], values in parentheses are the Stern-Volmer constant $k_q\cdot\tau$ determined using fluorescence quenching [9], values have the unit l/mol – [b] insoluble in this solvent.

Sensitized Crosslinking Kinetics

As indicated by Eq. 1, the crosslinking rate is related directly to the light intensity. In Fig. 4 an example is given for the crosslinking of VE_1 using the **DPA/I**$_{(1)}$ initiator system. The plot of R_P vs. the incident light intensity demonstrates the rate of polymerization is directly proportional to the absorbed ligth intensity.

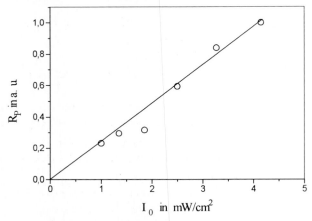

Fig. 4. Plot reaction rate vs incident light intensity (system: VE_1, initiator: **DPA/I**$_{(1)}$, [**DPA**] = $6\ 10^{-3}$ mol /l, [On] = $5\ 10^{-2}$ mol/l)

Thus, the light intensity exponent should be $\beta = 1$. Such a value is typical for a first-order chain termination reaction, found for many cationic induced polymerization processes.

The polymerization starts after an inhibition time t_i, which results from the presence of alkali used to stabilize the monomer. The values of the induction period depend on the inhibitor concentration and at constant inhibitor concentration on the incident light intensity, as schown by Eq. 5.

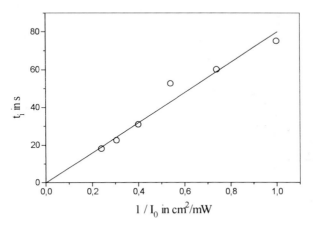

Fig. 5. Plot inhibition time vs. $1/I_o$ (system: VE$_1$, initiator: **DPA/I$_{(1)}$**, [DPA] = 6 10^{-3} mol/l, [On] = 5 10^{-2} mol/l).

In Fig. 5, a correlation between t_i and $1/I_o$ is given. The linearity of the plot demonstrates that Eq. 5 is fully obeyed. Under these conditions it is possible to determine the rate of acid formation (R_i) by means of Eq. 6. The given examples demonstrate that polymerization takes place under nearly ideal conditions. Under these circumstances one must find the kinetics of the sensitized proton formation. Using Eqs. 4 and 6 one can formulate:

$$R'_P = k''(x) \cdot [M]^{\alpha} \cdot (\eta_{sep} \cdot k_e [On]/(1/\tau + k_e [On]))$$

Eq. 8.

where k"(x) is a conversion dependent quantity, which gives in the inverted form a linear correlation between the reciprocal reaction rate and the reciprocal onium concentration:

$$1/R'_P = 1/(k''(x) \cdot [M]^{\alpha} \cdot \tau \cdot \eta_{sep} \cdot k_e [On]) + 1/(k''(x) \cdot [M]^{\alpha} \cdot \eta_{sep})$$

In Fig. 6 an example is given for the crosslinking of VE$_1$ using the **DPA/I$_{(1)}$** initiator system. The linearity of the plot demonstrates that Eq. 9 is fully obeyed. Moreover, it is possible to determine the $k_e \cdot \tau$-value by dividing and inverting the slope of the plot by the intercept. The values obtained (90) used the same order as the value determined using the sensitizer photolysis in DME/HMDS mixture (100). These kinetics directly reflects the generation of protons.

The inhibition time reflects the generation of protons in the system used, see Eq. 5. Usually in cationic crosslinking oxygen has no influence on the crosslinking kinetics. Nevertheless, our results show that the inhibition time was influenced by the air pressure (Fig. 7). Moreover, this influence is stronger at higher light intensities. Presumably, it is an effect of oxygen, which can inhibit a secondary radical-induced cation formation, see Scheme 5 and also [11].

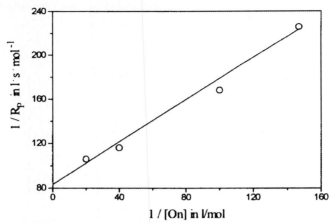

Fig. 6.   Plot inverse reaction rate vs inverse onium salt concentration (system: VE$_1$, initiator: **DPA/I$_{(1)}$**, [DPA] = $6 \cdot 10^{-3}$ mol/l, I$_o$ = 4,15 mW/cm^2)

This reaction starts only in that region of the layer in which both the oxygen initially present is consumed and the number of radicals is higher than the number of oxygen molecules approaching this area by diffusion, see also [7b, 10]. By means of this assumption one can explain the observed nonlinear t_i vs $1/I_o$ at higher light intensities under high intensity irradiation conditions, see also [6]. From the observed correlation one can deduce that at higher light intensities more cations were formed than at lower intensities. This complex relationship was observed in all systems and sensitizer used.

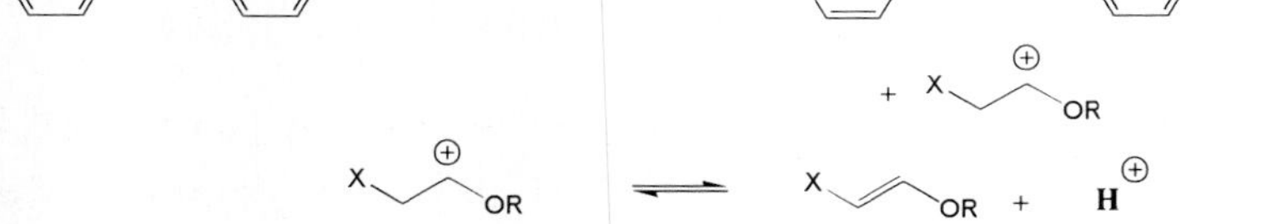

Scheme 5.

According to Eq. 4, it is also to be expected that the photoinitiator used influences the rate of crosslinking via Φ_{H^+} and η_{abs}. When using an iodonium salt/sensitizer as initiating system, Φ_{H^+} is a complex quantity and, among other things, will depend on the sensitizer used. Fig. 8 shows the effect of the sensitizer on R_P using a constant iodonium salt concentration. Surprisingly, there is no correlation between the reaction rate and quantum yield of proton formation. Moreover, the quantum yields were determined under strong monochromatic irradiation conditions with low light intensity. The crosslinking in this example is studied under high intensity irradiation and polychromatic conditions ($\lambda > 340$ nm). Under the conditions used, the measured results reflect the complex influence of the sensitizer of the product $\eta_{abs} \cdot \Phi_{H^+}$. Moreover, chain reactions are possible for proton formation under these conditions.

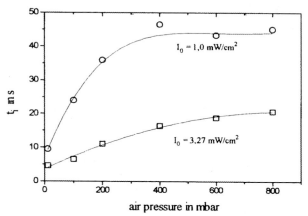

Fig. 7. Inhibition time vs air pressure (system: VE_1, sensitizer: anthracene ($6 \cdot 10^{-3}$ mol/l); $I_{(1)}$ ($5 \cdot 10^{-2}$ mol/l)).

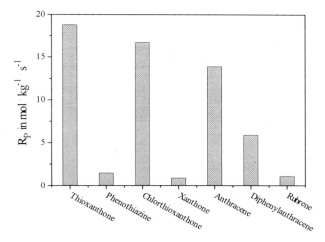

Fig. 8. Reaction of the crosslinking of $VE_{(1)}$ using several sensitizers and $I_{(1)}$ as onium salt ([Sens] = $6 \cdot 10^{-3}$ mol/l; [On] = $5 \cdot 10^{-2}$ mol/l, $\lambda > 340$ nm, $I_o = 57.5$ mW/cm², IR-detection).

Acknowledgement: Financial and material support by the *Deutsche Forschungsgemeinschaft* and by *Wacker-Chemie* (Burghausen) is grateful acknowledged.

References:

[1] R. P. Eckberg, K. D. Riding, *Radiation Curing of Polymeric Materials*, ACS Symposium Series no. 417, American Chemical Society, Washington D.C., **1990**, p. 382.

[2] Ch. Herzig, B. Deubzer, *Radtech '94 North America*, Orlando Florida, Conference Proceedings Vol I, **1994**, p. 635.

[3] a) J. V. Crivello, *Adv. Polym. Sci.* **1984**, *62*, 1.
b) R. J. DeVoe, P. M. Olofson, *Radiation Curing in Polymer Science and Technology, Vol. 2* (Eds.: J. P. Fouassier, J. F. Rabek), Elsevier, London, **1993**, p. 435.
c) N. P. Hacker, *Radiation Curing in Polymer Science and Technology, Vol. 2* (Eds.: J. P. Fouassier, J. F. Rabek), Elsevier, London, **1993**, p. 473.

[4] a) R. J. DeVoe, M. R. V. Sahyun, E. Schmidt, N. Serpone, D. K. Sharma, *Can. J.Chem.* **1988**, *66*.
b) R. J. DeVoe, M. R. V. Sahyun, E. Schmidt, *J. Imaging Sci.* **1989**, *33*, 39.

[5] S. Ulrich, D. Pfeifer, H.-J. Timpe, *J. Prakt. Chem.* **1990**, *332*, 563.

[6] U. Müller, A. Kunze. Ch. Decker, Ch. Herzig, J. Weis, *Radtech '95 Europe*, Maastricht Netherlands, Conference Proceedings of the Academic Day, **1995**, p. 251.

[7] a) H.-J. Timpe, B. Strehmel, *Makromol. Chem.*, **1992**, *192*, 779.
b) U. Müller, *J. Macromol. Sci., Pure Appl. Chem.* **1996**, *A33*, 33.

[8] U. Müller, *J. Macromol. Sci., Pure Appl. Chem.*, **1994**, *A31*, 1905.

[9] G. Eckert, M. Goez, B. Maiwald, U. Müller, *Ber. Bunsenges. Phys. Chem.* **1996**, *100*, 1191.

[10] U. Müller, C. Vallejos, *Angew. Makromol. Chem.*, **1993**, *206*, 171.

[11] P.-E. Sundell, S. Jönnson, A. Hult, *J. Pol. Sci., Part a, Polym. Chem.* **1991**, 1535.

Cationic Photoinitiators
for Curing Epoxy-Modified Silicones

C. Priou

Rhône-Poulenc Chimie
Silicone Division
1 rue des frères Perret, BP22, 69191 Saint Fons cedex, France
Tel.: Int. code + 72737671 – Fax: Int. code + 72736818

Keywords: Cationic Photopolymerization / Photoinitiators

Summary: In recent years, special attention has been given to cationic photopolymerization. The synthesis and characteristics of new cationic photoinitiators are presented. They are readily prepared by reacting a diaryliodonium halide with lithium tetrakis(pentafluorophenyl) borate. They exhibit a suprising high reactivity in apolar media such as silicone resins. The design of these new photoinitiators will be presented. An example of application in silicone release coatings will be given. This system offers advantages over thermal cure systems of lower temperature and faster cure permitting all types of plastic or paper substrates to be coated.

Introduction

During the past twenty years, development of compounds that efficiently initiate polymerization on irradiation have made possible the development of several new commercially important technologies based on these photoinitiators [1]. Their use in UV curable coatings is particularly notable. The most useful photoinitiators that have been explored to date are radical photoinitiators. Many applications today use this technology, in spite of important drawbacks [2]. The recent development of diaryliodonium, triarylsulfonium and ferrocenium salts as highly efficient photoinitiators for cationic polymerization has generated a new class of fast polymerizations.

According to recently published studies concerning the UV/EB curing market in north America [3], silicone release coatings represent one of the largest growth potentials for the coming years. The use of silicones as release coatings for pressure-sensitive adhesives has been well documented over the last 30 years. A typical construction (Fig. 1) consists of a paper or a film substrate upon which the silicone release coating has been cured (chemically crosslinked), an adhesive, and finally the adhesive laminate. The label and adhesive are removed from the release liner and used in subsequent applications. The force required to remove the label from the liner is called the release force.

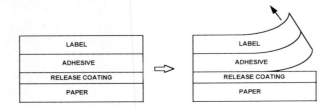

Fig. 1.

The silicone release coating is applied on the paper or film in a liquid form and is then cured, generally by heating.

Silicone release coated papers and films, when laminated to pressure-sensitive adhesive (PSA) coated materials, such as adhesive coated labels and tapes, provide both protection for the adhesive coated product (such as in preparation for slitting, die-cutting, and the like) and they provide subsequent easy release of the release coated paper or film from the adhesive surface immediately prior to its intended final use.

An approach to meet the growing demands of the converting industry has been the intensive development of radiation curable silicones over the past 5-10 years. Because of the relative safety and low cost of radiation curable technologies, new and better methods of converting such solventless silicone compositions have been particularly sought after. Electron Beam (EB) curable silicone release coatings, though very efficient and cost-effective, are very capital investment intensive in regard to purchasing the proper equipment.

A very elegant technology for curing silicone has been developed during the same period. Because silicone polymers themselves are transparent to UV light, radiation curing is particularly suitable for silicone polymers.

In 1983, radiation curable 100 % solvent-free silicone acrylates were introduced into the market [4]. This system provide the opportunity to be cured by either ultraviolet light (UV) or electron beam (EB). Similar in concept to peroxide initiation, silicone acrylate systems employ photoinitiators to generate free radicals and initiate cure, which is based on the polymerization of the acrylic C=C double bond via a radical chain reaction.

Although photoinitiated radical polymerization is widely employed, particularly for surface coatings, it suffers from several inherent problems. Among these deficiencies, the strong inhibiting effects of atmospheric oxygen should be mentioned. The growing radical chain is easily terminated by reaction with oxygen. This unwanted effect can be eliminated by curing in an inert atmosphere.

Over the past several years, considerable efforts have been directed towards the design of new, more efficient photoinitiators with the object of increasing the overall polymerization rates in the presence of air. Anionic polymerization can be immediatly ruled out since it is even more sensitive than radical chain polymerization to the presence of oxygen or impurities. In contrast, cationic polymerization does not suffer from the same limitations and, in fact, possesses certain highly desirable characteristics that makes it an attractive candidate in UV curing. This type of polymerization proceeds at very high rates even at a temperature as low as -100°C. These high reaction rates have been attributed to the lack of appreciable chain terminating side reactions. General Electric and 3M in the seventies introduced new cationic photoinitiators [5] that, by absorbing UV or visible light, undergo a fast photolysis, with formation of a Brφnsted acid capable of initiating the polymerization. This photoinduced cationic curing of epoxides and vinyl ethers, which involves the photogeneration of a Brönsted acid, is today well documented. The most popular photoinitiators are onium salts (iodonium and sulfonium), although the recently introduced ferrocenium salts are gaining popularity. The general structure of these three major classes of cationic photoinitiators is shown in Fig. 2. Diaryliodonium and triarylsulfonium salts can be employed for all known cationally polymerizable monomers, while the ferrocenium salts are best suited for the ring opening polymerization of epoxides.

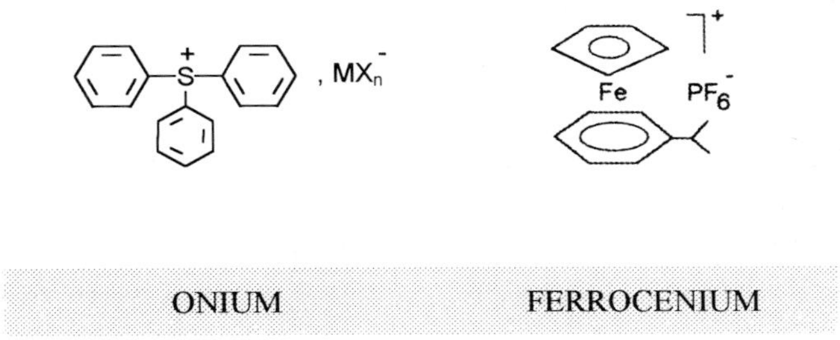

ONIUM   FERROCENIUM

Fig. 2.   Cationic photoinitiators.

Functional silicones used for release coatings applications are 100% solid silicone fluids. Generally, these polydimethylsiloxane polymers are functionalized by epoxy groups, as shown in Fig. 3.

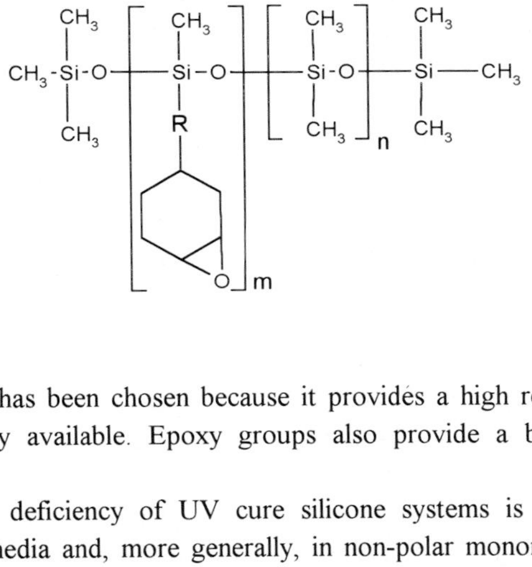

Fig. 3.   Epoxy silicone.

The epoxy functionality has been chosen because it provides a high reactivity and because the epoxy monomers are easily available. Epoxy groups also provide a better anchorage on any substrate.

The major performance deficiency of UV cure silicone systems is the limited solubility of photoinitiators in silicone media and, more generally, in non-polar monomers. Only silicones with high epoxide content can be used with common diaryliodonium photoinitiators. An additional problem associated with simple diaryliodonium salts is their toxicity. Diphenyliodonium hexafluoroantimonate has an oral LD 50 of 40 mg/kg (rats). Progress towards these goals has been reported by Eckberg and Larochelle [6] with the development of the photoinitiator bisdodecylphenyl iodonium hexafluoroantimonate. The long alkyl groups attached to phenyl rings greatly enhance its solubility and its toxicity is lower than for the diphenyl compound. However, this photoinitiator is not a pure compound, but rather a mixture of a large number of isomeric and related alkyl-substituted diaryliodonium salts.

Rhône-Poulenc innovate, in this promising field by proposing a new, patented system especially designed for paper and film release applications.

Design of the Cationic Photoinitiators

In a typical cationic system such as iodonium salts, the photodecomposition of the cationic part of the salt yields a Brφnsted acid [8] and can be represented by the equations shown in Scheme 1, which illustrate the chief photochemical processes involved.

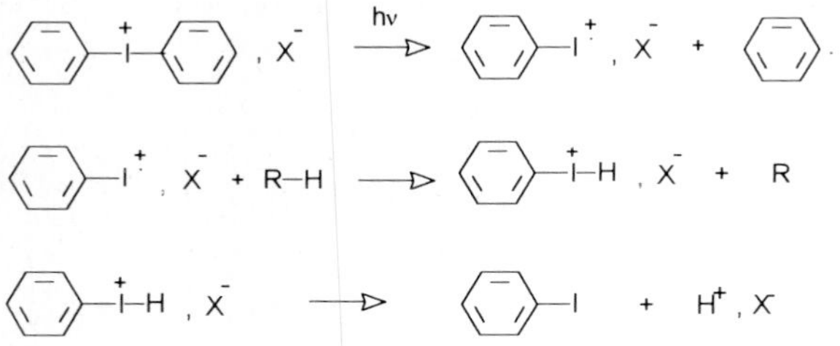

Scheme 1. Photodecomposition of a diaryliodonium salt.

Here, R–H represents a proton source, e.g., solvent or monomer. Considerable light has been shed on the mechanism of the photolyses of these compounds by Hacker and Decktar [9].

During this photochemical process, it has been shown that the anionic part of the salt, X⁻, does not play any role.

The Brφnsted acid, HX, generated attacks an epoxy group to yield a protonated species in the initiation step. The protonated epoxide can attack a second epoxide ring and propagation continues in this fashion until the growing chain is terminated. Termination may be envisioned to occur in several ways. First, the growing epoxy chain can cyclize, resulting in proton abstraction, or, alternatively, chain transfer agents can terminate chain growth and initiate a new chain. It is also possible that impurities can quench the acid.

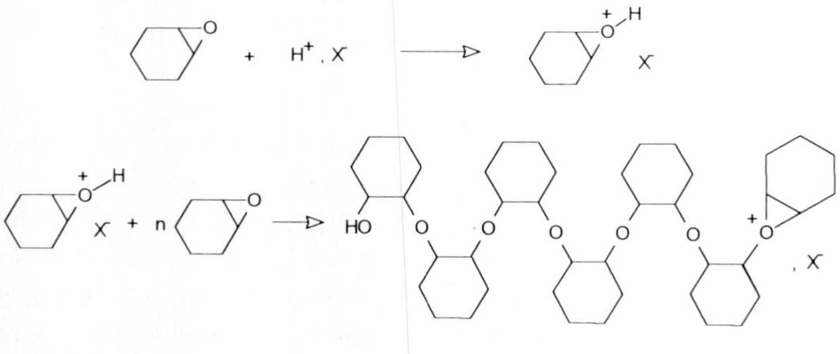

Scheme 2. Propagation steps.

The rate of the propagation step of a cationic polymerization is dependent on the reactivity of the monomer and the stability, or degree of separation, of the ion pair. The larger the negatively charged ion, the more loosely it is bound to the growing cationic chain end and the more active the propagating cationic species in the polymerization. It is for this reason that it is necessary to prepare diaryliodonium salts possessing counterions such as BF_4^-, PF_6^-, AsF_6^-, or SbF_6^-. These anions are called weakly coordinating anions. For this series, the SbF_6^- ion is the largest anion and the most loosely bound while BF_4^- is the smallest and therefore most tightly bound anion. From a practical standpoint, the most reactive onium salt photoinitiators that have yet been found for epoxy-based monomers are those containing the SbF_6^- anion.

On the basis of this analysis, we decided to develop new weakly coordinating anions that will be associated with various diaryliodonium salts, triarylsulfonium salts and ferrocenium salts.

Attention was given to the synthesis of bulky borate anions that seemed to display the required properties. In particular, the tetrakis(pentafluorophenyl)borate anion has focused our interest. This compound is a very stable, crystalline compound and is insensitive to air or moisture. We have found that certain of these salts, as will be described further in this paper, give excellent cationic photoinitiators when associated with a diaryliodonium cation, a triarylsulfonium cation or a ferrocenium cation.

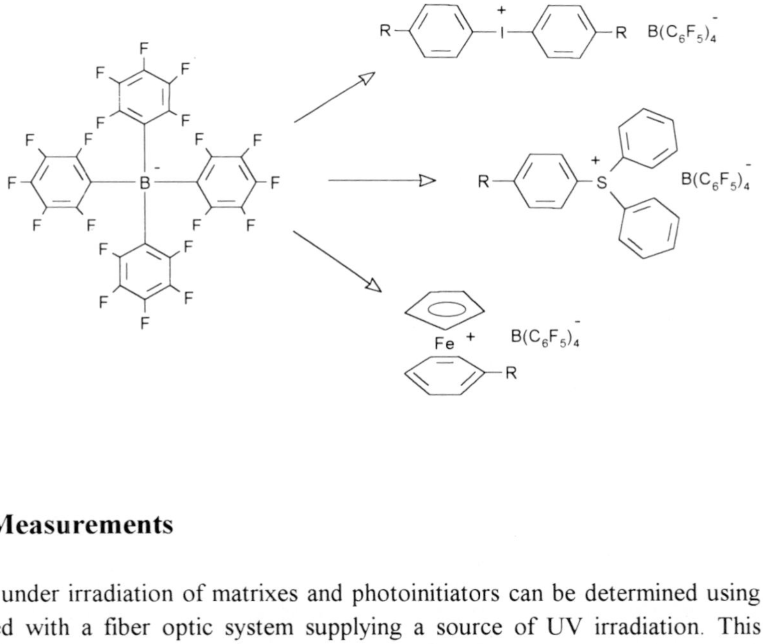

Scheme 3.

Reactivity Measurements

The reactivity under irradiation of matrixes and photoinitiators can be determined using a gel point apparatus fitted with a fiber optic system supplying a source of UV irradiation. This instrument operates by suspending a vibrating needle in the formulation to be monitored. The amplitude of vibration in air is around 0.1 mm. Any resistance to the movement of the needle (for example during curing) will result in a reduced amplitude of vibration. This amplitude attenuation provides the mechanism for monitoring the progress of cure. The cure rate can be monitored in small samples or even in thin films, such as in paint. This instrument has been modified in order to irradiate the bottom of the sample placed in a disposable UV cell. This is then irradiated with a portable exposure system

UVSPOTCURE 100SS (EFOS) equiped with a flexible wand that guides the UV light to the UV cell. A high pressure 100 W mercury short arc lamp is used in this method.

The needle is placed 2 mm above the bottom of the cell and the system is irradiated. Fig. 4 shows a typical result for a cure during which the liquid is transformed into a solid via the gel point. For each formulation, different concentrations of photoinitiator are evaluated in order to obtain the optimum concentration for a 2 mm thickness. In all the experiments, the gelling time t_{50} has been determined at this optimum concentration.

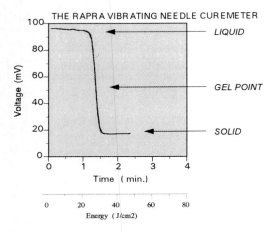

Fig. 4.   Typical gel point record.

The time required for gellation can be converted into energy by using a spot cure meter that measures and displays the intensity of the UV emitted within the bandwith 320-390 nm. Different formulations have been characterized using this apparatus.

Comparison of Photoinitiator Reactivity in Epoxy-silicones

Epoxysilicone UV200 is a functionalized polymer with cycloaliphatic epoxy groups, with a viscosity close to 350 mPa.s. Several photoinitiators have been evaluated in this polymer (Figs. 5 and 6).

The iodonium borate provides a very rapid cure compared with the corresponding antimony product. The difference in reactivity is even greater for the sulfonium salt system (Fig. 6).

Using only 500 ppm of sulfonium borate, curing is faster than for the iodonium salt. Under the same conditions the sulfonium hexafluoroantimonate produces no curing. The commercially available photoinitiator UVI 6974 (SbF_6^- anion), even when used at a concentration of 1.5 %, does not allow curing at such thicknesses.

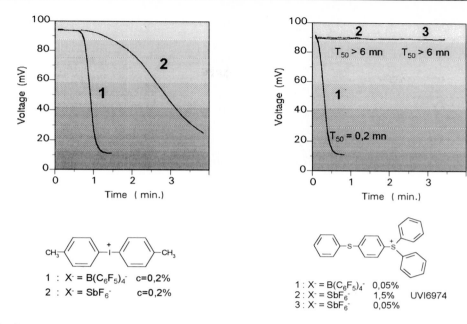

1 : X⁻ = B(C₆F₅)₄⁻   c=0,2%
2 : X⁻ = SbF₆⁻   c=0,2%

Fig. 5. Iodonium salts.

1 : X⁻ = B(C₆F₅)₄⁻   0,05%
2 : X⁻ = SbF₆⁻   1,5%   UVI6974
3 : X⁻ = SbF₆⁻   0,05%

Fig. 6. Sulfonium salts.

Photosensibilization

Several types of photosensitizers have been tested in the silicone epoxy polymer in association with the iodonium salt described above. Benzophenone was found to give poor performance (Fig. 7) whereas 2-ethyldimethoxyanthracene (Fig. 8), chlorothioxanthone (Fig. 9) and isopropyl-thioxanthone (Fig. 10) proved to be highly efficient. One of the major problems with epoxy-silicones is the poor solubility of these photosensibilizors. This problem is currently being studied.

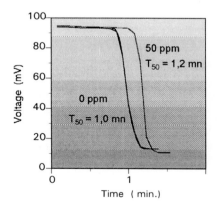

Fig. 7. Benzophenone.

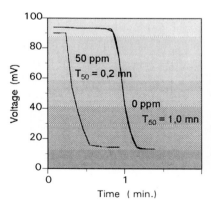

Fig. 8. 2-Ethyldimethoxyanthracene.

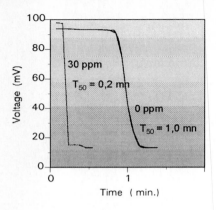

Fig. 9.   Chlorothioxanthone.

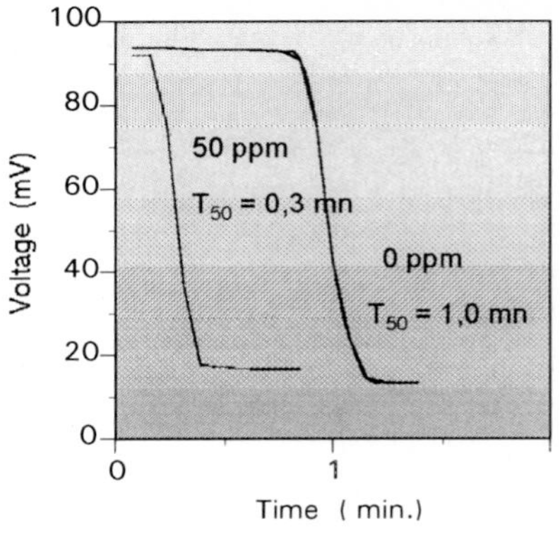

Fig. 10.   Isopropylthioxanthone.

Spectral Studies

The UV spectra of varions photoinitiators have been recorded and the absorbance converted into the molar extinction coefficient ε. Fig. 11 shows three products that differ in the nature of the anion associated with the iodonium cation.

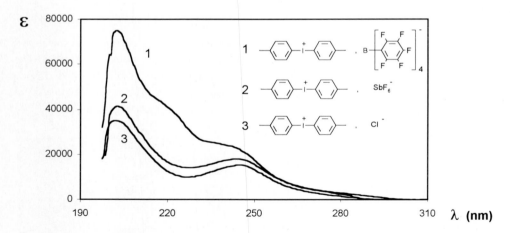

Fig. 11.   UV spectra of ditolyliodonium salts.

Ditolyliodonium chloride **3** and ditolyliodonium hexafluoroantimonate **2** have comparable UV spectra, characterized by a molar extinction coefficient $\alpha = 40000$ at $\lambda = 205$ nm and $\varepsilon = 18000$ at $\lambda = 245$ nm. The borate iodonium has a different spectrum due to the nature of the borate anion. Between 200 nm and 240 nm the molar extinction coefficient is twice that of the photoinitiator with chloride or hexafluoroantimonate anions. Between 240 and 290 nm, there is little difference between the spectra. The same phenomena have been observed with other iodonium cations.

In order to determine the influence of these differences in UV spectra, we used a "chemical filter" to absorb the radiation between 200 and 240 nm. This chemical filter A is a complex non-reactive molecule that has been especially designed for this purpose. It absorbs only between 200 and 240 nm with a molar extinction coefficient close to that of the photoinitiator. If the active range of ε is above 240 nm, there must be no difference in reactivity under UV whether the chemical filter A is present or not. Several formulations with the photoinitiator **1** and the filter A have been prepared with different molar ratios of the two components. These formulations have then been used to catalyse the epoxy-silicone UV200, and the gel times measured using the instrument described above.

Fig. 12 shows the UV spectra of the photoinitiator and different UV spectra corresponding to 1, 0.75, 0.5, and 0.25 equivalents of filter A.

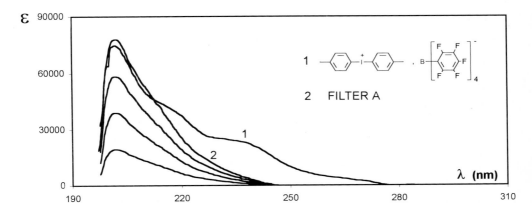

Fig. 12. Iodonium salt and filter A.

It is clear from this Fig. that only the range 200-240 nm is masked by the filter. The different gel times obtained are shown in Fig. 13 and are reported in Table 1.

Table 1. Gel times with iodonium salt.

eq A/1	0	0.25	0.5	0.75	1
t_{50} (mn)	0.6	1.4	4.4	8.8	no curing

The curing is clearly dependent on the UV absorption between 200 and 240 nm with this type of photoinitiator. We verified that there were no chemical reactions between the filter and the photoinitiator by restarting the experiment with a stoichiometric mixture of this photoinitiator and another comparable filter, characterized by an ε five times lower. No problems have been observed during curing under UV. The same spectral studies have been performed with the sulfonium salt (Fig. 14).

As with the iodonium salt, the differences between the two photoinitiators lie between 200 and 240 nm. The absorption at around 300 nm is the same when the antimonate anion is replaced by the borate anion.

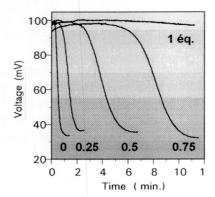

Fig. 13. Iodonium: gel point.

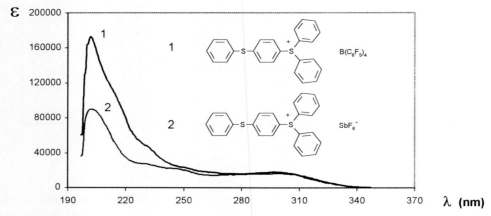

Fig. 14. Sulfonium salt.

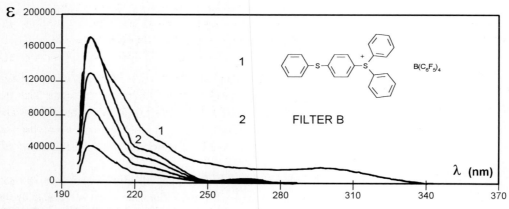

Fig. 15. Sulfonium salt and filter B.

As with the iodonium salt, the differences between the two photoinitiators lie between 200 and 240 nm. The absorption at around 300 nm is the same when the antimonate anion is replaced by the borate anion.

Another chemical filter B was used in order to mask the UV spectra between 200 and 240 nm as previously done with the iodonium salt. This filter has a molar extinction coefficient very close to that of the borate sulfonium salt. Several stoichiometric mixtures have been prepared, as shown in Fig. 15.

The varions gel times obtained with these mixtures are reported in Table 2 and shown in Fig.16.

Table 2. Gel times with sulfonium salt.

eq B/1	0	0.25	0.5	0.75	1
t_{50} (mn)	0.13	0.25	0.33	2.8	no curing

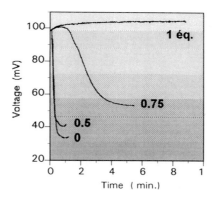

Fig. 16. Sulfonium: gel points.

As with the iodonium salt, this sulfonium salt is sensitive to the effect of the filter B. With one molar equivalent of filter B, there is no cure, even after a long period of irradiation. A noticable loss of reactivity is observed with 0.5 equivalents of filter, wheras 0.1 equivalents of filter A is enough to decrease dramatically the reactivity of the iodonium photoinitiator. The sulfonium photoinitiator seems to be less sensitive to the filter B than the iodonium salt. This difference can certainly be attributed to the relatively high absorption of the sulfonium salt above 280 nm (ε=18000). All these experiments with iodonium or sulfonium salts seem to show that absorption between 200 and 240 nm is responsible for the degradation of the photoinitiator and, therefore, UV lamps must emit at these wavelengths.

Reactivity by DPC

Photoinitiator reactivity studies were also performed with the aid of a Du Pont 930 differential scanning calorimeter. All the DPC experiments were conducted on 0,5-3 mg of the epoxy samples containing the photoinitiator. The samples were placed in aluminum pans and allowed to equilibrate for several minutes at 25°C in the DPC. The samples were then irradiated by opening the shutter and the course of the polymerization followed by recording the heat evolved.

The epoxy-silicone polymer used in these experiments is the same as previously used. DPC studies provide useful information that can be used to compare the chemical rates of the two systems. Some comparative results are reported in Table 3.

Table 3. DPC results.

	$\left[\langle\rangle\right]_2^+ I\ B(C_6F_5)_4^-$	$\left[\langle\rangle\right]_2^+ I\ SbF_6^-$
ENTHALPY	81 J/g	64.2 J/g
PEAK MAXIMUM	2.8 s	31.4 s
INDUCTION TIME	1.4 s	20.3 s

The total theoretical heat of polymerization assuming conversion of all the epoxide functions to polymer was calculated to be 85 J/g. This was calculated from heats of reaction for cycloaliphatic epoxide monomers of 94.5 KJ/mol.

With the borate photoinitiator, polymerization takes place much more rapidly than with the antimonate photoinitiator.

Acknowledgement: We would like to thank those in the *Rhône-Poulenc Release Coating Activity* who have contributed to producing this study, A. Soldat, S. Kerr, B. Benham, J. Richard.
The author expresses sincere thanks to Pr Abadie for recording DPC charts, Pr Fouassier for his contribution and to the *Rhône-Poulenc* company for permitting presentation of this work.

References:

[1]   S. P. Pappas (Ed.), *Radiation Curing Science and Technology*, Plenum, New York, **1992**.

[2]   Chang et al., *Handbook of Coatings Additives* **1992**, *2*, p.1.

[3]   K. Lawson, RADTECH'94 USA.

[4]   D. Vewers, RADTECH EUROPE Edinburgh, **1991**, p. 1.

[5]   J. V. Crivello et al., *J. Rad. Cur.* **1978**, *1*, 5.
      US Pat. 4,173,476 (3M).

[6]   US Pat. 4,279,717 (July 21, **1981**).
      US Pat. 4,421,904 (Dec.20, **1983**).

[7]   F. M. Beringer, R. A. Falk, M. Karniol, I. Lillien, G. Masullo, M. Mausner, E. Sommer, *J. Am. Chem. Soc.* **1959**, *81*, 342.

[8]   J. Knapczyk, J. J. Lubinkowski, W. E. Mc. Even, *Tet. Lett.* **1972**, *35*, 3739.

[9]   J. L. Dektar, N. P. Hacker, *J. Org. Chem.* **1990**, *55*, 639.

Photoconductivity in Polysilylenes: Doping with Electron Acceptors

Alexander Eckhardt, Volkmar Herden, Wolfram Schnabel*
Hahn–Meitner-Institut Berlin GmbH
Bereich Physikalische Chemie
Glienicker Str. 100, D-14109 Berlin, Germany

Keywords: Polysilylene / Photoconductivity / Time-of-Flight Method / Xerographic Method

Summary: Poly(methylphenyl silylene), PMPSi, was doped with compounds of the electron acceptor type. The quantum yield of charge carrier generation $\phi(cc)$ was significantly increased, relative to that of neat PMPSi. The extent of the increase in $\phi(cc)$ is correlated to the increase in electron affinity EA. On the other hand, the hole drift mobility μ is influenced by the dipole moment p of the dopant. μ is increased if $p \approx 0$ and decreased if $p > 0$. Additives of $p > 0$ are likely to interact (probably electrostatically) with holes passing through the polymer matrix. Requirements for dopants of the electron acceptor type regarding optimum improvement of the photoconductive properties of PMPSi are: low p values (close to zero) and high EA values.

Introduction

Polysilylenes of the general structure $-[R^1SiR^2]-$ exhibit electrical conductivity upon irradiation with UV light. The photoconductivity is based on the generation of charge carriers upon irradiation by UV light and the transport of holes [1-4]. Whereas the quantum yield of the charge carrier generation $\phi(cc) = 2.1 \times 10^{-3}$ charges/photon (at the electric field strength $F = 8 \times 10^7$ Vm^{-1}) is rather low, the drift mobility of the holes $\mu \approx 10^{-8}$ m²V^{-1}s^{-1} is rather high [5, 6]. Attempts in recent years to modify the photoconductive properties of polysilylene aimed at increasing $\phi(cc)$ without largely influencing μ. A frequently employed method to improve the photoconductivity of polymers consists in doping polymer films with appropriate additives. Doping of PMPSi with electron scavenging compounds generally resulted in soaring $\phi(cc)$ and plummeting μ values. The only exception was chloranil (2,3,5,6-tetrachloro-1,4-benzoquinone). In this case the addition of the dopant caused an increase in both $\phi(cc)$ and μ. It was realized that chloranil possesses a zero dipole moment ($p = 0$ D) and a high electron affinity. This finding prompted us to study in more detail and more systematically the role of the dipole moment and the electron affinity in influencing hole mobility and quantum efficiency in PMPSi.

Results

Charge carrier generation in undoped and doped PMPSi films

ϕ(cc) values were measured with the aid of the xerographic discharge method as described in the literature [7, 8].

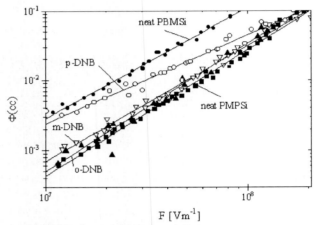

Fig. 1. Quantum yield of charge carrier generation vs. electrical field strength for neat PMPSi and PBMSi, and PMPSi doped with 3 mol% dinitrobenzene (DNB); λ_{inc} = 355 nm.

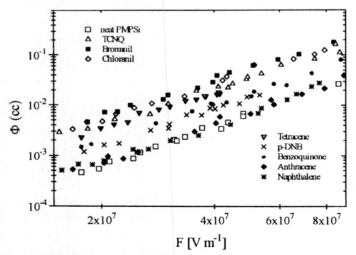

Fig. 2. Quantum yield of charge carrier generation vs. electrical field strength for neat PMPSi and PMPSi doped with 3 mol% electron acceptor; λ_{inc} = 339 nm.

Figs. 1 and 2 show double logarithmic plots of ϕ(cc) vs. the electric field strength F obtained at 295 K. In all cases ϕ(cc) increases with increasing field strength indicating competition between geminate recombination and separation of charge carriers. Notably, charge carrier generation occurs much more effectively in PBMSi than in PMPSi. If phenyl and biphenyl groups act as traps for photogenerated electrons the stabilization of geminate charge carrier pairs should be favored in the case of PBMSi. Biphenyl groups are more polarizable than phenyl groups and, therefore, more prone to electron delocalization. One can see from Table 1 and Fig. 2 that ϕ(cc) increases with increasing electron affinity upon the addition of electron acceptors to PMPSi. In the stage of charge carrier generation the electron can be captured by the additive. Thus the probability of charge carrier recombination is lowered and a higher ϕ(cc) value results. It is remarkable that the dipole moment of the additives has no influence on the quantum efficiency ϕ(cc).

Table 1. Electron affinity, dipole moments of additives, resulting mobility, and resulting quantum yield (3 mol% additive in PMPSi).

Additive	EA [eV][b]	p [Debye]	μ [m² V⁻¹ s⁻¹] at 295 K F = 3.6 x 10⁷Vm⁻¹ λ_{inc} = 347 nm	ϕ(cc) at 295 K F = 8 x 10⁷Vm⁻¹ λ_{inc} = 339 nm	ϕ(cc) at 295 K F = 8 x 10⁷Vm⁻¹ λ_{inc} = 355 nm
none			2.28×10^{-8}	2.10×10^{-2}	1.9×10^{-2}
o-DNB	0.0	6.0	5.02×10^{-9}		2.3×10^{-2}
m-DNB	0.3	3.8	1.42×10^{-8}		2.3×10^{-2}
p-DNB	0.7	0.0	3.10×10^{-8}	5.61×10^{-2}	3.4×10^{-2}
Naphthalene	-0.3	0.0	7.68×10^{-9}	2.85×10^{-2}	
Anthracene	0.5	0.0	2.57×10^{-8}	2.71×10^{-2}	
p-Benzoquinone	0.7	0.0	2.74×10^{-8}	7.81×10^{-2}	
Tetracene	1.0	0.0	3.06×10^{-8}	9.63×10^{-2}	
Chloranil	1.3	0.0	4.12×10^{-8}	1.25×10^{-1}	
Bromanil	1.4	0.0	4.80×10^{-8}	1.62×10^{-1}	
TCNQ[a]	1.7	0.0	5.71×10^{-8}	9.87×10^{-2}	

[a] TCNQ: tetracyanoquinone [b]: [14]

Charge Carrier Mobility in Undoped and Doped Polysilane Films

The hole drift mobility was determined with the aid of the time-of-flight method (TOF) [8] in conjunction with a frequency doubled ruby laser (λ_{inc} = 347 nm, flash length 20 ns).

Figs. 3 and 4 show that semilogarithmic plots of μ vs. the square root of the electric field strength recorded at T = 295 K yield straight lines in accord with the Poole-Frenkel equation [9]. A similar field strength dependence of μ was found in the whole temperature range examined in this work (295-388 K). Fig. 3 demonstrates that, in the case of PMPSi, μ is much larger than in the case of PBMSi. A longer hopping distance in PMPSi may serve for the explanation. Regarding the polaron model, the difference in μ values can be explained in terms of self trapping, which is more important in the case of PBMSi, since biphenyl groups are more polarizable than phenyl groups.

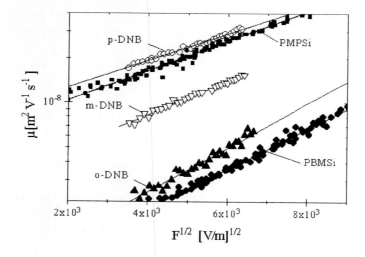

Fig. 3. Mobility vs. the square root of the electrical field strength for neat PMPSi and neat PBMSi, and for PMPSi doped with 3 mol% DNB. $\lambda_{inc} = 347$ nm.

Moreover, Fig. 3 demonstrates that the addition of *m*-DNB ($p = 3.8$ D) and *o*-DNB ($p = 6.0$ D) to PMPSi strongly reduces the charge carrier mobility μ. Similar effects were found in PBMSi. Dipolar compounds ($p > 0$) interact (probably electrostatically) with holes passing through the PMPSi matrix. Vanikov [10] interpreted the action of DNB in polycarbonates doped with aromatic amines on the basis of Marcus´polaron theory [11] and assumed a reorientation of the polar additives in the vicinity of the holes. On the other hand, Borsenberger and Bäßler introduced a disorder model [12, 13]. Accordingly, polar additives cause an enhancement of disorder concerning both the energy states and the position of the transporting sites that results in a diminution of the hole mobility. Dipolar reorientation is considered unimportant.

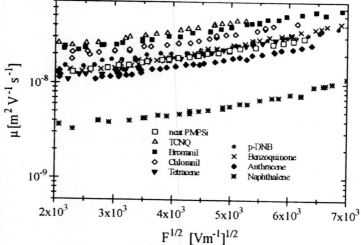

Fig. 4. Mobility vs. the square root of the electrical field strength; for neat PMPSi and PMPSi doped with electron acceptors (3 mol%). $\lambda_{inc} = 347$ nm.

When apolar electron acceptors ($p = 0$) were examined the hole drift mobility increased with increasing electron affinity (EA) as can be seen from Fig. 4 and Table 1. This trend applies to all compounds listed in Table 1. However, compounds having EA exceeding 1.4 eV turned out to be chemically unstable at the high electric fields afforded for these measurements. TCNQ is a typical compound behaving in that way.

Conclusions

The results of this work demonstrate that large, aromatic, and polarizable substituents attached to the polysilane backbone increase the quantum yield ϕ(cc), but decrease the mobility μ of charge carriers. Doping polysilylenes with additives of the electron acceptor type causes an increase in ϕ(cc), which is the more pronounced the higher the electron affinity. However, μ is strongly lowered if the dipole moment of the electron acceptors is larger than zero. Optimum photoconducting properties of polysilylenes (high ϕ(cc) and μ) are attainable with additives of high electron affinity EA and low dipole moment (p: close to zero).

References:

[1] K. A. Klingensmith, J. W. Downing, R. F. Miller, J. Michl, *J. Am. Chem. Soc.* **1986**, *108*, 7438.

[2] M. Fujino, *Chem. Phys. Lett.* **1986**, *135*, 451.

[3] M. A. Abkowitz, M. Stolka, R. J. Weagley, K. M. McGrane, F. E. Knier, *Adv. Chem. Series* **1990**, *224*, 467.

[4] M. Stolka, H. J. Yuh, K. M. McGrane, D. M. Pai, *J. Polym. Sci., Polym. Lett.* **1987**, *25*, 823.

[5] E. Brynda, S. Nespurek, W. Schnabel, *Chem. Phys.* **1993**, *175*, 459.

[6] A. Eckhardt, N. Yars, T. Wollny, S. Nespurek, W. Schnabel, *Ber. Bunsenges. Phys. Chem.* **1994**, *98*, 853.

[7] H. T. Li, P. J. Regensburger, *J. Appl. Phys.* **1963**, *34*, 1730.

[8] J. Mort, D. M. Pai, *Photoconductivity and Related Phenomena*, Elsevier, Amsterdam, **1976**.

[9] J. Frenkel, *Phys. Rev.* **1938**, *54*, 647.

[10] A. V. Vannikov, A. Yu. Kryukov, A. G.Tyurin, T. S. Zhuravleva, *Phys. Stat. Sol. (A)* **1989**, *115*, K47.

[11] R. A. Marcus, *Ann. Rev. Phys. Chem.* **1964**, *15*, 155.

[12] P. M. Borsenberger, H. Bäßler, *Phys. Stat. Sol. B* **1992**, *170*, 291.

[13] H. Bäßler, *Phys. Stat. Sol. B* **1993**, *175*, 15.

[14] G. Briegleb, *Angew. Chem.* **1964**, *76*, 326.

Functionalized Polycarbosilanes as Preceramic Materials

Stephan Back, Heinrich Lang*, Markus Weinmann, Wolfgang Frosch
Institut für Chemie, Lehrstuhl Anorganische Chemie
Technische Universität Chemnitz
Straße der Nationen 62, D-09107 Chemnitz, Germany
Fax: Int. code + (371)5311833
E-mail: heinrich.lang@chemie.tu-chemnitz.de

Keywords: Chlorinated Polycarbosilanes / Hydrido-Polycarbosilanes / Dehydro-genation / Preceramic Materials / Ceramic Materials

Summary: The stepwise synthesis of the polycarbosilanes $(Cl_2SiCH_2CH_2)_n$ **(5)** and $(H_2SiCH_2CH_2)_n$ **(6)** are described. On addition of catalytical amounts of transition metal complexes to polymer **6** dehydrogenation occurs and a further crosslinked carbosilane **(8)** is obtained by formation of new silicon-silicon bonds. Pyrolysis of carbosilane **8** produces a black ceramic material, containing β-SiC together with carbon. The ceramic yield after pyrolysis of **8** is approximately four times the yield obtained when **6** is employed as the starting material. From polymeric **8** preceramic fibers are accessible; subsequent pyrolysis yields ceramic fibers. Moreover, the carbosilane **8** can be utilized as a binder for ceramic powders.

Introduction

For the preparation of technically important metal carbide and metal nitride materials the application of organosilicon compounds as preceramic precursors is advantageous under certain conditions [1-5]. Compared with the conventional metallurgical powder process, one benefit is the utilization of very low process temperatures for the preparation of individual ceramic materials. Another improvement is the high purity of the ceramics obtained from tailor-made preceramic precursors. Usually, after pyrolysis organosilicon compounds afford silicon-containing ceramic powders. Likewise, they can also be used under certain conditions for the production of silicon carbide or silicon nitride fibers.

We here describe the synthesis of the polycarbosilane $(H_2SiCH_2CH_2)_n$ and its use as a preceramic compound for the preparation of silicon carbide in form of a powder or fibers.

Results and Discussion

Synthesis

The preparation of the carbosilanes $(Cl_2SiCH_2CH_2)_n$ (**5**) [6] and $(H_2SiCH_2CH_2)_n$ (**6**) consists of several steps, which are presented in Scheme 1 [7].

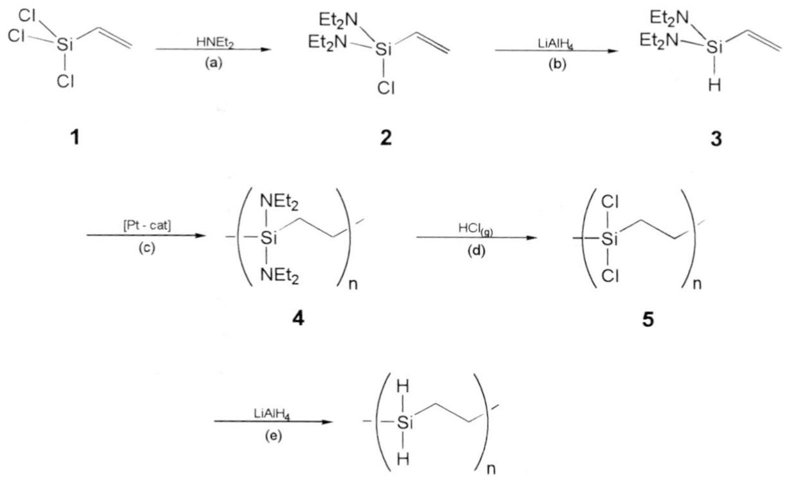

Scheme 1. Stepwise synthesis of the polycarbosilanes **5** and **6** [7].

Treatment of the vinyltrichlorosilane **1** with $HNEt_2$ in diethyl ether at $0°C$ produces the amino-functionalized chlorosilane $(Et_2N)_2Si(CH=CH_2)(Cl)$ (**2**) and $HNEt_2 \cdot HCl$ (reaction (a), Scheme 1). The addition of $LiAlH_4$ to a diethyl ether solution of **2** affords the silane $(Et_2N)_2Si(CH=CH_2)(H)$ (**3**) in high yield (reaction (b), Scheme 1). This monomer contains the constituents for the preparation of an amino-functionalized polycarbosilane. In the presence of catalytic amounts of hexachloroplatinic acid and 2-ethylhexanol (Speier's catalyst) hydrosilylation takes place and intermolecular addition of the Si–H unit to the carbon-carbon double bond entity is observed in a head-to-tail polymerization, delivering the polycarbosilane **4** (reaction (c), Scheme 1). On addition of $HCl_{(g)}$ to the polymer **4** in diethyl ether at $0°C$ spontaneous precipitation of $HNEt_2 \cdot HCl$ is observed and the chloro-functionalized polycarbosilane **5** is obtained as a colorless oil in quantitative yield (reaction (d), Scheme 1). Compound **5** serves as a suitable starting material for a great variety of silicon containing polymers [7b]. With regard to this aspect, the conversion of **5** into **6** is discussed: Treatment of diethyl ether solutions of chlorinated **5** with $LiAlH_4$ at ambient temperature smoothly produces the corresponding polycarbosilane **6** (reaction (e), Scheme 1), which shows high solubility in organic solvents.

In addition to the head-to-tail hydrosilylation described above we could show that monomeric vinyl-functionalized hypervalent silanes can be used for the preparation of polycarbosilanes containing silicon atoms in a trigonal-bipyramidal environment [7b, 8]. Treatment of

$(2\text{-}Me_2NCH_2C_6H_4)(H_2C=CH)SiH_2$ with catalytic amounts of Speier's catalyst at 200°C affords the polymeric carbosilane **7** in excellent yield.

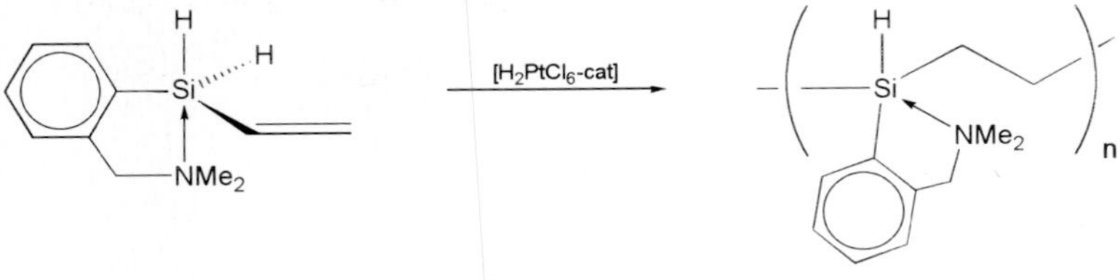

Scheme 2.

Polymeric **7** possesses a linear and regular structure with a $(2\text{-}Me_2NCH_2C_6H_4)SiH$ function. Compared to polymeric **5** and **6** hypervalent **7** is less soluble in common organic solvents.

The linear structure of polymers **5-7** was proved by ^1H, ^{13}C, and ^{29}Si NMR spectroscopic analysis.

Crosslinking

Dehydrogenative coupling of monomeric hydrosilanes is considered to be a potential route to oligo- or polysilanes circumventing the conventional halogenative coupling of halosilanes in presence of alkali metals or alkaline-earth metals [1g, 1k]. Dehydrogenative coupling of hydrosilanes occurs in the presence of transition metal complexes like titanocene, zirconocene or hafnocene derivatives [9].

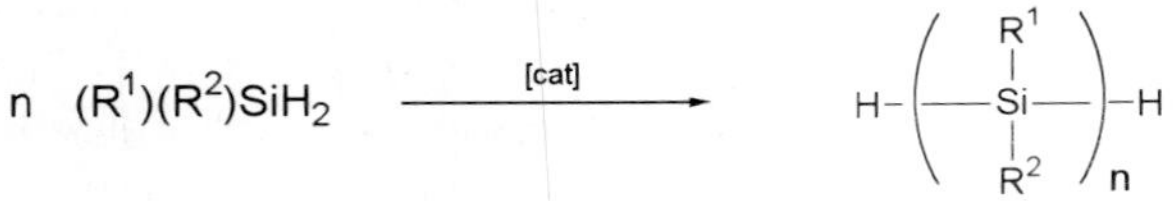

Scheme 3.

The carbosilane $(H_2SiCH_2CH_2)_n$ (**6**) contains silicon-hydrogen bonds, which should enable the desired dehydrogenative crosslinking in the presence of catalytic amounts of transition metal complexes [7b]. Improvement of the ceramic yield on subsequent pyrolysis of the organosilicon starting material utilized is expected. Therefore, we refluxed the organosilicon polymer **6** in the presence of a sufficient amount of for example, $Co_2(CO)_8$ in toluene for several hours. On elimination of dihydrogen the formation of new silicon-silicon bonds occurs and a further crosslinked polymer (**8**) is obtained.

From polymeric **8** preceramic fibers can be obtained by drawing them from saturated toluene-polymer solutions.

The crosslinking of polymer **6** already works with 0.01 weight percent of transition metal complexes applied. By raising the amount of suitable metal complexes in the polymer, hybrid polymers are obtained in which transition metal building blocks are incorporated into the polymeric hydridocarbosilane to form organosilicon-metal polymers [7b, 10].

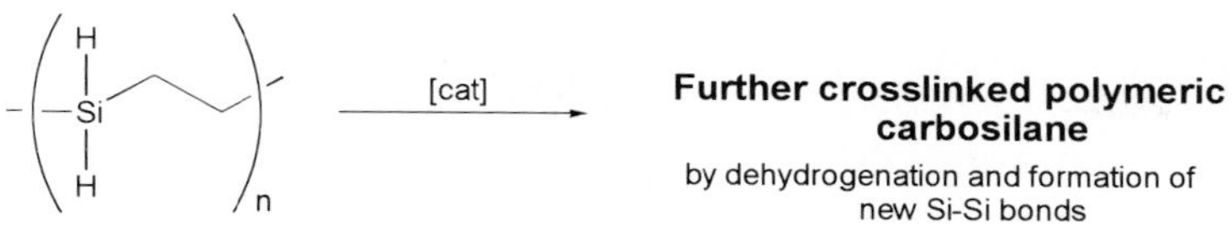

Scheme 4.

The transition metal initialized dehydrogenation of hydridocarbosilanes to yield further crosslinked organosilicon polymers can be monitored by IR- and ^{1}H NMR spectroscopic methods: Decreasing of the intensity of both the Si–H stretching vibration as well as of the Si–H resonance signal is observed.

Pyrolysis

Thermogravimetric studies have been carried out with samples of compounds **6** and **8**. The TGA yields of ceramic material obtained at 1000°C were 17 % for polymer **6** and 65 % for metal-modified polymer **8**, based on the weight of the samples loaded. This result shows that the metal-treated polycarbosilane **8** gives a ceramic yield, that is four times higher compared with that obtained when starting from hydridocarbosilane **6** [7].

Moreover, polymers **6** and **8** were pyrolyzed in bulk. These pyrolysis experiments were performed in a slow stream of nitrogen and the samples were heated to 1000°C at a rate of 10°C min^{-1}, remaining at this temperature for 30 minutes. Both of the ceramic products were black powders and in X-ray powder diffraction studies they showed only broad peaks of low intensity, indicating the presence of mainly amorphous material. To obtain crystalline materials, the ceramic products were heated slowly to 1400°C where they were held for 5 hours. The X-ray powder diffraction showed exclusively sharp peaks, characteristic of β-SiC, respectively. The increased ceramic yield obtained by pyrolysis of the metal modified carbosilane **8**, as compared with the polycarbosilane **6**, can be explained by an increased concentration of carbon as impurity, which was additionally evidenced by elemental analysis.

Acknowledgements: We thank the *Fonds der Chemischen Industrie* and the *Deutsche Forschungsgemeinschaft* for the financial support of this work and *Wacker-Chemie GmbH* (Burghausen) for providing us with a generous gift of various silicon compounds. We are grateful to Dr. C. Limberg and Miss S. Ahrens for many discussions.

References:

[1] For reviews see: a) *Organosilicon Chemistry II* (Eds. N. Auner, J. Weis), VCH, Weinheim, **1992**.

b) N. Auner, H. H. Karsch, J. Weis, *Nachr. Chem. Techn. Lab.* **1993**, *41*, 15.

c) D. W. Bruce, D. O´Hara, *Inorganic Materials*, Wiley, Chichester, **1992**.

d) D. Seyferth, H. Lang, Ch. A. Sobon, J. Borm, H. J. Tracy, N. Bryson, *J. Inorg. Organomet. Polymers* **1992**, *2*, 59.

e) H. Lange, G. Wötting, G. Winter, *Angew. Chem.* **1991**, *103*, 1606; *Angew. Chem. Int. Ed. Engl.* **1991**, *30*, 1579.

f) W. I. Ratzel, *Chem. Ing. Tech.* **1990**, *62*, 86.

g) R. D. Miller, J. Michl, *Chem. Rev.* **1989**, *89*, 1359.

h) J. Ackermann, V. Damrath, *Chem. unserer Zeit* **1989**, *23*, 86.

i) F. Aldinger, H. J. Kalz, *Angew. Chem.* **1987**, *99*, 381; *Angew. Chem. Int. Ed. Engl.* **1987**, *26*, 371.

j) D. Seyferth, *L´actualite chimique* **1986**, 71.

k) R. West, *L´actualite chimique* **1986**, 64.

[2] a) K. J. Thorne, S. E. Johnson, H. Zheng, J. D. Mackenzie, M. F. Hawthorne, *Chem. Mater.* **1994**, *6*, 110.

b) R. Corriu, P. Gerbier, Chr. Cuerin, B. Henner, *Adv. Mater.* **1993**, *5*, 380.

c) P. Sartori, W. Habel, A. Oelschläger, *J. Organomet. Chem.* **1993**, *463*, 47.

d) D. Seyferth, H. Lang, *Organometallics* **1991**, *10*, 551.

e) D. Seyferth, H. Lang, *Appl. Organomet. Chem.* **1991**, *5*, 463.

f) W. Habel, L. Mayer, P. Sartori, *Chem. Ztg.* **1991**, *115*, 301.

g) E. Bouillon, R. Pailler, R. Naslain, E. Bacque, J. P. Pillot, M. Birot, J. Dunogues, P. V. Huong, *Chem. Mater.* **1991**, *3*, 356.

h) M. Ishikawa, Y. Hasegawa, A. Kunai, T. Yamanaka, *J. Organomet. Chem.* **1990**, *381*, C57.

i) B. Boury, L. Carpenter, R. Corriu, H. Mutin, *New. J. Chem.* **1990**, *14*, 535.

j) R. Bortolin, B. Parbhoo, S. D. Brown, *J. Chem. Soc., Chem. Commun.* **1988**, 1079.

k) Y. Hasegawa, K. Okamura, *J. Mater. Sci.* **1986**, *21*, 321.

l) S. Yajima, K. Okumara, J. Hayashi, *J. Am. Ceram. Soc.* **1976**, *59*, 324.

[3] a) W. Uhlig, *Chem. Ber.* **1992**, *125*, 47.

b) K. Nate, M. Ishikawa, H. Ni, H. Watanabe, Y. Saheki, *Organometallics* **1987**, *6*, 1673.

c) R. West, *J. Organomet. Chem.* **1986**, *300*, 327.

[4] a) E. Werner, U. Klingebiel, F. Pauer, D. Stalke, R. Riedel, S. Schaible, *Z. Anorg. Allg. Chem.* **1991**, *596*, 35.

b) E. Bacque, J. P. Pillot, M. Birot, J. Dunogues, P. Lapouyade, E. Bouillon, R. Pailler, *Chem. Mater.* **1991**, *3*, 348.

c) M. Peuckert, T. Vaahs, M. Brück, *Adv. Mater.* **1990**, *2*, 398.

d) K. B. Schwartz, D. J. Rowcliffe, Y. D. Blum, R. M. Laine, *Mat. Res. Soc. Symp. Proc.* **1986**, *73*, 407.

e) D. Seyferth, G. H. Wiseman, *Am. Ceram. Soc.* **1984**, *67*, C132.

[5] *e.g.* a) D. Seyferth, C. Prud´homme, G. H. Wiseman, *Inorg. Chem.* **1983**, *22*, 2163.

b) R. R. Wills, R. A. Markle, S. P. Mukherjie, *Am. Ceram. Bull.* **1983**, *62*, 904.

[6] a) R. J. P. Corriu, W. E. Douglas, E. Layher, R. Shankar, *J. Inorg. Organomet. Polymers* **1993**, *3*, 129.

b) R. J. P. Corriu, D. Leclercq, P. H. Mutin, J M. Planeix, A. Vioux, *Organometallics* **1993**, *12*, 454.

[7] a) M. Weinmann, *Diploma thesis*, Universität Heidelberg, **1992**.

b) M. Weinmann, *Ph. D. thesis*, Universität Heidelberg, **1994**.

[8] H. Lang, M. Weinmann, A. Gehrig, B. Schiemenz, B. Nuber, *Organometallics*, submitted.

[9] a) T. Don Tilley, *Acc. Chem. Res.* **1993**, *26*, 22.

b) T. Nakano, H. Nakamura, Y. Nagai, *Chem. Lett.* **1989**, 83.

c) H. G. Woo, T. Don Tilley, *J. Am. Chem. Soc.* **1989**, *111*, 8043.

d) C. Aitken, J. F. Harrod, E. Samuel, *J. Organomet. Chem.* **1985**, *279*, C11.

[10] S. Blau, *Diploma thesis*, Universität Heidelberg, **1992**.

Novel Polyorganoborosilazanes for the Synthesis of Ultra-High Thermal Resistant Ceramics

Lutz Marc Ruwisch, Wolfgang Dressler, Silvia Reichert, Ralf Riedel*
Fachbereich Materialwissenschaft, Fachgebiet Disperse Feststoffe
Technische Hochschule Darmstadt
Petersenstraße 23, D-64287 Darmstadt, Germany
Fax: Int. code + (6151)166346
E-mail: dg9b@hrzpub.th-darmstadt.de

Keywords: Hydroboration of Vinylchlorosilanes / Preceramic Polymers / Polyborosilazane / Thermal Stability / Synthesis

Summary: Polyorganoborosilazanes possessing varying boron contents have been generated. The synthesized polymers reveal a good shape and thermal stability depending on the preparation conditions and on the state of cross-linking. Ceramics derived from polymers can be processed into additive free fibers and bulk materials. Layers can be produced by dip or spin-on coating. These polyorganoborosilazanes, Si_3BCN and Si_2BCN, are promising candidates for the synthesis of ultra-high thermally resistant ceramic materials, in particular for applications in power generating devices.

Introduction

The production of boron containing Si–C–N-ceramics provides an efficient system for materials with high thermal stability. Previously, the preparation of quaternary systems such as Si–B–C–N was realized by blending boron-containing compounds with polysilazanes. This processing route leads to an inhomogenous elemental distribution in the finally received ceramic. In the last few years several investigations of ceramics derived from different polymers have been performed [1-3]. Therefore, the stoichiometry and basic structure units determine the properties of the final non-oxide ceramic materials [4-5].

In our work we present a novel synthesis of boron containing polysilazanes as a single source precursor for ceramic materials such as fibers, layers and bulk materials in the system Si–B–C–N. We report on the thermal stability of these ceramics depending on their boron content and processing conditions of the molecular polymer before polycondensation. Therefore, we used the hydroboration reaction of dichloromethylvinylsilane with borane-dimethylsulfide and following polycondensation with ammonia [6]. This produces a polymer precursor with definite boron concentration in a Si/B ratio of 3:1. Co-condensation of the intermediate silylchloroborane with dichlorodimethylsilane supplies precursor with a higher Si/B ratio. On the other hand, further treatment with borane produces polymers with a lower Si/B ratio under formation of H_2 and borazine by bond cracking [3].

Preparation of Polyborosilazane

The hydroboration of dichloromethylvinylsilane in Eq. 1 was performed with borane-dimethylsulfide dissolved in toluene in a molar ratio of 3:1 at 0-10°C to exclude side reactions [7]. The first and second additions take place with high α-stereoselectivity, caused by a negative partial charge at the α-C-atom [8]. The third addition in Eq. 1 occurs in β-position due to steric hindrance at boron. In order to obtain polymer **2** we added monochloroborane-dimethylsulfide in toluene to dichloromethylvinylsilane under the same reaction conditions as for **1**, apart from a change in the molar ratio from 3:1 to 2:1 (Eq. 2).

$$3\ \underset{\underset{Cl}{|}}{\overset{\overset{Cl}{|}}{Me-Si}}\!\!-\!\!\diagup\!\!\diagup \quad +\ H_3B\!-\!SMe_2 \quad \xrightarrow[-\ SMe_2]{} \quad \left[Me\!-\!\underset{\underset{Cl}{|}}{\overset{\overset{Cl}{|}}{Si}}\!\!-\!\!\diagdown\!\!-\!\!B\overset{|}{\underset{|}{\!}}\!\!\overset{\bullet}{\diagup}\!\underset{\underset{Cl}{|}}{\overset{\overset{Cl}{|}}{Si}}\!-\!Me \right]_2$$

Eq. 1. Synthesis of **1**.

$$2\ \underset{\underset{Cl}{|}}{\overset{\overset{Cl}{|}}{Me-Si}}\!\!-\!\!\diagup\!\!\diagup \quad +\ H_2BCl\!-\!SMe_2 \quad \xrightarrow[-\ SMe_2]{} \quad \left[Me\!-\!\underset{\underset{Cl}{|}}{\overset{\overset{Cl}{|}}{Si}}\!\!\overset{\bullet}{\diagup}\!BCl \right]_2$$

Eq. 2. Synthesis of **2**.

As reported by Soderquist [9] in 1988 in Eq. 2 only α-hydroboration takes place at the vinyl group as determined by ^{13}C NMR. The reaction produces two chiral α-C-atoms (*) for compounds **1** and **2**. In comparison to ^{29}Si NMR spectra a *meso/RS* ratio of 8:10 was measured for compound **2**. The ^{11}B NMR spectra revealed broad peaks at 84 ppm (**1**) and 69 ppm (**2**). IR spectra attest C–B bonds at 1175 cm^{-1} for **1** and **2**. The complex spectra of **1** do not provide any information about the *meso/RS* ratio.

1

$$\begin{array}{c} SiCl_2Me \\ \diagup \\ B\!-\!\!\diagdown\!\!SiCl_2Me \\ \diagdown \\ SiCl_2Me \end{array}$$

$$\xrightarrow[-\ 6\ NH_4Cl]{\ [THF]\ \ +\ 9\ NH_3\ }$$

$$\left[\begin{array}{c} Me \\ | \\ -Si\!-\!NH- \\ | \\ \diagdown \\ B \\ \diagup\ \diagdown \\ R\ \ \ R \end{array} \right]_n$$

3

2

$$\begin{array}{c} SiCl_2Me \\ \diagup \\ B\!-\!Cl \\ \diagdown \\ SiCl_2Me \end{array}$$

$$\xrightarrow[-\ 10\ NH_4Cl]{\ [THF]\ \ +\ 15\ NH_3\ }$$

$$\left[\begin{array}{c} R \\ \diagdown \\ B\!-\!NH- \\ \diagup \\ R \end{array} \right]_n$$

4

R = -CH(Me)-Si(Me)-NH-

Eq. 3 Synthesis of **3** and **4**.

Subseqent dissolution in THF and condensation of **1** and **2** with ammonia produces polymers **3** and **4** with average molecular weight between 800 and 1100 g/mol. After separation of the byproduct, ammoniumchloride, by filtration and removal of the solvent at 180°C under vacuum, pure polymers in 85% yield were isolated as white solid powders (Eq. 3).

Thermolysis and Crosslinking

In a one-pot reaction polyorganoborosilazane **3** and **4** show a similar weight loss during the thermal gravimetric analysis (Fig. 1) TGA, between room temperature and 1000°C. The initial weight loss at 180°C corresponds to evaporation of oligomeric species. Up to 460°C further crosslinking takes place under evolution of ammonia. The resulting mass loss above 460°C is due to evolution of methane und H_2. As an essential requirement for the subsequent shaping we had to transform the polyorganoborosilazane which melts at 180°C into an infusible solid polymer by crosslinking.

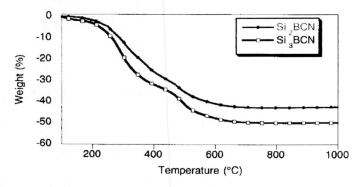

Fig. 1. Thermal gravimetric analysis (STA 429, Netzsch Gerätebau, Germany) of **3** and **4**. Heating rate 1.7 K/min, stationary He in graphite crucibles.

Thermal Stability Above 1000°C

The ultra-high thermal stability of polyorganoborosilazanes **3** and **4** is shown in Fig. 2. This provides a significantly increased thermal stability of about 500°C compared with Si_3N_4 and about 400°C enhanced stability as compared with SiCN. The continuous mass loss of about 2.4 % for **3** and 3.3 % for **4** analyzed in the temperature range 1400-2000°C is caused by the evolution of oxygen-containing gaseous species. Distillation of the reaction educts before ammonolysis results in a different thermal stability of the produced polymer **3a** as shown in the scattered line Fig. 2.

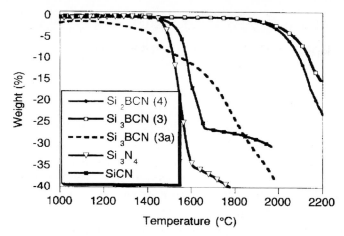

Fig. 2. Thermal gravimetric analysis (STA 429) of polycrystalline α-Si_3N_4 powdered (UBE E10), amorphous SiCN (NCP 200, Nichimen Corp., Japan) compared with **3** and **4**.

References:

[1] E. Werner, U. Klingebiel, S. Dielkus, R. Herbst-Irmer, *Z. Anorg. Allg. Chem.* **1994**, *620*, 1093.

[2] H.-P. Baldus, O. Wagner, M. Janssen, *Mat. Res. Soc. Symp. Proc.* **1992**, *271*, 821.

[3] D. Seyferth, H. Plenio, *J. Am. Ceram. Soc.* **1990**, *73*, 2131.

[4] R. Riedel, G. Passing, H. Schönfelder, R. J. Brook, *Nature* **1992**, *355*, 714.

[5] R. Riedel, *Naturwissenschaften* **1995**, *82*, 12.

[6] J. Bill, A. Kienzle, M. Sasaki, R. Riedel, in: *Ceramics: Charting the Future, Vol. 3B* (Ed.: P. Vincenzini), Faenza, **1995**, p. 1291.

[7] A. Kienzle, *Ph. D. thesis*, Universität Stuttgart, **1994**.

[8] H.-U. Reißig, *Chem. unserer Zeit* **1984**, *2*, 46.

[9] J. A. Soderquist, S.-J. Hwang Lee, *Tetrahedron* **1988**, *44*, 4033.

Precursors for Silicon-Alloyed Carbon Fibers

J. Dautel*, W. Schwarz
Institut für Anorganische Chemie
Universität Stuttgart
Pfaffenwaldring 55, D-70569 Stuttgart, Germany

Keywords: Polyaromatic Mesophase / Silicon-Alloyed Carbon Fibers / Co-pyrolysis / Monomeric and Polymeric Precursors

Summary: We have synthesized a few monomeric silicon compounds bearing benzyl, diphenylmethyl, 9,10-dihydroanthacenyl, 9-fluorenyl, 2,2-diphenylethyl, phenyl or *tert*-butyl groups, which are characterized by NMR data, and in several cases additionally by single crystal X-ray diffraction. Moreover, in some substances the silicon atoms are bridged by $-CH_2-CH_2-$ or $-C\equiv C-$ units. All synthesized compounds, which exclusively contain the elements silicon, carbon, and hydrogen, should be suitable as precursors for silicon-alloyed carbon fibers, while they could be co-pyrolised, if necessary after previous polymerization (catalytic dehydrogenative or by intermolecular hydrosilylation), in a mixture with the polyaromatic mesophase (PA-MP) to yield modified carbons. The latter investigation takes place in cooperation with the group of Prof. K. J. Hüttinger, Karlsruhe. The introduction of silicon improves strength, hardness, and oxidation resistance of these pyrolytically-prepared materials, so that in this way completely new applications are possible.

Introduction

The polyaromatic mesophase (PA-MP) is a nematic, discotic, chemotropic liquid crystal. Owing to its high density (about 1.5 gcm^{-3}), its high carbon yield of about 90 %, and its thermoplasticity, it is unique as a precursor of structure carbons. An important application is the manufacture of high modulus (HM) and ultra-high modulus (UHM) carbon fibers [1]. By alloying with silicon, physical and chemical properties of the materials, such as strength, hardness and oxidation resistance, can be improved. These modified carbons were available by chemical vapor deposition (CVD) processes only up to now. The preparation by liquid phase pyrolysis is novel, economic, and thus opens a completely new field of applications.

Principally, there are two ways of preparing silicon-alloyed PA-MPs: Co-pyrolysis of aromatics or mixtures of aromatics

a) with monomeric silicon-containing precursors,
b) with silicon-containing polymers (polysilanes, polycarbosilanes, and similar materials).

In both cases, all other elements except silicon, carbon, and hydrogen, must be absent in the precursors used in order to avoid impurities in the resulting material. The monomeric precursors should not be highly volatile, so that they do not distill off during MP formation. Moreover, it is favorable if the compounds already contain aromatic substituents. Thus we investigated compounds bearing benzyl, diphenylmethyl, 9,10-dihydro-9-anthracenyl, 9-fluorenyl, 2,2-diphenylethyl, or phenyl groups.

Silicon-containing polymers, which are used as starting materials, should be highly crosslinked in order to avoid thermal depolymerization during MP formation.

At present, in cooperation with the group of Prof. K. J. Hüttinger (Institut für Chemische Technik, Universität Karlsruhe), the applicability of the presented compounds as precursors for modified PA-MPs is being investigated.

Syntheses and NMR data

The benzyl- and diphenylmethyl-substituted compounds (numbering see Table 1) were prepared by reaction of organometallic reagents (Grignard or organo lithium compounds) with the appropriate chloro silanes. The hydrosilanes were obtained by reduction of the corresponding chlorosilanes with sodium dihydrobis(2-methoxyethoxy)aluminate ("Red-Al") in toluene. Whereas the chlorosilanes **14**, **18**, and **21** were synthesized by hydrosilylation using Speier's catalyst, we obtained trichlorodiphenylmethylsilane **5** by the Benkeser reaction [2] (Eq. 1).

$$(H_5C_6)_2C{=}O \quad \xrightarrow[\substack{- (nC_3H_7)_3NH^+Cl^- \\ - 1/n \ "[Cl_2SiO]_n"}]{\substack{+ \ 2 \ HSiCl_3 \\ + \ (nC_3H_7)_3N}} \quad (H_5C_6)_2CH{-}SiCl_3$$

Eq. 1.

The fluoro compounds could be obtained by halogen exchange reactions from the corresponding chloro compounds with zinc fluoride in diethylether. The acetylene-bridged compound **17** was prepared by reaction of di(*tert*-butyl)chlorosilane with sodium acetylide in THF (Eq. 2). Remarkably, during such reactions gaseous acetylene was formed, so that the acetylene bridged compounds arose in a one-step synthesis.

$$2 \ [(H_3C)_3C]_2SiHCl \quad \xrightarrow[\substack{- 2 \ NaCl \\ - \ HC{\equiv}CH}]{+2 \ Na{-}C{\equiv}C{-}H} \quad [(H_3C)_3C]_2SiH{-}C{\equiv}C{-}SiH[C(CH_3)_3]_2$$

Eq. 2.

In all synthesized compounds with Si–H bonds, the $\delta(^{29}Si)$ values exhibit significant dependence on the number of hydrogen bonds to silicon, whereas the $^1J_{(Si,H)}$ parameters appear in the narrow range of 190 up to 200 Hz. So the compounds R–SiH$_3$ show resonances around -50, $(R^1)(R^2)SiH_2$ around -30, and $(R^1)(R^2)(R^3)Si$–H around -10 ppm. Similar tendencies can be regarded in the ^{29}Si chemical shifts of the compounds R–SiF$_3$ (around -60 ppm) and $(R^1)(R^2)SiF_2$ (around -15 ppm). The

$\delta(^{19}F)$ values, however, occur almost independent of the number of Si–F bonds in the range of -140 ppm. $^1J_{Si-F}$ coupling constants can be observed at 280 up to 300 Hz. Compounds bearing Si–Cl bonds as well as compounds containing exclusively Si–C bonds are shifted distinctly to low field compared with Si–F and Si–H derivatives.

Table 1. Characteristic NMR data of the synthesized compounds. 25°C; [d_6]-benzene; ^{1}H: 250.133 MHz, standard Si(CH$_3$)$_4$; ^{19}F: 235.36 MHz, standard CFCl$_3$; ^{29}Si: 39.761 MHz, standard Si(CH$_3$)$_4$; δ, ppm; J, Hz.

No.	formula	$\delta(^{29}Si)$	$\delta(^1H)^{[a]}$	$\delta(^{19}F)$	$^1J_{Si-H}/$ $^1J_{Si-F}$
1	(C$_6$H$_5$–CH$_2$)$_2$SiH(CH$_3$) [3]	-7.86	4.08	—	191.4
2	(C$_6$H$_5$–CH$_2$)$_4$Si [4]	-0.99	—	—	—
3	9,10-Dihydro-9-anthracenyl-SiH(CH$_3$)$_2$ [5]	-11.08	3.96	—	190
4	9-Fluorenyl-SiH(CH$_3$)$_2$	-8.65	4.63	—	191
5	(C$_6$H$_5$)$_2$CH–SiCl$_3$ [2]	5.80	—	—	—
6	(C$_6$H$_5$)$_2$CH–SiF$_3$	-67.71	—	-139.34	289.8
7	(C$_6$H$_5$)$_2$CH–SiH$_3$	-50.95	3.92	—	200.5
8	(C$_6$H$_5$)$_2$CH–SiH(CH$_3$)$_2$ [6]	-9.22	4.41	—	188.8
9	(C$_6$H$_5$)$_2$CH–Si(CH$_3$)$_3$ [7]	3.08	—	—	—
10	[(C$_6$H$_5$)$_2$CH]$_2$SiH(CH$_3$)	-3.40	4.73	—	194.7
11	[(C$_6$H$_5$)$_2$CH]$_2$Si(CH$_3$)$_2$ [8]	4.63	—	—	—
12	[(C$_6$H$_5$)$_2$CH]$_2$Si(CH$_3$)(C$_6$H$_5$)	-3.35	—	—	—
13	[(C$_6$H$_5$)$_2$CH]$_3$Si–CH$_3$	3.06	—	—	—
14	(C$_6$H$_5$)$_2$CH–CH$_2$–SiCl$_3$ [9]	11.65	—	—	—
15	(C$_6$H$_5$)$_2$CH–CH$_2$–SiF$_3$	-59.51	—	-134.77	283.1
16	(C$_6$H$_5$)$_2$CH–CH$_2$–SiH$_3$	-60.90	3.93	—	195.8
17	[(H$_3$C)$_3$C]$_2$SiH–C≡C–SiH[C(CH$_3$)$_3$]$_2^{[b]}$	-9.66	3.85	—	196.1
18	C$_6$H$_5$–SiCl$_2$–CH$_2$–CH$_2$–SiCl$_2$–C$_6$H$_5$ [10]	19.31	—	—	—
19	C$_6$H$_5$–SiF$_2$–CH$_2$–CH$_2$–SiF$_2$–C$_6$H$_5$	-16.01	—	-142.76	299.1
20	C$_6$H$_5$–SiH$_2$–CH$_2$–CH$_2$–SiH$_2$–C$_6$H$_5$	-27.71	4.43	—	193.1
21	C$_6$H$_5$–SiCl$_2$–CH$_2$–CH$_2$–SiCl$_3$ [11]	12.79/18.42$^{[c]}$	—	—	—
22	C$_6$H$_5$–SiF$_2$–CH$_2$–CH$_2$–SiF$_3$	-59.51 (SiF$_3$) -16.97 (SiF$_2$)	—	-140.32 (SiF$_3$) -143.56 (SiF$_2$)	283.3 298.2
23	C$_6$H$_5$–SiH$_2$–CH$_2$–CH$_2$–SiH$_3$	-55.24 (SiH$_3$) -28.21 (SiH$_2$)	3.58 (SiH$_3$) 4.37 (SiH$_2$)	—	193.4 192.2

[a] Only SiH, SiH$_2$, and SiH$_3$ values are given – [b] δ(C≡C) 111.75 (62.896 MHz, standard Si(CH$_3$)$_4$) – [c] not assigned.

X-Ray Structures

Here we show molecule models of a few examples, which were investigated by single crystal X-ray diffraction.

2

10

11

12

13

14

17   **18**

Conclusions

While chlorosilanes are valuable starting compounds for preparing the corresponding hydrosilanes, the fluoro derivatives are only of poor preparative interest, but they are important comparison compounds, particularly in the assignment of the ^{29}Si resonances owing to the characteristic pattern of the Si–F couplings.

Compounds **1-4**, **7-13**, and **16** should be appropriate as monomeric precursors for modified PA-MPs, whereas the derivatives **17**, **20**, and **23** must be polymerized previously. The polymerization can be carried out either dehydrogenative (compounds **20** and **23**) by the use of transition metal complex catalysts or by intermolecular hydrosilylation of Si–H to the –C≡C– unit (compound **17**) to yield crosslinked polymeric precursors for Si-alloyed PA-MPs.

References:

[1] a) K. J. Hüttinger, *Chem.-Ztg.* **1988**, *112*, 355.

   b) M. Braun, A. Gschwindt, W. R. Hoffmann, K. J. Hüttinger, *"Influence of Powder Properties on the Sintering of Polyaromatic Mesophases to High-Strength Isotropic Graphite"*, in: *Ceramic Processing Science and Technology, Ceramic Transactions, Vol. 51* (Eds.: H. Hausner, G. L. Messing, S.-I. Hirano), The American Chemical Society, Westerville, Ohio, **1995**, p. 627.

[2] a) R. A. Benkeser, W. E. Smith, *J. Am. Chem. Soc.* **1969**, *91*, 1556.

   b) R. S. Davidson, R. Ellis, S. Tudor, S. A. Wilkinson, *Polymer* **1992**, *33*, 3031.

[3] a) A. V. Podol'skii, T. G. Cherezova, M. A. Bulatov, *Zh. Obshch. Khim.* **1977**, *47*, 1527; *J. Gen. Chem. (USSR)* **1977**, *47*, 1402; *Chem. Abstr.* **1977**, *87*, 151311z.

   b) L. G. Bell, W. A. Gustavson, S. Thanedar, M. D. Curtis, *Organometallics* **1983**, *2*, 740.

[4] a) H. V. Medoks, *J. Gen. Chem. (USSR)* **1938**, *8*, 291; *Chem. Abstr.* **1938**, *32*, 5392^3.

   b) E. M. Soshestvenskaya, *J. Gen. Chem. (USSR)* **1938**, *8*, 294; *Chem. Abstr.* **1938**, *32*, 5392^4.

[5] R. K. Dhar, A. Sygula, F. R. Fronczek, P. W. Rabideau, *Tetrahedron* **1992**, *48*, 9417.

[6] G. A. Olah, G. Rasul, L. Heiliger, J. Bausch, G. K. S. Prakash, *J. Am. Chem. Soc.* **1992**, *114*, 7737.

[7] a) A. D. Petrov, E. A. Chernyshev, M. E. Dolgaya, *Zh. Obshch. Khim.* **1955**, *25*, 2469; *J. Gen. Chem. (USSR)* **1955**, *25*, 2357; *Chem. Abstr.* **1956**, *50*, 9319d.

b) A. G. Brook, C. M. Warner, M. E. McGriskin, *J. Am. Chem. Soc.* **1959**, *81*, 981.

c) N. Duffant, C. Biran, J. Dunogues, R. Calas, P. Lapouyade, *J. Organomet. Chem.* **1970**, *24*, C51.

d) B. Glaser, H. Noeth, *Chem. Ber.* **1986**, *119*, 3253.

[8] a) J. V. Swisher, J. Perman, P. D. Weiss, J. R. Ropchan, *J. Organomet. Chem.* **1981**, *215*, 373.

b) C. Strohmann, *Chem. Ber.* **1995**, *128*, 167.

[9] A. I. Nogaideli, L. I. Nakaidze, V. S. Tskhovrebashvili, *Zh. Obshch. Khim.* **1974**, *44*, 1763; *J. Gen. Chem. (USSR)* **1974**, *44*, 1730; *Chem. Abstr.* **1974**, *81*, 169584b.

[10] a) G. F. Pavelko, E. G. Rozantsev, *Izv. Akad. Nauk SSSR, Ser. Khim.* **1967**, 2466; *Chem. Abstr.* **1968**, *69*, 77341u.

b) V. V. Semenov, N. F. Cherepennikova, E. Y. Ladilina, M. A. Lopatin, N. P. Makarenko, Y. A. Kurskii, *Metalloorg. Khim.* **1991**, *4*, 1329; *Chem. Abstr.* **1992**, *116*, 59472x.

[11] a) V. M. Vdovin, A. D. Petrov, *Zh. Obshch. Khim.* **1960**, *30*, 838; *J. Gen. Chem. (USSR)* **1960**, *30*, 852; *Chem. Abstr.* **1961**, *55*, 356h.

b) A. I. Petrashko, K. A. Andrianov, G. P. Korneeva, Z. M. Kuptsova, A. P. Moiseenko, L. Z. Asnovich, *Vysokomol. Soedin., Ser. A* **1967**, *9*, 2034; *Chem. Abstr.* **1967**, *67*, 117403h.

c) Z. V. Belyakova, L. K. Knyazeva, E. A. Chernyshev, *Zh. Obshch. Khim.* **1983**, *53*, 1591; *J. Gen. Chem. (USSR)* **1983**, *53*, 1435; *Chem. Abstr.* **1983**, *99*, 175875k.

One-Pot Syntheses of
Poly(diorganylsilylene-*co*-ethynylene)s

Wolfgang Habel, André Moll, Peter Sartori*

Anorganische Chemie
Gerhard Mercator Universität-Gesamthochschule-Duisburg
Lotharstr. 1, D-47048 Duisburg, Germany

Keywords: Silicon / Poly(carbosilanes) / Poly(diorganylsilylene-*co*-ethynylene)s / Di-Grignard

Summary: The synthesis of a series of symmetrically and asymmetrically substituted poly(diorganylsilylene-*co*-ethynylene)s of the general composition $[R^1R^2Si-C\equiv C-]_x$ (with R^1 and R^2 = alkyl-, alkenyl- or alkynyl-groups) was afforded by the reaction of diorganyl-dichlorosilanes with the di-Grignard reagent $BrMgC\equiv CMgBr$ in THF at room temperature. The products, obtained in good yields, have a regular alternating arrangement of ethynylene and silylene units. They were characterized by the usual spectroscopical methods and by gel permeation chromatography. The weight and number averages M_w and M_n reflect the oligomeric character of these compounds.

Introduction

Silicon organic, poly(carbosilane) analogous, unsaturated compounds with interesting optical, electronical, and precursor properties, can be formally constructed by an alternating arrangement of $-SiR_2-$ or $-SiR_2SiR_2-$ units with the organic bridging groups $-C\equiv C-$ or $-C\equiv C-C\equiv C-$ [1, 2]. The resulting products of the combinations are poly(silylene-*co*-ethynylene)s $[SiR_2-C\equiv C-]_x$, poly(disilylene-*co*-ethynylene)s $[(SiR_2)_2-C\equiv C-]_x$, poly(silylene-*co*-butadiynylene)s $[SiR_2-(C\equiv C)_2-]_x$ as well as poly(disilylene-*co*-butadiynylene)s $[(SiR_2)_2-(C\equiv C)_2-]_x$.

The poly(silylene-*co*-ethynylene)s, described in the literature, mainly contain SiPh- or SiMe-groups, rarely SiH- or SiVi-units. These polymers or oligomers are currently produced by the condensation of dihalosilanes with the di-Grignard reagent of acetylene [3, 4], dilithioacetylene [5-7] or disodiumacetylide [8].

$$
\begin{array}{lllll}
x\ BrMg-C\equiv C-MgBr & & & & 2x\ MgClBr \\
x\ Li-C\equiv C-Li & + & x\ Cl_2SiR_2 \longrightarrow [R_2Si-C\equiv C-]_x & + & 2x\ LiCl \\
x\ Na-C\equiv C-Na & & & & 2x\ NaCl \\
\end{array}
$$

Eq. 1. R_2 = Me₂, Ph₂, MeVi, MeH, or MePh

In this paper, we describe the preparation of a series of new poly(diorganylsilylene-*co*-ethynylene)s by the di-Grignard method.

Results and Discussion

The one step preparation of poly(diorganylsilylene-*co*-ethynylene)s was afforded by the reaction of the di-Grignard reagent of acetylene with various symmetrically or asymmetrically substituted dichlorodiorganylsilanes.

$$C_2H_5Br + Mg \longrightarrow C_2H_5MgBr$$

$$2\ C_2H_5MgBr + HC\equiv CH \longrightarrow BrMg–C\equiv C–MgBr + 2\ C_2H_6$$

Eq. 2.

$$x\ Cl_2SiR^1R^2 + x\ BrMg–C\equiv C–MgBr \longrightarrow [R^1R^2Si–C\equiv C–]_x + 2x\ MgClBr$$

Eq. 3. $R^1 = R^2 = CH_3$ (**I**), C_2H_5 (**II**), C_3H_7 (**III**), C_4H_9 (**IV**), C_5H_{11} (**V**), C_6H_{13} (**VI**) iC_3H_7 (**VII**), iC_4H_9 (**VIII**), tC_4H_9 (**IX**), *cyclo*-C_5H_{11} (**X**) or *cyclo*-C_6H_{13} (**XI**);
$R^1 = CH_3$ and $R^2 = CH=CH_2$ (**XII**), $CH_2CH=CH_2$ (**XIII**), $CH(CH_3)CH=CH_2$ (**XIV**) or $CH_2C(CH_3)=CH_2$ (**XV**);
$R^1 = CH=CH_2$ and $R^2 = CH_2CH=CH_2$ (**XVI**);
$R^1 = R^2 = CH=CH_2$ (**XVII**), $CH_2CH=CH_2$ (**XVIII**) or $CH_2C(CH_3)=CH_2$ (**XIX**);
$R^1 = CH_3$ and $R^2 = CH_3(CH_2)_2C\equiv C$ (**XX**), $CH_3(CH_2)_3C\equiv C$ (**XXI**), $CH_3(CH_2)_4C\equiv C$ (**XXII**) or $C_6H_5C\equiv C$ (**XXIII**);
$R^1 = R^2 = CH_3(CH_2)_3C\equiv C$ (**XXIV**).

The viscous products obtained in good yields were soluble in a variety of usual organic solvents. The number average molecular weights in the range M_n = 800-2440 g/mol clearly reflect the oligomeric character of the ethynylene compounds. The polydispersities of the compounds **I-XI** and **XX-XXIV** M_w/M_n = 1.2-7 correspond to a low degree of polymerization process. An increasing polymerization degree was observed for the alkenyl substituted products **XII-XIX**. The polydispersities in the range 20-50 indicate the high polymerization tendency of the alkenyl groups.

The strong absorption bands characteristic of the terminal ethynyl groups were observed in the regions 2038-2047 ($C\equiv C$-stretching) and 3273-3292 cm^{-1} ($\equiv C$–H-stretching). The very weak absorptions at 2107-2127 cm^{-1} indicate the bridging ethynylenes units. The IR spectra of the compounds **XII-XIX** exhibit strong bands at 1595 (**XII**) and 1594 cm^{-1} (**XVII**) for the ethenyl groups and at 1630-1641 cm^{-1} for the other alkenyl substituents. The alkynyl side groups lead to strong absorptions at 2165-2184 cm^{-1} ($C\equiv C$-stretching). The regular alternating arrangement of the ethynylene and silylene units is confirmed by the NMR spectroscopic investigations (Tables 1-4).

The behaviors of the unsaturated poly(diorganylsilylene-*co*-ethynylene)s **XII-XXIV** are dominated by the polymerization tendency, induced thermally or catalytically. These compounds are partially excellent precursors for the fabrication of oxygen-free SiC-fibers. Furthermore, subsequent reactions of the side groups lead to a number of new and interesting derivatives.

Table 1. ^{1}H, ^{13}C, and ^{29}Si chemical shifts δ [ppm, TMS] of poly(dialkylsilylene-*co*-ethynylene)s.

Compound	δ (^{1}H)	δ (^{13}C)	δ (^{29}Si)
I	0.36 (SiCH$_3$) 2.45 (C≡CH)	0.07 (SiCH$_3$), 86.12 (HC≡<u>C</u>Si) 94.53 (H<u>C</u>≡CSi), 110.71 (–C≡C–)	-40.15 (C–SiMe$_2$–C) -41.28 (–SiMe$_2$C≡CH)
II	0.65 (CH$_2$Si), 1.04 (CH$_3$) 2.42 (C≡CH)	5.86 (CH$_2$Si), 6.73 (CH$_3$), 84.10 (HC≡<u>C</u>Si), 95.04 (H<u>C</u>≡CSi) 109.72 (–C≡C–)	-31.51 (C–SiEt$_2$–C) -32.55 (–SiEt$_2$C≡CH)
III	0.75 (CH$_2$Si), 1.07 (CH$_3$) 1.55 (CCH$_2$C) 2.44 (C≡CH)	16.73 (CH$_2$Si), 17.30 (CH$_3$) 17.58 (β-CH$_2$[a], 84.92 (HC≡<u>C</u>Si) 94.86 (H<u>C</u>≡CSi), 110.57 (–C≡C–)	-35.87 (C–SiPr$_2$–C) -36.95 (–SiPr$_2$C≡CH)
IV	0.78 (CH$_2$Si), 0.95 (CH$_3$) 1.3-1.5 (CCH$_2$C) 2.41 (C≡CH)	13.61 (CH$_3$), 13.97 (CH$_2$Si), 25.71 (β-CH$_2$), 26.02 (γ-CH$_2$) 84.74 (HC≡<u>C</u>Si), 94.82 (H<u>C</u>≡CSi) 110.38 (–C≡C–)	-35.14 (C–SiBu$_2$–C) -36.23 (–SiBu$_2$C≡CH)
V	0.75 (CH$_2$Si), 0.94 (CH$_3$) 1.34 (CCH$_2$C) 2.42 (C≡CH)	14.11 (CH$_3$), 14.30 (CH$_2$Si), 22.55 (δ-CH$_2$), 23.18 (β-CH$_2$) 35.15 (γ-CH$_2$), 85.11 (HC≡<u>C</u>Si) 94.99 (H<u>C</u>≡CSi), 110.53 (–C≡C–)	-35.11 (C–SiPe$_2$–C) -36.24 (–SiPe$_2$C≡CH)
VI	0.65 (CH$_2$Si), 0.86 (CH$_3$) 1.30 (CCH$_2$C) 2.42 (C≡CH)	14.22 (CH$_3$), 15.21 (CH$_2$Si) 22.74 (ε-CH$_2$), 23.42 (β-CH$_2$) 31.58 (δ-CH$_2$), 33.10 (γ-CH$_2$) 84.76 (HC≡<u>C</u>Si), 94.69 (H<u>C</u>≡CSi) 110.51 (–C≡C–)	-35.23 (C–SiHx$_2$–C) -36.28 (–SiHx$_2$C≡CH)

[a] CH$_3$–CH$_2$(ε)–CH$_2$(δ)–CH$_2$(γ)–CH$_2$(β)–CH$_2$(α)–Si.

Table 2. ^{1}H, ^{13}C, and ^{29}Si chemical shifts δ [ppm, TMS] of branched and cyclic poly(dialkylsilylene-*co*-ethynylene)s.

Compound	δ (^{1}H)	δ (^{13}C)	δ (^{29}Si)
VII	1.05 (CH), 1.10 (CH$_3$) 2.52 (C≡CH)	13.86 (CH), 16.23 (CH$_3$) 81.26 (HC≡<u>C</u>Si), 95.84 (H<u>C</u>≡CSi) 106.97 (–C≡C–)	-37.06 (C–Si–iPr$_2$–C) -38.21 (–Si–iPr$_2$C≡CH)
VIII	0.70 (CH$_2$), 0.98 (CH$_3$) 1.94 (CH), 2.45 (C≡CH)	24.23 (CH$_2$), 25.04 (CH) 25.89 (CH$_3$), 85.76 (HC≡<u>C</u>Si) 95.07 (H<u>C</u>≡CSi), 111.53 (–C≡C–)	-38.57 (C–Si–iBu$_2$–C) -39.83 (–Si–iBu$_2$C≡CH)
IX	0.96 (CH$_3$), 2.57 (C≡CH)	25.46 (C), 30.18 (CH$_3$) 82.40 (HC≡<u>C</u>Si), 96.13 (H<u>C</u>≡CSi) 108.04 (–C≡C–)	-41.33 (C–Si–tBu$_2$–C) -42.45 (–Si–tBu$_2$C≡CH)
X	1.10 (CH), 1.55 (CH$_2$) 2.33 (C≡CH)	22.00 (CH), 26.85 (<u>C</u>H$_2$CH) 28.56 (CH$_2$<u>C</u>H$_2$CH$_2$) 84.61 (HC≡<u>C</u>Si), 94.80 (H<u>C</u>≡CSi) 110.52 (–C≡C–)	-22.60 (C–Si–*cy*Pe$_2$–C) -23.78 (–Si–*cy*Pe$_2$C≡CH)
XI	1.00 (CH), 1.25 (C<u>H</u>$_2$CH) 1.74 (CH$_2$C<u>H</u>$_2$CH$_2$) 2.42 (C≡CH)	23.09 (CH), 25.89-27.66 (CH$_2$) 83.32 (HC≡<u>C</u>Si), 95.45 (H<u>C</u>≡CSi) 109.64 (–C≡C–)	-29.05 (C–Si–*cy*Hx$_2$–C) -30.01 (–Si–*cy*Hx$_2$C≡CH)

Table 3. ^{1}H, ^{13}C and ^{29}Si chemical shifts δ [ppm, TMS] of poly(methylalkenylsilylene-co-ethynylene)s and poly(dialkenylsilylene-co-ethynylene)s.

Compound	δ (^{1}H)	δ (^{13}C)	δ (^{29}Si)
XII	0.41 (CH$_3$), 2.52 (C≡CH) 6.07 (<u>H</u>C=C<u>H</u>$_2$)	-1.38 (CH$_3$), 84.22 (H<u>C</u>≡CSi) 95.62 (H<u>C</u>≡CSi), 109.00 (–C≡C–) 132.00 (=CH$_2$), 135.96 (CH=)	-47.15 (C–MeSiVi–C) -48.10 (–MeSiViC≡CH)
XIII	0.31 (CH$_3$), 1.74 (CH$_2$) 2.50 (C≡CH) 4.91-5.00 (=CH$_2$) 5.73 (CH=)	-2.57 (CH$_3$), 22.93 (CH$_2$) 84.50 (HC≡CSi), 95.32 (H<u>C</u>≡CSi) 110.14 (–C≡C–), 115.50 (=CH$_2$) 131.72 (CH=)	-40,. 2 (C–MeSial–C) -41.60 (–MeSialC≡CH)
XIV	0.29 (SiCH$_3$), 1.17 (CH$_3$) 1.81 (CH), 2.47 (C≡CH) 4.92-4.98 (=CH$_2$) 5.85 (CH=)	-4.00 (SiCH$_3$), 12.72 (CH$_3$) 26.24 (CH), 84.09 (HC≡CSi) 95.60 (H<u>C</u>≡CSi), 109.88 (–C≡C–) 112.96 (=CH$_2$), 138.23 (CH=)	-35.37 (C–MeSi–αal–C) -36.49 (–MeSi–αalC≡ CH)
XV	0.02 (SiCH$_3$), 1.55 (CH$_2$) 1.70 (CH$_3$), 2.50 (C≡CH) 4.65, 4.91 (=CH$_2$)	0.88 (SiCH$_3$), 24.83 (CH$_2$) 25.29 (CH$_3$), 84.14 (HC≡CSi) 95.99 (H<u>C</u>≡CSi), 109.62 (–C≡C–) 109.90 (=CH$_2$), 145.45 (C=)	-40.52 (C–MeSi–βal–C) -41.48 (–MeSi–βalC≡CH)
XVI	1.73 (CH$_2$), 2.56 (C≡CH) 4.91-5.03 (=CH$_2$^{al}) 5.77 (CH=^{al}) 6.02-6.21 (SiC<u>H</u>=CH$_2$)	22.13 (CH$_2$), 82.57 (HC≡CSi) 96.47 (H<u>C</u>≡CSi), 109.31 (–C≡C–) 116.04 (=CH$_2$^{al}), 130.05 (CH=^{al}) 131.16 (=CH$_2$^{Vi}), 137.12 (CH=^{Vi})	-47.43 (C–ViSial–C) -48.32 (–ViSialC≡CH)
XVII	2.58 (C≡CH) 6.08-6.22 (<u>H</u>C=C<u>H</u>$_2$)	82.35 (H<u>C</u>≡CSi), 96.76 (H<u>C</u>≡CSi) 109.19 (–C≡C–), 130.40 (=CH$_2$) 137.49 (CH=)	-54.03 (C–SiVi$_2$–C) -54.62 (–SiVi$_2$C≡CH)
XVIII	1.89 (CH$_2$), 2.57 (C≡CH) 4.98-5.16 (=CH$_2$) 5.75 (CH=)	22.54 (CH$_2$), 80.93 (HC≡CSi) 96.45 (H<u>C</u>≡CSi), 107.05 (–C≡C–) 117.07 (=CH$_2$), 129.67 (CH=)	-40.41 (C–Sial$_2$–C) -1.20 (–Sial$_2$C≡CH)
XIX	1.77 (CH$_2$), 1.83 (CH$_3$) 2.52 (C≡CH) 4.67-4.73 (=CH$_2$)	24.38 (CH$_2$), 25.19 (CH$_3$) 83.93 (HC≡<u>C</u>Si), 95.93 (H<u>C</u>≡CSi) 110.67 (–C≡C–), 111.03 (=CH$_2$) 140.32 (C=)	-40.29 (C–Si–βal$_2$–C) -41.50 (–Si–βal$_2$C≡CH)

Table 4. ^{1}H, ^{13}C and ^{29}Si chemical shifts δ [ppm, TMS] of poly(methylalkynylsilylene-*co*-ethynylene)s and poly(dihexynylsilylene-*co*-ethynylene).

Compound	δ (^{1}H)	δ (^{13}C)	δ (^{29}Si)
XX	0.44 (SiCH$_3$), 0.96 (CH$_3$) 1.53 (CH$_2$), 2.21 (CH$_2$C≡) 2.50 (C≡CH)	0.88 (SiCH$_3$), 13.38 (CH$_3$) 21.50 (CH$_2$), 21.90 (C̲H$_2$C≡) 77.47 (CC≡C̲Si), 84.02 (HC≡C̲Si) 94.51 (HC̲≡CSi), 108.18 (CC̲≡CSi) 110.53 (–C≡C–)	-66.73 (C–MeSiPi–C) -67.37 (–MeSiPiC≡CH)
XXI	0.47 (SiCH$_3$), 0.90 (CH$_3$) 1.48 (CH$_2$), 2.26 (CH$_2$C≡) 2.52 (C≡CH)	0.71 (SiCH$_3$), 13.41 (CH$_3$) 19.50 (C̲H$_2$C≡), 21.71 (C̲H$_2$CH$_3$) 30.00 (CH$_2$), 78.23 (CC≡C̲Si) 83.74 (HC≡C̲Si), 95.53 (HC̲≡CSi) 108.02 (CC̲≡CSi), 110.41 (–C≡C–)	-66.80 (C–MeSiHi–C) -67.44 (–MeSiHiC≡CH)
XXII	0.48 (SiCH$_3$), 0.89 (CH$_3$) 1.41 (CH$_2$), 2.23 (CH$_2$C≡) 2.50 (C≡CH)	0.74 (SiCH$_3$), 13.81 (CH$_3$) 19.78 (C̲H$_2$C≡), 21.98 (C̲H$_2$CH$_3$) 27.54 (CH$_2$), 30.83 (C̲H$_2$ C$_2$H$_5$) 78.26 (CC≡C̲Si), 83.74 (HC≡C̲Si) 94.53 (HC̲≡CSi), 108.05 (CC̲≡CSi) 110.50 (–C≡C–)	-66.67 (C–MeSiHpi–C) -67.47 (–MeSiHpiC≡CH)
XXIII	1.03 (SiCH$_3$) 2.51 (C≡CH) 7.15, 7.71 (C$_6$H$_5$)	0.50 (SiCH$_3$), 83.31 (HC≡C̲Si) 86.10 (PhC≡C̲Si), 95.63 (HC̲≡CSi) 107.06 (PhC̲≡CSi), 108.14 (–C≡C–) 128.17-132.07 (C$_6$H$_5$)	-65.45 (C–MeSiEPh–C) -66.22 (–MeSiEPhC≡CH)
XXIV	0.89 (CH$_3$), 1.45 (CH$_2$) 2.24 (CH$_2$C≡) 2.51 (C≡CH)	13.21 (CH$_3$), 19.30 (C̲H$_2$C≡) 21.52 (C̲H$_2$CH$_3$), 29.58 (CH$_2$) 75.64 (CC≡C̲Si), 81.98 (HC≡C̲Si) 94.40 (HC̲≡CSi), 109.26 (CC̲≡CSi) 110.44 (–C≡C–)	-95.73 (C–SiHi$_2$–C) -96.23 (–SiHi$_2$C≡CH)

Acknowledgement: We gratefully acknowledge financal support by *Deutsche Forschungsgemeinschaft* and *Fonds der Chemischen Industrie*.

References:

[1]   R. M. Laine, F. Babonneau, *Chem. Mater.* **1993**, *5*, 260.

[2]   M. Birot, J.-P. Pillot, J. Dunguès, *Chem. Rev.* **1995**, *95*, 1443.

[3]   Korshak, A. M. Sladkov, L. K. Luneva, *Izv. Akad. Nauk SSSR, Otd. Khim Nauk* **1962**, 2251.

[4]   D. Seyferth, G. H. Wiseman, Y.-F. Yu, T. S. Targos, C. A. Sobon, T. G. Wood, G. E. Koppetsch, in: *Silicon Chemistry* (Eds.: J. Y. Corey, E. R. Corey, P. P. Gaspar), Ellis Horwood, Chichester, **1988**, 415.

[5]   S. Ijadi-Maghsoodi, T. J. Barton, *Macromolecules* **1990**, *23*, 4485.

[6]   T. J. Barton, *Front. Organosilicon Chem., [Proc. Int. Symp. Organosilicon Chem.]*, **1990**, *9th*, (Pub. **1991**) 3.

[7]   S. Ijadi-Maghsoodi, Y. Pang, T. J. Barton, *J. Polym. Sci., Part A, Polym. Chem.* **1990**, *28*, 955.

[8]   S.-M. Jo, W.-S. Lee, H.-S. Lyu, *Polym. Bull.* **1993**, *30*, 621.

Localization Phenomena of Photogenerated Charge Carriers in Silicon Structures: From Organosilicon Compounds to Bulk Silicon

Thomas Wirschem, Stan Veprek*
Institut für Chemie Anorganischer Materialien
Technische Universität München
Lichtenbergstraße 4, D-85747 Garching, Germany

Keywords: Luminescence / Nanocrystalline Silicon / Localization

Summary: The luminescence properties of a variety of silicon structures are presented. The sizes range from simple organosilicon compounds over nano- and microcrystalline silicon ,to bulk silicon. Time-resolved microwave absorption measurements have been used to show the importance of localization of the photogenerated charge carries for an efficient luminescence in systems with reduced effective dimensionality of the Si matrix.

Introduction

The desire for the integration of optoelectronic devices with silicon microelectronics has lead to the search for Si-based materials that emit light with a high quantum efficiency and fast response time. Crystalline silicon, however, does not show efficient light emission at room temperature, due to its indirect band gap of ≈ 1.1 eV. On the other hand, efficient light emission in the visible region can be observed from nanocrystalline Si separated and passivated with a silica tissue (nc-Si/a-SiO$_2$ composite) as a quasi zero-dimensional system even at room temperature. In order to elucidate the relation between luminescence properties and the spatial confinement of the photogenerated charge carriers, several low-dimensional Si nanostructures were investigated.

Luminescence from Si-Based Materials

Fig. 1 gives a survey of luminescence and optical absorption characteristics of some Si structures. For tetrakis(trimethylsilyl)silane (TTSS) sharp PL and absorption peaks are observed, underlining the molecular nature of this substance.

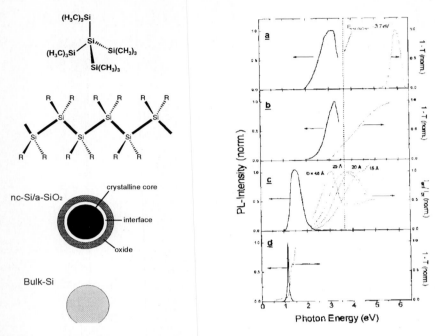

Fig. 1. Photoluminescence (PL) spectra (solid line) and absorption respectively excitation spectra (PLE) (dotted line) of a) tetrakis(trimethylsilyl)silane (TTSS) (soluted in hexane); b) polydimethylsiloxane (PDS); c) oxidized nanocrystalline silicon; d) bulk silicon [T: transmission; D: crystallite size].

In the case of polydimethylsiloxene (PDS) the PL peak is relatively narrow, whereas the absorption spectrum shows a broad onset. Time-resolved PL measurements show that the exciton is localized on 20 - 30 atoms in the chain, respectively on 2-3 atoms in branch structures [1].

The excitation spectra of nc-Si/a-SiO$_2$, prepared by plasma-CVD show a significant and systematic blue shift for decreasing crystallite sizes D, which reflects the increase of the band gap. Due to the distribution of crystallite sizes in the samples the evaluation of the spectra cannot follow conventional methods of extrapolation of $(I_{PLE} \cdot E_{h\nu})^{1/2}$ to zero, as expected for an indirect band gap semiconductor [2].

The luminescence of bulk silicon is orders of magnitude weaker than that of oxidized nc-Si and results from interband recombination. In this luminescence mechanism localization plays no role.

The influence of dimensionality on the optical band gap, especially the widening for chain-like σ-conjugated Si polymers [3] and for small Si nanocrystallites, is demonstrated in Fig. 2. Theoretical predictions from Delerue et al. [4] are also included. Our experimental data agree fairly well with this theory.

The spectral distribution and the position of the maximum of the photoluminescence (PL) for the nc-Si/a-SiO$_2$ composites remain, however, unchanged at 1.5 eV, indicating that the processes of absorption and recombination must be regarded separately.

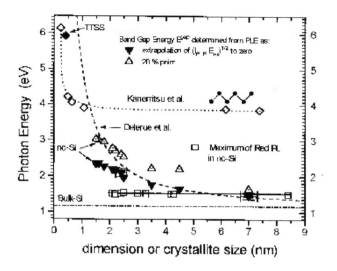

Fig. 2. Band gap widening as a function of dimensionality.

Time-Resolved Microwave Absorption

These measurements are based on the principle that scattering of the free charge carriers in a microwave cavity causes a change of the impedance Q, which can be detected. Unlike the absorption in the infra-red, where the energy but not the momentum of the photogenerated carrier is changed, in this experiment only the momentum of the carriers is changed, but not their energy (less than 50 µeV).

The apparatus is shown schematically in Fig. 3: The samples on silica glass or undoped Si wafer are introduced into the microwave resonator driven by a Gunn oscillator at a frequency of about 10.5 GHz. The cavity with the sample is then tuned to resonance. By choosing an appropriate attenuation a constant signal level is monitored by a Schottky diode. The sample is irradiated by a 3 ns short pulse nitrogen laser and the transient mw-absorption signal is registered by a digitizing oscilloscope.

In Figs 4a + 4b the relationship between efficient luminescence and localization of charge carriers is clearly demonstrated. The sample shows no luminescence but a strong microwave absorption signal due to the delocalized electrons, *as prepared. After oxidation* and treatment in forming gas (FG) there is efficient photoluminescence (PL) but the microwave absorption has vanished due to the localization of the electrons in states at the Si–SiO$_2$ interface (see Fig. 5).

The dependence of the microwave absorption on the size of the Si structures is shown in Fig. 4c (the crystallite size of nc-Si was determined by XRD [5]). TTSS and PDS show *no* signal. Nanocrystalline Si shows absorption only for crystallite sizes (D) above about 5-6 nm. The obvious localization of the excitons for smaller values can either be attributed to the extension of the exciton itself or to the fact that the band gap widening is big enough to cause the localization in surface states (see Fig. 5). The roughly linear increase of the mw signal with increasing crystallite sizes D

demonstrates that the absorption of energy from the microwave field is limited in this regime only by scattering of the charge carriers at the grain boundaries.

Fig. 4b shows that the localization for small crystallites can be overcome by applying a substrate bias during the deposition of the nc-Si films, which induces a compressive stress in the films [6].

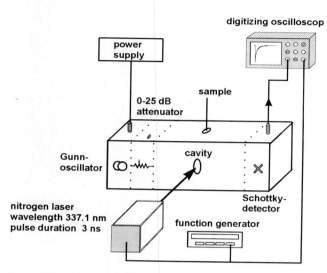

Fig. 3.   Scheme for time-resolved microwave absorption measurements.

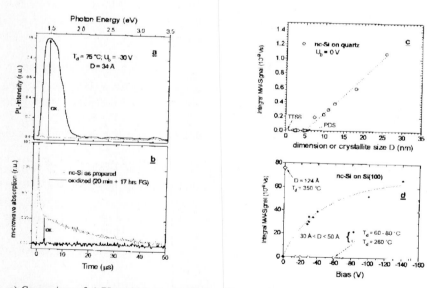

Fig. 4.   a) Comparison of a) PL and b) mw signal of nc-Si; as prepared and after oxidation; c) dependence of mw signal for nc-Si on quarz substrate, PDS and TTSS; d) dependence of mw signal from substrate bias U_b for nc-Si on Si subtrates deposited at different temperatures (T_d).

For low deposition temperature (T_d = 60-80°C) the films deposited at floating potential have tensile stress. As the compressive stress grows in the films with the applied substrate bias the bond dilation within the grain boundaries is reduced and vanishes at a compressive stress of ≈ 30 kbar. In this way the charge carriers can move more easily through the relaxed grain boundaries without being scattered.

Conclusions

For oxidized nc-Si there exist surface states at the Si/SiO$_2$ interface in which photogenerated electron-hole pairs can be localized when the optical band gap has increased enough in small crystallites (Figs. 5a + 5b). From these states radiative recombination of the excitons can occur on a time scale of tens of microseconds. *Ab initio* calculations for one sided oxidized Si planar sheets show that there is a *direct-allowed* transition of 1.66 eV at the Γ point [7]. In TTSS and PDS the sharp PL- and absorption bands together with the results from the microwave absorption indicate that the origin of the luminescence is of a molecular nature caused by localization in the Si backbone.

Those molecular systems, small clusters, nanocrystallites or strongly confined artificial systems may have a potential for a fully Si-based optoelectronic in the future.

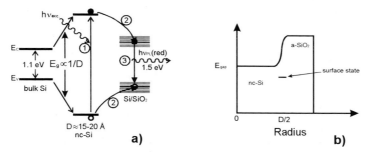

Fig. 5.   a) scenario for PL from nc-Si/a-SiO$_2$; band gap widening; absorption of UV light in the crystallite (1) followed by localization in surface states (2) and recombination (3); b) band gap scheme as function of radius (with surface states in the interfacial region).

Acknowledgement: We would like to thank the *Deutsche Forschungsgemeinschaft* (DFG) for financial support of this work.

References:

[1] Y. Kanemitsu, K. Suzuki, Y. Masumoto, T. Komatsu, K. Sato, S. Kyushin, H. Matsumoto, *Solid State Commun.* **1993**, *86*, 545.

[2] E. J. Johnson, in: *Semiconductors and Semimetals* (Eds.: R. K. Willardson, A. C. Beer), Academic, New York, **1967**, *3*, 153.

[3] Y. Kanemitsu, K. Suzuki, Y. Nakayoshi, Y. Masumoto, *Phys. Rev.* **1992**, *B46*, 3916.

[4] C. Delerue, M. Lannoo, G. Allan, *J. Lumin.* **1993**, *57*, 249.

[5] S. Veprek, F.-A. Sarott, M. Rückschloß; *J. of Non-crystalline Solids*, **1991**, *137-138*, 733.

[6] S. Veprek, F.-A. Sarott, Z. Iqbal, *Phys. Rev.* **1987**, *B36*, 3344.

[7] K. Takeda, K. Shiraishi, *Solid State Commun.* **1993**, *85*, 301.

Functionalized Structure-Directing Agents for the Direct Synthesis of Nanostructured Materials

Peter Behrens

Institut für Anorganische Chemie
Ludwig-Maximilians-Universität München
Meiserstraße 1, D-80333 München, Germany
Tel.: Int. code + (89)5902356 – Fax: Int. code + (89)5902578
E-mail: pbe@anorg.chemie.uni-muenchen.de

Keywords: Materials / Microporous Solids / Organometallic complexes / Porosils / Structure-directed Synthesis

Summary: Organometallic complexes like cobalticinium cations act as structure-directing agents in the synthesis of microporous solids with open frameworks composed of silica. During the hydrothermal synthesis they become occluded by the crystallizing framework. In contrast to the post-synthetic methods that have been used before to prepare organometallic/silica composites, this direct synthesis gives homogeneous, stoichiometric and highly crystalline compounds. In the nanostructured materials, the occluded complexes are spatially organized, and often they are aligned. Also, they are stabilized and protected by the surrounding framework. Organometallic/silica composites prepared by direct synthesis in which the complex acts as structure-directing agent thus offer the possibility to combine the advantages of solid-state and molecular materials.

Introduction

The term "organosilicon chemistry" normally refers to compounds in which silicon atoms are connected to organic groups. Prime examples are the widely used silicones. Their properties are strongly influenced by the nature of the organic groups. Another important class of materials that contain silicon and whose properties are determined by organic species are the zeolite-type microporous solids ("zeotypes") [1]. In these oxidic materials, Si occurs as $[SiO_{4/2}]$ tetrahedra. The organic molecules used in their synthesis [2-4] influence the connectivity of the framework of corner-sharing $[SiO_{4/2}]$ tetrahedra and the type of pores it forms. They are thus designated "structure-directing agents" (SDAs) [5-9]. For example, the silica sodalite structure [10] shown in Fig. 1 was formed by using 1,3,5-trioxane molecules as SDAs [11]. These molecules have been occluded in the cages of the sodalite structure, in which they are disordered over four equivalent sites.

The use of organic molecules in the synthesis of zeotype solids is an especially interesting preparative method for extended inorganic solid materials. Organic molecules cannot survive the harsh conditions of the classical high-temperature route involving reaction of the components in the solid state. Structure-directed synthesis thus belongs to the "soft chemistry" routes for the preparation of solid-state compounds [12, 13].

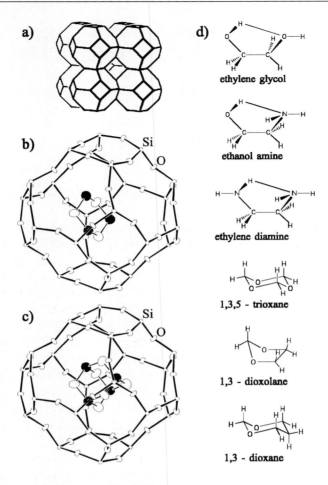

a)

b) Si
O

c) Si
O

d)
ethylene glycol

ethanol amine

ethylene diamine

1,3,5 - trioxane

1,3 - dioxolane

1,3 - dioxane

Fig. 1.   Silica sodalites: a) The topology of the sodalite framework (oxygen atoms omitted); b) position of a 1,3,5-trioxane molecule in the cage of the silica sodalite structure. This corresponds to a snapshot because the trioxane molecule is disordered in the sodalite cage; c) disorder cluster of the trioxane molecule in the sodalite cage; d) organic molecules known to direct the silica sodalite structure.

Silica can react under relatively mild conditions in a solvent like water if hydrothermal conditions are employed. Under these conditions, reactive amorphous silica can be dissolved. If appropriate organic molecules are present during (re-)crystallization, they can be occluded into the growing silica crystals, and the space they occupy cannot be filled by silica. The organic molecules thus inhibit the formation of dense silica phases such as quartz. The inclusion of the organic molecules explains (at least in part) their structure-directing effect. The fact that the silica frameworks formed are metastable and the relatively mild conditions under which these reactions are carried out (temperatures below 200°C, pressures below 500 bar) clearly show that these reactions are kinetically controlled. Fig. 2 compares the thermodynamic stability of microporous silica with that of dense crystalline phases. This compilation is based on the experimental determination of formation enthalpies by Navrotsky and Davis [14]. It shows that the microporous modifications are metastable, but also that their formation enthalpies are only slightly higher than those of the dense phases, so that they can be obtained in kinetically controlled reactions [5].

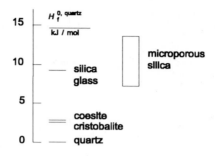

Fig. 2. Enthalpies of formation of silica phases and modifications. Left: dense crystalline phases, right: microporous substances. This compilation is based on the experimental work of Petrovic *et al.* [14].

The relation between SDAs and the zeotype structures they direct is not simply bijective. Usually, the formation of a certain zeotype structure can be directed by more than one SDA [5]. For example, the sodalite structure shown in Fig. 1 can also be directed by molecules other than trioxane, namely, 1,3-dioxolane [15], 1,3-dioxane [16], ethylene glycol [17], ethanolamine [18], and ethylenediamine [18]. Also, depending upon the reaction conditions, an organic molecule may direct the formation of different frameworks [8, 9]. It has therefore been proposed that the term "template", formerly used instead of SDA, should now be reserved for cases of strong structure direction and strong host-guest relationships, in line with its usage in other areas of the natural sciences [5]. A strict condition for a true "template" is that it should not exhibit disorder in the voids of the silica host.

Zeotypes comprise a variety of different framework compositions. The most prominent group are the zeolites, crystalline aluminosilicates containing channel-like voids with apertures from 4 to ca. 15 Å. Zeolites have important uses as catalysts and catalyst supports, as ion-exchangers and in separation processes. Aluminosilicates with cage-like voids are called clathralites in analogy to clathrate hydrates [19].

In analogy to zeolites and clathralites, microporous frameworks formed from pure SiO_2 are designated zeosils and clathrasils, respectively [19]. The term "porosils" comprises these two different kinds of silica modifications [8, 9]. Compared to zeolites, porosils have so far only found limited applications, e.g., as strongly hydrophobic sorbents for organic molecules from wastewaters. However, they represent an important model case for the synthesis of zeotypes, as the chemistry of the synthesis gel or solution is simpler than that of syntheses with two framework atoms (e.g., Al and Si in the case of zeolites). They also are an interesting host material for the organization of advanced materials based on zeotype solids [20].

More than one hundred different zeotype structures, based on different topologies of the linkage of corner-sharing $[TO_{4/2}]$ tetrahedra (T = Al, Si, P ...), have been discovered so far. The various topologies are compiled in the "Zeolite Atlas" [21] under their respective three-letter code (e.g., LTA for "Linde Type A" or zeolite A, MFI for "ZSM-5" or "Mobil Five"). Apart from the framework composition, the high variability of framework structures is related to the fact that all of these compounds are metastable [14] and are formed by kinetic control under the influence of various SDAs.

In this paper, we first discuss the role of SDAs in the hydrothermal syntheses of zeotypes (part 2) and then present some examples of the preparation of advanced materials based on zeolites (part 3). In part 4, recent work from our group that uses functional organometallic complexes to directly synthesize nanostructured materials will be presented.

Structure-Directed Synthesis of Zeotypes

Similar to the porosils, the dense, thermodynamically stable SiO_2 modification α-quartz is also prepared under hydrothermal conditions. However, in the industrial process for the production of quartz, the temperatures are rather high (around 400°C). In this process, NaOH is added as a mineralizer to the aqueous solution to promote dissolution of the silica precursor. The reaction mixtures for the preparation of porosils and other zeotype materials also generally contain a mineralizer, but the reaction conditions are much milder. Synthesis temperatures are below 200°C, typically between 140 and 180°C. Some zeolites can even be prepared from aqueous solutions under reflux at normal pressure. These mild synthesis conditions provide the kinetic control necessary to form metastable products [5-9].

In the synthesis of many zeolites, hydrated alkali or alkaline earth metal cations and water are occluded in the crystallizing framework. To a certain degree, these inorganic cations influence the formation of particular zeolite structures. For example, from Na^+-containing solutions, the aluminosilicate frameworks of zeolite A, zeolite X or sodalite may crystallize, depending on the other conditions of synthesis. The crystallization of another zeolite structure, called Rho, requires the presence of Cs^+ cations. Thus, the (hydrated) inorganic cations may be regarded as SDAs in zeolite syntheses. For these inorganic cations, the mechanism for structure direction is still largely unknown, but the formation of zeolites can also be directed by organic cations or molecules, e.g., by alkylammonium ions. In such cases, the relationship between the SDA and the crystallizing framework is usually more obvious [5].

Organic SDAs are also generally applied in the synthesis of porosils. The most obvious relationship between the crystallizing host framework and the structure-directing guest molecule is that the size of the SDA determines the minimum size of the voids created in the framework [8, 9]. To some extent, there is also correspondence between the shape of the SDA and the shape of the pore system. For example, spherical SDAs usually lead to spherical voids, i.e., they direct the crystallization of clathrasils. SDAs extended along one direction often produce one-dimensional channel systems, and branched SDAs generally give three-dimensionally interconnected pore systems [5-9]. However, crystal structures of porosils determined at room temperature exhibit severe orientational and/or rotational disorder of the guest species in most cases (see Fig. 1c for the disorder in trioxane silica sodalite). At lower temperatures, the guest molecule may choose a preferred conformation and configuration with regard to the framework such that its disorder is reduced. At higher temperatures during the synthesis, however, the orientational and rotational disorder of the SDA molecules will be even more pronounced [5]. This is one of the reasons why the relationships between the different host frameworks and the SDAs is not a simple bijective one. Another reason is that the silica framework is not able to fully adapt its structure to the size and the shape of the molecules it occludes during crystallization, because its own arrangement has to fulfil certain requirements with respect to bond lengths, bond angles, and connectivity [22].

The prime requirement for a molecule to act as an SDA in a porosil synthesis is stability under the hydrothermal synthesis conditions. Although these are very mild compared to conventional high-temperature solid-state syntheses, they are harsh enough to destroy labile organic compounds. In the usual procedure, a reaction mixture consists of a reactive silica source (amorphous silica, *e.g*, fumed or precipitated silica, silica gel), a mineralizer (e.g. NaOH), the SDA and water. Alkaline mineralizers such as NaOH raise the *pH* to 12 or higher. These strongly alkaline solutions are treated at elevated temperatures (up to 200°C) for several days.

In a milder variant, fluoride anions are used as mineralizers [23]. A typical synthesis system then consists of reactive silica, the SDA, water and NH_4F or HF as fluoride source. The *pH* can be adjusted to neutral or to slightly acidic. It is presumed that the silica is mobilized as fluoride or oxofluoride complexes. Fluoride-mediated syntheses usually yield compounds of high crystallinity, but the crystallization times are longer than for alkaline preparations.

The fluoride route also works under "dry" conditions, i.e., in systems to which no water has been added [24, 25]. A typical synthesis system then consists only of silica, the SDA and NH_4F. This solid mixture is filled into the autoclave and heated. Of course, the reaction of silica with NH_4F to form fluoride-containing Si complexes liberates water, but the amount formed is very small. This may be important for SDAs that are sensitive to water.

To act as an SDA, a molecule must fulfil certain other prerequisites besides stability [5-7]. A critical factor is its solubility in the solvent used (usually water). The potential SDA must possess at least a limited solubility to take part in the reaction. However, if the solubility is too high or if the SDA forms strong hydrogen bonds with water, then its tendency to co-crystallize with the silica will be low and it will prefer to stay in solution. For this reason, alcohols are only weak SDAs, for instance. The fact that ethylene glycol and ethanolamine are grouped with other SDAs for the synthesis of silica sodalite in Fig. 1d seems to contradict this conclusion. However, these molecules act as SDAs only in solvothermal synthesis, in which the SDA simultaneously acts as the solvent.

The question of solubility is coupled to the charge and the polarity of the SDA. Anionic molecules cannot act as SDAs because the silicate ions that aggregate around the SDA during crystallization are also negatively charged. Neutral molecules can act as SDAs, but must possess a certain minimum polarity. Primary, secondary, and tertiary amines are usually good SDAs, as are cationic species, e.g., tetraalkylammonium ions. For cationic SDAs, the solubility can be influenced by means of the counteranion (e.g., OH^-, F^-, Cl^-, Br^-, PF_6^-).

Conformational stability of an SDA reduces the disorder of the molecules during the formation of the framework. This may lead to stronger structure-directing effects [8, 9].

Our knowledge on SDAs summarized above has mainly been gathered by indirect, empirical synthetic investigations. In the last few years, more direct studies of the structure-directing effects have been performed [4, 26-28]. For example, Burkett and Davis have shown by $^1H^{29}Si$ CP MAS NMR that in the synthesis of the porosil silicalite, short-range intermolecular interactions of the van der Waals type are established before crystallization starts [26]. The impetus for all these studies is the development of synthetic methods that allow the precise prediction of the outcome of solid-state syntheses, i.e., the idea to design solid-state structures [29]. This goal has so far not been achieved, in contrast to molecular organic and inorganic chemistry, where well-developed synthetic concepts for planning multi-step syntheses are available.

Functional Structure-Directing Agents

A novel approach in the materials chemistry of zeotypes uses their highly organized pore structure for the preparation of advanced materials with potentially interesting magnetic, optical or electronic properties [20]. An illustrative example is the work of Caro, Marlow and coworkers, who prepared a material that exhibits strong second harmonic generation (SHG), a phenomenon in non-linear optics (NLO) that is important for laser applications. The material is based on the aromatic molecule *p*-nitroaniline (*p*NA) [30-32]. This polar molecule also exhibits strong hyperpolarizability, an important precondition for strong SHG. The crystal structure of pure *p*NA, however, is centrosymmetric. Therefore, it cannot produce a macroscopic SHG effect. Caro and co-workers used AlPO$_4$-5, a zeotype with an aluminophosphate framework, as host material for *p*NA. AlPO$_4$-5 was first crystallized hydrothermally with, e.g., tetrapropylammonium hydroxide as SDA. The hexagonal crystals formed were then calcined (heated to 500°C in air) to empty the pores, which are parallel one-dimensional channels with a diameter of about 7 Å. Then *p*NA was inserted into these empty pores from the vapour phase. The resulting inorganic-organic host-guest composite showed strong SHG with a clear dependence on crystal orientation (Fig. 3). This behaviour is in agreement with a structural arrangement in which the *p*NA molecules are aligned in the channels and all their dipoles possess the same orientation [31-32].

The alignment of the molecules is also clear from the visual inspection of *p*NA-loaded AlPO$_4$-5 crystals under polarized light (Fig. 3) [30]. The crystals appear yellow only when the electric field vector E of the linearly polarized light oscillates in the direction of the transition dipole moment of the blue absorption of the *p*NA molecules (which is responsible for their yellow colour). When E and these dipole moments are perpendicular to each other, no absorption occurs in the visible region so that the *p*NA/AlPO$_4$-5 crystals appear colourless [30].

In another example for the synthesis of novel zeotype-based materials, Herron *et al.* prepared small clusters of compound semiconductors in zeolite Y [33]. They first exchanged the Na$^+$ cations of the zeolite (which counterbalance the charge of the framework in the as-synthesized state) against Cd^{2+} cations and then treated these samples with H$_2$S. The authors concluded that they had prepared zeolite Y samples containing ultra-small cubane-like [Cd$_4$S$_4$] clusters (with some of the S replaced by O), further growth of which was prevented by the surrounding zeolite framework. In the optical absorption spectra, these samples exhibited a strong blue-shift compared to bulk CdS. This was ascribed to the quantum confinement of the electronic states of the small cadmium sulfide clusters [33].

Both these examples demonstrate the organizing power of the pore systems of zeotypes on the geometric arrangement of the guest species. Porous frameworks can in addition provide protection of occluded molecules and thus stabilize them thermally and against chemical and photochemical attack [20]. For example, radical molecules that were created inside the void structure of zeolites are stabilized [34], and the stability of organic dye molecules against photobleaching increases dramatically on insertion into various zeotypes [35]. The strongest stabilization is obtained, when the guest species are encapsulated in isolated voids of clathrate-like compounds. These voids are not readily accessible for reactive molecules, not even for small ones (e.g. O$_2$), and the molecules are isolated from each other and cannot form the oligomeric or polymeric products which are typical of their decomposition in the pure bulk state. Of course, the guest species to be protected in these cage-like voids cannot be introduced post-synthetically, but have to be occluded during the synthesis process. For 1,3-dioxolane molecules encapsulated in the isolated cages of a silica sodalite matrix (the molecules had acted as SDAs during the synthesis of this host-guest compound), significant thermal stabilization was observed: The decomposition temperature rises from 730 K for the free molecule to 800 K for the encapsulated one [15].

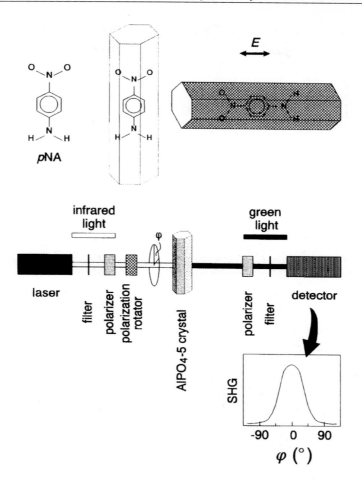

Fig. 3. Nanostructured composite of *p*-nitroaniline (*p*NA, top left) and AlPO$_4$-5. Top right: Anisotropic absorption of the composite: The crystals appear yellow only when the electric field vector E of the linearly polarized light oscillates in the direction of the transition dipole moment of the blue absorption of the *p*NA molecules. When E and these dipole moments are perpendicular to each other, no absorption occurs in the visible region, so that the *p*NA/AlPO$_4$-5 crystals appear colourless. Center: Experiment to measure the second harmonic generation (SHG) of the *p*NA/AlPO$_4$-5 composite. Bottom right: The measured SHG effect is a function of cos φ^4, where φ is the rotation angle of the polarization rotator. This result is in agreement with the expectation if it is assumed that the pNA molecules are aligned in the channels and their dipoles all possess the same orientation. After the work of Caro, Marlow and co-workers [30-32].

The properties of zeotype host-guest composites described above – i.e., spatial organization, protection and stabilization of guest species – will become more important as molecules exhibit further properties that are essential for their use as materials. The well-developed synthetic methods for molecular compounds allow the preparation of designed molecular entities that possess predictable properties. However, no such thoroughly elaborated synthetic methods are available for the construction of organized arrays of functional molecules in their solid structures [36]. This is cumbersome since often the arrangement of the molecules in their solid compounds is detrimental to the effects (e.g. non-linear optical, ferroelectric, electro-optical) that are to be exploited in materials. For example, many structures of molecules in the solid state are centrosymmetric. Also, molecular materials are often sensitive to thermal, chemical or photochemical attack. The spatial organization, protection and stabilization provided by zeotype hosts promise solid-state materials that may help to remedy the disadvantages of molecular materials. The problem then is how to prepare zeotype/molecule (inorganic/organic) host-guest arrangements so as to optimize the material properties of the composite [20].

The synthesis of the pNA/AlPO$_4$-5 and that of the cadmium sulfide/zeolite Y composites are typical of the preparation methods used to generate zeotype-based host-guest materials. These usually involve multi-step syntheses. Typical reaction sequences are:

- preparation of the zeotype with an organic SDA $\rightarrow$ calcination to empty the pore system $\rightarrow$ insertion of functional molecules $\rightarrow$ possibly additional synthesis steps to modify the host-guest arrangement,
  or
- preparation of a zeolite with alkali metal cations $\rightarrow$ ion-exchange against other metal cations $\rightarrow$ further treatment of the material to produce, e.g., metal or semiconductor clusters.

The disadvantage of these multi-step procedures is that each synthesis step after the initial formation of the zeotype is potentially able to damage the zeotype framework. Also, the additional steps may not give quantitative yields, so that a host-guest composite with non-optimum loading and, therefore, non-optimum properties results. For example, in the case of the pNA/AlPO$_4$-5 composite discussed above, the AlPO$_4$-5 crystals are not fully loaded by pNA, probably due to pore blockage of the one-dimensional channels.

In view of the fact that molecules are used as SDAs in the preparation of zeotypes, a possible solution to these synthetic problems is the use of SDAs that already contain the desired molecular function. Highly crystalline composite materials could then be formed directly by hydrothermal syntheses. In an ideal case, no further post-synthesis steps would be necessary, and such materials would possess optimum filling of the pores. As with the materials prepared by multi-step procedures, the zeotype framework would serve to spatially organize, stabilize and protect the guest molecules it occludes during the crystallization.

Classical SDAs such as alkylamines or alkylammonium cations, which were mentioned in the second part of this article, do not possess special properties which would make them of interest for the construction of materials. Aromatic molecules, in contrast, can be equipped with many different properties. For example, they can act as chromophores, may possess large hyperpolarizabilities (for NLO applications such as SHG, cf. the pNA molecule) or stabilize special electronic structures like radicals. However, aromatic molecules are notoriously poor SDAs. One of the few exceptions is pyridine. The hydrothermal treatment of a pyridine – HF – H$_2$O silica solution at 190°C, for instance, results in the formation of large crystals of dodecasil 3C, a clathrasil with small cages. These crystals, which can grow to millimeter size, are acentric at room temperature and exhibit SHG, although the effect is only weak [37].

We are investigating organometallic complexes as potential functional SDAs for the direct synthesis of nanostructured zeotype/molecule host-guest composites. Organometallic complexes can be designed and prepared with a variety of different properties. They too can act as chromophores and possess high hyperpolarizabilities for NLO applications [38]. In addition, metal complexes can be paramagnetic, redox-active or can perform switching functions [39]. Supralattices, *i.e.,* ordered arrangements of oriented functional molecules, should exhibit interesting macroscopic properties. So far, such ordered arrangements have mainly been produced by using Langmuir-Blodgett-type arrangements or by the crystal engineering approach. The inclusion of functionalized molecules into silica hosts in addition promises a stabilization of the functional organometallic SDA.

The use of organometallic molecules as SDAs is promising due to several other interesting properties. Bein [40] has recently reviewed the chemistry and applications of zeotype/organometallic composites that were prepared by conventional post-synthesis modifications. In most cases, the results apply to directly synthesized composite structures as well. For example, organometallic molecules can serve as valuable precursors for catalytic systems that are immobilized inside the zeotype framework and thus profit from the shape-selectivity that makes zeolites themselves important catalytic systems.

Organometallics as a novel class of SDAs also offer the chance to generate new framework topologies, and this has in fact been accomplished in one case (see below). In particular, it might also be possible to discover the "Holy Grail" of synthetic zeolite chemistry, namely, the preparation of a chiral zeotype structure by using metal complexes as SDAs with a chiral shape.

Organometallic Complexes as Structure-Directing Agents

Syntheses

Fig. 4 shows the porosil structures that have so far been synthesized with organometallic complexes as SDAs [25, 41-47]. Among them are three clathrasils, namely, nonasil (NON), octadecasil (AST) and dodecasil 1H (DOH) [41, 42]. All these frameworks possess a large cage and additional smaller cages. The volume of the main cages increases from the concave $[5^8 6^{12}]$ cage of the NON structure through the spherical $[4^6 6^{12}]$ void of the AST structure to the barrel-shaped $[5^{12} 6^8]$ cage of the DOH framework [8, 9, 42]. The smaller cages supplement the large cages to fill the space between them (the sodalite structure shown in Fig. 1 is the only porosil structure that can be built from a single type of cage).

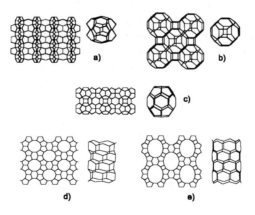

Fig. 4.   Porosil structures (oxygen atoms omitted) that have so far been synthesized with organometallic complexes as SDAs: a) NON topology, b) AST topology, c) DOH topology, d) ZSM-48, e) UTD-1.

In addition to the clathrasils, two zeosils, namely, ZSM-48 [47] and UTD-1, [45, 46] have been prepared (Fig. 4). ZSM-48 has channels lined by ten [SiO$_{4/2}$] tetrahedra, i.e., it contains ten-membered rings. All the clathrasils and ZSM-48 had been synthesized before with organic SDAs. The UTD-1 framework, however, is novel. It is the first one that possesses channels lined by fourteen-membered rings [45, 46]. This is the largest aperture that has so far been found among the porosils. Therefore, its synthesis represents an important breakthrough in the synthesis of porosils, and it is noteworthy that this breakthrough was achieved by using an organometallic complex as SDA.

The first organometallic SDA that was successfully used for the crystallization of zeotypes was the cobalticinium cation. After an early patent report, Shepelev and Balkus and our group independently reported the synthesis of cobalticinium nonasils [41, 42]. In our syntheses, we observed in addition the formation of octadecasil, [42] dodecasil 1H [42] and ZSM-48 [47]. The conditions under which these different products are formed are summarized in Table 1.

In the synthesis system with fluoride ions as mineralizer at 140-160°C, we obtained a NON compound of formula [Co(C$_5$H$_5$)$_2$]F[SiO$_2$]$_{22}$, designated as cobalticinium fluoride nonasil and abbreviated as [Co(cp)$_2$]F-NON. The presence of fluoride ions has been unambiguously shown by single-crystal structural analysis [42] and from ^{19}F MAS NMR spectroscopy [50, 51]. Variation of the synthesis temperature provided access to two other compounds: cobalticinium fluoride octadecasil, [Co(C$_5$H$_5$)$_2$]F[SiO$_2$]$_{10}$ = [Co(cp)$_2$]F-AST, and cobalticinium fluoride dodecasil 1H, [Co(C$_5$H$_5$)$_2$]F[SiO$_2$]$_{34}$ = [Co(cp)$_2$]F-DOH. Between 160 and 180°C, pure AST is produced. Syntheses carried out between 180 and 190°C yield a mixture of crystals of AST and DOH. In the fluoride system, this dodecasil compound can only be obtained as a by-product [42]. A further increase of the synthesis temperature might lead to formation of pure cobalticinium fluoride dodecasil 1H. This route is however, not possible, because the cobalticinium cation decomposes in the reaction mixture above 200°C. The sequence of these compounds with temperature reflects the rule that in clathrasil syntheses, structures with larger maximum cage volumes crystallize preferentially with increasing synthesis temperature [8, 9]. This rule can be traced back to the increasing dynamic motion of the SDA molecules in the synthesis solution, giving rise to increased space requirements in the structure-determining step.

The DOH framework with its voluminous cage can also be stabilized by increasing the size of the SDA [25]. On simply adding one methyl group to each cyclopentadienyl ring of the cobalticinium cation, DOH becomes the only porosil compound that forms over a wide temperature range. The formula of the 1,1'-dimethylcobalticinium fluoride dodecasil 1H is [Co(C$_5$H$_4$(CH$_3$))$_2$]F[SiO$_2$]$_{34}$ = [Co(cpme)$_2$]F-DOH [21]. Similarly, replacing one cyclopentadienyl ring by a benzene ligand (and exchange of Co^{3+} against Fe^{2+} to preserve the +1 charge of the complex) directs the synthesis to the large-cage compound benzene cyclopentadienyl iron fluoride dodecasil 1H, [Fe(C$_5$H$_5$)(C$_6$H$_6$)]F[SiO$_2$]$_{34}$ = [Fe(bz)(cp)]F-DOH [25].

These two DOH compounds have only been obtained by the special "dry-synthesis" method described above [24, 25]. However, the reasons for the use of this method are different in the two cases. The benzene cyclopentadienyl iron complex decomposes quickly in aqueous synthesis systems at higher temperatures. Colourless needles are produced then. Under the conditions of the dry synthesis the complex is reasonably stable, although the synthesis is not easily reproduced. It works best when a reactive silica source such as fumed or precipitated silica is used and when the autoclaves in which the reactions take place have a large free volume. Such a dependence of the product on the starting material or the reaction vessel used is typical of reactions carried out under kinetic control [25].

Table 1. Synthesis conditions and crystallographic characterization of porosils synthesized in our group by using organometallic cations as SDAs.

SDA	synthesis temp. [°C]	frame-work type	designation	a [Å]	b [Å]	c [Å]	α, β, γ [°] if $\neq 90°$
$[Co(C_5H_5)_2]^+$			as PF_6^- salt in the fluoride system system				
	140-160	NON	$[Co(cp)_2]F$-NON	22.170(1)	14.953(1)	13.628(1)	–
	160-190	AST	$[Co(cp)_2]F$-AST	13.144(2)	9.534(1)	9.350(1)	$\beta = 90.36(4)$
	180-190	DOH	$[Co(cp)_2]F$-DOH	13.762(5)	13.762(5)	11.115(4)	$\gamma = 120.00$
			as PF_6^- salt in the dry synthesis system				
	150-180	AST	$[Co(cp)_2]F$-AST	13.139(2)	9.532(1)	9.346(1)	$\beta = 90.27(4)$
			as Cl^- salt in the basic synthesis system				
	140-180	NON	$[Co(cp)_2]$-NON	22.181(9)	14.872(4)	13.830(5)	–
	160-180	ZSM-48	$[Co(cp)_2]$-ZSM-48	14.30(3)	20.36(6)	8.36(1)	–
	175	DOH	$[Co(cp)_2]$-DOH	13.799(2)	13.799(2)	11.151(2)	$\gamma = 120.00$
$[Co(C_5H_4CH_3)_2]^+$			as PF_6^- salt in the dry synthesis system				
	140-180	DOH	$[Co(cpme)_2]F$-DOH	13.798(1)	13.798(1)	11.200(1)	$\gamma = 120.00$
$[Fe(C_6H_6)(C_5H_5)]^+$			as PF_6^- salt in the dry synthesis system				
	150-170	DOH	$[Fe(bz)(cp)]F$-DOH	13.813(2)	13.813(2)	11.197(2)	$\gamma = 120.00$

In contrast to the iron complex, the 1,1'-dimethylcobalticinium cation is stable under the conditions of the hydrothermal fluoride synthesis in aqueous solution. However, no solid product is formed in such a system. Possibly, the dimethylcobalticinium cation is only a weak SDA and is not able to form a porosil when diluted in an aqueous solution. Under dry-synthesis conditions, its concentration attains the maximum value possible [25]. When the dry-synthesis method is applied with the cobalticinium cation as SDA, a $[Co(cp)_2]F$-AST compound is the only product. It is in all respects similar to that obtained in aqueous fluoride syntheses.

As is also typical of kinetically controlled syntheses, variations in the reaction conditions can change the product distribution. This is exemplified by syntheses with NaOH as mineralizer. The thermal stability of the cobalticinium cation in basic media is somewhat lower than in fluoride-containing solutions. Three products can be obtained by using the standard basic method (Table 1). Two of them have similar framework structures to the products obtained by the fluoride method. These are $[Co(cp)_2]$-NON, which was first synthesized by Balkus and Shepelev, [41] and $[Co(cp)_2]$-DOH [47]. The properties of these compounds are clearly different from those produced by the fluoride method. This is already evident from the lattice constants listed in Table 1. The differences probably reflect different mechanisms for compensating the positive charge of the SDA. In highly crystalline $[Co(cp)_2]F$-NON (synthesized in the fluoride system), the charge of the cobalticinium cation is compensated by a fluoride anion bonded to a framework Si atom, which thus becomes five-coordinate [42]. $[Co(cp)_2]$-NON (synthesized in basic medium) possibly contains negatively charged defects in the silica framework, consisting of Si–O⁻...H–O–Si groups. Such defects have been described by Koller et al. [52] for the zeosil silicalite-1 obtained with the tetrapropylammonium cation as SDA.

In the structures of several other microporous solids, including porosils, AlPOs and GaPOs, that were synthesized by the fluoride method, the fluoride anion was found in a different location than in [Co(cp)₂]-NON, i.e., not directly bonded to one of the framework atoms. Instead, it occupies the center of a double four-ring unit. An example is another porosil of AST structure, namely quinuclidinium-AST [53]. Therefore, it has been supposed that the F⁻ anion acts as a co-template, stabilizing structures that contain double four-ring units [23]. In agreement with this idea, we find that AST is only formed in fluoride-containing syntheses, and not in fluoride-free preparations.

In contrast, ZSM-48 forms only in basic synthesis media. It is the only zeosil that has so far been obtained with the simple cobalticinium cation as SDA. The depiction of the channel structure in Fig. 4, in which one-dimensional channels are lined by ten [SiO₄/₂] tetrahedra, is an idealized one, however. ZSM-48 type zeosils and zeolites typically exhibit intergrowth disorder. This is also true of the compound synthesized with the cobalticinium cation. An X-ray powder diffraction pattern clearly shows disorder-induced broadening of some reflections. The synthesis, which is not easily reproduced, gives ZSM-48 as bundles of tiny yellow needles [47].

The influence of temperature on syntheses conducted in the basic medium is not as clear as in the fluoride system. [Co(cp)₂]-NON may crystallize in the temperature range of 140-180°C. Pure samples are routinely obtained in the temperature range of 140-160°C. At higher temperatures (160-180°C), [Co(cp)₂]-ZSM-48 or [Co(cp)₂]-DOH is usually found as by-products. It has not yet been possible to find conditions that reliably and reproducibly lead to pure ZSM-48. Temperatures at the level of decomposition of the cobalticinium cation in basic solution (175-180°C) and large autoclave volumes (50 ml instead of the usual 10 ml) seem to favour the formation of [Co(cp)₂]-DOH. In this way, a pure cobalticinium-containing DOH compound can be obtained. This was not possible in the fluoride-mediated reaction (see above), which gave [Co(cp)₂]F-DOH only as a by-product in syntheses with [Co(cp)₂]F-AST as the main product.

In addition to the results on porosils given above, the cobalticinium cation has also been applied to the synthesis of microporous solids of other host compositions. Balkus et al. have described their results for the synthesis of open-framework aluminophosphates, [54] and Kallus et al. have determined the structure of a cobalticinium-containing gallium phosphate. This compound, however, is not a true zeotype since it does not possess a three-dimensional framework structure [55].

Identity and Contents of Guest Species

A first indication of successful incorporation of intact organometallic SDAs is the colour of the porosil crystals. All the organometallic SDAs described in this work are yellow. Correspondingly, in normal light, porosil crystals from successful syntheses also appear yellow (the behaviour under polarized light is more peculiar, see below). UV-Vis spectra quantify this visual impression of colour [42, 56].

In addition, various other methods have been used to certify that the molecules that have become occluded by the growing silica crystals correspond to the intact organometallic cations employed as SDAs. Among them are IR spectroscopy, ¹H, and ¹³C solid-state MAS NMR spectroscopy and CoK XANES (X-ray Absorption Near Edge Structure) and EXAFS (Extended X-ray Absorption Fine

Structure) analyses [42, 43]. When spectra from the organometallic/porosil composites are compared to the corresponding spectra of the pure hexafluorophosphate salts of the organometallic cations or their solutions, certain minor deviations are often observed. These can be explained by interactions of the organometallic cations with the silica host. All spectral data thus agree with the presence of intact organometallic SDAs in the pores.

As an example for such deviations, Fig. 5 shows the C–H stretching region of the IR spectra of cobalticinium cations in various surroundings. Cobalticinium cations typically show one ν_{C-H} band in the region from 3120 to 3130 cm^{-1}, as exemplified in Fig. 5 by the hexafluorophosphate salt and several clathrasil compounds. Remarkably, the IR spectrum of [Co(cp)$_2$]F-NON exhibits an additional band at 3100 cm^{-1}. The shift to lower wavenumbers corresponds to a decrease in bond strength and has been ascribed to the occurrence of C–H···O–Si hydrogen bonds between host framework and guest cation in this compound, which in turn weakens certain C–H bonds. The existence of these unusual C–H···O–Si hydrogen bonds was also proved by X-ray structural analysis [43]. Interestingly, [Co(cp)$_2$]-NON synthesized in basic medium does not show this extra absorption. Although a full explanation of these differences cannot yet be given, they are probably related to the different mechanisms of charge compensation described above.

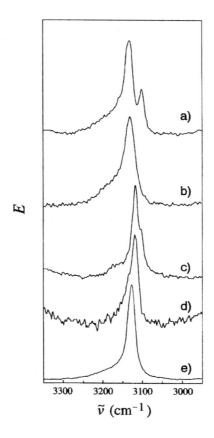

Fig. 5. C–H stretching region of IR spectra of the cobalticinium cation in various surroundings: a) [Co(cp)$_2$]F-NON, b) [Co(cp)$_2$]-NON, c) [Co(cp)$_2$]-AST, d) [Co(cp)$_2$]-DOH, e) [Co(cp)$_2$] PF$_6$.

Data on the chemical composition of the compounds have mainly been obtained by thermogravimetry. This is a usual procedure for determining the guest content of organic/inorganic host-guest composites. After heating to temperatures that ensure complete thermal decomposition of the complexes, the mass loss in percent is a measure for the ratio of organometallic guest species to inorganic framework component. We assume that metal ions of the complex are not volatalized during decomposition and that their charge is compensated by oxide and/or fluoride ions. For the clathrasils, we then find that the experimental mass loss agrees well with the value calculated on the basis that each large cage contains one organometallic cation, i.e., there are no empty large cages [44].

Structure and Organization

In this part, we will review some structural studies on organometallic/porosil composites. Our focus is on the arrangement and spatial organization of the organometallic chromophores, as these are important features for potential applications of these materials, as described in part 3 of this review. Apart from X-ray structure determinations on single crystals or powders, the observation of the crystals under a polarizing microscope is a simple and valuable method for obtaining a first impression of the spatial organization of the organometallic complexes.

Fig. 6 summarizes schematically the observations made with the polarizing microscope on porosils containing organometallic complexes. For the organometallic sandwich complexes used in this work, the absorptions in the visible region have their transition dipole moments parallel to their principal symmetry axis, i.e. along the line that runs through the centroids of the coordinated π-ligands and the metal atom. These absorptions give rise to a yellow colour. When the principal symmetry axes are aligned, anisotropic absorption is observed, as shown in Fig. 3 for the *p*NA/AlPO$_4$-5 composite. When viewed under polarized light, porosil crystals with different orientations will exhibit different colours or different colour intensities. Crystals of cobalticinium/porosil composites appear yellow when the principal axes of the complexes are oriented parallel to the electric field vector *E* of the polarized light; they are colourless when these two directions are orthogonal. Oblique angles between the symmetry axis and the field vector produce a yellow colour of non-maximum intensity. When the complexes are not aligned, the crystals are yellow, independent of the orientation between the complexes and the electric field.

Fig. 6a depicts schematically the absorption behaviour of crystals of [Co(cp)$_2$]F-NON. When the electric field vector is parallel to the *c* axis, the orthorhombic crystals appear yellow. If the light is polarized perpendicular to *c* axis, i.e., along the *b* axis, they are colourless (colour photographs of these crystals have appeared in refs. [20, 44]). With the knowledge of the relationship between the crystal morphology and the crystal structure it is possible to deduce the orientation of the complex in the large cage of the NON structure (Fig. 6a, right). As the crystal structure analysis shows, this orientation provides a perfect fit between the organometallic SDA and the [5^{8}6^{12}] cage (see below). The anisotropic absorption behaviour of [Co(cp)$_2$]F-NON has been quantified by polarized UV/Vis spectroscopy [44]. Also, in the IR spectrum, the C–H stretching bands of the occluded cobalticinium cation, when measured with polarized light, show a corresponding absorption behaviour [44].

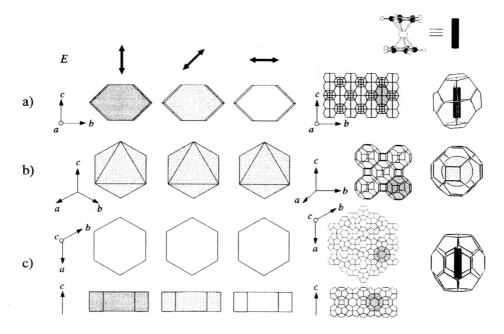

Fig. 6. Anisotropic absorption behaviour under the polarizing microscope of crystals of (a) [Co(cp)₂]F-NON, (b) [Co(cp)₂]F-AST and [Co(cp)₂]F-DOH. The shading of the crystals indicates yellow colours of different intensities.

No polarization effects are observed in crystals of [Co(cp)₂]F-AST, which have an octahedral morphology (the crystal system is monoclinic but pseudo-cubic). Independent of the direction of the electric field vector, they always show a strong yellow colour, indicating that there is no alignment of the organometallic cations in different cages. This result is in agreement with the almost spherical shape of the $[4^6 6^{12}]$ cage of the AST structure, which does not seem to offer a preferred orientation for the cobalticinium cation.

The optical behaviour of [Co(cp)₂]F-DOH is very interesting (Fig. 6c). The crystals are simple hexagonal prisms. Viewed parallel to the c-axis, they appear colourless, independent of the direction of E. When placed on one of their basal (100) faces, their colour changes with the direction of the electric field vector. From this behaviour, it may be deduced that the symmetry axes of the organometallic complexes are aligned along the c axis of the hexagonal crystals. Transferring this result to the crystal structure, we see that these axes are aligned with the sixfold axes of the barrel-shaped $[5^{12} 6^8]$ cages. Polarization microscopy can only probe the direction of the transition dipole moments, and is insensitive to the rotation of the organometallic complexes around their principal axes. It can therefore not detect rotational disorder around these axes. Structural analyses (see below) of [Co(cp)₂]F-NON and [Co(cp)₂]F-DOH show that the degree of rotational disorder in these two compounds is different, quite in contrast to their similar behaviour in polarizing microscope experiments (Figs. 6a and 6c).

Crystals of [Co(cpme)₂]F-DOH have a similar shape to those of [Co(cp)₂]F-DOH, but they are much smaller [25]. Anisotropic absorption is thus very difficult to observe. In fact, we were not able to distinguish any colour changes during rotation of the electric field vector with the crystals in various orientations. As we will see below, this fits very well to the structure of [Co(cpme)$_2$]F-DOH as determined by diffraction methods. We will now turn to the results obtained by these methods.

The crystals of [Co(cp)$_2$]F-NON were amenable to a single-crystal structural investigation [42]. The main results are shown in Fig. 7. The highest space group symmetry that the nonasil framework can attain, i.e., its topological symmetry, is *Fmmm*. The symmetry of the cobalticinium-containing nonasil is lower. It crystallizes in space group *Pccn*, a non-maximal subgroup of *Fmmm*. Correspondingly, the point symmetry of the large [5^{8}6^{12}] cage of the NON structure is lowered from *mmm* to $\bar{1}$. The cobalticinium cation in its staggered conformation also possesses this symmetry operation. In fact, in the structure of [Co(cp)$_2$]F-NON, the organometallic complex adopts this conformation and is inserted into the [5^{8}6^{12}] cages in such a way that the symmetry of host and guest agree with one another. Fig. 7 shows this arrangement from two different perspectives. [Co(cp)$_2$]F-NON is thus one of the rare cases among zeotype structures in which there is no disorder of the guest species. Remember that orientational and/or rotational disorder is otherwise rather common in these host-guest compounds (cf. trioxane silica sodalite in Fig. 1c); furthermore, sandwich complexes such as the cobalticinium cation often exhibit rotational disorder in their crystal structures. The fact that the cobalticinium cation occupies a fixed position in the structure of the nonasil has been attributed to the occurrence of the C–H$\cdots$O–Si hydrogen bonds discussed above [43]. The crystal structure of [Co(cp)$_2$]F-NON has been determined at 220 K, but further investigations by differential scanning calorimetry, temperature-dependent X-ray diffraction and IR spectroscopy, as well as thermodynamical arguments, have shown that rotational disorder of the organometallic complex cation begins only at 485 K [42, 43].

The fact that all the [5^{8}6^{12}] cages of the NON structure have the same orientation (Fig. 4a) and the absence of rotational and orientational disorder of the cobalticinium cation are responsible for the strict alignment of all organometallic cations in the clathrasil (Fig. 7). This uniform supralattice of aligned cobalticinium chromophores is an impressive example of the organizing power that porous hosts exert on their guests. This structural picture also explains the observations made with the polarizing microscope (see above and Fig. 6a) [44].

No detailed structural investigations have so far been performed on [Co(cp)$_2$]F-AST. The picture that has evolved from the investigations with the polarizing microscope (Fig. 6b) suggests that refinement of the heavily disordered guest species in the nearly spherical cage would be very cumbersome. In earlier structure determinations on microporous compounds with the AST topology, namely, the porosil quinuclidinium-AST [53] and the aristotype, the aluminophosphate AlPO$_4$-16 containing quinuclidine, [57] the guest species were also found to be disordered.

Similar problems occur in the structural investigation of [Co(cp)$_2$]F-DOH, although the disorder is less pronounced here (Fig. 6c). The framework atoms and the cobalt atom of the complex can easily be localized, but the carbon atoms of the guest species only yield a spread-out intra-cage electron density. In fact, modelling investigations show that the cobalticinium cation can occupy five different positions in the barrel-shaped [5^{12}6^8] cage of the DOH structure that are structurally different but energetically practically equivalent (the symmetry misfit between the [5^{12}6^8] cage with topological symmetry 6/*mmm* and the cobalticinium cation with fivefold symmetry in both its high-symmetry conformations additionally causes an sixfold orientational disorder of each of these positions) [58].

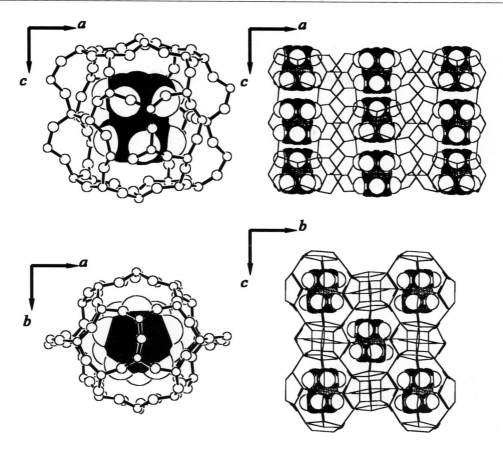

Fig. 7. Structure of [Co(cp)$_2$]F-NON: Different perspectives showing the cobalticinium cation in the [$5^8 6^{12}$] cage and their alignment in the structure.

In all these positions, the principal axis of the cobalticinium cation is aligned with the sixfold axis of the cage, although its conformation may vary. This alignment of all the cations is thus not disturbed by the rotational disorder. The structural picture that emerges from the modelling investigation and the necessarily incomplete structure determination thus is in agreement with the polarization-microscopic study (Fig. 6c).

The structural arrangement of [Co(cpme)$_2$]F-DOH has been determined by a combination of molecular modelling and Rietveld refinement techniques [59]. Fig. 8 shows the arrangement of the 1,1'-dimethylcobalticinium cation in the DOH cage. After introduction of the methyl groups it is no longer possible for the organometallic complex to adopt a position in which its principal axis (which connects the centroids of the cyclopentadienyl ligands and the metal atom) is parallel to the c axis of the crystals. Instead, these two axes are now inclined, due to the increased space requirements of the organometallic guest cation. The fact that no polarization effects have been observed for [Co(cpme)$_2$]F-DOH crystals under the polarizing microscope agrees with such a structural arrangement if we take into account the orientational disorder present due to the host-guest symmetry misfit.

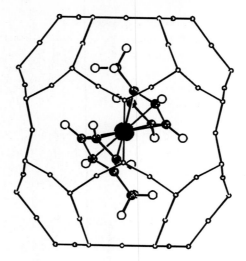

Fig. 8. Structural arrangement of the 1,1'-dimethylcobalticinium cation in the DOH cage of [Co(cpme)₂]F-DOH.

Protection and Stabilization

With regard to thermal stabilization of organometallic complexes occluded in clathrasils, we have obtained some spectacular results. The decomposition temperatures (defined as the onset of mass loss in thermogravimetric experiments) of the cobalticinium cations encapsulated in certain clathrasil host structures show a large increase in thermal stability: compared to the simple hexafluorophosphate salt of the cobalticinium cation, which decomposes at 650 K (even this is a remarkable high stability for an organometallic compound), the decomposition temperatures rise to 900 K in the NON and to 880 K in the AST compound. These values are independent of the atmosphere (nitrogen or oxygen) in which the experiment takes place [42]. According to these results, [Co(cp)₂]F-NON may be considered as the thermally most stable organometallic compound. In this case the fit between the [5⁸6¹²] NON cage and the complex cation is so tight that it may be presumed that the thermal stabilization is to a certain degree due to the suppression of vibrational modes of the complex cations that might ultimately lead to their pyrolytic decomposition.

Conclusions

The idea of using organometallic SDAs that exhibit an additional property (in the cases treated so far, that of a chromophore) to directly synthesize host-guest composites with potential materials applications clearly seems to be working. The resulting nanostructures exhibit different degrees of spatial organization, and the occluded molecules are stabilized and protected by encapsulation. In our continuing work we intend to use organometallic molecules that have other properties (i.e., magnetic properties, high hyperpolarizabilities, redox properties, chirality, switchability) to provide access to further interesting materials. In our accompanying research we aim to develop even milder synthetic methods (e.g., microwave heating) in order to perform reactions with more sensitive organometallic complexes. Also, we will extend our research to other host compositions (e.g. alumino- and zincophosphates) and to inorganic complex compounds (i.e., complexes that do not contain metal-carbon bonds). As a first result, we have recently obtained a new porosil structure type that is related to that of the zeolite chabazite. This compound (named LMU-2) was prepared by using the cationic aza crown ether complex *bis*(1,4,7-triazacyclononane)-cobalt(III) as SDA [60].

Acknowledgement: The research described in this review benefitted from the efforts of Gianpietro van de Goor, Clemens C. Freyhardt, Benedikt Lindlar, Christian Panz, Andreas M. Schneider, and Volker Hufnagel. Part of it was carried out in the laboratories of Professor Felsche (Universität Konstanz), and special thanks are due to Gerhard Wildermuth (Universität Konstanz) for thermoanalytical investigations. Financial support from the *DFG*, the *Fonds der Chemischen Industrie*, the *Gesellschaft der Freunde der Universität München* and from the *Riedel-de Haen-Stiftung* is gratefully acknowledged.

References:

[1]   A comprehensive treatise on zeolite chemistry is available in: *"Comprehensive Supramolecular Chemistry"* (Series Eds.: J. L. Atwood, J. E. D. Davies, D. D. MacNicol, F. Vögtle), Vol. 7: *"Solid State Supramolecular Chemistry: Two- and Three-Dimensional Inorganic Networks"* (Volume Eds.: G. Alberti, T. Bein), Pergamon, Oxford, **1996**. Valuable introductions to the various aspects of zeotype science, written by leading experts, may be found in the books that publish the lectures held during the „Zeolite Summer Schools" which proceed the International Zeolite Conferences: H. van Bekkum, E. M. Flanigen, J. C. Jansen (Eds.): *Introduction to Zeolite Science and Practice*, *Stud. Surf. Sci. Catal. 58*, Elsevier, Amsterdam, **1991**; J. C. Jansen, M. Stöcker, H. G. Karge, J. Weitkamp (Eds.): *Advanced Zeolite Science and Applications*, *Stud. Surf. Sci. Catal. 85*, Elsevier, Amsterdam, **1994**; H. Chon, S. I. Woo, S.-E. Park (Eds.): *Recent Advances and New Horizons in Zeolite Science and Technology, Stud. Surf. Sci. Catal. 102*, Elsevier, Amsterdam, **1996**.

[2]   M. L. Occelli, H. Kessler (Eds.): *Synthesis of Porous Materials*, Dekker, New York, **1996**.

[3]   H. Kessler: *"Syntheses of Molecular Sieves"*, in: *Comprehensive Supramolecular Chemistry* (Series Eds.: J. L. Atwood, D. D. MacNicol, J. E. D. Davies, F. Vögtle), Vol. 7 (Volume eds.: G. Alberti, T. Bein), Pergamon, Oxford, **1996**, p. 425.

[4]   S. L. Burkett, M. E. Davis, *"Synthetic Mechanisms and Strategies for Zeolite Synthesis"*, in: *Comprehensive Supramolecular Chemistry* (Series Eds.: J. L. Atwood, D. D. MacNicol, J. E. D. Davies, F. Vögtle), Vol. 7 (Volume Eds.: G. Alberti, T. Bein), Pergamon, Oxford, **1996**, p. 465.

[5]   M. E. Davis, R. F. Lobo, *Chem. Mater.* **1992**, *4*, 756.

[6]   R. F. Lobo, S. I. Zones, M. E. Davis, *J. Incl. Phenom. Mol. Recogn.* **1995**, *21*, 47.

[7]   Y. Kubota, M. M. Helmkamp, S. I. Zones, M. E. Davis, *Microporous Mater.* **1996**, *6*, 213.

[8]   H. Gies, in: *Inclusion Compounds* (Eds.: J. L. Atwood, J. E. D. Davies, D. D. MacNicol), Vol. 5, Oxford University Press, Oxford, **1991**, p. 1.

[9]   H. Gies, B. Marler, *Zeolites* **1992**, *12*, 42.

[10] P. Behrens, G. van de Goor, M. Wiebcke, C. Braunbarth, A. M. Schneider, J. Felsche, G. Engelhardt, P. Fischer, K. Fütterer, W. Depmeier, Poster contribution to the 8th International Symposium on Molecular Recognition and Inclusion, Ottawa, 31. July - 5. August 1994; and to be published.

[11] J. Keijsper, C. J. J. den Ouden, M. F. M. Post, *Stud. Surf. Sci. Catal.* **1989**, *49B*, 237.

[12] A. Stein, S. W. Keller, T. E. Mallouk, *Science* **1993**, *259*, 1558.

[13] R. Roy: *"Low-Temperature Materials Syntheses"*, in: *Reactivity of Solids* (Ed.: V. V. Boldyrev), Blackwell, Oxford, **1996**, p. 253.

[14] I. Petrovic, A. Navrotsky, M. E. Davis, S. I. Zones, *Chem. Mater.* **1993**, *5*, 1805.

[15] G. van de Goor, P. Behrens, J. Felsche, *Microporous Mater.* **1994**, *2*, 493.

[16] C. M. Braunbarth, *Ph. D. thesis*, Konstanz, **1997**.

[17] D. M. Bibby, M. P. Dale, *Nature* **1985**, *317*, 157; D. M. Bibby, N. I. Baxter, D. Grant-Taylor, L. M. Parker, *ACS Symp. Ser.* **1989**, *398*, 209.

[18] C. M. Braunbarth, P. Behrens, J. Felsche, G. van de Goor, G. Wildermuth, G. Engelhardt, *Zeolites* **1996**, *16*, 207.

[19] F. Liebau, H. Gies, R. P. Gunawardane, B. Marler, *Zeolites* **1986**, *6*, 373.

[20] P. Behrens, G. D. Stucky: *"Novel Materials Based on Zeolites"*, in: *Comprehensive Supramolecular Chemistry* (Series Eds.: J. L. Atwood, D. D. MacNicol, J. E. D. Davies, F. Vögtle), Vol. 7 (Volume Eds.: G. Alberti, T. Bein), Pergamon, Oxford, **1996**, p. 721.

[21] W. M. Meier, D. H. Olson, Ch. Baerlocher, Atlas of Zeolite Structure Types, 4th ed., Butterworth-Heinemann, **1996**; W. M. Meier, D. H. Olson, Ch. Baerlocher, *Zeolites* **1996**, *17*, 1.

[22] G. Brunner, *Zeolites* **1993**, *13*, 592.

[23] H. Kessler, *Stud. Surf. Sci. Catal.* **1989**, *52*, 17; H. Kessler, *Mater. Res. Soc. Symp. Proc.* **1991**, *233*, 47; E. Klock, L. Delmotte, M. Soulard, J. L. Guth, in: *Proceedings of the 9th International Zeolite Conference* (Eds.: R. Van Ballmoos *et al.*), Butterworth-Heinemann, **1993**, p. 611.

[24] R. Althoff, K. Unger, F. Schüth, *Microporous Mater.* **1994**, *2*, 557; R. Althoff, B. Sellegren, B. Zibrowius, K. Unger, F. Schüth, *Stud. Surf. Sci. Catal.* **1995**, *98*, 36.

[25] G. van de Goor, B. Lindlar, J. Felsche, P. Behrens, *J. Chem. Soc., Chem. Commun.* **1995**, 2559.

[26] S. L. Burkett, M. E. Davis, *J. Phys. Chem.* **1994**, *98*, 4647.

[27] S. L. Burkett, M. E. Davis, *Chem. Mater.* **1995**, *7*, 920.

[28] S. L. Burkett, M. E. Davis, *Chem. Mater.* **1995**, *7*, 1453.

[29] M. E. Davis, *CHEMTECH* **1994**, 22; M. E. Davis, *Nature* **1996**, *382*, 583; D. W. Lewis, D. J. Willock, C. R. A. Catlow, J. M. Thomas, G. J. Hutchings, *Nature* **1996**, *382*, 605; M. E. Davis, S. I. Zones: *"A Perspective on Zeolite Syntheses: How Do You Know What You'll Get?"*, in: *Synthesis of Porous Materials* (Eds.: M. L. Occelli, H. Kessler), Dekker, New York, **1996**, p. 1.

[30] F. Marlow, J. Caro, *Zeolites* **1992**, *12*, 433.

[31] L. Werner, J. Caro, G. Finger, J. Kornatowski, *Zeolites* **1992**, *12*, 658.

[32] J. Caro, G. Finger, J. Kornatowski, J. Richter-Mendau, L. Werner, B. Zibrowius, *Adv. Mater.* **1992**, *4*, 273.

[33] N. Herron, Y. Wang, M. M. Eddy, G. D. Stucky, D. E. Cox, K. Moller, T. Bein, *J. Am. Chem. Soc.* **1989**, *111*, 530.

[34] V. Ramamurthy, J. V. Caspar, D. R. Corbin, *J. Am. Chem. Soc.*, **1991**, *113*, 594, 600; V. Ramamurthy, D. F. Eaton, *Chem. Mater.* **1994**, *6*, 1128.

[35] R. Hoppe, G. Schulz-Ekloff, D. Wöhrle, Ch. Kirschhock, H. Fuess, *Stud. Surf. Sci. Catal.* **1994**, *84*, 821; D. Wöhrle, G. Schulz-Ekloff, *Adv. Mater.* **1994**, *6*, 875.

[36] J.-M. Lehn, *Science* **1993**, *260*, 1762; J.-M. Lehn, *Top. Curr. Chem.* **1993**, *165*; D. T. Seto, G. M. Whitesides, *J. Am. Chem. Soc.* **1993**, *115*, 905.

[37] H. K. Chae, W. G. Klemperer, D. A. Payne, C. T. A. Suchicital, D. R. Wake, S. R. Wilson, *ACS Symp. Ser.* **1991**, *127*, 528.

[38] N. J. Long, *Angew. Chem., Int. Ed. Engl.* **1995**, *34*, 37.

[39] P. Gütlich, A. Hauser, H. Spiering, *Angew. Chem., Int. Ed. Engl.* **1994**, *33*, 2024.

[40] T. Bein: *"Inclusion Chemistry of Organometallics in Zeolites"*, in: *Comprehensive Supramolecular Chemistry* (Series Eds.: J. L. Atwood, D. D. MacNicol, J. E. D. Davies, F. Vögtle), Vol. 7 (Volume Eds.: G. Alberti, T. Bein), Pergamon, Oxford, **1996**, p. 579.

[41] K. J. Balkus, S. Shepelev, *Microporous Mater.* **1993**, *1*, 383.

[42] G. van de Goor, C. C. Freyhardt, P. Behrens, *Z. Anorg. Allg. Chem.* **1995**, *621*, 311.

[43] P. Behrens, G. van de Goor, C. C. Freyhardt, *Angew. Chem., Int. Ed. Engl.* **1995**, *34*, 2680.

[44] G. van de Goor, K. Hoffmann, S. Kallus, F. Marlow, F. Schüth, P. Behrens, *Adv. Mater.* **1996**, *8*, 65.

[45] K. J. Balkus, A. G. Gabrielov, N. Sandler, *Mater. Res. Soc. Symp. Proc.* **1995**, *368*, 369.

[46] C. C. Freyhardt, M. Tsapatsis, R. F. Lobo, K. J. Balkus, M. E. Davis, *Nature* **1996**, *381*, 295.

[47] P. Behrens, Ch. Panz, V. Hufnagel, B. Lindlar, C. C. Freyhardt, G. van de Goor, *Solid State Ionics*, in press.

[48] K. J. Balkus, A. G. Gabrielov, S. Shepelev, *Microporous Mater.* **1995**, *3*, 489.

[49] S. Kallus, J. Patarin, B. Marler, *Microporous Mater.* **1996**, *7*, 89.

[50] G. van de Goor, Ph. D. thesis, Universität Konstanz, **1995**.

[51] E. Klock-Châtelain, H. Kessler, personal communication.

[52] H. Koller, R. F. Lobo, S. L. Burkett, M. E. Davis, *J. Phys. Chem.* **1995**, *99*, 12588.

[53] P. Caullet, J. L. Guth, J. Hazm, J. M. Lamblin, H. Gies, *Eur. J. Solid State Inorg. Chem.* **1991**, *28*, 345.

[54] K. J. Balkus, A. G. Gabrielov, S. Shepelev, *Microporous Mater.* **1995**, *3*, 489.

[55] S. Kallus, J. Patarin, B. Marler, *Microporous Mater.* **1996**, *7*, 89.

[56] P. Behrens, G. van de Goor, M. Wark, A. Trnoska, A. Popitsch, *J. Mol. Str.* **1995**, *348*, 85.

[57] J. M. Bennett, R. N. Kirchner, *Zeolites* **1991**, *11*, 502.

[58] A. M. Schneider, P. Behrens, unpublished results.

[59] A. M. Schneider, P. Behrens, *Chem. Mater.*, submitted.

[60] P. Behrens, V. Hufnagel, C. C. Freyhardt, A. M. Schneider, H. Koller, K. Polborn, submitted.

Novel Aspects of the Chemical Modification of Silica Surface

Valentin Tertykh
Department of Chemisorption
Institute of Surface Chemistry, National Academy of Sciences
Prospekt Nauki 31, 252022 Kiev, Ukraine
Tel.: Int. code + (38044)2656520 – Fax: Int. code + (38044)2640446
E-mail: tertykh@public.ua.net

Keywords: Chemisorption / Modification / Silica / Surface

Summary: Types of heterolytic reactions with the participation of the silica surface sites are examined. Chemisorption regularities of the organosilicon compounds, cotaining trimethylsilyl group (trimethylhalo- and trimethylpseudohalosilanes), in reaction with surface silanols of fumed silica are analyzed. Peculiarities of surface reactions with the participation of some chlorides and oxochlorides are discussed. Some examples of the addition reactions are considered.

Introduction

By chemical modification of the silica surface it has become possible to design new highly-selective adsorbents and catalysts, active polymer fillers, efficient thickeners of dispersive media. Interest in the modified silicas, in particular, in the activated matrices based on functional organosilicas has quickened in the past few years as a result of the favorable prospects for their application for various kinds of chromatographic separation, preparation of grafted metal complex catalysts, immobilized enzymes and other biologically active compounds [1].

The present work gives analytical treatment to the main types of heterolytic processes with the participation of silica surface sites, the specific features for reactions of the electrophilic proton substitution in the structural silanol groups at interaction with some halo-and pseudohalosilanes, halogenides and oxohalogenides of various elements. Some possibilities of the addition reactions in the synthesis of the surface chemical compounds are considered.

Types of Heterolytic Reactions with the Participation of Silica Surface Sites

The surface of highly-dispersed pyrogenic silica prepared at 400-600°C is dominated by the isolated silanol groups spaced about 0.7 nm apart, as previously determined [2]. Such a magnitude was obtained by a chemical method consisting of successive introduction of first the trichlorisilyl groups ($\equiv$ SiOSiCl$_3$) with silicon tetrachloride and then also the adequate alkoxysilyl groups by treatment of chlorinated silica with alcohols into the surface layer. Varying a radical length for *n*-aliphatic alcohol in the reaction involving the grafted trichlorisilyl groups it is possible to estimate an average distance between the silicon atoms in silanol groups on silica surface. It was shown [2] that in case of methanol it was possible to introduce $-SiCl_3$ groups into the chemical reaction completely. For ethanol and propanol the achieved chemisorption values (as recalculated for $-Si(OR)_3$ groups) were equal to 90 and 80% respectively, and in the case of *n*-butanol about 2/3 of the surface Si–Cl bonds are involved into reaction. Surface of dehydrated fumed silica may be represented in the form of the patches exhibiting the structure as in Scheme 1.

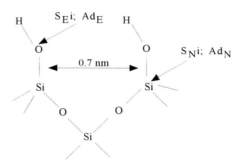

Scheme 1. Arrangement of silanol groups on the dehydrated fumed silica surface and types of some heterolytic reactions with its participation.

Thus, it is possible to use the dehydrated fumed silica surface for studying of chemisorption of bi- and tri- functional molecules. So, such a matrix may be considered as adequately geometrically homogeneous, i.e., containing only locally arranged fixed sites of the same type. The silica surface prepared at more moderate temperatures contains strongly bound water in different quantities exerting a considerable influence on the chemical reactions which proceed in the surface layer.

In the majority of cases the reactions realized with the participation of silica surface sites are classified as the heterolytic processes of substitution, addition, elimination or rearrangement [1]. In order to classify the reactions involving silica surfaces it is expedient to adopt the concepts of electrophilic and nucleophilic mechanisms of attack [1, 3-5] as developed by C. K. Ingold [6] and L. .H. Sommer [7] for the heterolytic reactions of carbon and silicon compounds. Direction of attack will be governed by the reagent nature and the character of asymmetric distribution of electron density on atoms of an active surface site. In such a case asymmetry in electron density distribution may either exist in an initial site or to arise as a result of reciprocal polarization when a reagent molecule approaches a surface site. As rightly emphasized [4], in the case of a silica surface it seems impossible for an attacking reagent to approach in the rear, and so the reaction mechanism should be realized similar to the processes of intramolecular substitution (S_Ei or S_Ni).

It is possible to set aside a wide group of reactions at which the attack is realized by an electrophilic reagent in respect to an oxygen atom of a surface silanol group. It includes the reaction of electrophilic proton substitution (S_Ei) at interaction with different chloro- and alkoxysilanes, organosilazanes, various organometallic compounds, halogenides of different elements, and also the processes of electrophilic addition (Ad_E). For instance, the interaction of trimethylmethoxysilane and ethyleneimine with surface silanol groups represents a typical example of a reaction proceeding according to mechanisms S_Ei and Ad_E, respectively:

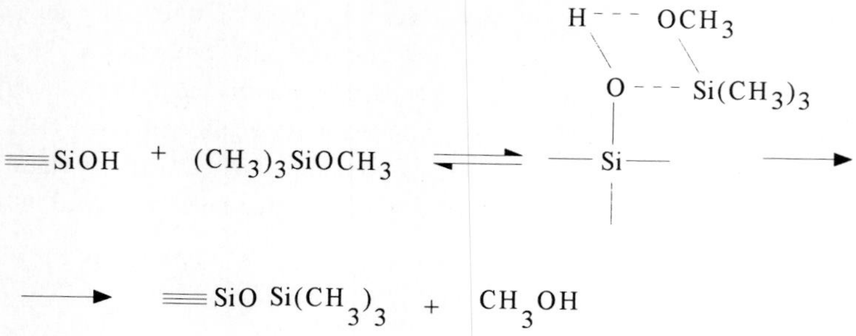

$$\equiv SiOH \;+\; (CH_3)_3SiOCH_3 \;\rightleftharpoons\;$$

(structure showing Si with H---OCH₃ and O---Si(CH₃)₃)

Scheme 2.

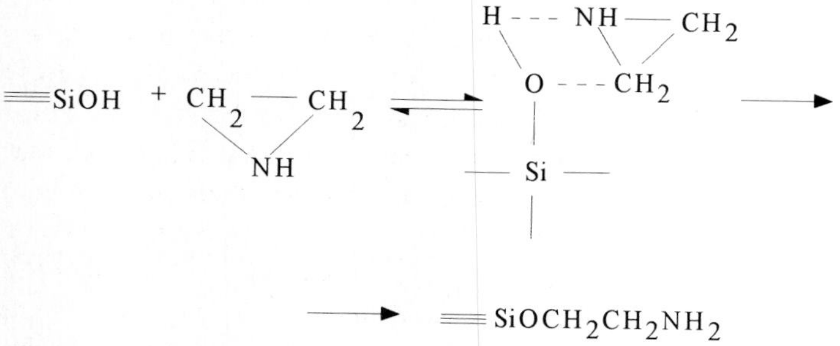

$$\longrightarrow \;\equiv SiO\,Si(CH_3)_3 \;+\; CH_3OH$$

$$\equiv SiOH \;+\; CH_2\!-\!CH_2 \;\rightleftharpoons\;$$

(structure with NH bridge; H---NH—CH₂ and O---CH₂ bound to Si)

$$\longrightarrow \;\equiv SiOCH_2CH_2NH_2$$

Scheme 3.

The other group of reactions proceeds with attack of a nucleophilic reagent at a silicon atom of silica surface. They include the processes of nucleophilic substitution or nucleophilic addition ($S_N i$ or Ad_N). For example, methanolysis of chlorinated silica is a typical process for $S_N i$ mechanism:

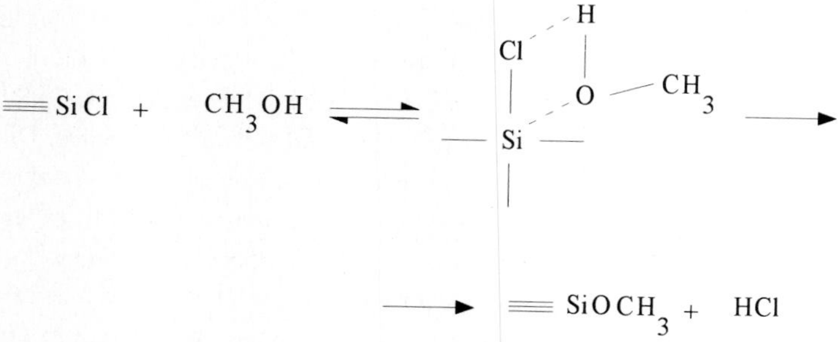

$$\equiv SiCl \;+\; CH_3OH \;\rightleftharpoons\;$$

(structure showing Si with Cl, H, O—CH₃)

$$\longrightarrow \;\equiv SiOCH_3 \;+\; HCl$$

Scheme 4.

Interaction of product grafted on the silica surface $\equiv$SiH groups with olefins corresponds to nucleophilic addition (Ad_N):

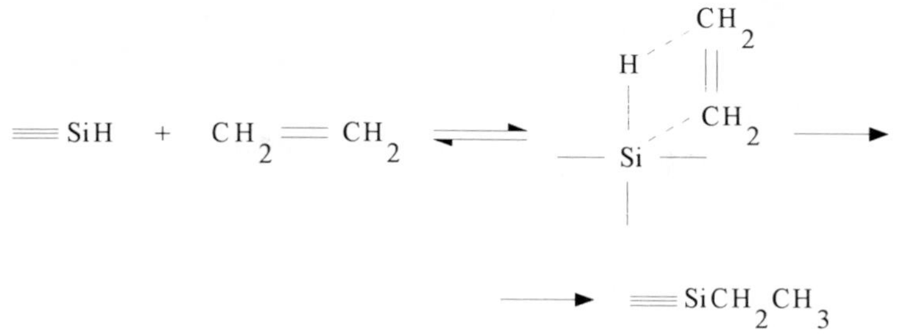

$$\equiv SiH \; + \; CH_2 = CH_2 \; \rightleftharpoons \; \begin{array}{c} H \quad CH_2 \\ | \quad || \\ | \quad CH_2 \\ -Si- \\ | \end{array} \; \longrightarrow$$

$$\longrightarrow \; \equiv SiCH_2CH_3$$

Scheme 5.

In some cases both the silicon and oxygen atoms of the siloxane bonds may simultaneously act as the reactive surface sites (the processes according to $Ad_{N,E}$ mechanism).

For the majority of surface reactions it is postulated that the rate-determining step is the formation of quasicyclic (predominately four-centered) transient complexes. This process is followed by cleavage of the existing bonds and by formation of new compounds. Formation and rebuilding of four-centered transient complexes with donor-acceptor bonds may bring in an essential contribution into the height of activation barrier. Quantum chemical computations of sections of potential energy surfaces along the reaction pathway [8-11] indicate that 50-70 % of the activation energy magnitude for reactions with participation of surface silanol groups is determined by expenditures for proton transfer in the cyclic transient complexes. From calculations [12] two minima on the potential curves for interaction of modifying reagent molecules with surface silanol groups are due to formation of linear hydrogen-bonded and following cyclic donor-acceptor complexes prior to the transition state for proton transfer.

Interaction of Surface Silanols with Organosilicon Compounds Containing the Trimethylsilyl Group

The interaction of the isolated silanol group of dehydrated fumed silica with methylchlorosilanes of the $Cl_nSi(CH_3)_{4-n}$ series, where $n = 1-4$, proceeds in the main monofunctionally (1:1) according to the scheme:

$$\equiv SiOH + Cl_nSi(CH_3)_{4-n} \rightarrow \equiv SiOSi(CH_3)_{4-n}Cl_{n-1} + HCl$$

Scheme 6.

As this take takes place, the activation energy of the process decreases with increase in n, being equal to 159, 126, 105, and 80 kJ/mol of the grafted groups for $n = 1-4$, respectively [13]. As to the given reaction series for the simplest case a quantitative relationship between activity and structure may be described with the Hammet equation, taking advantage of the principle of linearity in the free energy changes (LFE-principle). A plot of , lg k/k_o vs. $\Sigma_i\sigma_i^*$ (Fig. 1, a) is linear (k_o and k–are the rate constants in the process of chemisorption of trimethylchlorosilane, taken as a standard in series under study, and of the given chlorosilane; $\Sigma_i\sigma_i^*$–inductive constants of the substituents). It is essential that the activation energy vs. $\Sigma_i\sigma_i^*$ relation also corresponds to the LFE-principle (Fig. 1, b), thus giving evidence of its applicability not only for free energies of reaction, but also for activation energies. It is evident that a relation between the activation energies and the relative rate constants is expressed as a linear dependence. The value of the reactivity constant ρ^*, estimated from the slope of lg k/k_o vs. $\Sigma_i\sigma_i^*$ for the given reaction series is equal to $+0.53$. It is known that a positive value of ρ^* is characteristic of the reactions favored by the electronegative substituents. And actually, as was noted above, the rate constants for the reactions of electrophilic substitution of proton in a silanol group exhibit a sequential rise in the succession from $(CH_3)_3SiCl$ to $SiCl_4$ at a given temperature. Low absolute value of the Taft reaction constant as compared to the ρ^* values for similar reactions of the organosilicon compounds in solutions [7] results from the fact that the structural silanol group on the silica surface acts in this process as a more weak nucleophilic reagent compared with the hydroxyls of water and alcohols.

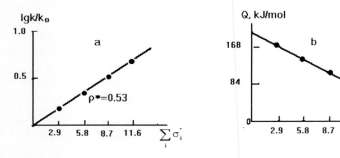

Fig. 1.   Dependence of the relative rate constants (a) and the activation energies (b) vs. the total inductive effect of the substituents in the reagent for chemisorption of methylchlorosilanes $Cl_nSi(CH_3)_{4-n}$ (n = 0-4) by isolated silanol groups of the silica surface.

When comparing the reactivities of the trimethylsilyl-sustituted chloro- and alkoxysilanes, activities of the corresponding organosiloxanes and aminosilanes in the reactions of electrophilic proton substitution in the isolated silanol groups of the silica surface the following sequence was obtained: $Si-N(H,R) > Si-O(H,R) > Si-Cl > Si-O(Si) > Si-C(H,R)$

For the first time this sequence has been described in a paper [5]. For the series of compounds studied there is a linear correlation between the activation energies of the chemisorption process and proton-accepting properties of the atoms bound to silicon within a reactive functional group of the modifying reagent (Fig. 2). Actually, as was shown in [5], there is a quite satisfactory correlation between the calculated values of activation energy for the process and those of proton affinity for the carbon, chlorine, oxygen, and nitrogen atoms. At the same time, direct experimental data, characterizing proton affinity of the same atoms within certain organosilicon compounds, are absent. The accepting properties of these atoms characterize to some degree the values of the 3750 cm^{-1} band shifts, $\Delta\nu_{OH}$ – the band which corresponds to the stretching vibrations of O–H in the isolated silanol surface groups at formation of hydrogen bonds with the molecules of organosilicon compounds. It was found that there is a linear relationship between the $\Delta\nu_{OH}$ values and those for the process activation energies.

For the explanation of differences in reactivities of molecules, interacting with isolated silanol groups according to mechanism S_{Ei}, Yu.I.Gorlov et al. [14] made the supposition that the activation barrier of such reactions is determinated only by the deformation energy of the attacking molecule, as its distance from the surface fragments carrying the OH-group relatively large. On the other hand, for reactions taking place ty mechanism S_{Ni}–Si, it is supposed that the deformation energy of the corresponding surface fragment makes the main contribution to the activation barrier. Supposed approach (deformation model) to explain differences in reactivities of molecules attacking the same surface site is found the most suitable only for case when the leaving group and structure of the corresponding four-centerd transient complex are identical. Really, this model explained rather well the differencies in reactivities of chlorosilanes in the series from $SiCl_4$ to $(CH_3)_3SiCl$ (Table 1).

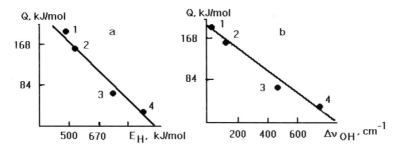

Fig. 2. Dependence of the activation energies for reaction of electrophilic substitution of proton in surface silanol groups *vs.* the proton-accepting properties of atoms bound with silicon (a) and the values of $3750\ cm^{-1}$ shifts (b) after adsorption of organosilicon compounds: $1 - Si(CH_3)_4$; $2 - (CH_3)_3SiCl$; $3 - (CH_3)_3SiOCH_3$; $4 - (CH_3)_3SiN(CH_3)_2$.

Table 1. Optimum reaction temperature and activation energy of the reaction of methylchlorosilanes with iso–lated silanol groups of the silica surface.

Modifying Reagent	Deformation energy [kJ/mol][14]	Optimum Reaction Temperature [°C][1, 13]	Activation Energy [kJ/mol][1, 13]
$SiCl_4$	67	200	80
CH_3SiCl_3	100	250	105
$(CH_3)_3SiCl_2$	113	300	126
$(CH_3)_3SiCl$	165	380	159

But consideration of only the deformation energies of the electrophilic reagents is evidently insufficient in cases when the leaving groups are different (for example, for the other trimethylsilyl-substituting compounds $(CH_3)_3SiN(CH_3)_2$, $(CH_3)_3SiOCH_3$, etc.). Besides the application of deformation model to different trimethylhalosilanes, for which the following sequence of activities in relation to surface silanol groups was obtained [15, 16]: $(CH_3)_3SiI > (CH_3)_3SiBr \gg (CH_3)_3SiCl$, also results in a less confident explanation. Computations of the deformation energies for molecules $(CH_3)_3SiI$, $(CH_3)_3SiBr$ and $(CH_3)_3SiCl$ gave [17] not strongly differing values: 128, 144, and 165 kJ/mol, respectively. At the same time the reactivities of trimethylhalosilanes differ strikingly [15, 16]. Whereas a rather high temperature (above 300°C) is necessary for chemisorption of trimethylchlorosilane with isolated silanol groups of the silica surface, the reaction with trimethylbromosilane proceeds at 50°C at a high rate and the complete substitution of protons of silanol groups in the reaction with trimethyliodosilane occurs already at room temperature. Maximum achieved concentrations of the grafted trimethylsilyl groups are equal for all trimethylhalosilanes studied, corresponding to the concentration of the isolated silanol groups on the surface of initial silica.

Interaction of trimethylhalosilanes with the silica surface may be realized in several ways, i.e., either by the mechanism of electrophilic proton substitution in the isolated silanol groups by the trimethylsilyl ones or by addition of organosilanes to the siloxane bonds (mechanism $Ad_{N,E}$). Consideration must be given to the homolytic reactions to proceed, especially in the case of $(CH_3)_3SiI$, although such reactions are not characteristic for silicon compounds. Really, the results obtained show that the radical reactions and addition of $(CH_3)_3SiI$ and $(CH_3)_3SiBr$ to the siloxane bonds do not occur at the given temperatures. First of all, the halogens (in the form of $\equiv SiI$ or $\equiv SiBr$ groups) were not found in the surface compounds, suggesting that the chemisorption with the participation of siloxane bonds is absent. Secondly, the grafting of trimethylbromo- and trimethyliodosilane does not occur on methoxysilica at moderate temperatures, and consequently, radical processes do not occur. The fact that the concentration of trimethylsilyl groups in the reaction products is equal to concentration of the hydroxyl groups on the surface of initial fumed silica, gives evidence that the reaction proceedes between surface sites and trimethylhalosilanes by the mechanism of electrophilic proton substitution in the structural silanol group (S_Ei). Actually, elimination of a proton from the reaction site (due to reaction of methoxylation on silica surface) results in the fact that the reaction with $(CH_3)_3SiI$ and $CH_3)_3SiBr$ does not proceed under usual conditions. It is important that the bond length decreases and the silicon-halogen bond energy increases in the same succession as the sequence of activities of trimethylhalosilanes in relation to surface silanol groups. When explaining the differences in reactivity of trimethylhalosilanes, account should be taken of the easier polarizability of the large-sized atoms (and the appropriate silicon-halogen bonds), as well as the differences in the proton-accepting properties of the atoms related to silicon.

For trimethylsilyl-containing compounds with Si–N bond, for example, for trimethylpseudo-halosilanes having a of common formula $(CH_3)_3SiX$, where X = N_3, NCO, NCS, $NCNSi(CH_3)_3$, it was found that the given compounds are chemisorbed by the surface of fumed silica, prepared at 600°C, even at room temperature (Fig. 3).

Absence of distinct changes in the intensity of the band of surface silanol groups (3750 cm^{-1}) together with simultaneous appearance of the absorption bands characteristic of the chemisorbed trimethylsilyl and nitrogen-containing ($\equiv SiN_3$, $\equiv SiNCO$, $\equiv SiNCS$ or $\equiv SiNCNSi(CH_3)_3$) groups is attributed to the participation of the siloxane bridges in the reaction (mechanism $Ad_{N,E}$):

$$\equiv SiOSi\equiv + XSi(CH_3)_3 \rightarrow \equiv SiOSi(CH_3)_3 + \equiv SiX$$

Scheme 7.

Registration of the absorption bands in IR-spectra, characteristic of the chemisorbed $\equiv SiX$ groups on methoxysilica surface after its contact with the vapor of N-containing trimethyl-substituted silanes serves as an additional argument in favour of direct addition of the $(CH_3)_3SiX$ molecules to the siloxane bonds. The $\equiv SiN_3$, $\equiv SiNCO$, $\equiv SiNCS$ or $\equiv SiNCNSi(CH_3)_3$ groups fixed on the surface are easily subjected to hydrolysis or methanolysis [16].

T, %

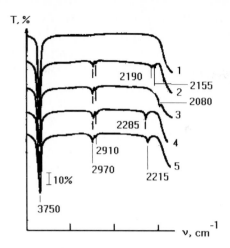

Fig. 3. Infrared absorption spectra of the fumed silica, prepared at 600°C (1) and modified 1h at 20°C by vapor of (CH$_3$)$_3$SiN$_3$ (2), (CH$_3$)$_3$SiNCS (3), (CH$_3$)$_3$SiNCO (4) or [(CH$_3$)$_3$Si]$_2$NCN (5).

As temperature increases, the process of electrophilic proton substitution in the isolated silanol surface groups with formation of the grafted trimethylsilyl groups proceeds also. However, as distinct from other Si–N-containing compounds, e.g., hexamethyldisilazane, trimethyl(dimethyl-amino)silane, dimethylaminotriethoxysilane, trimethylsilylpiperidine, and trimethylsilylimidazole, which react with the ≡SiOH groups even at room temperature, the complete involvement of the structural silanols in reaction with trimethylasidosilane requires heating up to 400°C (up to 500°C in the case of trimethylthioisocyanatesilane), and up to 550°C when using trimethylisocyanatesilane and bis(trimethlsilyl)carbodiimide. Reactivity of trimethylpseudohalosilane with Si–N bonds in respect to –OH surface groups decreases in succession:

(CH$_3$)$_3$SiN$_3$ > (CH$_3$)$_3$SiNCS > (CH$_3$)$_3$SiNCO, [(CH$_3$)$_3$Si]$_2$NCN

Characteristically, the activity of the same reagents in the reactions of siloxane bond cleavage changes in the reverse order.

For the purpose of explaining the differences in reactivity of the trimethylpseudohalosilanes in the reaction with the silica surface sites of the same type the analytical treatment was given to geometric and electronic configuration of the attacking molecules. Calculations have shown that the length of an Si–N bond depends on the composition of a pseudohalogenide group and decreases in the sequence SiN$_3$ > SiNCS > SiNCO. Stability of the molecular system, as governed by enthalpy of formation of the corresponding trimethylpseudohalosilane, increases with the decrease of the Si–N bond length. Thus, reduction in activity of (CH$_3$)$_3$SiX compounds in the sequence X= N$_3$, NCS, NCO correlates with calculations, concerning the rise in stability of the reagent molecules and decrease in the equilibrium length of the Si–N bond. The calculations are also giving evidence of a distinct charge alteration on the atoms of the X-group trimethylpseudohalosilanes, which may favour dissociation of the molecules on the corresponding surface sites.

As to reactions of heterolytic cleavage, the Si–C bond usually exhibits a rather low activity. This fact also concerns the reactions involving silica surface sites. For instance, according to [13], tetramethylsilane enters into reaction with the free hydroxyl surface groups at temperatures above 550°C. The activation energy of the process, according to the S$_{Ei}$ mechanism, accounts for around

185 kJ/mol of the grafted trimethylsilyl groups. Trimethylcyanosilane, however, is the exception. As was shown in [18], the $(CH_3)_3SiCN$ molecules may rather easily react with the OH-groups of the silica surface. Even at room temperature about 70 % of the isolated silanol groups are substituted by the trimethylsilyl ones within 1 hour. Reaction with trimethylcyanosilane does not proceed if a proton is excluded from the reaction site by methoxylation of the surface. These experimental data form a foundation allowing to attribute the chemisorption of the $(CH_3)_3SiCN$ molecules to the processes of electrophilic proton substitution in the silanol groups. When analyzing the reasons of sharp differences in activity of the $(CH_3)_3SiCH_3$ and $(CH_3)_3SiCN$ molecules in the reaction with the $\equiv SiOH$ surface groups account should be taken of a higher electronegativity of a CN-group as compared to the methyl one, which may result in increase in the effective positive charge on a silicon atom in the trimethylcyanosilane molecule. It is significant also that the proton-accepting properties of a CN-groups are higher than those of the methyl one. Thus, according to [18] shift of the 3750 cm^{-1} band on adsorption of $(CH_3)_3SiCN$ vapor is equal to 330 cm^{-1}, whereas $\Delta\nu_{OH}$ at adsorption of tetramethylsilane amounts to only 40 cm^{-1}.

Chemisorption of bis(trimethylsilyl)sulfate $[(CH_3)_3SiO]_2SO_2$ on silica surface begins even at room temperature, and the complete involvement of the OH-groups in reaction occurs at $100°C$ [19]. Reaction is realized by S_Ei mechanism without any sulfonation on silica surface. When comparing the reactivities of different trimethylsilyl-containing compounds with the Si–O bond in interaction with the OH surface groups, it was noted [1] that activity of a modifying reagent exhibits a strong dependence on the nature of the oxygen atom substituent:

$(CH_3)_3SiOSO_2OSi(CH_3)_3 > (CH_3)_3SiOCH_3 > (CH_3)_3SiOSi(CH_3)_3$ hexamethyldisiloxane above $380°C$. Characteristically that high activity in respect to the structural silanol groups on the surface is typical also of the other sulfur-containing organosilicon compounds, i.e., trimethylsilylfluoromethyl-sulfonate $(CH_3)_3SiOSO_2CF_3$ and hexamethyldisylthione $(CH_3)_3SiSSi(CH_3)_3$ [20].

Bogillo [21] offered some linear relationships between quantum chemical indexes of modifying reagent molecules (their maximum charges and energies of the frontier orbitals) and their empirical donor-acceptor parameters as well as heats of the hydrogen-bonded complexes formation with silica surface OH groups or the chemisorption activation energies. These relationships may be used for estimation of the complexes stability and reactivity of various organic and organosilicon compounds towards active sites of silica surface. The electron-donating ability of surface sites and electron-accepting ability of reagent molecules play a leading part in the rate of electrophilic substitution of proton in the surface silanol groups. Just as the opposite properties of reagent molecules affect the stability of hydrogen-bonded precursors in these reactions. The overall activation reaction barrier reduces the increase of preliminary adsorption complex stability and the overall reaction rate grows.

Some Peculiarities of Surface Chemical Reactions with Chlorides and Oxo-chlorides of Different Elements (Elimination Reactions).

Formerly the data were obtained [22] that thionyl chloride enters into interaction with the hydroxyl groups of silica surface by electrophilic proton substitution, and after that the forming surface compounds transforms (elimination process), and the ultimate result of such transformation corresponds to nucleophilic substitution on silicon atom.

We have suggested that such transformations of surface chemical compounds, simulating to certain degree some of the catalytic transformations on the surface of oxides, may be considered as a more general class of reactions, which unites the processes that proceed according to the S_Ei and S_Ni mechanisms. Based on the analytical treatment it was shown [1] that a similar type of reactions involving the structural hydroxyl groups on the surface of dispersed silica is realized also in the other cases. In particular, chemisorption of WCl_6 [23] by the isolated silanol groups the forming surface chemical compounds ($\equiv SiOWCl_5$) will decompose on heating with elimination of tungsten oxychloride:

$$\equiv SiOWCl_5 \ \rightarrow \ \equiv SiCl + WOCl_4$$

Scheme 8.

The surface compounds $\equiv SiOWOCl_3$ formed on reaction of the silanol groups with $WOCl_4$ experience transformations with the formation of $\equiv SiCl$ groups on the surface and WO_2Cl_2 release [24]. Manifestation of the elimination processes (E) was also detected on chemisorption of halogenides and oxyhalogenides of the other elements (Table 2).

Table 2. Structure of intermediate surface compounds and the products of elimination in reactions of some chlorides and oxochlorides with isolated silanol groups.

Modifying Reagent	Intermediate Surface Compounds	Product of Elimination Reaction	Ref.
CCl_4	$\equiv SiOCCl_3$	$COCl_2$	[22]
$SOCl_2$	$\equiv SiOSOCl$	SO_2	[22]
$TiCl_4$	$\equiv SiOTiCl_3$	$TiOCl_2$	[22]
WCl_6	$\equiv SiOWCl_5$	$WOCl_4$	[23]
$WOCl_4$	$\equiv SiOWOCl_2$	WO_2Cl_2	[24]
WCl_5	$\equiv SiOWCl_4$	$WOCl_3$	[25]
$MoOCl_4$	$\equiv SiOMoOCl_3$	MoO_2Cl_2	[26]
$MoCl_5$	$\equiv SiOMoCl_4$	$MoOCl_3$	[27]
PCl_5	$\equiv SiOPCl_4$	$POCl_3$	[28]

In all these reactions the formation of the $\equiv SiCl$ groups and elimination of the appropriate relatively volatile oxochlorides were discovered. It would be interesting in the future to determine the factors, which govern stability of the chemical compounds formed in such a case.

Addition Reactions in the Synthesis of some Surface Compounds

Reaction of phenols with the $\equiv SiNH_2$ groups, which were previously introduced into the surface layer of chlorinated silica, was used for preparation of the grafted phenol derivatives [29]. It was shown that on silica, bearing the $\equiv SiNH_2$ groups, the quaternary ammonium compounds are forming at the initial stage (the process of electrophilic addition Ad_E), which after destruction at 100-140°C, results in chemisorption of the adequate ether groups (elimination reaction).

Recently the ample possibilities of the use of solid-phase hydrosilylation reactions for the formation of dense modifying coverings has been established [30-36]. In these works it has been shown that ≡Si–H groups, grafted on the silica surface, may be practically quantitatively involved into chemical reactions with alk-1-enes both in the presence of Speier's catalyst and in the absence of catalyst at high temperature. Modifying silicas with grafted hydrocarbon radicals, containing 6, 8, 10, 14, 16, or 18 carbon atoms, were synthesized by such method. The achieved concentrations of grafted groups is decreasing from 3.70 $\mu mol/m^2$ for $n = 6$ to 0.71 $\mu mol/m^2$ for $n = 18$. However in spite of the decrease of grafted hydrocarbon group concentration with increasing n, the hydrophobic nature of modified silicas is enhanced. It is obvious that hydrocarbon radicals with the high content of carbon atoms do not have a brush-like structure, rather these groups are inclined and they pave the silica matrix surface. Catalytic addition of functional olefins (acetyl acetone, vinyl acetate, acrylamide) to Si–H groups on the silica surface proceeds with the lesser yield (from 75 to 30%). For example, in the case of styrene addition by use of solid-phase hydrosilylation reaction the content of grafted phenyl ethyl groups amounts to 1.53 $\mu mol/m^2$ [36]. The decrease in the reactivity of functional olefins in the process indicated may be related to the decrease of the electron density on carbon-carbon double bonds of reagent molecules since in the case of unsaturated compounds this density is determined by the mesomeric effect of substituents.

The attachment of one of the participants of the hydrosilylation process to solid surface provides more convenient and favorable standpoints and approaches to researches into reaction mechanisms. The experimental data [37] give evidence that solid-phase catalytic hydrosilylation reactions involve the formation of intermediate complexes containing both reagents and Speier's catalyst. The possibility has also been established of effecting a non-catalytic reaction of solid-phase hydrosilylation of olefins at high pressure and temperature. The modifying coatings obtained using solid-phase hydrosilylation reactions have a rather high hydrolytic stability and reproducible properties. This allows [38] to regard the reaction of solid-phase hydrosilylation as to the basis for one of promising methods of producing surface chemical compounds with Si–C bonds.

References:

[1]   V. A. Tertykh, L. A. Belyakova, *Chemical Reactions with the Silica Surface Participation*, Naukova Dumka, Kiev, **1991** (in Russian).

[2]   V. A. Tertykh, V. V. Pavlov, K. I. Tkachenko, A. A. Chuiko, *Teoret. Eksperim. Khimiya* **1975**, *11*, 415.

[3]   S. M. Budd, *Phys. Chem. Glasses* **1961**, *2*, 111.

[4]   V. V. Strelko, V. A. Kanibolotskii, *Kolloid. Zhurn.* **1971**, *33*, 750.

[5]   V. A. Tertykh, V. V. Pavlov, *Adsorbtsiya i Adsorbenty* **1978**, *6*, 67.

[6]   C. K. Ingold, "*Structure and Mechanism in Organic Chemistry*", Cornell University Press, Ithaca, London, **1969**.

[7]   L. H. Sommer, "*Stereochemistry, Mechanism, and Silicon*", McGraw-Hill, **1965**.

[8]   V. M. Gun'ko, A. A. Chuiko, *Kinetika i Kataliz* **1991**, *32*, 322.

[9]   V. M. Gun'ko, *Kinetika i Kataliz* **1991**, *32*, 576.

[10] V. M. Gun'ko, *Kinetika i Kataliz* **1993**, *34*, 691.

[11] V. M. Gun'ko, E. F. Voronin, E. M. Pakhlov, A. A. Chuiko, *Langmuir* **1993**, *9*, 716.

[12] V .I. Bogillo, V. M. Gun'ko, *Langmuir* **1996**, *12*, 115.

[13] V. A. Tertykh, V. V. Pavlov, K. I. Tkachenko, A. A. Chuiko, *Teoret. Eksperim. Khimiya* **1975**, *11*, 174.

[14] Yu. I. Gorlov, V. A. Zayats, A. A. Chuiko, *Teoret. Eksperim. Khimiya* **1989**, *25*, 756.

[15] V. A. Tertykh, L. A. Belyakova, A. M. Varvarin, L. A. Lazukina, V. P. Kukhar', *Teoret. Eksperim. Khimiya* **1982**,*18*, 717.

[16] V. A. Tertykh, L. A. Belyakova, A. M. Varvarin, *React. Kinet. Catal. Lett.* **1989**,*40*, 151.

[17] Yu. I. Gorlov, *React. Kinet. Catal. Lett.* **1993**, *50*, 89.

[18] A. M. Varvarin, L. A. Belyakova, V. A. Tertykh, L. A. Lazukina, V. P. Kukhar' *Teoret. Eksperim. Khimiya* **1987**, *23*, 117.

[19] A. M. Varvarin, L. A. Belyakova, V. A. Tertykh, L. A. Lazukina, V. P. Kukhar' *Teoret. Eksperim. Khimiya* **1989**, *25*, 377.

[20] J. Chmielowiec, B. A. Morrow, *J.Colloid Interface Sci.* **1983**, *94*, 319.

[21] V. I. Bogillo, *"Kinetics of organic compounds chemisorption from the gas phase on oxides surface"*, in: *Adsorption on New and Modified Inorganic Sorbents* (Eds.: A.Dabrowski, V.A.Tertykh), Elsevier, Amsterdam, **1996**, p.237.

[22] V. V. Pavlov, V. A. Tertykh, A. A. Chuiko, K. P. Kazakov, *Adsorbtsiya i Adsorbenty* **1976**, *4*, 67.

[23] I. V. Babich, Yu. V. Plyuto, A. A. Chuiko, *Zhurn. Fiz. Khimii* **1988**, *62*, 516.

[24] I. V. Babich, Yu. V. Plyuto, A. A. Chuiko, *Dokl. AN UkrSSR* **1987**, *N4*, 39.

[25] P. A. Mutovkin, I. V. Babich, Yu. V. Plyuto, A. A. Chuiko, *Ukr. Khim. Zhurn.* **1993**, *59*, 727.

[26] A. A. Gomenyuk, I. V. Babich, Yu. V. Plyuto, A .A. Chuiko, *Zhurn. Fiz. Khimii* **1990**, *64*, 1662.

[27] Yu. V. Plyuto, A. A. Gomenyuk, I. V. Babich, A. A. Chuiko, *Kolloid. Zhurn.* **1993**, *55*, *N6*, 85.

[28] P. A. Mutovkin, Yu. V. Plyuto, I. V. Babich, A. A. Chuiko, *Ukr. Khim. Zhurn.* **1991**, *57*, 367.

[29] E. F. Voronin, V. I. Bogomaz, V. M. Ogenko, A. A. Chuiko, *Teoret. Eksperim. Khimiya* **1980**, *16*, 801.

[30] V. A. Tertykh, L. A. Belyakova, A. V. Simurov, *Zhurn. Fiz. Khimii* **1990**, *64*, 1410.

[31] V. A. Tertykh, L. A. Belyakova, A. V. Simurov, *Dokl. AN SSSR* **1991**, *318*, 647.

[32] J. E. Sandoval, J. J. Pesek, *Anal. Chem.* **1991**, *63*, 2634.

[33] V. A. Tertykh, L. A. Belyakova, A. V. Simurov, *Mendeleev Commun.* **1992**, 46.

[34] V. A. Tertykh, L. A. Belyakova, *Zhurn. Fiz. Khimii* **1993**, *67*, 2116.

[35] A. V. Simurov, L. A. Belyakova, V. A. Tertykh, *Functional Materials* **1995**, *2*, 51.

[36] V. A. Tertykh, S. N. Tomachinsky, *Functional Materials* **1995**, *2*, 58.

[37] L. A. Belyakova, A. V. Simurov, *Ukr. Khim. Zhurn.* **1994**, *60*, 1400.

[38] V. A. Tertykh, L. A. Belyakova, *"Solid-phase hydrosilylation with participation of modified silica surface"*, in: *Adsorption on New and Modified Inorganic Sorbents* (Eds.: A.Dabrowski, V.A.Tertykh), Elsevier, Amsterdam, **1996**, p.147.

Microporous Thermal Insulation:
Theory, Properties, Applications

Hans Katzer*, Johann Weis
Wacker-Chemie GmbH
Geschäftsbereich S – Werk Kempten
Max-Schaidhauf-Str. 25, D-87437 Kempten, Germany

Keywords: Fumed Silica / Aerogels / Thermal Insulation

Summary: The physical requirements for a high performance insulation material can best be fulfilled with pressed mixtures consisting of fumed silica and opacifiers or with aerogels. No other commercially available non-flammable system can compete with these, either by heat flow transferred through gases, or through solid-state transmission. The high infrared permeability of thermal radiation can be reduced with the addition of opacifiers. Since commencement of commercial production at the beginning of the 1950s, the interest in microporous thermal insulation materials has grown continuously. Therefore, for example, aerogels are used in the insulation of lighting elements, refrigerators and water boilers. Products made from fumed silica are largely used as insulation for radiant heaters in glass ceramic hobs, in night storage heaters and in numerous industrial applications in high temperature areas.

Introduction

An increased awareness of the general public regarding environmental and ecological issues has also become a driving force for new developments in the area of thermal insulation.

Whereas for thousands of years, thermal insulation was selected mainly on subjective experience, today thermal insulation can only be developed based on accurate knowledge of the fundamental principles of heat transfer.

Heat Flow Components

For heat transfer through any particular medium, having a temperature T on the one side and a temperature $T + \Delta T$ on the other, Fourier's empirical law applies:

$$\dot{Q} = \lambda \cdot f \cdot \frac{\Delta T}{\Delta s}$$

$\dot{Q}$ = total heat flow through a solid
λ = total thermal conductivity
f = surface area
ΔT = temperature gradient
Δs = thickness

In a porous thermal insulation, the total heat flow $\dot{Q}$ is normally composed of a gaseous heat flow $\dot{Q}_{Gas}$, which can consist of a conductive and a convective component, a solid conduction heat flow $\dot{Q}_S$ and a radiative heat flow $\dot{Q}_{Rad}$:

$$\dot{Q} = \dot{Q}_{Gas} + \dot{Q}_S + \dot{Q}_{Rad}$$

Corresponding to the above, the following formula is also valid: $\lambda = \lambda_{Gas} + \lambda_S + \lambda_{Rad}$

In a good thermal insulation material, the individual portions of λ_{Gas}, λ_S, and λ_{Rad} should be as small as possible.
Let us take a closer look at the individual addends:

Heat Flow Through a Gas

For heat transfer through a gas, the following formula applies:

$$\lambda_{Gas0} = \frac{1}{3} \cdot c_V \cdot \rho_{Gas} \cdot l_{Gas} \cdot v_{Gas}$$

λ_{Gas0} = thermal conductivity of the gas
c_v = specific heat capacity at constant volume
ρ_{Gas} = density
l_{Gas} = mean free path of the gas molecules
v_{Gas} = mean velocity of the gas molecules

For a porous insulation material this formula applies under the condition that the pore diameter is much larger than the mean free path of the gas molecule.
If this condition is not met, that is, if the pore diameter has a similar magnitude as to that of the mean free path, for example, if a very fine powder is compacted, then the following formula according to Kaganer [1] applies:

$$\lambda_{Gas} = \frac{\lambda_{Gas0}}{1 + 2\beta \cdot Kn}$$

λ_{Gas0}	=	thermal conductivity of the gas
λ_{Gas}	=	thermal conductivity of the gas in pores
β	=	weight factor
Kn	=	Knudsen number as with

$$Kn = \frac{l_{Gas}}{\delta}$$

l_{Gas}	=	mean free path of the gas molecules
δ	=	pore diameter

In a porous system the thermal conductivity component λ_{Gas} is reduced when there is an increase in the mean free path. This can be achieved by evacuation of the system, for example.

Since the mean free path of O_2- or N_2-molecules lies at approx. 70 nm under normal conditions, thermal insulation materials with pore diameters in this range would have a great technical importance. Such systems were first achieved during the 1950s.

If volatile silicon compounds, such as $SiCl_4$, are burned in a hydrogen / oxygen flame, the first results are primary particles with a diameter of approx. 10 nm.

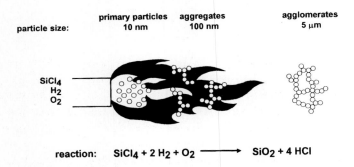

Fig. 1. Production of fumed silica in a flame process.

These spheres, which are "sticky" at high temperature, adhere together when they collide in the gas stream and build aggregates with lengths of approx. 100 nm. At lower temperature they agglomerate into larger groups. A very fluffy white powder with an apparent density of 0.02 - 0.04 g/cm^3 develops. On the other hand, the basic amorphous SiO_2 has a density of 2.2 g/cm^3. When this powder is compressed to a panel with a density of approx. 0.22 g/cm^3, the result is a material with a porosity of 0.9.

According to Kaganer [1], the pore diameter δ of such a material can be rated using the following formula:

$$\delta = \frac{2}{3} \cdot \frac{\Pi}{1-\Pi} \cdot d$$

δ = pore diameter
Π = porosity
d = diameter of the particle

For fumed silica ($d = 10$ nm, $\Pi = 0.9$) the results at normal pressure are:
$\delta = 60$ nm and $Kn = 70$ nm/60 nm = 1.2

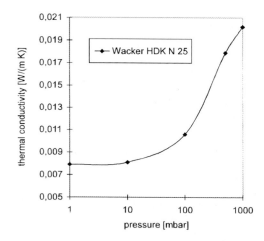

Fig. 2.   Thermal conductivity versus residual gas pressure (compressed fumed silica BET 250 m^2 g^{-1}).

The sharp decline in the thermal conductivity in the range 1000-100 mbar shows that the pore diameter of this material is of the same order as the mean free path of the gas molecules. Based on the above, fumed silica is excellent for use in the production of the so-called "microporous" thermal insulation.

Since the 1980s a new microporous insulation material based on silicic aerogels is in a test phase. About 60 years ago Kistler from Stanford University developed a method to produce aerogels from silicic material without shrinking them. He reacted waterglass with hydrochloric acid, washed out the sodium and chloride ions from the gel, substituted the water with alcohol and dried the gel in an autoclave above the critical point. Other methods begin with tetramethoxysilane for the production of monolithic aerogels.

The structure of aerogel is similar to the structure of a compressed fumed silica. The diameter of the SiO$_2$ sphere is about 5 nm, the diameter of the pores in the aerogel is about 50 nm. Aerogels are, therefore, also excellent microporous insulation materials.

Heat Flow Through Solid Matter

For thermal conductivity in inorganic, non-metallic solid matters, the following formula applies:

$$\lambda_{SM} = \frac{1}{3} c_v \cdot v \cdot l_P \cdot \rho_P$$

λ_{SM}	=	thermal conductivity of the solid matter
c_v	=	specific heat capacity (based on volume)
v	=	velocity of sound
l_P	=	mean free path of the phonons
ρ_P	=	density of the phonons

The phonons can interact amongst themselves or in reciprocal action with other particles, same as the gas molecules. Therefore, heat transfer is dependent on the structure of the solid matter. Since phonons are distributed with every disturbance of the crystal lattice, the value of λ_{SM} drops as the number of crystal lattice defects, impurities, grain boundaries or amorphous areas rises.

In a thermal insulation made of moulded powder, heat transfer in the solid is achieved exclusively through the contact points between the particles.

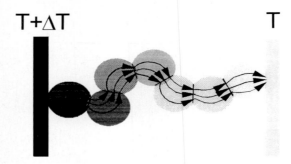

T+ΔT   T

Fig. 3.   Thermal conductivity by way of the contact points.

The following applies according to Kaganer [1]:

$$\lambda_{Cont} = N \cdot a \cdot \frac{(1-\Pi)^2 \cdot \lambda_{SM}}{r} \approx \lambda_S$$

λ_{Cont}	=	thermal conductivity by way of contact points
N	=	number of particle contact points with their neighbours
a	=	contact radius
Π	=	porosity
λ_{SM}	=	thermal conductivity of the solid matter
r	=	radius of particles
λ_S	=	thermal conductivity of the solid porous material

From this follows that along with the thermal conductivity, the hardness of the solid material is also significant, as this is responsible for the contact radius.

When the values for fumed silica (λ_{Quartz} = 1.6 $Wm^{-1}K^{-1}$, r = 5 nm, $N \approx 1.16$ (according to Kiselev [2]), a $\approx$ 2 nm (estimated from SEM and TEM) are inserted, the result is a value of about 0.007 $Wm^{-1}K^{-1}$.

Table 1. Thermal conductivity of solids.

Material	λ_S in $Wm^{-1}K^{-1}$	
Aerogel (0.07 g/cm³)	0.0002	Kaganer [1]
Aerogel (0.457 g/cm³)	0.001	Kaganer [1]
Fumed Silica + 16% Opacifier	0.006	Büttner et al. [3]

Experimentally, λ_{Cont} can be determined when the total thermal conductivity of the system is determined in vacuum as a function of the temperature.

Under these conditions, λ_{Gas} is zero and $\lambda_{Rad} \sim T^3$, so that all measurement values lie on a straight line with the equation $\lambda = \alpha + \beta T^3$. The point of intersection of the straight line with the y-axis (T^3 = 0) yields the value for λ_S. These measurements have been described in several publications.

Radiative Heat Flow

Each body sends out thermal radiation at a temperature above 0 K. The radiated power which is emitted increases extraordinarily at a higher temperature. The Stefan-Boltzmann law applies here:

$$E = \sigma \cdot T^4$$

E	=	radiated power
σ	=	Stefan-Boltzmann constant
T	=	absolute temperature

With the help of the "Wien's law": $\Lambda_{max} \cdot T$ = const. = 2897,8 µm K the wavelength Λ_{max} of the maximum emissions can be calculated.

At a temperature between 0 and 1000°C it lies approximately between 10 and 2 µm. Based on considerations from the area of astrophysics, the energy transfer by radiation within a fine-grained material can be treated like a heat transfer through photons with a certain mean free path:

$$\lambda_{Rad} = \frac{16}{3} \cdot \frac{n^2 \cdot \sigma \cdot T^3}{[e(T) \cdot \rho]}$$

σ	=	Stefan-Boltzmann constant
n	=	mean refractive index
T	=	temperature in degrees absolute
$e(T)$ =		specific extinction coefficient
ρ	=	density

Since microporous thermal insulation made from fumed silica is used in higher temperature areas and SiO₂ is radiation permeable in the infrared, the addition of a so-called opacifying agent is needed to reduce the radiation.

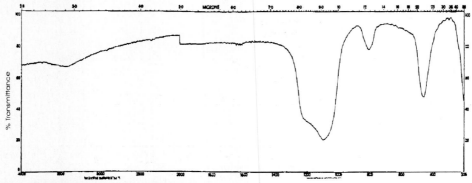

Fig. 4. IR spectrum of fumed silica (Wacker HDK T30).

The physicist Mie created the theoretical basis for choosing the appropriate opacifying agent. Through his work the extinction profile of ball-shaped dielectric particles and, therefore, the optimum particle diameter of an opacifier dependent on wavelength, has become calculable.

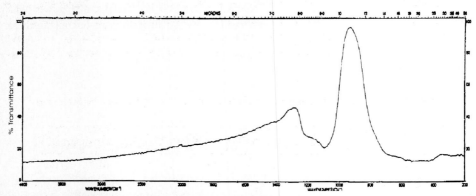

Fig. 5. IR spectrum of fumed silica + opacifier (Ilmenite FeTiO₃).

The opacifier should be distributed as evenly as possible in the fumed silica matrix. An optimal method is to mix the opacifying agent in the agglomeration phase during production.

For a long time the technical application of this microporous insulation was limited, as the material made from pressed powder was hardly workable and, for example, had to be protected through a covering made from glass fiber.

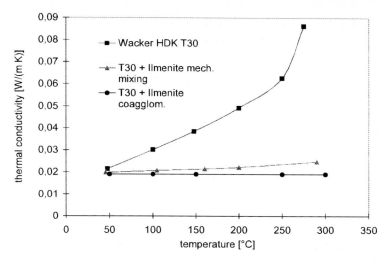

Fig. 6. Thermal conductivity versus temperature.

A raw material containing good mechanical properties could be attained through the addition of so-called hardening agents. Compounds such as B_4C, CaB_6, TiB_2 in a finely distributed form are suitable as hardening agents. These are added in quantities of approx. 0.5 to 1.5 wt.% to the original materials–fumed silica and opacifier–before pressing. Subsequently, the formed piece is heated to approx. 700-800°C. With the use of, for example, a fine B_4C-powder as a hardening agent, an extremely strong reaction results at approx. 550-600°C, which leads promptly to highly localized spot sintering and through further reaction of the resulting boroxide also to glazing. The important issue in this process is to steer the reaction in such a way that the solid conductivity λ_S does not increase significantly.

Table 2. Properties of microporous thermal insulation materials before and after hardening reaction.

	before reaction	after reaction
λ [Wm^{-1} K^{-1}]	0.021	0.022
BET-surface [m^2 g^{-1}]	142	140
Bending strength [N mm^{-2}]	0.08	0.35
Compression strength [N mm^{-2}]	0.94	1.26

Applications for Aerogels

Unfortunately, aerogels are not yet produced on a large industrial scale, so only some feasibility studies with material from pilot plants have been performed. Refrigerators and water boilers insulated with aerogels showed a remarkable reduction of energy consumption. Aerogel-filled windows can reduce energy losses in buildings significantly and have a high savings potential for heating materials.

Applications for Microporous Thermal Insulation Materials

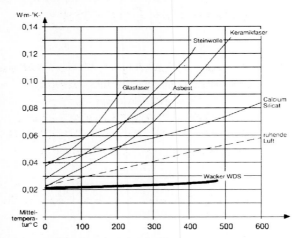

Fig. 7. Thermal conductivity of different insulation materials versus temperature.

The uses of microporous insulation materials are almost unlimited. This material can be used to protect specific areas from high temperature. It can increase the effective volume and the productivity of heated devices for the same energy consumption.

In household appliances it helps to achieve slimline units and to increase thermal efficiency, for example, in night storage heaters and radiant heaters in glass ceramic hobs. In industrial applications the main fields are thermal insulation of ceramic kilns, glazing kilns, laboratory furnaces, and industrial baking furnaces. In the aluminum and glass industry microporous material offers the advantage of a reduction in the amount of heat loss through the walls of containers and feeder pipes. There are also many other applications, for example, in the automotive industry, fire protection, shipbuilding, environmental technology and so forth. The advantages of a microporous insulation material become apparent when compared with conventional insulations, as can be seen in Fig. 7.

Acknowledgement: The author thanks Dr. H. Reiss, *ABB Heidelberg*, for many stimulating discussions and for reading and correcting the German manuscript.

References:

[1]   M. G. Kaganer, *Thermal Insulation in Cryogenic Engineering* (transl. by A. Moscona), Israel Progr. Sci. Transl., Jerusalem, **1969**.

[2]   A. V. Kiselev, *"Korpuskulyarnaya struktura adsorbentov-gelei"*, in: Sbornik *Metody issledovaniya struktury vysokodispersnykh i poristykh tel*, Moskva, Izd. AN SSSR., **1958**.

[3]   D. Büttner, J. Fricke, R. Krapf, H. Reiss, *"Measurement of the thermal conductivity of evacuated load-bearing, high-temperature powder and glass board insulations with a 700x700 mm² guarded hot-plate device"*, in: *Proc. 8th Eur. Conf. Thermophys. Prop.*, Baden-Baden, **1982**.

Nitridosilicates – High Temperature Materials with Interesting Properties

Wolfgang Schnick*, Hubert Huppertz, Thomas Schlieper
Laboratorium für Anorganische Chemie
Universität Bayreuth
Universitätsstraße 30, D-95440 Bayreuth, Germany
Tel.: Int. code + (921)552530 – Fax: Int. code + (921)552788
E-mail: wolfgang.schnick@uni-bayreuth.de

Keywords: Nitridosilicates / High Temperature Synthesis / Network Structures / Materials

Summary: Systematic investigations enabled a broad synthetic access to the nitridosilicate class of compounds. Typical building blocks in these compounds are SiN_4 tetrahedra, which may be linked via common vertices forming complex network structures. The first results suggest that the structural variabilities of nitridosilicates are more versatile as compared to the oxosilicates. Highly cross-linked network structures are accessible, which cause the remarkable stability of nitridosilicates.

Introduction

Aiming at the development of new inorganic materials, we focussed our work on the synthesis of new nitrides [1], which are compounds containing nitrogen as the electronegative element. In selecting the electropositive elements to be bound to nitrogen, the following aspects seemed to be important: with respect to the demanded properties of the materials, the constituting elements should exhibit a relatively low specific weight. Furthermore, rather stable covalent chemical bonds should occur in combination with nitrogen. At the same time, polymeric cross-linked structures should be possible in the solid. And, last but not least, the respective chemical elements should possess a sufficient availability. Particularly, the following light main-group elements meet these requirements: boron, aluminum, carbon, silicon, and phosphorus.

In the last few years we began our efforts to develop new inorganic solid materials with the synthesis and structure determination of the phosphorus nitrides [2]. Based on the synthesis of pure, defined, and crystalline P_3N_5 [3] we started a systematic preparative and structural investigation of ternary and multinary phosphorus nitrides.

In all phosphorus nitrides known so far PN_4 tetrahedra occur as typical building units of the P–N partial structures. Depending on the degree of condensation these PN_4 tetrahedra may be linked via common vertices forming rings, chains, cages, or three-dimensional network structures (Fig. 1). Accordingly, the phosphorus nitrides exhibit strong analogies to the silicates and phosphates.

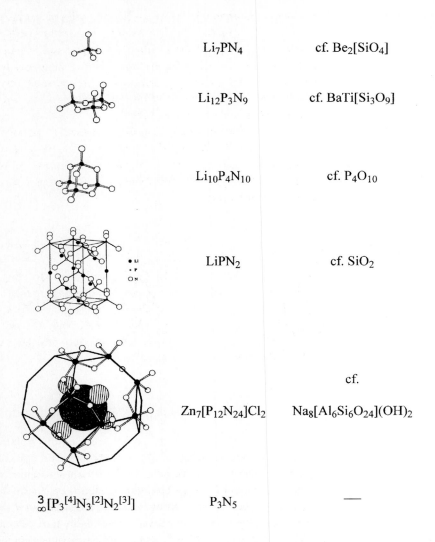

Li$_7$PN$_4$ cf. Be$_2$[SiO$_4$]

Li$_{12}$P$_3$N$_9$ cf. BaTi[Si$_3$O$_9$]

Li$_{10}$P$_4$N$_{10}$ cf. P$_4$O$_{10}$

LiPN$_2$ cf. SiO$_2$

cf.

Zn$_7$[P$_{12}$N$_{24}$]Cl$_2$ Na$_8$[Al$_6$Si$_6$O$_{24}$](OH)$_2$

$\frac{3}{\infty}$[P$_3^{[4]}$N$_3^{[2]}$N$_2^{[3]}$]   P$_3$N$_5$ —

Fig. 1. In phosphorus nitrides PN$_4$ tetrahedra occur as typical building blocks, which may be linked via common vertices exhibiting structural analogies to silicates and phosphates.

Due to their structures and properties some of the phosphorus nitrides may have the potential for various ceramic applications, e.g., ionic conductors [4], sintering additives [5], pigments [1], or microporous materials [6]. The highly condensed phosphorus nitrides LiPN$_2$, HPN$_2$, M$_{(7-x)}$H$_{2x}$[P$_{12}$N$_{24}$]X$_2$, and P$_3$N$_5$ exhibit a high chemical stability towards hydrolysis by acids and bases, which is comparable to the behavior of silicon nitride Si$_3$N$_4$. However, with respect to the development of high performance materials the phosphorus nitrides show a significant disadvantage as their thermal stability seems to be limited to a maximum temperature of 1000°C [1].

This limitation is the result of two effects concerning the P–N network structures occurring in the phosphorus nitrides: on the one hand, the P–N bonds do not exhibit a sufficient bonding stability and on the other the cross-linking of corner sharing PN_4 tetrahedra normally is limited to a situation in which the nitrogen atoms are linked only to two neighboring P atoms ($N^{[2]}$) analogously to the situation of oxygen in phosphates and silicates. Only in binary phosphorus nitride and some very few ternary compounds like $K_3P_6N_{11}$ the cross-linking is enhanced and a small portion of the nitrogen atoms each connect three phosphorus atoms ($N^{[3]}$) while the other ones form ($N^{[2]}$) bridges.

In order to raise the stability of nonmetal nitride network structures the element phosphorus has to be exchanged against silicon. In phosphorus nitrides the relative amount of ($N^{[3]}$) never seems to exceed a value of 2/5 and in most examples it is zero. In contrast, in silicon nitrides even all of the nitrogen atoms each may connect three silicon atoms ($N^{[3]}$). This topological situation is realized in the two modifications of binary silicon nitride Si_3N_4 (Fig. 2).

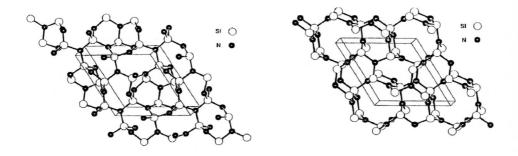

Fig. 2.   Crystal structures of α-(left) and β-Si_3N_4 (right). All of the nitrogen atoms form ($N^{[3]}$) bridges between three neighboring silicon atoms.

As compared to phosphorus nitrides the cross-linking of the TN_4 tetrahedra (T = P, Si) significantly is enhanced in silicon nitrides while at the same time the T–N bond energy increases from phosphorus (290 kJ/mol) to silicon (335 kJ/mol) [1]. Accordingly, silicon nitride exhibits a thermal stability up to 1900°C, which is much higher compared with the value of phosphorus nitride [1]. Due to its high thermal and chemical stability in combination with a low density (ρ = 3.2 gcm^{-3}), a remarkable hardness (Vickers hardness = 1400-1700 MNm^{-2}), and a high mechanical strength (up to 1400°C) silicon nitride has become one of the most important nonoxidic materials. The applications of Si_3N_4 range from ceramic turbochargers, valve tappets, and cutting tools to integrated semiconductor modules [7, 8].

However, a systematic investigation of the chemistry of ternary and multinary silicon nitrides in analogy to the phosphorus nitrides has been prevented by the extraordinary chemical stability of Si_3N_4. For this reason, Si_3N_4 has only been used in the past by way of exception as an educt in chemical reactions.

Synthesis of Ternary and Multinary Nitridosilicates

Analogous to the formation of ternary oxosilicates by reaction of an alkaline metal oxide with an acidic nonmetal oxide the synthesis of a ternary phosphorus nitride is possible (Scheme 1).

$$CaO \quad + \quad SiO_2 \quad \longrightarrow \quad CaSiO_3$$

alkaline   acidic

$$2\, Ca_3N_2 \quad + \quad P_3N_5 \quad \longrightarrow \quad 3\, Ca_2PN_3$$

Scheme 1.   Synthesis of a ternary compound by reaction of the respective metal oxide/nitride and the nonmetal oxide/nitride.

In several cases it does not seem to be possible to apply this simple synthetic approach to ternary nitridosilicates. For example, the formation of $CaSiN_2$ by reaction of Ca_3N_2 and Si_3N_4 seems to be rather unlikely because below 1400°C Si_3N_4 exhibits no marked reactivity whereas Ca_3N_2 surely is not stable above 1000°C. Therefore we have developed a novel synthetic approach to ternary and multinary nitridosilicates abandoning the use of the less stable metal nitrides. Instead of that we found that the reaction of the pure metals with silicon diimide $Si(NH)_2$ leads to the formation of pure and single phase ternary and quaternary nitridosilicates (Scheme 2) [9].

$$2\, M \quad + \quad 5\, Si(NH)_2 \quad \xrightarrow[\text{HF furnace}]{1500\text{-}1650°C} \quad M_2[Si_5N_8] \quad + \quad N_2 + \quad 5\, H_2$$

$$M = Ca,\, Sr,\, Ba$$

$$3\, M' \quad + \quad 6\, Si(NH)_2 \quad \xrightarrow[\text{HF furnace}]{1650°C} \quad M'_3\,[Si_6N_{11}] \quad + \quad \tfrac{1}{2}\, N_2 + 6\, H_2$$

$$M' = Ce,\, Pr$$

Scheme 2.   Synthesis of ternary nitridosilicates by reaction of the respective metals with silicon diimide $Si(NH)_2$.

A specific advantage of this method is the possibility for the variation of the metals and the fact that preparative amounts of the products are accessible as coarse crystalline and single phase products in short reaction times. The reactions may be interpreted as the dissolution of an electropositive metal in a nitrido analogous, polymeric acid ($Si(NH)_2$) by the evolution of hydrogen.

A special high-temperature furnace has been developed in our laboratory based on a high-frequency generator (200 kHz, 0-12 kW) used for the inductive heating. The tungsten crucible containing the reaction mixture is positioned in a water-cooled quartz-reactor and is heated under pure nitrogen atmosphere by inductive coupling through a water-cooled induction coil, which is connected to the high-frequency generator. The experimental setup of the furnace is shown in Fig. 3. Advantageously this furnace makes possible very fast heating rates as well as quenching of the reaction mixture.

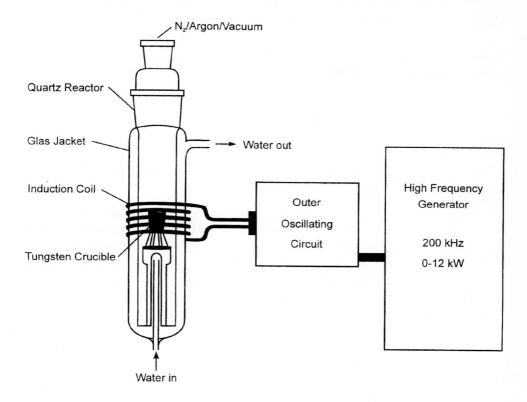

Fig. 3. High-frequency furnace for the synthesis of nitridosilicates.

Alkaline Earth Metal Nitridosilicates

According to Scheme 3 the alkaline earth metal nitridosilicates $Ca_2Si_5N_8$, $Sr_2Si_5N_8$, and $Ba_2Si_5N_8$ have been obtained as colorless solids. All three compounds are chemically inert towards hydrolysis by acids and bases and have a thermal stability up to 1600°C. In the solid $Ca_2Si_5N_8$ is built up by a three-dimensional network structure ${}_{\infty}^{3}\left[\left(Si_5^{[4]}N_4^{[2]}N_4^{[3]}\right)^{4-}\right]$ of corner-sharing SiN_4 tetrahedra [9]. Half of the nitrogen atoms each are connecting two silicon atoms ($N^{[2]}$) the rest of the N atoms are bridging three silicon ($N^{[3]}$). These ($N^{[3]}$) are arranged nearly coplanar in sheets vertical to [100] resulting in layers of highly condensed Dreier rings (Fig. 4). As expected, the bonds $Si-N^{[2]}$ are markedly shorter (167 to 171 pm) as compared to the bonds $Si-N^{[3]}$ (173 to 180 pm). Only the simply bridging ($N^{[2]}$) atoms seem to have a negative charge and they coordinate the Ca^{2+} cations. The distances $Ca^{2+}-N^{[2]}$ (232 to 284 pm) correspond to the sum of the ionic radii.

Fig. 4.   Crystal structure of $Ca_2Si_5N_8$, view along [021] (left). Sheets of highly condensed Dreier rings occur vertical to [100].

$Sr_2Si_5N_8$ and $Ba_2Si_5N_8$ are isotypic and have a network structure of corner-sharing SiN_4 tetrahedra similar to $Ca_2Si_5N_8$. However both structure types are topologically different [10] as the distribution of (Si_nN_n) ring sizes differ markedly (see below). According to $_\infty^3\left[\left(Si_5^{[4]}N_4^{[2]}N_4^{[3]}\right)^{4-}\right]$ half of the nitrogen atoms are each connecting two or three silicon, respectively. And again the $(N^{[3]})$ are arranged in sheets vertical to [100], however in contrast to $Ca_2Si_5N_8$, these sheets are significantly corrugated (Fig. 5).

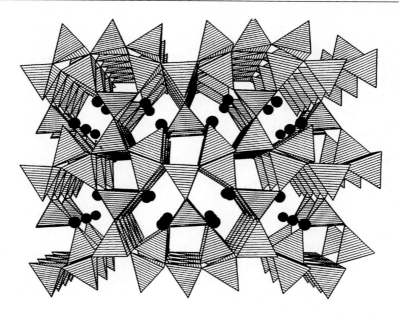

Fig. 5.   Crystal Structure of $Sr_2Si_5N_8$ and $Ba_2Si_5N_8$, view along [100].

The metal cations Sr^{2+} and Ba^{2+} are situated in channels along [100] formed by Sechser rings. Analogously to $Ca_2Si_5N_8$ the metal ions Sr^{2+} and Ba^{2+} are mainly coordinated by ($N^{[2]}$) atoms. The distances Sr^{2+}–$N^{[2]}$ (254 to 296 pm) and Ba^{2+}–$N^{[2]}$ (268 to 300 pm) nearly correspond to the sum of the respective ionic radii.

Rare Earth Metal Nitridosilicates

The general procedure for the synthesis of nitridosilicates by reaction of the corresponding metal with silicon diimide $Si(NH)_2$ also is applicable to the rare earth metals. We obtained yellow $Ce_3Si_6N_{11}$ and greenish $Pr_3Si_6N_{11}$ [11, 12]. According to $^3_\infty\left[\left(Si_6^{[4]}N_9^{[2]}N_2^{[3]}\right)^{9-}\right]$ a three-dimensional network structure of corner sharing SiN_4 tetrahedra occurs besides Ce^{3+} and Pr^{3+}, respectively In these two compounds only a minor portion of the nitrogen atoms each connect three silicon ($N^{[3]}$) while most of them form simple Si–$N^{[2]}$–Si bridges.

Fig. 6. Crystal structure of $Ce_3Si_6N_{11}$ and $Pr_3Si_6N_{11}$, view along [001].

The SiN_4 tetrahedra form layers built up by Vierer and Achter rings. These layers are stacked along [001] and are connected by double tetrahedra bridging the Achter rings (Fig. 6). The contact distances M^{3+}–N correspond to the sum of the ionic radii.

Meanwhile some rare earth metal nitridosilicates $Ln_3Si_6N_{11}$ (Ln = La, Ce, Pr, Nd, Sm) and $LnSi_3N_5$ (Ln = La, Ce, Pr, Nd) have been synthesized by the reaction of the respective silicides $LnSi_2$ and "$LnSi_3$" with N_2. However, a problem of this procedure is the avoidance of metallic impurities of the products [13].

Classification of the Si–N Networks in Nitridosilicates

For the classification and differentiation of the complex Si–N network structures in the nitridosilicates we calculated the cycle-class sequences using an algorithm presented by Klee et al. [14, 15]. Substantially, this algorithm counts the number of (Si_nN_n) rings per unit cell. As summarized in Table 1 the highly condensed nitridosilicates, in contrast to the modifications of SiO_2, show a tendency to form odd ring sizes, while in the SiO_2 modifications only even ring sizes (Si_nN_n) with $n = 4, 6, 8 \ldots$ occurs.

As discussed above the topological differences between the Si–N networks in $Ca_2Si_5N_8$ and $Sr_2Si_5N_8/Ba_2Si_5N_8$ become obvious. In the same way the ring size distribution in α- and β-Si_3N_4 significantly differs.

Table 1. Ring size distribution in SiO_2 modifications and nitridosilicates.

Ring Size	3	4	5	6	7	8
Quartz	–	–	–	3	–	21
Cristobalite	–	–	–	4	–	6
Coesite p = 35000 bar	–	4	–	6	–	18
Zeolite Rho	–	18	–	28	–	78
$MgSiN_2$	–	–	–	2	–	3
$Ce_3Si_6N_{11}$	2	1	–	12	20	55
$CeSi_3N_5$	2	1	6	10	32	97
$Ca_2Si_5N_8$	6	2	16	20	83	219
$Sr_2Si_5N_8$ $Ba_2Si_5N_8$	5	3	16	25	90	239
α-Si_3N_4	6	15	33	137	540	1911
β-Si_3N_4	3	6	21	67	264	972

Characteristic structural elements in oxo- and nitridosilicates are SiO_4 and SiN_4 tetrahedra, respectively, which are connected by corner sharing. In oxosilicates a maximum cross-linkage is reached in silicon dioxide, wherein the molar ratio of tetrahedral centers to bridging atoms is $Si:O = 1:2$. All bridging atoms ($O^{[2]}$) each are bonded to two Si. In nitridosilicates which are built up by SiN_4 tetrahedra not before a molar ratio of $Si:N > 1:2$ bridging atoms ($N^{[3]}$) were found, which connect three silicon (e.g., $^3_\infty\left[\left(Si_6^{[4]}N_9^{[2]}N_2^{[3]}\right)^{9-}\right]$ and $^3_\infty\left[\left(Si_5^{[4]}N_4^{[2]}N_4^{[3]}\right)^{4-}\right]$). Accordingly nitridosilicates extend the structural possibilities of silicates and lead with a molar ratio $Si:N > 1:2$ to higher condensed network structures. The final point of increasing cross-linking is reached in Si_3N_4, when according to $^3_\infty\left[Si_3^{[4]}N_4^{[3]}\right]$ all nitrogen atoms ($N^{[3]}$) are connecting three silicon atoms.

Unexpected Structural Possibilities in Nitridosilicates

Consequently in Si–N networks we do not expect to find any nitrogen atoms which connect four Si tetrahedral centers simultaneously ($N^{[4]}$). In condensed tetrahedral structures such a feature can only be expected to occur when the molar ratio of tetrahedral centers T to bridging atoms X has a value of $1.0 \geq T:X \geq 0.75$. This would require partial cationic Si–N structures which have not been observed so far. Recently, we obtained $BaYbSi_4N_7$ [16] which is the first compound with nitrogen atoms ($N^{[4]}$) each connecting four silicon atoms (Fig. 7).

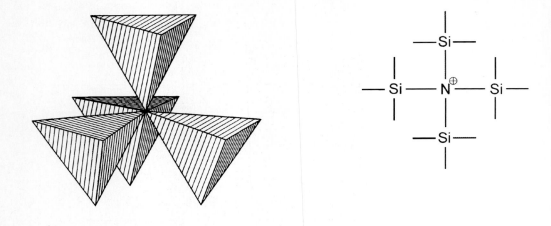

Fig. 7.   In BaYbSi₄N₇ for the first time (N$^{[4]}$) have been found each of them connecting four silicon atoms.

BaYbSi₄N₇ containes a network structure of corner sharing SiN₄ tetrahedra $\frac{3}{\infty}\left[\left(Si_4^{[4]}N_6^{[2]}N^{[4]}\right)^{5-}\right]$.

Though a molar ratio of Si:N = 4:7 should lead to $\frac{3}{\infty}\left[\left(Si_4^{[4]}N_5^{[2]}N_2^{[3]}\right)^{5-}\right]$ no N$^{[3]}$ was found. In contrast to that there is a corresponding number (N$^{[4]}$) each connecting four Si tetrahedral centers. As expected the bond lengths to (N$^{[4]}$) in BaYbSi₄N₇ are significantly longer (Si–N: 188-196 pm) than those to (N$^{[2]}$) (Si–N: 170-172 pm). The Si–N network structure in BaYbSi₄N₇ is built up by [N(SiN₃)₄] building blocks with a star-like shape as shown in Fig. 7. By connecting these groups through common (N$^{[2]}$) atoms a stacking variant of the wurtzite analogous AlN structure type is formed (Fig. 8). By systematic elimination of tetrahedra from this arrangement along [100] channels are formed by Sechser rings. In these channels the Ba^{2+} and Yb^{3+} are positioned. The metal ions show anticuboctahedral (Ba^{2+}) and octahedral (Yb^{3+}) coordination by nitrogen atoms of the Si₄N₇ network, respectively. The contact distances Ba^{2+}–N and Yb^{3+}–N each correspond to the sum of the ionic radii.

In BaYbSi₄N₇ for the first (N$^{[4]}$) connections between four Si tetrahedral centers have been found which extend the structural possibilities in nitridosilicates by a surprising dimension. Particularly in highly condensed nitridosilicates (Si:N > 1:2) thoroughly cross-linked Si–N network structures should be possible, for which attractive material properties (high mechanical, thermal, and chemical stability) can be predicted. While the structural chemistry of oxosilicates is limited to terminal oxygen atoms and simple bridging (O$^{[2]}$), the nitridosilicates show extended structural possibilities with (N$^{[2]}$) (N$^{[3]}$), and (N$^{[4]}$) connections of Si tetrahedral centers. These variations have not as yet been found in oxosilicates and may not be possible with silicon and oxygen.

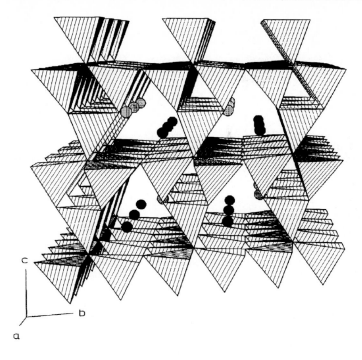

Fig. 8.   Crystal structure of $BaYbSi_4N_7$.

Molecular Preorganization

For the development of new high-performance materials it is particularly interesting to search for new covalent nitrides with highly cross-linked structures. The binary nitrides P_3N_5 and Si_3N_4 are now well known but what about ternary nitrides in the system Si–P–N?

In the past, attempts to prepare such ternary nitrides by reaction of the respective binary nitrides always have failed because the binary nitrides do not melt congruently and also because of the low self-diffusion coefficients of these materials. However, for the synthesis of $SiPN_3$ a molecular precursor $Cl_3SiNPCl_3$ has been proven to be specifically useful [5]. In this compound the required structural element of two vertex sharing tetrahedra centered by phosphorus and silicon and connected via a common nitrogen atom is pre-organized on a molecular level. The precursor compound is obtained (Scheme 3) in a three-step synthesis starting from $((CH_3)_3Si)_2NH$ which is commercially available.

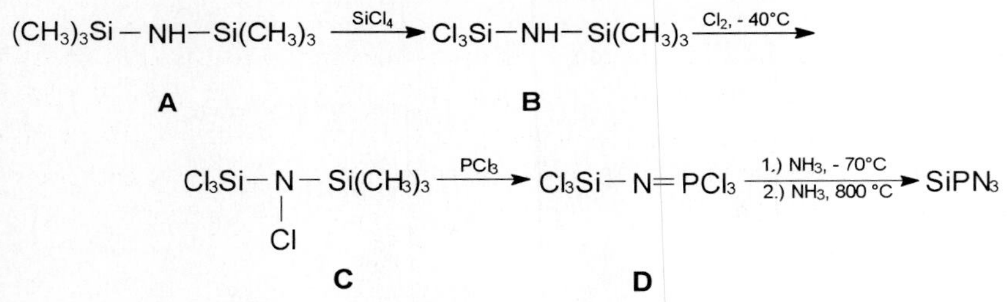

$$(CH_3)_3Si-NH-Si(CH_3)_3 \xrightarrow{\text{SiCl}_4} Cl_3Si-NH-Si(CH_3)_3 \xrightarrow{\text{Cl}_2, -40°C}$$

A   **B**

$$Cl_3Si-\underset{\underset{Cl}{|}}{N}-Si(CH_3)_3 \xrightarrow{\text{PCl}_5} Cl_3Si-N=PCl_3 \xrightarrow[\text{2.) NH}_3, 800°C]{\text{1.) NH}_3, -70°C} SiPN_3$$

C   **D**

Scheme 3.   Molecular preorganization of SiPN₃.

Low temperature ammonolysis, followed by removal of the ammonium chloride formed and pyrolysis in a stream of ammonia, leads to SiPN₃ [5]. During the reaction starting from molecules and leading to the SiPN₃ solid the structural element of two corner sharing PN₄ and SiN₄ tetrahedra is being preserved. Accordingly, silicon phosphorus nitride SiPN₃ has a three-dimensional network structure of corner-sharing PN₄ and SiN₄ tetrahedra (Fig. 9).

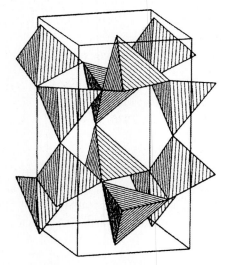

Fig. 9.   Crystal stucture of SiPN₃ which is homeotypic to Si₂N₂NH and Si₂N₂O.

With respect to the structure and some properties the ternary nitride SiPN₃ is a hybrid of both binary nitrides P₃N₅ and Si₃N₄. The thermal stability of SiPN₃ is significantly higher when compared with P₃N₅. However, it does not reach the rather high stability of Si₃N₄. The reason for this behavior is the possibility to form gaseous phosphorus during thermal decomposition of SiPN₃ above 1000°C. After several hours at 1400°C the decomposition product is pure α-Si₃N₄, which acts as a nucleus for crystallization. Calcination of commercially available amorphous Si₃N₄ by the addition of small amounts of SiPN₃ gives pure crystalline Si₃N₄ with a low oxygen content and a large amount of α-Si₃N₄, which is preferred for sintering processes. For this reason SiPN₃ might be a valuable sinter additive [17].

Acknowledgement: The work summarized in this review has been supported by the *Fonds der Chemischen Industrie*, the *Ministerium für Wissenschaft und Forschung des Landes Nordrhein-Westfalen*, the *Bundesministerium für Bildung, Wissenschaft, Forschung und Technologie (BMBF)*, the *Bayerisches Staatsministerium für Unterricht, Kultus, Wissenschaft und Kunst* and exceptionally by the *Deutsche Forschungsgemeinschaft*.

References:

[1]   W. Schnick, *Angew. Chem. Int. Ed. Engl.* **1993**, *32*, 806.

[2]   W. Schnick, *Comments Inorg. Chem.* **1995**, *17*, 189.

[3]   W. Schnick, J. Lücke, F. Krumeich, *Chem. Mater.* **1996**, *8*, 281.

[4]   W. Schnick, J. Lücke, *Solid State Ionics* **1990**, *38*, 271.

[5]   H. P. Baldus, W. Schnick, J. Lücke, U. Wannagat, G. Bogedain, *Chem. Mater.* **1993**, *5*, 845.

[6]   W. Schnick, *Stud. Surf. Sci. Catal.* **1994**, *84*, 2221.

[7]   C. Boberski, R. Hamminger, M. Peuckert, F. Aldinger, R. Dillinger, J. Heinrich, J. Huber, *Angew. Chem. Int. Ed. Engl. Adv. Mater.* **1989**, *28*, 1560; *Adv. Mater.* **1989**, *1*, 378.

[8]   G. Petzow, *Ber. Bunsenges. Phys. Chem.* **1989**, *93*, 1173.

[9]   T. Schlieper, W. Schnick, *Z. Anorg. Allg. Chem.* **1995**, *621*, 1037.

[10] T. Schlieper, W. Milius, W. Schnick, *Z. Anorg. Allg. Chem.* **1995**, *621*, 1380.

[11] T. Schlieper, W. Schnick, *Z. Anorg. Allg. Chem.* **1995**, *621*, 1535.

[12] T. Schlieper, W. Schnick, *Z. Kristallogr.* **1996**, *211*, 254.

[13] M. Woike, W. Jeitschko, *Inorg. Chem.* **1995**, *34*, 5105.

[14] W. E. Klee, *Z. Kristallogr.* **1987**, *179*, 67.

[15] A. Beukemann, W. E. Klee, *Z. Kristallogr.* **1994**, *209*, 709.

[16] H. Huppertz, W. Schnick, *Angew. Chem. Int. Ed. Engl.* **1996**, *35*, 1983.

[17] H. P. Baldus, W. Schnick, 21.01.92 DE 4201484 (patent).

Author Index

A

Abele B. C. 206
Abele S. 342
Abraham W. 587
Amelunxen K. 152
Apeloig Y. 144
Appel A. 152
Auer H. 296
Auner N. 1, 101, 106, 113, 471

B

Back S. 622
Barton T. J. 17
Becker G. 342
Beckmann J. 403
Behrens P. 649
Belzner J. 58, 418, 429
Benninghoven A. 520
Bertrand G. 223
Böhme U. 262
Born D. 484
Brendler E. 307
Bruchertseifer Ch. 531
Brüggerhoff S. 531
Busse K. 496

D

Dautel J. 632
Dehnert U. 418

Digeser M. 157
Dittmar U. 395
Dressler W. 628
Driess M. 126
Drost Ch. 358
du Mont W.-W. 286

E

Eberle U. 342
Eckhardt A. 617
Egenolf H. 31
Ehlend A. 364
Ehrich H. 484
Eikenberg D. 76
Ernst M. 241

F

Fabis R. 526
Fearon G. 1, 471
Findeis B. 172
Finger Ch. M. M. 296
Frank C. W. 587
Frenzel A. 120
Frey R. 189
Frosch W. 622

G

Gade L. H. 172

Garrison M. 587

Gehrhus B. 44

Gilges H. 271

Girreser U. 195

Goetze B. 106

Gordon M. S. 17

Götze H.-J. 531

Graschy S. 317

Graßhoff J. 500

Grobe J. 515, 526, 531, 538

Grogger C. 317

Gross T. 178

Gunzelmann N. 407

H

Habel W. 638

Häberle N. 566

Harkness B. R. 550

Hartmann M. 157

Hassler K. 241, 248, 301

Heermann J. 466

Heikenwälder C.-R. 101

Heine A. 162

Heinicke J. 36, 50, 254

Hengge E. 257, 317, 322, 327, 333

Herden V. 617

Hermann U. 322

Herrschaft B. 106

Herzig Ch. 594

Herzog U. 312

Hitchcock P. B. 44

Hornung F. M. 364

Huppertz H. 691

I

Ijadi-Maghsoodi S. 17

J

Jähn A. 241

Jehle H. 267

Jeschke E. 178

Jeske J. 286

Jones P. G. 286, 466

Jung J. 39

Junge K. 353

Jurkschat K. 403

Jursch M. 531

Jutzi P. 76, 275

K

Kaim W. 364

Kairies S. 543

Kalikhman I. 435

Karch R. 271

Karsch H. H. 53, 65, 237, 460

Katzer H. 682

Kelling H. 496

Kleewein A. 327

Klein K.-D. 504, 510

Klingebiel U. 120, 348, 358

Klinkhammer K. W. 82, 322

Kneifel S. 211

Knorr M. 211

Köll W. 301

Köppen K. 504

Kost D. 435

Köstler W. 182

Kreuzer F.-H. 566

Krivonos S. 435

Kroke E. 95

Kunze A. 594

L

Lameyer L. 162

Lang H. 423, 622

Lange D. 496

Lange T. 291

Lankat R. 412

Lappert M. F. 44

Lassacher P. 257

Leigeber H. 566

Leising G. 327

Lickiss P. D. 369

Lieske H. 484

Lin J. 17

Linti G. 182, 189

M

Ma Z. 17

Maercker A. 195

Maier G. 31, 39

Malisch W. 267, 407, 412, 415

Malkina O. 25

Mantey S. 254

Marschner Ch. 333

Marsmann H. C. 395

Maurer R. 566

Meichel E. 423

Meinel S. 36

Melter M. 423

Memmler H. 172

Meudt A. 39

Meyer K. 520

Mita I. 550

Mitzel N. W. 248

Moll A. 638

Möller S. 267

Motz G. 342

Mucha F. 400

Müller B. 25

Müller G. 452

Müller L.-P. 286

Müller T. 144

Müller U. 594

Murugavel R. 376

N

Neumayer M. 407

Niesmann J. 120

Nöth H. 152

Notheis Ch. 307

O

Oehme H. 178

Oprea A. 50

P

Pachaly B. 478

Pacl H. 39

Palitzsch W. 262

Pape A. 452

Patyk A. 86

Pätzold U. 327

Perneker K. 407

Petri S. H. A. 275

Peulecke N. 353

Pillong F. 281

Polborn K. 152, 296

Popowski E. 353

Pöschl U. 301

Power M. D. 17

Priou C. 605

Pritzkow H. 126

R

Rademacher B. 157

Rankin D. W. H. 248

Reichert S. 628

Reider K. 195

Reiners J. 510

Reinke H. 178, 353

Reisenauer H. P. 31, 39

Reising J. 412, 415

Rell S. 126

Richter L. 510

Richter Robin 291

Richter Roland 460

Richter-Mendau J. 484

Riedel R. 628

Rikowski E. 395

Roesky H. W. 376

Roewer G. 262, 291, 312, 400

Roschmann K. 407

Rose K. 543

Rutz W. 496

Ruwisch L. M. 628

S

Sander W. 86

Sartori P. 638

Schaefer D. 510

Schär D. 429

Schenzel K. 241

Schlieper T. 691

Schlüter P. 53

Schmidt M. 152

Schnabel W. 617

Schneider M. 415

Schnick W. 691

Schollmeyer D. 403

Schreiber K. A. 237

Schubart M. 172

Schubert U. 271

Schulze N. 291

Schütt O. 281

Schwarz W. 342, 632

Seelbach W. 267

Shimizu H. 17

Siehl H.-U. 25

Sixt T. 364

Smart B. A. 248

Sohn H. 144

Sperveslage G. 515

Stadtmüller S. 504, 510

Stalke D. 162

Steinberger H.-U. 113
Sternberg K. 353
Stohrer J. 566
Stoppek-Langner K. 515, 520, 531
Strehmel B. 587
Strohmann C. 206, 211, 217, 281
Stüger H. 257

T

Tachikawa M. 550
Tacke M. 70
Tacke R. 466
Takahashi M. 555
Tasch S. 327
Tertykh V. 670
Thomas B. 307
Trommer M. 86

U

Uhlig F. 322
Uhlig W. 337
Urban H. 189

V

Venzmer J. 504
Veprek S. 643
Voigt A. 376

W

Wack E. 217
Wagner R. 510
Wagner S. 117

Walawalkar M. G. 376
Waldkircher M. 452
Weidenbruch M. 95
Weinmann M. 423, 622
Weis J. 1, 471, 478, 566, 594, 682
Weißmüller J. 510
Wendler Ch. 496
Wessels M. 538
West R. 144
Westerhausen M. 157
Wiberg N. 117, 152, 296
Wieneke M. 157
Winkler U. 126
Wirschem T. 643
Witt E. 65, 460
Witte-Abel H. 358
Wolfgramm R. 348
Wu Y. 510

Z

Zanin A. 286
Zeine Ch. 526
Zhang X. 17
Zink R. 248

Subject Index

A

ab initio calculations 39, 70, 144, 241, 248
Active copper species 484
Addition reactions 120
Aerogels 682
Aldehyde-functionalized silanes 538
Alkali metal ion complexes 395
Alkaline earth metal 157
Alkenylsilanes 515
(3-Alkoxypropyl)diethoxymethylsilanes 526
(3-Alkoxypropyl)triethoxysilanes 526
Alkylalkoxysilanes 531
Allyl analogues 126
Aminodisilanes 254
Aminomethylation 237
(Aminomethyl)silanes 206
Aminosilanes 257
Antiaromaticity 223
Aromaticity 144
Aryllithium compounds 418
Azomethine compounds 400

B

Bis(amino)silylene 44
Bis(hypersilyl)germylene 82
Bis(hypersilyl)silylene 82
Bis(lithiomethyl)silanes 217
(Bromomethyl)silanes 281
Bromosilanes 296

C

Carbene 17, 223
Carbocations 25, 281
Carbodiimides 223
Catalysis 376
Cation formation 594
Cationic photopolymerization 505
Ceramic materials 622
Chelate 211
Chelate assistance 271
Chelating substituents 418
Chemisorption 670
Chlorinated polycarbosilanes 622
Chloromethane 478
Chloromethylphenyloligosilanes 307
Cholestric reflectors 566
^{13}C NMR 25
Coat 555
Cocondensation 31
Condensation 415, 496
Conductivity 555
Contact angle ....
Copolysilane 510
Copper catalyst 478
Co-pyrolysis 632
Crosslinking 566
Crystal structure 120, 275, 342, 348,
    358, 364, 403, 466

Crystal structure analyses

    see Crystal structure

Cycloaddition reactions 36, 53, 101,106, 117

[n+m] Cycloadditions

    see cycloaddition reactions

Cycloadditions

    see cycloaddition reactions

Cyclodisilazanes 353

Cyclopentasilanes 301

Cyclosilanes 317, 322, 327

Cyclotetrasilanes 296

Cyclotrisilanes 296

D

Decamethylsilicocene 76

Dehydrogenation 622

Dendrimers 333

DFT calculations 25

1,4-Diaza-1,3-butadienes 53

Diastereoselective synthesis 407

Diazasilaheterocycles 237

Diazomethane 223

Dichlorodivinylsilane 106

Dichlorosilanes 50

Diffuse reflectance infrared 520

Diffuse reflectance infrared Fourier

transformation 515

Digallanes 182

1,3-Digermacyclobutanes 217

Di-Grignard 638

Diisocyanates 364

Dilithiovinylsilanes 195

Diorganosilicon oxide 403

Diphosphene 286

Direct process 478

1,3-Disilacyclobutanes 217

Disilagermirane 162

Disilanes 257, 271, 281, 291

Disilenes 95

Distribution 496

Di-*tert*-butyldisilane 248

Donor-Acceptor complexes 452

E

EDX 484

Electrical 555

Electrochemistry 317

Electron diffraction 248

Ene reactions 117

Epoxysilanes 515

Exchange reactions 275

Extraction constants 395

F

Ferrio-silanes 407

Fluorescence probes 587

Fluorescence spectra 327

Fluorosilylhydroxylamines 348

Fumed silica 682

G

Gallium silyls 182

Gallium(I) compounds 182

1-Germa-3-silacyclobutanes 217

Germaethenes 117

Germanium 466

Germaromatics 144

Germole dianion 144

Group 1 metals 152

Group 12 metals 152

Group 13 metals 152

H

Heating of chloromethyloligosilanes 307

Heterobimetallics 211

Heterobutadienes 423

Heterocycles 364

Heterogeneously catalyzed disproportionation 291

Heterosiloxanes 376

Hexacoordinate silicon 435

Hexacoordination 460

High temperature synthesis 691

Highly coordinated silicon compounds 418

Hybrid materials 543

Hydrido-Polycarbosilanes 622

Hydrido-Silyl complexes 271

Hydridosilylamides 353

Hydroboration of vinylchlorosilanes 628

Hydrochlorodisilanes 257

Hydroformylation 538

Hydrogenation 113, 312

Hydrolysis 504

Hydrophobation 526

Hydrophobization 531

Hypersilyl compounds 178

Hypersilyl gallanes 189

Hypersilyl gallates 189

Hypervalent silanes 423

I

Iminosilanes 120

Insertion 364

IR spectroscopy 241, 248

Iron 262

Isonitriles 95

K

Kinetics 496, 594

L

Lewis acid 291

Lewis base 291

Lithiation 342

(Lithiomethyl)silanes 206

Lithium hydride elimination 195

Lithium methylamines 237

Lithium silyl cuprates 162

Lithium derivatives 358

Localization 643

Luminescence 643

M

Materials 649, 691

Matrix isolation 31, 39, 86

Mesophase 550

Metallasiloxanes 376

Metallo-Disilanes 267

Metallo-Siloxanes 412, 415

Methoxysilanes 520

Methylchlorosilanes 478

Methylhalodisilanes 241

Microporous solids 649

Mixed diorganotin 403

Mobility 587

MOCVD 275

Model compounds 376

Modification 670

Molybdenum 262

Molybdenum compounds 275

Monomeric precursors 632

Monoterpenes 101

Multinuclear NMR 435, 446

N

Nanocrystalline silicon 643

Network structures 691

Networks 587

Nitridosilicates 691

Nitrilimines 223

NMR spectroscopy 322

Nucleophilic substitution 429

Nucleophilicity 58

O

Octylsilane 531

Oligomers 423, 550

Oligosilanes 262, 312

Organometallic complexes 649

Organophosphorus halides 286

Organosilicon coating 531

O-Silylpyrazolones 358

Oxidation 86

Oxidative addition 271

P

Pentacoordinate

   see pentacoordination

Pentacoordination 446, 460

Phosphanide 157

Phosphinoalkylsilanes 271

Phosphinomethanides 65, 460

Phosphorus cation (PIII) 65

Phosphorus coordination 452

Photoconductivity 617

Photocrosslinking 594

Photoinduced cyclization 412

Photoinduced metallation 412

Photoinitiators 605

Photoisomerization 31

Platinum 211

Polarity 587

Polyaromatic mesophase 632

Polyborosilazane 628

Poly(carbosilanes) 638

Poly(diorganylsilylene-co-ethynylene)s 638

Poly(diphenylsiloxane) 550

Polymeric precursors 632

Polymers 327, 423, 566

Polysilanes 178, 291, 333

Polysilylene 617

Polysilyne 337

Porosils 649

Potassium-Graphite 337

Preceramic materials 622

Preceramic polymers 628

Precursors 543

Preservation of historical monuments 526

Promoter 484

Propylsilane 531

R

Raman spectroscopy 241, 248

Reaction behavior 353

Reaction intermediates 162

Rearrangement 17, 82

Reductive metalation 195

Reductive silylation ....

Regioselectivity 538

Regiospecific chlorination 267

Regiospecific oxofunctionalization 267

Rhenium 211

Rochow reaction 484

S

SA-layer 520

SEM 484

2-Silaallene 106

Silaamidides 120

Silaaromatics 144

Silacycles 281

Silacyclobutanes 101

Silacyclopentenes 53

Silaethenes 117

Silaheterocycles 95, 101, 126, 460

Silanediols 496

Silanes 312, 504

Silanetriols 376

Silanols 369

Silanone 76, 369

Silanorbornanes 113

Silanorbornenes 113

Silaspirocycles 106

Silathione 76

Silazanes 342

Silene 86, 101

Silica 670

λ^5Si-Silicates 466

Silicon 17, 400, 478, 638

Silicon compounds 275

Silicon halides 429

Silicon-alloyed carbon fibers 632

Silicon-arsenic compounds 126

Silicon-functionalized olefins 113

Siliconium compounds 429

Silicon-nitrogen bonds 364

Silicon-phosphorus Compounds 126

Silicone elastomers 555

Siloxane-1,1-diols 496

Siloxanes 566

Siloxene 327

Silsesquioxanes 395

Silyl complexes 211

Silyl effect 25

Silyl triflate 337

Silylene 36, 39, 44, 50, 58, 76, 86, 95

Silylhydrazines 358

Silylidenearsanes 126

Silylidenephosphanes 126

Silylphosphanes 286

Si-surfactants 510

^{29}Si NMR 307, 312, 403

^{119}Sn NMR 403

Solvent effect 195

Spirocyclic E^{II} compounds 65

Spreading 510

Stannaethenes 117

Stannanes 312

Stannylsilanes 322

Stereodynamics 435, 446

Steric hindrance 369

Sterically hindered silicon centers 418

Steriogenic metal centers 407

Stone consolidants 526

Structure 369, 452

Structure-directed synthesis 649

Supersilyl 152

Supersilyldisilanes 296

Surface 670

Surface activity 504

Surface modification 515

Surface Tension 510

Surfactants 504

Synthesis 369, 628

T

Tetrasilatetrahedrane 296

Thermal rearrangements 348

Thermal stability 628

Thermolysis 36, 95

Three-membered rings 82

Time-of-flight method 617

Time-of-flight SIMS 520

Transition metal hydride 76

Transition metal silicon compounds 262

Transparent 555

Trialkylsilyl 157

Triazines 543

Trichlorosilane 53

Tripodal amido ligands 172

Tris(silyl)hydroxylamines 348

Tris(trimethylsilyl)silyl derivatives 178

Tris(trimethylsilyl)silyllithium 162

Trisilanes 301

Trisiloxanes 504

Trisilylmethanes 172

Trisilylsilanes 172

V

Vinyllithium dimerization 195

W

Wetting tension 510

X

Xerographic method 617

X-Ray analysis
    see Crystal structure

X-Ray crystal structures
    see Crystal structure

Z

Zinc 157, 484

Zwitterions 466

t